# Biosolids Engineering and Management

Humana Press
Handbook of Environmental Engineering Series

Volume 1: Air Pollution Control Engineering. L. K. Wang, N. C. Pereira, and Y. T. Hung (eds.) 504 pp. (2004)

Volume 2: Advanced Air and Noise Pollution Control. L. K. Wang, N. C. Pereira, and Y. T. Hung (eds.) 526 pp. (2005)

Volume 3: Physicochemical Treatment Processes. L. K. Wang, Y. T. Hung, and N. K. Shammas (eds.) 723 pp. (2005)

Volume 4: Advanced Physicochemical Treatment Processes. L. K. Wang, Y. T. Hung, and N. K. Shammas (eds.) 690 pp. (2006)

Volume 5: Advanced Physicochemical Treatment Technologies. L. K. Wang, Y. T. Hung, and N. K. Shammas (eds.) 710 pp. (2007)

Volume 6: Biosolids Treatment Processes. L. K. Wang, N. K. Shammas, and Y. T. Hung (eds.) 820 pp. (2007)

Volume 7: Biosolids Engineering and Management. L. K. Wang, N. K. Shammas, and Y. T. Hung (eds.) 800 pp. (2008)

Volume 8: Biological Treatment Processes. L. K. Wang, N. C. Pereira, Y. T. Hung, and N. K. Shammas (eds.) (2008)

VOLUME 7
HANDBOOK OF ENVIRONMENTAL ENGINEERING

# Biosolids Engineering and Management

Edited by

**Lawrence K. Wang,** PhD, PE, DEE
*Lenox Institute of Water Technology, Lenox, MA
Krofta Engineering Corporation, Lenox, MA
Zorex Corporation, Newtonville, NY*

**Nazih K. Shammas,** PhD
*Lenox Institute of Water Technology, Lenox, MA
Krofta Engineering Corporation, Lenox, MA*

**Yung-Tse Hung,** PhD, PE, DEE
*Department of Civil and Environmental Engineering
Cleveland State University, Cleveland, OH*

*Editors*
**Lawrence K. Wang**
Dean & Director (retired), Lenox Institute of Water Technology
Assistant to the President, Krofta Engineering Corporation
Vice President, Zorex Corporation
1 Dawn Drive, Latham, NY 12110 USA
larrykwang@juno.com
lawrencekwang@gmail.com

**Nazih K. Shammas**
Professor and Environmental Engineering Consultant
Ex-Dean and Director, Lenox Institute of Water Technology
Advisor, Krofta Engineering Corporation
35 Flintstone Drive, Pittsfield, MA 01201, USA
n.shammas@shammasconsult.com
nazih@n-shammas.org

**Yung-Tse Hung**
Professor, Department of Civil and Environmental Engineering
Cleveland State University
16945 Deerfield Drive, Strongsville, OH 44136, USA
y.hung@csuohio.edu

ISBN 978-1-58829-861-4      e-ISBN 978-1-59745-174-1

Library of Congress Control Number: 2008922724

© 2008 Humana Press, a part of Springer Science+Business Media, LLC

All rights reserved. This work may not be translated or copied in whole or in part without the written permission of the publisher (Humana Press, 999 Riverview Drive, Suite 208, Totowa, NJ 07512 USA), except for brief excerpts in connection with reviews or scholarly analysis. Use in connection with any form of information storage and retrieval, electronic adaptation, computer software, or by similar or dissimilar methodology now known or hereafter developed is forbidden.
The use in this publication of trade names, trademarks, service marks, and similar terms, even if they are not identified as such, is not to be taken as an expression of opinion as to whether or not they are subject to proprietary rights.
While the advice and information in this book are believed to be true and accurate at the date of going to press, neither the authors nor the editors nor the publisher can accept any legal responsibility for any errors or omissions that may be made. The publisher makes no warranty, express or implied, with respect to the material contained herein.

Printed on acid-free paper

9 8 7 6 5 4 3 2 1

springer.com

The Editors of the Handbook of Environmental Engineering series dedicate this volume and all subsequent volumes to Thomas L. Lanigan (1938–2006), the founder and President of Humana Press, who encourged and vigorously supported the editors and many contributors around the world to embark on this ambitious, life-long handbook project (1978–2009) for the sole purpose of protecting our environment, in turn, benefiting our entire mankind.

# Preface

The past thirty years have seen a growing desire worldwide that positive actions be taken to restore and protect the environment from the degrading effects of all forms of pollution—air, water, soil, and noise. Since pollution is a direct or indirect consequence of waste, the seemingly idealistic demand for "zero discharge" can be construed as an unrealistic demand for zero waste. However, as long as waste continues to exist, we can only attempt to abate the subsequent pollution by converting it to a less noxious form. Three major questions usually arise when a particular type of pollution has been identified: (1) How serious is the pollution? (2) Is the technology to abate it available? (3) Do the costs of abatement justify the degree of abatement achieved? This book is one of the volumes of the *Handbook of Environmental Engineering* series. The principal intention of this series is to help readers formulate answers to the above three questions.

The traditional approach of applying tried-and-true solutions to specific pollution problems has been a major contributor to the success of environmental engineering and has accounted in large measure for the establishment of a "methodology of pollution control." However, the realization of the ever-increasing complexity and interrelated nature of current environmental problems renders it imperative that intelligent planning of pollution abatement systems be undertaken. Prerequisite to such planning is an understanding of the performance, potential, and limitations of the various methods of pollution abatement available for environmental scientists and engineers. This series of handbooks reviews at a tutorial level a broad spectrum of engineering systems (processes, operations, and methods) currently being utilized, or of potential utility, for pollution abatement. We believe that the unified interdisciplinary approach presented in these handbooks is a logical step in the evolution of environmental engineering.

Discussion of the various engineering systems presented shows how an engineering formulation of the subject flows naturally from the fundamental principles and theories of chemistry, microbiology, physics, and mathematics. This emphasis on fundamental science recognizes that engineering practice has in recent years become more firmly based on scientific principles rather than on its earlier dependency on empirical accumulation of facts. It is not intended, though, to neglect empiricism where such data lead quickly to the most economic design; certain engineering systems are not readily amenable to fundamental scientific analysis, and in these instances we have resorted to less science in favor of more art and empiricism.

Since an environmental engineer must understand science within the context of application, we first present the development of the scientific basis of a particular subject, followed by exposition of the pertinent design concepts and operations, and detailed explanations of their applications to environmental quality control or remediation. Throughout the series, methods of practical design and calculation are illustrated by numerical examples. These examples clearly demonstrate how organized, analytical reasoning leads to the most direct and clear solutions. Wherever possible, pertinent cost data have been provided.

Our treatment of pollution-abatement engineering is offered in the belief that the trained engineer should more firmly understand fundamental principles, be more aware of the similarities and differences among many of the engineering systems, and exhibit greater flexibility and originality in the definition and innovative solution of environmental pollution problems. In short, the environmental engineer should by conviction and practice be more readily adaptable to change and progress.

Coverage of the unusually broad field of environmental engineering has demanded an expertise that could only be provided through multiple authors. The authors use their customary personal style in organizing and presenting their topics; consequently, the topics are not discussed in a homogeneous manner. Moreover, owing to limitations of space, some of the authors' topics could not be discussed in great detail, and many less important topics had to be merely mentioned or commented on briefly. All authors have provided an excellent list of references at the end of each chapter for the benefit of the interested readers. As each chapter is meant to be self-contained, some mild repetition among the various texts was unavoidable. In each case, all omissions or repetitions are the responsibility of the editors and not the individual authors. With the current trend toward metrication, the question of using a consistent system of units has been a problem. Wherever possible, the authors have used the British system (fps) along with the metric equivalent (mks, cgs, or SIU) or vice versa. Conversion factors for environmental engineers are attached as an appendix in this handbook for the convenience of international readers. The editors sincerely hope that this duplication of units will prove to be useful to the reader.

The goals of the *Handbook of Environmental Engineering* series are (1) to cover entire environmental fields, including air and noise pollution control, solid waste processing and resource recovery, physicochemical treatment processes, biological treatment processes, biosolids management, water resources, natural control processes, radioactive waste disposal, and thermal pollution control; and (2) to employ a multimedia approach to environmental pollution control since air, water, soil, and energy are all interrelated.

As can be seen from the above handbook coverage, no consideration is given to pollution by type of industry, or to the abatement of specific pollutants. Rather, the organization of the handbook series has been based on the three basic forms in which pollutants and waste are manifested: gas, solid, and liquid. In addition, noise pollution control is included in the handbook series.

This book, volume 7, *Biosolids Engineering and Management*, is a sister book to volume 6, *Biosolids Treatment Processes*. Both biosolids books have been designed to serve as basic biosolids treatment textbooks as well as comprehensive reference books. We hope and expect they will prove of equally high value to advanced undergraduate and graduate students, to designers of wastewater, biosolids, and sludge treatment systems, and to scientists and researchers. The editors welcome comments from readers in all of these categories. It is our hope that both books will not only provide information on the physical, chemical, and biological treatment technologies, but also serve as a basis for advanced study or specialized investigation of the theory and practice of individual biosolids management systems.

This book (Volume 7) covers the topics of sludge and biosolids transport, pumping and storage, sludge conversion to biosolids, waste chlorination for stabilization, regulatory requirements, cost estimation, beneficial utilization, agricultural land application, biosolids landfill engineering, ocean disposal technology assessment, combustion and incineration, and process selection for biosolids management systems. The sister book (Volume 6) covers topics on biosolids characteristics and quantity, gravity thickening, flotation thickening, centrifugation, anaerobic digestion, aerobic digestion, lime stabilization, low-temperature thermal processes, high-temperature thermal processes, chemical conditioning, stabilization, elutriation, polymer conditioning, drying, belt filter, composting, vertical shaft digestion, flotation, biofiltration, pressurized ozonation, evaporation, pressure filtration, vacuum filtration, anaerobic lagoons, vermicomposting, irradiation, and land application.

The editors are pleased to acknowledge the encouragement and support received from their colleagues and the publisher during the conceptual stages of this endeavor. We wish to thank the contributing authors for their time and effort, and for having patiently borne our reviews and numerous queries and comments. We are very grateful to our respective families for their patience and understanding during some rather trying times.

*Lawrence K. Wang, Lenox, MA*
*Nazih K. Shammas, Lenox, MA*
*Yung-Tse Hung, Cleveland, OH*

# Contents

Preface ............................................................................................................... vii
Contributors ...................................................................................................... xxi

1. **Transport and Pumping of Sewage Sludge and Biosolids**
   *Nazih K. Shammas and Lawrence K. Wang* ............................................... 1

   1. Introduction .................................................................................................. 1
      1.1. Sewage Sludge and Biosolids ................................................................ 1
      1.2. Biosolids Applications .......................................................................... 2
      1.3. Transport and Pumping of Sewage Sludge and Biosolids ................... 2
   2. Pumping ...................................................................................................... 2
      2.1. Types of Sludge and Biosolids Pumps ................................................. 3
      2.2. Application and Performance Evaluation of Sludge and Sludge/Biosolids Pumps ............... 12
      2.3. Control Considerations ........................................................................ 14
   3. Pipelines .................................................................................................... 18
      3.1. Pipe, Fittings, and Valves .................................................................... 18
      3.2. Long-Distance Transport .................................................................... 18
      3.3. Headloss Calculations ......................................................................... 21
      3.4. Design Guidance ................................................................................. 22
      3.5. In-Line Grinding ................................................................................. 26
      3.6. Cost ..................................................................................................... 26
   4. Dewatered Wastewater Solids Conveyance ............................................... 28
      4.1. Manual Transport of Screenings and Grit ........................................... 29
      4.2. Belt Conveyors ................................................................................... 29
      4.3. Screw Conveyors ................................................................................ 32
      4.4. Positive-Displacement–Type Conveyors ............................................ 33
      4.5. Pneumatic Conveyors ......................................................................... 33
      4.6. Chutes and Inclined Planes ................................................................. 36
      4.7. Odors ................................................................................................... 36
   5. Long-Distance Wastewater Solids Hauling ............................................... 36
      5.1. Truck Transportation .......................................................................... 37
      5.2. Rail Transportation ............................................................................. 42
      5.3. Barge Transportation .......................................................................... 47
      5.4. Design of Sludge/Biosolids Hauling ................................................... 51
      5.5. Example .............................................................................................. 54
   6. Potential Risk to Biosolids Exposure ......................................................... 55
      6.1. Biosolids Constituents that Require Control of Worker Exposure ..... 56
      6.2. Steps to Be Taken for Protection of Workers ..................................... 57
   Nomenclature ................................................................................................. 59
   References ..................................................................................................... 60
   Appendix ....................................................................................................... 64

2. **Conversion of Sewage Sludge to Biosolids**
   *Omotayo S. Amuda, An Deng, Abbas O. Alade,
   and Yung-Tse Hung* ..................................................................................... 65

   1. Introduction ................................................................................................ 65
      1.1. Sewage and Sewage Sludge Generation .............................................. 65
      1.2. Composition and Characteristics of Sewage ....................................... 66
      1.3. Sewage and Sewage Sludge Treatment ............................................... 68
      1.4. Biosolids Regulations ......................................................................... 70

- 2. Sewage Clarification ............................................................................................................ 72
  - 2.1. Sedimentation Clarification ........................................................................................ 72
  - 2.2. Flotation Clarification ................................................................................................. 72
  - 2.3. Membrane Clarification .............................................................................................. 73
- 3. Sewage Sludge Stabilization ................................................................................................ 73
  - 3.1. Aerobic Stabilization .................................................................................................. 74
  - 3.2. Alkaline Stabilization ................................................................................................. 75
  - 3.3. Advanced Alkaline Stabilization ................................................................................ 77
  - 3.4. Anaerobic Digestion ................................................................................................... 77
  - 3.5. Composting ................................................................................................................. 84
  - 3.6. Pasteurization .............................................................................................................. 86
  - 3.7. Deep-Shaft Digestion ................................................................................................. 87
- 4. Conditioning ......................................................................................................................... 87
  - 4.1. Chemical Conditioning ............................................................................................... 87
  - 4.2. Heat Conditioning ....................................................................................................... 88
  - 4.3. Cell Destruction .......................................................................................................... 89
  - 4.4. Odor Conditioning ...................................................................................................... 90
  - 4.5. Electrocoagulation ...................................................................................................... 91
  - 4.6. Enzyme Conditioning ................................................................................................. 92
  - 4.7. Freezing ...................................................................................................................... 92
- 5. Thickening ............................................................................................................................ 93
  - 5.1. Gravity Thickening ..................................................................................................... 93
  - 5.2. Centrifugation Thickening .......................................................................................... 95
  - 5.3. Gravity Belt Thickening ............................................................................................. 97
  - 5.4. Flotation Thickening .................................................................................................. 97
  - 5.5. Rotary Drum Thickening ............................................................................................ 97
  - 5.6. Anoxic Gas Flotation Thickening ............................................................................... 97
  - 5.7. Membrane Thickening ................................................................................................ 99
  - 5.8. Recuperative Thickening .......................................................................................... 100
  - 5.9. Metal Screen Thickening .......................................................................................... 100
- 6. Dewatering and Drying ...................................................................................................... 100
  - 6.1. Belt Filter Press ........................................................................................................ 100
  - 6.2. Recessed-Plate Filter Press ....................................................................................... 101
  - 6.3. Centrifuges ................................................................................................................ 103
  - 6.4. Drying Beds .............................................................................................................. 104
  - 6.5. Vacuum Filtration ..................................................................................................... 106
  - 6.6. Electro-Dewatering ................................................................................................... 107
  - 6.7. Metal Screen Filtration ............................................................................................. 107
  - 6.8. Textile Media Filtration ............................................................................................ 108
  - 6.9. Membrane Filter Press .............................................................................................. 109
  - 6.10. Thermal Conditioning and Dewatering .................................................................. 109
  - 6.11. Drying ..................................................................................................................... 109
- 7. Other Processes .................................................................................................................. 113
  - 7.1. Focused Electrode Leak Locator (FELL) Electroscanning ...................................... 113
  - 7.2. Lystek Thermal/Chemical Process ........................................................................... 113
  - 7.3. Kiln Injection ............................................................................................................ 113
- 8. Case Study .......................................................................................................................... 114
- 9. Summary ............................................................................................................................ 114
- Acronyms ............................................................................................................................... 114
- References .............................................................................................................................. 115

## 3. Biosolids Thickening-Dewatering and Septage Treatment
*Nazih K. Shammas, Azni Idris, Katayon Saed, Yung-Tse Hung, and Lawrence K. Wang* ............................................................................................... *121*

- 1. Introduction ........................................................................................................................ 122
- 2. Expressor Press .................................................................................................................. 123
- 3. Som-A-System .................................................................................................................... 125
- 4. CentriPress ......................................................................................................................... 127

|     | 5. Hollin Iron Works Screw Press | 128 |
|-----|---|---|
|     | 6. Sun Sludge System | 132 |
|     | 7. Wedgewater Bed | 134 |
|     | 8. Vacuum-Assisted Bed | 136 |
|     | 9. Reed Bed | 137 |
|     | 10. Sludge-Freezing Bed | 139 |
|     | 11. Biological Flotation | 140 |
|     | 12. Septage Treatment | 140 |
|     |     12.1. Receiving Station (Dumping Station/Storage Facilities) | 140 |
|     |     12.2. Receiving Station (Dumping Station, Pretreatment, Equalization) | 141 |
|     |     12.3. Land Application of Septage | 142 |
|     |     12.4. Lagoon Disposal | 144 |
|     |     12.5. Composting | 145 |
|     |     12.6. Odor Control | 146 |
|     | References | 147 |

## 4. Waste Chlorination and Stabilization
### *Lawrence K. Wang* .................................................................................................. *151*

|     | 1. Introduction | 151 |
|-----|---|---|
|     |     1.1. Process Introduction | 151 |
|     |     1.2. Glossary | 152 |
|     | 2. Wastewater Chlorination | 153 |
|     |     2.1. Process Description | 153 |
|     |     2.2. Design and Operation Considerations | 154 |
|     |     2.3. Process Equipment and Control | 157 |
|     |     2.4. Design Example—Design of a Wastewater Chlorine Contact Chamber | 158 |
|     |     2.5. Application Example—Coxsackie Sewage Treatment Plant, Coxsackie, NY, USA | 165 |
|     | 3. Sludge Chlorination and Stabilization | 167 |
|     |     3.1. Process Description | 167 |
|     |     3.2. Design and Operation Considerations | 169 |
|     |     3.3. Process Equipment and Control | 171 |
|     |     3.4. Application Example—Coxsackie Sewage Treatment Plant, Coxsackie, NY, USA | 178 |
|     | 4. Septage Chlorination and Stabilization | 183 |
|     |     4.1. Process Description | 183 |
|     |     4.2. Design and Operation Considerations | 184 |
|     |     4.3. Process Equipment and Control | 186 |
|     |     4.4. Design Criteria | 186 |
|     | 5. Safety Considerations of Chlorination Processes | 187 |
|     | 6. Recent Advances in Waste Disinfection | 188 |
|     | Nomenclature | 189 |
|     | Acknowledgments | 189 |
|     | References | 190 |

## 5. Storage of Sewage Sludge and Biosolids
### *Nazih K. Shammas and Lawrence K. Wang* ........................................................... *193*

|     | 1. Introduction | 193 |
|-----|---|---|
|     |     1.1. Need for Storage | 194 |
|     |     1.2. Risks and Benefits of Solids Storage Within Wastewater Treatment Systems | 194 |
|     |     1.3. Storage Within Wastewater Sludge Treatment Processes | 194 |
|     |     1.4. Field Storage of Biosolids | 195 |
|     |     1.5. Effects of Storage on Wastewater Solids | 195 |
|     |     1.6. Types of Storage | 196 |
|     | 2. Wastewater Treatment Storage | 197 |
|     |     2.1. Storage Within Wastewater Treatment Processes | 197 |
|     |     2.2. Storage Within Wastewater Sludge Treatment Processes | 206 |

3. Facilities Dedicated to Storage of Liquid Sludge .................................................................. 208
　　　　3.1. Holding Tanks .......................................................................................................... 208
　　　　3.2. Facultative Sludge Lagoons ..................................................................................... 213
　　　　3.3. Anaerobic Liquid Sludge Lagoons ........................................................................... 229
　　　　3.4. Aerated Storage Basins ............................................................................................. 232
　　4. Facilities Dedicated to Storage of Dewatered Sludge .............................................................. 233
　　　　4.1. Drying Sludge Lagoons ............................................................................................ 234
　　　　4.2. Confined Hoppers or Bins ........................................................................................ 237
　　　　4.3. Unconfined Stockpiles .............................................................................................. 241
　　5. Field Storage of Biosolids ........................................................................................................ 242
　　　　5.1. Management of Storage ............................................................................................ 243
　　　　5.2. Odors ........................................................................................................................ 245
　　　　5.3. Water Quality ........................................................................................................... 250
　　　　5.4. Pathogens .................................................................................................................. 255
　　6. Design Examples ...................................................................................................................... 261
　　Nomenclature ................................................................................................................................ 267
　　References .................................................................................................................................... 267
　　Appendix ....................................................................................................................................... 272

6. **Regulations and Costs of Biosolids Disposal and Reuse**
   *Nazih K. Shammas and Lawrence K. Wang* ............................................................ *273*

　　1. Introduction ............................................................................................................................. 274
　　　　1.1. Historical Background ............................................................................................. 274
　　　　1.2. Background of the Part 503 Rule ............................................................................. 275
　　　　1.3. Risk Assessment Basis of the Part 503 Rule ............................................................ 276
　　　　1.4. Overview of the Rule ............................................................................................... 276
　　2. Land Application of Biosolids .................................................................................................. 277
　　　　2.1. Pollutant Limits, and Pathogen and Vector Attraction Reduction Requirements ........ 280
　　　　2.2. Options for Meeting Land Application Requirements ............................................. 280
　　　　2.3. General Requirements and Management Practices .................................................. 290
　　　　2.4. Frequency of Monitoring Requirements .................................................................. 292
　　　　2.5. Record-Keeping and Reporting Requirements ........................................................ 292
　　　　2.6. Domestic Septage ..................................................................................................... 293
　　　　2.7. Liability Issues and Enforcement Oversight ............................................................ 293
　　3. Surface Disposal of Biosolids ................................................................................................... 294
　　　　3.1. General Requirements for Surface Disposal Sites ................................................... 295
　　　　3.2. Pollutant Limits for Biosolids Placed on Surface Disposal Sites ............................. 296
　　　　3.3. Management Practices for Surface Disposal of Biosolids ....................................... 297
　　　　3.4. Pathogen and Vector Attraction Reduction Requirements for Surface
　　　　　　 Disposal Sites ............................................................................................................ 302
　　　　3.5. Frequency of Monitoring Requirements for Surface Disposal Sites ........................ 303
　　　　3.6. Record-Keeping and Reporting Requirements for Surface Disposal Sites .............. 305
　　　　3.7. Regulatory Requirements for Surface Disposal of Domestic Septage ..................... 305
　　4. Incineration of Biosolids .......................................................................................................... 305
　　　　4.1. Pollutant Limits for Biosolids Fired in a Biosolids Incinerator ............................... 306
　　　　4.2. Total Hydrocarbons .................................................................................................. 314
　　　　4.3. Management Practices for Biosolids Incineration ................................................... 316
　　　　4.4. Frequency of Monitoring Requirements for Biosolids Incineration ........................ 317
　　　　4.5. Record-Keeping and Reporting Requirements for Biosolids Incineration .............. 320
　　5. Pathogen and Vector Attraction Reduction Requirements ....................................................... 320
　　　　5.1. Pathogen Reduction Alternatives ............................................................................. 320
　　　　5.2. Requirements for Reducing Vector Attraction ......................................................... 328
　　6. Costs .......................................................................................................................................... 332
　　　　6.1. Description of Alternatives ...................................................................................... 333
　　　　6.2. Cost Relationships ................................................................................................... 336
　　　　6.3. Sludge Disposal Cost Curves ................................................................................... 336
　　　　6.4. Procedure for Using the Diagram ............................................................................. 337

Acronyms ................................................................................................................................337
Nomenclature ........................................................................................................................338
References .............................................................................................................................338
Appendix ...............................................................................................................................342

## 7. Engineering and Management of Agricultural Land Application
### *Lawrence K. Wang, Nazih K. Shammas, and Gregory Evanylo* ............................ 343

   1. Introduction ................................................................................................................344
      1.1. Biosolids ............................................................................................................344
      1.2. Biosolids Production and Pretreatment Before Land Application ....................344
      1.3. Biosolids Characteristics ...................................................................................345
      1.4. Agricultural Land Application for Beneficial Use ............................................347
      1.5. U.S. Federal and State Regulations ...................................................................348
   2. Agricultural Land Application ...................................................................................353
      2.1. Land Application Process .................................................................................353
      2.2. Agricultural Land Application Concepts and Terminologies ...........................355
   3. Planning and Management of Agricultural Land Application ...................................361
      3.1. Planning .............................................................................................................361
      3.2. Nutrient Management .......................................................................................361
   4. Design of Land Application Process ..........................................................................364
      4.1. Biosolids Application Rate Scenario ................................................................364
      4.2. Step-by-Step Procedures for Biosolids Application Rate Determination .........366
      4.3. Simplified Sludge Application Rate Determination .........................................372
   5. Operation and Maintenance .......................................................................................373
      5.1. Operation and Maintenance Process Considerations ........................................373
      5.2. Process Control Considerations ........................................................................373
      5.3. Maintenance Requirements and Safety Issues ..................................................373
   6. Normal Operating Procedures ....................................................................................374
      6.1. Startup Procedures ............................................................................................374
      6.2. Routine Land Application Procedures ..............................................................374
      6.3. Shutdown Procedures .......................................................................................374
   7. Emergency Operating Procedures ..............................................................................374
      7.1. Loss of Power or Fuel .......................................................................................374
      7.2. Loss of Other Biosolids Treatment Units ..........................................................374
   8. Environmental Impacts ...............................................................................................375
   9. Land Application Costs ..............................................................................................376
 10. Practical Applications and Design Examples ............................................................376
     10.1. Biosolids Pretreatment Before Agricultural Land Application ........................376
     10.2. Advantages and Disadvantages of Biosolids Land Application ......................377
     10.3. Design Worksheet for Determining the Agronomic Rate ................................378
     10.4. Calculation for Available Mineralized Organic Nitrogen .................................378
     10.5. Risk Assessment Approach Versus Alternative Regulatory Approach to Land Application of Biosolids ........................................................................................................378
     10.6. Tracking Cumulative Pollutant Loading Rates on Land Application Sites ......383
     10.7. Management of Nitrogen in the Soils and Biosolids .......................................383
     10.8. Converting Dry Tons of Biosolids per Acre to Pound of Nutrient per Acre ....386
     10.9. Converting Percent Content to Pound per Dry Ton .........................................387
     10.10. Calculating Net Primary Nutrient Crop Need .................................................387
     10.11. Calculating the Components of Plant Available Nitrogen in Biosolids .........388
     10.12. Calculating the First Year $PAN_{0-1}$ from Biosolids .....................................389
     10.13. Calculating Biosolids Carryover Plant Available Nitrogen ...........................390
     10.14. Calculating Nitrogen-Based Agronomic Rate ................................................391
     10.15. Calculating the Required Land for Biosolids Application .............................394
     10.16. Calculating the Nitrogen-Based and the Phosphorus-Based Agronomic Rates for Agricultural Land Application ..........................................................................................394
     10.17. Calculating the Lime-Based Agronomic Rate for Agricultural Land Application ................396

  10.18. Calculating Potassium Fertilizer Needs ...................................................................397
  10.19. Biosolids Land Application Costs and Cost Adjustment ............................................398
 11. Glossary of Land Application Terms ........................................................................400
 Nomenclature ..................................................................................................................404
 References .......................................................................................................................406
 Appendix A ......................................................................................................................410
 Appendix B ......................................................................................................................412
 Appendix C ......................................................................................................................413

## 8. Landfilling Engineering and Management
### *Puangrat Kajitvichyanukul, Jirapat Ananpattarachai, Omotayo S. Amuda, Abbas O. Alade, Yung-Tse Hung, and Lawrence K. Wang* ......................................................... *415*

 1. Introduction ..................................................................................................................415
 2. Regulations and Pollutant Standards for Biosolids Landfilling ...................................416
 3. Types of Biosolids for Landfilling ...............................................................................419
 4. Requirements of Biosolids Characteristics for Landfilling ..........................................421
  4.1. Class A Pathogen Requirements ........................................................................421
  4.2. Class B Pathogen Requirements ........................................................................423
  4.3. Other Biosolids Characteristics for Landfilling .................................................423
  4.4. Analytical Methods in Determining Biosolids Characteristics ..........................427
 5. Biosolids Treatment for Landfilling .............................................................................427
  5.1. Conditioning ......................................................................................................428
  5.2. Thickening .........................................................................................................428
  5.3. Stabilization .......................................................................................................429
  5.4. Dewatering ........................................................................................................431
 6. Design of Biosolids Landfilling ...................................................................................432
  6.1. Landfilling Application for Biosolids ................................................................432
  6.2. Biosolids Monofill .............................................................................................433
  6.3. Design Criteria ...................................................................................................436
 7. Case Study and Example ..............................................................................................438
  7.1. Future Trends in Biosolids Landfilling ..............................................................438
  7.2. Calculation Examples ........................................................................................439
 References .......................................................................................................................441

## 9. Ocean Disposal Technology and Assessment
### *Kok-Leng Tay, James Osborne and Lawrence K. Wang* ......................... *443*

 1. Introduction ..................................................................................................................444
 2. Convention on the Prevention of Marine Pollution by Dumping of Wastes and Other Matter—London Convention 1972 .........................................................446
 3. Waste Assessment Guidance .......................................................................................446
 4. Waste Assessment Audit .............................................................................................447
 5. Waste Characterization Process and Disposal Permit System ....................................449
  5.1. Assessment of Material for Disposal ................................................................449
  5.2. Chemical Screening ..........................................................................................450
  5.3. Biological Testing .............................................................................................451
  5.4. Ecological and Human Health Risk Assessment ..............................................454
  5.5. Water Quality Issues .........................................................................................457
 6. Disposal Site Selection ................................................................................................457
 7. Disposal Site Monitoring .............................................................................................458
  7.1. Acoustic Geophysical Surveys ..........................................................................459
  7.2. Currents and Sediment Transport Survey .........................................................460
  7.3. Chemical and Biological Sampling ..................................................................460
  7.4. Case Studies ......................................................................................................461

8. Land-Based Discharges of Wastes to the Sea: Engineering Design Considerations .........................................463
    8.1. Ocean Outfall System ..............................................................................464
    8.2. Initial Dilution .........................................................................................466
    8.3. Dispersion Dilution .................................................................................466
    8.4. Decay Dilution .......................................................................................466
    8.5. Outfall Design Criteria ............................................................................467
    8.6. Design Example .....................................................................................468
 9. Marine Pollution Prevention (The City of Los Angeles Biosolids Environmental Management System) ......................................................................................469
10. Ocean Disposal Technology Assessment and Conclusions .................................471
Nomenclature ...........................................................................................................472
References ................................................................................................................473

## 10. Combustion and Incineration Engineering
### *Walter R. Niessen* .................................................................................................479

 1. Introduction to Incineration ..................................................................................479
 2. Process Analysis of Incineration Systems .............................................................480
    2.1. Stoichiometry .........................................................................................480
    2.2. Thermal Decomposition (Pyrolysis) .......................................................494
    2.3. Mass Burning .........................................................................................499
    2.4. Suspension Burning ................................................................................502
    2.5. Air Pollution from Incineration ..............................................................502
    2.6. Fluid Mechanics in Furnace Systems .....................................................510
 3. Incineration Systems for Municipal Solid Waste ..................................................515
    3.1. Receipt and Storage ................................................................................519
    3.2. Charging .................................................................................................520
    3.3. Enclosures ..............................................................................................522
    3.4. Grates and Hearths .................................................................................524
    3.5. Combustion Air ......................................................................................530
    3.6. Flue Gas Conditioning ...........................................................................531
    3.7. Air Pollution Control .............................................................................533
    3.8. Special Topics ........................................................................................538
 4. Thermal Processing Systems for Biosolids ...........................................................560
    4.1. Introduction ............................................................................................560
    4.2. Objectives and General Approach ..........................................................562
    4.3. Low-Range (Ambient, 100°C) Drying Processes ...................................566
    4.4. Mid-Range (250° to 1000°C or 300° to 1800°F) Combustion Processes ...........576
    4.5. High-Range (>1100°C or >2000°F) Combustion Processes ..................588
    4.6. Discussion ..............................................................................................589
 5. Economics of Incineration ....................................................................................590
    5.1. General ...................................................................................................592
    5.2. Capital Investment .................................................................................594
    5.3. Operating Costs ......................................................................................594
 6. An Approach to Design .........................................................................................594
    6.1. Characterize the Waste ...........................................................................594
    6.2. Lay Out the System in Blocks ................................................................597
    6.3. Establish Performance Objectives ..........................................................597
    6.4. Develop Heat and Material Balances .....................................................597
    6.5. Develop Incinerator Envelope ................................................................597
    6.6. Evaluate Incinerator Dynamics ...............................................................599
    6.7. Develop the Design of Auxiliary Equipment .........................................599
    6.8. Review Heat and Material Balances .......................................................599
    6.9. Build and Operate ..................................................................................599
Appendix: Waste Thermochemical Data .................................................................599
    A.1. Refuse Composition ...............................................................................600
    A.2. Solid Waste Properties ...........................................................................601
    A.3. Ash Composition ...................................................................................601

Nomenclature.................................................................................................................601
References....................................................................................................................602

## 11. Combustion and Incineration Management
*Mingming Lu and Yu-Ming Zheng*.................................................................. 607

    1. Introduction.............................................................................................................607
        1.1. Overview of Biosolids Incineration..............................................................607
        1.2. Overview of the Dewatering Process............................................................608
        1.3. Overview of Air Pollution Control Devices.................................................609
        1.4. Overview of the Ash-Handling System.........................................................611
        1.5. U.S. Federal and State Regulations...............................................................613
    2. Operation and Management of the Multiple Hearth Furnace................................621
        2.1. Process Description.......................................................................................621
        2.2. Design and Operating Parameters.................................................................623
        2.3. Performance Evaluation, Management, and Troubleshooting of the Multiple Hearth Furnace................626
    3. Operation and Management of the Fluidized Bed Furnace...................................633
        3.1. Process Description.......................................................................................633
        3.2. Design and Operating Parameters.................................................................634
        3.3. Performance Evaluation, Management, and Troubleshooting of the Fluidized Bed Furnace..................635
        3.4. Fluidized Bed Incinerator with Improved Design.........................................637
        3.5. Comparison Between Multiple Hearth and Fluidized Bed Furnaces............639
    4. Other Incineration Processes..................................................................................640
        4.1. Electric Infrared Incinerators........................................................................640
        4.2. Co-Incineration..............................................................................................640
        4.3. Other Sludge Incineration Techniques..........................................................643
    Nomenclature.................................................................................................................644
    References....................................................................................................................644

## 12. Beneficial Utilization of Biosolids
*Nazih K. Shammas and Lawrence K. Wang*..................................................... 647

    1. Introduction.............................................................................................................647
    2. Federal Biosolids Regulations................................................................................649
        2.1. Background....................................................................................................649
        2.2. Risk Assessment Basis of Part 503...............................................................650
        2.3. Overview of Part 503.....................................................................................651
        2.4. Requirements for Land Application..............................................................651
        2.5. Requirements for Biosolids Placed on a Surface Disposal Site...................653
        2.6. Requirements for Pathogen and Vector Attraction Reduction......................653
        2.7. Requirements for Biosolids Fired in Incinerators.........................................653
        2.8. Enforcement of Part 503 and Reporting Requirements................................655
        2.9. Relationship of the Federal Requirements to State Requirements...............655
    3. Land Application of Biosolids................................................................................656
        3.1. Perspective.....................................................................................................656
        3.2. Principles and Design Criteria......................................................................658
        3.3. Options for Meeting Land Application Requirements.................................659
        3.4. Site Restrictions, General Requirements, and Management Practices........668
        3.5. Process Design...............................................................................................668
        3.6. Facilities Design............................................................................................669
        3.7. Facility Management, Operations, and Monitoring......................................670
    4. Surface Disposal of Biosolids.................................................................................670
        4.1. Perspective.....................................................................................................670
        4.2. Differentiation Among Surface Disposal, Storage, and Land Application..671
        4.3. Pollutant Limits for Biosolids.......................................................................671
        4.4. Pathogens and Vector Attraction Reduction Requirements..........................672
        4.5. Frequency of Monitoring Requirements.......................................................673
        4.6. Regulatory Requirements for Surface Disposal of Domestic Septage.........674

Contents xix

  5. Incineration of Biosolids as an Energy Source ..................................................................675
    5.1. Perspective..................................................................................................................675
    5.2. Recovery of Energy from Biosolids............................................................................676
    5.3. Factors Affecting Heat Recovery................................................................................679
    5.4. Pollutant Limits for Biosolids Fired in Incinerators....................................................680
  6. Other Uses of Wastewater Solids and Solid By-Products ....................................................684
  7. Examples ...............................................................................................................................685
    7.1. Example 1: Determination of the Annual Whole Sludge (Biosolids) Application Rate (AWSAR)....................................................................................................................685
    7.2. Example 2: Determination of the Amount of Nitrogen Provided by the AWSAR Relative to the Agronomic Rate .................................................................................................685
 Nomenclature................................................................................................................................686
 References....................................................................................................................................687

## 13. Process Selection of Biosolids Management Systems
### *Nazih K. Shammas and Lawrence K. Wang*..................................................................... *691*

  1. Introduction ...........................................................................................................................691
  2. The Logic of Process Selection.............................................................................................692
    2.1. Identification of Relevant Criteria ...............................................................................693
    2.2. Identification of System Options .................................................................................693
    2.3. System Selection Procedure ........................................................................................693
    2.4. Parallel Elements .........................................................................................................701
    2.5. Example of Process Selection at Eugene, Oregon.......................................................704
  3. Sizing of Equipment ..............................................................................................................707
  4. Approaches to Sidestream Management ...............................................................................710
    4.1. Sidestream Production.................................................................................................710
    4.2. Sidestream Quality and Potential Problems................................................................711
    4.3. General Approaches to Sidestream Problems .............................................................712
    4.4. Elimination of Sidestream............................................................................................712
    4.5. Modification of Upstream Solids Processing Steps.....................................................712
    4.6. Change in Timing, Return Rate, or Return Point ........................................................713
    4.7. Modification of Wastewater Treatment Facilities........................................................714
    4.8. Separate Treatment of Sidestreams..............................................................................715
  5. Contingency Planning............................................................................................................721
    5.1. Contingency Problems and Their Solutions................................................................721
    5.2. Example of Contingency Planning for Breakdowns ...................................................722
  6. Site Variations........................................................................................................................725
  7. Energy Conservation.............................................................................................................725
  8. Cost-Effective Analyses ........................................................................................................726
  9. Checklists...............................................................................................................................727
  10. U.S. Practices in Managing Biosolids ..................................................................................729
    10.1. Primary Biosolids Processing Trains.........................................................................729
    10.2. Secondary Biosolids Processing Trains.....................................................................734
    10.3. Combined Biosolids Processing Trains.....................................................................735
    10.4. Types of Unit Processes.............................................................................................737
 References....................................................................................................................................739

## Appendix: Conversion Factors for Environmental Engineers
### *Lawrence K. Wang*........................................................................................................... *745*

## Index................................................................................................................................789

# Contributors

ABBAS O. ALADE, MSc • *Assistant Lecturer, Department of Chemical Engineering, Ladoke Akintola University of Technology, Ogbomoso, Nigeria*

OMOTAYO S. AMUDA, PhD • *Senior Lecturer, Department of Pure and Applied Chemistry, Ladoke Akintola University of Technology, Ogbomoso, Nigeria*

JIRAPAT ANANPATTARACHAI, MSc • *Researcher, Department of Environmental Engineering, King Mongkut's University of Technology Thonburi, Bangkok Thailand*

AN DENG, PhD • *Associate Professor, College of Civil Engineering, Hohai University, Nanjing, China*

GREGORY EVANYLO, PhD • *Professor and Extension Specialist, Crop and Soil Environmental Sciences, Virginia Tech, Blacksburg, VA*

YUNG-TSE HUNG, PhD, PE, DEE • *Professor, Department of Civil and Environmental Engineering, Cleveland State University, Cleveland, OH*

AZNI IDRIS, PhD • *Professor, Department of Chemical & Environmental Engineering, Universiti Putra Malaysia, Serdang, Selangor, Malaysia*

PUANGRAT KAJITVICHYANUKUL, PhD • *Associate Professor, Department of Environmental Engineering, Mongkut's University of Technology Thonburi, Bangkok Thailand*

MINGMING LU, PhD • *Associate Professor, Department of Civil and Environmental Engineering, University of Cincinnati, Cincinnati, OH*

WALTER R. NIESSEN, MSc, PE, DEE • *President, Niessen Consultants, S. P., Andover, MA*

JAMES OSBORNE, BS • *Senior Manager, Jim Osborne Environmental Consultants, Chelsea, Quebec, Canada*

KATAYON SAED, PhD • *Assistant Professor, Department of Civil Engineering, Universiti Putra Malaysia, Serdang, Selangor, Malaysia*

NAZIH K. SHAMMAS, PhD • *Professor, Ex-Dean and Director, Lenox Institute of Water Technology, Lenox, MA; Advisor, Krofta Engineering Corporation, Lenox, MA; Environmental Engineering Consultant*

KOK-LENG TAY, PhD • *Head, Contaminated Sites and Wastes Management Unit, Environment Canada, Atlantic Region, Dartmouth, Nova Scotia, Canada*

LAWRENCE K. WANG, PhD, PE, DEE • *Dean and Director (retired), Lenox Institute of Water Technology, Lenox, MA; Assistant to the President (retired), Krofta Engineering Corporation, Lenox, MA; and Vice President (retired), Zorex Corporation, Newtonville, NY*

YU-MING ZHENG, PhD • *Research Fellow, Division of Environmental Science & Engineering, National University of Singapore, Singapore*

# 1
# Transport and Pumping of Sewage Sludge and Biosolids

## Nazih K. Shammas and Lawrence K. Wang

**CONTENTS**

INTRODUCTION
PUMPING
PIPELINES
DEWATERED WASTEWATER SOLIDS CONVEYANCE
LONG-DISTANCE WASTEWATER SOLIDS HAULING
POTENTIAL RISK TO BIOSOLIDS EXPOSURE
NOMENCLATURE
REFERENCES
APPENDIX

**Abstract** The fundamental objective of all wastewater treatment operations is to remove undesirable constituents present in wastewater and consolidate these materials for further processing and disposal. Solids removed by wastewater treatment processes include screenings and grit, naturally floating materials called scum, and the removed solids from primary and secondary clarifiers called sewage sludge. This chapter discusses the transportation of solids or the movement of sewage sludge, treated sludge (biosolids), scum, or other miscellaneous solids from point to point for treatment, storage, or disposal. Transportation includes movement of solids by pumping and pipelines, conveyors, or hauling equipment.

**Key Words** Sewage sludge • biosolids • transport • pumping • pipelines • headloss • conveyors • hauling • trucks • trains • barges • risk to exposure.

## 1. INTRODUCTION

### 1.1. Sewage Sludge and Biosolids

Solids removed by wastewater treatment processes include screenings and grit, naturally floating materials called scum, and the removed solids from primary and secondary clarifiers called sewage sludge. The term *biosolids* is the new name for what had previously been referred to as stabilized sewage sludge. Biosolids are primarily organic treated wastewater residues from municipal wastewater treatment plants—with the emphasis on the word *treated*—that are suitable for recycling as a soil amendment. *Sewage sludge* is now the term used to refer to untreated primary and secondary organic solids. This usage of terminology differentiates between biosolids, which refer to the organic solids that have received stabilization treatment at a municipal wastewater treatment plant, and the many other types of sludges (such as industrial oil and gas field wastes) that cannot be beneficially recycled as soil amendment.

## *1.2. Biosolids Applications*

Biosolids can be used as a slow release nitrogen fertilizer with low concentrations of other plant nutrients. In addition to significant amounts of nitrogen, biosolids also contain phosphorus, potassium, and essential micronutrients such as zinc and iron. Many soils in the western United States are deficient in micronutrients. Biosolids are rich in organic matter that can improve soil quality by improving water-holding capacity, soil structure, and air and water transport. Proper use of biosolids can ultimately decrease topsoil erosion.

Moreover, biosolids may provide an economic benefit in addition to their environmental advantages. Continuous application of three dry tons per acre every other year to dry land planted with wheat may produce comparable yields, higher protein content, and larger economic returns compared with the use of 50 to 60 pounds per acre of commercial nitrogen fertilizer.

## *1.3. Transport and Pumping of Sewage Sludge and Biosolids*

The fundamental objective of all wastewater treatment operations is to remove undesirable constituents present in wastewater and consolidate these materials for further processing, utilization, or disposal. This chapter discusses the transportation of solids removed by the wastewater treatment processes or the movement of scum, sewage sludge, biosolids, or other miscellaneous solids from point to point for treatment, storage, utilization, or disposal. Transportation includes movement of solids by pumping and pipelines, conveyors, or hauling equipment.

## 2. PUMPING

Biosolids pumps have many uses in a municipal wastewater treatment plant. Settled primary sludge must be moved regularly; activated sludge must be returned continuously to aeration tanks, with the extra biosolids wasted; scum must be pumped to digestion tanks; and biosolids must be recirculated and transferred within the plant in processes such as digestion, trickling filter operation, and final disposal. The type of pumping station used at the plant depends on the characteristics of the sludge itself.

Unless biosolids have been dewatered, they can be transported most efficiently and economically by pumping through pipelines. Biosolids are subject to the same physical laws as other fluids. Simply stated, work put into a fluid by a pump alters velocity, elevation, and pressure, and overcomes friction loss. The unique flow characteristics of biosolids create special problems and constraints. Nevertheless, biosolids have been successfully pumped through short pipelines at up to 20% solids by weight, as well as in pipelines of over 10 miles (16 km) long at up to 8% solids concentrations (1).

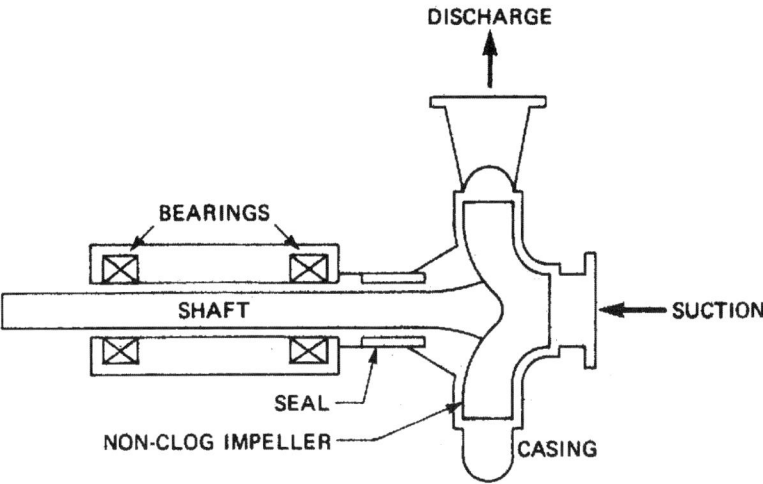

**Fig. 1.1.** Centrifugal pump. *Source*: US EPA (1).

## 2.1. Types of Sludge and Biosolids Pumps

Wastewater sludge and biosolids can range in consistency from a watery scum to thick paste-like slurry. A different type of pump may be required for each type of solids. Pumps that are currently utilized for sludge and biosolids transport include centrifugal, torque flow, plunger, piston, piston/hydraulic diaphragm, progressive cavity, rotary, diaphragm, ejector, and air lift types. Water eductor pumps are sometimes used to pump grit from aerated grit removal tanks.

### 2.1.1. Centrifugal Pumps

A centrifugal pump (Figure 1.1) consists of a set of rotating vanes in a housing or casing. The vanes may be either open or enclosed. The vanes impart energy to a fluid through centrifugal force. The nonclog centrifugal pump for wastewater or biosolids, in comparison to a centrifugal pump designed to handle clean water, has fewer but larger and less obstructed vane passageways in the impeller; has greater clearances between impeller and casing; and has sturdier bearings, shafts, and seals. Such nonclog centrifugal pumps may be used to circulate digester contents and transfer sludges with lower solids concentrations, such as waste activated sludge. The larger passageways and greater clearances result in increased reliability at a cost of lower efficiency.

The basic problem with using any form of centrifugal pump on sludge/biosolids is choosing the correct size. At any given speed, centrifugal pumps operate well only if the pumping head is within a relatively narrow range; the variable nature of sludge/biosolids, however, causes pumping heads to vary. The selected pumps must be large enough to pass solids without clogging of the impellers and yet small enough to avoid the problem of diluting the sludge/biosolids by drawing in large quantities of overlying wastewater. Throttling the discharge to reduce the capacity of a centrifugal pump is impractical both because of energy inefficiency and because frequent clogging of the throttling valve will occur. It is recommended that centrifugal pumps requiring capacity adjustment be equipped with variable-speed drives. Fixed capacity in multiple pump applications is achieved by equipping each pump with a discharge flow meter and using the flow meter signal in conjunction with the variable speed drive to control the speed of the pump. Seals last longer if back suction pumps are used. Utilizing the back of the impeller for suction removes areas of high pressure inside the pump casing from the location of the seal and prolongs seal life.

Propeller or mixed flow centrifugal pumps are sometimes used for low head applications because of higher efficiencies; a typical application is return activated sludge pumping. When being considered for this type of application, such pumps must be of sufficient size (usually at least 12 inches (in) [300 mm] in suction diameter) to provide internal clearances capable of passing the type of debris normally found within the activated sludge system. Such pumps should not be used in activated sludge systems that are not preceded with primary sedimentation facilities.

### 2.1.2. Torque Flow Pumps

A torque flow pump (Figure 1.2), also known as a recessed impeller or vortex pump, is a centrifugal pump in which the impeller is open faced and recessed well back into the pump casing. The size of particles that can be handled by this type of pump is limited only by the diameter of the suction or discharge openings. The rotating impeller imparts a spiraling motion to the fluid passing through the pump. Most of the fluid does not actually pass through the vanes of the impeller, thereby minimizing abrasive contact with it and reducing the chance of clogging. Because there are no close tolerances between the impeller and casing, the chances for abrasive wear within the pump are further reduced. The price paid for increased pump longevity and reliability is that the pumps are relatively inefficient compared with other nonclog centrifugals; 45% versus 65% efficiency is typical. Torque flow pumps for sludge/biosolids service should always

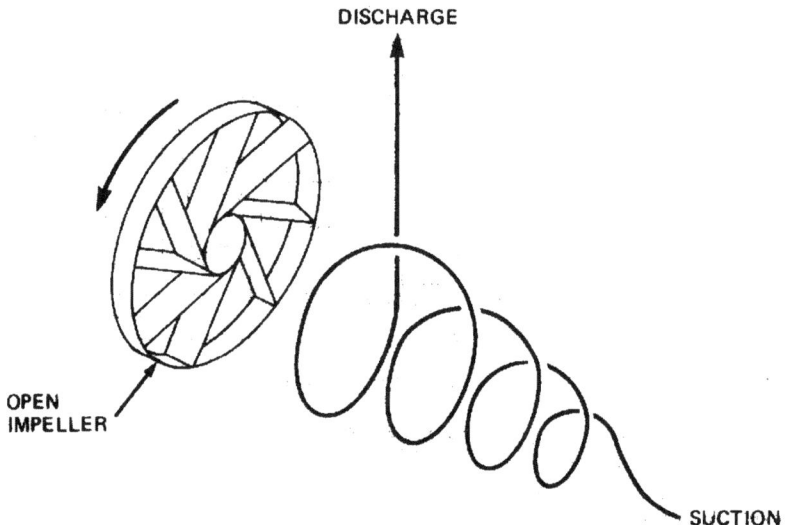

**Fig. 1.2.** Torque flow pump. *Source:* US EPA (1).

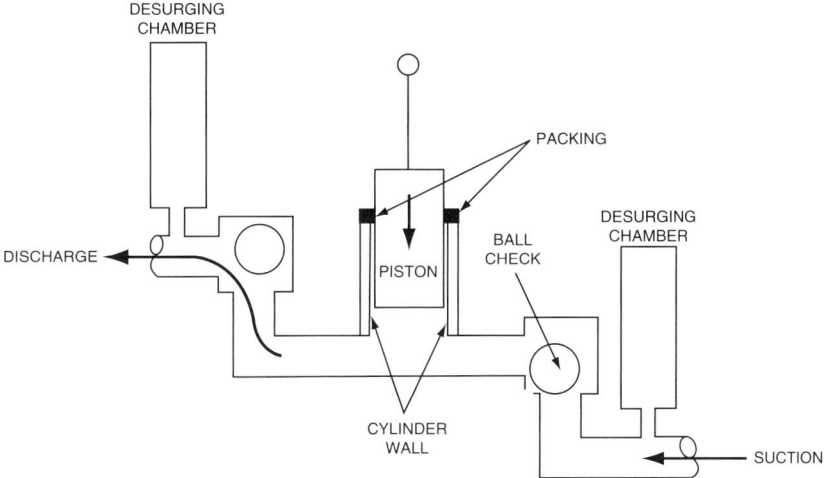

**Fig. 1.3.** Plunger pump. *Source*: US EPA (1).

have nickel or chrome abrasion-resistant volute and impellers. The pumps must be sized accurately so that excessive recirculation does not occur at any condition at operating head. Capacity adjustment and control is achieved in the same manner as for other centrifugal pumps.

### 2.1.3. Plunger Pumps

Plunger pumps (Figure 1.3) consist of pistons driven by an exposed drive crank. The eccentricity of the drive crank is adjustable, offering a variable stroke length and hence a variable positive displacement pumping action. The check valves, ball or flap, are usually paired in tandem before and after the pump. Plunger pumps have constant capacity regardless of large variations in pumping head, and can handle sludges up to 15% solids if designed specifically for such service. Plunger pumps are cost-effective where the installation requirements do not exceed 500 gallons per minute (gpm) (32 L/s), a 200 ft (61 m) discharge head, or 15% sludge solids (1). Plunger pumps require daily routine servicing by the operator, but overhaul maintenance effort and cost are low.

The plunger pump's internal mechanism is visible. The pump's connecting rod attaches to the piston inside its hollow interior, and this "bowl" is filled with oil for lubrication of the journal bearing. Either the piston exterior or the cylinder interior houses the packing, which must be kept moist at all times. Water for this purpose is usually supplied from an annular pool located above the packing; the pool receives a constant trickle of clean water. If the packing fails, sludge may be sprayed over the surrounding area.

Plunger pumps may operate with up to 10 ft (3 m) of suction lift; however, suction lifts may reduce the solids concentration that can be pumped. The use of the pump with the suction pressure higher than the discharge is not practical because flow will be forced past the check valves. The use of special intake and discharge air chambers reduces noise and vibration. These chambers also smooth out pulsations of intermittent flow. Pulsation dampening air chambers, if used, should be glass lined to avoid destruction by hydrogen

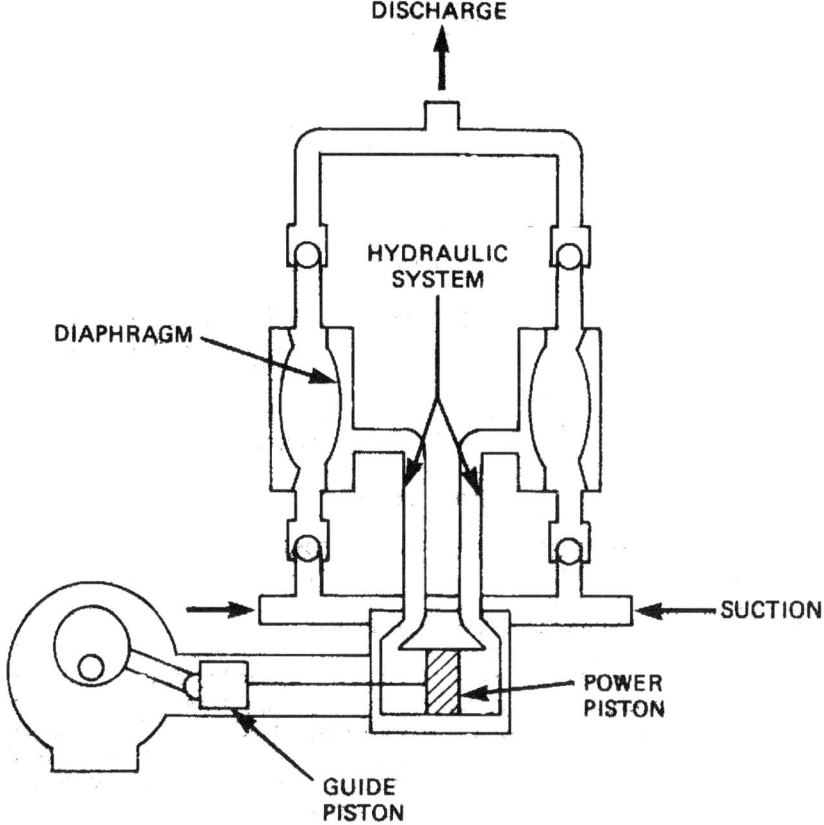

**Fig. 1.4.** Piston pump. *Source*: US EPA (1).

sulfide corrosion. If the pump is operated when the discharge pipeline is obstructed, serious damage may occur to the pump, motor, or pipeline; this problem can be avoided by a simple shear pin arrangement.

*2.1.4. Piston Pumps*

Piston pumps are similar in action to the plunger pumps, but consist of a guide piston and a fluid power piston (Figure 1.4). Piston pumps are capable of generating high pressures at low flows. These pumps are more expensive than other types of positive displacement sludge pumps and are usually used in special applications such as feed pumps for heat treatment systems. As with other types of positive displacement pumps, shear pins or other devices must be used to prevent damage due to obstructed pipelines.

A variation of the piston pump has been developed for use where reliability and close control are needed. The pump utilizes a fluid power piston driving an intermediate hydraulic fluid (clean water), which in turn pumps the sludge/biosolids in a diaphragm chamber (Figure 1.5). The speed of the hydraulic fluid drive piston can be controlled to provide pump discharge conditions ranging from constant flow rate to constant pressure. This pump is used primarily as a feed pump for filter presses. This special pump has the

# Transport and Pumping

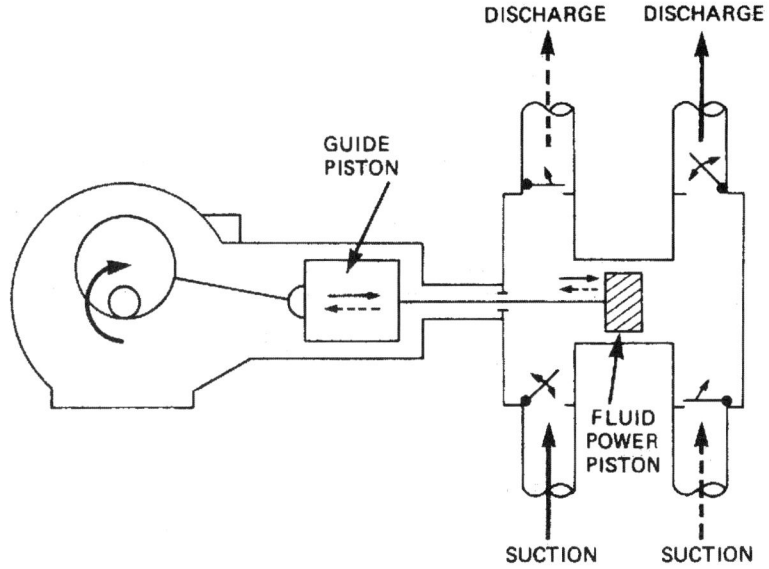

**Fig. 1.5.** Combination piston-hydraulic diaphragm pump. *Source*: US EPA (1).

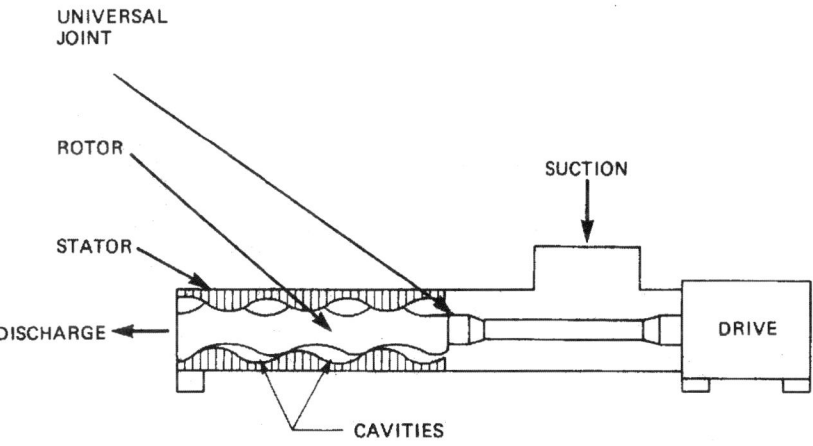

**Fig. 1.6.** Progressive cavity pump. *Source*: US EPA (1).

greatest initial cost of any piston pump, but the cost is usually offset by low maintenance and high reliability.

### 2.1.5. Progressive Cavity Pumps

The progressive cavity pump (Figure 1.6) has been used successfully on almost all types of sludge/biosolids. This pump comprises a single-threaded rotor that operates with an interference clearance in a double-threaded helix stator made of rubber. A volume or "cavity" moves "progressively" from suction to discharge when the rotor is rotating, hence the name "progressive cavity." The progressive cavity pump may be operated at discharge heads of 450 ft (137 m) on sludge/biosolids. Capacities are available to 1200 gpm (75 L/s). Some progressive cavity pumps will pass solids up to 1.125 in (2.9 cm) in diameter. Rags or stringy material should be ground up before entering this pump. The rotor is inherently self-locking in the stator housing when not in operation, and acts as a check valve for the sludge/biosolids pumping line. An auxiliary motor brake may be specified to enhance this operational feature.

The total head produced by the progressive cavity pump is divided equally between the number of cavities created by the threaded rotor and helix stator. The differential pressure between cavities directly relates to the wear of the rotor and stator because of the slight "blow by" caused by this pressure difference. Because wear on the rotor and stator is high, the maintenance cost for this type of pump is the highest of any sludge/biosolids pump. Maintenance costs are reduced by specifying the pump for one class higher pressure service (one extra stage) than would be used for clean fluids. This creates many extra cavities, reduces the differential pressure between cavities, and consequently reduces rotor and stator wear. Also, speeds should not exceed 325 revolutions per minute (rpm) in sludge/biosolids service, and grit concentrations should be minimized.

Since the rotor shaft has an eccentric motion, universal joints are required between the motor shaft and the rotor. The design of the universal joint varies greatly among different manufacturers. Continuous duty, trouble-free operation of these universal joints is best achieved by using the best quality (and usually most expensive) universal gear joint design. Discharge pressure safety shutdown devices are required on the pump discharge to prevent rupture of blocked discharge lines. No-flow safety shutdown devices are often used to prevent the rotor and stator from becoming fused due to dry operation. As previously mentioned, these pumps are expensive to maintain. However, flow rates are easily controlled, pulsation is minimal, and operation is clean. Therefore, progressive cavity pumps are widely used for pumping sludge/biosolids.

### 2.1.6. Diaphragm Pumps

Diaphragm pumps (Figure 1.7) utilize a flexible membrane that is pushed or pulled to contract or enlarge an enclosed cavity. Flow is directed through this cavity by check valves, which may be either ball or flap type. The capacity of a diaphragm pump is altered by changing either the length of the diaphragm stroke or the number of strokes per minute. Pump capacity can be increased and flow pulsations smoothed out by providing two pump-chambers and utilizing both strokes of the diaphragm for pumping. Diaphragm pumps are relatively low-head and low-capacity units; the largest available air-operated diaphragm pump delivers 220 gpm (14 L/s) against 50 ft (15 m) of head (1). The distinct advantage of the diaphragm pumps is their simplicity. Their needs for operator attention and maintenance are minimal. There are no seals, shafts, rotors, stators, or packing in contact with the fluid; also, diaphragm pumps can run in a dry condition indefinitely.

Flexure of the diaphragm may be accomplished mechanically (push rod or spring) or hydraulically (air or water). Diaphragm life is more a function of the discharge head and the total number of flexures than the abrasiveness or viscosity of the pumped fluid. Power to drive air driven diaphragm pumps is typically double that required to operate a mechanically driven pump of similar capacity. However, hydraulically operated (air or

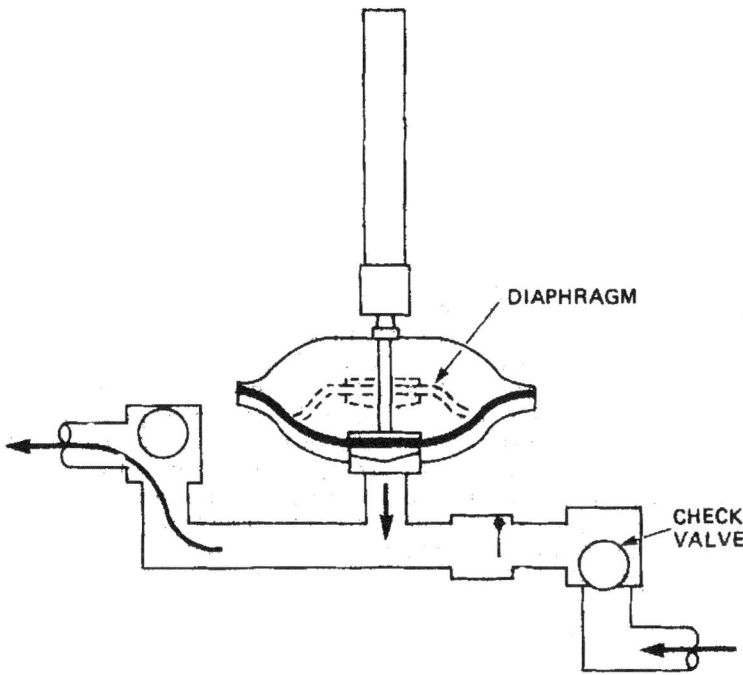

**Fig. 1.7.** Diaphragm pump. *Source*: US EPA (1).

water) diaphragms generally outwear mechanically driven diaphragms by a considerable amount. Hydraulically driven diaphragm pumps are suitable for operation in hazardous explosion-prone areas; also a pressure release means in the hydraulic system provides protection against obstructed pipelines. Typical repairs to a diaphragm pump usually cost less than USD 172 (2007 basis) for parts and require approximately 2 hours of labor (1). In some locations, high humidity intake air causes icing problems to develop at the air release valve and muffler on an air-driven diaphragm pump. A compressed air dryer should be used in the air supply system when such a condition exists.

The overall construction of some diaphragm pumps, the common "trash pump," is such that abrasion may cause the lightweight casings to fail before the diaphragms, since the pumps are not designed for continuous service. For wastewater treatment applications, the mechanical diaphragm "walking beam" pumps are more appropriate. These pumps are dependable, have quick cleanout ball or flap check valves, and are presently used to handle scum and sludge at numerous small plants throughout the country.

One air-driven diaphragm pump is sold in a package expressly intended for pumping sludge from primary sedimentation tanks and gravity thickeners. The basic pump package consists of a single-chambered, spring return diaphragm pump, an air pressure regulator, a solenoid valve, a gauge, a muffler, and an electronic transistorized timer. This unit pumps a single 3.8-gallon (14.4-L) stroke after an interval of time. The interval is readily adjusted to match the pumping rate to the rate of formation of the sludge blanket in the sedimentation tank or thickener. The large single stroke capacity of this pump has several maintenance advantages. Not only is total flexure count reduced, but ball valve flushing is improved, so large particles cause less difficulty. The maximum recommended solids size is 7/8 in (2.2 cm). Pump stroke speed is constant regardless of the selected pump flow so that minimum scouring velocities are always maintained in the discharge piping during the pumping surge.

The traditional sequence of intermittent pumping for primary sedimentation tanks has been to thicken for an interval without pumping and then draw the sludge blanket down. A relatively long interval is required by pump motors, since frequent motor starts can cause overheating. Theoretically, if the sludge concentration is 10% on the bottom and decreases to 8% at the top of the pumped sludge zone, then the pumped average is 9%. However, by using air drive, a diaphragm pump can operate with starts every few seconds instead of every several minutes or longer. The manufacturer claims its system will draw single intermittent pulses from the 10% bottom layer since the sludge blanket depth is maintained at a virtually constant height. Downstream sludge treatment processes can have greater solids capacity because more concentrated sludges can be obtained.

The City of San Francisco ran independent pump evaluation tests (2) and concluded that proper use of air-driven diaphragm pumps increases the sedimentation tanks' ability to concentrate sludges. The sludge collection system in the sedimentation tanks and the sludge pumping equipment had to be controlled together to give optimum thickening. Savings in operations and maintenance as well as improved thickening were accomplished by lowering the overall average rate of sludge withdrawal and making the sludge collectors work continuously at a reduced rate instead of intermittently. When considering such a pump installation, the capacity requirement is based on the maximum rate at which the sludge blanket forms in the tank and not the capacity required to maintain minimum pipe velocities.

*2.1.7. Rotary Pumps*

Rotary pumps (Figure 1.8) are positive displacement pumps in which two rotating synchronous lobes essentially push the fluid through the pump. Because rotary pump lobe configurations can be designed for a specific application, rotary pumps are suitable for jobs ranging from air compressor duty to wastewater sludge/biosolids pumping. Rotational speed and shearing stresses are low. Wastewater pumping lobes are non-contact and clearances are factory changed according to the abrasive content of the slurry. It is not recommended that the pumps be considered self-priming or suction lift pumps, although they are advertised as such. Experience at one plant indicates that the pump operates best with a bottom suction and top discharge. Only very limited operational data are available for rotary pumps used on sludge/biosolids. Several manufacturers advertise hard metal two-lobed pumps for sludge/biosolids usage. Lobe replacement for these pumps appears to be less costly than rotor and stator replacement on progressive cavity pumps. Some manufacturers are offering hard rubber three-lobed rotary pumps, which are used successfully for sludge/biosolids pumping. Rotary pumps, like other positive displacement pumps, must be protected against pipeline obstructions.

# Transport and Pumping

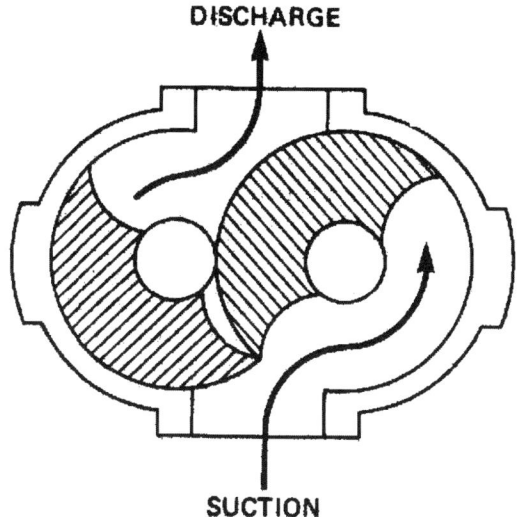

**Fig. 1.8.** Rotary pump. *Source*: US EPA (1).

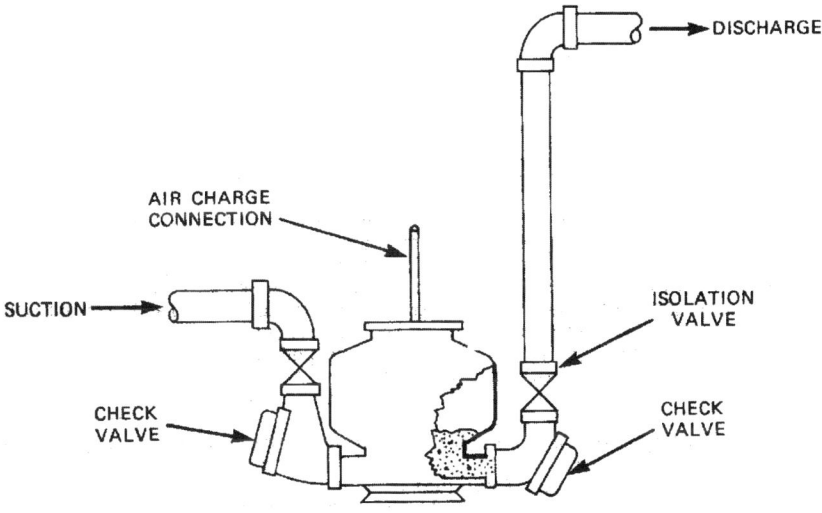

**Fig. 1.9.** Ejection pump. *Source*: US EPA (1).

### 2.1.8. Ejector Pumps

Wastewater ejectors use a charging pot, which is intermittently discharged by compressed air supply (Figure 1.9). Ejectors are most applicable for incoming average flow rates less than 150 gpm (9 L/s). These pumps require a positive suction and usually discharge to a vented holding tank or basin. Scum and sludge/biosolids can incapacitate the standard mechanical or electronic probe-type level sensors offered by most manufacturers to sequence pot discharge; custom instrumentation may be necessary. Large flushing and cleanout connections should be provided. If ejectors are to be used to discharge sludge/biosolids to an anaerobic digester where the air could produce an explosive mixture, special precautions should be taken to see that the units cannot bleed excessive quantities of air into the digester. Ejector pumps have been used in some installations to pump thickened waste-activated sludge produced by the dissolved-air flotation (DAF) process.

### 2.1.9. Gas Lift Pumps

Gas lift pumps use low-pressure gas released within a confined riser pipe with an open top and bottom. The released gas bubbles rise, dragging the liquid up and out of the riser pipe. Air is commonly used, in which case the pump is called an air-lift pump.

Air-lift pumps are used for return activated sludge and similar applications; gas-lift pumps using digester gas are used to circulate the contents of anaerobic digesters. The main advantage of these relatively inefficient pumps is the complete absence of moving parts. Gas-lift sludge/biosolids pumps are usually limited to lifts of less than 10 ft. The capacity of a lift pump can be varied by changing its buoyant gas supply. Reliable gas-lift pumping requires the gas supply to be completely independent of outside flow or pressure variables. Gas-lift pumps with an external gas supply and circumferential diffuser can pass solids of a size equivalent to the internal diameter of the confining riser pipe without clogging. When the gas is supplied by a separate inserted pipe, the obstruction created negates this nonclog feature. Gas-lift pumps, because of their low lifting capability, are very sensitive to suction and discharge head variations, and to variations in the depth of buoyant gas release. Special discharge heads are usually required to enhance the complete separation of diffused air once the discharge elevation has been reached.

### 2.1.10. Water Eductors

Water eductors use the suction force (vacuum) created when a high-pressure water stream is passed through a streamlined confining tube (Venturi). Like the air-lift pump, water eductors have no moving parts. When water is required to transport a solid material, the water eductor becomes a very convenient pump. Most water eductors with reasonable water demands cannot pump solids the size of a golf ball. However, they have been successfully used to remove grit from aerated grit removal tanks and discharge the grit into dewatering classifiers.

## 2.2. Application and Performance Evaluation of Sludge and Sludge/Biosolids Pumps

This section describes appropriate applications for the pumps and identifies some limitations and constraints. It covers screenings, grit, scum, sewage sludge, as well as sludge/biosolids.

Centrifugal pumps are used to handle large volumes of flow that have low solids content, and when precise control of the flow rate is not required. Centrifugal pumps often are used to return activated sludge and waste unthickened solids from primary and secondary treatment processes. This pump also is used for recirculation of digester contents (with less than 4% or 5% total suspended solids [TSS], and for scum and skimmings removal (3).

The most common types of positive displacement pumps used for sludge and sludge/biosolids include the plunger, rotary pump, and diaphragm pump. The plunger pump is the most popular pump for handling high viscosity sludge/biosolids containing large and abrasive materials. Of the rotary positive displacement pumps, only the progressing cavity pump has widespread use in pumping sludge/biosolids. This pump is self-priming and delivers a smooth flow in contrast to the plunger-type pump. Not only is the rotary displacement pump able to handle very thick sludge/biosolids, but it also may be used to transport sludge cake; it can pump centrifuge and filter cakes having 15% to 40% TSS. When the progressing cavity pump is made of the right materials, it also may be used for handling chemical slurries. The diaphragm-type pump may be used for the same applications as a plunger pump, except that it has no problems with abrasion. This pump also may be used for handling strong or toxic chemicals when leakage of the chemicals is a major concern.

There are two main types of sludge/biosolids grinding pumps. The first type is a comminuting device, which produces only enough head to pass solids through the grinder itself. This unit is mostly used for comminuting thickened sludge/biosolids, scum, and screenings that may cause clogging in dewatering systems. The second type of pumps combines both grinding of solids and pumping of the liquid and comminuted solids. The unit may be used to grind scum and screenings, handle sludge/sludge/biosolids flows, and break up relatively large trash particles.

Table 1.1 lists various types of sludge/biosolids pumps, their capacities, and delivered pressure (3). This table may be used as a general guide to evaluating the performance of sludge/biosolids pumps at a treatment plant. For a very precise evaluation, the actual operating characteristics of the pump should be checked against manufacturers design data for the pump. Pumps cannot be expected to operate beyond their designed capacity and intended use.

Suction conditions require special attention when pumping sludge/sludge/biosolids. When pumping water or other newtonian fluids, calculations, of net positive suction head (NPSH) can be used to determine permissible suction piping arrangements. However, sludge/biosolids are non-newtonian fluids, especially at high solids concentrations. This behavior may drastically reduce the available NPSH. Consequently, long suction pipelines should be avoided, and the sludge/biosolids pump should be several feet below

Table 1.1
Type, capacity, and delivered pressure of sludge/biosolids pumps

| Type of pump | Capacity (gpm) | Delivered pressure (psi) |
| --- | --- | --- |
| Plunger pump | up to 500 | 100–150 |
| Rotary positive displacement (progressing cavity pump) | up to 400 | up to 500 |
| Diaphragm | up to 100 | up to 100 |
| Sludge grinding pumps (comminuting and pumping type) | 25–300 | — |
| Screw lift pump | up to 80,000 | — |

*Source*: US EPA (3).

the liquid level in the tank from which the sludge is to be pumped. If these conditions are not met, a pump will not be able to handle sludge/biosolids at high concentrations.

Special precautions are usually required to reliably pump screenings and grit. Screenings should be ground up and pumped by pumps with the ability to pass large material. Torque flow pumps are ideal for this application. Grit pumping requires special abrasion and nonclogging considerations. Both screenings and grit pumps should be easy to disassemble with quick access to the volute and impeller.

Table 1.2 presents an application matrix that identifies the various types of sludge/sludge/biosolids normally encountered in wastewater applications, and provides a guide for the suitability of each type of pump in that service (1).

## 2.3. Control Considerations

To be effective, sludge/sludge/biosolids pumping systems must be flexible under different plant operating conditions. The overall piping, valves, and pumping system must be set up to allow bypassing and provide standby pumping capacity when problems occur. The most important control considerations that must be understood by the operator are as follows (3):

1. The total quantity of sludge/sludge/biosolids per day to be handled
2. The rate at which solids build up and must be removed

Unless the system and the operator are prepared to handle grit and other solids during times of heavy solids inflow, the operator may find all the sludge/biosolids lines plugged and overloaded. The operator also must be aware of the effects of overpumping and underpumping from different unit operations. For example, removing solids at too high a rate results in thin sludge/biosolids and overpumping of the thickener.

Sludge removal rates may depend on downstream operations such as dewatering and combustion. For these processes, a uniform rate of solids delivery is necessary. On the other hand, sludge/biosolids flow is not so critical in downstream units like aerobic or anaerobic digestion. If a sludge/biosolids concentrator is used, time is needed to accumulate the solids (known as a "solids inventory"). If the removal rate is higher than the stocking rate of the inventory, then the solids concentration will be lowered. Whenever possible, the concentration of the withdrawn sludge/biosolids should be used to determine the sludge/biosolids pumping rate.

When multiple units are used, it is best to withdraw thickened sludge/biosolids from two or more units at the same time using a multiple pump arrangement. This practice results in much more uniform and highly concentrated sludge/biosolids than doubling the pumping rate of a single unit. Another way is to pump at a higher rate but for a much shorter time.

Sludge removal from an aerobic or anaerobic digester is more efficient when the pumping schedule is based on solids accumulation. The pumping program should tend to underpump the thickener or secondary settling tank so that a daily manual check on the sludge/biosolids inventory can be made, and the pumping schedule adjusted to remove any accumulated inventory.

Table 1.3 shows the problems that could be encountered in the operation of sludge/biosolids pumps, their probable causes, and their solutions (3).

**Table 1.2**
**Applications for sludge/biosolids pumps**

| Pump type | Miscellaneous solids | | | | Primary sludge | | Secondary sludge (Biosolids) | | | | Digested sludge, percent | | Lagooned sludge, percent | |
|---|---|---|---|---|---|---|---|---|---|---|---|---|---|---|
| | Ground screenings | Grit | Scum | Septage | Settled sludge | Thickened sludge | Trickling filter | Activated sludge | Thickened sludge | | Mixed <6 | Thickened ≥6 | Wet <10 | Dry >15 |
| | | | | | | | | | Float | Gravity | | | | |
| Centrifugal | 0 | 0 | 0 | 0 | 3 | 2 | 4 | 4 | 0 | 3 | 4 | 3 | 4 | 1 |
| Torque flow | 5 | 4 | 0 | 5 | 4 | 3 | 4 | 4 | 0 | 4 | 4 | 3 | 3 | 0 |
| Plunger | 0 | 0 | 4 | 4 | 4 | 4 | 4 | 1 | 1 | 4 | 4 | 4 | 4 | 0 |
| Piston | 0 | 0 | 0 | 3 | 3 | 3 | 3 | 3 | 3 | 3 | 3 | 3 | 3 | 0 |
| Progressive cavity | 4 | 1 | 5 | 4 | 5 | 5 | 5 | 5 | 5 | 5 | 4 | 5 | 5 | 5 |
| Piston/hydraulic diaphragm | 0 | 0 | 0 | 2 | 3 | 3 | 3 | 3 | 3 | 3 | 3 | 3 | 3 | 0 |
| Diaphragm | 4 | 0 | 4 | 4 | 5 | 5 | 5 | 5 | 5 | 5 | 5 | 4 | 3 | 0 |
| Rotary | 0 | 0 | 0 | 0 | 3 | 3 | 3 | 3 | 3 | 3 | 3 | 3 | 3 | 0 |
| Pneumatic ejector | 4 | 3 | 3 | 3 | 3 | 3 | 3 | 3 | 3 | 3 | 3 | 3 | 3 | 0 |
| Air lift | 0 | 2 | 0 | 0 | 4 | 0 | 4 | 4 | 0 | 0 | 0 | 0 | 0 | 0 |
| Water eductor | 0 | 3 | 0 | 0 | 0 | 0 | 0 | 0 | 0 | 0 | 0 | 0 | 0 | 0 |

Key:
0 - Unsuitable
1 - Use only under special circumstances
2 - Use with caution
3 - Suitable with limitations
4 - Suitable
5 - Best type to use
*Source*: US EPA (1).

**Table 1.3**
**Troubleshooting of pumps' operational problems**

| Indicators/Observations | Probable Cause | Check or Monitor | Solutions |
|---|---|---|---|
| 1. Overpumping. | 1a. Long suction line with high head loss on suction side of pump. | 1a. Sludge dilutes as it breaks through sludge blanket if pump is operating at too high a rate. | 1a. (1) Pump sludge more frequently. (2) Reduce speed of pump. |
| 2. Unwanted (dilute) flow of sludge through pump. | 2a. Improper location of pump. 2b. Ball valve too light or ball hung up on trash accumulations. | 2b. Visual inspection. | 2a. Relocate pump. 2b. Change the weight of the ball check to prevent it from lifting and allowing dilute sludge to flow through the pump. |
| 3. Water hammer. | 3a. High suction head and high discharge pressure. | 3a. Check pressures. | 3a. (1) Be sure suction and discharge air chambers are filled with air. (2) Change ball checks and seating arrangement. (3) Modify pumping rate. |
| 4. Pump inefficiency at high suction. | 4a. Air leakage through pump seals or valve stem seals. | 4a. Pour water around seal and visibly inspect sealing check; you may also hear the leak. | 4a. Check seating and seals on valves, valve covers, valve stems, and piston on plunger pump (repair or replace damaged and worn parts). |

| | | |
|---|---|---|
| 5. Grease build-up in raw sludge line. | 5a. Sludge characteristics. | 5a. (1) Fill raw sludge line with digested sludge and let sit overnite.<br>(2) Recirculate warm sludge through raw sludge line to "melt" grease. |
| 6. Excessive wear on pumps. | 6a. Plunger pump operating on a short stroke for long periods of time. | 6a. Run pump on a longer stroke at slower speed. |
| | 6b. Improper clearance adjustment on grinder pump. | 6b. Properly adjust clearance of cutters. |
| | 6b. Cutter clearance. | |
| 7. Excessive leakage around seals on shafts and plungers. | 7a. Excessive wear on shaft or cylinder. | 7a. Replace shaft or plunger, replace mechanical seals with water-lubricated seals. |
| | 7a. Packing rapidly destroyed. | |
| 8. Progressive cavity pump unable to transport sludge. | 8a. Slippage occurring in pump due to wear on stators and rotors. | 8a. Replace stator and/or rotor. |
| | 8a. Stator and rotor condition. | |
| | 8b. Pump operating at excessive speeds. | 8b. Reduce operation to 200 to 300 rpm. |
| | 8b. Excessive wear. | |

*Source*: US EPA (3).

## 3. PIPELINES

### 3.1. Pipe, Fittings, and Valves

Materials for wastewater solids pipelines include steel, cast and ductile iron, pretensioned concrete cylinder pipe, thermoplastic, fiberglass reinforced plastic, and others. Steel and iron are most common. With steel or iron, external corrosion may occur in unprotected buried lines; corrosion may be adequately controlled under most conditions by coatings and, where needed, cathodic protection. Inside the pipe, a lining of cement, plastic, or glass may be used to protect the pipe from internal corrosion and abrasion. With raw sludge and scum, linings have an additional function: they provide a smooth surface that greatly retards accumulations of grease on the pipe wall (4, 5). With anaerobically digested biosolids, linings may be useful to prevent crystals of struvite from growing on the pipe wall. Smooth linings are especially valuable in pump suction piping and in key portions of piping (header pipes and the like) where maintenance shutdowns would cause process difficulties.

Fittings and appurtenances must be compatible with sludge/biosolids and pipe. Long sweep elbows are preferred over short radius elbows. Grit piping may be provided with elbows and tees made of special erosion-resistant materials. Valves of the nonlubricated eccentric plug type have proven reliable in sludge/biosolids pipeline service. Care must be taken if a cleaning tool is to pass through the valves. Grit pipelines are usually equipped with tapered lubricated plug valves.

Wastewater solids piping should be designed for reasonably convenient maintenance. Even under good conditions, pipe may occasionally have erosive wear, grease deposits, or other difficulties. Pipe in tunnels or galleries is more accessible than buried pipe. An adequate number of flanged joints, mechanical couplings, and takedown fittings should be provided. It is recommended that 4 to 6 in (100 to 150 mm) be considered the minimum diameter for wastewater solids pipelines to minimize grease clogging or particle blockage and facilitate maintenance. Blind flanges and cleanouts should be provided for ease of line maintenance. Gas formation by wastewater solids left for long periods in confined pipe or equipment can create explosive pressures; therefore, provision should be made for flushing and draining all pipes, pumps, and equipment. The pressure rating of wastewater solids pipelines should be adequate for unusual as well as routine operating pressures. Unusual pressures occasionally occur due to high solids concentrations, pipe obstructions, gas formation, water hammer, and cleaning operations.

Temperature changes may cause stress in the pipe. Temperatures are changed by heated material as it enters cold pipe, by flushing, and by the use of hot fluids during cleaning to remove grease. Pipe should be designed to accommodate such stresses.

### 3.2. Long-Distance Transport

Sludge/biosolids may be pumped for miles. A pipeline is frequently less expensive than the alternatives of trucks, rail cars, or barging (6), especially if, by pipelining, mechanical dewatering can be avoided. Pipelines may have less environmental impact along their routes than trucks.

Tables 1.4 and 1.5 describe some typical pipelines for unstabilized and digested sludge/biosolids. An examination of these tables shows the following (1):

Table 1.4
Long pipeline carrying unstabilized sludge/biosolids

| Characteristics | Cleveland, OH Easterly to Southerly | Indianapolis, IN Southport to Belmont | Jacksonville, FL District II to Buckman | Kansas City, MO West Side to Big Blue River | Philadelphia, PA Southeast to Southwest |
|---|---|---|---|---|---|
| Length, mile | 13.2 | 7.5 | 7 | 6.6 | 5 |
| Diameter, inches | 12 | Twin 14 | 8 | 12 | 8 |
| Pipe material | Cast iron, unlined | Ductile iron | — | Ductile iron | Ductile iron |
| Sludge type | Primary, waste-activated | Primary, waste-activated | Primary, waste-activated | Primary | Primary, excluding scum |
| Percent solids | 3–3.5 | 0.75–1.75 | 3 | 0.4–1.0 | 2.5–5 |
| Percent volatile | 65 | 68 | 80 | 50–70 | 50 |
| Flow rate, gpm | 350 | 1,000 minimum | 500 normal | 1,000 | 500 |
| Velocity, ft/s | 1.0 | 2 minimum | 3 | 2.8 | 3 |
| Total pressure, psig | 150–175 | 90 normal | 90 normal | 65 | 90 normal |
| Pump type | Centrifugal, three in series | Centrifugal | Centrifugal, two in series | Centrifugal | Centrifugal |
| Operating schedule | Continuous | Continuous | 30–60 minutes every two hours | Continuous | Continuous |
| Use of cleaning tool | Every 4–6 weeks | None | Possible, not needed | Weekly | Every 1 to 2 weeks |
| Septicity of sludge | — | Yes | Yes | Some chlorine used | Not much odor |
| Comments | Difficulty with solids accumulation at receiving plant | Thickeners do not work as well on sludge that has been pumped from Southport | Heat treatment dewatering less | Good thickening at receiving plant | Good thickening at receiving plant |

Source: US EPA (1).

**Table 1.5**
Long pipeline carrying digested sludge/biosolids

| Characteristics | Chicago, IL lagoon | Denver, CO Northside to Metro | Fort Wayne, IN | Rahway Valley Sanitary Authority, NJ | San Diego, CA Point Loma |
|---|---|---|---|---|---|
| Length, mile | 1.7 | 2 | 3 | 3 | 7.5 |
| Diameter, inches | 16 | Twin 8 | 12, some 10 | 8 | 8 |
| Material | Steel | Cast iron | Unlined cast iron | — | Fiber reinforced plastic |
| Sludge type | Lagooned | Anaerobically digested primary | Digested | Anaerobically digested primary and waste-activated | Anaerobically digested primary |
| Percent solids | 13 average 15 maximum | 4–7 | 5 maximum | 3–4 | Up to 7.5 |
| Percent volatile | 40 | 49 | 35–40 | — | 57 |
| Flow rate, gpm | 1300 | 700 | 600 | 500 | 550–600 |
| Velocity, ft/s | 2.1 | 2 | 1.6 | 3 | 3.5 |
| Total pressure, psig | 87 | 40–60 | 20–30 | 80 | 155 |
| Pumps | Centrifugal with mixers | Centrifugal | Centrifugal | Two-stage centrifugal, formerly reciprocating | Torque flow |
| Operating schedule | Intermittent | 1–2 h/d, not flushed | 3 h/d, can flush but not needed | 4 h/d, not flushed | 5 times/week, flushed before and after use |
| Use of cleaning tool | None | None | None | Not needed | None |

*Source*: US EPA (1).

*Transport and Pumping*                                                                                      21

1. Centrifugal pumps are widely used, even on unstabilized sludge/biosolids.
2. Operating pressures are usually below 125 pounds per square inch gauge (psig) (860 kN/m² gauge).
3. Velocities are usually below 3.5 ft/s (1.1 m/s).
4. If the volatile solids content of the sludge/biosolids is low, the sludge/biosolids can be pumped at a high total solids concentration. This is well illustrated by the lagoon sludge pipelines, which have operated at up to 18% solids; lagooned sludge has a very low volatile content.

In some cases, sludge/biosolids thickening at the receiving location was adversely affected by the shearing or the septicity that occurred in the pipelines. Special flushing practices after pipeline use, or use of a pipe cleaning device, was not done in several cases. The need for these techniques depends on the nature of the sludge/biosolids being pumped.

## 3.3. Headloss Calculations

The literature indicates that the hydraulic characteristics of wastewater sludge/biosolids have not been well defined because of their indefinite nature, and that finite predictions of headlosses are impossible to make. They are not available in standard tables. The approach has been to provide an adequate safety factor when designing sludge/biosolids pump and piping systems (7).

Head requirements for elevation change and velocity are the same as for water. However, friction losses may be much higher than friction losses in water pipelines. Relatively simple procedures are often used in design work; such a procedure is described below. The accuracy of these procedures is often adequate, especially at solids contents below 3% by weight. However, as the pipe length, percent total solids, and percent volatile solids increase, these simple procedures may give imprecise or misleading results. In water piping, flow is almost always turbulent. Formulas for friction loss with water, such as Hazen-Williams and Darcy-Weisbach, are based on turbulent flow. Sludge/biosolids also may flow turbulently, in which case the friction loss may be roughly that of water. Sludge/biosolids, however, are unlike water in that laminar flow also is common. When laminar flow occurs, the friction loss may be much greater than for water. Furthermore, laminar flow laws for ordinary newtonian fluids, such as water, cannot be used for laminar flow of sludge/biosolids because sludge/biosolids are not a newtonian fluid; it follows different flow laws.

Figure 1.10 may be used to provide good estimates of friction loss under laminar flow conditions (1). Sludge/biosolids have been successfully and reliably pumped in the laminar flow range. Most of the installations operate in this range. Figure 1.10 should be used in he following situations (1, 7):

1. Velocities are at least 2.5 ft/s (0.8 m/s). At lower velocities, the difference between sludge/biosolids and water may greatly increase.
2. Velocities do not exceed 8 ft/s (2.4 m/s). Higher velocities are not commonly used because of high friction loss and abrasion problems.
3. Thixotropic behavior is not considered. Thixotropy implies time-dependent change in viscosity that drops with time of shearing, followed by a gradual recovery when shearing

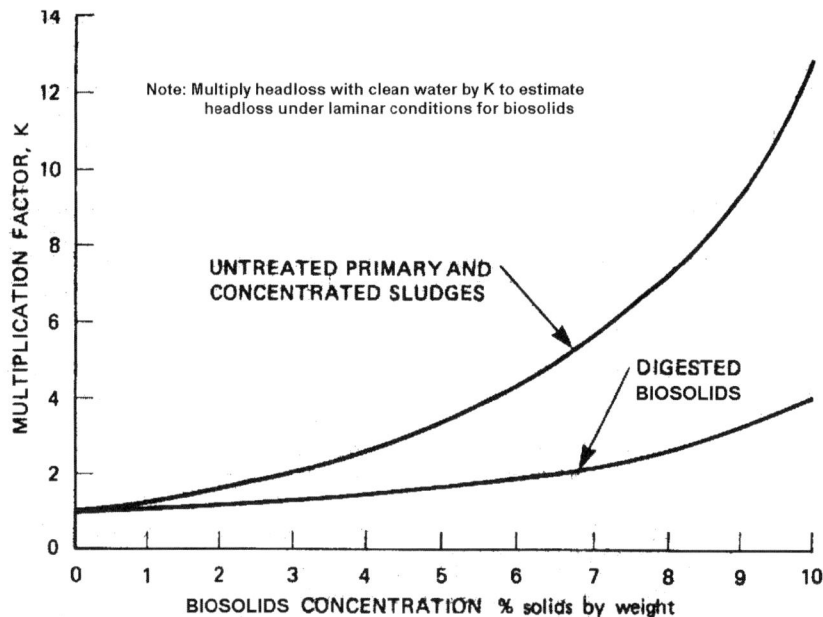

**Fig. 1.10.** Friction headloss factor for laminar flow of solids/biosolids. *Source*: US EPA (1).

   is stopped. Friction losses may be much higher in suction piping. Also, when starting a pipeline that has been shut down for over a day, unusually high pressures may be needed.
4. The pipe is not seriously obstructed by grease or other materials.

In particular cases where high velocities, turbulent flow, or thixotropic conditions prevail and more precise calculations of friction losses are desired, the readers are referred to the literature (8–17).

As an example, consider a pipe carrying unstabilized primary sludge or biosolids. The pipe is 500 ft (152 m) long and 6 in (150 mm) in diameter; flow rate is 300 gpm (19 L/s). Assume that the solids concentration may be up to 7% solids on occasion. Using the Hazen-Williams formula with a C (Hazen-Williams coefficient) of 100, a friction loss of 6.5 ft (2.0 m) would apply. If laminar sludge flow occurs, Figure 1.10 gives a multiplication factor of 5.8, so a friction loss of 38 ft (12 m) might occur. The friction loss could easily vary from 6.5 to 38 ft (2.0 to 12 m) in actual operation due to changes in sludge properties and factors not considered in Figure 1.10 such as higher velocities and turbulent flows.

Grit slurries are usually dilute; also, grit particles do not stick to each other. Therefore, ordinary friction formulas for water are usually adequate. A velocity of about 5 ft/s (1.5 m/s) is typically used. Low velocities may cause deposition of grit within the pipe; high velocities may cause erosion.

### 3.4. Design Guidance

Proper preplanning of a pipeline installation is of great importance. For example, a pump breakdown or a plugged pipeline has a great impact on plant operation, and its likelihood can be greatly minimized by good initial design and equipment selection (18).

If digestion is to be part of the system, the digesters may be located either before or after the long sludge/biosolids pipeline. However, sludge/biosolids are much easier to pump after they have been digested. In addition, raw sludge may cause problems related to thickening, odors, and corrosion at the receiving point, since septic conditions may develop in the pipeline. If raw sludge is to be pumped long distances, the least environmental impact will result if the pipeline contents are discharged directly into anaerobic digesters.

Sludge/biosolids that have been piped for a long distance may experience floc breakdown. If this occurs, thickening and dewatering may be impaired. Chemical conditioning may require a higher chemical dose; thermal conditioning may produce sludge/biosolids with poorer dewatering properties.

Sludge/biosolids are thixotropic. The most economical sludge/biosolids pumping occurs at the critical velocity where turbulent flow begins and where the mixing and agitation reduce the viscosity (and headloss). Critical velocity is a function of solids content as well as pipe size. Some critical velocities at 4% solids for various pipe diameters are as follows (19):

| Pipe diameter, in. D | Critical velocity, ft/s | |
|---|---|---|
| | Lower | Higher |
| 8 | 3.58 | 4.52 |
| 10 | 3.55 | 4.42 |
| 14 | 3.45 | 4.31 |
| 20 | 3.40 | 4.23 |

Increasing the velocity is one method for causing turbulence, viscosity reduction, and self-cleaning. However, velocities much above the critical will involve an excessive headloss because of friction. Headlosses attributable to the sludge/biosolids characteristics increase in the following situations (20, 21):

1. The solids concentration increases
2. Size of the coarse sludge/biosolids particles increases
3. The volatile content increases
4. The temperature decreases
5. The velocities are too high or too low

Effective grit removal is necessary for economical pumping. Grit increases the sludge/biosolids viscosity and settles out during periods of little or no flow causing a temporary increase in pipe roughness and headloss. Pumping anaerobically digested sludge/biosolids results in lower headloss as a result of friction than pumping raw primary sludge of the same solids content (dry basis) and flow condition. Turbulent flow tends to prevent deposition of grease.

Pipeline materials and linings influence headlosses as a result of differing friction flow factors. Mechanical and chemical aids such as macerators, in-line mixers, and polymers are sometimes used to reduce viscosity and headloss. Operating controls usually are float control, density gauges, flowmeters, pressure switches, and pressure gauges (19–21). Brief characteristics of pipeline and pumping stations are shown in Table 1.6.

**Table 1.6**
**Brief characteristics of pipeline and pumping stations**

| | | Characteristics | | |
|---|---|---|---|---|
| Type | Material | Application | Advantages | Disadvantages |
| Pipe | Cast or ductile iron, unlined | General use; high pressures | Less expensive than most | Undergoes corrosion; grease buildup |
| Pipe | Cast or ductile iron, cement-lined | General use; high pressures | No corrosion Slow grease buildup | More expensive than unlined |
| Pipe | Cast iron glass lined | Not in general use | No corrosion Slow grease buildup | Most expensive |
| Pipe | Steel, unlined | General use; high pressures | Less expensive than most | Undergoes corrosion |
| Pipe | Steel, cement-lined | General use; high pressures | No corrosion Slow grease buildup | More expensive than unlined steel |
| Pipe | Fiberglass-reinforced epoxy pipe | Not in general use; moderate pressures | No corrosion Slow grease buildup | As expensive as glass-lined pipe |
| Pipe | Plastic | Not in general use; lower pressures | No corrosion Slow grease buildup | Expensive |
| Pumping station packaged | Various | In general use | Less costly than field constructed | 10,000 gpm max. 200 ft head max. |
| Pumping station field constructed | Various | In general use | High capacity High heads | Field constructed More expensive |

*Source*: US EPA (19).

The following special design features should be considered for long distance pipelines (1):

1. Provide two pipes unless a single pipe can be shut down for several days without causing problems in wastewater treatment system.
2. Consider external corrosion and pipe loads just as for any other utility pipeline, for example, water or natural gas. External corrosion has been a problem on some long sludge/biosolids pipelines. Electrical return currents, acid soils, saline groundwater, and other factors may cause serious difficulty unless special corrosion control measures are used. Advice of specialists on the need for cathodic protection is advised.
3. Provide for adding controlled amounts of water to dilute the sludge/biosolids or flush the line. Primary effluent may be used in raw sludge pipelines; disinfected final effluent may be preferred for digested biosolids pipelines. The water connection should have a flow rate indicator. The flushing water should flow at about 3 ft/s (0.9 m/s).
4. Provide for inserting and removing a cleaning tool ("pig," "go-devil") that can be sent through the line if needed (8, 22, 23). Such cleaning may be frequently required if unstabilized sludge is pumped, even if scum is handled separately. If tool cleaning is to be used, some additional recommendations apply:
    a. Valves must provide an unobstructed waterway to pass the tool.
    b. Flushing water pressure should be sufficient to push the tool through the full length of pipeline.
    c. Pipe bend fittings should be 45° or, if possible, $22\frac{1}{2}°$. Some cleaning tools will pass 90° bends, but such bends are likely to be trouble spots. Length/radius of bends should be checked with the tool supplier.
    d. A recording or totalizing flowmeter should be provided. If the tool gets stuck in the line, the flow record can be used to compute the number of gallons pumped since the tool was inserted. Thus, the tool can be located and retrieved.
5. The pipeline route should be selected for ease of maintenance.
6. At high points, air or gas relief valves should be provided. With care, automatic relief valves can be made reliable on digested sludge/biosolids lines; however, in unstabilized sludge lines, grease and debris generally cause automatic valves to be unreliable. Simple manual blow-off valves are generally better for unstabilized sludge. Air and gases from sludge/biosolids pipelines may be odorous. In confined spaces, the air or gas may also be toxic, flammable, explosive, and corrosive.
7. If sludge/biosolids are to be pumped at more than about 3% solids, the pumps and pipeline should be designed for high and variable friction headlosses. Sludge/biosolids may flow more like a Bingham plastic than an ordinary newtonian fluid. A multiplication factor, such as those on Figure 1.10, should not be used. A more accurate design method should be used.
8. If centrifugal pumps are used, flow rates will be somewhat unpredictable because of the varying flow resistance properties of the sludge/biosolids. Storage provisions should be made for these variations. Pumps should be capable of operating at shutoff head (The head generated by the pump at closed valve) with very low flow during pipeline startup.
9. Positive displacement pumps may experience difficulty when starting a long sludge/biosolids pipeline. The thixotropic nature of sludge/biosolids may cause very high resistance to flow during startup. Consequently, excessive pressures may be generated by positive displacement pumps. To avoid this problem, variable speed drives should be provided and the pumps should be started at low speeds. An air chamber may be installed on the discharge side of the pumps; the chamber will assist in startup, as well as dampen

pulsations. With digested sludge/biosolids, a relief valve piped back to the digesters may be used near the pumps.
10. For very long lines, a booster pumping station may be required. If positive displacement booster pumps are used, a holding tank should be provided. It is practically impossible to match booster pumping rates to the sludge/biosolids flow reaching the booster station unless centrifugal pumps are used.
11. Waterhammer is best controlled by limiting velocity. Unless a special evaluation is made, velocities should not exceed about 3 ft/s (0.9 m/s). Even lower velocities may be required in some cases.

### 3.5. In-Line Grinding

In-line grinders are used to reduce the size of solids to prevent problems with the operation of downstream processes. Grinders require high maintenance; therefore, they should not be installed unless shown to be absolutely necessary. For locations where a grinder may be installed in the future, removable spool pieces should be inserted into the pipeline to facilitate the later installation of a grinder. Grinders may be applicable to streams carrying debris, rags, or stringy materials, but are usually not needed for streams carrying only secondary sludge. Grinders have often been installed preceding equipment with ball or flapper check valves. However, utilizing dual check valves, proper stroke seating can be obtained and the grinders can often be eliminated. Grinders remain a necessity upstream from small-diameter, high-pressure positive displacement pumps.

Sophisticated, slow-speed, hydraulic or electric grinders that can sense blockages and clear themselves by reverse operation are now available. Special combination centrifugal pump-grinders are available for use as digester circulation pumps, and are effective in preventing rag balls. Experience indicates such pumps require as much maintenance as grinders.

### 3.6. Cost

Each pipeline is highly site specific due to static head and the dynamic head requirements dictated by the pipe material and size and the characteristics of the sludge/biosolids being pumped. The following points were considered in evaluating the costs for pipelines and pumping stations (24):

1. Depending on the length and pressure drop of the line, intermediate lift stations are provided.
2. Depending on the sludge/biosolids utilization procedure to be followed at the site, dewatering at the site may be necessary.
3. Construction cost includes pipeline and pumping stations, one major highway crossing per mile, one single rail crossing per 5 miles, nominal number of driveways, and minor road crossings.
4. Pipeline is buried 3 to 6 ft (add 15% to cost for 6 to 10 ft), with no elevation change in the pipeline.
5. Pipeline is cement-lined cast or ductile iron, 4-inch minimum pipe size.
6. Pumps are dry-pit, horizontal or vertical, nonclog centrifugal (1780 rpm).
7. Construction cost does not include rock excavation or major unusual problems; add 70% to cost for rock.

8. Operation cost includes labor, supplies, and electrical power for pump stations, 12 h pumping/d, flow velocity of 4 ft/s.
9. Flushing or "pigging" of entire line may become necessary, requiring shutdown of line.

The approximate energy requirements for a pipeline can be computed from the following equation when assuming a wire to water efficiency of 60% for the pumping station (19):

$$E = \text{kwh/yr} = 1900 \, (\text{MGD/yr} \times \text{ft of total head}) \qquad (1)$$

where

$E$ = energy requirement, kwh/yr
kwh = kilowatt hours
MGD = million gallons per day

The construction costs for pipelines and pumping stations are shown in Figure 1.11 and the operation and maintenance costs in Figure 1.12 (24). The costs (2007 basis) can be calculated by using the U.S. Army Corps of Engineers Civil Works Construction

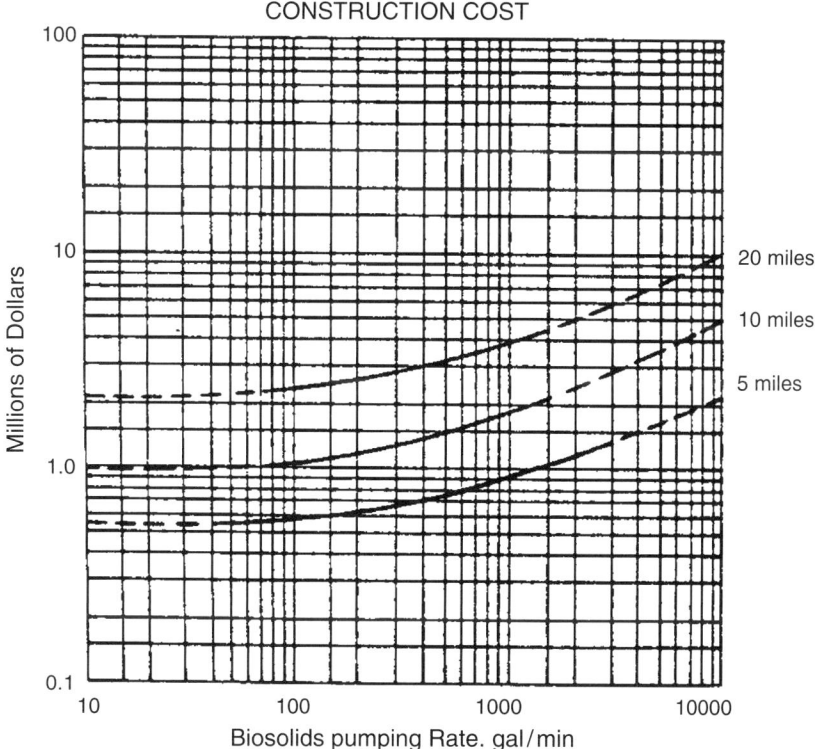

**Fig. 1.11.** Pipeline and pumping station construction cost. *Note*: Cost (2007 basis) = $1.94 × Cost from figure. *Source*: US EPA (19).

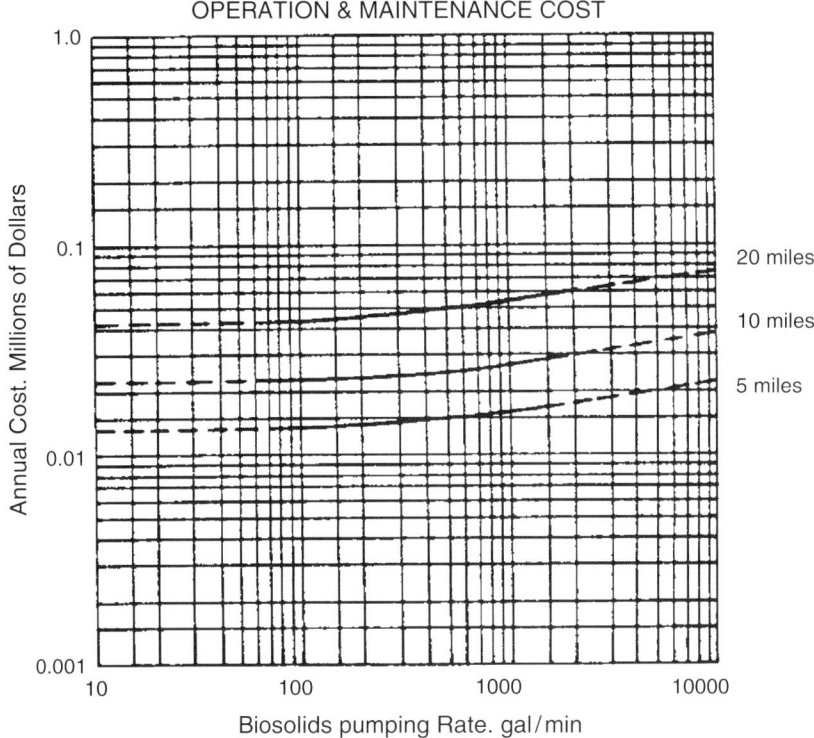

**Fig. 1.12.** Pipeline and pumping station operation and maintenance cost. *Note*: Cost (2007 basis) = \$1.94 × Cost from figure. *Source*: US EPA (19).

Yearly Average Cost Index for Utilities (25) as shown in the appendix to this chapter.

$$\text{Cost (2007 basis)} = \frac{2007\ \text{Index}}{1980\ \text{Index}} \times \text{Cost from Figs. 1.11 and 1.12} \quad (2)$$

$$= \frac{539.74}{277.60} \times \text{Cost from Figs. 1.11 and 1.12}$$

$$= 1.94 \times \text{Cost from Figs. 1.11 and 1.12}$$

## 4. DEWATERED WASTEWATER SOLIDS CONVEYANCE

Dewatered or dried sludge/biosolids, screenings, ash, and grit can be conveyed by most forms of industrial materials handling equipment, including belt, tubular, and screw conveyors; slides and inclines; elevators; and pneumatic systems. Each method may be used to advantage in certain applications. Because the consistency of wastewater solids is highly variable, and because the solids are often difficult to move and may tend to flow, the design of this equipment must consider the most severe conditions that may be expected. The U.S. Environmental Protection Agency (US EPA) report (26) discusses emerging technologies, assessment, and research needs for conveyance systems.

Transport and Pumping

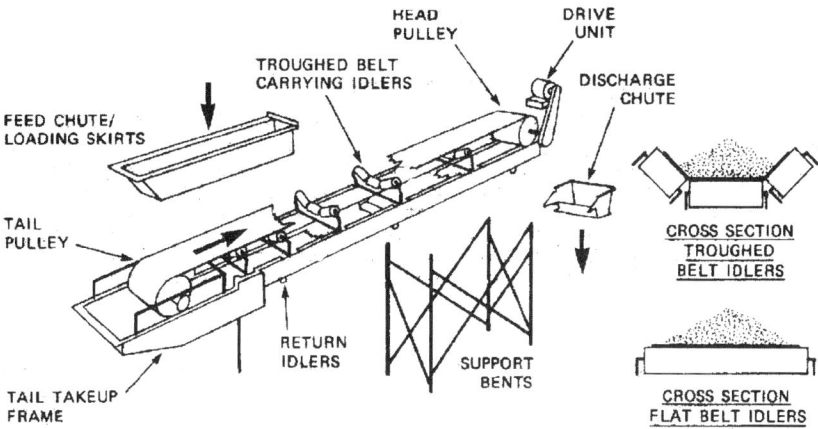

**Fig. 1.13.** Belt conveyer. *Source*: US EPA (1).

### 4.1. Manual Transport of Screenings and Grit

A common method of handling screenings or grit is simply to place a mobile container (27) beneath the discharge point and to periodically empty the mobile container into a larger container to be hauled away to a landfill. The mobile container may have wheels for ease of movement or it may be maneuvered by an overhead crane. The principal disadvantage of this approach is the amount of manual labor required. However, for small or isolated operations, this may be the most appropriate method.

### 4.2. Belt Conveyors

Troughed belt conveyors are simple and reliable (Figure 1.13). They may be equipped with load-cell weight-bridge sections for totalization of conveyed solids weight. Totalization is useful when an accurate solids balance must be calculated for a dewatering facility or treatment plant. Sludge/biosolids concentrated enough to maintain a semisolid shape (15%) can be conveyed at about 18° maximum inclination on troughed belt conveyors. Sludge/biosolids with higher solids content can be moved up steeper slopes. Where wash sprays are utilized, splash pans should be provided on the underside of belts to direct the used washwater to a proper disposal point. Such splash protection assists in keeping the area dry and preventing head and tail pulley slippage. Head and tail pulley lagging (grooving), crowning, and other auxiliary ways of maintaining belt guidance should be thoroughly reviewed with conveyor manufacturers before specifying a troughed belt installation. Most troughed belt installations for sludge/biosolids currently utilize steel idlers and pulleys with lubricated antifriction bearings. The fisheries industry, which also uses conveyors in constantly wet applications, is successfully using lubricated thermoplastic idler bearings with Schedule 80 polyvinyl chloride (PVC) pipe rollers; these provide longer service life than is achieved with all-steel construction.

In sludge/biosolids applications, belt failures usually occur first at the zipper-like mechanical belt seams. Endless belts with field vulcanized seams may be specified to eliminate this mode of failure. Belt material must be resistant to dilute sulfuric acid, formed by the reaction of hydrogen sulfide and moisture. Material selection must also consider oil, grease, and a multitude of other elements found in sludge/biosolids.

Belt conveyors have been successfully used to transport coarse solids removed from mechanically cleaned bar racks, and can be used to transport grit. Special consideration should be given to the type of belt design, construction materials, bearings, type of drive, and controls. Since screenings are heavily laden with water, the belt must be designed to contain and direct draining water to a point of disposal. A means of changing belt speeds should be provided so that a range of loads can be accommodated.

The handbook on belt conveyors for bulk materials published by the Conveyor Equipment Manufacturers Association (28) is a good reference for general design of belt conveyors. However, there is little specific information available relating to the special problems associated with the cohesive, nonuniform properties of dewatered sludge/biosolids. Experience at existing facilities using this type of conveying equipment and transporting sludge/biosolids with similar characteristics provides the most useful design information.

The experience of the County Sanitation Districts of Los Angeles County in the operation of a two-stage digested biosolids dewatering station provides useful guidance for conveying centrifuge-dewatered digested biosolids (29). The facility includes solid bowl centrifuges as a first stage, after which the centrate is screened and then dewatered using basket centrifuges. The system uses belt conveyors to transport dewatered biosolids between production, storage, and truck loading. The system has 44 belt conveyors totaling approximately one-half mile in length. Troughed conveyor belts carry both first-stage centrifuge cake at 32% solids and second-stage centrifuge cake at 17% solids. Dewatered biosolids are usually stored in the 12 storage bins at 22% to 24% solids and then transported to trucks by additional belt conveyors. Helpful guidelines resulting from startup of this facility include the following (1):

1. *Reduction of splashing at transfer points*: The dump point should be enclosed and the drop distance minimized. Skirtboards (stationary sidewalls at edges of belts) should be used at critical areas and covered if necessary. Rubber gaskets from hoppers to skirtboards and on the bottom of skirtboards may be required to reduce splashing or spillage. Where long drops cannot be avoided, transfer chutes should have interior impact baffles to dissipate the momentum of falling biosolids.
2. *Removal of biosolids from returning belts*: Counterweighted rubber-bladed scrapers at head pulleys are not effective in scraping biosolids off return belts and are a maintenance problem. The use of adjustable tension finger-type scrapers is recommended. To avoid problems with idler roller vibration and irregularities, and to ensure continuous contact, scrapers should be installed beyond the idler on the flattened portion of the belt.
3. *Assuring minimum pulley slippage*: Appurtenances that contact the dirty side of the belt should be avoided. Figure 1.14 illustrates both the undesirable and the recommended design features of inclined belt conveyors. Snubber pulleys and trippers (devices that remove the moving material from the belt) cannot be successfully used for biosolids. Gravity counterweight take-ups should be avoided, and screw take-ups should be used instead. Where long lifts are required, multiple short belts should be used instead of one long belt to avoid the need for gravity take-ups.
4. *Importance of housekeeping facilities*: Notwithstanding the care taken to avoid spillage or splashing, biosolids handling facilities are dirty, and must be designed to facilitate cleanup.

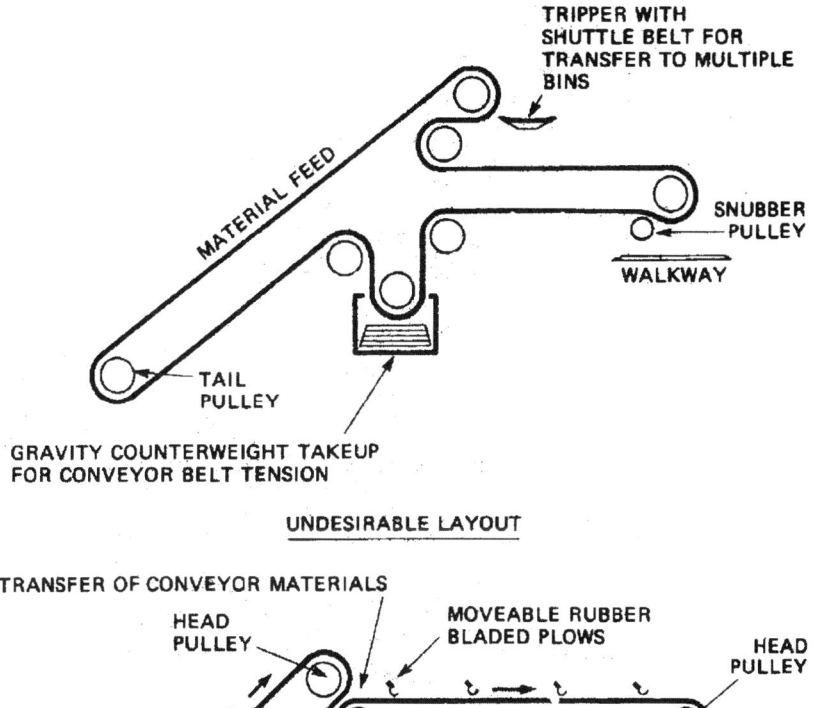

**Fig. 1.14.** Inclined belt conveyer features. *Source*: US EPA (1).

Nonskid cover plates, rather than grating, should be used for all access areas except those immediately over storage hoppers. Convenient hose stations should be located to serve all areas. Floors and slabs should be provided with exaggerated drainage slopes (up to 1 in/ft) and should drain to liberally distributed drain sumps. Special care should be used at all transfer points, take-up pulleys, and dump points to minimize biosolids spillage or splashing, or to provide surroundings that are easily cleaned.

Flexible conveyors are now available in styles with integral pockets, sidewalls, and cleats that allow steep, high capacity operations on almost all materials (Figure 1.15). The belts may change inclination at several points in their run. They are best cleaned by a combination brush and spray cleaner. Except for belt pockets, sidewalls, and cleats,

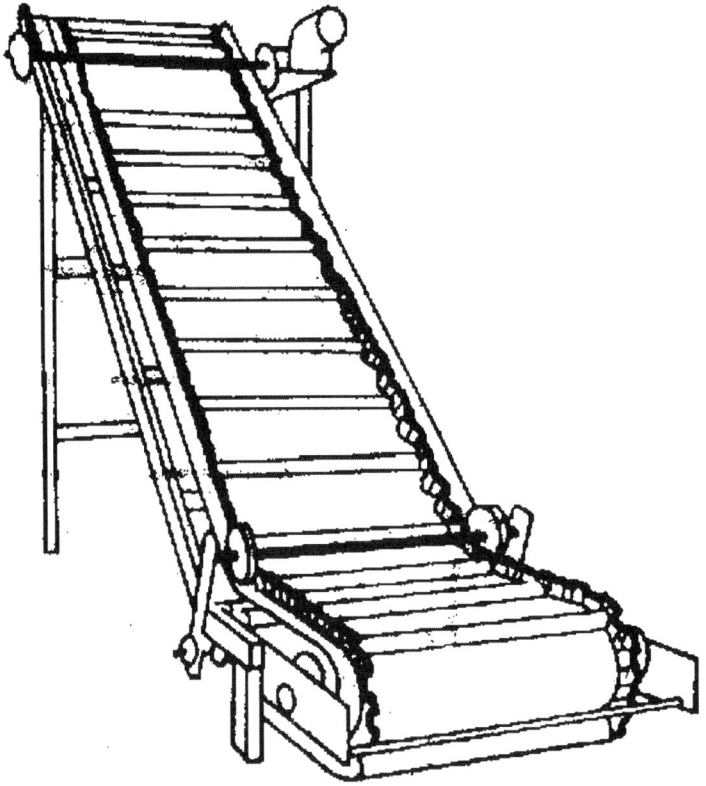

**Fig. 1.15.** Flexibly cleated and side walled flat belt conveyor. *Source*: US EPA (1).

their mechanical components are similar to those on troughed belts; maintenance costs for mechanical drives and rollers are also similar.

There are patented flexible conveyors that not only can change inclination but also can change direction or even spiral vertically upward. One unit may replace several straight-line belts. These units are not actually belts but segmental-chain and sprocket-driven mechanisms with interlocked, pleated rubber trough sections. Drive mechanism wear and corrosion is high in comparison with flat belt conveyors. These conveyors are not recommended where there is sufficient room to allow installation of multiple conventional troughed or pocketed conveyors.

*4.3. Screw Conveyors*

Screw conveyors (Figure 1.16) are silent, reliable, and economical (30). They are used for horizontal movement of grit or biosolids, or may be used to convey dewatered biosolids up inclines. (The degree of incline depends on biosolids moisture content and consistency.) Conservative sizing, abrasion-resistant construction materials, and integral wash-down systems within enclosed housings are recommended for solids handling facilities. All enclosed housings should have numerous quick opening access plates for maintenance and observation. Screw conveyors for dewatered biosolids should not

# Transport and Pumping

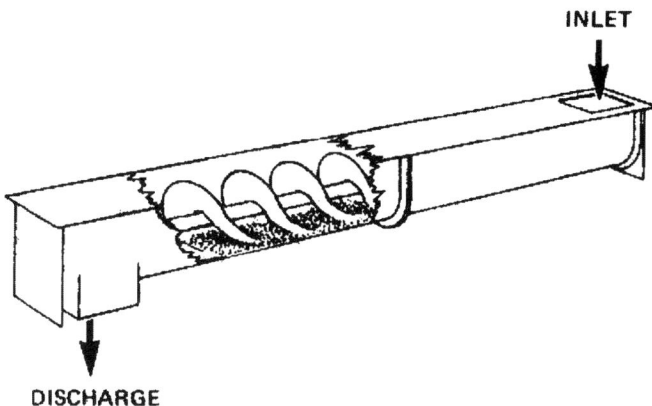

**Fig. 1.16.** Screw conveyor. *Source*: US EPA (1).

have internal intermediate bearings because biosolids can pile up on the bearing and restrict or prevent flow. For this reason, screw conveyor lengths should be limited to 20 ft. Screw conveyors with reversible direction, or with several slide gate controlled discharge openings in the bottom of the conveyor housing, allow the point of conveyor discharge to be changed as appropriate, providing flexibility of operation.

Screw conveyors have been successfully used for transporting grit, but their application to screenings is questionable because rags may become entangled on the conveyor shaft. Oversized objects, such as sticks, can jam the screw or fall out of the conveyor, creating housekeeping problems. To reduce wear, open or ribbon-type screw conveyors are sometimes used for grit.

## 4.4. Positive-Displacement–Type Conveyors

Positive-displacement–type conveyors include tubular conveyors and bucket elevators. Tubular conveyors (Figure 1.17) are tubular conduits through which circular flights are pulled by chains. They may be used for the horizontal transportation of dry solids such as incinerator ash or semi-dry grit. They are several times as expensive as flat belts per linear foot, but require much less room, are fully enclosed and airtight, and can be routed anywhere a conduit will fit. Maintenance is high. Most plants utilizing these conveyors routinely replace the chain elements at least once a month.

Bucket elevators (Figure 1.18) incorporate chain-and-sprocket–driven buckets in a manner similar to the tubular conveyors, except that the chain flights are not in continual contact with the product. As a result, mechanical longevity is greatly increased. They are usually restricted to vertical lifts with limited horizontal displacement.

## 4.5. Pneumatic Conveyors

Pneumatic conveyors are usually not appropriate for dewatered biosolids, but can effectively handle screenings, grit, and dry finely divided materials such as incinerator ash. Screenings and grit can be easily transported, even over long distances, through the use of a batch pneumatic ejector system (Figure 1.19). Such pneumatic ejector systems

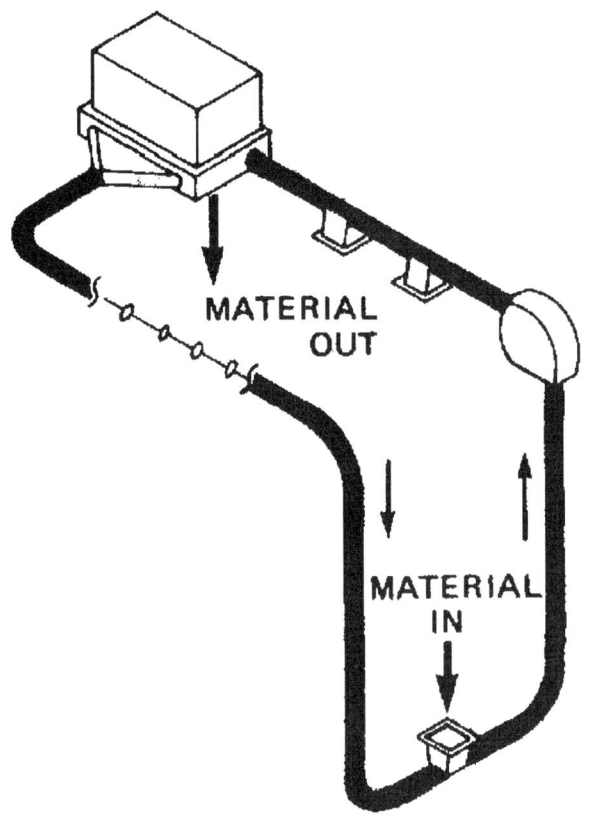

**Fig. 1.17.** Tubular conveyor. *Source*: US EPA (1).

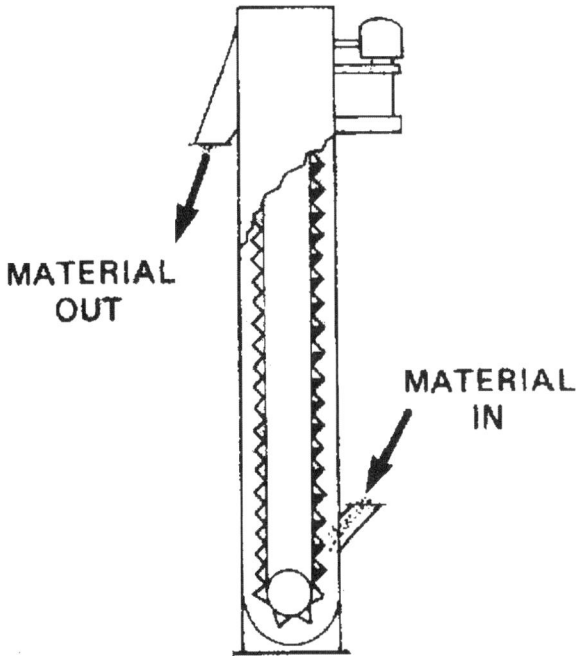

**Fig. 1.18.** Bucket elevator. *Source*: US EPA (1).

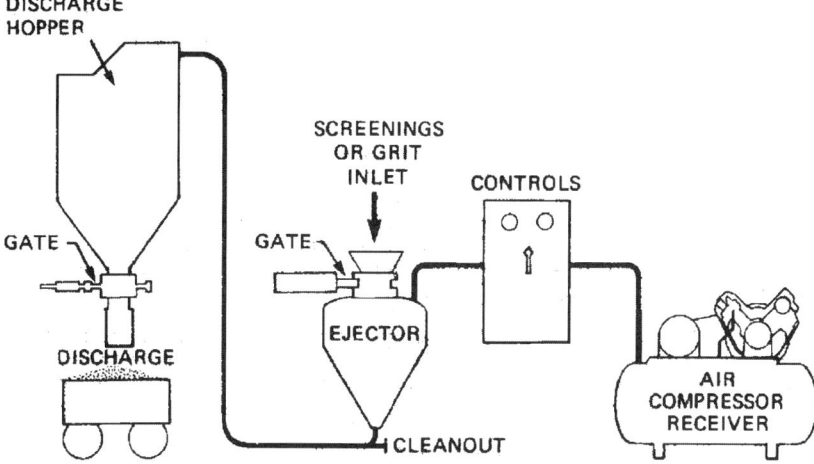

**Fig. 1.19.** Pneumatic ejector. *Source*: US EPA (1).

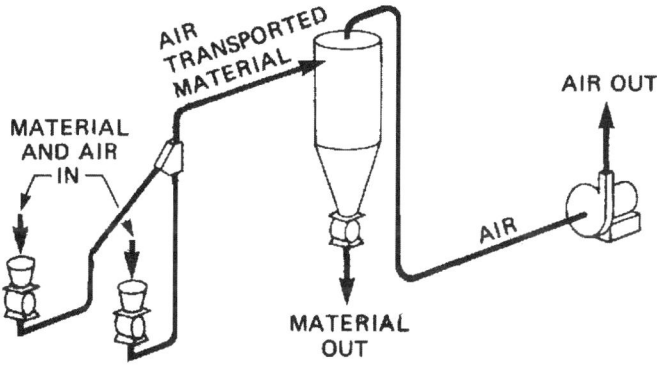

**Fig. 1.20.** Pneumatic conveyor. *Source*: US EPA (1).

have provided good service for distances up to one-half mile and up to 100 ft of lift. The transport system between the points of loading and discharge is a totally enclosed pipe, which is clean and odor-free and can be easily routed along available passages. The entire system utilizes a minimum of moving parts. Consideration must be given to the use of abrasion resistant materials, especially at pipe bends, and an air pressure system consistent with the distance and lift to be traversed.

Continuous pneumatic conveying systems (Figure 1.20), either pressure or vacuum type, are widely used where dry, particulate materials are to be transported. Their use in biosolids transport is limited to materials such as incinerator ash. Where long distances or complex routings are involved, pneumatic conveyor systems are especially well suited to ash transport. Ash is an extremely abrasive material, and rotary valves and elbow segments in particular must be carefully specified to provide maximum abrasion resistance. The blowers may require noise shielding.

## 4.6. Chutes and Inclined Planes

Chutes and inclined planes for sludge/biosolids, screenings, ash, and grit should be tested for minimum inclination on the specific transported product whenever possible. In general, inclinations for dewatered sludge/biosolids should be greater than 60° from the horizontal. For dry bulk materials, such as ash, the inclinations should at least be greater than the material's natural angle of repose.

## 4.7. Odors

Open sludge/biosolids conveyance can be a source of odors. All solids transporting facilities should be well ventilated and, if necessary, provided with odor control for the vented air. Even with stabilized biosolids, if large holding or equalization tanks are required for the pumping system, floating covers or special odor control facilities for venting tank air should be provided when the detention time is greater than several hours.

## 5. LONG-DISTANCE WASTEWATER SOLIDS HAULING

Because biosolids must be hauled from a biosolids production facility to an application site, transportation is an important part of biosolids management. It is often necessary to transport the wastewater solids for long distances, that is, beyond the boundaries of the wastewater treatment plant site. This may be done by pumping if the material is biosolids or scum or by other methods, which shall be termed long-distance hauling. Long-distance hauling includes trucking, rail transport, and barging.

Ettlich (31), in developing cost formulas for transport of wastewater sludge/biosolids, made the following general observations about the comparative economics of the long-distance transportation of dewatered sludge/biosolids methods:

1. Total annual cost for railroad is less than truck for all annual sludge volumes (7500 to 750,000 yd$^3$ [5730 to 573,450 m$^3$]) and distances (20 to 320 miles [32 to 515 km]) studied with and without terminal facilities for loading and unloading sludge to the transport vehicle.
2. Railroad facilities are more capital intensive than truck facilities.
3. Transport equipment can be leased for both truck and railroad transport.

Similarly, Ettlich made the following general observations about the comparative economics of the long-distance transportation of liquid sludge/biosolids methods:

1. Trucking is the least expensive mode for one-way distances of 20 miles (30 km) or less and sludge/biosolids volumes less than 10 to 15 MG (million gallons)/yr (38,000 to 57,000 m$^3$/yr).
2. Pipeline is the least expensive mode for all cases when the annual sludge/biosolids volume is greater than approximately 30 to 70 MG (110,000 to 260,000 m$^3$).
3. Pipeline is not economically attractive for annual sludge/biosolids volumes of 10 MG (38,000 m$^3$) or less because of the high capital investment.
4. Pipeline is capital intensive and the terminal points are not easily changed. Pipeline is ideal for large volumes of sludge/biosolids transported between two fixed points.
5. Rail and barge are comparable over the 7 to 700 MG (30,000 to 2,600,000 m$^3$) volume range for long-haul distances.

6. Barge is more economical than rail for short to medium distances for annual sludge/biosolids volumes greater than 30 MG (110,000 m$^3$).

While much information is available on costs of transporting sludge/biosolids in specific situations (32–34), there is a wide disparity in reported costs since there are so many variables in each situation. Consequently, it is much more accurate to utilize an approach such as Ettlich's than to rely upon cost estimates from other treatment plants where conditions may be quite different (31).

## 5.1. Truck Transportation

For most small plants and some large plants, the use of trucks is the best approach. Trucking provides a viable option for transport of both liquid and dewatered sludge/biosolids. Trucking provides flexibility not found in other modes of transport since terminal points and route can be changed readily at low cost (31). Provided trucks are leased rather than purchased, a truck hauling option is not capital intensive and allows more flexibility than pumping or other transport modes. This flexibility is valuable since locations of reuse or disposal may change.

### 5.1.1. Types of Trucks

Sludge/biosolids hauling trucks are similar to standard highway trucks because both types of trucks must use public roads and comply with their overall vehicle width, height, and gross weight restrictions. Most of the variability can be seen in sludge/biosolids containment body configuration. For the majority of cases, which involve comparatively short distances with one-way travel times less than 1 hour, ease and speed of loading and unloading are of paramount importance. The larger trucks are the most economical except for one-way haul distances less than 10 miles and annual sludge/biosolids volumes less than 3000 yd$^3$ for dewatered sludge/biosolids and for less than 1 MGD/yr for liquid sludge/biosolids. Diesel engine power is preferred because of cheaper fuel and lower maintenance costs (19). Generally, diesel engines are used in the larger trucks and are the economical choice for small trucks that are operated at high annual mileage (31). Depending on state road laws, the type of vehicle would vary, as follows (25, 35):

- Two-axle and three-axle trucks
- Two-axle tractor with one-axle semitrailer
- Two-axle tractor with two-axle semitrailer
- Three-axle tractor with two-axle semitrailer
- Two- or three-axle truck with a two- or three-axle trailer

There are some limitations on the use of trucks for transport of sludge/biosolids (12, 35, 36):

1. State road laws limit the load of vehicles. In Ohio, for example, the maximum payloads would be as follows:

    | Vehicle | Payload, tons |
    | --- | --- |
    | 3-axle tractor, 2-axle semitrailer | 22 |
    | 3-axle truck | 10 |
    | 2-axle truck | 7.5 |

2. Generally, highway and road loadings are in accordance with the American Association of State Highway Officials class standards H10, H15, and H20.
3. Load limits may be restricted by practical on-site road conditions.
4. While truck transport generally has a lower initial investment cost, it will have higher operating cost relative to rail for most levels of design volume.
5. Vehicles carrying sludge/biosolids should be able to reach the site without passing through heavily populated areas or business districts.

Where it is determined that economic, environmental, and institutional considerations allow direct land application of liquid digested sludge/biosolids, special tank trucks are available equipped with specially designed spreaders, auger beaters, or special application apparatus. Some manufacturers equip their trucks with subsoil injectors for subsurface treatment. Use of such dual-purpose trucks allows transport and ultimate disposal without an intermediate storage/pumping step. Specialized tanks or trucking equipment can be custom built for specific applications (37–39). Some companies produce flexible tanks designed to fit on a flatbed truck (40).

Vehicles or containers used for transportation of sludge/biosolids should be loaded in a manner such that the contents will not fall, leak, or spill during transportation (41). Spillage or leakages from sludge/biosolids hauling operations are unacceptable because of aesthetic and health considerations. This has meant a shift away from belly-dump vehicles, even for a very well dewatered sludge/biosolids cake. There is increased concern for covering the top of the sludge/biosolids to minimize both odor release during transit and the chance of spillage due to sudden stops or accidents. Note that any open loads transported to a site must be covered to comply with the U.S. Department of Transportation (DOT) requirements. Consequently, tank-type bodies are gradually becoming the most common, even for mechanically dewatered sludge/biosolids (37–39). These vehicles require unusually large hatch openings for loading purposes, and well-designed watertight hatches or tailgates for unloading. Tanks for liquid sludge/biosolids transport are of more standard design, but the provision of internal baffles to minimize load shifting is recommended for highway transport.

### 5.1.2. Owned Equipment vs. Contract Hauling

The foregoing concerns apply equally whether or not the wastewater treatment management agency contracts out its sludge/biosolids hauling or uses its own vehicles. The choice between utilizing agency personnel and contracting for private companies to drive sludge/biosolids trucks is often decided not on the basis of cost, but on the size of the plant. Smaller plants favor the use of both their own vehicles and staff (42, 43).

The choice of contract hauling can be limited to the provision of tractor units and driver services, with the trailers owned by the agency. This has two major benefits. First, treatment plant staff assigned to sludge/biosolids handling or dewatering operations are working in the immediate vicinity of the trailers, and can therefore re-spot the trailers under a conveyor belt at the best times. Second, with most contracts awarded for only 1- to 3-year terms, the contractor would otherwise need to figure in his bid price a very rapid amortization of custom trailers, which may be of no further use to him if he is not re-awarded the contract at a later date, even though the trailers may have a useful life far

# Transport and Pumping

in excess of the contract period. Since it is economically sensible to operate with more trailers than tractor units, trailer cost depreciation can be a significant overall cost factor.

## 5.1.3. Haul Scheduling

A common problem that is usually not recognized is the need to properly schedule trucking operations. In general, the total cost of truck transport will be decreased (per unit of material hauled) if the daily period of truck operation is increased, because capital-intensive equipment is better utilized. However, restrictions may be placed on any significant truck operations, such as requiring specific routes or limiting operations to daylight hours (31). Such haul scheduling may require the provision of some form of temporary sludge/biosolids storage at the plant. Whenever intermittent operations are possible, however, mechanically dewatered sludge/biosolids is usually loaded directly from a conveyor belt. Using trucks or trailer bodies as temporary storage may not be the most economical method when drivers' work hours, overtime pay, and the cost of re-spotting trailers under a belt are considered.

## 5.1.4. Design Criteria

In designing sludge/biosolids handling facilities, it is desirable to provide several dump points with the capability to quickly shift from one to another. If trailers are used, the ability to fill several units before the tractor unit returns adds flexibility to scheduling and reduces storage requirements. If the receiving vessel for dewatered sludge/biosolids is not self-powered (such as a trailer), consideration should be given to movable dump conveyors in order to allow the load to be distributed uniformly within the vessel. Dewatered sludge/biosolids will mound when loaded from a single point. This may prevent effective utilization of the transport vessel (44–46). Design criteria for truck hauling are as follows (20, 24, 47):

1. Dewatered sludge/biosolids should have a minimum of 12% total solids.
2. Vehicle should be loaded by gravity from a storage tank and gravity unloaded at the site.
3. Loading equipment should be sized to fill the vehicle in 20 minutes maximum.
4. Size of vehicle should be selected so that density and solid contents of sludge/biosolids achieve approximate payload of the vehicle. For a 25% solid content, vehicle sizes of 10 yd$^3$ and 30 yd$^3$ are most cost-effective.
5. Loading tank should be sized to fill at least one vehicle.
6. Highway vehicles offer flexibility of movement to various sites when compared to rail.
7. Transport to the site should be one element of an integrated design for the production and ultimate disposal of the sludge/biosolids.
8. Other important elements of this design are the methods and equipment to be used for unloading, storage, and distributing the sludge/biosolids over the site.

## 5.1.5. Costs of Sludge/biosolids Trucking

When considering sludge/biosolids trucking, it is worthwhile to remember that pumping equipment can handle digested sludge/biosolids at least up to 20% solids concentration, and to note that the layout and design of loading and unloading facilities can contribute markedly to cost savings (Figure 1.21).

The following points were considered in evaluating the costs for trucking of sludge/biosolids (24, 48):

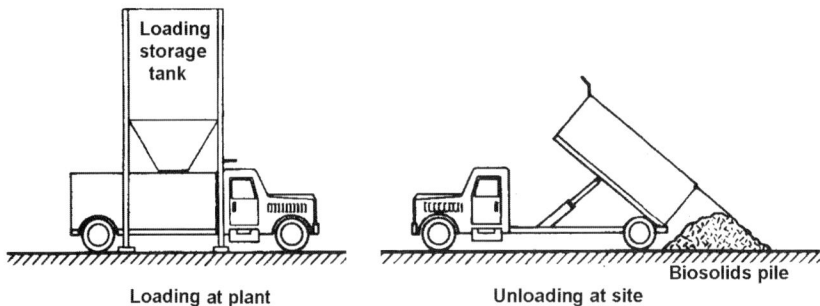

**Fig. 1.21.** Loading and unloading of sludge/biosolids in truck transport. *Source*: US EPA (19).

1. Construction cost includes truck purchase and load/unload facilities. Truck sizes of 10 yd$^3$ and 30 yd$^3$ are the most cost-effective for the volume of 25% total solids sludge/biosolids transported. Storage at loading site is not included.
2. Operation and maintenance cost includes truck maintenance, truck operation, and fuel; labor for truck and loading facility operation; and electric power and supplies for the loading facility.
3. Loading storage tank is sized to fill either a 10- or 30-yd$^3$ truck in 20 minutes maximum.
4. Travel distances are 10, 20, or 40 miles one-way to the site.
5. Operation is for 8 hours a day, 360 days a year.

The energy requirement for hauling 10,000 yd$^3$/yr a one-way hauling distance of 1 mile is approximately $6.2 \times 10$ BTU and $2.8 \times 10$ BTU for 10 yd$^3$ and 30 yd$^3$ trucks, respectively (49).

The construction costs for sludge/biosolids transport by trucking are shown in Figure 1.22, and the operation and maintenance costs are shown in Figure 1.23 (24, 48). The costs (2007 basis) can be calculated by using the U.S. Army Corps of Engineers Civil Works Construction Yearly Average Cost Index for Utilities (25) as shown in the appendix:

$$\text{Cost (2007 basis)} = 1.94 \times \text{Cost from Figs. 1.22 and 1.23}$$

### 5.1.6. Training of Workers

State and local regulations are an important component of sludge/biosolids management. Proactive procedures, training, and plans to prevent transportation incidents should be stressed. A basic knowledge of sludge/biosolids quality and how to interact with the public is essential to designing a spill response plan. Workers need to know the following (50–53):

1. The rules of the road
2. The name of the state agency that licenses truck drivers
3. The name of the state or local agency that designates vehicle weight, height, and width restrictions for highways
4. The name of the state or local agency that administers seasonal weight restrictions (e.g., restrictions that apply to unpaved roads during spring thaw in northern states)
5. The characteristics of acceptable/unacceptable routes for hauling biosolids from treatment plant to the application site

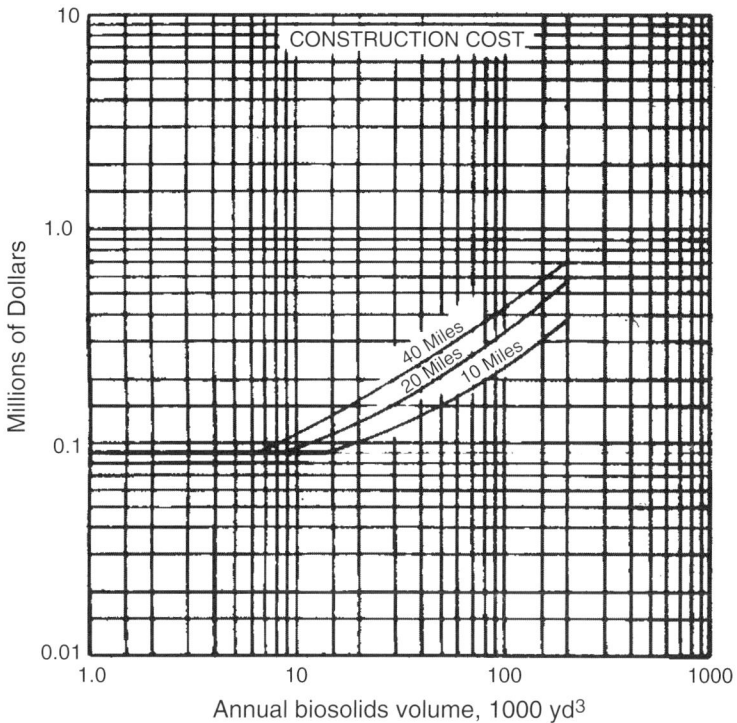

**Fig. 1.22.** Construction cost for truck transport of dewatered sludge/biosolids. *Note*: Cost (2007 basis) = $1.94 × Cost from figure. *Source*: US EPA (19).

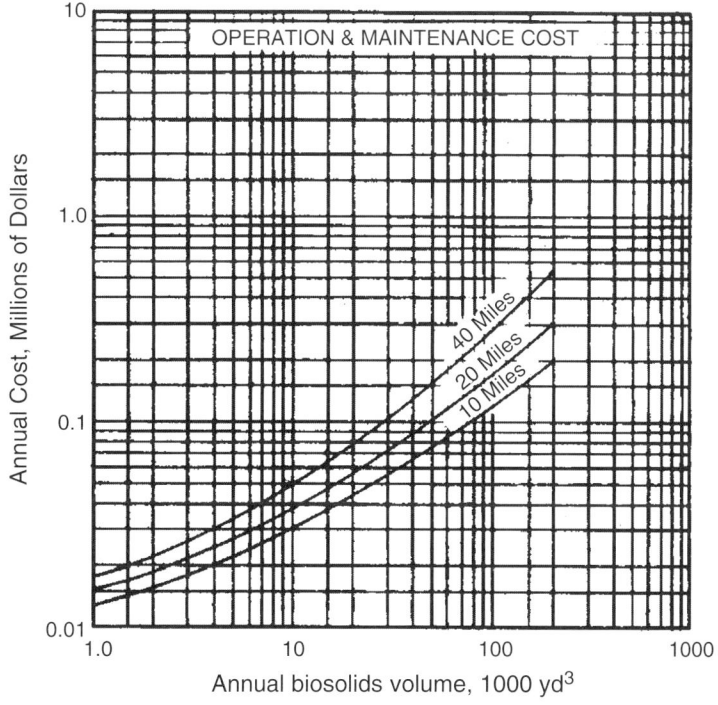

**Fig. 1.23.** Operation and maintenance cost for truck transport of dewatered sludge/biosolids. *Note*: Cost (2007 basis) = $1.94 × Cost from figure. *Source*: US EPA (19).

Workers should learn the rules of the road that govern biosolids hauling to a land application site. Truck drivers should be trained and get acquainted with the following:

1. Appropriate cleanup procedures to be performed on biosolids application equipment before it leaves the application site for the public roads
2. Alternatives to reduce dust from truck traffic on rural roadways
3. Improper trucking practices
4. A list of what should be included in a trucking contractor's standard operating plan
5. Specifications for contracts with outside trucking companies (vendors)

Workers should be trained to know what actions to take to prepare for a biosolids spill. Workers need to know what is to be done in response to A biosolids spill:

1. Identify what items must be included in a spill response plan, and where written copies of the plan should be kept.
2. Know when regulatory agency, law enforcement, or transportation department personnel must be notified of a biosolids spill.
3. Know the appropriate response actions for a spill of a large or small quantity of biosolids.
4. Know the written information the driver should provide to the first responder in the case of a large spill.
5. Know that it is important that spill responders receive written information on biosolids stating that biosolids are not a hazardous waste.
6. Know the trucking contractor's standard operating plan.
7. A one-page information sheet should be kept in all vehicles outlining spill response actions.

### 5.2. Rail Transportation

Rail transportation is suitable for biosolids of any concentration. It is, however, not a common method of transporting biosolids in the United States (50).

#### 5.2.1. Advantages and Disadvantages

Rail transportation has a lower energy cost per unit volume of sludge/biosolids than pipelining and truck hauling, and once found to be feasible has a right-of-way already established, which is not the case with a pipeline. Rail transportation can suffer from many of the same problems as pipelines, such as large unrecoverable capital expenditures and fixed terminal points. In addition, it has some of the same problems associated with trucking, such as an ongoing administrative burden, vulnerability to labor disputes and strikes, risk of spills, and, because of the labor requirements, an operational cost that will rise continually (19). However, special circumstances may favor rail hauling. For example, if sludge/biosolids are to be used to rehabilitate strip-mined lands, a rail line may have been built for hauling out the coal. That line would still be available for the transport of sludge/biosolids.

#### 5.2.2. Routes

The construction of a new railroad line may not be cost-effective or even possible for the sole purpose of transporting wastewater sludge/biosolids. New construction is normally limited to a short spur from a mainline railroad or the provision or expansion of small switching yards on a large treatment plant site in conjunction with chemical delivery facilities. Any attempt at longer new lines is impractical. This limits the overall route selection, generally between the treatment plant site and the final sludge disposal point, to railroad lines already in existence. In turn, this limits either the selection of rail for sludge transport or severely limits the choice of or subsequent change in disposal site location.

*5.2.3. Haul Contracts*

Railroad cars must be hauled by a railroad company, except possibly for switching. Therefore, a contract must be obtained with the railroad. Since this contract hauling is a major cost element, and since the railroad often cannot provide rapid and realistic cost estimates, some time and consideration will be required.

Railroads are a regulated utility; this complicates the rate quotation process. Rates are of two general types: a class rate and a special commodity rate. The class rates are readily obtained, but are usually prohibitively expensive for sludge/biosolids. To obtain a special commodity rate, the following procedure is necessary (1):

1. An application is submitted to the railroad, including a complete description of what is to be shipped; how it is to be shipped (type of material, liquid or solid); precisely where it is to be shipped; the frequency of shipping (how much per day, per week); the approximate loading and unloading time; what other types of materials are similar in form, concentration, and makeup to the material being shipped; and a statement of the price the shipper would be willing to pay in cents/100 lb net weight (45.4 kg).
2. The local railroad—the carrier—reviews the load, distance, terrain, switching requirements, and competition, and calculates a rate.
3. The rate is published by the local freight bureau for a notice period of 30 days for review by other, possibly competing, carriers, and by one of the five regional freight bureaus: western, southwestern, central, southern, or northeastern. The regional freight bureaus are conglomerations of the local ones, and they regulate and control prices between bureau jurisdictions.
4. Comments and appeals of rates can be made to the Interstate Commerce Commission (ICC). An appeal of a proposed rate causes that rate to be suspended for a 7-month period for the case to be heard by the suspension board of ICC and for the carrier to justify that rate. Historically, appeals have caused proposed rates to be eliminated from the carriers' tariffs. This effectively eliminates the option of rail transport of sludge/biosolids for this locality.

Generally speaking, railroads are interested in providing sludge/biosolids transportation. However, many railroads are unfamiliar with sludge/biosolids hauling; similarly, many environmental engineers are unfamiliar with railroad procedures (54).

*5.2.4. Railcar Supply*

There are three methods of ensuring railcar equipment adequacy: by leasing, by outright purchase, or through placement of the required number of cars in "assigned service" by the carrier under the terms of the haul contract (1). Generally, an assigned service option is only available for a solid (dry) or semisolid (mechanically dewatered) sludge/biosolids that can be transported in hopper cars. Liquid sludge/biosolids must be carried in tank cars, which are not normally available "free" from the railroad. As a generalization, the amortization of the purchase of either type of car at approximately USD 200,000 to USD 300,000 (2007 basis) will be at considerably higher cost than the rental or lease fee (1). Consequently, it is expected that the assigned service option would be selected for hopper cars, and a lease arrangement negotiated with a private tank car rental company for tank cars.

Railroad hopper car use is subject to minimum shipment fees per car and certain demurrage criteria. For example, a single hopper car minimum shipment is 180,000 lb (82,000 kg) and demurrage criteria are that the car must be loaded within 48 hours and unloaded within 24 hours. Reference time is 7 a.m. If a car is delivered between midnight and 8 a.m., the time begins at 7 a.m. the same day. If a car is delivered between 8 a.m. and midnight, the time begins at 7 a.m. the following day. Typical hopper car capacities are 2600, 3215, and 4000 ft$^3$ (75, 91, and 113 m$^3$), with the smallest size being typically the most readily available (1).

Tank cars are normally rented by the month from private tank car rental companies with a minimum 5-year commitment. A large noninsulated coiled car (coiled to prevent freezing during the winter months) will rent for approximately USD 1000 per month (2007 prices). Tank car capacities are typically 10,000 to 20,000 gallons (37,850 to 75,700 L). The selection of rail transport, with its high transit times, for more putrescible sludge/biosolids without special gas venting and control equipment, should be avoided. Typical minimum tank and hopper car requirements are shown in Table 1.7.

The exact calculation of car requirements is very site- and area-specific and should be checked directly for any given situation. It should be recognized that the speed of railroad transport will depend in part on the track conditions and on the railroad's normal traffic schedule; the track conditions may also limit the loads carried per car, and hence the size and number of cars required. As a guide only, typical transit times are shown in Table 1.8.

### 5.2.5. Ancillary Facilities

Railroad transport of sludge/biosolids requires loading storage and equipment (tanks, pumps, and piping for liquid sludge/biosolids and hoppers and conveyors for dewatered sludge/biosolids), railroad sidings, and unloading equipment. Unloading is ordinarily accomplished by gravity. Car maintenance and storage will be undertaken by the owner of the cars—not normally the wastewater treatment authority—but car cleaning and wash-down facilities may be required.

### 5.2.6. Design Criteria

The fixed position of a railroad limits disposal site locations. Generally, a minimum of 40 miles one way for each trip and a total load of 1000 ton/day are needed to compete with truck transportation. Reduced costs are possible if cars are owned by the shipper, which is justifiable for larger plants.

Transport to the site should be one element of an integrated design for the production and ultimate disposal of the sludge/biosolids. Other important elements of this design are the methods and equipment to be used for unloading, storing, and distributing the

Transport and Pumping

**Table 1.7**
**Minimum tank car requirements**

| Approximate secondary treatment plant size, MGD | Annual biosolids volume, MG | One-way distance, mile | Car loads[a] | | Cars required[b] |
| --- | --- | --- | --- | --- | --- |
| | | | Per year | Per day | |
| 5 | 7.5 | 20 | 375 | 1 | 5 |
| | | 40 | 375 | 1 | 5 |
| | | 80 | 375 | 1 | 7 |
| | | 160 | 375 | 1 | 8 |
| | | 320 | 375 | 1 | 9 |
| 10 | 15 | 20 | 750 | 2 | 9 |
| | | 40 | 750 | 2 | 9 |
| | | 80 | 750 | 2 | 13 |
| | | 160 | 750 | 2 | 15 |
| | | 320 | 750 | 2 | 17 |
| 50 | 75 | 20 | 3750 | 10 | 47 |
| | | 40 | 3750 | 10 | 47 |
| | | 80 | 3750 | 10 | 68 |
| | | 160 | 3750 | 10 | 78 |
| | | 320 | 3750 | 10 | 89 |
| 100 | 150 | 20 | 7500 | 21 | 97 |
| | | 40 | 7500 | 21 | 97 |
| | | 80 | 7500 | 21 | 139 |
| | | 160 | 7500 | 21 | 160 |
| | | 320 | 7500 | 21 | 181 |

[a]Car size 20,000 gal (76 m$^3$).
[b]Estimate assumes that ample storage is available so that extra cars are not required for peak biosolids production or scheduling problems.
*Source*: US EPA (1).

**Table 1.8**
**Transit times for railroad transportation**

| One-way distance, mile | Round-trip transit time, day,[a] |
| --- | --- |
| 20 | 4 |
| 40 | 4 |
| 80 | 6 |
| 160 | 7 |
| 320 | 8 |

[a]For estimating rail car demand, an allowance of 25% to 50% should be added to accommodate scheduling and car holdup problems. Also, the transit time does not include time for loading and unloading, which must be estimated separately.
*Source*: US EPA (1).

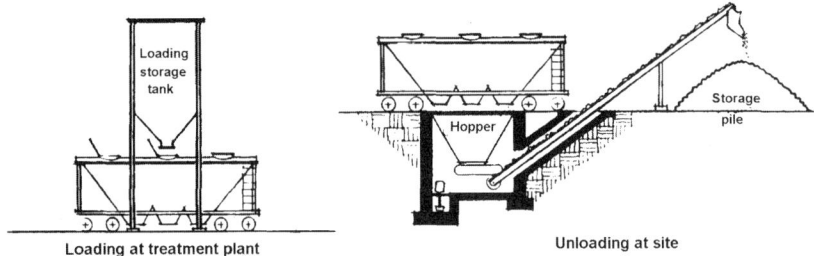

**Fig. 1.24.** Loading and unloading of sludge/biosolids in rail transport. *Source*: US EPA (19).

biosolids over the site. By virtue of its existing right of way, the railroad in many instances can provide the opportunity to use marginal or poor land of the type that can be reclaimed in some way by application of biosolids. The following design criteria for truck hauling should be taken into consideration (24, 36, 50):

1. The rail line should be near or next to the site to reduce the length of the spur or siding.
2. Sludge/biosolids should have a density to achieve the approximate payload of the rail car.
3. Cars should be covered to avoid the impact of odors.
4. Cars should be gravity loaded from a storage tank above the car at treatment plant (Figure 1.24).
5. Cars should be gravity unloaded into a hopper below the car at the site (Figure 1.24).
6. Movement of up to 74,000 yd$^3$ of sludge/biosolids should be done with 50-yd$^3$ (approx. 50 ton) cars.
7. Movement of greater than 74,000 yd$^3$ of sludge/biosolids should be done with 100 yd$^3$ (approx. 100 ton) cars.

### 5.2.7. Manpower, Energy Requirements, and Costs

The wastewater authority will have labor requirements for loading and unloading railroad cars and for associated maintenance; estimates are given in Table 1.9. Data on energy demands associated with railroad transportation are not readily available, but energy demands are relatively low compared with other transportation modes. The fuel consumed in transporting the sludge/biosolids nevertheless should be estimated for inclusion in the sludge/biosolids management program's energy effectiveness analysis. Rail transportation can be considered to require approximately 25% of the energy in BTU/ton mile when compared to truck transportation (51).

The following points were considered in evaluating the costs for rail transportation of sludge/biosolids (24, 48, 51, 55):

1. Construction cost includes construction of loading facilities; the loading storage tank is sized for one carload (cars are gravity loaded).
2. The railroad provides hopper cars.
3. Construction cost does not include any construction work and equipment at site.
4. Operation and maintenance costs include rail haul charges, labor, electric power, and supplies for the loading facilities; 50 yd$^3$ cars for 0 to 74,000 yd$^3$ of sludge/biosolids; 100 yd$^3$ cars for greater than 74,000 yd$^3$ of sludge/biosolids.
5. Rail haul charges are based on travel distances of 40, 80, or 160 miles one way in the central and north central areas of the country. Adjustments for other areas of the country are as follows:

*Transport and Pumping* 47

**Table 1.9**
**Manpower requirements for railroad transport**

| | Liquid biosolids | | | Dewatered biosolids | |
|---|---|---|---|---|---|
| Annual volume, MG | Labor, man hours/yr | | Annual volume, thousand yd$^3$ | Labor, man hours/yr | |
| | Operation | Maintenance | | Operation | Maintenance |
| 7.5 | 4124 | 130 | 7.5 | 1650 | 130 |
| 15 | 4124 | 260 | 15 | 3300 | 260 |
| 150 | 10,500 | 500 | 150 | 4125 | 500 |
| 750 | 28,500 | 1200 | 750 | 10,000 | 1200 |

*Source*: US EPA (1).

| Area | Approximate RR rate variation |
|---|---|
| Northeast | 25% higher than average |
| Southeast | 25% lower than average |
| Southwest | 10% lower than average |
| West Coast | 10% higher than average |

6. Costs are based on 8 hours of operation a day.
7. Unloading costs not included.

The construction costs for sludge/biosolids transportation by rail are shown in Figure 1.25 and the operation and maintenance costs are shown in Figure 1.26 (19). The costs (2007 basis) can be calculated by using the U.S. Army Corps of Engineers Civil Works Construction Yearly Average Cost Index for Utilities (25) as shown in the appendix:

$$\text{Cost (2007 basis)} = 1.94 \times \text{Cost from Figs. 1.22 and 1.23}$$

### 5.3. Barge Transportation

Barge transportation for the ocean dumping of sludge/biosolids has been practiced for many decades around the world. Recent decisions to limit ocean dumping combined with rapidly escalating costs for dewatering or drying sludge/biosolids have led to more consideration of barge transport of liquid sludge/biosolids between the wastewater treatment plant or plants and land disposal sites many miles distant. Barge transportation of sludge/biosolids is generally only feasible for liquid sludge/biosolids (to the solids concentration limit at which it may be pumped) and over longer distances, generally over 30 miles. Additional information is available elsewhere (31, 34, 56).

#### 5.3.1. Routes and Transit Times

It is evident that the key feature in consideration of barge transportation is the proximity to a suitable waterway. However, in planning a barge transport system, the transit time also plays a critical role. The traffic on the waterway, physical features such as drawbridges, locks, and height limitations, and natural characteristics such as currents,

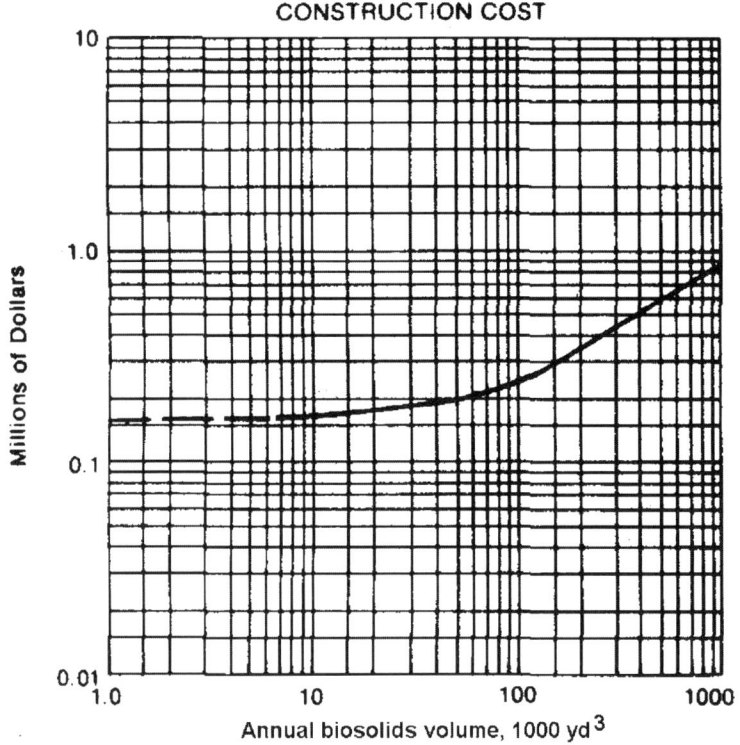

**Fig. 1.25.** Construction cost for rail transport of dewatered sludge/biosolids. *Note*: Cost (2007 basis) = $1.94 × Cost from figure. *Source*: US EPA (19).

tides, and even wave heights will all affect the transit time. Local operators familiar with the waterway should be contacted for information, and a conservative safety factor should be applied. Loading and unloading times then must be added to estimate the overall turnaround time—the key feature when contracting for towing service. Towing speeds and cost estimates are given in Table 1.10.

*5.3.2. Haul or System Contracting*

Only for very large plants should ownership of the motive power unit(s) (tug or powered barge) be considered. Self-propelled barges are no longer generally considered cost-effective when initiating a new system, although the specifics of any particular case could modify this conclusion. This means the choice for most wastewater treatment authorities narrows down to contracting for either complete barge transport services or for tug service alone. Full-service contracts may prove the best for small operations with intermittent transport requirements. Moderate to large plants will generally favor contract towing only, with the barge(s) owned by the authority. Contractual agreements should clearly define in detail all services to be provided and include a barging schedule. In certain cases it may be possible for two or more wastewater treatment authorities to

Transport and Pumping

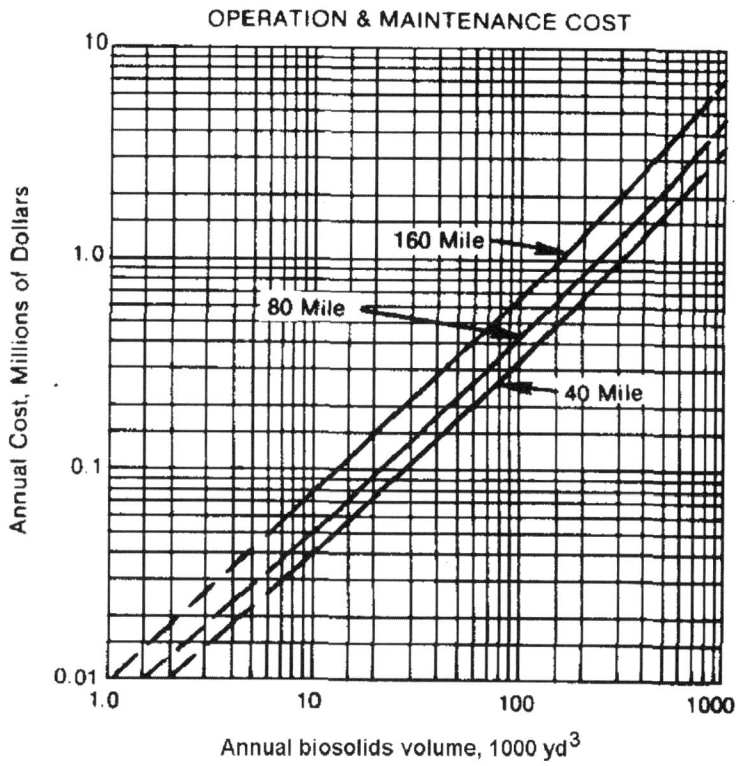

**Fig. 1.26.** Operation and maintenance cost for rail transport of dewatered sludge/biosolids. *Note*: Cost (2007 basis) = $1.94 × Cost from figure. *Source*: US EPA (19).

join in a common contractual agreement whereby sludge/biosolids from two or more plants are picked up in tandem by the one haul contractor.

*5.3.3. Barge Selection and Acquisition*

Both the useful life and salvage value of barges tend to be high. This often leads to a decision to purchase rather than lease equipment. The size and number of barges depend on plant size and the specific sludge/biosolids processing system.

Some data on typical barge sizes and costs are given in Table 1.11. Physical dimensions of barges are not standardized, since they are usually custom built within certain dimensions set by some waterway constriction, such as lockage limitations. Lead times on construction are about 2 years. Barge proportions are commonly a length to breadth 4 or 5 to 1, and a breadth to depth 3 or 4 to 1. For inland waterways, about 2 ft (0.6 m) of freeboard under the maximum loaded condition is usually adequate. Barges are very common in the 20,000 to 25,000 barrel (3200 to 4000 m$^3$) capacity range. Construction costs in 2007 prices were about USD 16/ft$^3$ ($565/m$^3$) for a 25,000 barrel (4000 m$^3$) barge, with only a slight reduction in unit costs as size increases, to about USD 15/ft$^3$ ($517/m$^3$) at the 100,000 barrel (16,000 m$^3$) size. Greater flexibility in operations usually dictates the choice

**Table 1.10**
Tug costs for various barge capacity

| Barge Capacity, barrels | Average velocity, knots | | Tug costs, 2007 (USD/hour) |
|---|---|---|---|
| | Loaded | Unloaded | |
| 25,000 | 6 | 7 | 320 |
| 50,000 | 7 | 8 | 400 |
| 100,000 | 8 | 10 | 520 |

*Source*: US EPA (1).

**Table 1.11**
Typical barge sizes and costs[a]

| Capacity,[b] barrels | Dimensions, ft | | | | Cost, 2007 (USD) | |
|---|---|---|---|---|---|---|
| | Length | Breadth | Depth | Draft | New | Used |
| 14,000 | — | — | — | — | — | 600,000 |
| 20,000 | 240 | 52 | 15 | 13.5 | 2,930,000 | — |
| 23,000 | 240 | 60 | 13.5 | — | — | — |
| 27,000 | — | — | — | — | — | 1,730,000 |
| 33,000 | — | — | — | — | — | 1,660,000 |
| 35,000 | 286 | 62 | 18 | 16 | 4,660,000 | — |
| 50,000 | 320 | 70 | 20 | 18 | 6,120,000 | — |

[a]Examples are for barges custom built for liquid sludges but do not include pumps necessary for unloading.
[b]One barrel equals 42 gal (159 L).
*Source*: US EPA (1).

of smaller barges, unless distances are about 200 miles (330 km) or more and the number of waterway restrictions low. Then the increased speed offered by a larger tug/barge combination will substantially cut transit time and thus reduce towing fees (1).

*5.3.4. Ancillary Facilities*

A critical factor in determining the feasibility of barging sludge/biosolids lies in the cost of facilities for loading and offloading, and receiving the sludge/biosolids. If the treatment plant is not close to the waterway, it may be desirable to locate sludge/biosolids storage tank or lagoon near the barge loading dock. For a tank, design would need to be similar to an unheated digester because of continued anaerobic decomposition. Lagoons should be operated as facultative sludge/biosolids lagoons. In either case, the costs of the tank or lagoon should be included in the barge system costs.

In most cases, it is desirable to load and meter the flow from a fixed pumping station located on a fixed wharf. Offloading is often accomplished by a pump mounted on the barge itself. The disposal site should be located near a dock capable of mooring a suitably sized barge. Floating docks are usually more expensive in both marine and freshwater environments than fixed wharfs, due to the complexity of anchoring devices capable of sustaining the loads imposed by a large barge. In certain instances, however, a floating dock may be more acceptable from an environmental standpoint. Unloading to a land pipeline typically takes about 6 hours. If a tug must remain with the barge during the unloading period, rapid unloading becomes economically important.

### 5.3.5. Spill Prevention and Cleanup

One important element of a barge transportation system is a well-developed spill prevention and cleanup program. Spills resulting from accidents during transport can result in serious water pollution and associated health problems. Sludge/biosolids spills should be contained immediately and transferred to storage tanks or another barge as quickly as possible to reduce risks. The risk of spills during loading and unloading can be minimized by careful attention to design and operator training.

## 5.4. Design of Sludge/Biosolids Hauling

### 5.4.1. Background

Sludge/biosolids hauling and landfilling may be approached in a manner similar to that for a typical solid waste disposal problem. Most solid waste disposal systems have at least four definable components: storage, collection, haul, and disposal.

Storage costs are site-specific and depend largely on the method selected in the sludge/biosolids handling train. They may be simply the costs associated with the purchase of bins for storage of waste activated sludge/biosolids or primary sludge, and a dump truck for storage of digested sludge/biosolids that have been centrifuged or vacuum filtered.

Collection costs are dependent on a time-labor relationship to transfer the sludge/biosolids from storage to the transporting vehicle, as a dump or tank truck. There may not be a collection cost associated with labor; however, a cost would be incurred to provide a vehicle during the loading period. Larger facilities may require that a driver be assigned to the vehicle during loading periods. Collection costs may be significant when it is necessary to shovel sludge/biosolids from drying beds into trucks for transportation to the landfill.

Transportation costs are associated with such parameters as truck cost, truck size, haul time, labor, and operating costs per unit time for items such as depreciation, fuel, insurance, maintenance, etc. Operating costs may be estimated from the manufacturer's rating information and used in conjunction with estimates of sludge/biosolids production from various wastewater treatment processes.

The lowest possible moisture content attainable at a reasonable cost should be produced for economical sludge/biosolids hauling and landfill operations. A reduction of moisture content will produce savings in storage, initial equipment, operating, and labor costs.

**Table 1.12**
**Normal quantities of sludge/biosolids produced by different treatment processes**

| Wastewater Treatment Process | gal biosolids/MG Treated | Solids % | biosolids Specific Gravity |
|---|---|---|---|
| Primary sedimentation | | | |
|   Undigested | 2950 | 5.0 | 1.02 |
|   Digested in separate tanks | 1450 | 6.0 | 1.03 |
| Trickling filter | 745 | 7.5 | 1.025 |
| Chemical precipitation | 5120 | 7.5 | 1.03 |
| Primary sedimentation and activated sludge | | | |
|   Undigested | 6900 | 4.0 | 1.02 |
|   Digested in separate tanks | 2700 | 6.0 | 1.03 |
| Activated sludge | | | |
|   Waste sludge | 19,400 | 1.5 | 1.005 |
| Septic tanks, digested | 900 | 10.0 | 1.04 |
| Imhoff tanks, digested | 500 | 15.0 | 1.04 |

*Source*: US EPA (57).

**Table 1.13**
**Process efficiencies for dewatering of wastewater sludge/biosolids**

| Unit process | Solids capture, % | Cake Solids, % |
|---|---|---|
| Centrifugation | | |
|   Solid bowl | 80–90 | 5–13 |
|   Disk-nozzle | 80–97 | 5–7 |
|   Basket | 70–90 | 9–10 |
| Dissolved air flotation | 95 | 4–6 |
| Drying beds | 85–99 | 8–25 |
| Filter press | 99 | 40–60 |
| Gravity thickener | 90–95 | 5–12 |
| Vacuum filter | 90+ | 28–35 |

*Source*: US EPA (57).

The normal quantities of sludge/biosolids produced by different treatment processes are given in Table 1.12, while Table 1.13 shows the process efficiencies for dewatering of the wastewater sludge/biosolids (57, 58).

*5.4.2. Input Data*

Input data include the following:

1. Average wastewater flow, MGD
2. Sludge volume, gal/MG
3. Raw sludge solids concentration, percent

# Transport and Pumping

4. Dewatered sludge concentration, percent
5. Vehicle loading time, hours
6. Round-trip haul time, hours
7. Vehicle capacity, yd$^3$
8. Solids capture in dewatering process, percent
9. Distance to disposal site, miles

### 5.4.3. Design Parameters

Design parameters include the following:

1. Sludge volume/MG treated (Table 1.12)
2. Raw sludge concentration, percent (1.5–15%) (Table 1.12)
3. Concentrated solids, percent (4–60%) (Table 1.13)
4. Vehicle capacity, yd$^3$/truck
5. Truck loading time, hours (0.5–2.0 hours)
6. Haul time, hours (local conditions)
7. Daily work schedule, hours (6–8 hours)
8. Solids capture, percent (70–99%) (Table 1.13)

### 5.4.4. Design Procedure

The design procedure is as follows:

1. Compute the sludge volume hauled, yd$^3$/d (57–60):

$$BV = \frac{(Q_{avg})(BF)(SS)}{(7.48)(27)(CSS)} \quad (3)$$

where
$BV$ = biosolids/sludge volume hauled, yd$^3$/d
$Q_{avg}$ = average wastewater flow, MGD
$BF$ = biosolids/sludge flow, gal/MG (Table 1.12)
$SS$ = suspended solids concentration in sludge/biosolids flow, percent (Table 1.12)
$CSS$ = concentrated suspended solids from final treatment process, percent (Table 1.13)

2. Calculate the number of vehicles for collection and hauling of the biosolids/sludge:

$$NTR = \frac{(BV)(LT + HT)}{(HPD)(CAP)} \quad (4)$$

where
$NTR$ = number of trucks required
$LT$ = loading time, hours (0.5–2.0 hours)
$HT$ = round-trip haul time, hours (local conditions)
$HPD$ = work schedule, hours/day (6–8 hours)
$CAP$ = vehicle capacity, yd$^3$/truck (3–12 yd$^3$)

3. Compute the tons of biosolids hauled per day

$$TBH = \frac{(Q_{avg})(BF)(SS)(SCAP)(8.34)}{(100)(CSS)(2000)} \quad (5)$$

where
$TBH$ = tons of biosolids/sludge hauled per day (ton/d)
$SCAP$ = solids capture, percent (Table 1.13)

## 5.4.5. Output Data

The output data are as follows:

1. Volume of biosolids/sludge to be dewatered, gallons per day (gpd)
2. Initial moisture content, percent
3. Final moisture content, percent
4. Volume of biosolids/sludge hauled, yd$^3$/day
5. Truck capacity, yd$^3$
6. Time to make one load, hours
7. Work schedule, hours/day
8. Number of trucks required
9. Tons of biosolids/sludge hauled per day (ton/d)
10. Distance to disposal site, miles

## 5.5. Example

Given:

1. Wastewater treatment plant $Q_{avg} = 1.0$ MGD
2. Sludge flow $= 2700$ gal/MG
3. Suspended solids in sludge flow $= 6\%$
4. Solids capture $= 95\%$
5. Concentrated suspended solids from final treatment process $= 12\%$
6. Truck capacity $= 4$ yd$^3$/truck
7. Truck loading time $= 1.5$ hours
8. Round-trip haul time $= 2.5$ hours
9. Work schedule $= 8$ hours

Compute:

1. Sludge volume hauled
2. Number of vehicles required
3. Tons of sludge hauled per day

   1. Compute the sludge volume hauled from Eq. 3:

   $$BV = \frac{(Q_{avg})(BF)(SS)}{(7.48)(27)(CSS)} \quad (3)$$

   where
   $BV =$ sludge volume hauled, yd$^3$/d
   $Q_{avg} =$ average daily flow, 1.0 MGD
   $BF =$ sludge flow, 2700 gal/MG
   $SS =$ suspended solids in sludge flow, 6%
   $CSS =$ concentrated suspended solids from final treatment process, 12%

   $$BV = \frac{(1.0)(2700)(6)}{(7.48)(27)(12)}$$

   $$BV = 6.7 \text{ yd}^3/\text{d}$$

   2. Calculate number of vehicles required from Eq. 4:

   $$NTR = \frac{(BV)(LT+HT)}{(HPD)(CAP)} \quad (4)$$

where
$NTR$ = number of trucks required
$BV$ = sludge volume hauled, 6.7 yd³/d
$LT$ = loading time, 1.5 hours
$HT$ = round-trip haul time, 2.5 hours
$HPD$ = work schedule, 8 hours
$CAP$ = vehicle capacity, 4 yd³/truck

$$\text{NTR} = \frac{(6.7)(1.5 + 2.5)}{(8)(4)}$$

$$\text{NTR} = 1 \text{ truck at 2 trips/d}$$

3. Compute tons of sludge hauled per day from Eq. 5:

$$\text{TBH} = \frac{(Q_{avg})(BF)(SS)(SCAP)(8.34)}{(100)(CSS)(2000)} \tag{5}$$

where
$TBH$ = tons of sludge hauled per day, ton/d
$Q_{avg}$ = average flow, 1.0 MGD
$BF$ = sludge flow, 2700 gal/MG
$SS$ = suspended solids in sludge flow, 6%
$SCAP$ = solids capture, 95%
$CSS$ = cake suspended solids, 12%

$$\text{TBH} = \frac{(1.0)(2700)(6)(95)(8.34)}{(100)(12)(2000)}$$

$$\text{TBH} = 5.3 \, \text{t/d}$$

## 6. POTENTIAL RISK TO BIOSOLIDS EXPOSURE

This section discusses controlling health risks to workers from Class B biosolids during handling and land application. One purpose of biosolids treatment is to eliminate disease-causing organisms, the pathogens. Treatment also reduces the attractiveness of the residues to insects, birds, and rodents. The product, Class A biosolids, is a material that can be recycled for uses with no restrictions.

Class B biosolids have undergone treatment that has reduced but not eliminated pathogens. Class B biosolids may contain pathogens. As a result, federal regulations for use of Class B biosolids require additional measures to restrict public access and to limit livestock grazing for specified time periods after land application. This allows time for the natural die-off of pathogens in the soil. Whereas US EPA rules (Title 40 of the Code of Federal Regulations [CFR], Part 503) restrict public access to lands treated with Class B biosolids in order to protect public health, these rules do not apply to workers involved with Class B biosolids handling and land application.

Workers may come into either direct or indirect contact with biosolids during any phase of the treatment, transport, or application process, or after they are land applied. Workers and employers may be well aware of the need for precautions when contacting untreated wastewater but less aware of the need for basic precautions when using Class B biosolids. This section provides information, guidance, and recommendations to employers and employees working with Class B biosolids to minimize occupational risks from pathogens (61). It does not address other potential safety and health issues such as injuries or exposures to chemicals.

## 6.1. Biosolids Constituents that Require Control of Worker Exposure

There are four major types of human disease-causing organisms (pathogens) that can be found in sewage: (1) bacteria, (2) viruses, (3) protozoa, and (4) helminths (parasitic worms). Class B biosolids may contain the same types of pathogens as the source sewage, but at reduced concentrations. Both Class A and Class B biosolids may also contain chemicals (including metals) and allergens. The US EPA recognizes that occupational exposure can occur, and states that workers exposed to Class B biosolids might benefit from several additional precautions such as the use of dust masks when spreading dry materials, the use of gloves when touching biosolids, and routine hand washing before eating, drinking, smoking, or using the bathroom (62, 63).

The association among poor hygiene, raw sewage, and infectious disease is well established. Most of the pathogenic bacteria, viruses, and parasites in biosolids are enteric, which means they are present in the intestinal tracts of humans and animals. Enteric organisms that may be found in biosolids include, but are not limited to, *Escherichia coli, Salmonella, Shigella, Campylobacter, Cryptosporidium, Giardia*, Norwalk virus, and enteroviruses. Exposure may potentially result in disease (e.g., gastroenteritis) or in a carrier state in which an infection does not clinically manifest itself in the individual but can be spread to others. These enteric organisms are usually associated with self-limited gastrointestinal illness but can develop into more serious diseases in sensitive populations such as immune-compromised individuals, infants, young children, and especially the elderly (64).

The disease risk is a function of the number and types of pathogens in the Class B biosolids relative to the exposure levels and infective dose. Because data are sparse on what constitutes an infective dose, it is prudent public health practice to minimize workers' contact with Class B biosolids and soil or dusts containing Class B biosolids during production, transport, and application, and at land application sites during the period when public access is restricted. Class A biosolids may also present some health risk to workers, since some chemicals and biological constituents in Class A biosolids are not regulated by the US EPA.

Workers could be exposed to pathogens and irritants when working with Class B biosolids during the period when public access is restricted. A National Institute for Occupational Safety and Health (NIOSH) field investigation at one biosolids land application and storage site that did not comply with the US EPA requirements reported the following (65):

1. NIOSH interviewed employees who worked in all phases of the biosolids operation. Some employees reported repeated episodes of gastrointestinal illness after working with the biosolids at the treatment plant, during transport or during land application.

2. NIOSH observed among workers an inconsistent awareness, provision, and use of protective equipment and hygiene practices appropriate for handling Class B biosolids (or biosolids that do not comply with US EPA standards).
3. NIOSH collected bulk samples from different locations within the biosolids storage site and found measurable concentrations of fecal coliforms. Fecal coliforms are used as an indicator for the presence of other enteric microorganisms. Enteric bacteria were detected in air samples collected at the land application site.
4. The local department of environmental services recently informed NIOSH that biosolids applied at this site intermittently exceeded (by up to 4.5 times) the US EPA fecal coliform upper limit for Class B biosolids prior to the NIOSH survey.
5. The substandard biosolids were applied at the agricultural site before the monitoring results were received from the laboratory.

The US EPA reports that high-pressure spray applications may result in some aerosolization of pathogens, and that application or incorporation of dewatered biosolids may cause localized fine particulate/dusty conditions. Also, farm workers may be exposed to biosolids after application and during the restricted period. Ancillary workers (for example, laborers hired to clean trucks that were used to haul biosolids) can be exposed to biosolids. Exposures to substandard biosolids can occur when these materials are loaded and hauled to approved landfills or incinerators for disposal.

Additional study of worker exposures to pathogens and other toxics possibly present in Class B biosolids is needed. This will reduce scientific uncertainty about these issues and allow further refinement of worker precautions.

To protect workers who have direct contact with Class B biosolids and thus are likely to have an exposure to pathogens, employers should provide a basic level of protection, including appropriate measures from those listed below. While the measures are worded to refer to Class B biosolids, most also apply to tasks involving contact with sewage, untreated or partially treated sludge, or substandard biosolids.

## 6.2. Steps to Be Taken for Protection of Workers

### 6.2.1. Provision of Basic Hygiene Recommendations for Workers

Basic hygiene precautions are important for workers handling biosolids. The following list, originally developed by the US EPA (66), provides a good set of hygiene recommendations:

1. Wash hands thoroughly with soap and water after contact with biosolids.
2. Avoid touching the face, mouth, eyes, nose, genitalia, or open sores and cuts while working with biosolids.
3. Wash hands before eating, drinking, or smoking, and before and after using the bathroom.
4. Eat in designated areas away from biosolids-handling activities.
5. Do not smoke or chew tobacco or gum while working with biosolids.
6. Use barriers between skin and surfaces exposed to biosolids.
7. Remove excess biosolids from footgear prior to entering a vehicle or a building.
8. Keep wounds covered with clean, dry bandages.
9. Thoroughly but gently flush eyes with water if biosolids contact eyes.
10. Change into clean work clothing on a daily basis and reserve footgear for use at worksite or during biosolids transport.

11. Do not wear work clothes home or outside the work environment.
12. Use gloves to prevent skin abrasion.

In addition, NIOSH recommends extra steps (see next sections) to provide a more comprehensive set of precautions for use by employers and employees (67).

*6.2.2. Provision of Appropriate Protective Equipment, Hygiene Stations, and Training*

6.2.2.1. PERSONAL PROTECTIVE EQUIPMENT

Appropriate personal protective equipment (PPE) should be provided for all workers likely to have exposure to biosolids. The choices of PPE include goggles, splash-proof face shields, respirators, liquid-repellent coveralls, and gloves. Face shields should be made available for all jobs in which there is a potential for exposure to spray or high-pressure leaks, or aerosolized biosolids during land application (68). Management and employee representatives should work together to determine which job duties are likely to result in this type of exposure, to conduct appropriate on-site monitoring, and to determine which type of PPE is needed in conjunction with a qualified safety and health professional. If respirators are needed, a comprehensive program would include respirator fit-testing and training or retraining.

6.2.2.2. HYGIENE AND SANITATION

Hand-washing stations with clean water and mild soap should be readily available whenever contact with biosolids occurs. In the case of workers in the field, portable sanitation equipment, including clean water and soap, should be provided. Cabs should be wiped down and cleaned of residual mud (or settled dust) frequently to reduce potential for exposure to biosolids.

6.2.2.3. TRAINING

Periodic training on standard hygiene practices for biosolids workers should be conducted by qualified safety and health professionals to cover issues such as the following:

1. Frequent and routine hand washing (the most valuable safeguard in preventing infection by agents present in biosolids), especially before eating or smoking
2. The proper use of appropriate PPE, such as coveralls, boots, gloves, goggles, respirators, and face shields
3. The removal of contaminated PPE and the use of available on-site showers, lockers, and laundry services
4. Proper storage, cleaning, or disposal of contaminated PPE
5. Instructions that work clothes and boots should not be worn home or outside the immediate work environment
6. Prohibition of eating, drinking, or smoking while working in or around biosolids
7. Procedures for controlling exposures to chemical agents that may be in biosolids

6.2.2.4. REPORTING

Workers should be trained to report potentially work-related illnesses or symptoms to the appropriate supervisory or health care staff. This may aid in the early detection of work-related health effects.

### 6.2.2.5. IMMUNIZATIONS

Ensure that all employees are up to date on tetanus-diphtheria immunizations, since employees are at risk of soil-contaminated injuries. Current Centers for Diseases Control and Prevention (CDC) recommendations do not support hepatitis A vaccination for sewage workers (69, 70).

### 6.2.3. Good Environmental Practices to Prevent and Minimize Occupational Exposures

The good environmental practices to prevent and minimize occupational exposures include the following (71):

1. Where feasible, substituting Class A biosolids could reduce the pathogen exposure risks during land application compared to applying Class B biosolids. Feasibility may be affected by local customer preferences, since the two types of biosolids vary in the nutrient value they provide to end-users.
2. Monitor the source material coming from the wastewater treatment facility. Check monitoring results to assure they meet specified Class B or Class A standards prior to land application operations.
3. Monitor stored biosolids prior to application to ensure that the biosolids are properly stabilized and that unacceptable regrowth or cross-contamination from substandard material has not occurred.
4. Where local conditions permit, inject biosolids below the soil or incorporate (thoroughly mix) into tilled soil. This will minimize postapplication worker contact with applied biosolids and prevent resuspension into the air during periods of dryness.
5. On windy days, avoid spreading or disturbing dry biosolids (e.g., compost) that would create dust. On windy days, avoid spreading biosolids by high-pressure spray.
6. Avoid unnecessary mechanical disturbance and contact with land-applied Class B biosolids during the period when public access is restricted.
7. Equip heavy equipment used at storage and application facilities with sealed, positive-pressure, air-conditioned cabs that contain filtered air-recirculation units.
8. Monitor worker exposures when adjusting precautions to address site-specific issues.

## NOMENCLATURE

$BF$ = biosolids/sludge flow, gal/MG
$BV$ = biosolids/sludge volume hauled, yd$^3$/d
$CAP$ = vehicle capacity, yd$^3$/truck (3–12 yd$^3$)
$CSS$ = concentrated suspended solids from final treatment process, %
$HPD$ = work schedule, h/d
$HT$ = round-trip haul time, h
$K$ = multiplication factor
$Q_{avg}$ = average wastewater flow, MGD (million gallons per day)
$LT$ = loading time, h
$NTR$ = number of trucks required
$SCAP$ = solids capture, %
$SS$ = suspended solids concentration in biosolids/sludge flow, %
$TBH$ = tons of biosolids/sludge hauled per day, ton/d

## REFERENCES

1. U.S. Environmental Protection Agency, *Process Design Manual, Sludge Treatment and Disposal*, US EPA 625/1-79-011, Cincinnati, OH, September (1979).
2. City of San Francisco, *Primary Sludge Pump Evaluation*, City of San Francisco, Division of Sanitary Engineering, San Francisco, CA, October (1975).
3. G. L. Culp and N. F. Heim, *Field Manual for Performance Evaluation and Troubleshooting at Municipal Wastewater Treatment Facilities*, U.S. Environmental Protection Agency, Washington, DC, January (1978).
4. A. E. Sparr, Pumping sludge long distances, *Journal of Water Pollution Control Federation*, 43, 1702, August (1971).
5. M. L. Williams, A guide to the specification of glass lined pipe, *Water and Sewage Works*, 124, 10, 76, October (1977).
6. U.S. Environmental Protection Agency, *Transport of Sewage Sludge*, US EPA 600/2-77-216, Washington, DC, December (1977).
7. M. J. McFarland, *Biosolids Engineering, Chapter 5, Transport, Storage and Facilities Design*, McGraw-Hill, New York (2000).
8. J. R. Wolfs, Factors affecting sludge force mains, *Sewage and Industrial Wastes*, 22, p. 1, January (1950).
9. R. R. Rimkus and R. W. Heil, The rheology of plastic sewage sludge. In: *Proceedings of the 2nd National Conference on Complete Water Reuse*, Chicago, 4–8 May 1975, American Institute of Chemical Engineers, L. K. Cecil (ed.) p. 722 (1975).
10. C. E. Pehl, Comments on transport of biosolids in waste-amended soils, A. D. Karathanasis, D. M. C. Johnson and C. J. Matocha, *Journal of Environmental Quality*, 35, 1–2 (2006).
11. R. R. Rimkus and R. W. Heil, Breaking the viscosity barrier. In: *Proceedings of the Second National Conference on Complete Water Reuse*, Chicago, 4–8 May 1975, American Institute of Chemical Engineers, L. K. Cecil (ed.) p. 716 (1975).
12. Metcalf and Eddy, Inc., *Wastewater Engineering Treatment and Reuse*, 4th ed., McGraw-Hill, New York (2003).
13. B. O. A. Hedstrom, Flow of plastic materials in pipes, *Industrial Engineering Chemistry*, 44, 651 (1952).
14. G. E. Alves, D. F. Boucher and R. L. Pigford, Pipeline design for non-Newtonian solutions and suspensions, *Chemical Engineering Progress*, 48, 385 (1952).
15. R. W. Hanks and D. R. Pratt, On the flow of Bingham plastic slurries on pipes and between parallel plates, *Journal of Society of Petroleum Engineers*, p. 342, December (1967).
16. T. L. Chou, Flow of concentrated raw sewage sludges in pipes, *Journal of Sanitary Engineering Division ASCE*, 84, SA1, 1557, February (1958).
17. P. A. Vesilind, *Treatment and Disposal of Wastewater Sludges*, Ann Arbor Science, Chapter 4, 2nd ed. (1979).
18. Carollo, *Biosolids Management*, Planners & Designers, http://www.carollo.com/ 231/section.aspx/download/44 (2007).
19. U.S. Environmental Protection Agency, *Innovative and Alternative Technology Assessment Manual*, US EPA 430978009, available from National Technical Information Service (NTIS) #PB81–103277, 117 pp., February (1980).
20. Water Pollution Control Federation, *Wastewater Treatment Plant Design, Manual of Practice*, Lancaster Press, Inc., Lancaster, PA (1977).

21. A. E. Sparr, Pumping sludge long distances, *Journal of Water Pollution Control Federation*, 43, 8, 1702–1711, August (1971).
22. L. W. Weller, Pipeline transport and incineration, *Water Works and Wastes Engineering*, Kansas City, MO, September (1965).
23. J. J. Wirts, Tips and quips—contribution from Cleveland, *Sewage Works Journal*, 20, 3, 571, May (1948).
24. U.S. Environmental Protection Agency, *Areawide Assessment Procedures Manual*, Vol. III, Report No. 600/9–76–014, Washington, DC, July (1976).
25. U.S. Army Corps of Engineers, *Civil Works Construction Cost Index System Manual*, 110–2–1304, Washington, DC, 44 pp., http://www.nww.usace.army.mil/cost (2007).
26. U.S. Environmental Protection Agency, *Emerging Technologies for Conveyance Systems: New Installations and Rehabilitation Methods*, US EPA 832–R-06–004, Prepared by the Parsons Corporation, Fairfax, VA, for US EPA, Washington, DC, July (2006).
27. G. Tchobanoglous, H. Theisen and R. Eliassen, *Solid Wastes*, Chapter 5, McGraw-Hill, New York (1977).
28. Conveyor Equipment Manufacturers Association, *Belt Conveyors for Bulk Materials*, Cahners Publishing, Newton, MA (1966).
29. B. E. Hansen, D. L. Smith, and W. E. Garrison, Start-up problems of sludge dewatering facility. In: *Proceedings of the 51st Annual Water Pollution Control Federation Conference*, Anaheim, CA, October (1978).
30. Conveyor Equipment Manufacturers Association, *Screw Conveyers*, Book No. 350, Naples, FL (1971).
31. W. F. Ettlich, Economics of transport methods of sludge. In: *Proceedings of the Third National Conference on Sludge Management: Disposal and Utilization*, Information Transfer Inc., p. 7, Miami Beach, FL, December 14–16 (1976).
32. F. E. Dallon and R. R. Murphy, Land disposal IV: reclamation and recycle, *Journal of Water Pollution Control Federation*, 45, 7, 1489, July (1973).
33. U.S. Environmental Protection Agency, *Cost of Landspreading and Hauling Sludge from Municipal Wastewater Treatment Plants*, US EPA/530/SW-6l9, Cincinnati, OH, October (1977).
34. C. F. Guarino, M. D. Nelson, S. A. Townsend, T. E. Wilson and E. F. Ballotti, Land and sea solids management alternatives in Philadelphia, *Journal of Water Pollution Control Federation*, 47, 11, 255l, November (1975).
35. U.S. Environmental Protection Agency, *Costs of Hauling and Land Spreading of Domestic Sewage Treatment Plant Sludge*, Report No. 670/2–74–010, Washington, DC, February (1974).
36. U.S. Environmental Protection Agency, *Process Design Manual for Sludge Treatment and Disposal*, Report No. 625/1–74–006, Washington, DC, October (1974).
37. Liberty Transport, *Biosolids Management Service—Transportation Equipment*, McCarthy Family Farms, Inc. http://www.mccarthyfarms.com/liberty.htm (2007).
38. Sumas Transport, Inc *Biosolids Management*, http://www.sumas.com/biosolids.htm (2007).
39. C. Fulhage, *Equipment for Off-Site Application of Biosolids*, County and Regional Extension Centers, University of Missouri-Columbia, http://muextension.missouri.edu/xplor/envqual/wq0432.htm (2006).
40. C. H., Billings, S. H. Conner, J. R. Kircher and G. M. Scales, *1979 Public Works Manual*, Public Works Journal Corp., p. D-49 (1979).

41. Georgia State Environmental Protection Division, *Guidelines for Land Application of Sewage Sludge (Biosolids) at Agronomic Rates*, Atlanta, GA, June (2006).
42. M. Lang, National manual of good practice for biosolids. In: *Water Environment Federation Biosolids Specialty Conference*, San Diego, CA, February 21 (2001).
43. Department of Environmental Quality, *Western Australian Guidelines for Direct Land Application of Biosolids and Biosolids Products*, Western Australia, February (2002).
44. Bureau of Sanitation, *Critical Control Points that Influence the Management of the Biosolids Program*, City of Los Angeles, CA, May (2006).
45. Department of Primary Industries and Water, *Biosolids application methods*, Information Sheet No. 4, Tasmania, Australia, www.dpiw.tas.gov.au/inter.nsf/Attachments/CDAT-5S3UNR/$FILE/Info%20Sheet%20No%204.pdf (2007).
46. California State Water Resources Control Board, *General Waste Discharge Requirements for Biosolids Land Application, Chapter 9, Traffic*, Sacramento, CA, February (2004).
47. U.S. Environmental Protection Agency, *Pretreatment and Ultimate Disposal of Wastewater Solids*, Report No. 902/9–74–002, Washington, DC (1974).
48. U.S. Environmental Protection Agency, *Transport of Sewage Sludge*, US EPA-600/2–77–216, Washington, DC, December (1977).
49. U.S. Environmental Protection Agency, *Alternatives for Small Wastewater Treatment Systems, On-Site Disposal/Septage Treatment and Disposal*, Report No. 625/4–77–011, Washington, DC, October (1977).
50. U.S. Environmental Protection Agency, *Sanitary Landfilling*, Report on a Joint Conference Sponsored by the National Solid Waste Management Association and US Environmental Protection Agency, November 14–15, 1972, Kansas City, MO (1973).
51. Ford Foundation, *Exploring Energy Choices*, Detroit, MI (1974).
52. Toward Quality Biosolids Management: A Trainer's Manual, Transportation to Land Application, http://cropandsoil.oregonstate.edu/News/Publicat/Sullivan/TQBM/tm4.html (2007).
53. P. N. Kane, A biosolids technician training course with a "hands on" team approach using professionals from the field, *Journal of Extension*, 40, 2, April (2002).
54. N. Heller, Working with the railroad. In: *Proceedings of the 3$^{rd}$ National Conference on Sludge Management: Disposal and Utilization*, Information Transfer Inc., p. 50, Miami Beach, FL, December 14–16 (1976).
55. U.S. Environmental Protection Agency, *Summary Report: Pilot Plant Studies on Dewatering Primary Digested Sludge*, Report No. 670/2–73– 043, Washington, DC, August (1973).
56. U.S. Environmental Protection Agency, *Evaluation of Sludge Management Systems: Evaluation Checklist and Supporting Commentary*, Washington, DC, April (1979).
57. U.S. Environmental Protection Agency, *Design of Wastewater Treatment Facilities—Major Systems*, US EPA/43019–79–008, Reprint of Department of the Army, EM 1110–2–501, Washington, DC, September (1978).
58. U.S. Environmental Protection Agency, General design considerations, Chapter 14. In: *Process Design Manual, Land Application of Sewage Sludge and Domestic Septage*, US EPA-625–R-95–001, Washington, DC (1995).
59. B. M. Clark and J. I. Gillean, Systems analysis and solid waste planning, *Journal of Environmental Engineering Division, ASCE*, 100, 7 (1971).
60. C. F. Guarino, Land and sea solids management alternatives in Philadelphia, *Journal of Water Pollution Control Federation*, 47, 2551 (1975).

61. National Institute for Occupational Safety and Health, *Guidance for Controlling Potential Risks to Workers Exposed to Class B Biosolids*, Publication Number 2002–149, Washington, DC, July (2002).
62. U.S. Environmental Protection Agency, *Environmental Regulations and Technology—Control of Pathogens and Vector Attraction in Sewage Sludge*, US EPA/625/R–92/013, Washington, DC (1999).
63. U.S. Environmental Protection Agency, *Office of Inspector General Status Report: Land Application of Biosolids*, 2002–S-000004, Washington, DC (2002).
64. J. M Scarlett-Kranz, J. G. Babish, D. Strickland, and D. J. Lisk, Health among municipal sewage and water treatment plant workers, *Toxicology and Industrial Health, 3*, 3, 311–319 (1987).
65. National Institute for Occupational Safety and Health, *Biosolids Land Application*, www.cdc.gov/niosh, Washington, DC (2007).
66. U.S. Environmental Protection Agency, *Biosolids*, Washington, DC, www.epa.gov/owm/bio.htm (2007).
67. National Institute for Occupational Safety and Health, *Guidance for Controlling Potential Risks to Workers Exposed to Class B Biosolids*, Publication No. 2002–149, Washington, DC July (2002).
68. S. Laitinen, J. Kangas, M. Kotimaa, J. Liesivouri, P. J. Martikainen, A. Nevalainen, R. Sarantila, and K. Husman, Workers' exposure to airborne bacteria and endotoxins at industrial wastewater treatment plants, *American Industrial Hygiene Association Journal, 55*, 11, 1055–1060 (1994).
69. Centers for Disease Control and Prevention, *Prevention of Hepatitis A Through Active or Passive Immunization*, MMWR 48 (RR–12), Washington, DC (1999).
70. D. Trout, M. S. Mueller, L. Venczel, A. Krake, Evaluation of occupational transmission of Hepatitis A virus among wastewater workers, *Journal of Occupational and Environmental Med*icine, 42, 1, 83–87 (2000).
71. U.S. Environmental Protection Agency, *Biosolids Management Program*, Region 8. www.epa.gov/region08/water/wastewater/biohome/biohome.html (2007).

## APPENDIX

**U.S. Army Corps of Engineers Civil Works Construction Yearly Average Cost Index for Utilities (25)**

| Year | Index | Year | Index |
|---|---|---|---|
| 1967 | 100 | 1988 | 369.45 |
| 1968 | 104.83 | 1989 | 383.14 |
| 1969 | 112.17 | 1990 | 386.75 |
| 1970 | 119.75 | 1991 | 392.35 |
| 1971 | 131.73 | 1992 | 399.07 |
| 1972 | 141.94 | 1993 | 410.63 |
| 1973 | 149.36 | 1994 | 424.91 |
| 1974 | 170.45 | 1995 | 439.72 |
| 1975 | 190.49 | 1996 | 445.58 |
| 1976 | 202.61 | 1997 | 454.99 |
| 1977 | 215.84 | 1998 | 459.40 |
| 1978 | 235.78 | 1999 | 460.16 |
| 1979 | 257.20 | 2000 | 468.05 |
| 1980 | 277.60 | 2001 | 472.18 |
| 1981 | 302.25 | 2002 | 484.41 |
| 1982 | 320.13 | 2003 | 495.72 |
| 1983 | 330.82 | 2004 | 506.13 |
| 1984 | 341.06 | 2005 | 516.75 |
| 1985 | 346.12 | 2006 | 528.12 |
| 1986 | 347.33 | 2007 | 539.74 |
| 1987 | 353.35 | | |

# 2
# Conversion of Sewage Sludge to Biosolids

## Omotayo S. Amuda, An Deng, Abass O. Alade, and Yung-Tse Hung

**CONTENTS**

INTRODUCTION
SEWAGE CLARIFICATION
SEWAGE SLUDGE STABILIZATION
CONDITIONING
THICKENING
DEWATERING AND DRYING
OTHER PROCESSES
CASE STUDY
SUMMARY
ACRONYMS
REFERENCES

**Abstract** With the increasing generation of sewage sludge, its proper handling and disposal play a critical role in protecting our environment. Sustainable sludge treatment and management depend on controlling the quantity, quality, and characteristics of biosolids in favor of efficient sludge handling and biosolids beneficial use. One of the ultimate goals in the treatment is to convert sludge into biosolids or a product subject to subsequent treatment, in simplified operations that are cost-effective and environmentally safe. This chapter discusses the main treatment processes—clarification, stabilization, conditioning, thickening, dewatering, and drying—of sludge. The principle, operation, relevant diagrams, and criteria of each treatment method are presented. A case study of sludge management operation is also included.

**Key Words** Biosolids • sewage sludge conversion • treatment processes • stabilization • conditioning • thickening • dewatering • drying • beneficial use.

## 1. INTRODUCTION

### 1.1. Sewage and Sewage Sludge Generation

Sewage sludge can be described as a complex mixture of suspended and dissolved inorganic and organic materials. According to the United States Environmental

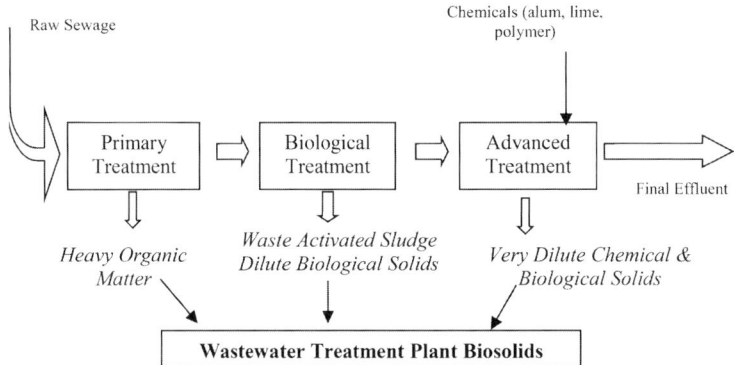

**Fig. 2.1.** Schematic flow of generated sludge.

Protection Agency (US EPA), sewage sludge means solids, semisolids, or liquid residue generated by the processes of domestic treatment works. Sewage sludge includes, but is not limited to, domestic septage; scum or solids removed in primary, secondary, or advanced wastewater treatment processes; and any material derived from sewage sludge.

The concern of scientists and engineers with respect to sewage sludges is to have these residuals treated before being discharged and render them suitable for beneficial purposes. Treatment processes yield less harmful effluents and aggregates of solids residual, which still require further treatment before being declared environmentally safe. This chapter provides a summary of sludge treatment processes. For detailed discussion and design of sewage sludge conversion methods into biosolids, the reader is referred to the book *Biosolids Treatment Processes* (1), which was edited by the same editors of this book and published by Humana in 2007.

### 1.2. Composition and Characteristics of Sewage

Sewage sludge is a heterogeneous medium largely consisting of water (>90%) and solids (<10%). About two thirds of the impurities in sewage are organic, predominantly, nitrogenous compounds, carbohydrates, and cellulose. Inorganic solids and organic colloids are major constituents of a typical sludge. The inorganic components include metallic salts and other dissolved elements in the sludge.

Before sewage sludge is completely converted into biosolids, its composition experiences great variations in terms of treatment methods. Figure 2.1 depicts the generated sludge flow diagram in a wastewater facility. In the primary and secondary multistage settlings/sedimentations, the solids generated are much different. The primary sludge represents settled solids from raw sewage sludge, and the secondary sludge is mainly the organic, microbial mass loaded with nutrients, particularly nitrogen. The primary sludge is more granular in nature and concentrated than the secondary sludge, whereas about 70% of both primary and secondary sludge contents are organic (2).

The mass of solids in the primary sludge can be estimated using this expression:

$$Mp = \gamma \times S \times Q \qquad (1)$$

where

$M_p$ = mass of primary solids, kg/s
$\gamma$ = efficiency of primary clarifier
$S$ = total suspended solid in influent to primary clarifier, kg/m³
$Q$ = flow rate, m³/s

The volume of wet sludge can be estimated with the following equation:

$$V = (M/S)/1000 \qquad (2)$$

where

$V$ = volume of sludge produced, m³/s
$M$ = mass of dry solids, kg/s
$S$ = solids content expressed as decimal fraction
$1000$ = density of water, kg/m³

It is inferred that the increase in the percentage of solids content in the sludge results in reduction in the sludge volume. For example, a uniform cylindrical tank (22 m in diameter and 100 m in height) is filled with sludge of 4% solids. (1) Calculate the dry mass of the solids. (2) When the liquid is decanted to about 50% of its volume and the sludge has settled, calculate the percent of solids in sludge.

To calculate volume of sludge,

$$V = \pi r^2 h$$
$$= 3.14 \times (21/2)^2 \times 100$$
$$= 34{,}650 \, \text{m}^3$$

where '$r$' and '$h$' are the radius and height of the cylindrical tank, respectively

Half volume is

$$V_{\text{half}} = V/2$$
$$= 34{,}650/2$$
$$= 17{,}325 \, \text{m}^3.$$

(1) Dry mass of solids. According to Eq. 2,

$$M = 1000 \times V \times S$$
$$= 1000 \times 34{,}650 \times 0.04$$
$$= 1{,}386{,}000 \, \text{kg/s}$$

(2) To calculate the percent of solids in sludge, when the sludge has settled, and the liquid is decanted to about 50% of its volume

$$S = (M/1000 V_{\text{half}})$$
$$= (1{,}386{,}000/1000 \times 17{,}325)$$
$$= 0.08, \text{ or } 8\%$$

## 1.3. Sewage and Sewage Sludge Treatment

Principally, in wastewater treatment, the impurities therein are concentrated into solids residue and then separated from the bulk liquid for further treatment. The concentrated solids residue is referred to as sludge. The sludge after treatment is referred to as biosolids because of the nature of its components, which are principally biomass (3).

### 1.3.1. Objectives of Sewage Treatment

The primary objective of sewage sludge treatment is to separate water phase from the solid phase, eliminate offensive matters, and avoid associated pathogens and odors (4). In the treatment, separation of solid and liquid phases is accomplished via various physical, chemical, and biological methods. The portion of trapped organisms and microbes or pathogens that have the potential of causing diseases are destroyed or treated in a way to incapacitate them and their "nefarious" activities. The second objective is to prevent the rivers, underground water, and other sources of potable water supplies from being contaminated, thereby eliminating water-borne diseases. The final objective is to prevent indiscriminate dumping of sewage sludge as a pollutant into the ecosystem.

### 1.3.2. Sewage Treatment Processes

There are three stages involved in sewage treatment: screening (or preliminary), primary and secondary treatment processes. One stage leads to the next, depending on the choice of the nature of the end product of the treatment.

#### 1.3.2.1. SCREENING (OR PRELIMINARY) TREATMENT

The main objective of the screening treatment is to remove or reduce coarse solids in the form of large floating and suspended materials that are likely to damage processing equipment. The sewage is forced through the bar screen with openings ($\geq 15$ mm) between the bars. The accumulated or screened solids on the bar racks may be cleaned manually or mechanically with automated rakes. The types and sizes of screens as well as the type of wastewater determine the quantity and characteristics of screened solids collected for disposal. The product of the screening are often incinerated, buried, or used as landfills.

Preliminary wastewater or sewage treatment may at times involve comminution as an alternative to the use of racks in the screening process. This process cuts up the coarse solids into smaller sizes, which facilitate further the processing of the influents. This process, however, is not safe from the presence of grit, which usually accompanies the sewage. The grit is removed in the grit chamber. With the use of differential settlement in the grit chamber or channels, the heavier grit particles settle out while the lighter organic matters remain in suspension (5).

The quantity of grit collected depends on the amount of household garbage and sandy soil in contact with the sewer or wastewater in question. Since larger percentage of grit consists of sand gravel and cinders, it is often disposed of in landfills in conformity with the appropriate environmental regulations. The removal of organic matter adhering to the grit is a major challenge in this process.

#### 1.3.2.2. PRIMARY TREATMENT

The raw sewage or wastewater from the preliminary tank or screening process enters the primary sedimentation stage where it undergoes further treatment (Figure 2.1). The purpose of sedimentation in the treatment of wastewater or sewage is to reduce the amount of suspended solids in the influent. The settling characteristics, such as density, size, ability to flocculate, retention time, surface loading, and weir overflow rate, are the relevant factors that affect the removal of particles during sedimentation. More than half of the suspended solids and more than one third of the biochemical oxygen demand (BOD) are removed.

The associated problem with this process is the floating scum, which must be removed and combined with the sludge. The concentrated solids residue may be removed under the influence of hydrostatic pressure. The concentrated solids are collected for further processing in the sludge treatment plant.

1.3.2.3. SECONDARY TREATMENT

The secondary sewage or wastewater treatment referred to as biological treatment (Figure 2.1), can be accomplished either through biological filtration or sludge activation. The two processes are similar in the number of reactors used and the application of microorganisms to oxidize the BOD in the flow (6).

Biological wastewater treatment is basically employed to convert the dissolved organic finely divided matters in the wastewater into settleable biological or inorganic solids, which are then removed in secondary clarifiers.

### 1.3.3. Biosolids Treatment

Biosolids refer to treated sewage sludge that meets the US EPA's pollutant and pathogen requirements for land application and surface disposal. Conversion of sewage sludge to biosolids is basically accomplished physically through the removal of substantial water volume, and biologically through the elimination or incapacitation of pathogens. As a result, generated biosolids that are rich in nutrients (such as nitrogen and phosphorus) and valuable micronutrients, are qualified for uses in home gardening, commercial agriculture, silviculture, greenways, recreational areas, and reclamation of drastically disturbed sites, such as those subjected to surface mining.

Facilities employ a number of both simple and complex technologies and processes to convert sludge to biosolids. One report divides these technologies, based on their stages of developments, into three categories: embryonic, innovative, and established (3). These processes, in terms of their treatment effects and generated products, involve mainly five steps: clarification, stabilization, conditioning, thickening (concentration), dewatering, and drying. Following the treatment sequence, sewage sludge is eventually converted into biosolids. Based on treatment tools and principles, each stepwise process includes various methods. The ultimate goal of any technology or technique is to use or combine physical, chemical, and biological methods to eliminate water, organic matter, and pathogens from the solids residues (7).

### 1.3.4. Biosolids Applications

Biosolids may be put to many beneficial uses, besides occasional use in landfilling. The most common method is to use biosolids in land applications. Biosolids are used on agricultural land, forests, range lands, or disturbed land in need of reclamation. They are added to soil to supply nutrients and replenish the soil organic matter. They improve soil structure and tilth, enhance moisture retention, and reduce soil erosion, which make conditions more favorable for root growth and increase the drought tolerance of

**Table 2.1**
**Biosolids application to various vegetation sites**

| Type of site/vegetation | Schedule | Application frequency | Application rate (dry tons per acre) |
|---|---|---|---|
| Agricultural land | | | |
|   Corn | April, May, after harvest | Annually | 5–10 |
|   Small grains | March–June, August, fall | Up to 3 times per year | 2–5 |
|   Soybeans | April–June, fall | Annually | 5–20 |
|   Hay | After each cutting | Up to 3 times per year | 2–5 |
| Forest land | Year round | Once every 2–5 years | 5–100 |
| Range land | Year round | Once every 1–2 years | 2–60 |
| Reclamation sites | Year round | Once | 60–100 |

*Source*: US EPA (3).

vegetation. They also supply the nutrients that are important to plant growth, for example nitrogen and phosphorus, as well as some essential micronutrients such as nickel, zinc, and copper. Biosolids can also serve as an alternative for expensive chemical fertilizers. The nutrients in the biosolids offer several advantages over those in inorganic fertilizers because they are organics and are released slowly to growing plants. These organic forms of nutrients are less water-soluble and therefore less likely to leach into groundwater and surface waters.

The amount of biosolids that may be applied to a site is a function of the amount of nutrients required by the vegetation and the amount of metals found in the biosolids. Table 2.1 summarizes the application frequency, timing, and rates for various types of sites.

## *1.4. Biosolids Regulations*

The US EPA's State Sludge Management Program (Title 40 of the Code of Federal Regulations [CFR], Part 501) and Standards for the Use and Disposal of Sewage Sludge (Part 503) are two regulations for the use or disposal of sewage sludge. The Part 503 rule establishes standards, which consist of general requirements, pollutant limits, management practices, and operational standards, for the final use or disposal of sewage sludge generated during the treatment of domestic sewage in treatment plants. Standards are included in this part for sewage sludge applied to the land, placed on a surface disposal site, or fired in a sewage sludge incinerator. Also included in this part are pathogen and alternative vector attraction reduction requirements for sewage sludge applied to the land or placed on a surface disposal site.

The Part 503 rule governs the treatment and monitoring requirements that wastewater utilities must meet prior to disposal of their biosolids. The US EPA developed the rule in its belief that beneficial reuse of the biosolids is the best ultimate disposal method. So, while the rule covers requirements for nonbeneficial methods, such as incineration or landfilling, its principal focus is on the requirements that wastewater systems must follow for land application of biosolids.

Biosolids treated to a very high standard such that they are devoid of pathogens and pollutants are referred to as "exceptional quality" (EQ) biosolids. There are no restrictions associated with the application of these materials, and some wastewater utilities even package and sell this material as a compost or soil amendment. To be classified as EQ, the biosolids must meet strict limits for pollutant concentrations (mainly heavy metals), pathogens, and attractiveness to vectors (i.e., insects that will carry possible diseases from the material).

The Part 503 rule defines two types of biosolids with respect to pathogen reduction: Class A (no detectable pathogens) and Class B (a reduced level of pathogens). Both classes are safe, but additional requirements are necessary with Class B materials. These requirements are detailed in the rule and include such things as limiting public access to the site of application, limiting livestock grazing, and controlling crop harvesting schedules. Class A biosolids are not subject to these use restrictions and can generally be used like any commercial fertilizer.

The rule allows wastewater treatment facilities greater latitude in the use of Class A biosolids than Class B, but producing this level entails greater, more sophisticated and intensive treatment. It is in this difference that the rule gets very complicated. For instance, if the biosolids are land applied, certain restrictions on application rates, land use, and contact times are set for Class B biosolids. On the other hand, certain Class A biosolids have virtually no restrictions and, in several cities, the material is packaged and sold as a fertilizer or soil amendment.

Class A requirements can be met in a number of ways, but most often sludge is held at high temperatures (i.e., heating processes) or high pH (i.e., alkaline stabilization) and high temperature for specified periods of time. Methods for obtaining Class B biosolids are more varied and less intensive.

In addition to biosolids performance standards regarding pathogen levels, the Part 503 rule also established maximum concentrations for nine metals that cannot be exceeded in biosolids products. These are known as ceiling concentrations. These federal maximum allowable metals concentrations are provided in Table 2.2. The rule also established more stringent pollutant concentrations, which is used to define EQ biosolids. Biosolids products that do not exceed pollutant concentrations meet Class A pathogen reduction

**Table 2.2**
**Maximum metal concentrations in biosolids**

| Metal | Ceiling concentration (mg/kg) | Pollutant concentrations (mg/kg) |
|---|---|---|
| Arsenic | 75 | 41 |
| Cadmium | 85 | 39 |
| Copper | 4300 | 1500 |
| Lead | 840 | 300 |
| Mercury | 57 | 17 |
| Molybdenum | 75 | No established limit |
| Nickel | 420 | 420 |
| Selenium | 100 | 100 |
| Zinc | 7500 | 2800 |

*Source*: US EPA (3).

requirements, and are often referred to as EQ products. Products meeting these requirements may be freely distributed for a variety of uses. Further details on regulations are provided in Chapter 6 of this book.

## 2. SEWAGE CLARIFICATION

Clarification is a popular unit operation in wastewater treatment and involves separation of suspended particles that are heavier than water. The aim of this process, in all the steps involving treatment of wastewater or sewage, is to produce a clarified effluent. Clarification can be accomplished by sedimentation, flotation, and membrane clarifications.

### 2.1. Sedimentation Clarification

Basically, sedimentation clarification is employed to produce a clarified effluent and concentrated sewage sludge that can be handled and further processed. Sewage is forced to flow at a low rate to be able to accomplish the slow sedimentation process in a basin. This process has other applications, particularly in the removal of unwanted solid particles in wastewater and also in sludge thickening.

The four main types of settling processes that occur in sedimentation tanks are discrete particle, flocculant, hindered, and compression settling. In the settling of discrete particles, the sedimentation basin is designed by setting a terminal velocity equal to or greater than the terminal velocity of likely particles that are present in the wastewater. At a high flow rate, the particles in relatively dilute solutions coalesce or flocculate and settle faster with an increase in size. It is obvious that the concentration of the particles, range of particle sizes, the depth of the sedimentation tank, as well as the opportunity for constant velocity can affect the extent to which flocculation occurs (10, 11).

The hindered settling involves high concentrations of suspended solids. Due to the high concentration of the contacting particles, they tend to settle at a position that withstands their densities, while the liquid in which they are found moves through their interstices. However, the concentration and the characteristics of the solids usually affect the settling rate. Compression sedimentation involves mechanical stirring efforts that help compact sludge by breaking up the flocculants and allowing water to escape. While in the settling tank, the sludge is further compressed mechanically to make it more compact.

### 2.2. Flotation Clarification

Flotation is another unit operation in the process of wastewater treatment that employs the use of fine gas bubbles to separate solid particles from a liquid phase. Concentration of biological sludge and the removal of suspended matter are achieved through this process.

In the process, the bubbled air or gas is attached to the particulate matters in the wastewater. Basically lighter than water, these particulates rise to the water surface and are collected by a skimming operation. The process helps to remove slow-settling particles, large or small, more completely and faster compared to the sedimentation process, which results in shortening the clarification time.

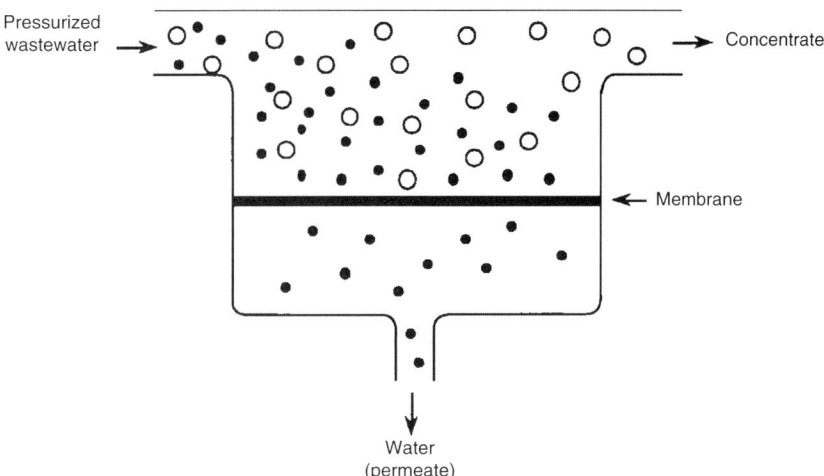

**Fig. 2.2.** Schematic diagram of a membrane clarification unit.

The chemicals used in aiding the flotation process help to create a surface or structure that can adsorb or entrap air bubbles. Inorganic chemicals like aluminum and ferric salts as well as organic polymers can be used independently to achieve the desired effects in the wastewater (12). The extent of removal depends on the method of producing the gas bubbles. Flotation can be accomplished by dissolved-air flotation (DAF), air flotation, and vacuum flotation.

## 2.3. Membrane Clarification

This is a form of filtration of effluent to remove fine suspended solids and precipitated chemicals through the use of membranes or filters (Figure 2.2). However, the tendency for the membranes to clog or go blind with gelatinous metal acid precipitates rendered this process unpopular and lead to the development of semicontinuous and continuous membrane clarifications.

In semicontinuous clarification, the filtration and unclogging occur sequentially. The filtration takes its normal course, but the removal of the suspended solids is often achieved through a complex process involving straining, interception, sedimentation, flocculation, and adsorption mechanisms.

Backwashing is the common method for cleaning and unclogging the membranes. When a predesignated level of headloss in the water leaving the membrane is reached, the operation is aborted. The flow is reversed with water sufficient to fluidize the membrane and thus the accumulated materials are washed away. Both filtering and clearing take place simultaneously in the continuous filtration operation of membrane clarification.

## 3. SEWAGE SLUDGE STABILIZATION

Stabilization is a chemical and physical process applied to sewage sludge to reduce or destroy pathogens, minimize or eliminate the potential of odor generation, and reduce the material vector attraction potential (13).

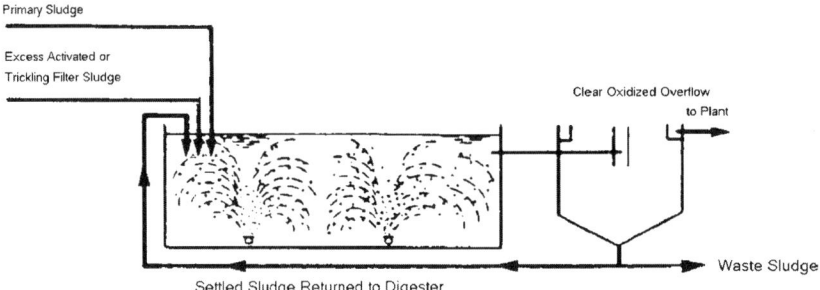

**Fig. 2.3.** Schematic diagram of aerobic digestion for sludge stabilization.

The quantity of the sludge to be treated, the complementary process, and the application of the treated sludge in landfill disposal and land applications for beneficial uses are the important factors to be considered when selecting or designing a sludge stabilization process. Therefore, in the application of stabilization process, chemicals are added and the mixture is heated in the treatment process of the sludge in order to reduce the volatile contents of the sludge and render the sludge unsuitable for the survival of microorganisms.

### 3.1. Aerobic Stabilization

Aerobic stabilization is commonly applied to treat organic sludge produced from various treatment operations. It has wide application in the treatment of waste-activated sludge (WAS), mixture of primary and waste-activated sludge, waste sludge from operation plants, and waste sludge from activated-sludge plant without primary settling.

As the available substrates depleted in a given sludge, the microorganisms undergo endogenous phase where about 80% of their cell tissue is oxidized aerobically to carbon dioxide, water and ammonia. Therefore, the pH of the sludge drops as the ammonia further oxidized to nitrate. In a conventional aerobic digester (Figure 2.3), the sludge is aerated for an extended period of time in open tank using aeration equipment. The digester can be operated as a batch or continuous-flow reactor.

The following factors are to be considered in the design of aerobic digesters:
- The need to maintain constant temperature since the tanks are subjected to varying atmospheric weather condition
- Destruction and biodegradation of the organic content of the sludge to bring about solids reduction
- Required oxygen that is sufficient for aerobic digestion, particularly for cell tissue and BOD in sludge
- Constant energy supply to facilitate a thorough mix of the contents of the aerobic digester
- Periodic check and adjustment of the pH level of the digester to facilitate effective process operation
- The process efficiency varies in terms of sludge age and sludge characteristics. As a result, a pilot work is required to make an effective design.

Some modifications have been introduced into the conventional aerobic digestion. The common one is using high-purity oxygen aerobic digestion, in which oxygen is used instead of air. It facilitates the use of closed tanks, unlike the conventional open tanks, and it helps to overcome variation in atmospheric conditions that often affect the temperature of conventional aerobic digestion. Its solids concentration ranges between 2% and 4%. However, the need to generate pure oxygen renders this process less cost-effective (14).

### 3.1.1. Autothermal Thermophilic Aerobic Digestion (ATAD)

Thermophilic aerobic digestion is another refined modification of aerobic digestion. As a temperature self-regulating process, it is set to correct the associated temperature problem with conventional aerobic digestion. About 70% of biodegradable organics are removed at a very short detention time of 3 to 4 days in a large pilot scale experimentation of this process. In most European nations, it has been incorporated into dual sludge digestion processes.

### 3.1.2. Anoxic Aerobic Digestion

This is an innovative aerobic process of digesting sludge that aims at improving denitrification, thus enhancing aerobic digestion. The embedded aerator works intermittently—on and off—to facilitate denitrification of the sludge under anoxic condition when the aerator is turned off. The denitrification process brings the pH of the sludge approximately to neutral thereby enhancing aerobic digestion, nitrogen removal, and effective pathogen destruction. The product has improved dewaterability, filtrate quality, and pH control compared to conventional aerobic digestion.

### 3.1.3. Simultaneous Sludge Digestion and Metal Leaching (SSDML)

This is an embryonic modification of aerobic digestion designed to increase pathogen reduction as well as metals solubility during the digestion process. It reduces odors and bacteria in the sludge (15). It is a simultaneous process of sludge digestion and metal leaching, involving the addition of elemental sulfur to the sludge during digestion. After a few days, the solubility of the toxic metals within the sludge increases due to the lower pH of the mixture.

## 3.2. Alkaline Stabilization

This process (Figure 2.4) involves the addition of lime or alkali to untreated sludge material to raise the pH to or greater than 12, in order to make the conditions unfavorable for the growth of pathogens and other microorganisms. At such a pH level, the sludge cannot undergo putrefaction, and so odor and other related nuisances are eliminated; but a pH below 9.5 can mar these achievements (13). This process can achieve the minimum requirements for both Class A and Class B biosolids with respect to pathogens, depending on the amount of alkaline material added and other processes employed. Generally, alkaline stabilization meets the Class B requirements when the pH of the mixture of wastewater solids and alkaline material is at 12 or above after 2 hours of contact.

The process involves the use of materials like hydrated lime, quicklime, fly ash, cement kiln dust, and carbide lime. Alkaline stabilized sludge is suitable mostly for land applications like landscaping, agriculture, and mine reclamation. In agriculture, it serves as fertilizers, which favor conditions for vegetative growth by improving soil properties

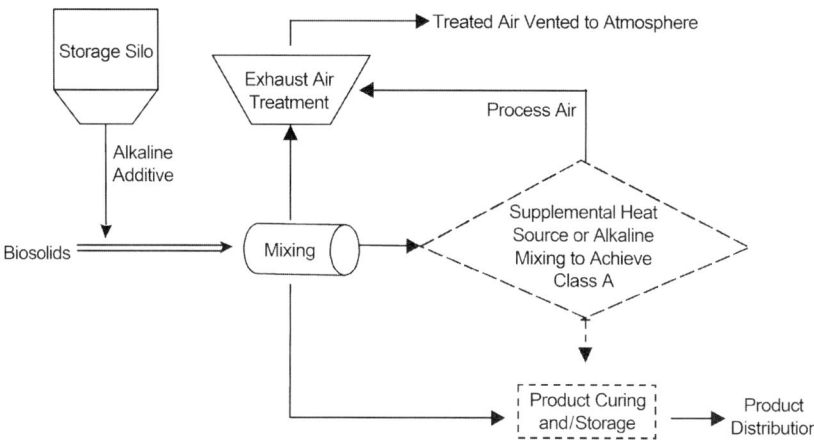

**Fig. 2.4.** Schematic diagram of alkaline stabilization operation.

like pH, texture, tilth, and water-holding capacity (13). In practice, alkaline stabilization has been found to be less expensive than composting or thermal drying (13). This process can be achieved in the following two steps.

### 3.2.1. Alkaline Pretreatment

This step involves the application or addition of a selected and measured alkali to sludge prior to dewatering. A high pH value requires higher dose of alkali, and there must be sufficient contact time in order to effectively eliminate the pathogens present. Table 2.3 shows the design criteria for Class B alkaline stabilization.

### 3.2.2. Lime Stabilization

The minimum US EPA criterion for lime stabilization is a pH of 12 for a period of 2 hours (13). The required dosage varies with the type of sludge and solids concentration. Lime stabilization requires more lime than conditioning processes and does not cause destruction of organics. However, it provides a single-treatment process that can be easily controlled and produces sludge suitable for land application. Representative lime dosage requirements as a function of percent sludge solids are shown in Table 2.4.

According to the Part 503 rule, Class A requirements can be achieved when the pH of the mixture is maintained at or above 12 for at least 72 hours, with a temperature of 52°C maintained for at least 12 hours during this time. In one process, the mixture is air dried to over 50% solids after the 72-hour period of elevated pH. Alternatively, the process may be manipulated to maintain temperatures at or above 70°C for $\geq$30 minutes, while maintaining the pH requirement of 12. This higher temperature can be achieved by overdosing with lime (that is, adding more than is needed to reach a pH of 12), by using a supplemental heat source, or by using a combination of the two. Monitoring for fecal coliforms or *Salmonella* sp. is required prior to release by the generator for use.

## Table 2.3
### Typical design criteria for Class B alkaline stabilization

| Parameter | Design criterion |
|---|---|
| Alkaline dose | 0.25 pound per pound of wastewater solids at 20% solids |
| Retention time in mixer | 1 minute |
| Retention time in curing vessel | 30 minute |

*Source*: US EPA (13).

## Table 2.4
### Lime requirements in chemical stabilization

| Lime dosage to obtain desired pH | Sludge, % solids | | | | |
|---|---|---|---|---|---|
| | 1 | 2 | 3 | 3.5 | 4.4 |
| pH = 11 $Ca(OH)_2$, mg/L | 1400 | 2500 | 3700 | 6000 | 8200 |
| pH = 12 $Ca(OH)_2$, mg/L | 2600 | 4300 | 5000 | 9000 | 9500 |

*Source*: US EPA (16).

### 3.3. Advanced Alkaline Stabilization

This process involves the addition of a selected and weighed alkaline compound to dewatered-sludge, often in the processing machines such as a paddle mixer or screw conveyor, in order to raise the pH of the mixture.

The commonly used alkaline compound for this process is quicklime. It generates a temperature high enough to raise the temperature of the mixture above 122°F (50°C) when in contact with water, and this is enough to inactive worm eggs. To meet up with the aim of stabilization, there must be a good mixing that will facilitate contact between the applied lime and small particles of the sludge.

Application of dry lime to dewatered sludge prevents scaling and maintenance problems associated with lime-sludge dewatering equipment. This is an important advantage over the alkaline pretreatment method of stabilization (13). Table 2.5 presents typical lime stabilization facilities.

### 3.4. Anaerobic Digestion

Anaerobic digestion is historically one of the oldest forms of biological wastewater treatment that is still in use. The process, carried out in an airtight reactor (Figure 2.5), converts biologically organic materials in mixtures of primary and biological sludges to biogas (methane) and carbon dioxide. The stabilized sludge, often with reduced organic and pathogen contents, may be accomplished in a batch or continuous process. This process involves hydrolysis, acidogenesis, and methanogenesis (17). During hydrolysis, the enzymes present in the sludge convert high-molecular-mass compounds into compounds suitable for use as a source of energy and cellular carbon. Acidogenesis involves conversion of the product of hydrolysis into intermediate compounds having

Table 2.5
Representative lime stabilization facilities

| Location | Solids Production (dry tons/day) | Disposal/ end use | Pathogen reduction | Process employed |
|---|---|---|---|---|
| Hampton Road, VA | 8 | Land application | Class A | Bio*Fix |
| Reedsville, WI | 1–2 | Land application | Class A | EnVessel Pasteurization™ |
| Lancaster, OH | 3 | Land application | Class A & B | Bio*Fix |
| Raleigh, NC | 20 | Land application | Class A | N-Viro AASAD |
| Howard Co, MD | 20 | Land application | Class A & B | Bio*Fix |
| Cookeville, TN | <1 | Land application | Class A | EnVessel |
| Charlotte, NC | 20 | Agricultural liming agent | Class A | Bio*Fix |
| Washington, DC | 270 | Land application | Class B | Post-Lime Stabilization |
| Middlesex, NJ | 146 | Landfill cover and agricultural liming agent | Class A | N-Viro AASAD |
| Toledo, OH | 63 | Agricultural liming agent | Class A | N-Viro AASAD |
| Greenville, SC | 10 | Landfill cover | Class A | N-Viro AASAD |
| Tarpon Springs, FL | 3 | Land application | Class A | N-Viro AASAD |
| Syracuse, NY | 30 | Agricultural liming agent | Class A | N-Viro AASAD |
| Urbana Champagne, IL | 5 | Land application | Class B | N-Viro |
| Urbana Gwynedd, PA | 2–3 | Land application | Class B | Generic |
| Bergen County, NJ | 45 | Landfill cover | Class A | Bio*Fix |

AASAD, Advanced Alkaline Stabilization Anaerobic Digestion.
*Source*: US EPA (13).

intermediate molecular masses, while methanogenesis is the conversion of product of acidogenesis to methane and carbon dioxide. The microorganisms, predominantly bacteria, are involved at each step of the reactions. A few examples include *Clostridium* spp., *Actinomyces, Escherichia coli* (nonmethanogenic) *Methanobacterium, Methanococcus*, and *Methanosarcia* (methanogenic). Though they are all very specific in their activities, they complement one another in the two pathways that lead to formation of methane. End products of fermentation like hydrogen and acetate are converted to methane and carbon dioxide by the methanogens, while the acidogens produce the hydrogen (18).

A dynamic equilibrium state between nonmethanogenic and methanogenic bacteria, absence of dissolved oxygen, pH value between 6.6 and 7.6, sufficient amount of nutrients like nitrogen and phosphorus, and optimum temperature ranges are important conditions to maintain an anaerobic treatment process that will stabilize sludge efficiently (19). The primary factor in determining whether a multistage anaerobic digestion process is feasible for a system is the feed solids concentration. Because a multistage process can be sensitive to changes in the feed solids, it might not be feasible if the characteristics

**Table 2.6**
**Substances with potential to cause biological inhibition in anaerobic digestion**

| Substance | Moderately inhibitive (mg/L) | Strongly inhibitive (mg/L) |
|---|---|---|
| Calcium | 1500–4500 | 8000 |
| Magnesium | 1000–1500 | 3000 |
| Sodium | 3500–5500 | 8000 |
| Potassium | 2500–4500 | 12,000 |
| Ammonia nitrogen | 1500–3000 | 3000 |
| Copper | NA | 50–70 (total) |
| Chromium VI | NA | 200–250 (total) |
| Chromium | NA | 180–420 (total) |
| Nickel | NA | 30 (total) |
| Zinc | NA | 1.0 (soluble) |

*Source*: US EPA (17).

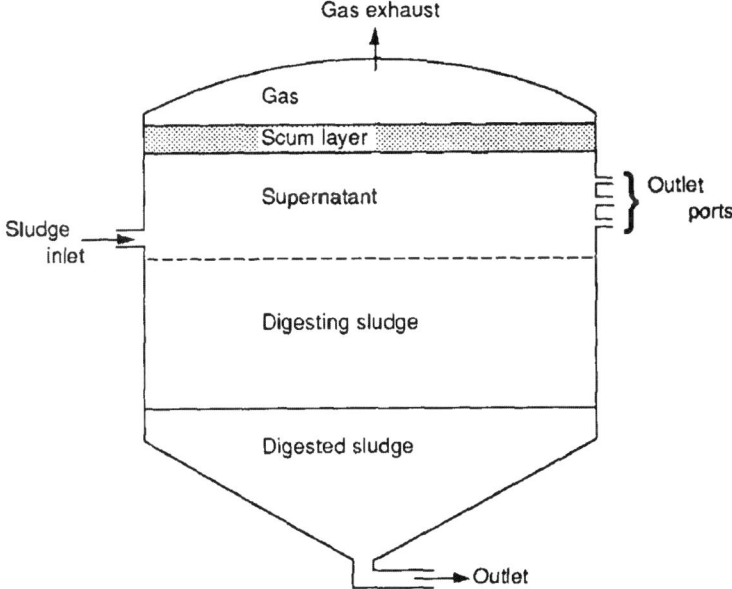

**Fig. 2.5.** Schematic diagram of anaerobic digestion system.

of the feed solids concentrations vary significantly. The volatile solids (VSs) content in the feed should preferably be at least 50% and the feed should not contain substances at levels that may inhibit the biological processes associated with anaerobic digestion (Table 2.6). Wastewater residuals containing lime, alum, iron, and other substances can be successfully digested as long as the VS content remains high enough to support the growth of microorganisms.

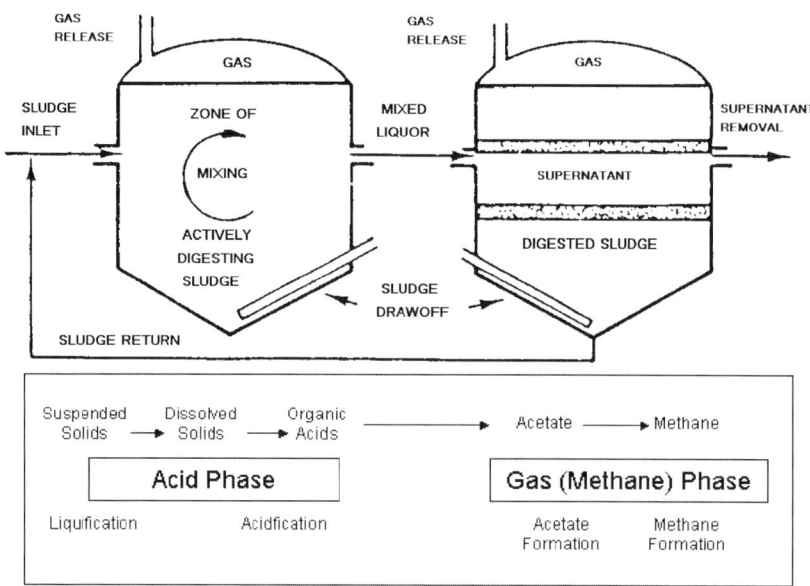

**Fig. 2.6.** Schematic diagram of two-stage anaerobic digestion of sludge.

### 3.4.1. Two-Stage Digesters

To improve the efficiency of high-rate digesters in the treatment of sludge, two digesters are connected in series (Figure 2.6). The first tank, often equipped with mixing and heating facilities, is primarily for digestion of the sludge. The second, closed tank is primarily designed to store, concentrate, and generate biogas. The tank is neither heated nor mixed, which facilitates the formation and collection of clear supernatant from the treated sludge, thus assisting dewatering of the sludge (17).

Typically, gas production and collection are the problems associated with this method. The gas prodution is not stable volume-wise and is dependent on the content of the volatile solids and biological activities present in the digester (17). Gas yield in the secondary stage is more than that in the primary stage. Gas collection requires attention because of its potential explosive nature particularly when air is allowed to mix with it. Pressure relief valves are required in order to release the pressure buildup in the reactor. The two common ways of collecting gas are the fixed and floating cover systems. These methods are cost sensitive; the capital cost is basically twice the cost of a single-stage digester. Interfering substances, such as excessive quantities of heavy metals, sulfides, and chlorinated hydrocarbons, may degrade its digestion performance. However, this can be mitigated by periodic cleaning.

### 3.4.2. Anaerobic-Baffled Reactor (ABR)

This is an innovative anaerobic digestion process designed to reduce sludge production. The reactor is compartmentalized with alternating hanging and standing baffles. This design facilitates flow of forced liquid up and down the compartments. The solids retention time, which also facilitates anaerobic digestion in a short retention time between 4 and 10 hours, is well separated from the hydraulic retention time. This process is often accompanied with lower sludge yield and higher BOD than in other high rate anaerobic treatment systems. This method is very appropriate for on-site primary treatment in low-income communities as sited in South Africa (20).

### 3.4.3. Columbus Biosolids Flow-Through Thermophilic Treatment (CBFT3)

This is an innovative low-cost anaerobic process for converting Class B to Class A type biosolids (21). It perhaps takes its name from the Columbus Water Works of Georgia, where pilot-scale production of Class A biosolids at a digestion temperature of 127°F (53°C) was achieved in less than 6 days (22). The reactor consists of continuous-feed thermophilic anaerobic digester connected to a long narrow plug-flow reactor. The product is later subjected to mesophilic digestion in order to minimize odors in the final product (3).

### 3.4.4. High Rate Plug Flow

This process, which uses a plug flow insulated tank, is designed to minimize the total suspended solids of feed sludge effectively (about 85%) within a short time of one day. Generation of "recoverable" methane gas for energy use and preheating of the sludge is one of the advantages of the system (23). Preheated sludge (at about 95°F [35°C]), which may be further screened and degritted, is often fed into the rectangular-shaped tank of 10,000 gallon/d (38 m$^3$/d) capacity for processing. The associated problems of carbon-to-nitrogen ratio and maintenance of pH are corrected by the addition of a carbon supplement and sodium bicarbonate into the mix, respectively. This process requires shorter retention time and creates a product of higher total solids, compared to established anaerobic digestion. Pilot scale demonstration of this process has been carried out in some U.S. cities (3).

### 3.4.5. Temperature-Phased Anaerobic Digestion

Temperature-phased anaerobic digestion is an innovative digestion process that combines thermophilic and mesophilic anaerobic digestion to improve the quality of biosolids, particularly the odor. The first phase of digestion is thermophilic at a temperature greater than 131°F (55°C), and the second phase is mesophilic achieved at a temperature of 95°F (35°C). The combination of thermophilic and mesophilic digestion processes gives this technology an advantage over either process alone (24, 25).

### 3.4.6. Thermal Hydrolysis

In this process, dewatered sludge (15–20% solids) is fed through a hydrolysis vessel in order to reduce biosolids mass and increase biogas production. The process facilitates the oxidation of sludge under a high temperature and pressure of about 320°F (160°C) and 100 psi (690 kN/m$^2$), which eventually destroys the pathogens present in the sludge and break down the cell structures, to release energy-rich compounds. These compounds are hydrolyzed and later digested anaerobically to further destroy the high-volatility solids present at about 65%. The process enhances increased biogas production. Unfortunately, the presence of odor in the product is an impending nuisance to this process. However, it has received a wide application in Europe, Australia, and Japan (26).

*3.4.7. Thermophilic Anaerobic Digestion Fermentation*

This is another digestion process that is designed to convert sewage sludge and organic waste to useful biosolids used as a fertilizer at a temperature that facilitates thermophilic digestion. The temperature is usually between 120° and 135°F (49° and 57°C). The process involves auto-heated anaerobic digestion of influent waste material, as a substrate, at a short residence time of 30 hours to produce single-cell proteins (solids), which are then dried to pellet. This process is much faster than mesophilic digestion and yields a 20% increase in protein content than do other similar established technologies. In Canada, this process is popular with the conversion of high-protein animal feed supplement and wastewater treatment sludge into fertilizer materials (3).

*3.4.8. Three-Phase Anaerobic Digestion*

This is another innovative anaerobic digestion process designed to produce Class A biosolids that are fit for direct land applications. It facilitates increasing dewaterability, biogas production, and odor reduction. The process involves three phases. The first phase is digestion of volatile fatty acid at a temperature of 95°F (35°C). The second phase is a thermophilic condition that facilitates digestion at a temperature ranging between 122° and 133°F (50° to 56°C). There is no heat application in the third phase, but successful digestion is accomplished at a steady temperature of 95°F (35°C). Very similar to single-stage anaerobic digestion, however, this process has improved pathogen destruction, very high volatile solid reduction, and gas production. A full-scale application of this process was reported in California in 2000 (6, 27).

*3.4.9. Two-Phase Anaerobic Digestion*

This process, also called two-phase acid/gas anaerobic digestion, facilitates increased production of methane gas and shortens the time of biosolids digestion. The acid and gas phases are employed to break down the organic materials in the sludge. The acid phase digestion, favored at a pH range of 5.5 to 6.0, involves hydrolysis, acidification, and acetification, while the gas stage favors high-quality methane gas production (17). Basically, this process involves physical separation of two different groups of bacteria into two separate reactors where they are cultured under the optimum conditions that favorably support their growth (28).

Acidogenic bacteria are grown in the acid phase digester at pH range of 5.5 to 6.0 to facilitate fermentation of volatile fatty acid in the sludge. They cannot thrive well in the methane gas production phase because there are no substrates for them to feed on. The methanogenic bacteria in the gas phase is grown at a pH greater than 6.0 to produce high-quality methane. The characteristics of the constituents of the sludge determine the residence time in the reactor, which is between 7 and 10 days. The methanogenic bacteria will not survive in the acid phase because the pH value as well as the retention time of the acid-phase reactor are very low.

The process has overcome sensitivity of methanogenic bacteria to overloads, which is peculiar to fermentation processes and has also minimized and eliminated foaming problems. This treatment process, put in place in Illinois, has generated sufficient methane gas to power a 1.5-megawatt generator (3).

### 3.4.10. Anaerobic Digestion with Ozone Treatment

This embryonic process basically involves the use of ozone to break down organic matter in sludge in order to increase the effectiveness of anaerobic digestion. The anaerobically digested sludge from the digester is diverted to reaction tank where it is exposed to low levels of ozone. The ozonized sludge is then thickened in the thickening tank and later sent back to the digester, where it is mixed with the nonozonized sludge. The product of this stage is subjected to further processing like dewatering or recycled to the ozone reaction. The source of ozone is corona discharges capable of transforming molecular oxygen into ozone (29). Experimental results showed that 0.06 g of ozone per gram of dissolved solids was necessary to destroy the biological activity of the digested biosolids (30).

### 3.4.11. Ferrate Addition

In this embryonic process, ferrate, a powerful liquid oxidizing chemical, is injected into the process stream of sludge to stabilize and disinfect the sludge. It generally enhances the sludge production for beneficial use. With this process, microorganisms like *E. coli* are reported to be inactivated to as low as 99.9% (31, 32). Also, odor-producing compounds, such as sulfides, mercaptans, and alkyl amines in the sludge, are eliminated due to their affinity for ferrate. The resulting treated sludge has a pH ranging between 12 and 13 (33).

### 3.4.12. Irradiation

Irradiation employs the use of gamma and beta rays to destroy microorganisms in the sludge. With irradiation, the colloidal nature of the microorganism cell protoplasm is altered, leading to their death. Gamma rays often produced from radioactive isotopes such as cobalt 60 and cesium 137 (at 1 megarad at 68°F [20°C]) can penetrate substantial thickness and thus destroy the pathogens in the sludge. The beta rays dose-base radiations have limited penetration ability, and thus thin layers of the sludge are allowed to pass under the source of the ray. The process helps to disinfect the sludge; however, it does not produce sludge that satisfies the vector attraction reduction (VAR) requirements of the US EPA (3). The product of this process may further be enhanced by addition of lime. Its low space requirement has made it more attractive to users.

### 3.4.13. Acidification

This embryonic process of disinfecting wastewater solids is operated in batch manner. The sludge is fed into a tank and mixed continuously for 2 hours with chlorine dioxide. Sulfuric acid is added to acidify the solids to a pH between 2.3 and 3.0. Sodium nitrite is then added to the mixture in the tank in order to convert nitrous acid at that pH value. As a result of acidification and generation of nitrous acid gas, pressure between 15 and 35 pounds per square inch gauge (psig) (103 and 241 $kN/m^2$ gauge) is built up in the tank. The materials in the reactor are processed with continuous mixing for 2 hours, after which the pH may be adjusted to produce biosolids for desired use (34).

A research report indicates that fecal coliforms and viral densities were reduced to a level below detectable limits, and the viability of helminths was 0%. Very similar to the

Synox process, however, the use of less expensive and more reliable chlorine dioxide and nitrous acid instead of ozone gives this process an edge (3).

## 3.5. Composting

Composting is the biological degradation of organic materials under controlled aerobic condition to stable end products that may be put to beneficial use (35). The end product is a virtually pathogens-free and odor-free humus-like material designated as Class A biosolids that can be applied as soil conditioner and fertilizer (35). During the decomposition of the organic materials in the sludge, the compost temperature ranges between 122° and 158°F (50° and 70°C), which eventually destroys the pathogenic organisms present in the compost.

Composting can take place under aerobic or anaerobic conditions. Aerobic composting favors a high rate of material decomposition, leading to a rise in temperature necessary for pathogen destruction and consequently odor reduction. The composting process involves three stages: mesophilic, thermophilic, and cooling. At the mesophilic stage, the presence of fungi and acid-producing bacteria increases the temperature of the compost to ambient temperature (104°F [40°C]). The mesophilic stage organisms are soon replaced by thermophilic microorganisms like actinomycetes and fungi. The temperature rise (about 104° to 158°F [40° to 70°C]) introduces the thermophilic stage, which enhances maximum degradation and stabilization of organic materials in the compost. During the cooling stage, the microbial activity is reduced and the condition favors the existence of mesophilic microorganisms. More water is evaporated at this stage, while formation of humic acid is completed and the pH is stabilized (35).

The length of time for composting at a specific temperature is important in determining the eventual use of the compost end product. The time and temperature requirements for producing classes A and B biosolids by composting are presented in Table 2.7.

Composting is characterized by the use of bulking agents. Since the sludge is often rich in nitrogen, there is a need for mixing it with higher carbon-containing materials like wood chips, sawdust, and newsprint as bulking agents, to increase the porosity of the mixture. Porosity facilitates adequate oxygen enrichment of the composting mass.

There are four general methods for composting biosolids: aerated static pile, windrow, in-vessel, and vermicomposting. All methods use the same scientific principles but vary in their procedures and equipment needs.

Table 2.7
Time and temperature requirements for biosolids composting

| Production | Regulatory requirement |
|---|---|
| Class A | Aerated static pile or in-vessel: 55°C for least 3 days |
| | Windrow: 55°C for at least 15 days with 5 turns |
| Class B | 40°C or higher for 5 days during which temperature exceed 55°C for at least 4 hours |

*Source*: US EPA (36).

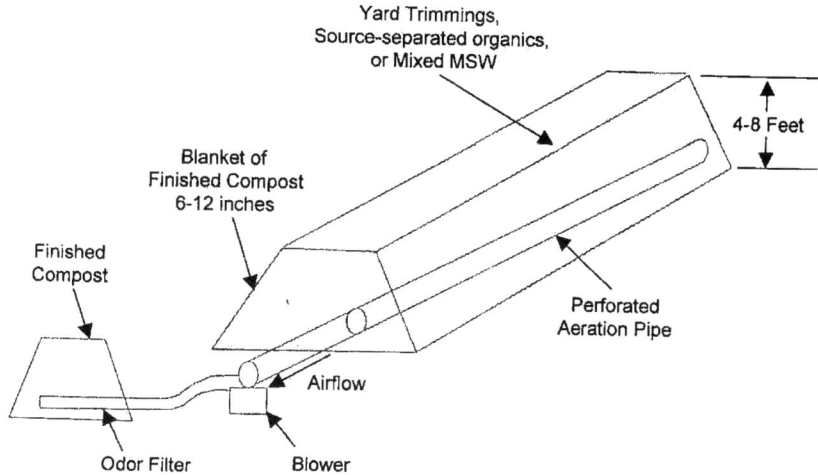

**Fig. 2.7.** Schematic diagram of static pile of sludge composting. MSW, Municipal Solid Wastes.

### 3.5.1. Aerated Static Pile

Under this system (Figure 2.7), a pretreated dewatered sludge is mixed with a bulking agent and placed on pipes through which air is drawn. The height of the sludge pile is between 2 and 2.5 m. The pile is decomposed by thermophilic organisms, which raise the operating temperature to 140°F (60°C) for 21 to 28 days, after which the rate of decomposition and temperature drops. The pile is further cured for another 30 days or dried depending on weather conditions. The drying (about 40–43% moisture) facilitates clean separation of compost from the bulking agent while the curing helps to remove the offensive odor and completes stabilization. The screening step helps to recover the bulking agents that are recycled for further composting.

This method of composting, particularly during the drying state, is weather dependent and, as such, two dry and sunny days are needed to maximize quality. It might not be effective for sludge containing heavy metals and industrial chemicals, because they are not completely removed in the process (36).

### 3.5.2. Windrow

As in the case of static pile, dewatered sludge undergoing the windrow method is mixed with bulking agent and piled in long rows (36). The windrow is 3 to 6 ft high and 6 to 14 ft wide at the base. The sludge pile is mechanically turned and mixed to increase the amount of oxygen needed for composting. The periodic mixing, which is done about five times, allows the outer materials to be subjected to higher temperature (about 131°F [55°C]) deeper in the pile and this leads to the release of offensive odor. The composting takes about 21 to 28 days, after which the materials are moved into curing piles. The drums and belts, mechanically powered by moving devices and self-propelled models that straddle the compost piles, are major mechanical devices employed in this method. This process requires large area and poses particulate and odor pollution to the environment (36).

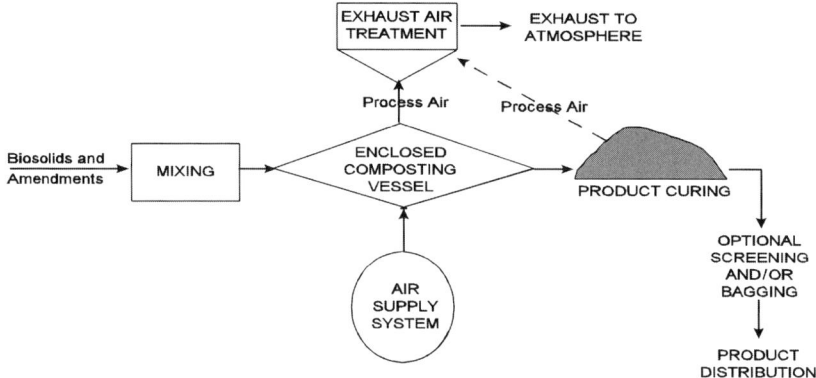

**Fig. 2.8.** Schematic diagram of in-vessel process unit.

*3.5.3. In-Vessel*

The in-vessel method of composting usually takes place in a covered vessel (silo-like) and enables effective control and monitoring of the composting materials. Generally, a mixture of sludge and bulking agent is fed into the designed vessel where it is aerated and mixed with augers and other devices; it is later moved to the discharge point as processed biosolids (Figure 2.8). The product is then stored in a pile for curing before being subjected to any beneficial use. Three common types of in-vessel composting reactors are vertical plug-flow, horizontal plug-flow, and agitated bin. They differ mainly in the aeration and loading/unloading systems (35).

This method uses a smaller area; it generates little particulate matter into the environment, compared to other methods. However, being an enclosed reactor, there is a need to control the potential odor in the vessel. Similarly, being a mechanical device, associated breakdown or mechanical fault of the main device or part of the device has resulted in the system's minimal acceptance (36).

*3.5.4. Vermicomposting*

This is an innovative approach to composting that involves the use of earthworms alongside microorganisms present in the sludge to produce qualitative compost (i.e., Class A) fit for land applications. The earthworms are grown within the bed of the composting system, and their population is adjusted according to the available substrates in the feed; as such, there is no need to restock. The earthworms decompose the organic materials to produce a fine-grained casting that contains plant growth regulators and other substances that validate them as biofertilizer and biopest control agents (3). This method demands effective monitoring of compost moisture content and temperature, but does not involve intensive labor and thermophilic stage for stabilization.

## 3.6. Pasteurization

Pasteurization is a process of disinfecting mostly wet sludge by inactivating the microorganisms' eggs and cysts in the sludge at a temperature greater than 158°F (70°C). The direct steam injection and indirect heat exchanger are the two methods involved in pasteurization of wet sludge; however, the latter has not been much embraced because of fouling of the exchanger by the organic matter (14).

Untreated or digested sludge at 68°F (20°C) is preheated to a temperature of 268°F (131°C), and superheated steam at a temperature of 347°F (175°C) is injected into the material before being sent into the pasteurization reactor. The pasteurized sludge (at 158°F [70°C]) is then pumped out for storage or utilization, as the case may be. Moreover, both methods of pasteurization have gained application in Europe and the U.S. (14).

### 3.7. Deep-Shaft Digestion

Deep-shaft wet air oxidation is an autothermophilic aerobic digestion process that is employed to increase energy efficiency and produce Class A biosolids. It involves treatment of sludge in a subsurface that is 250 to 350 ft (76 to 107 m) deep. The three reactor zones of the system are oxidation, mixing, and saturation zones.

The compressed air that is forced into the deep reactor facilitates thorough mixing and supplies oxygen, which promotes rapid digestion of sludge. The whole process takes 4 days to produce dewatered biosolids (30–35% solids). The offgas is collected in a fixed-film Biofilter, where it is separated and treated. The process involves the use of some polymers and is still in its embryonic state of development (3)

## 4. CONDITIONING

Conditioning is the process of enhancing the characteristics of sewage sludge for subsequent processing to meet the required standard for beneficial use or disposal. The sludge, freshly produced from wastewater, possesses some characteristics that do not aid handling and processing. There is the need to recondition it in such a way as to facilitate further processing, such as thickening, dewatering, and drying. Conditioning is a crucial step to attaining a thicker waste stream. Furthermore, different conditioning methods are employed to achieve different goals. Conditioning methods can be physical, chemical, or biological. Besides the conventional methods, such as the use of polymers, inorganic chemicals, or heat treatment, emerging methods of conditioning such as freeze-thaw cycles, acoustics, microwave pretreatment, and physical lyses are also being employed. The conditioning or reconditioning methods and their purposes are presented in Table 2.8.

### 4.1. Chemical Conditioning

This is a process of using chemicals to condition sludge primarily for dewatering. It involves coagulation of the solids and loss of the absorbed water, which may reduce sludge moisture contents between 65% and 85%. Ferric chloride, lime, alum, and organic polymers are some of the chemicals used in the conditioning of sludge. Sludge properties, such as source, type, concentration, age, and alkalinity, are important factors that affect the choice of types and dosage of sludge conditioning chemicals (14). Table 2.9 lists the dosages of typical chemicals (e.g., $FeCl_3$, $CaO$, and $KMnO_4$) for various sludge solids in the chemical conditioning process.

Table 2.8
Sludge conditioning methods and purposes

| Conditioning method | Affected process | Purpose |
| --- | --- | --- |
| Polymer addition | Thickening | Improve loading rate, degree of concentration, and solids capture |
| Polymer addition | Dewatering | Improve production rate, cake solids content, and solids capture |
| Inorganic chemical addition | Dewatering | Improve production rate, cake solids content, and solids capture |
| Elutriation | Dewatering | Decrease acidic chemical conditioner demand and increase degree of concentration |
| Heat treatment | Dewatering | Eliminate or decrease chemical use, improve production rate, cake solids content, and stabilization; some conversion may also occur |
| Ash addition | Dewatering | Provide improved cake release from belt-type vacuum filters and facilitate filter pressing; it can also result in higher filter yields and reduced chemical requirements |

*Source*: US EPA (37).

Table 2.9
Chemical dose rates

| Types of solids | $FeCl_3$ (mg/L) | $CaO$ (mg/L) | $KMnO_4$ (mg/L) |
| --- | --- | --- | --- |
| Primary | 1–2 | 6–8 | 10–40 |
| Primary + trickling filter | 2–3 | 6–8 | 10–40 |
| Primary + WAS | 1.5–2.5 | 7–9 | 10–40 |

WAS, waste-activated sludge.
*Source*: US EPA (3).

While maintaining the flocs and minimum retention time, the sludge and the coagulants are properly mixed and immediately transferred to the dewatering unit as soon as it reaches its conditioning state. The appropriate chemicals, in liquid form, are often measured and mixed with the sludge in the corrosion-resistant reaction vessel.

### 4.2. Heat Conditioning

Heat conditioning (Figure 2.9) is a process of heating biosolids to a temperature between 290° and 410°F (144° and 210°C) for a short period under a pressure of 150 to 400 psig (1034 to 2758 kN/m² gauge). It prepares conditioned sludge for thickening and dewatering processes without the use of chemicals.

Wet sludge is passed through a heat exchanger into a reactor vessel, where the temperature and pressure are reconditioned by injecting steam directly into the sludge. This process takes about 30 minutes; thereafter, the sludge is discharged into the

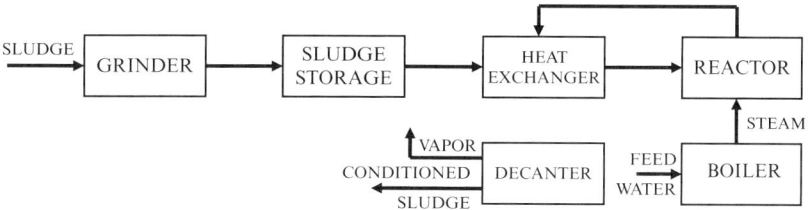

**Fig. 2.9.** Schematic diagram of heating conditioning.

thickener-decant tank for further processing. The merits of heat conditioning (14) are as follows:

1. It serves as an alternative superior to chemical conditioning in improving dewatering characteristics of treated sludge.
2. It renders the biosolids fairly suitable for disposals through various methods, like land application.
3. It helps reduce the offensive odor characteristics of biosolids.
4. It is a better option for many types of sludge that cannot be stabilized biologically.
5. It reduces the amount of heat needed to incinerate the biosolids.

Limitations of heat conditioning include the following:

1. It is capital intensive to run and maintain.
2. Operating conditions require specialized supervision and maintenance.
3. Some of the equipment is prone to corrosion.
4. The treated sludge is not free of heavy metals and as such not applicable where the presence of metals is offensive.
5. There is the need to pretreat the biosolids supernatant liquor before returning it to the treatment plant because it contains very high concentration of soluble organic compounds and ammonia nitrogen.

## 4.3. Cell Destruction

Cell destruction is one of the innovative technologies recently introduced to facilitate the conditioning of sludge. The external aids prompting cell conditioning include chemical additives, physical forces (i.e., ultrasonic exposures), and biological treatment.

### 4.3.1. Chemical Cell Destruction

This conditioning process is designed to destroy the cell membrane of microbes in the sludge in order to increase the performance of anaerobic digestion of sludge. This further increases the amount of biogas generation. Sodium hydroxide (NaOH) is added to the sludge, leaving the secondary clarifiers in another tank, where a high-pressure homogenizer or "cell-disrupter" is further employed to provide an enormous and sudden pressure drop. The process leads to lyses of the bacterial cells in the sludge. The sludge in this unit impinges on an impact ring, disrupting the cell membranes and liquefying the sludge. This may be mixed with primary and aerobically digested sludges to produce stabilized biosolids and biogas (38).

In this process, the chemical additives break down the cellular bonds, and the caustic soda used is comparatively cheaper than ferrous and aluminum salts. The accompanying biogas generation reduces the volume of biosolids for further processing, though there is a need to provide facilities that will collect the generated biogas.

*4.3.2. Ultrasonic Cell Destruction*

The ultrasonic cell destruction is primarily employed to increase the rate at which anaerobic digestion of the sludge occurs. It also improves sludge settling, facilitates denitrification, and promotes the recovery of biogas.

Extremely high pressures and temperatures are needed in order to generate gas bubbles that produce shear stresses that break up surfaces of cellular matters within the sludge. Acoustic waves are applied to achieve this prior to digestion. High- or low-frequency waves are applied based on the disintegration demands (3).

The process enhances anaerobic digestion by increasing the rate of flocculation and cell disintegration. It also increases the digestion rate, volatile suspended solids concentration, and gas production compared to the established processes. However, it is not totally absolved of problems associated with hydrolysis in anaerobic digestion and is a relatively expensive process (3).

*4.3.3. Biological Cell Destruction*

This innovative conditioning process promotes biological disintegration of organic matters in organic sludge to carbon dioxide and water. It invariably reduces the volume of the sludge. This process is particularly applicable to organic sludge of food processing industries (39).

Biological destruction of cells is typically applicable to activated sludge. This is achieved by the addition of strong chemical oxidizing agents that weaken and break down the bacterial cell walls in the sludge. The processed sludge is returned to the activated sludge unit, where it decomposes into carbon dioxide and water.

This technology is restricted to Japan (39). It facilitates the disintegration of organic matters in organic sludge and reduces the volume of the sludge to facilitate transportation.

## *4.4. Odor Conditioning*

Unprocessed sludge and the processing sludges are often accompanied with characteristically nuisance odors. These offensive odors are generated as a result of the biodegradation of the abundant proteins, amino acids, and carbohydrates present in the sludge (40). They are simply related to the decomposed organic and inorganic matters of sulfur, mercaptans, ammonia, amines, and organic fatty acids. Their release is expedited as a result of heating, aerating, and digesting effects (41). Odor generation is common at all sludge treatment processes, and may be present at the point source or in ambient air.

The following operation and maintenance practices at processing facilities are to be examined in a holistic way for controlling the odor (42, 43):

1. Siting of processing plants away from populated neighborhoods
2. Minimizing the length of time biosolids are stored
3. Avoiding land application when wind conditions transport odors to residential areas

**Table 2.10**
Odor removal efficiencies of conditioning systems

| System | $H_2S$ | $NH_3$ | Odor units (D/T) |
|---|---|---|---|
| Biofilter | >98% | >80% | >95% |
| Activated sludge (coarse bubble) | <85%–92% | >90% | 90–95% |
| Activated sludge (fine bubble) | >99.5% | N/A | >99.5% |
| Wet scrubbers | >95% | >95% | <80–99% |
| Regenerative thermal oxidizers | N/A | N/A | >95% |
| Chemical oxidants | >99% | N/A | Up to 99% |
| Counteractants and neutralizing agents | 30% | 30% | N/A |

*Source*: US EPA (40).

4. Upholding the value of air dispersion modeling, manual or computerized, which will give credible information about the necessary adjustments to make to control the odor
5. Preventing anaerobic conditions and employing oxidizing agents to prevent formation of hydrogen sulfide (44)
6. Using an equation of the impacts of blending different types of solids and storage (45)

Table 2.10 presents the reported removal efficiencies of odor conditioning systems for typical odors ($H_2S$ and $NH_3$).

As an advantage, controlling odor facilitates a good working environment for the workers within the plants and facilitates the acceptance of technologies of processing sludge in near or remote sites by the community (46).

The disadvantages of odor in processing biosolids include the difficulty of finding personnel willing to work in the plant sites, and the detrimental effects on the aesthetics and the quality of life in neighboring communities (46).

### 4.5. Electrocoagulation

Electrocoagulation is an embryonic technology applicable to biosolids conditioning. It has been used for applications in the mining and metals industries since the beginning of the 20th century, and has been employed in wastewater treatment plant in Vancouver, Canada, to remove heavy metals and emulsified oils (47). Electrocoagulation facilitates the rate at which anaerobic digestion of biosolids occurs, improves sludge settling, and facilitates denitrification and the recovery of biogas for energy production, as in the case of ultrasonic cell destruction process.

The underlying principle of the electrocoagulation process is electrolysis. Electrical current is used to dissolve the anode, which introduces chemically reactive aluminum into the wastewater stream. The very fine negatively charged ions and particles in suspension attract this positively charged aluminum to agglomerates of particles. These agglomerates increase in size until they no longer remain in suspension. However, accompanying gas bubbles, as a result of hydrolysis, make the agglomerates float on the surface of the wastewater, which is then separated by flotation (3). Continued supply of electricity facilitates maximum removal of several wastewater constituents including heavy metals, insoluble BOD and total coliforms. However, antifreeze-solvents and soluble BOD are not removed by this process.

## 4.6. Enzyme Conditioning

Enzyme conditioning is employed to degrade the sludge organic material in order to improve the following qualities:

1. Increasing dewaterability of biosolids
2. Reducing odors
3. Improving digestion processes

Enzymes have proved effective in degradation of fats and oils in meat industry wastewater treatment (4), as well as in reducing odors and chemical oxygen demand (COD) in aeration tanks and biodigesters. Its novel application to biosolids conditioning is equally promising (47).

Specially cultured bacteria and enzymes are often added to attack (feed on) organic materials, thereby converting them into carbon dioxide and water. Unlike chemical conditioning, this process offers limited or no damage to biological treatment systems. It saves cost since the employed enzymes will favor an enzymic reaction lasting for days. Also, a small amount of the enzymes is needed to accomplish the treatment of a large volume of sludge. However, specialized nutrients like humic acids and amino acids are often required to culture the enzymes.

## 4.7. Freezing

Freezing is a method of conditioning sludge and improving the dewaterability of waste treatment plant solids. It aids the separation of the solids from the sludge, thus reducing the volume of final materials for disposal.

Research has indicated that the freezing process is highly effective for conditioning sewage sludge (48). It also indicated that the freezing rate, solid contents, and the length of time the sludge remains frozen (cure time) are the most important variables to consider when analyzing the freezing process (49).

When sludge freezes, free water that surrounds the sludge flocs begins to freeze until it crystallizes. Once the free water is frozen, the interstitial water held by capillary forces between the particles is extracted by diffusion and added to the growing crystallized structure (50).

During the process of ice formation, sludge is separated and concentrated before the growing ice front. The well-arranged growing ice crystalline structure cannot accept other atoms without intense local strain due to its symmetry. Therefore, almost every solute in the water is separated, at optimum freezing rate, by the growing ice front. The freezing rate is an important variable in the freeze conditioning process. An optimum freezing rate is defined as one that allows complete dewatering of the floc particles (49). A freezing speed of 6 cm/h may be an effective rate of freezing; otherwise, the freezing process may not produce the beneficial effect of sludge dewaterability due to entrapped flock particles.

The curing time is the time at which the ice block is kept under subfreezing conditions and is closely related to the freezing time. It facilitates extra freezing time to ascertain that the ice that was frozen last had enough time to dehydrate completely and, hence, to ascertain the optimum dewatering conditions. It was recommended that at least 1 hour

Conversion of Sewage Sludge to Biosolids

of storage time at below-freezing temperatures should be employed in order to improve the sludge dewaterability (50). Sludge concentration is also an important variable in the optimization of freezing process determining a facility's ultimate size and operating costs (51).

Energy consumption and size of the conditioning facility can be reduced by 50% if the sludge is thickened, from 1% to 2%, before treatment (52). Several researchers have obtained dewatered solids content of 15% to 20% for alum sludge, 30% for iron hydroxide sludge, and between 15% and 25% solids for biological (i.e., digested wastewater) sludge.

The Electric Power Research Institute (EPRI) program sponsored the construction and operation of a mobile pilot facility to test the use of mechanically freeze/thaw conditioning on a variety of water treatment residuals and wastewater plant biosolids. It was suggested that freeze/thaw conditioning would be economical only in cases with extremely high tipping costs (greater than USD 60/ton) and low electricity costs (less than USD 0.07/kwh) (51, 53).

## 5. THICKENING

Once the sludge is conditioned, it needs to be thickened. Thickening is a process used in increasing the concentration of the sludge in order to facilitate subsequent processes like stabilization, dewatering, drying, digestion, and incineration (8). The thickening of the sludge invariably facilitates lower costs for transportation in terms of volume and pipe size, and lower pumping costs. For example, thickening liquid-solids (slurry) from 3% to 6% reduces the volume by 50%. This process is commonly used at most treatment plants, thereby enhancing effective management of the solids. This process spans established, innovative, and embryonic states of developments. Usually, thickening relies on physical methods, such as gravity thickeners (in combination with a coagulant or polymer), dissolved air flotation thickeners, and occasionally gravity belt thickeners (54). Thickening reduces the water content, and higher solids content can be achieved. However, once the solids content exceeds 10%, the thickened material is considered to be dewatered. The following are the types of thickening processes available.

### 5.1. Gravity Thickening

This usually takes place in sedimentation tanks, where the feed sludge is allowed to settle and compact, and later withdrawn from the bottom of the tank. The principle used herein is based on the tendency of dense solids to settle out of liquid and then concentrate the solids (8). Sedimentation tanks used in the thickening process are usually circular ones with conical bottoms (Figure 2.10). Tanks consist of two truss-type steel scraper arms, spanning the tanks and fixed to the shaft of a motorized mechanism. Sludge is introduced at the middle of the tank and the solids settle at the bottom, where they are pumped to the digesters or other required processes. The supernatant liquid flows out over the effluent weir to be returned to the preliminary tank or to the head works of the treatment plant. To prevent occasional odor and solidification, a high flow of dilution water is used, and chlorine is added (8). Relevant performance parameters are presented in Table 2.11. The thickening efficiency varies with the solids type and feed content. The

**Table 2.11**
Performance of conventional gravity thickening

| Types of solids | Feed (% TS) | Thickened solids (% TS) |
|---|---|---|
| Primary (PRI) | 2–7 | 5–10 |
| Tricking filter (TF) | 1–4 | 3–6 |
| Rotating biological contactor (RBC) | 1–3.5 | 2–5 |
| Waste-activated sludge (WAS) | 0.2–1 | 2–3 |
| PRI + WAS | 0.5–4 | 4–7 |
| PRI + TF | 2–6 | 5–9 |
| PRI + RBC | 2–6 | 5–8 |
| PRI + lime | 3–4.5 | 10–15 |
| PRI + (WAS + iron) | 1.5 | 3 |
| PRI + (WAS + aluminum salts) | 0.2–0.4 | 4.5–6.5 |
| Anaerobic digested PRI + WAS | 4 | 8 |

TS, total solids.
*Source*: US EPA (8).

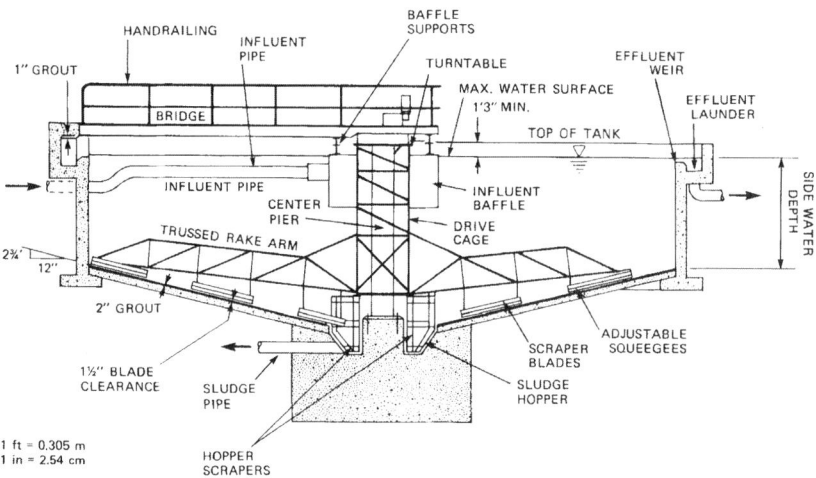

**Fig. 2.10.** Schematic diagram of gravity thickener.

solids types include primary, trickling filter (TF), waste activated sludge (WAS), and others.

Thickening in a sedimentation tank is achieved through three different settling processes: gravity settling, hindered settling, and compaction settling. Gravity settling occurs when the solids settles down in the tank under gravity by virtue of their weight. Hindered settling occurs near the bottom of the tank where the settlement rate reduces as the solids progressively concentrate. Compaction of underlying solids by the newly settled solids leads to compaction settling (8). This thickening process yields over 10% solids in the sludge and produces a well-clarified supernatant with low concentration

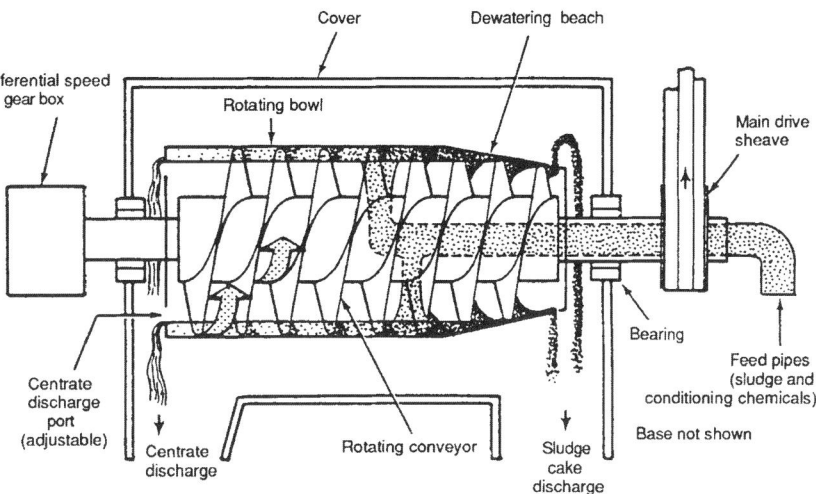

**Fig. 2.11.** Schematic diagram of countercurrent solid bowl centrifuge for sludge thickening.

of suspended solids (8). However, this process does not perform satisfactorily on most waste activated, mixed primary-waste activated, and alum or iron sludges. It is highly dependent on the dewaterability of the sludge being treated.

## 5.2. Centrifugation Thickening

This process employs the use of centrifugal forces to concentrate the solid particles within the volume of sludge. It has two advantages: thickening and dewatering. It is a high-speed process that forcefully separates wastewater solids from liquid through the use of high speed cylindrical bowl. In general, the solid produced is near "dried cake" in the cylindrical bowl (Figure 2.11) before being discharged for storage, transportation. or other end uses. The imperforate basket centrifuge (Figure 2.12) is another common type of centrifuges used for sludge thickening. The solid bowl centrifuge is conical in shape with a tapered end and is usually mounted horizontally for effective functioning.

Operating as continuous feed units, they remove solids by a scroll conveyor and discharge liquid over the weir. The imperforate basket, a batch-wise operating unit, works well for soft or fine solid particles that are difficult to filter. The solids accumulate against the wall of the bowl and the concentrated liquor is decanted primarily for disposal or recycling. The thickening process is economically viable for fairly large facilities. Table 2.12 presents the ranges of expected centrifuge performance for various types of wastewater solids. The solids content for a blend of primary and WAS vary depending on the percentage of each type of solids. Wastewater solids with a higher percentage of primary sludge can be dewatered to the higher end of the range of total solids cake. Wastewater solids with a higher percentage of WAS will probably dewater to the lower end of the range and require polymer to reach the higher end of the range.

Centrifugation requires sturdy foundations to protect against the vibrations and noise that are always associated with centrifuge operations. Adequate electric power should be furnished to run the large motors. The centrate, concentrated with suspended and

**Table 2.12**
**Range of expected centrifuge performance**

| Types of wastewater solids | Feed (%TS) | Polymer (g/kg) | Cake (%TS) |
| --- | --- | --- | --- |
| Primary undigested | 4–8 | 5–30 | 25–40 |
| WAS undigested | 1–4 | 15–30 | 16–25 |
| Primary + WAS undigested | 2–4 | 5–16 | 25–35 |
| Primary + WAS aerobic digested | 1.5–3 | 15–30 | 16–25 |
| Primary + WAS anaerobic digested | 2–4 | 15–30 | 22–32 |
| Primary anaerobic digested | 2–4 | 8–12 | 25–35 |
| WAS aerobic digested | 1–4 | 20 | 18–21 |
| Hi-temp aerobic | 4–6 | 20–40 | 20–25 |
| Hi-temp anaerobic | 3–6 | 20–30 | 22–28 |
| Lime stabilized | 4–6 | 15–25 | 20–28 |

*Source*: US EPA (55, 56).

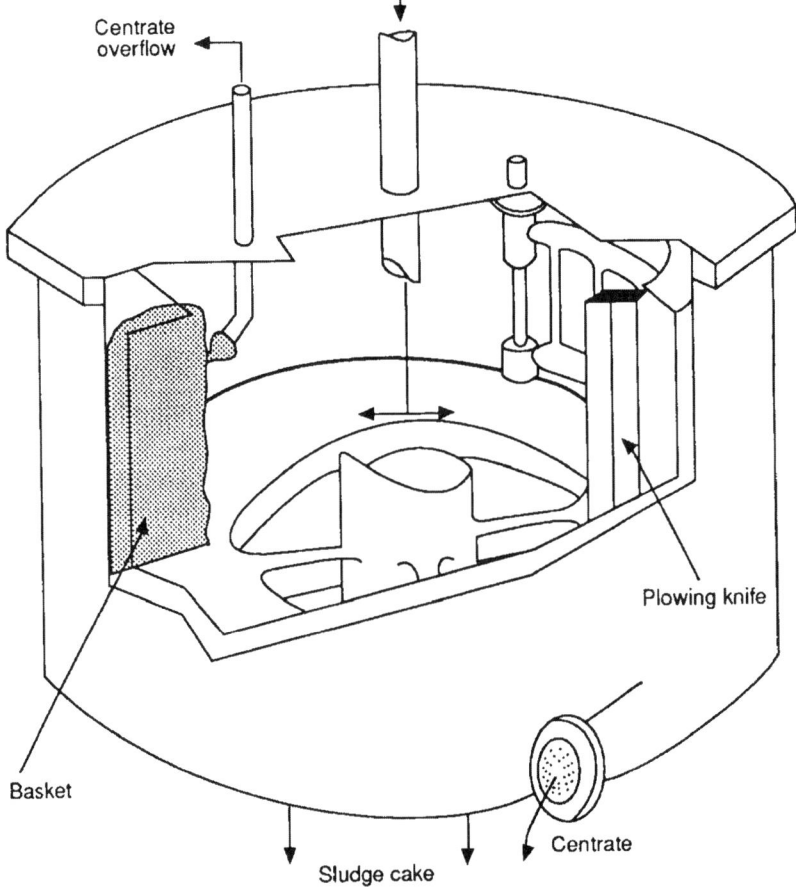

**Fig. 2.12.** Schematic diagram of basket centrifuge for sludge thickening.

## Conversion of Sewage Sludge to Biosolids

nonsettling solids, should be disposed of safely. Feeds are often subjected to degritting and screening processes to prevent blockage of discharge orifices.

### 5.3. Gravity Belt Thickening

The polymer-conditioned sludge is fed into the feed box attached to one end, which then evenly distributes the sludge across the width of a gravity belt moving over rollers driven by a variable speed motor. As the belt moves on, water drains out and the thickened sludge is discharged at the other end. Depending on its efficiency, often some features like plow blades are placed along the length of the belt, which further facilitate dewatering of the sludge. About 3% of solids thickening is achieved in this process. The relevant performance is presented in Table 2.13.

### 5.4. Flotation Thickening

Flotation thickening is widely used for waste sludge from suspended growth biological treatment processes. Applicable sludges include primary sludge, aerobically digested sludge, chemical containing sludge, and others generated in the suspended-growth process. Common sedimentation tanks used for waste activated sludge can be used for flotation thickening. A pressurized air-wastewater inlet is introduced at about half the height of the tank. The solution is depressurized and the dissolved air is released as finely divided bubbles carrying the sludge to the top as float solids, where it is removed.

It is widely accepted that float solids concentration via flotation thickening is highly influenced by the ratio of air to the solids in sludge, sludge characteristics, the rate of solids loading, and the polymer added. This process has an effectiveness of thickening the sludge to 6% solids, whereas gravity thickening yields only 3% solids (3). It also requires less land than gravity thickening; however, the stripping of volatile organic materials from the sludge has been a strong environmental concern for the choice of this process. Table 2.14 presents relevant parameters for flotation thickening processes.

### 5.5. Rotary Drum Thickening

The rotary drum thickening process (Figure 2.13) is so named because it employs the use of a rotary drum to thicken the in-fed sludge. The system contains two processing parts: a waste-activated sludge conditioning part and a screening part. The conditioning part of the drum is where the polymer and other desired conditioning additives are mixed with incoming sludge. This conditioned sludge then moves to the screening part, where as it passes through the rotating screen drum it thickens and rolls out of the drum, while the separated water also decants out through the screen.

The efficiency of this process ranges from 3% to 4% for solids. Polymer addition in this process is necessary; however, its low maintenance and small space for operation make it more useful than other methods.

### 5.6. Anoxic Gas Flotation Thickening

Also called biological flotation, this type of thickening process is one of the innovative thickening processes emerging in wastewater treatment technology. It is primarily designed to thicken contents of an anaerobic digester in order to increase digester

**Table 2.13**
Factors affecting gravity thickening performance

| Factor | Effect |
| --- | --- |
| Nature of the solids feed | Affect the thickening process because some solids thicken more easily than others |
| Freshness of feed solids | High solids age can result in septic conditions |
| High volatile solids concentrations | Hamper gravity setting due to reduced particle specific gravity |
| High hydraulic loading rates | Increase velocity and cause turbulence that will disrupt setting and carry the lighter solids past the weirs |
| Solid loading rate | If rates are high, there will be insufficient detention time for setting; if rates are too low, septic conditions may arise |
| Temperature and variation in temperature of thickener contents | High temperatures will result in septic conditions; extremely low temperatures will result in lower setting velocities if temperature varies, setting decreases due to stratification |
| High solids blanket depth | Increase the performance of the setting by causing compaction of the lower layers, but it may result in solids being carried over the weir |
| Solids residence time | An increase may result in septic conditions; a decrease may result in only partial setting |
| Mechanism and rate of solids withdrawal | Must be maintained to produce a smooth and continuous flow otherwise, turbulence, septic conditions, altered setting, and other anomalies may occur |
| Chemical treatment | Chemicals, such as potassium permanganate, polymers, or ferric chloride, may improve setting and/or supernatant quality |
| Presence of bacteriostatic agents of oxidizing agents | Allows for longer detention timers before anaerobic conditions create gas buckle and floating solids |
| Cationic polymer addition | Helps thicken waste-activated solids and clarify the supernatant |
| Use of metal salt coagulants | Improve overflow clarity but may have little impact on underflow concentration |

*Source*: US EPA (8).

capacity. It simply involves the use of digester gas to separate and thicken solids removed during anaerobic digestion. These solids are removed and returned to the digester while the gases are also discharged to the digester. The process is very similar to dissolved air flotation thickening; however, digester gas is used in anoxic gas flotation thickening, while ordinary air is used in air flotation. The efficiency of this process ranges between 6% and 10% (3). Reports from a few plants using this process in the U.S. show that odor and the amount of polymer required are relatively reduced compared to the established processes (3).

**Table 2.14**
**Thickening by dissolved air flotation**

| Sludge type | Feed solids concentration (%) | Typical loading rate without polymer (1b/ft²/day) | Typical loading rate without polymer (1b/ft²/day) | Float solids concentration (%) |
|---|---|---|---|---|
| Primary + WAS | 2.0 | 20 | 60 | 5.5 |
| Primary + (WAS + FeCl$_3$) | 1.5 | 15 | 45 | 3.5 |
| (Primary + FeCl$_3$) + WAS | 1.8 | 15 | 45 | 4.0 |
| WAS | 1.0 | 10 | 30 | 3.0 |
| WAS + +FeCl$_3$ | 1.0 | 10 | 30 | 2.5 |
| Digested primary + WAS | 4.0 | 20 | 60 | 10.0 |
| Digested primary + (WAS + FeCl$_3$) | 4.0 | 15 | 45 | 8.0 |
| Tertiary, alum | 1.0 | 8 | 24 | 2.0 |

*Source*: US EPA (8).

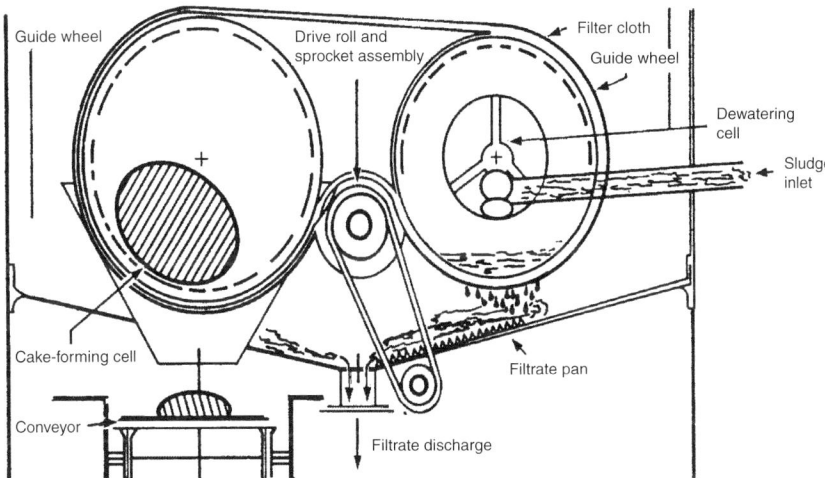

**Fig. 2.13.** Schematic diagram of rotary drum thickening unit.

## 5.7. Membrane Thickening

This innovative thickening process is mostly applicable to waste-activated sludge. It employs the use of a basin with a suspended biomass and a membrane system to thicken out the solids, mostly biomass, from the water, particularly in an aerobic environment. The process, therefore, requires continuous oxygen mixing to function well. The types of membranes available for this process include tubular, hollow-fiber, spiral wound, plate and frame, and pleated cartridge filters. The membrane is of a smaller footprint than the membranes used in the established thickening technologies. Efficiency of about 4% for solids has been reported of this process (38). This process is being employed in many U.S. cities, such as Dundee, Michigan, and cities in Georgia (38).

## 5.8. Recuperative Thickening

The recuperative thickening process facilitates the removal and thickening of digested biosolids from the anaerobic digestion process. They are later recycled to the digester. It is an innovative wastewater treatment technology that, besides thickening, enhances biosolids reduction, destruction, gas production, and dewatering, and it has other beneficial uses. It is similar to the aforementioned anoxic gas filtration. Moreover, established biosolids processing equipment can be applied to this process. The process has not gained wide use in the U.S.; a full-scale test at the Spokane, Washington, Advanced Wastewater Treatment Plant has had an encouraging success (3).

## 5.9. Metal Screen Thickening

This is an embryonic thickening technology that facilitates conditioning and thickening of sludge in one basin. It employs the use of a set of slit screens with 1-mm openings in a mixing tank to thicken the sludge, which is filtered by cross-flowing through the screens. The system is set to overcome the usual clogging associated with membrane problems. There has not been any reported field application of this process, but the pilot test indicated a sludge treatment of about 200 kg solids per hour (3).

## 6. DEWATERING AND DRYING

Dewatering and drying are physical processes employed in sludge treatment to remove or reduce a substantial amount of the moisture content of sewage sludge for further processing and beneficial uses. This process is able to transform sludge into solids with contents ranging between 10% and 40% total solids. Once the solids content exceeds about 25%, the material can be formed into a ball that will hold its shape, and it becomes difficult to squeeze any excess water from the material. Besides improving treated sludge handling and reducing the cost of transportation, dewatering and drying enhance the sludge for further processing, like incineration, composting, and landfilling, depending on the solids content. Mostly, the process renders the sludge odorless and nonputrescible, since the absence of moisture content and high temperature will deactivate most microorganisms or further anaerobic digestion of the sludge.

The principle of dewatering is based on natural evaporation and percolation as well as mechanical dewatering using physical means like filtration, squeezing, and compaction. The types of sludge to be dewatered, the nature and purpose of the dewatered product, and the available space have been identified as the criteria for selecting the dewatering process. Drying, on the other hand, is accomplished naturally by vaporization or mechanically by supplying external heat to provide latent heat necessary for evaporation. To effectively remove the moisture contents, sludge treatment is best preceded by dewatering. Studies clearly show that the two cannot be used interchangeably because one complements the other.

## 6.1. Belt Filter Press

The belt filter press is a mechanical device employed to dewater sludge and biosolids to a moist "cake" with the use of pressure (57). The belt filter press is like a conveyor belt running over guided rollers (Figure 2.14) incorporating the following basic features:

# Conversion of Sewage Sludge to Biosolids

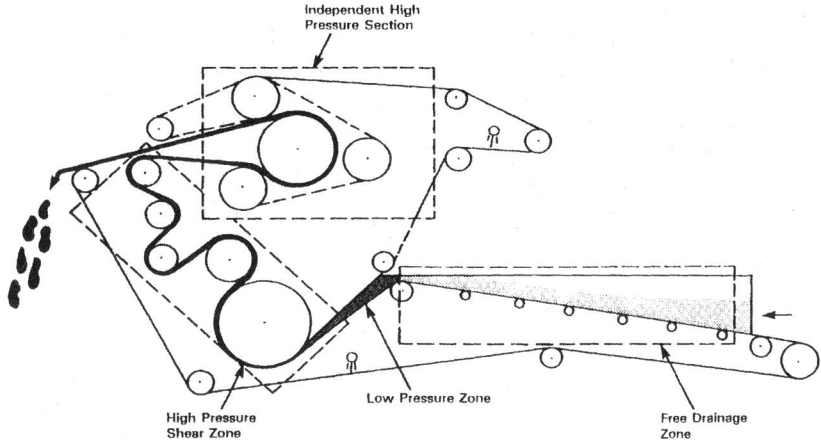

**Fig. 2.14.** Belt filter press for dewatering.

a polymer conditioning zone, a gravity drainage zone, a low-pressure squeezing zone, and a high-pressure squeezing zone. The basic principle of operation is that another belt running the length of the filter belt in the same direction is pressed on the filter belt by means of pressure rollers to dewater the moist or wet sludge that is continuously fed onto the filter belt. In another unit, a low-pressure section, the sludge is subjected to shear forces as the belts move through a number of rollers where it is further dewatered. The resulting cake is scrapped off the surface of the belt and the system works on a continuous basis.

The process is free of auxiliary problems like vacuum system and sludge-pickup problems that are associated with rotary vacuum filters. However, the penetration of the filter belt by the sludge, typical of filter belt units, can be corrected, often as the system proceeds, by coagulating the sludge with polymeric flocculants. Odor problems and the need to wash the belt at the end of each operational shift are among the disadvantages of this process (57). Belt presses require more operator attention if the feed solids vary in their solids concentration or organic matter. This should not be a problem if the belt presses are fed from well-mixed digesters (58, 59). Typical data for various types of sludge dewatered on belt filter presses are presented in Table 2.15.

## 6.2. Recessed-Plate Filter Press

A recessed plate filter (Figure 2.15) is a mechanical device used in dewatering sludge and biosolids to produce a high-grade cake of solids concentration (60). The two common types of recessed-plate filter press in use are the fixed-volume and variable-volume recessed plate filter presses.

The fixed-volume recessed plate filter press consists of series of polypropylene square concave plates with a hole in the middle. They are recessed on both sides and arranged in a vertical position facing each other, thereby creating a chamber to pressurize solids and squeeze out liquid through a filter cloth lining the chamber. Mostly chemically conditioned sludge is pumped into the center hole to fill each chamber; then high pressure (about $960\,kN/m^2$) is applied for 1 to 3 hours to force the liquid out through the

**Table 2.15**
**Typical data for various types of sludge dewatered on belt filter presses**

| Types of wastewater sludge | Total feed solids (%) | Polymer (g/kg) | Total cake solids (%) |
|---|---|---|---|
| Raw primary | 3 to 10 | 1 to 5 | 28 to 44 |
| Raw WAS | 0.5 to 4 | 1 to 10 | 20 to 35 |
| Raw Primary + WAS | 3 to 6 | 1 to 10 | 20 to 35 |
| Anaerobically digested primary | 3 to 10 | 1 to 5 | 25 to 36 |
| Anaerobically digested WAS | 3 to 4 | 2 to 10 | 12 to 22 |
| Anaerobically digested primary + WAS | 3 to 9 | 2 to 8 | 18 to 44 |
| Aerobically digested primary + WAS | 1 to 3 | 2 to 8 | 12 to 22 |
| Oxygen activated WAS | 1 to 3 | 4 to 10 | 15 to 23 |
| Thermally conditioned primary + WAS | 4 to 8 | 0 | 25 to 50 |

*Source*: US EPA (57).

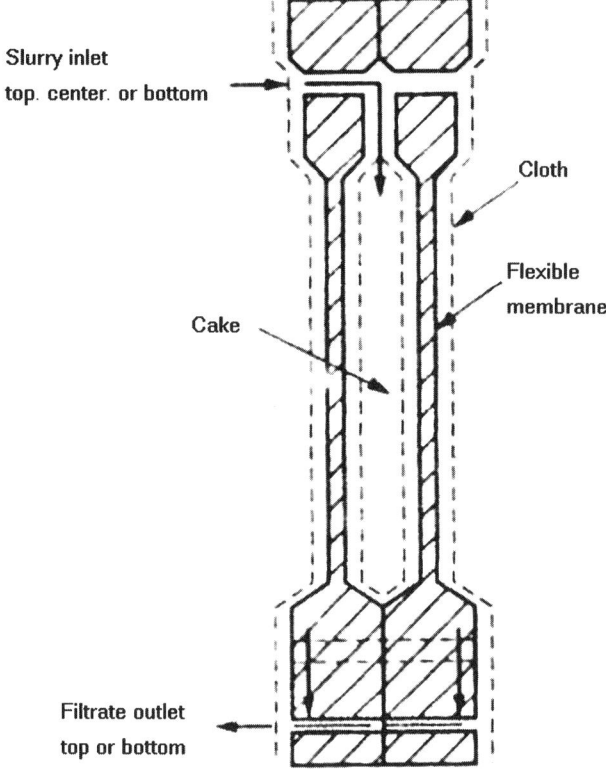

**Fig. 2.15.** Cross section of recessed plate filter press.

filter cloth and holes. After this, the plates are separated to release the caked biosolids with moisture content varying from 48% to 70%. Unlike the belt filter press, this process can be operated only in batches in a cycle of 2 to 5 hours, though it could be automated to make it faster (60).

**Table 2.16**
**Performance of plate filter press for various types of domestic wastewater sludges**

| Types of wastewater sludge | Feed TS (%) | Typical cycle time (hours) | Cake TS (%) |
|---|---|---|---|
| Primary + WAS | 3–8 | 2–2.5 | 45–50 |
| Primary + WAS + trickling filter | 6–8 | 1.5–3 | 35–50 |
| Primary + WAS + $FeCl_3$ | 5–8 | 3–4 | 40–45 |
| Primary + WAS + $FeCl_3$ (digested) | 6–8 | 3 | 40 |
| Tertiary with lime | 8 | 1.5 | 55 |
| Tertiary with aluminum | 4–6 | 6 | 36 |

*Source*: US EPA (60).

The variable-volume recessed plate filter press is similar to fixed-volume press particularly in design and setup, except for the presence of a rubber diaphragm. When pressure is applied as in the fixed-volume press, the diaphragm expands to achieve maximum squeeze of the sludge to produce the cake. The process steps, include filling, compression with applied pressure ranging from 100 to 300 psi (690 to 2070 $kN/m^2$), opening the press, discharging the cake, and closing the press per unit batch takes less than 1 hour. Though it saves time and is applicable to a wider variety of sludges, this type of filter press requires considerable maintenance (60). Its performance data for various domestic wastewater solids are presented in Table 2.16.

## 6.3. Centrifuges

Although centrifugation has been an old process applied to wastewater sludge to thicken the slurry or remove settleable solids in sludge, it has been extended to the dewatering of biosolids. The following are the centrifugal devices often employed.

### 6.3.1. Solid-Bowl Centrifuge

Solid-bowl centrifuge is a continuously fed system that removes solids by scroll conveyor and discharge liquid over a weir. The conical-shaped bowls lift solids out of the sludge feed and allow them to dry to cake that contains approximately 70% to 80% moisture, before being discharged (56). This system may be used to dewater sludge without chemical conditioning, though the solids capture improves when the solids are conditioned with polymer that is added in-line or in the bowl of the centrifuge.

### 6.3.2. Imperforate Basket Centrifuge

This process is mostly applicable to activated sludge with no chemical conditioning. It concentrates and dewaters the feed sludge at a 90% solid capture rate. The feed sludge is first skimmed to remove soft sludge from the inner wall of the basket before the initiation of plowing. However, the solid capture rate of the system can be improved by longer residence time and chemical conditioning.

## 6.4. Drying Beds

Sludge drying beds are used to dewater mostly digested biosolids by evaporation due to exposure to the air and drainage through the sludge mass. The dried sludge is often disposed off as landfill or as soil conditioner. The drying beds are the most widely used low-cost process for dewatering biosolids. The common types of drying beds include the following.

### 6.4.1. Conventional Drying Beds

Typical sand drying beds are made up of layers of gravel (8 to 18 in [20 to 46 cm]) covered with a fine sand layer (4 to 9 in [10 to 23 cm]), and they have underdrains that are spaced 8 to 20 ft (2.4 to 6.1 m) apart. The underdrains consist of lateral drainage lines made of vitrified clay pipes laid with open joints with a minimum slope of about 1% (14).

The drying area is partitioned to individual beds of 20 ft (6.1 m) wide and 20 to 100 ft (6.1 to 30.5 m) long, convenient to be filled in a normal loading cycle. Two or three creosoted planks mounted on top of each other to a height of 15 to 18 in (38 to 46 cm) are commonly used to partition the interior of the bed. The outer boundaries of covered beds and open drying beds are made up of concrete walls and creosote or earthen embankment, respectively. Piping to the sludge beds is often made of cast iron or plastic pipes and is designed for a velocity of at least 2.5 ft/s (0.76 m/s). An important concern of the operators of such beds is draining, which is addressed by making provision to flush the lines and to prevent their freezing in cold weather. Furthermore, distribution devices are put in place to control the flow of sludge into the desired beds. Splash plates are placed in front of the sludge outlet to spread the product over the beds and to prevent erosion of the bed sand (14).

The dried sludge is often coarse and dark brown in color and has about 60% moisture content. It can be removed mechanically using a shovel or another electric form of scrapers; however, there should be provision for driving a truck that negotiates the length or side of the beds to facilitate loading. Moreover, open beds must be sited away from residential areas because of the occasional odor. Weather conditions, sludge characteristics, land values, and the proximity of residential areas are a few factors that affect the design and use of drying beds. Furthermore, drying beds are generally restricted to well-digested or stabilized sludge because raw sludge is odorous, attracts insects, and does not dry well at reasonable depths. Sand bed pores are often clogged with oil and grease, thereby retarding the drainage rate (14).

### 6.4.2. Paved Drying Beds

The paved version of drying beds is of two types: drainage and decanting. The drainage type, which functions like the conventional drying bed, is usually rectangular in shape, 6 to 15 m wide by 69 to 151 ft (21 to 46 m) long, and with vertical side walls of 7 to 10 ft (2 to 3 m). The base of drying beds are lined with concrete or bituminous concrete. The lining should have a minimum of 1.5% slope to the center of the unpaved drainage area. Paved beds allow the use of mechanical equipment for cleaning, and as such facilitate removal of a higher moisture content of sludge than in the case of hand cleaning in conventional beds.

The decanting type of paved bed is designed to enhance decanting of the supernatant and mixing of the drying sludge for enhanced evaporation. About 20% to 30% of water can be decanted by this system and can produce 40% to 50% solids concentration in 30 to 40 days' drying time of continuous sunny days as in an arid climate (3, 61).

### 6.4.3. Vacuum-Assisted Drying Beds

This process is employed to accelerate dewatering and drying of biosolids in drying beds. It is achieved by connecting a vacuum system to the underside of the porous filter bed. This drying bed is often fed with polymer-conditioned sludge, which is dewatered by gravity drainage followed immediately by application of vacuum. The treated sludge is allowed to air dry for to 2 days before being removed, and the surface of the porous plates are washed with a high-pressure hose to remove the remaining sludge residue. This process helps to reduce the cycle time required for sludge dewatering (14).

### 6.4.4. Artificial Media Drying Beds

This type of drying bed employs a porous medium underlying the beds to drain off the moisture content of the sludge. The artificial media used most often are stainless steel, wedge wire, and high-density polyurethane formed into panels. The wedge wires with 0.8 ft (0.25 m) slotted openings between the bars have their flat part on top and are arranged into panels to form a false floor for control and effective drainage. The advantages of artificial media drying beds over conventional sand beds include the following:

1. Nonclogging
2. Constant and rapid drainage
3. Easy maintenance
4. Drying of aerobically digested sludge
5. Higher throughput

The high-density polyurethane media system employs the use of interlocking panels of about 40 ft (12.3 m) square, placed on a sloped slab or in prefabricated steel, and self-dumping trays for draining of the moisture content of sludge. Advantages of this the system include the following (14):

1. Can be used for opened or covered beds
2. Ability to dewater dilute and anaerobically digested waste-activated sludge
3. Very low amount of suspended solids
4. Easy cleaning of the fixed units of the setup

### 6.4.5. Quick Drying Beds

This is an innovative form of drying bed that enhances the drying of biosolids to a higher solids percentage (30% to 60% dry cake). The system consists of a series of pipes underlying the base of a sand bed (0.8–1 in [2–2.5 cm] thick), which drains water that enters the bed before the sludge is applied. A honeycomb grid is placed on the base filled with 10- to 15-mm rocks and then covered with a final layer of sand to make the bed. The system has an auxiliary in-line polymer preparation subsystem that injects polymer into the flocculation device. Sludge fed onto the bed is dried by gravity drainage and natural evaporation processes. The sludge flows evenly across the bed surface until the bed is saturated and the entrapped air is forced out of the filter media. When the underdrain is opened, a siphoning effect is created that causes rapid dewatering of the sludge by cracking the sludge, thereby allowing air to circulate around the cake, and hence further increases drying. A 45% to 60% dry cake can be produced within 5 to 7 days, and more moisture content can be removed as long as the sludge remains on the bed (62).

### 6.5. Vacuum Filtration

Vacuum filtration is a process of dewatering sludge to produce a cake with a moisture content that facilitates handling and subsequent processing using a rotary vacuum filter. The rotary vacuum filter consists of a partially submerged cylindrical drum in a sludge vat. The radial sections of the drum are connected to ports in a valve plate at the hub through internal piping technique. These plates rotate in contact with fixed valve plates that have similar ports, which are connected to a vacuum supply, a compressed-air supply, and an atmospheric vent. The applied vacuum draws liquid through the filter media to form a partially dewatered sludge cake that emerges from the liquid sludge pool; this is further dewatered by suction. The cake rotates alongside the drum until the scrapping point, when it is removed. The belt and drum types use cloth media, while the coil type uses stainless steel springs. Overall, the machines and sludge processing require high operating skill (14, 63).

The vacuum filter design and performance data are listed in Table 2.17. The sludge type and conditioning procedure may interrupt operation. Poor release of the filter cake from the belt is occasionally encountered. Sometimes the operation cost may be extremely high if the sludge is hard to dewater.

Table 2.17
Vacuum filter design and performance data

| Type of sludge | Loading rate, lb/ft$^2$/h (kg/m$^2$/h) | Sludge cake, % total solids |
|---|---|---|
| Undigested sludges | | |
|   Primary | 5–10 (24–49) | 25–32 |
|   Primary and tickling filter | 3–6 (15–29) | 20–28 |
|   Primary and activated sludge | 2–5 (10–24) | 16–26 |
|   Activated sludge | 1–2 (5–10) | 12–20 |
| Digested sludges | | |
|   Primary and trickling filter | 4–6 (20–29) | 20–28 |
|   Primary and activated sludge | 3–5 (15–24) | 2–24 |
| Lime (raw primary) sludge | | |
|   Low lime | 3–6 (15–29) | 25–30 |
|   High lime | 5–10 (24–49) | 30–40 |
|   Polymer (raw primary) sludge | 8–10 (39–49) | 25–38 |

*Source*: US EPA (54).

## 6.6. Electro-Dewatering

This is an innovative process of dewatering biosolids with the use of direct current. The voltage applied to the biosolids mixture in the initial stage of dewatering leads to electrophoresis, which is the migration of particles to electrodes of opposite charge.

As the sludge turns to cake, electrophoresis is replaced with electro-osmosis, where ions migrate to the appropriate electrode to compensate for particle charges. Electro-dewatering can be combined with conventional filter presses. Recent studies indicate that the technique can be applied to dewater a wide range of sludge, particularly those with high conductivities (5, 64).

### 6.6.1. Electroacoustic Dewatering

This type of electro-dewatering process combines the use of electrical field and ultrasound waves to enhance effective dewatering of biosolids. The electric field, as in the case of electro-dewatering, facilitates electrophoresis and electro-osmosis, while the ultrasound waves produce acoustical force, which helps to maintain a steady flow of electricity throughout the biosolids. The introduction of the ultrasonic waves led to a decrease in specific energy consumption, increase in filtration rate, and cleanliness of the cathode. Studies indicated that this process produces cakes that have a 3.4% to 10.4% increase in solids content compared to conventional dewatering (3, 65). The technique is in its embryonic state of development.

### 6.6.2. Electro-Osmotic Dewatering

This process, also in its embryonic state of development, has been successfully used in the ceramics and construction industries for dewatering of ceramic products and soil at building foundations, respectively. In sludge dewatering, the process involves the use of imposed electric field to force ionic particles in a biosolids mixture to migrate to their attractive electrodes, which is followed by electro-dewatering of the sludge (3).

## 6.7. Metal Screen Filtration

The inclined screw press is a form of metal screen filtration for dewatering of sludge. It is cost-effective with simplified operations and lower polymer usage. Liquid sludge fed into a flocculation reactor via a pump is dosed with polymer in a static inline mixer. This flocculated and conditioned liquid sludge overflow into a stainless steel vat containing rotating inclined screws (approx. 20°) and wedge wire screen (200 μm) where the moisture content of the sludge is filtered out. The dewatering is achieved in the system due to the frictional force at the sludge/screen interface as well as the increased pressure caused by the outlet restriction. The screen basket design accounts for the system's effectiveness. Its lower and wider sections function as a pre-dewatering zone where free water drains by gravity. The second section, which is the pressure zone, has baskets with reduced diameter that enhance sludge compression between the narrow flights of the screw. The dewatered sludge cake (20% to 25% solids) driven through a gap between the cone and the basket is allowed to drop on conveyors or other forms of containers. This system with less noise, low vibration, and overall low wear due to its slow rotational speed has shown advantages over other technologies. It has been installed in some cities in Utah and Maine (3, 66).

## 6.8. Textile Media Filtration

### 6.8.1. Bucher Hydraulic Press

The Bucher hydraulic press is an innovative form of textile media filtration that uses a hydraulic device to increase the sludge cake's solids content. The press consists of a cylinder, a moving piston, and several layers of cloth filters. The moving piston squeezes the sludge in the cylinder and the water moves through the layer of filters and is collected and recycled into the wastewater treatment system. The "dry" cake in the cylinder is then removed. This procedure indicates that the system's operation is semicontinuous (67). Though very similar to a belt filter press, it has shown a 25% solids content improvement. It is widely used in Europe and North America, particularly in the food and beverage processing industries (3, 68).

### 6.8.2. Drainer System

This process, in its embryonic state of development, is a form of textile media filtration (3). It is primarily a low-maintenance dewatering device that produces, at low cost, a dewatered sludge cake with a solids concentration of up to 90%. The conical gravity flow-filtration-drainer mechanism of the system's dehydration unit removes a maximum quantity of water from the sludge. The mechanism is immersed in flocculated sludge, and the filtrate flows by gravity in between the spaces of the double-walled cylinder of a fine mesh filter medium on a stainless steel frame. The weight of additional sludge to the cone compresses the sludge to increase its solids concentration at the bottom. The system is equipped with an internal high-pressure jet mechanism that cleans the filter medium at set intervals. It is very similar to the Imhoff Ian vacuum filtration system and has been embraced widely in Scotland and Sweden, where the technology is believed to have originated (69).

### 6.8.3. Geotextile Tube Container

The geotextile tube container is another innovative version of textile media filtration process for dewatering of biosolids. It has been used in the past to dewater dredge materials from the stripping of harbors, but has been improved by its application to the dewatering of fine-grained biosolids. Its high-strength polypropylene fabric coupled with a geosynthetic tube allows effluent to permeate through the tube wall, while retaining the fine grain solids of the feed. The efficiency of the system can be improved with the use of chemically activated sludge, particularly to give a clarified effluent worth recycling into the plant. At the end of each batch, the dewatered biosolids gain more dryness by desiccation due to the escape of water vapors through the geotextile, and the cake is finally discharged having the required dryness (70). Advantages of the system compared with established filter press include odor reduction, less potential for spills, and improved biosolids handling (71).

### 6.8.4. Simon Moos

This is a dewatering process that is also in its embryonic state of development. It is often used as an on-site dewatering technique for sludge from septic tanks, geese traps, small wastewater treatment plants, and some industrial sludge. It has three distinct parts: the dosing plant, an attached or separate pump, and a dewatering container. The polymer-treated sludge is pumped from the dosing plant into the dewatering container, where the water flows through a special set of filter nets installed inside the container and outdrain ports located on each side of the container (3).

*6.8.5. Tubular Filter Press*

The tubular filter press is another dewatering technique in its embryonic state of development. It facilitates dewatering and thickening of inorganic sludge. It consists of tube-shaped filter presses made of propriety fabric. Sludge pumped at high velocity is dewatered in the press to form a cake around the tube walls. The cake is then dislodged by a roller, drained further, and collected in a hopper. This process is widely used for dewatering of mining wastewater. It is also employed in South Africa to treat drinking water (58, 72).

## 6.9. Membrane Filter Press

This is a new version of a filter press designed to increase the percent of solids in biosolids cake. It contains chambers and plates darned with filter cloth. Liquid biosolids are pumped into the chamber, where the clear water or filtrate passes through the filter just as in the convention filter press. On completion, the formed filter cake is further compressed by inflating the filter membrane with pressurized fluid, thereby reducing its volume. The final cake is then discharged for beneficial uses. The membrane filter press has been employed in chemical processing industries, pharmaceuticals, food product and ingredient manufacturing, and industrial waste dewatering and recycling (3).

## 6.10. Thermal Conditioning and Dewatering

The thermal conditioning process is used to coagulate the biosolids, break down the gel structure, and invariably dewater the biosolids by reducing the water affinity of the sludge solids as a result of heat treatment. In most cases, biosolids that are difficult to stabilize or condition are subjected to thermal conditioning for further processing of the biosolids (14).

## 6.11. Drying

Sludge drying is a process of dewatering sludge by naturally or mechanically vaporizing of its water content to the air, in order to increase its incinerating efficiency or processing into fertilizers. Drying enhances volume and weight reductions for the ease of transportation, deactivation of biological processes, and grinding of the sludge (3). The evaporation of the water content in the sludge is accomplished by the difference in the atmospheric pressure. With the use of mechanical means, auxiliary heat is used to heat the ambient air in the surrounding of the sludge, thus providing the latent heat necessary for evaporation (14).

Evaporation rate of biosolids is given as

$$W = k_g (H_s - H_a) A \tag{3}$$

where

$W$ = evaporation rate, kg/h
$k_g$ = mass-transfer coefficient of gas phase (kg/m$^2$ · h) per unit of humidity difference ($\Delta H$)

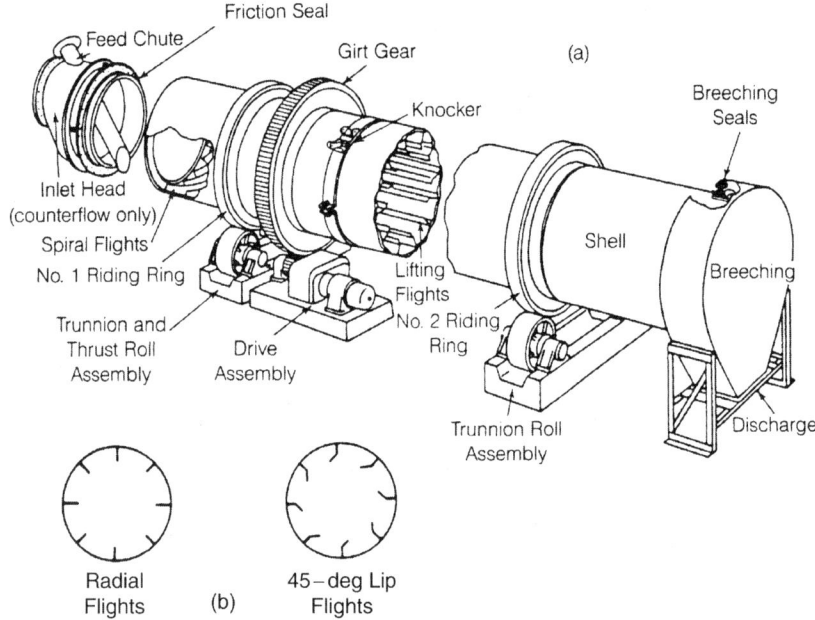

**Fig. 2.16.** Rotary dryer. (a) Isometric view. (b) Alternative flight arrangements.

$H_s$ = saturation humidity of air at sludge-air interface, mass water vapor/mass of dry air, kg/kg

$H_a$ = humidity of drying air, mass of wet vapor/mass of dry air, kg/kg.

$A$ = area of wetted surface exposed to drying medium, m$^2$

Factors that can affect drying are increase in surface area, exposure of new area, and maximum contact between dry air and wet sludge.

### 6.11.1. Direct Drying

Direct drying is the process of allowing biosolids to come in contact with hot gases, in order to evaporate the moisture content in it. It is often applied to biosolids that are marketed as agricultural product. During processing, biosolids are mixed with the feed in order to raise the solid content of the feed mixture and to avoid the "plastic" phase, which often renders the material difficult to mix or move inside the dryer (Figure 2.16). Products of this process are durable and are usually in pellets having uniform size and texture (73).

### 6.11.2. Flash Drying

Flash drying is a form of the direct drying process where a turbulent flow of hot gases dries the pulverized sludge. The hot gases and the sludge are forced up a duct, in which most of the drying takes place. This is followed by a cyclone that separates the vapor and solids. This technique yields dry biosolids with 8% moisture content that can be used as Class A biosolids.

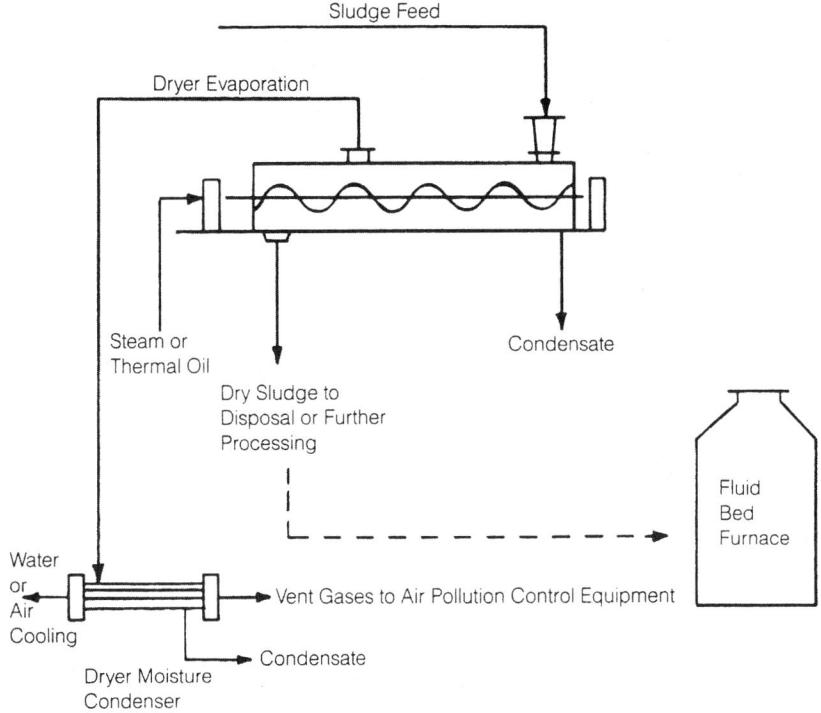

**Fig. 2.17.** Schematic diagram of hollow-flight dryer system.

An innovative form of this technique dewaters biosolids to a minimum of 25% solids after which the biosolids are discharged into the thermal stage as fine-grained spray. This technique introduces the use of sweep gas, pneumatic conveyor, rotary valve, and venture scrubber as an improvement over the old flash drying technology. It does not require biosolids to be dewatered prior to entering the unit (74).

*6.11.3. Indirect Drying*

In indirect drying process, the biosolids are not in direct contact with the heating medium or source. The heating medium is usually metal walls of varying designs. Moisture content of biosolids evaporates when it comes in contact with the surfaces of the metal walls. The heat transfer surfaces have a series of hollow metal paddles mounted on a rotating shaft. The shaft, through which the heat flows, agitates the solids in order to improve maximum heat transfer and facilitates the movement of the biosolids through the dryer. The products of this technique are oversized pellets, which require further processing like granulation and compaction to increase its uniformity, consistency, and durability. A hollow-flight dryer for indirect drying process is schematically presented in Figure 2.17.

In comparison with direct dryers, indirect drying produces less odor; thus, it requires less odor control equipment and has a higher thermal efficiency. With less dust during the drying process, it has a lower risk of explosion than do direct dryers. Examples of indirect dryers include steam dryers, hollow-flight dryers, and tray dryers (73).

### 6.11.4. Belt Drying

Belt drying is a process designed to achieve solids concentrations of over 90%. It consists of two or more slow-moving belts in series with air supplied through or around the belts. Dewatered solids are thinly spread on the first belt to increase its surface area as it comes in contact with preheated air that is either blown through the belt or pumped into the area surrounding the belts. The commercially available dryers of this nature are operated at a low pressure to minimize odor. Dust releases are often associated with the biosolids drying technologies (75).

### 6.11.5. Direct Microwave Drying

In direct microwave drying, a high efficiency and multimode microwave is designed to remove excess water from waste activated sludge and reduce or destroy pathogens present in the sludge. The microwave is specifically designed to remove moisture by vibrating water. A final product of 10% moisture content (of biosolids of initial moisture content of 85%) and about 100% pathogen kill without any change in nutrient content is achieved through the application of this technique. In comparison with other drying technologies, which use natural gas as fuel, and rely on convention current to heat solid from the outside, microwave heats the material simultaneously and the heat is generated from within the material. This technology is in use in Ireland and the U.S. (3, 73).

### 6.11.6. Fluidized Bed Drying

This process offers a safer and more reliable flexible technology for converting digested or undigested municipal sludge to Class A biosolids. The process involves formation of larger granules of biosolids by pumping mechanically dewatered wet cake from storage into fluidized bed dryer. Here it is thoroughly mixed with already dried granules. The surface of the heat exchanger is immersed in the fluidized layer of the solids and the system is heated indirectly. The whole process of drying takes place within the fluid bed itself and yields a dry product that is over 90% solids, dust-free, and mechanically stable. These products can be used for land application for beneficial use as Class A biosolids or as fuel and mineral source in cement kilns. The process is widely used in chemical and pharmaceutical industries where it is required to dry pellets, powders, and granules. This drying process is in its innovative state of development and is being applied in Europe and the U.S. (3).

### 6.11.7. Chemical Drying

This drying process does not involve the use of any heating medium, direct or indirect. In this process biosolids are dried through chemical reactions, using ammonium salts (or anhydrous ammonia and concentrated organic acids), which are mixed with the biosolids. The organic acids and ammonia react exothermically to produce sulfates and phosphates. The biosolids, as a result of the heat and pressure from the reaction, become dried and sterilized. Where ammonium salts are used, the product is in the form of a hard granular material that can be combined with plant nutrients to further raise its nutritive value (3).

# 7. OTHER PROCESSES

Other sludge processing methods to achieve improved biosolids, besides common processes like drying, thickening, and composting, include the following technologies.

## 7.1. Focused Electrode Leak Locator (FELL) Electroscanning

This is an innovative method of processing biosolids, primarily to reduce its volume without digestion, thickening, dewatering, or polymer addition. It is often operated as an auxiliary process to the main treatment plants. Pumped sludge from a main treatment plant is mixed in a side-stream bioreactor to convert an aerobic dominant bacterial population to a facultative dominant bacterial population. The process enhances the selective destruction of aerobic bacteria that break down and use the remains of the aerobic bacteria and their by-products (76). These facultative bacteria are controlled by recycling them to the treatment plant where the environment favors their destruction. The technology, which reduces the volume of biosolids, employs the use of recycling process to maximize the use of microbes that process the biosolids to the desired state (77). This process has recorded a remarkable success in the treatment of biosolids in Ohio.

## 7.2. Lystek Thermal/Chemical Process

Lystek is a treatment and processing technology for the production of Class A biosolids that are free of pathogens. The process is a batch operation where the feed solids are treated with heat and the addition of a chemical under controlled conditions. The product is about 12% to 15% solids with a viscosity of less than 1500 cP. The product retains its pumping ability, which is required to minimize the cost of biosolids production, storage, transport, and land application (3). Since the retention times are very short, the system may be automated to control pH, temperature, time, and other relevant parameters to meet the product standard. The product is high-quality Class A biosolids that can be put to beneficial use in land applications. The process is employed at the Guelph Wastewater Treatment Plant in Ontario, Canada (78).

## 7.3. Kiln Injection

This process of biosolids treatment is designed to remove or reduce emission of nitrogen oxides ($NO_X$) in kiln stacks and to produce a solids blend for fuel usage. Nitrogen oxides and other related poisonous gases (VOCs, $SO_X$) are often emitted, during incineration, into the environment; they are totally unfriendly even in small amounts, which tend to accumulate with time. Biosolids can be used to reduce $NO_X$ emission and serve as an alternate fuel source.

Dewatered biosolids are injected into the kiln, particularly, at a point where the exhaust gases leave. The operating temperature is between 1600°F (870°C) and 1700°F (930°C), and the $NO_X$ emitting is reduced (or consumed) by the reaction of ammonia present in the biosolids with oxygen to form nitrogen and water. The design of the kiln varies, and the temperature and points of injection differ from one another. For biosolids to be used to augment fossil fuel, 80% dry solids are fed into the calciner combustion zone of the cement manufacturing process. Production and release of gaseous reactions are typical problems associated with this process. Emissions of carbon monoxide, sulfur dioxide, total hydrocarbons, and hazardous air pollutants (HAPs) are still an environmental challenge, though for now there has not been a record of output beyond acceptable levels. However, residual products, particularly ashes, are bound in the cement and have been found to be good additives in cement manufacturing. Despite being a new technology in the management of biosolids, it has received wide application in the U.S. and Europe (3).

## 8. CASE STUDY

A case study is cited to demonstrate the success of the conversion of sludge to biosolids and its beneficial applications (79). The City of Milwaukee, Wisconsin, has marketed and sold its treated biosolids nationwide as a fertilizer since the 1920s under the brand name Milorganite. The city blends a mixture of digested solids from one wastewater treatment plant and waste-activated sludge from a second wastewater plant. The blended waste stream is thickened using gravity belt thickeners, and then dewatered on belt filter presses. The dewatered solids stream is dried in a rotary drum dryer, and then either packaged in 40-lb (18-kg) bags for retail purchase, 50-lb (22.5-kg) bags for professional purchase, or shipped in bulk for larger consumers. The city adds ferric chloride prior to the thickener to improve its operation and to provide the iron content guaranteed in the final product. The product's development grew from the recognition by city engineers in the 1920s that they needed to find a way to dispose of the solids generated during wastewater treatment. Working with researchers at the University of Wisconsin–Milwaukee, they found that the solids nutrient content was remarkably similar to commercially available fertilizers. Subsequent field tests showed that the material was an excellent fertilizer even when overapplied, which was a distinct advantage compared to other types of fertilizers.

## 9. SUMMARY

According to the Part 503 rule, which governs the management, treatment and disposal of sewage sludge, the sludge can be converted to Class A or B biosolids, which is a valuable product applied mainly for land applications. To implement the phase transition from a waste to a beneficial resource, a series of complex and crucial solids processing works need to be executed on raw and processed sludges. Over decades of practice and research, a great number of treatment methods (physical, chemical, and biological) are used to eventually improve the offensive properties of sludge (e.g., high water content, contaminants, and pathogens), making it legally, technically, environmentally, economically, and socially acceptable biosolids. The merits of conversion include, but not limited to, the sewage sludge cleanup, the protection of surface and underground water, and the reuse of the product as a soil amendment.

## ACRONYMS

AASAD = Advanced Alkaline Stabilization Anaerobic Digestion.
BOD = biochemical oxygen demand
COD = chemical oxygen demand
DTS = dry ton solids

EPRI = Electric Power Research Institute
HAPs = hazardous air pollutants
MSW = Municipal Solid Wastes
TF = trickling filter
TS = total solids
US EPA = U.S. Environmental Protection Agency
VS = volatile solids
VOC = volatile organic compounds
WAS = waste-activated sludge


## REFERENCES

1. L. K. Wang, N. K. Shammas and Y. T. Hung (eds.) *Biosolids Treatment Processes*, 820 pp, Humana Press, Totowa, NJ (2007).
2. U.S. Environmental Protection Agency, *Multi-stage Anaerobic Digestion*, US EPA/832/F–06/031, Washington, DC (2006).
3. U.S. Environmental Protection Agency, *Emerging Technologies for Biosolids Management*, US EPA/832/R–06/005, Washington, DC (2006).
4. W. E. Toffey and M. Higgins, Results of trials and chemicals, enzymes and biological agents for reducing odorant intensity of biosolids. In: *Proceedings of the WEF Residuals and Biosolids Management Conference 2006: Bridging to the Future*, Cincinnati, OH, March 12–14 (2006).
5. B. Rabinowitz and R. Stephen, Full-scale demonstration of waste-activated sludge homogenization at the Los Angeles Country Joint Water Pollution Control Plant. In: *Proceedings of the WEF/AWWA Joint Residuals and Biosolids Management Conference*, Cincinnati, OH, March 12–15 (2006).
6. H. R. Salsali and S. Sattar, Influence of staged operation of mesophilic anaerobic digestion on microbial reduction. Presented at WEFTEC, Washington, DC (2005).
7. A. C. Robert, *Standard Handbook of Environmental Engineering*, McGraw-Hill, New York (1989).
8. U.S. Environmental Protection Agency, *Biosolids Technology Fact Sheet Gravity Thickening*, US EPA/832/F–03/022, Washington, DC (2003).
9. 40 CFR, Part 503 standards for the use and disposal of sewage sludge, *Electronic Code of Federal Regulations*, Washington, DC, June 20 (2007).
10. R. E. Weiner and R. A. Matthews, *Environmental Engineering*, 4$^{th}$ ed., pp. 510, Butterworth-Heineman, New York (2003).
11. O. S. Amuda and A. Alade, Coagulation/flocculation process in the treatment of abattoir wastewater, *Desalination*, 196, 1–3, 22–31, September (2002).
12. O. S. Amuda and I. A. Amoo, Coagulation/flocculation and sludge conditioning in the treatment of industrial wastewater, *Journal of Hazardous Materials*, 141, 3, 778–783, March (2007).
13. U.S. Environmental Protection Agency, *Biosolids Technology Fact Sheet Alkaline Stabilization of Biosolids*, US EPA/832/F–00/052, Washington, DC (2000).
14. G. Tchobanoglous and F. L. Burton, *Wastewater Engineering Treatment, Disposal and Reuse*, 3$^{rd}$ ed. Tata, McGraw-Hill, New Delhi, India (1995).
15. J. Blais, N. Meunier, G. Mercier, P. Drogui and R. D. Tyagi, Pilot plant study of simultaneous sewage sludge digestion and metal leaching, *Journal of Environmental Engineering-ASCE*, 130, 5, 516–525 (2004).

16. U.S. Environmental Protection Agency, *Innovative and Alternative Technology Assessment Manual*, US EPA/430/9-78/009, Cincinnati, OH (1978).
17. U.S. Environmental Protection Agency, *Multi-Stage Anaerobic Digestion*, US EPA/832/F–06/031, Washington, DC (2006).
18. T. Wilson and L. Potts, *An Update on Full Scale 2-Phase AG Systems-2003*, presented at WEFTEC (2003).
19. F. Chang, L. Otten, E. Lepange and V. Opstal, *Is 100% Diversion from Land Fill an Achievable Good?* A Report to New and Emerging Engineering, Policies and Practices Advisory Group, Canada (2004).
20. K. M. Foxo, S. Pillay, T. Lalbahdur, N. Rodda, F. Holder and C. A. Buckley, The anaerobic baffled reactor (ABR): an appropriate technology for on-site sanitation, *Water*, 30, 5, 44–50 (2004).
21. I. Willis and P. Schafer, Upgrading to Class A anaerobic digestion: is your biosolids program ready to make the move? *Public Works Magazine* (2006).
22. K. Esters, Columbus water works honored techniques that turns human waste into safe fertilizer, *Water Industries News* (2005).
23. C. Burnett, Pilot test result of the bio terminator high rate plug flow anaerobic digester improvements in dewaterability. In: *Proceedings of the WEF/AWWA Joint Residuals and Biosolids Management Conference*, Nashville, TN (2005).
24. J. L. Richards and Associate Ltd. *Ravens View WPCP Secondary Treatment and Capacity Upgrades Class A Update*. Technical Memo No. 5–Biosolids. Management Upgrades. Prepared for Utilities Kingson (2004).
25. D. C. Inman, S. Murthy, P. Schafer, J. Schlegel, J. Webb and J. T. Novak, A comparative study of two-state thermophilic single-sage mesophilic and temperature-phased anaerobic digestion. In: *Proceedings of the WEF/AWWA Joint Residuals and Biosolids Management Conference*, Nashville, TN (2005).
26. D. Steven, J. Kelly, C. Liston and D. Oemete, Biosolids management in England and France, *Water*, 29, 1, 56–61, (2002).
27. S. A. Lee, D. D. Drury, C. A. Baker, J. S. Bowers and R. H. Nienhuis, Three-phase thermophilic digestion disinfects biosolids, *Water Environment Research Federation Biosolids Technical Bulletin*, 8, 6, 5–7, (2003).
28. H. G. Kelly, *Engineering Technologies in Biosolids Treatment*, West Vancouver, Canada, (2003).
29. European Environmental Press, *Europe: Using Ozone to Reduce Sludge*. http://www.waterandwastewater.com/www_services/news_center/publish/article_00540.shtml (2005).
30. R. Vranitsky and J. Lahnsteiner. Sewage sludge disintegration using ozone—a method of enhancing the anaerobic stabilization of sewage sludge. In: *Proceedings of the European Biosolids and Organic Residuals Workshop, Conference and Exhibition* (2002).
31. R. S. Reimers, V. K. Sharma, S. D. Pilliai and D. R. Reinhard, Application of ferrates in biosolids and manure management with respect to disinfection and stabilization. In: *Proceedings of the WEF/AWWA Joint Residuals and Biosolids Management Conference*, Nashville, TN (2005).
32. H. Kim, P. Millner, V. Sharma, L. McConnell, A. Torrents, M. Ramirez and C. Peot, Ferrate: nature's most powerful oxidizer: it's potential as disinfection, *Treatment for Thickened Sludge*, http://www.ars.usda.gov/research/publications/Publications.htm?seq_no_115=190364 (2006).

33. A. Chao, Quality improvement of biosolids by ferrate (VI) oxidation of offensive odor compounds, *IWA Publishing Journal*, http://www.iwapouline. com/wst033030119.htm, August 8 (2006).
34. R. S. Reimers, L. S. Pratt-Ward, H. B. Bradford, F. P. Jussari and W. Schmitz, Development of neutralizer process for disinfection and stabilization of municipal wastewater residues. In: *Proceedings of the WEF Residuals and Biosolids Management Conference 2006, Bridging to the Future*, Cincinnati, OH, March 12–14 (2006).
35. U.S. Environmental Protection Agency, *In Vessel Composting of Biosolids*, US EPA/332/R–00/061, Washington, DC (2000).
36. U.S. Environmental Protection Agency, *Biosolids Technology Fact Sheet Use of Composting for Biosolids Management*, US EPA/832/F–02/024, Washington, DC (2002).
37. U.S. Environmental Protection Agency, *Process Design Manual for Sludge Treatment and Disposal*, US EPA/625/1-74/006, Washington, DC (1974).
38. Metcalf and Eddy, Inc., *Wastewater Engineering Treatment and Reuse*, 4$^{th}$ ed., McGraw-Hill, New York (2003).
39. Japanese Advanced Environmental Equipment, *BIODIET®. Global Environmental Technology Database NETT21* (2002).
40. U.S. Environmental Protection Agency, *Biosolids and Residuals Management Fact Sheet Odor Control in Biosolids Management*, US EPA/832/F–00/067, Washington, DC (2000).
41. S. S. Schiffman, J. M. Walker, P. Dalton, T. S. Lorrig, J. H. Raymer, D. Shusterman and C. M. Williams, Potential health effects of odor from animal operation, wastewater treatment and recycling of byproducts, *Journal of Agromedicine*, 7, 1, 7–82 (2000).
42. Metcalf and Eddy, Inc., *Wastewater Engineering Treatment Disposal Reuse*, 2$^{nd}$ ed., McGraw-Hill, New York (1979).
43. R. W. Moncrieff, *The Chemical Series*, 3$^{rd}$ ed., Leonard Hill, London (1967).
44. U.S. Environmental Protection Agency, and U.S. Department of Agriculture, *Guide to Field Storage of Biosolids and Other Organic By-Products Used in Agriculture and for Soil Resource Management*, US EPA/832/B–00/007, Washington, DC (2000).
45. J. R. Hentz, H. C. Lawrence and R. Alan, Separating solids solves odor emission problems, *Biosolids Technical Bulletin*, July/August (2000).
46. U.S. Environmental Protection Agency, *Biosolids Generation, Use, and Disposal in the United States*, US EPA/530/R-99/009, Washington, DC (1999).
47. D. Dursun, A. Ayol and S. K. Dentel. Pretreatment of biosolids by multi-enzyme mixtures leads to dramatic improvements in dewaterability. In: *Proceedings of the WEF/AWWA Joint Residuals and Biosolids Management Conference*, Cincinnati, OH, March 12–15 (2006).
48. T. E. Vik, *Anaerobic Digester Methane to Energy: A statewide Assessment*, A report to Focus on Energy, McM. No. W0937-920459, Wisconsin (2003).
49. U.S. Environmental Protection Agency, *Clean Sheds Needs Survey 2000*, US EPA/832/R–03/001, Washington, DC (2000).
50. Electric Power Research Institute, *Mechanical Freeze/Thaw and Freeze Concentration of Water and Wastewater Residuals*, WO-671002, Palo Alto, CA (2001).
51. J. Sheridan and B. Cutis, Case book: revolutionary technology cuts biosolids production and costs, *Pollution Engineering*, 36, 5, (2004).
52. Global Energy Partners LLC, *Solid Handling for Water and Wastewater Treatment: Tech Review*, Lafayette, CA (2006).
53. Electric Power Research Institute, *Engineering Environmental Technologies: An Analysis of New Treatment Technologies for the California Energy Commission*, Palo Alto, CA (2003).

54. U.S. Environmental Protection Agency, *Process Design Manual for Dewatering Municipal Wastewater Sludges*, US EPA/625/1-82/014, Cincinnati, OH (1982).
55. N. P. Cheremisinoff, *Handbook of Water and Wastewater Treatment Technologies*, pp. 654, Butterworth-Heinemann, New York (2002).
56. U.S. Environmental Protection Agency, *Biosolids Technology Fact Sheet Centrifuge Thickening and Dewatering*, US EPA/832/F–00/053, Washington, DC (2000).
57. U.S. Environmental Protection Agency, *Biosolids Technology Fact Sheet Belt Filter Press*, US EPA 832-F-00-057, Washington, DC (2000).
58. E. P. A. Coopman, H. P. Schivarz and M. J. Pryor, The dewatering of a mining sludge containing hexavalent chromium using a tubular filter press, *A South African Development Water Supply* 1, 5–6, 371–376 (2001).
59. R. T. Henderson and S. T. Schultz, Centrifuges versus belt presses in San Bernardino, California. In: *Proceedings of WEF/AWWA Joint Residuals and Biosolids Management Conference: Strategic Networking for the 21st Century*, Water Environment Federation, Arlington, VA (1999).
60. U.S. Environmental Protection Agency, *Biosolids Technology Fact Sheet Recessed-Plate Filter Press*, US EPA/832/F–00/058, Washington, DC (2000).
61. M. McFarland, *Biosolids Engineering*, McGraw-Hill, New York (2000).
62. A. Evans, Biosolids reduction and the Deskin quick dry filter bed. In: *Proceedings of the 65th Annual Water Industry Engineers and Operators' Conference*, pp. 67–75, Geelong, September 4–5 (2002).
63. American Society of Civil Engineers, *Wastewater Treatment Plant Design*, Water Pollution Control Federation Manual of Practice No. 8, PA (1977).
64. WERF, Demystifying the dewatering process new techniques and technologies shed light on a complex process, *WERF Progress Newsletter*. http://www.werf.us/press/springo6/dewatering.cfm, WERF Alexandria, VA (2006).
65. M. Abu-Orf, C. D. Miller, C. Park and J. J. Novak J.J., Innovative technologies to reduce water content of dewatered municipal residuals, *Science and Technology*, 1, 2, 83–91, (2001).
66. P. S. Atherton, R. Stetson, T. McGovern and D. Smith, Innovative biosolids dewatering system proved a successful part of the upgrade to the old town Maine Water Pollution Control Facility. In: *Proceeding of the 2005 WEFTEC. The Water Quality Event*, pp. 6050–6665, Washington, DC (2005).
67. G. Kolisch, M. Boehler, F. C. Arancibia, D. Pinnow and K. V. Kraus, A new approach to improve sludge dewatering using a semicontinuous hydraulic press system, *Water Science Technology*, 52, 10–11, 211–218 (2005).
68. F. Soroustian, Y. Shang, E. J. Whitman and R. Roxburgh, Biosolids and manure dewatering with a hydraulic de-juicing press. In: *Proceedings of the WEF Residuals and Biosolids Management Conference 2006, Bridging to the future*, Cincinnati, OH, March 12–14 (2006).
69. C. Rand, East of Scotland water cuts treatment costs. http://www.entingeringtalk.com/news/sim/sim102html (2000).
70. WaterSolve LLC. http://www.gowatersolve.com/geotubes (2006).
71. B. J. Mastin and G. E. Lebster, Dewatering with Geotube® containers: a good fit for a midwest wastewater facility? In: *Proceedings of the WEF Residuals and Biosolids Management Conference 2006: Bridging to the Future*, Cincinnati, OH, March 12–14 (2006).
72. R. Bastian, The biosolids (sludge) treatment, beneficial use, and disposal situation in the USA, *European Water Pollution Control Journal*, 7, 2, 62–79, March (1997).

73. U.S. Environmental Protection Agency, *Biosolids Technology Fact Sheet Heat Drying*, US EPA/832/F–06/029, Washington, DC (2006).
74. King County Department of Natural Resources, *Regional Wastewater Service Plan*, Annual Report Wastewater Treatment Division, King County, WA (2001).
75. B. Frewerd, *Harnessing the Power of Biosolids*. Krugerine Cary, NC (2006).
76. J. T. Novak, D. H. Chon, B. A. Curtis and M. Doyle, Reduction of sludge generation using the cannibal process: mechanism and performance. In: *Proceedings of the WEF Residuals and Biosolids Management Conference 2006, Bridging to the future*, Cincinnati, OH, March 12–14 (2006).
77. J. T. Novak, D. H. Chon, B. A. Curtis and M. Doyle, Reduction of sludge generation using the cannibal process: mechanism and performance. In: *Proceedings of the WEF Residuals and Biosolids Management Conference 2006, Bridging to the future*, Cincinnati, OH, March 12–14 (2006).
78. A. Singh, F. Mosher, O. P. Ward and W. Key, An advanced biosolids treatment and processing technology for beneficial application of high solids and pathogen free product. In: *Proceeding of the 3$^{rd}$ Canadian Organic Residues and Biosolids Management Conference*, Calgary, Canada, June 1–4 (2005).
79. Water and Energy Technology Application Service, *Solids Handling for Water and Wastewater Treatment*. http://www.nyserda.org/Programs/Environment/SolidsHandlingTechReview.pdf. (2006).

# 3
# Biosolids Thickening-Dewatering and Septage Treatment

## Nazih K. Shammas, Azni Idris, Katayon Saed, Yung-Tse Hung, and Lawrence K. Wang

**CONTENTS**

    INTRODUCTION
    EXPRESSOR PRESS
    SOM-A-SYSTEM
    CENTRIPRESS
    HOLLIN IRON WORKS SCREW PRESS
    SUN SLUDGE SYSTEM
    WEDGEWATER BED
    VACUUM-ASSISTED BED
    REED BED
    SLUDGE-FREEZING BED
    BIOLOGICAL FLOTATION
    SEPTAGE TREATMENT
    REFERENCES

**Abstract** This chapter deals with processes for biosolids dewatering and septage treatment. Septage is the liquid and solid material pumped from a septic tank or cesspool when it is cleaned. A selection of recent methods that show promising applications is presented. These include Expressor Press, Som-A-System, CentriPress, screw press, Sun Sludge System, wedgewater bed, vacuum-assisted bed, reed bed, biosolids-freezing bed, biological flotation, and septage management systems.

When septage is to be ultimately treated at a wastewater treatment plant or independent septage-treatment facility, a receiving station is required in order to provide preliminary treatment and equalization. Other management options discussed include land application of septage, lagoon disposal, and composting and odor control.

**Key Words** Septage • management systems • dewatering • land application • lagoons • composting • odor control

From: *Handbook of Environmental Engineering, Volume 7: Biosolids Engineering and Management*
Edited by: L. K. Wang, N. K. Shammas and Y. T. Hung © The Humana Press, Totowa, NJ

## 1. INTRODUCTION

This chapter discusses the current processes for biosolids dewatering and septage treatment. Septage is the liquid and solid material pumped from a septic tank or cesspool when it is cleaned (1). The main objective of biosolids dewatering is to remove water or moisture content, thereby reducing the residual volume (2). The end product is a sludge cake or powder that possesses solid characteristics and is no longer considered a liquid. This treatment leads to a substantial reduction of the costs of subsequent treatment and disposal. In most applications, the ultimate percent solid content of dewatered biosolids is set by the requirements for subsequent treatment and disposal options. The percent solid content for dewatered biosolids is always significantly higher than the percent solid content of thickened biosolids.

The combination of processes used for biosolids treatment prior to dewatering, transport, and disposal varies widely in many countries and from plant to plant. Generally, the dewatering process is preceded by one of the following stabilization processes: anaerobic or aerobic digestion; and thickening by gravity, centrifugation, air floatation, chemical (alkaline treatment), or heat treatment (2). In some cases, raw biosolids, particularly raw primary sludge, may be dewatered directly without prior thickening, although the handling and the method of ultimate disposal would have to be considered carefully (3). It is a common practice to have further treatment of dewatered biosolids by means of stabilization using methods such as composting. If volume and organic reduction is the target, technologies of incineration and gasification are becoming a popular choice; otherwise the dewatered biosolids ultimately may be reused by spreading on agricultural or landscaped areas or disposed off by trucks to either a landfill or designated area for land spreading (2).

A number of biosolids dewatering techniques are currently being used by many wastewater treatment plant operators. The selection of any biosolids dewatering system depends on the characteristics of biosolids to be dewatered, available space, and moisture content requirements of the biosolids cake for ultimate disposal. When land is available and the biosolids quantity is small, natural dewatering systems are more attractive. These include drying lagoons and drying beds (4). The mechanical dewatering systems are generally selected where land is not available. Common mechanical biosolids dewatering systems are more appropriate for larger plants to maximize space requirement and also to ease handling operation. The mechanical systems include vacuum filter (5), centrifuge (6), filter press (7), and belt filter press (8).

Some biosolids, particularly those that are aerobically digested, are not readily amenable to mechanical dewatering. These biosolids can be dewatered on sand beds with good results. When particular biosolids must be dewatered mechanically, it is often difficult or impossible to select the optimum dewatering device without conducting bench-scale or pilot studies. Trailer mounted, full-size equipment is available from several manufacturers for field testing purposes (9, 10). Advanced biosolids treatment processes using thermal and thermochemical processes or chemical oxidation have been developed to improve biosolids dewatering and to facilitate handling and ultimate disposal (2, 4).

When evaluating or selecting a dewatering process, one must keep in mind the affect of the prior wastewater and biosolids treatment processes as well as the subsequent use or disposal practices. The choice of a reuse strategy or disposal process is in turn strongly influenced by local, state, and federal regulations. A dewatering process cannot be evaluated without considering the other processes involved in the overall wastewater/solids handling system. This evaluation or selection can be a complex procedure because of the large number of possible combinations of unit processes available for wastewater treatment and biosolids thickening, stabilization, conditioning, dewatering, and ultimate use/disposal.

Above all, the design engineer must ensure that capacity limitations in the biosolids processing system are not the direct cost of impaired effluent quality. That is, the design should provide for sufficient standby capacity or an alternative mode of biosolids handling, whereby solids can be removed from the wet end-processing in an orderly manner, even if the primary means of biosolids disposal is unavailable or has failed in some manner. This criterion applies equally well to small and large plants, whether utilizing mechanical or nonmechanical means of biosolids disposal (2, 11, 12).

There are many new developments in the technology of biosolids dewatering and septage management. A selection of recent methods that show promising applications is presented in the following sections. This review highlights some of the systems used by municipalities, including Expressor Press, Som-A-System, CentriPress, screw press, Sun Sludge System, wedgewater bed, vacuum-assisted bed, reed bed, biosolids-freezing bed, biological flotation, and septage management systems.

## 2. EXPRESSOR PRESS

Eimco Water Technologies, Ltd., Aubum, Australia, which is one of the leading manufacturers of dewatering equipment developed a belt press using some modification to incorporate a twin belt mechanism for use primarily in the industrial market. Substantial tests have been conducted with municipal biosolids and various kinds of fibrous industrial waste sludges. The device, named the Expressor® or Expressor Press, consists of, in its basic form, two or three S-shaped rolls (wraparound) and a series of five pressure (P) rolls (direct) on which the pressure can be individually varied. An Expressor Press with this configuration is shown in Figure 3.1 (12).

In a second configuration, a unit called the Hybrid Expressor Press contains a gravity drainage section, four or five S rolls, and the five variable-pressure P rolls. Depending on the model being considered, the P roll pressure can be varied from zero above the belt tension up to 200 kg/cm (1000 lb/inch [in]). This new unit is capable of producing a very dry cake from the most difficult sludges with the use of press aids. A variety of press aids have been employed, but the most widely investigated material has been sawdust. The unit can produce an autogenous cake from waste-activated sludge using between 50% and 125% sawdust by dry weight, based on the content of sludge solids. The water displacement by the press aid varies from slightly over 1 kg $H_2O$/kg to as much as 3 kg $H_2O$/kg press aid added. The water displacement is based on the kg $H_2O$/kg sludge solids with and without press aid. The cake produced varies from 30% to 40% solids and, in some instances, runs somewhat higher than 40% (12).

Other press aids have been tested, including sand, soil, finely divided paper, fly ash, and coal fines. All work to some degree to increase the cake resistance to shear in the P rolls and hence permit higher pressure and, in turn, higher solids content. Press aids in the 30- to 80-mesh region seem to be the most effective. With materials not particularly

**Fig. 3.1.** Photograph of a typical expressor belt press.

resistant to shear, such as paper and fiber, the particle size seems to have little impact on final sludge solids.

The press has also been tested on primary sludges and on a mixture of primary and waste-activated sludge from a pulp and paper manufacturing facility. Typical cake concentrations varied from 40% to 47% solids without a press aid. Wastes from the manufacture of pulp or paper would seem to work particularly well with this equipment because of the fibrous nature of the primary sludge. Also of interest is the ability of the press to produce an alum sludge cake of 40% to 60% solids using soil as a press aid in one test and sawdust in another. In each case, the press aid used was approximately 100% of the weight of dry solids of the alum sludge.

Determination of the pressure profile is a function of the biosolids, the biosolids blend, and the quantity and nature of the press aid used. On primary and waste-activated sludge in the normal proportions (i.e., approximately 50-50) and the pure waste-activated sludge, the P roll pressures are usually tapered and vary from 10 kg/cm (56 lb/in) on the first roll to 60 to 250 kg/cm (336 to 1401 lb/in) on the last roll (12).

The biosolids in the Portland Columbia River Wastewater Treatment Plant were dewatered during a demonstration study at that facility. The activated sludge feed varied from 2.5% to 3.5% solids, and each test was run at approximately 100% of the press aid by weight. Sawdust additive yielded a cake in the range of 30% to 40% solids, while the paper press aid produced a cake from 35% to somewhat over 40% solids. The solids capacity of the press varies from 225 to 600 kg/m/h (102 to 272 lb/ft/h), and an acceptable hydraulic feed rate ranges from 1.6 to 3.2 L/s (25 to 51 gallons per minute [gpm]) on a 1-m-wide (39-in) machine. The basic press has been investigated for further dewatering of cake derived from other dewatering equipment (12).

## 3. SOM-A-SYSTEM

The Som-A-System Screw Press, which is manufactured by Somat Corp., Coatesville, PA, consists of a vertical, rotating screw enclosed by dual stainless steel screens. The screens and screw are encased in stainless steel housing with a removal cover on each side. Tiny perforations in the inner screen allow only water to escape. The outer screen has larger holes and easily collects the pressate, which sprays inside the housing and drains into a receptacle. Brushes are located along the edge of the screw to sweep the cake that builds up on the screen, allowing a clear opening for the pressate to escape (12).

The feed enters at the bottom of the screw press. A buildup of biosolids cake on the screw is recommended to obtain good dewatering. As the pressate drains, the cake becomes progressively drier and is pushed to the top, where it is discharged into a waiting dump truck or hopper. A back-pressure system is located below the discharge chute and gives the cake a final squeeze before discharge. One plant, however, removes this cone, which collected hairballs, with no adverse effect on its operation or dewatering results. The Som-A-Press is shown in Figure 3.2.

Biosolids that floc easily and are fibrous are the most conducive to a screw press operation. Feed concentration is critical to achieving high cake solids. The higher the feed concentration, the higher the cake solids and the unit capacity (kg/h). Table 3.1 reports feed solids, cake solids, and solids recovery from several different plants using the Som-A-System. The key to the action of the unit is bridging of the holes in the screen, as the bulk of the particles in the biosolids will be finer than the holes in the screen. Consequently, a proper biosolids conditioning is essential using this system. Table 3.2 presents the polymer usage of several plants using the Som-A-System.

A slow screw speed yields a better cake, although it will also decrease the throughput. High flow rates and screw speed generally result in a discharge of wet sludge. A variable speed pump regulates the feed rate to the screw press. At Pinetop, Arizona, the plant generally keeps the feed rate at 2.5 L/s (40 gpm), which is near the maximum. Biosolids that have been aerobically digested at a 20-day detention easily yield an acceptable 12% to 15% cake at a feed rate of 2.5 L/s (40 gpm). However, if the biosolids had a lower detention time, a feed rate of 2.5 L/s (40 gpm) would produce a wetter cake (12).

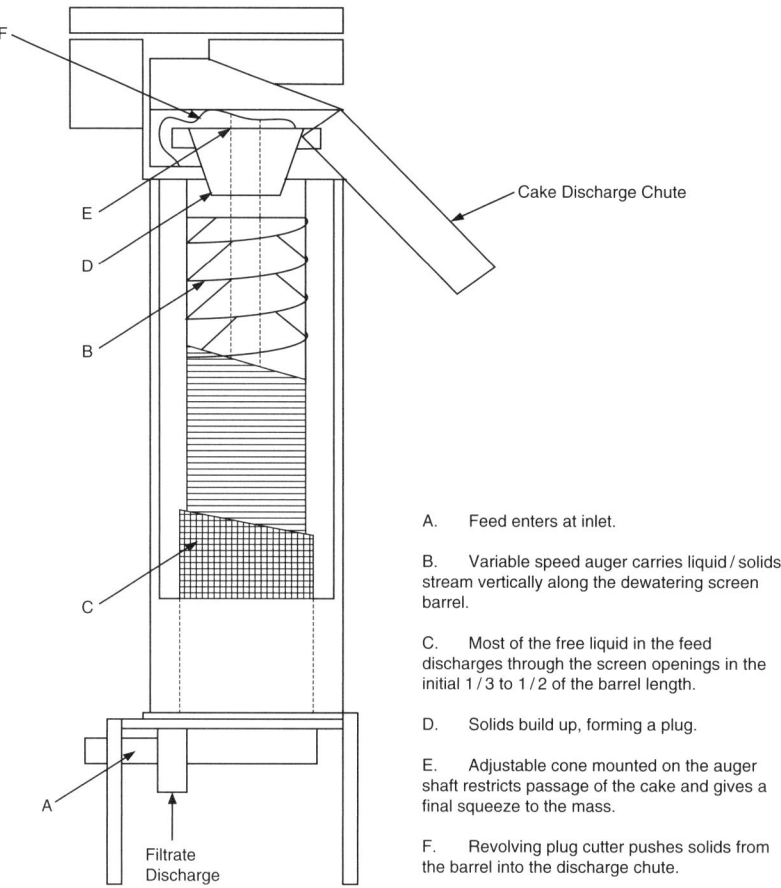

**Fig. 3.2.** Functional schematic of Som-A-Press (12).

The Som-A-System is very simple to operate and maintain. Depending on the biosolids, the press can normally be operated with only periodic checks. Many plants simply turn the machine on in the morning and periodically check the feed solids, the cake solids, and the level of biosolids in the waiting dump truck. Some operations require more attention to the feed biosolids to ensure that the proper concentration is achieved and no water is being fed to the press. The unit is also relatively easy to disassemble. General maintenance involves routine lubrication and washing the screens to prevent buildup of biosolids, which can prematurely wear the brushes. Repairs reported by plants have been limited to replacement of inner and outer screens and brushes after approximately 1500 hours (12).

The low capital cost of this screw press is a primary attraction, and comparative economic evaluation points favorably to it. It is ideal for operations with limited space requirements, since the system occupies, at a maximum, approximately 3 m$^2$ (32 ft$^2$) of floor space. On the other hand, potential drawbacks include low unit capacity, higher polymer dosage, and lower cake solids. The capacity of the presses can be a deterrent because the small throughput demands a multiplicity of units, which can be more

Biosolids and Septage Treatment

Table 3.1
Som-A-System operating data (12)

| Plant | Sludge | Average plant flow, MGD | Feed solids TSS, % | Feed rate, gpm | Cake solids, % | Solids recovery, % |
|---|---|---|---|---|---|---|
| Camden, NY | Aerob. digested | 0.6 | 1–2 | 10–24 | 10 | 85 |
| Churchville, NY | Aerob. digested | 0.11 | 2.5–3.5 | 10.5 | 12.3 | — |
| New Canaan, CT | Aerob. digested | 0.25 | 1.0–1.5 | 30–40 | 12–17 | 84–94 |
| Danville, VA | WAS/stab. | 16.2 | 5–6 | 15 | 21–23 | 86 |
|  | scum from DAF |  | 8 | 30–40 | 28–30 | 90 |
| Pinetop, AZ | Aerob. digested | 0.4 | 2 | 40 | 12–15 | 88–90 |
| Sunriver, OR | Aerob. digested | 0.5 | 0.5–0.75 | 35–40 | 7–12 | 85[1] |
| Frisco, CO | Aerob. digested | 1.0 | 2 | 15–18[2] 40[3] | 11 | — |
| Provo, UT | Anaer. digested | 1.5 | 2 | 30 | 7–15 | 87–94 |

DAF, dissolved-air flotation; gpm, gallons per minute; MGD, million gallons per day; TSS, total suspended solids; WAS, waste-activated sludge.
[1] Normally the solids recovery runs 90% to 94%.
[2] Undersized polymer pump limits feed rate 15–18 gpm; new pump ordered.
[3] With larger pump, expect to run presses at 40 gpm.

Table 3.2
Som-A-System chemical conditioning data (12)

| Plant | Polymer Name | Dosage | USD/lb | USD/ton DS |
|---|---|---|---|---|
| Camden, NY | Percol 767 | 5–20 | — | — |
| Churchville, NY | Percol 757 | 1.57 | 2.80 | 4.39 |
| New Canaan, CT | Percol 757 | 6–9 | 3.25 | 24.38 |
| Danville, VA | Cationic | 21 | 0.92 | 19.32 |
| Pinetop, AZ | Percol 757 | 8–10 | — | 27.00 |
| Sunriver, OR | Allied CC4450 | 21 | 1.65 | 34.65 |
| Frisco, CO | Percol 757 | 12 | 2.70 | 32.40 |
| Provo, UT | Percol 763 | 21 | — | — |

DS, digested sludge.

difficult to control. One of the plants expressed disappointment on the amount of polymer required, and were attempting some experiments in an effort to reduce the quantity.

## 4. CENTRIPRESS

Based on observed field demonstrations, there have been significant improvements in the capabilities of a newly designed solid bowl continuous flow centrifuge. The improvements were in the area of cake solids concentration. During testing, the centrifuge was operated in parallel with a filter press system. The new centrifuge design, called the

CentriPress, produced cake solids that were as high as those produced by the filter press system (12).

A model, 45-cm-diameter by 135-cm-long (18-in by 53-in), centrifuge at the Marienfelde sewage treatment plant (STP) Berlin, Germany was operating on digested primary and waste activated sludge. These same biosolids were fed to 91.5-cm-diameter by 274-cm-long (36-in by 108-in) centrifuge, which was dewatering the plant biosolids to a cake product of approximately 22% (total solids) TS. The CentriPress was producing a granular cake of 30% to 32% TS. The "standard" centrifuge produced a cake with a 60% higher moisture content. Both centrifuge installations were recovering in excess of 90% of the feed solids.

A larger unit, 91.5-cm diameter by 274-cm long (36-in by 108-in), is operating at Vienna, Austria, wastewater treatment plant (WWTP). This unit is dewatering a heated primary and waste-activated sludge to a cake solids content of 40% to 42% TS. Results from the centrifuge are comparable to those produced by a recessed plate filter press. The manufacturer had taken orders in Europe for the new machine.

Test runs using a Humboldt-Wedag CentriPress Koeln, Germany were performed by the Metropolitan Sanitary District of Greater Chicago (MSDGC) at the West-Southwest STP. This plant employs high-speed centrifuges for dewatering digested primary sludge (PS) and waste-activated sludge (WAS) that has an original solids ratio of 0.21 PS to 0.79 WAS. The existing centrifuges produce a cake of 14% to 16% TS. The tests were conducted using two types of cationic polymers as shown in Table 3.3.

One of the polymers was not cost-effective for the plant's digested biosolids. The tests used different feed rates and differential speeds, with the polymer adjusted to maintain the total suspended solids (TSS) recovery in the range of 85% to 95%. The key results of Table 3.3, using American Cyanamid 2540C polymer, are as follows (12):

|                       | Average | Range       |
|-----------------------|---------|-------------|
| Cake solids, %        | 29.4    | 26.2–33.9   |
| Solids recovery, %    | 92.7    | 78.4–97.9   |
| Polymer dosage, kg/MG | 7.45    | 3.23–15.93  |
| USD/MG                | 3.37    | 5.86–29.22  |

Figure 3.3 shows the effects of polymer dosage on the solids recovery of the CentriPress. About 5 kg/metric ton (10 lb/ton) of cationic polymer was required to maintain the solids recovery in excess of 90% TS. Table 3.3 does not indicate that higher dosages of polymer were beneficial to improve cake solids, although recoveries above 95% were achieved.

The use of low differential speeds appears to be the key to achieving good cake solids. As shown in Figure 3.4, there was a good correlation between cake solids and centrifugal force at about 2600 g (12).

## 5. HOLLIN IRON WORKS SCREW PRESS

This Korean screw press, which was manufactured in Seoul, Korea, was evaluated for dewatering biosplids from liquid to cake, and second-stage (cake to drier cake) operations have also been evaluated. The Hollin Iron Works (HIW) screw press, shown in Figure 3.5, is continuously fed with biosolids being conditioned by polymer.

**Table 3.3**
**Results of Chicago WSW CentriPress Study (12)**

| | Machine data | | | Sludge data | | | | Polymer data | | Performance | | | |
|---|---|---|---|---|---|---|---|---|---|---|---|---|---|
| Run | G-force, g's | Diff. speed | Feed rate, gpm | Feed conc., % | Feed solids, ton/d | Volatiles % | Flow rate, gpm | Polymer (dry), lb/ton | Cake solids, % | Centrate solids, %TSS | Capture, % | Cost, USD/ton |
| 1[1] | 2300 | 3 | 27 | 4.18 | 6.78 | 48.3 | 3.45 | 14.67 | 29.1 | 4200 | 91.27 | 11.24 |
| 2 | 2300 | 2 | 27 | 4.19 | 6.79 | 48.0 | 3.09 | 13.12 | 26.3 | 1700 | 96.57 | 10.06 |
| 3 | 2300 | 1.8 | 25 | 4.00 | 6.01 | 47.0 | 3.28 | 15.73 | 29.6 | 5000 | 89.00 | 12.06 |
| 4 | 2300 | 2 | 31 | 4.09 | 7.61 | 46.7 | 4.10 | 8.41 | 29.2 | 1000 | 97.89 | 6.45 |
| 5 | 2600 | 2 | 26.5 | 4.16 | 6.62 | 48.4 | 3.9 | 10.61 | 33.2 | 3200 | 93.21 | 8.74 |
| 6 | 2600 | 2 | 26.5 | 4.10 | 6.52 | 48.9 | 3.9 | 10.28 | 33.9 | 2700 | 94.17 | 8.88 |
| 7 | 2600 | 3.5 | 32 | 3.96 | 7.61 | 48.7 | 3.73 | 9.42 | 29.7 | 7200 | 83.85 | 7.76 |
| 8 | 2600 | 5.5 | 32 | 4.07 | 7.82 | 49.0 | 2.63 | 6.46 | 26.2 | 10,000 | 78.42 | 5.33 |
| 9 | 2600 | 5.5 | 32 | 4.10 | 7.89 | 49.2 | 4.58 | 11.15 | 27.4 | 4200 | 91.15 | 9.19 |
| 10 | 2600 | 2.2 | 16 | 4.12 | 3.96 | 48.6 | 3.27 | 19.83 | 30.2 | 1900 | 95.99 | 16.34 |
| 11 | 2600 | 2.8 | 16 | 3.80 | 3.65 | 50.1 | 2.67 | 17.57 | 31.1 | 2400 | 94.41 | 14.48 |
| 12 | 2600 | 7.5 | 50 | 4.13 | 12.43 | 48.8 | 5.57 | 10.76 | 29.8 | 1400 | 97.07 | 8.87 |
| 13 | 2600 | 5.8 | 50 | 4.11 | 12.34 | 48.1 | 5.57 | 11.93 | 28.4 | 3700 | 92.20 | 9.83 |
| 14 | 2600 | 2.5 | 22 | 4.23 | 5.59 | 50.7 | 3.91 | 21.84 | 28.5 | 2900 | 94.10 | 18.20 |
| 15 | 2600 | 2.5 | 22 | 4.25 | 5.62 | 50.5 | 4.4 | 24.45 | 28.4 | 1500 | 96.98 | 20.37 |
| 16 | 2600 | 2.7 | 22 | 4.20 | 5.55 | 50.1 | 3.2 | 31.85 | 28.9 | 1300 | 97.34 | 26.56 |
| 17[2] | 2600 | 2.5 | 22 | 4.25 | 5.62 | 49.4 | 2.73 | 33.25 | 29.2 | 2900 | 94.11 | 76.48 |
| 18 | 2600 | 2.8 | 22 | 4.25 | 5.62 | 49.4 | 2.73 | 33.25 | 32.0 | 1700 | 96.51 | 76.48 |
| 19 | 2600 | 2.9 | 22 | 4.00 | 5.28 | 49.8 | 2.4 | 31.14 | 29.4 | 1100 | 97.61 | 71.58 |
| 20 | 2600 | 2.2 | 22 | 4.10 | 5.42 | 49.2 | 2.4 | 30.31 | 31.3 | 2200 | 95.30 | 69.71 |
| 21 | 2600 | 2 | 22 | 4.08 | 5.39 | 50.6 | 3.57 | 45.34 | 32.8 | 1300 | 99.20 | 104.28 |
| 22 | 2600 | 2 | 22 | 4.16 | 5.50 | 50.5 | 3.50 | 42.8 | 34.8 | 2200 | 95.31 | 98.44 |
| 23 | 2600 | 5 | 26.5 | 4.09 | 6.51 | 49.3 | 3.50 | 36.16 | 30.0 | 2400 | 94.84 | 83.17 |
| 24 | 2600 | 1.5 | 18 | 4.05 | 4.38 | 49.1 | 2.5 | 38.39 | 30.8 | 1000 | 97.85 | 88.30 |
| 25 | 2600 | 1.2 | 18 | 4.08 | 4.41 | 49.4 | 2.9 | 44.23 | 29.7 | 1200 | 97.45 | 101.73 |

[1] Tests 1 through 16 used American Cyanamid 2540C polymer.
[2] Tests 17 through 25 used Allied Chemical Percol 778F525.

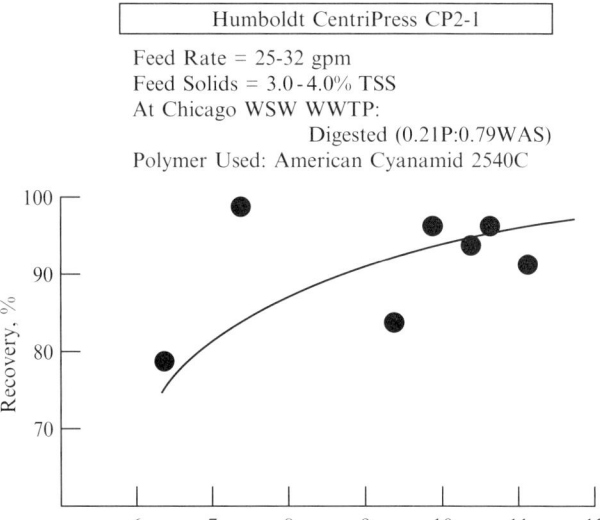

**Fig. 3.3.** Effect of polymer dose on solids recovery (12).

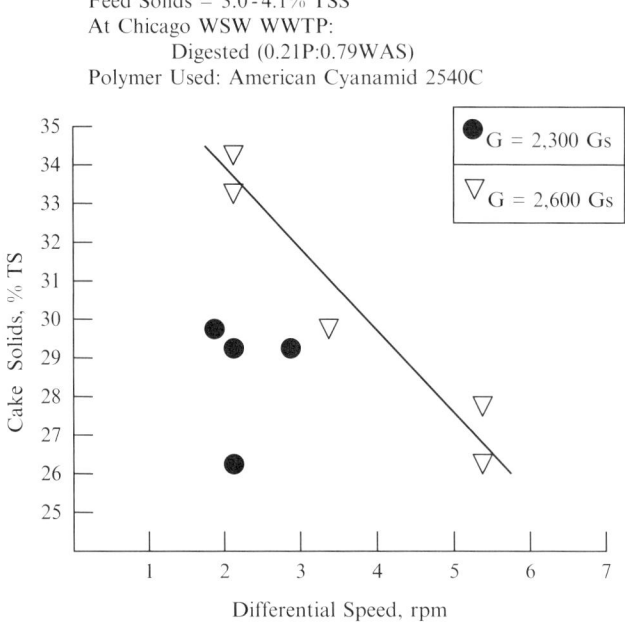

**Fig. 3.4.** Cake solids vs. differential speed (12).

# Biosolids and Septage Treatment

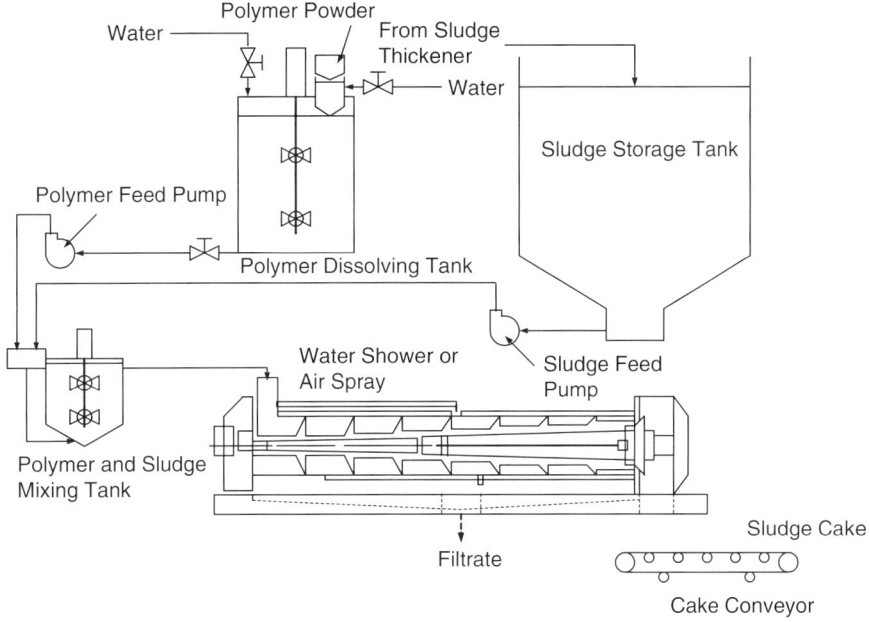

**Fig. 3.5.** Schematic of the screw press dewatering system (courtesy of Hollin Iron Works) (12).

Once inside the unit, the biosolids are subjected to gradually increasing pressure as they progress through the screw press. The maximum pressure before discharge may exceed 10 kg/cm$^2$ (147 lb/in$^2$). In some instances the dewatering may be enhanced by heating (a normal experience with screw presses) prior to the dewatering. This screw press is said to be relatively simple and easy to maintain. Also, the low operating speed helps keep repair costs to a minimum. The HIW reports that there are over 100 units in operations (or installed) for various types of wastewater treatment and are providing satisfactory service. Some results reported are shown in Table 3.4.

More data and a better definition of the feed biosolids are required to fully evaluate the possibility of the screw press replacing conventional dewatering equipment. Past excessive secondary solids losses must be evaluated as a function of the cake solids content produced.

The MSDGC tested a pilot HIW screw press. The unit was tested on primary and anaerobically digested biosolids at the West-Southwest STP. The test was performed over a period of 2 days, and approximately 12 separated runs were undertaken. Biosolids flow rate, dilute polymer concentration, and polymer flow were varied. With an average biosolids feed concentration of 4.5%, the test unit attained the following average results:

Cake concentration:     17.5%
Solids recovery:        94.5%
Pressate concentration: 3720 mg/L (0.43 lb/gal)
Polymer usage:          8 dry kg polymer/dry metric ton solids (16 lb/dry ton)

Table 3.4
Test results for Hollin Iron Works screw press (12)

| Sludge | Feed solids, % | Cake solids, % | Solids recovery, % | Polymer dosage, kg/MG |
|---|---|---|---|---|
| Digested A | 4.65 | 20.9 | 93.0 | 9.1 |
| Digested B | 4.93 | 25.3 | 97.8 | 7.6 |
| Primary A | 2.85 | 20.5 | 95.0 | 13.4 |
| Primary B | 2.37 | 21.2 | 95.9 | 16.0 |
| Paper Mill 1 | 3.45 | 48.6 | 99.0 | 1.0 |
| Paper Mill 2 | 4.08 | 44.6 | 98.9 | NR |
| Paper Mill 3 | 2.95 | 42.3 | 98.9 | NR |
| Paper Mill 4 | 2.4 | 23.0 | 95.4 | NR |

MG, million gallons.

Based on this pilot test results, the MSDGC decided to purchase a full-sized screw press. It anticipates that a full-size screw press may be a cost-effective alternative to centrifugation due to the following considerations (12):

- Low initial cost
- Lower electric power consumption
- Equal or higher cake concentrations
- Slow operating speed (low G force)
- Lower maintenance cost
- Comparable polymer cost

Krofta and Wang (48, 49) invented a combined dissolved-air flotation (DAF) and screw press (SP) process equipment for simultaneous sludge thickening and dewatering. The combined DAF-SP process unit was installed at the Lenox Water Treatment Plant, Lenox, Massachusetts, and proved to be a successful operation (48, 49). In the Lenox plant's full-scale operation, the DAF chamber at the bottom floats the sludge to the water surface and thickens it. The SP that sits on the top of the DAF chamber picks up the thickened sludge and further dewaters it. The theory, principles, and operational data of DAF-SP sludge thickening-dewatering can be found elsewhere (18, 48–52).

## 6. SUN SLUDGE SYSTEM

The Sun Sludge System (Hi-Compact) for biosolids dewatering was developed in Tokyo, Japan, and has been licensed for marketing and manufacture for Europe and the United States. The principle of the process is to develop a structured material from a cake of poor dewatering characteristics, and to form liquid channels. The cake is then subjected to high pressures. To that end, dewatered wastewater biosolids are reduced to pellets that are subsequently coated by a powdery layer such as ash or pulverized coal. Compressing a stack of these pellets results in a compact block interwoven with a network of drainage layers; the water being removed by pressing flows through a line of least resistance to the nearest drainage layer, as shown in Figure 3.6.

In the system, biosolids are first dewatered by conventional dewatering equipment such as vacuum filters, centrifuges, or continuous belt filters to a 20% to 25% solids

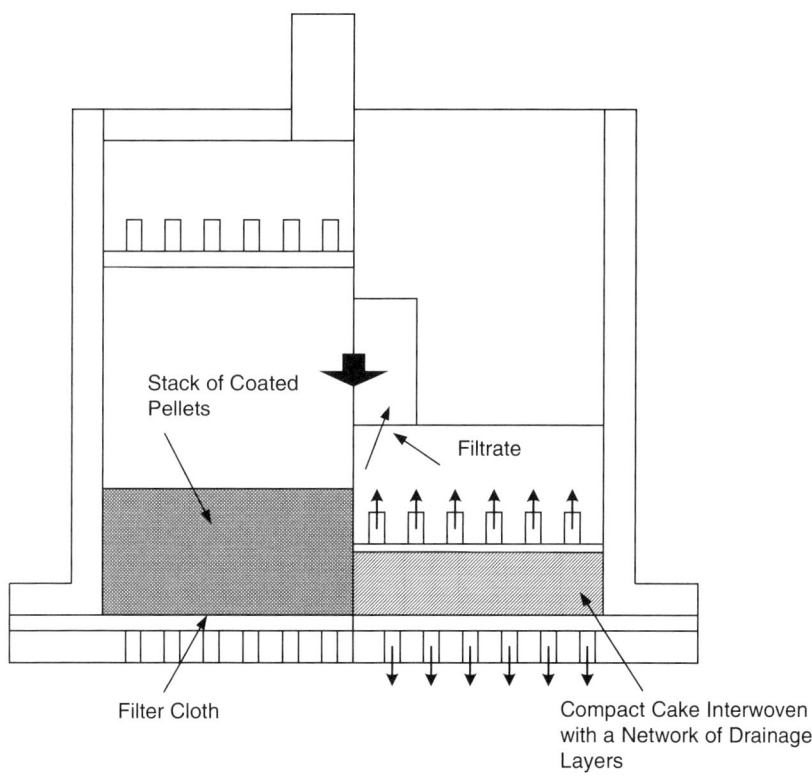

**Fig. 3.6.** Sludge press function of high-compact method (courtesy of Humboldt-Wedag) (12).

concentration. This material is then conditioned in a unit called a disintegrating palletizer, which first breaks and forms the biosolids cake into small particles with a dry powder, forming biosolids-like pellets. The dry additive used should be mostly water insoluble and should not break up at the high pressures used. Materials such as diatomaceous earth, gypsum, calcium carbonate, incinerator ash, coal powder, bone meal, dried pulp, sawdust, and soil have been used, either alone or in combination. The conditioners should be added in the ratio of 40% to 60% by weight per unit dry weight of the original biosolids cake. The effective biosolids pellet's diameter should not be greater than 20 mm (0.8 in). Best performance occurs when the effective diameter of the pellets is between 3 and 5 mm (0.1 and 0.2 in). Also, the conditioning agents should coat only the surface and should not be kneaded into the biosolids pellets for maximum effectiveness (12).

The conditioned biosolids cake particles are conveyed to a hydraulic press where additional water is removed, and a cake of greater than 40% solids is produced. The palletized biosolids are pressed between two sheets of filter cloth that cover thick plates that have a number of perforations of 2 to 10 mm (0.1 to 0.4 in) in diameter. Compression is carried out in two steps. The initial compression step is usually at 15 to 25 kg/cm$^2$ (210 lb/in$^2$) for 45 seconds followed by a pressure of 30 kg/cm$^2$ (430 lb/in$^2$) for 5 minutes (12). In practice, the compression has occurred at 15 kg/cm$^2$ (210 lb/ft$^2$) for 45 seconds, followed by a pressure of 30 kg/cm$^2$ (430 lb/in$^2$) for 1 minute. The palletized biosolids cake is compressed by hydraulic cylinder to form a disk-shaped solid with a 40% to 55% solids concentration. As an example, a mixture of primary and waste-activated sludge having a 2% solid concentration could first be dewatered by a belt filter press to a solids concentration of 25% and then, with the Sun Sludge System, could be further dewatered to a solids concentration of 55%.

The Ashigara Works of Japan has successfully been using this system for waste-activated sludge treatment since 1982. The excess biosolids are dewatered by a belt press to a water content of 80%, then palletized and conditioned with incinerator ash and further dewatered to a water content of 50% or less. Ash is added in the ratio of 10% to 15% by weight of the amount of belt press cake (or 50% to 75% of the dry solids). Biosolids cakes are incinerated and heat is recovered (12).

A field demonstration of the process produced a cake of 55% from a 32% cake mixed with biosolids ash (50% by dry solids weight) every 3 minutes. The unit was a pilot scale producing in excess of 1000 kg cake/h (2204 lb/h). This would be equivalent to 370 kg/h (816 lb/h) of wastewater sludge solids. In this demonstration, pressures up to 60 kg/cm$^2$ (853 lb/in$^2$) were employed and the press time was shortened to about 3 minutes.

While the product from the press is very hard, it is also quite friable. It can be easily fragmented into particles that are dry to the touch and can easily be transported pneumatically. The palletizing/pressing operation reduced the moisture content from 3.5 kg H$_2$O/kg TS to 1.3 kg H$_2$O/kg TS (biosolids-only basis). The feed would have been suitable for boiler feed and would produce an equilibrium temperature of about 1090°C (2000°F). While the process mechanics look favorable, the machine design capable of long-term operation at 50 to 60 kg/cm$^2$ (710 to 850 lb/in$^2$) needs further evaluation.

## 7. WEDGEWATER BED

Wedgewater, or wedgewire beds, are often constructed with an interlocking synthetic filter media placed on a concrete basin with an underdrain system. Polymer is always added to the biosolids before placement on the media surface. Wedgewater bed systems can produce biosolids with a final solids content of about 8% to 12% in 24 hours and up to 20% given additional drying time. Beds are usually uncovered, but may be covered to protect biosolids from excessive precipitation. The process is best suited for smaller treatment plants, 1893 m$^3$/d (0.5 MGD) or less, and in moderate climates. The United States Army Construction Engineering Research Laboratory (ACERL) has found that wedgewater systems have been used successfully in plants with flows up to 28,387 m$^3$/d (7.5 MGD). A typical facility consists of the following (12–14):

- An outdoor, concrete construction with synthetic media plates
- Filtrate collection and drainage system (outlets)
- Polymer feed system
- Sludge distribution system (inlets)
- Washwater system

Figure 3.7 shows a typical section of a wedgewater drying bed. The main structure should consist of a concrete floor with a drainage system; sidewalls, approximately 0.6 m (2 ft) high; biosolids distribution piping; supernatant decanting system; and vehicle

# Biosolids and Septage Treatment

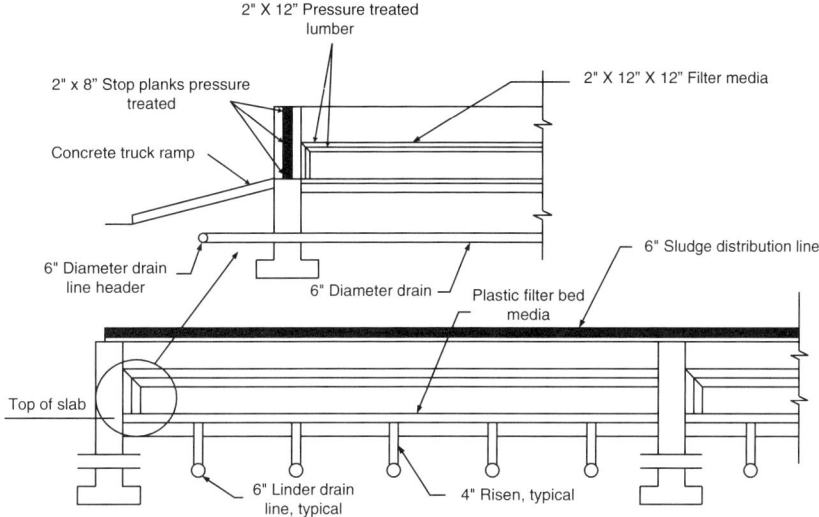

**Fig. 3.7.** Typical sections of a wedgewater drying bed (11).

entrance for biosolids removal. Manually removable wooden planks are to be installed at the vehicle entrance. Although most wedgewater operations are uncovered, use of a translucent roof or canopy is recommended in areas with significant precipitation. Filtrate is drained by gravity through the wedgewater media over the concrete floor. The floor should be designed with a slope of 0.5% to 1.0% to facilitate gravity drainage and avoid solids buildup under the media. Additional pipe drainage system may be installed. As a rule of thumb, there should be drainage outlet for each 2.25 m$^3$ (25 ft$^3$) of media (11).

The media manufacturer's recommendations should be considered for design of a synthetic media dewatering system. The basis for design is the plant's average annual biosolids production rate in dry solids (i.e., kg/yr or lb/yr) and the number of cycles per week that can conveniently be performed. For reliability, a minimum of two beds should be constructed. The U.S. Environmental Protection Agency (US EPA) states that solids loading rates of 2 to 5 kg/m$^3$/cycle (0.4 to 1.0 lb/ft$^3$/cycle) are typical (12). Adjustments, based on the expected efficiency and effectiveness of the operation, may also be considered. The number of operational cycles per year will vary. While the literature suggests that 24-hour cycle times are acceptable, it is recommended that the design be based on two cycles per week. The design shall allow for downtime for cleaning of beds. Biosolids loading rates should be approximately 9.4 L/s (150 gpm). The general dimensions of each bed should be limited to approximately 7.6 m wide by 15.25 m long (25 ft by 50 ft). This will avoid problems with thermal expansion of the media and with the separation of solids before even distribution of biosolids can occur. Additional biosolids distribution inlets are also required as compared to conventional sand drying beds. Supernatant decanter devices are recommended to simultaneously remove water from the surface of the bed. High-pressure washwater systems using treated effluent are recommended for tile cleaning. The supernatant and filtrate shall be routed back to the headworks, primary clarifier, or aeration basin for additional treatment (11).

Problems associated with these systems are inadequate media cleaning, front-end loader damage, and engineering errors. If the beds are properly designed, constructed, operated, and maintained, the beds will have a long life, and underdrain cleaning will be required only once or twice a year. A polyurethane blade should be used on the front-end loader bucket, avoiding the use of skid-steering loaders to prevent media damage. Wedgewater beds have less media clogging if high-pressure hoses are used to clean the tiles (11).

## 8. VACUUM-ASSISTED BED

The vacuum-assisted dewatering bed (VADB) uses commercially available equipment to apply a vacuum to the underside of a rigid, porous, media bed on which conditioned biosolids have been applied. The theory is that gravity, assisted by the vacuum, draws the water through the media, leaving the dry solids on top. The VADB systems are capable of dewatering biosolids to final solids content of about 14% in 24 hours and 18% or higher in an additional 24 hours. The primary elements of a typical facility are as follows (12):

- An outdoor, concrete structure with synthetic media plates
- Filtrate collection and drainage system
- Polymer feed system
- Biosolids distribution system
- Vacuum system
- Washwater system
- Controls

Figure 3.8 shows a schematic view of a vacuum-assisted biosolids drying bed.

The VADBs are generally used for smaller treatment plants, that is, up to 7579 $m^3$/d (2.0 MGD). Biosolids are seldom as dry as those removed from sand drying beds. Total solids content varies from site to site and depends on several factors including the basic type of treatment process, biosolids conditioning, biosolids feed rates, and cycle times. The bed design is similar to that of a wedgewater bed described above. In cases of adverse weather conditions, the bed should be covered. The system equipment manufacturer's recommendations should be considered for any design of a VADB system. The basis for design is the plant's average annual biosolids production rate in dry solids (i.e., pounds or kilograms per year) and the number of cycles per week that can conveniently be performed. For reliability, a minimum of two beds should be constructed. The US EPA recommends a solids loading rate of 10 $kg/m^2$/cycle (2 $lb/ft^2$/cycle). Adjustments, based on the expected efficiency and effectiveness of the operation, may be considered. Most VADB designs are based on a 24-hour cycle time. Biosolids loading rates should be approximately 9.4 L/s (150 gpm). Supernatant decanter devices should be installed to simultaneously remove water from the surface of the beds (11).

A common complaint of VADB operators is that the biosolids require long drying cycles. This problem is due mainly to inadequate drainage caused by media or underdrain clogging and to media destruction caused by front-end loaders or epoxy failure. Plant operators recommended that a polyurethane blade be used on the front-end loader bucket to prevent damage. Skid-steering loaders are also inappropriate for this system. Tile cleaning is more difficult than for wedgewater beds. Media blinding

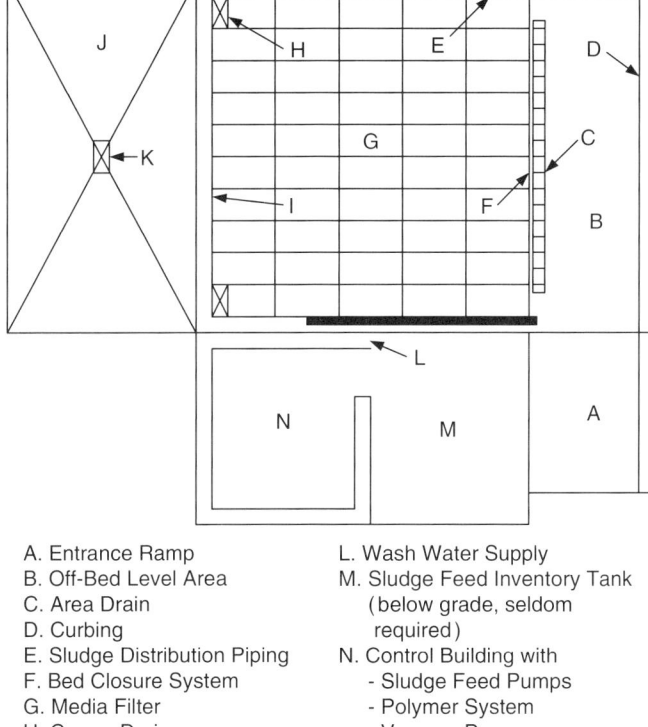

A. Entrance Ramp
B. Off-Bed Level Area
C. Area Drain
D. Curbing
E. Sludge Distribution Piping
F. Bed Closure System
G. Media Filter
H. Corner Drain
I. Bed Containment Wall
J. Truck Loading Area
K. Area Drain
L. Wash Water Supply
M. Sludge Feed Inventory Tank
   (below grade, seldom
   required)
N. Control Building with
   - Sludge Feed Pumps
   - Polymer System
   - Vacuum Pumps
   - Control Panel
   - Filtrate Receiver / Pumps
     (below grade)

**Fig. 3.8.** Schematic plan view of vacuum assisted sludge drying bed (11).

was reported as a major problem with a few existing VADB systems. The VADBs produce a faster turnover rate than sand beds, and VADB systems can be operated year-round (11).

## 9. REED BED

An emerging and popular technique for biosolids dewatering in the U.S. for the past few years employs a reed bed (sometimes referred to as wetland). Since biosolids are applied to a predesigned stand or growth, essentially a bed of reeds, this treatment method is popularly called the "reed bed" process. The most common reed species utilized is from the genus *Phragmites*.

The Max-Planck Institute of West Germany originally conducted research in the late 1960s and early 1970s on the use of the reed bed system to process and dewater wastewater biosolids from small wastewater treatment plants. Although the process was originally used for wastewater treatment, it was extended to biosolids dewatering in 1980. Using the reed bed system, biosolids from wastewater treatment plants are applied to an actively growing stand of a common reed under controlled conditions. The growing reeds derive moisture and nutrients from the biosolids, and with time the rooted plants and the accompanying root ecosystem alter the characteristics of the biosolids, resulting in dewatering and improved biosolids characteristics. In addition to evapotranspiration, natural environmental processes, such as evaporation and drainage, contribute to the moisture loss and dewatering as with conventional drying beds. Wastewater treatment plants in the northeastern U.S. have been using reed bed technology successfully for dewatering biosolids since the early 1980s (13).

The primary elements and characteristics of the reed bed process are as follows (12):

- Bed construction is similar to that of sand drying beds. Often retrofitted sand drying beds are used.
- Excavated trenches are lined with an impermeable material and filled with two sizes of gravel and a top layer of filter sand.
- Reed root stock or small plants are planted in the sand layer, and the trenches are flooded to promote reed growth.
- A 1-m freeboard above the sand layer is provided to allow for long-term biosolids storage.
- After plants are well established, stabilized, thickened biosolids (3% to 4% solids) are applied to the bed in 10-cm (4-in) layers at regular intervals.
- Annual harvesting of reeds and their disposal by landfilling, burning, or composting is required.
- Biosolids are not removed regularly. Biosolids removal cycle time is 6 to 10 years.

A comparison between biosolids dewatering with conventional sand beds and the reed bed method shows that reed beds can provide adequate or marginal dewatering for both aerobically and anaerobically digested biosolids, if all the existing and drying beds are converted to reed beds. The most obvious advantage of reed beds is the elimination of labor for regular biosolids removal from sand drying beds. The process also offers many distinct advantages with respect to reduced costs, labor, and maintenance. Reed beds can also be constructed using existing biosolids drying beds. Greenhouses with reed bed should be used with caution because greenhouse environments may generate severe heat and drought stress on the reeds. Higher volumes of aerobically digested biosolids may be dewatered than that of the anaerobically digested biosolids.

Figure 3.9 shows a typical section view of a reed bed for dewatering of biosolids.

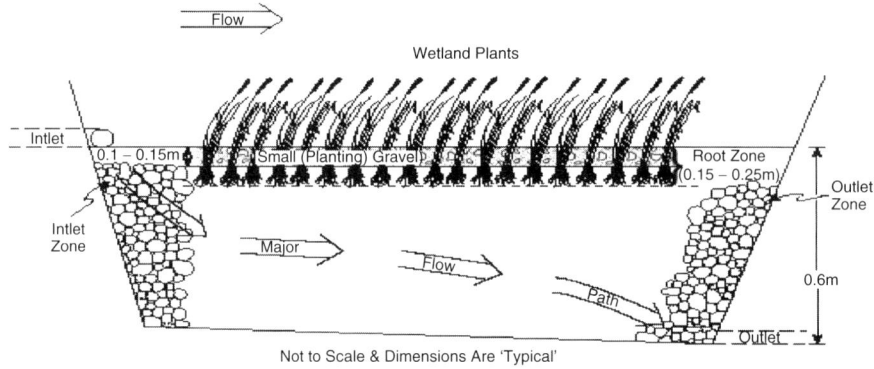

**Fig. 3.9.** Schematic view of reed bed (14).

Suggested solids loading rates are as follows (12):

- For aerobic biosolids: 9 to 95 kg/m$^2$/yr (2 to 20 lb/ft$^2$/yr). The US EPA indicates that operational systems in the northeastern U.S. average loading rates of 81 kg/m$^2$/yr (17 lb/ft$^2$/yr). The ACERL studies indicate an average loading rate of 52 kg/m$^2$/yr (10.9 lb/ft$^2$/yr) for systems in the U.S.
- For anaerobic biosolids: 9 to 57 kg/m$^2$/yr (2 to 12 lb/ft$^2$/yr). The ACERL studies indicate an average loading rate of 22 kg/m$^2$/yr (4.7 lb/ft$^2$/yr) for systems in the U.S.

It is recommended that multiple beds be provided to allow for biosolids removal and maintenance of beds (13).

The US EPA recommends that thickened biosolids (3% to 4% solids) be applied to the beds. The reeds must be harvested annually and subsequently disposed of in an acceptable manner. Operational problems include aphids and weed growth for younger reeds. Labor for weeding operations should be estimated from 1 to 10 man-days/yr. Cost estimates vary depending on the size of the operation. Salinity affects the reed's height and growth. The maximum recommended salinity is 4.5%. During freezing months, biosolids application is normally stopped and the reeds are harvested (13, 14).

## 10. SLUDGE-FREEZING BED

A sludge-freezing bed is a unit operation that uses natural freeze-thaw cycles to condition the biosolids for dewatering. It is most applicable in regions having 3 months per year or more of temperatures at or below 0°C (subfreezing temperatures). Freezing beds can be used with conventional drying beds to provide year-round biosolids dewatering.

The design incorporates a covered, in-ground containment structure with drainage and ramp access. Drainage may be similar to that of conventional sand drying beds or synthetic media (wedgewire) systems. During the winter months, the biosolids are added to the bed in layers. Successive layers are added as the previous layer freezes. At the end of the cold season, the bed is allowed to thaw and drain. Dewatering occurs by the removal of the melt water by the underdrain system. Once the desired solid/liquid content is achieved, the dewatered biosolids are removed by mechanical means. The bed may be used as a conventional covered drying bed during the warmer months (15, 16).

The Cold Regions Research and Engineering Laboratory (CRREL) was involved in a demonstration project at Fort Greely, Alaska, and assisted in developing a freezing bed design for projects constructed at Eielson Air Force Base, Alaska, and Fort McCoy, Wisconsin.

The primary features and characteristics of the freezing bed dewatering system are as follows (15, 16):

- The facility consists of a basin with an underdrain system where biosolids are deposited in layers and allowed to freeze.
- Basins are usually covered to keep out precipitation.
- The process requires no chemical addition; that is, polymers are not required.
- The operation requires no special skills to operate.

Freezing is dependent on natural climatic conditions at the proposed site. Any location that has 3 months per year or more of temperatures at or below 0°C may be considered.

Biosolids freezing is a reliable dewatering method for most of the northern U.S. Any type of biosolids will benefit from the freeze-thaw cycle. However, it is recommended that stabilized and thickened biosolids (3% to 7% solids content) be applied to avoid odor problems, maximize effectiveness, and reduce cost. The system is designed for a worst-case condition, such as the warmest winter, to ensure successful performance. A pre-engineered metal roof will be considered as part of the facility design to protect the area from snow.

The size and capacity of the freezing beds depends on the depth of biosolids that can be frozen and subsequently thawed in a season. In very cold climates, the depth of biosolids that can be frozen may be greater than the depth that can be thawed. In that case, the thawing depth will be limiting and should be used for design. The freezing depth ranges from less than 30 cm (12 in) to more than 180 cm (70 in) for most of the northern U.S. (15–17).

## 11. BIOLOGICAL FLOTATION

In a biological flotation system, fermentations take place in the presence of anaerobic bacteria, nitrates, and substrates under anaerobic environment. Anaerobic bacteria in the biosolids convert nitrate and the organic substrate as a carbon source (such as methanol) to nitrite, water, and carbon dioxide fine bubbles. Nitrite further reacts with a substrate (such as methanol) in the same biosolids, producing fine nitrogen bubbles, more fine carbon dioxide bubbles, water, and hydroxide ions. The biological biosolids, such as activated sludge, can then be floated to the surface by the fine nitrogen and carbon dioxide bubbles. While the energy consumption of this process is low, its detention time is long, in the range of 1 or 2 days (2, 18, 50–52). Wang's chemical reactions in a biological flotation reactor for thickening of secondary activated sludge, under anaerobic conditions, assuming nitrate ($NO_3^-$) is present in the biosolids, and methanol ($CH_3OH$) is added as the substrate, are as follows:

$$\begin{array}{r} 6\,NO_3^- + 2\,CH_3OH \rightarrow 6\,NO_2^- + 2\,CO_2 + 4\,H_2O \\ 6\,NO_2^- + 3\,CH_3OH \rightarrow 3\,N_2 + 3\,CO_2 + 3\,H_2O + 6\,OH^- \\ \hline 6\,NO_3^- + 5\,CH_3OH \rightarrow 5\,CO_2 + 3\,N_2 + 7\,H_2O + 6\,OH^- \end{array}$$

It should be noted that the secondary activated sludge usually contains residual soluble BOD = biological oxygen demand, COD = chemical oxygen demand, or TOC = total organic carbon, which may avoid the necessity of adding an external organic substrate such as $CH_3OH$.

## 12. SEPTAGE TREATMENT

### 12.1. Receiving Station (Dumping Station/Storage Facilities)

A septage receiving station with dumping station and storage facilities (Figure 3.10) provides for the transfer of septage from hauler trucks to a temporary holding tank from which it can be drawn at a controlled rate. With such a facility, septage can be discharged to an interceptor sewer or directly to the headworks of a treatment plant. The

# Biosolids and Septage Treatment

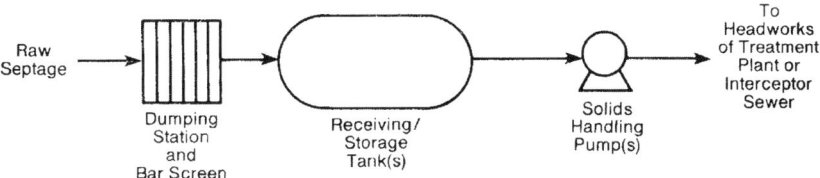

**Fig. 3.10.** Flow diagram of septage receiving station (23).

dumping station should provide for both direct hose connections (preferred) and open pit discharges. The dumping pit should be equipped with a coarse bar screen, and should be covered and preferably locked when not in use. A manual-controlled or timer-controlled pump discharge facilitates feeding septage at a predetermined rate over specific periods of time in order to maximize the dilution of septage by wastewater (19–22).

Where septage is to be transferred from haulers' trucks to other vehicles (e.g., large tanker trucks for transport to centralized treatment facilities, or specialized land application equipment), the same basic facilities as described above could be used, with the exception that tanker trucks or trailers would replace the permanent storage tanks. Where land application is involved, longer-term storage may be required during adverse weather conditions, and lagoon storage facilities should be considered in such cases. If septage is to be discharged to an interceptor sewer where flows are high, storage facilities might not be required. Odor control may be required depending on station location.

Grit and solids residuals that may accumulate in holding tank must be cleaned out periodically. This can be accomplished by removing the solids using vacuum truck equipment, or by flushing the solids out of the tank using high-pressure water. Periodic removal of screenings is also required.

The recommended design criteria are as follows:

1. Bar Screen: 0.5-in × 1.5-in bar stock and 0.5-in to 0.75-in spacing
2. Hauler truck hose connection: 4-in diameter
3. Piping and valves: 8-in diameter
4. Holding tank capacity: 1 day peak flow (not including supplemental storage requirements associated with land application systems, etc.)

## 12.2. Receiving Station (Dumping Station, Pretreatment, Equalization)

When septage is to be ultimately treated at a wastewater treatment plant or independent septage treatment facility, a receiving station is required in order to provide preliminary treatment and equalization. This normally consists of a dumping pit with screening, grit removal, and flow equalization (Figure 3.11). Features that should be provided include a sloped ramp and hose-down facilities at the unloading location, a channel in front of a bar screen for more uniform flow and to avoid direct discharge of septage onto screen, manually or mechanically cleaned bar screens, solids handling pumps, sampling/monitoring capability, ventilation system, and odor control (19–24).

Grit removal can either precede storage and equalization or follow it. If a grit chamber precedes equalization, it must be designed to handle the discharge of individual or multiple truckloads of septage as they come. If storage and equalization precede grit removal, the grit removal process can be designed to handle the average flow, and can be operated

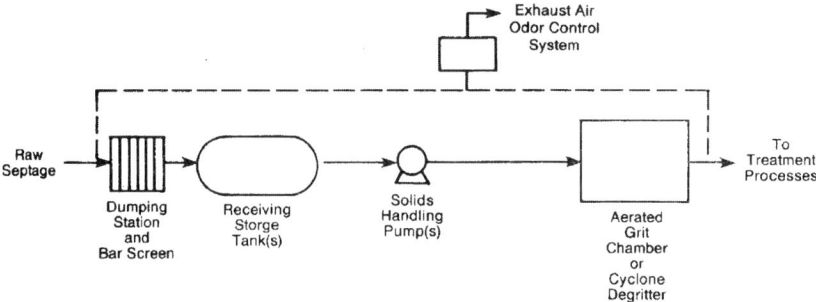

**Fig. 3.11.** Flow diagram of septage receiving station with pretreatment (23).

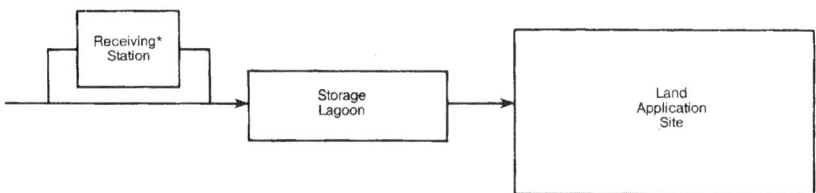

**Fig. 3.12.** Flow diagram of land application of septage (23).

according to a set schedule coinciding with subsequent treatment operations. Cyclone degritters may be substituted for aerated grit chambers if septage solids concentration is less than 2%.

Provisions must be made for removal and disposal of screenings and grit residuals, plus accumulated solids that settle out in flow equalization tanks. Landfilling is the most common method of disposal.

The recommended design criteria are as follows (23):

1. Bar screen: 0.5-in × 1.5-in bar stock and 0.5-in to 0.75-in spacing
2. Hauler truck hose connection: 4-in diameter
3. Piping and valves: 8-in diameter
4. Degritting equipment: as per manufacturer's specifications for design flow
5. Equalization tanks: multiple tanks, total capacity twice peak daily flow
6. Pumps: sized according to average design flow and operational schedule

### 12.3. Land Application of Septage

Raw septage and septage solids may be spread on the surface of the land or incorporated into the subsurface topsoil layers (Figure 3.12). Surface spreading includes spreading from septage hauler trucks or transfer vehicles such as tank wagons, spray irrigation, ridge and furrow practices, and overland flow (25, 26). Application by the hauler trucks is the most common method practiced. Spray irrigation of septage requires the use of high-pressure large nozzle systems to prevent clogging. Ridge and furrow methods involve spreading septage in the furrows and planting crops on the ridges. Overland flow methods are best suited to lands with a slope of 2% to 6%.

Subsurface application techniques include plow furrow cover (PFC) and subsurface injection (SSI). The PFC method of application applies septage in a narrow furrow created by the plow shear and is immediately covered by the plow moldboard. The SSI method of application applies septage in a narrow band behind a sweep that opens a cavity 10 to 15 cm (6 to 8 in) deep. This kind of device opens a mole-type hole with an oscillating chisel point and injects the septage into the hole (27–31).

Federal criteria (Title 40 of the Code of Federal Regulations [CFR], Part 257) specify that septage applied to the land or incorporated into the soil must be treated by a process to significantly reduce pathogens (PSRP) prior to application or incorporation, unless public access to the facility is restricted for at least 12 months after application has ceased, and unless grazing by animals whose products are consumed by humans is prevented for at least 1 month after application. The PSRPs include aerobic digestion, air drying, anaerobic digestion, composting, lime stabilization, and other techniques that provide equivalent pathogen reduction (2).

The criteria also require septage to be treated by a process to further reduce pathogens (PFRP) prior to application or incorporation, if crops for direct human consumption are grown within 18 months subsequent to application or incorporation, and if contact between the septage applied and edible portion of the crop is possible. The PFRPs include composting, heat drying, heat treatment, thermophilic aerobic digestion, and other techniques that provide equivalent pathogen reduction (2).

Constituents of the septage may limit the acceptable rate of application, the crop that can be grown, or the management or location of the site. Nitrogen requirements of the crop normally dictates the annual septage application rates. It is also required that soil pH be maintained at 6.5 or above to minimize the uptake of the trace elements.

The potential for contaminated runoff, soil compaction, crop damage, or trucks getting stuck preclude the application of septage during periods when soil moisture is too high. Therefore, septage application is limited to only a part of the year. For the period of the year when septage cannot be applied, storage facilities must be provided. Many states regulate the total volume of septage that can be applied as a function of soil drainage characteristics.

Septage contains all of the essential plant nutrients. It can be applied at rates that will supply all the nitrogen and phosphorus needed by most crops. Application rates depend on septage composition, soil characteristics, and cropping practices. Annual application rates have varied from 282 $m^3$/hectare [ha] (30,000 gallon/acre) to 1880 $m^3$/ha (200,000 gallon/acre). Applying septage at a rate to support the nitrogen needs of a crop avoids problems with overloading the soil (23).

There is a potential for heavy metals and pathogens to contaminate soil, water, air, vegetation, and animal life, which ultimately become hazardous to humans. Accumulations of metals in the soil may cause phytoxic effects, the degree of which varies with the tolerance level of the particular crop (32). Toxic substances such as cadmium that accumulate in plant tissues can subsequently enter the food chain, reaching human beings directly by ingestion or indirectly through animals. If available nitrogen exceeds plant requirements, it can be expected to reach groundwater in the nitrate form. Toxic

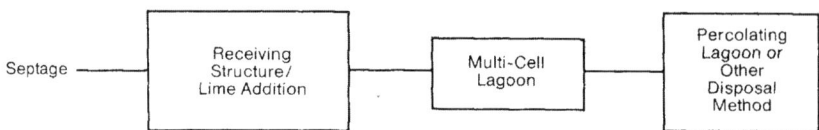

**Fig. 3.13.** Flow diagram of septage disposal by lagoons (23).

materials can contaminate groundwater supplies or can be transported by runoff or erosion to surface waters if improper loading occurs. Aerosols that contain pathogenic organisms may be present in the air over a landspreading site, especially where spray irrigation is the means of septage application. Other potential impacts include public acceptance and odor.

### 12.4. Lagoon Disposal

The use of lagoons for the disposal of septage is a common alternative in rural areas. The design and operation of lagoons vary from simple septage pits to sealed basins with separate percolation beds. Most lagoons are operated in the unheated anaerobic or facultative phase.

A typical lagoon system consists of two earthen basins arranged in series (Figure 3.13). The first or primary lagoon receives the raw septage via a vertical discharge chamber entering under the surface of the liquid near the lagoon bottom to minimize odors. It may be lined or unlined, depending on the geological conditions of the site. The supernatant from the primary lagoon, which has undergone some clarification and possibly anaerobic digestion, is drawn off into the second lagoon or percolation bed, where it is allowed to percolate into the ground. Once the solids have accumulated in the primary lagoon until the point where no further clarification occurs, the lagoon is drained and the solids are allowed to dry. The dried solids are then removed, sometimes further dewatered, and disposed of at a landfill or buried (33–35).

Aeration may be applied to supplement the supply of oxygen to the system and for mixing. Lagoons may be lined with various impervious materials such as rubber, plastic, or clay as required by geological conditions. Where groundwater quality is of concern, the effluent from septage lagoons can be applied to the land or treated and discharged to surface water, rather than using percolation beds.

The pH of the lagoon must be maintained at 8.0 or greater to control odors. This may be accomplished with the use of hydrated lime added each time a truckload is discharged to the receiving chamber. Lagoon effluent can be disposed of by applying spray irrigation or overland flow. If the effluent is to be discharged to a surface water it should be further treated using either polishing ponds or sand filters, and disinfected as required.

In very cold climates, reduced biological activity occurs and ice may form on the surface. Overloading may create potential odor problems. Potential exists for groundwater contamination with percolation beds and seepage pits or lagoons. Hence, extensive site evaluation is recommended and groundwater should be monitored near the lagoon site. Odor and vector problems are possible in immediate vicinity of lagoons.

Settled solids from primary lagoon have to be removed and properly disposed of periodically, every few months to once every 5 or 10 years depending on size of lagoon.

The recommended design criteria for lagoons are as follows (23):

1. Detention time: 20 to 30 days for settling alone; 1 to 2 years for stabilization (i.e., 80% to 90% removal of BOD and volatile solids [VS])
2. Area loading rate: 20 lb VS/d/1000 ft$^2$ (facultative sludge lagoon)
3. pH: 8.0 using lime
4. Minimum depth: 0.9 m (3 ft) (plus additional depth for sludge storage and anaerobic zone)
5. Minimum separation distance from high groundwater level: 1.3 m (4 ft)

## 12.5. Composting

Composting is the stabilization of organic material through the process of aerobic, thermophylic decomposition. It is a disposal technique that offers good bactericidal action and up to 25% reduction in organic carbon. Septage is transformed into a humus-like material that can be used as a soil conditioner.

Composting is classified into three types of operations, which differ principally by the aeration mechanism they employ. They are windrow, aerated static pile, and mechanical composting (2, 35). Although all three methods may be applied for composting septage, the method that appears to offer the greatest potential as a septage treatment alternative is the aerated static pile method because it permits more uniform composting and minimizes land requirements.

Septage is usually first dewatered and then mixed with bulking agents (e.g., wood-chips, sawdust, bark chips, leaves, etc.) prior to composting to decrease the moisture content of the mixture, increase the porosity of the septage, and ensure aerobic conditions during composting. The mixture is then constructed into a pile as shown in Figure 14. A blanket of finished compost completely surrounds the composting mixture in order to reduce heat loss and minimize odors.

The aerated pile undergoes decomposition by thermophylic organisms, whose activity generates a concomitant elevation in temperature to 60°C (140°F) or more. Aerated

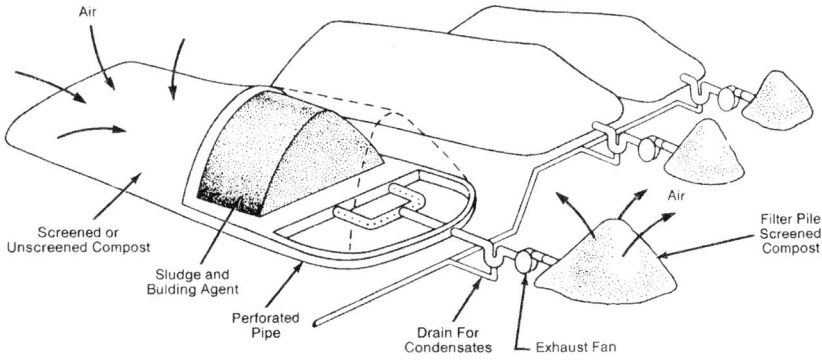

**Fig. 3.14.** Aerated static piles for septage composting (23).

conditions in the pile are maintained by drawing air through the pile at a predetermined rate. Exhaust air is forced through a small pile of screened finished compost for odor control. The composting period normally lasts 3 weeks (36–43).

Following the composting period, the aerated piles are taken down, moved, and stored in piles for 4 weeks or longer to ensure that no offensive odors remain and to complete stabilization. The composted material can be separated from the bulking agent, which is generally recycled for further usage. The finished compost material is then ready for utilization as a low-grade organic fertilizer or soil conditioner, or for land reclamation.

Windrow and mechanical composting are commonly used to stabilize wastewater biosolids and can be adapted to treat septage. The Lebo process, which is a variation of windrow composting, treats raw septage without dewatering, by first aerating the septage in a reactor and then mixing it with sawdust before composting, which takes up to 6 months. The aerated static pile method can also be used to compost raw septage; however, excessive quantities of bulking agent are required to maintain the desired moisture content.

Dewatering of septage is recommended prior to composting to minimize the amount of bulking agent required. However, if large quantities of bulking agent are available at reasonable cost, raw septage can be treated.

In areas of significant rainfall it may be necessary to provide a cover for the pile. A drainage and collection system is generally required to control storm water runoff and leachate from the pile.

Composting represents the combined activity of succession of mixed populations of bacteria, *Actinomycetes,* and other fungi. The principal factors that affect the biology of composting are moisture, temperature, pH, nutrient concentration, and availability and concentration of oxygen. A summary of pertinent design parameters follows (23, 36–42):

1. Moisture content: 40–60%
2. Oxygen: 5–15%
3. Temperature peak: 55°–65°C (130°–150°F)
4. pH: 5–8
5. Carbon/nitrogen (C/N) ratio: 20:1–30:1
6. Land requirement: 0.2–0.3 acre/dry ton septage solids/d (0.09–0.13 ha/dry metric ton/d)
7. Blower size: 1/4 kW (1/3 horsepower [hp])
8. Septage pile dimensions:
    2.7 m (9 ft) high
    4.6 m (15 ft) diameter
    0.3 m (1 ft) base
    0.5 m (1.5 ft) blanket
    0.75 m$^3$ (1 yd$^3$) filter pile

## 12.6. *Odor Control*

Soil filters provide breakdown of malodorous compounds by both chemical and biological means. This is accomplished by collection and forcing air from contained process units through networks of perforated pipes buried in the soil, or through a mixture of iron oxide and woodchips (Figure 3.15).

# Biosolids and Septage Treatment

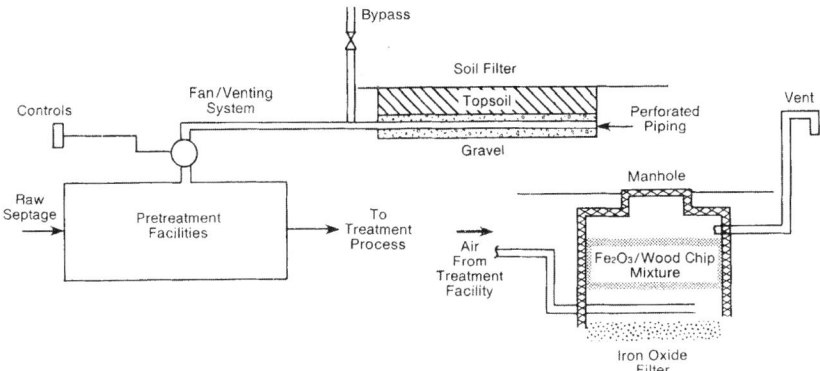

**Fig. 3.15.** Septage odor control (23).

Common modifications include use of compost rather than soil as filter media; above ground, enclosed filters for smaller volumes of gas; use of rooted vegetation to maintain loose soil; and enhanced biological activity. Alternative odor control methods include exhaust gas scrubbing in aeration basins and incineration in biosolids combustion units. Chemical scrubbers and activated carbon filters have also been used with mixed success.

Odorous gases are contained and vented to the soil filter area via a piping network. Given sustained biological activity, filters may regenerate during periods when no gases are passing through. Pilot- and full-scale studies have demonstrated complete elimination of odors by use of soil filters (i.e., no detectable odors in vicinity of soil filter). Gases with $H_2S$ concentrations greater than 100 mg/L have been deodorized ($H_2S$ <1 mg/L) by this method (44–47).

The recommended design criteria for odor control are as follows (23, 44, 45):

1. Minimum soil depth: 0.5 m (20 in)
2. Air loading rate: 60 $m^3/m^2$/hr (200 $ft^3/ft^2$/hr) for soil filters
3. Detention time: not less than 30 seconds at peak air flow
4. Soil type: moist loam, sandy loam, compost
5. Soil temperature: above 3°C (38°F)
6. Soil moisture: sprinkling may be required in dry summer periods; proper drainage must be provided to prevent saturation of the soil

## REFERENCES

1. N. K. Shammas and L. K. Wang, Emerging natural biological treatment processes. In: *Advanced Biological Treatment Processes,* L. K. Wang, N. K. Shammas, and Y. T. Hung (eds.) Humana Press, Inc., Totowa, NJ (2008).
2. L. K. Wang, N. K. Shammas, and Y. T. Hung (eds.) In: *Biosolids Treatment Processes,* 820 pp. L. K. Wang, N. K. Shammas, and Y. T. Hung (eds.). Humana Press, Inc., Totowa, NJ (2007).
3. O. E. Albertson, E. B. Burris, S. C. Reed, J. A. Semon, J. E. Smith, Jr., and A. T. Wallace, Dewatering municipal wastewater sludges, *Pollution Technology Review* No. 202. Noyes Data Corporation, Park Ridge, NJ (1991).
4. L. K. Wang, L., Yan, N. K. Shammas, and G. P. Sakellaropoulos, Drying beds. In: *Biosolids Treatment Processes,* pp. 403–430. L. K. Wang, N. K. Shammas, and Y. T. Hung (eds.) Humana Press, Inc., Totowa, NJ (2007).

5. N. K. Shammas and L. K. Wang, Vacuum filtration. In: *Biosolids Treatment Processes,* pp. 495–518. L. K. Wang, N. K. Shammas, and Y. T. Hung (eds.), Humana Press, Inc., Totowa, NJ (2007).
6. L. K. Wang, S. Y. Chang, Y. T. Hung, and S. H. Muralidhara, Centrifugation clarification and thickening. In: *Biosolids Treatment Processes,* pp. 101–134. L. K. Wang, N. K. Shammas, and Y. T. Hung (eds.). Humana Press, Inc., Totowa, NJ (2007).
7. N. K. Shammas and L. K. Wang, Pressure filtration. In: *Biosolids Treatment Processes,* pp. 541–582. L. K. Wang, N. K. Shammas, and Y. T. Hung (eds.). Humana Press, Inc., Totowa, NJ (2007).
8. N. K. Shammas and L. K. Wang, Belt filter presses. In: *Biosolids Treatment Processes,* pp. 519–540. L. K. Wang, N. K. Shammas, and Y. T. Hung (eds.). Humana Press, Inc., Totowa, NJ (2007).
9. S. R. Qasim, *Wastewater Treatment Plant, Planning, Design and Operation,* 2nd ed., Technomic Publishing Co., Inc. Lanchester, PA (1999).
10. G. Tchobanoglous, F. L. Burton, and H. D. Stensel, *Wastewater Engineering, Treatment and Reuse,* 4th ed., Metcalf & Eddy, Inc. McGraw-Hill, New York (2003).
11. E. Neyens, J. Baeyens, R. Dewil, and B. De Heyder, Advanced sludge treatment affects extracellular polymeric substances to improve activated sludge dewatering, *Journal of Hazardous Material*, 106, 2–3, 83–92 (2004).
12. U.S. Environmental Protection Agency, *Design Manual: Dewatering Municipal Wastewater Sludges,* US EPA/625/1–87/014, Cincinnati, OH, September (1987).
13. B. J. Kim, R. R. Cardenas, C. S. Gee, and J. T. Brandy, *Performance Evaluation of Existing Wedgewater and Vacuum-Assisted Bed Dewatering System*, U.S. Army Construction Engineering Research Laboratory (US ACERL) (1992).
14. B. J. Kim, R. R. Cardenas, and S. P. Chennupati, *An Evaluation of Reed Bed Technology to Dewater Army Wastewater Treatment Plant Sludge*, U.S. Army Construction Engineering Research Laboratory (US ACERL) (1993).
15. U.S. Environmental Protection Agency, *Manual on Constructed Wetlands Treatment of Municipal Wastewaters,* Cincinnati, OH (2000).
16. C. J. Martel, Development and design of sludge freezing beds, *ASCE Journal of Environmental Engineering*, 15, 4 (1989).
17. C. J. Martel, Operation and performance of sludge freezing bed at Fort McCoy, Wisconsin, In: *Proceedings of the 7th International Cold Regions Engineering Specialty Conference*, Edmonton, Alberta, Canada (1994).
18. L. K. Wang, *Theory and Applications of Flotation Process*, Lenox Institute of Water Technology (formerly Lenox Institute for Research), Lenox, MA, Technical Report No. LIR/11–85/l58, 1985. U. S. Department of Commerce, National Technical Information Service, Springfield, VA, NTIS-PB86–194198/AS (1985).
19. A. S. Eikum, *Treatment of Septage—European Practice*, Norwegian Institute for Water Research, Report No. 0–80040, February (1983).
20. Roy F. Weston, Inc., *Concept Engineering Report—Septage Management Facilities for Ocean County Utilities Authority*, October (1980).
21. J. J. Kolega, A. W. Dewey, and C. S. Shu, Streamline septage receiving stations, *Water and Wastes Engineering,* 8, July (1971).
22. Whitman and Howard, Inc., *A Study of Waste Septic Tank Sludge Disposal in Massachusetts,* Division of Water Pollution Control, Water Resources Commission, Boston (1976).
23. U.S. Environmental Protection Agency, *Septage Treatment and Disposal Handbook,* US EPA-625/6–84–009, Municipal Environmental Research Laboratory, Cincinnati, OH, September (1984).

24. A. J. Condren, *Pilot-Scale Evaluations of Septage Treatment Alternatives*, U.S. Environmental Protection Agency, Washington, DC, Report No. 600/2–78–164, NTIS No. PB-288415/AS, September (1978).
25. N. K. Shammas and L. K. Wang, Land application of biosolids. In: *Biosolids Treatment Processes,* pp. 705–745. L. K. Wang, N. K. Shammas, and Y. T. Hung (eds.). Humana Press, Inc., Totowa, NJ (2007).
26. L. K. Wang, N. K. Shammas, and Y. T. Hung (eds.) *Advanced Biological Treatment Processes,* Humana Press, Inc., Totowa, NJ (2008).
27. J. W. Rezek and I. A. Cooper, *Septage Management*, U.S. Environmental Protection Agency, Washington, DC, Report No. 600/8–80–032, NTIS No. PB-8l-l4248l, August (1980).
28. U.S. Environmental Protection Agency, *Applications of Sludges and Wastewaters on Agricultural Land: A Planning and Educational Guide,* Washington, DC, Report No. MCD-35, March (1978).
29. U.S. Congress. Criteria for classification of solid waste disposal facilities and practices, *Federal Register*, 44, 53438–53468, 13 September (1979).
30. U.S. Environmental Protection Agency, *Process Design Manual for Land Application of Municipal Sludge*, Washington, DC, Report No. 625/1–83–016, October (1983).
31. E. L. Stone, Microelement nutrition of forest trees: a review. In: *Forest Fertilization—Theory and Practice*, Tennessee Valley Authority, Muscle Shoals, AL (1968).
32. D. R. Keeney, K. W. Lee, and L. M. Walsh, *Guidelines for the Application of Wastewater Sludge to Agricultural Land in Wisconsin*, Technical Bulletin No. 88, Wisconsin Department of Natural Resources (1975).
33. M. A. Vivona and W. Herzig, The use of septage lagoons in New England, *Sludge*, March–April (1980).
34. New England Interstate Water Pollution Control Commission, *Evaluation of Acton's Septage Disposal Facility* (1980).
35. N. K. Shammas and L. K. Wang, Biosolids composting. In: *Biological Treatment Processes,* L. K. Wang, N. C. Pereira and Y. T. Hung (eds.) Humana Press, Inc., Totowa, NJ (2008).
36. E. Epstein, G. B. Wilison, W. D. Gurge, R. Mullen, and L. D. Enkiri, A Forced aeration system for composting wastewater sludge, *Journal of Water Pollution Control Federation*, 48, 4, April (1976).
37. D. Mosher and R. K. Anderson, *Composting Sewage Sludge by High- Rate Suction Aeration Techniques*, U.S. Environmental Protection Agency, Interim Report No. SW-6l4d (1977).
38. R. Wolf, Mechanized sludge composting at Durham, New Hampshire, *Compost Science Journal of Waste Recycling*, November-December (1977).
39. J. Heaman, Windrow composting—a commercial possibility for sewage sludge disposal, *Water Pollution Control*, January (1975).
40. R. P. Poincelot, The biochemistry of composting process, *National Conference on Composting Municipal Residues and Sludges*, Information Transfer, Inc., Rockville, MD, August (1977).
41. G. M. Wesner, *Sewage Sludge Composting*, U.S. Environmental Protection Agency, Technology Seminar Publication on Sludge Treatment and Disposal, Cincinnati, OH, September (1978).
42. Stearns and Wheeler, Inc., *Draft Interim Septage Management Plan*, Sussex County, NJ, Municipal Utilities Authority, April (1980).
43. K. D. White and L. K. Wang, Natural treatment and on-site processes, *Water Environment Research*, 72, 5, 67–71 (2000).

44. The Calgon Corporation, *Effective Odor Control with Calgon Granular Activated Carbon Systems*, Pittsburgh, PA (1981).
45. D. A. Carison and C. P. Leiser, Soil beds for the control of sewage odors, *Journal of Water Pollution Control Federation*, 34 (1966).
46. G. T. Kleinheinz and P. C. Wright, Biological odor and VOC control process. In: *Biological Treatment Processes,* L. K. Wang, N. C. Pereira and Y. T. Hung (eds.) Humana Press, Inc., Totowa, NJ (2008).
47. L. K. Wang, N. C. Pereira, and Y. T. Hung (eds.) *Biological Treatment Processes,* Humana Press, Inc., Totowa, NJ (2008).
48. M. Krofta and L. K. Wang, Application of dissolved air flotation to the Lenox Massachusetts water supply: water purification by flotation. *Journal of New England Water Works Association*, pp. 249–264 (1985).
49. M. Krofta and L. K. Wang, Application of dissolved air flotation to the Lenox Massachusetts water supply: sludge thickening by flotation and lagoon. *Journal of New England Water Works Association,* pp. 265–284 (1985).
50. L. K. Wang, Y. T. Hung, and N. K. Shammas (eds.) *Physicochemical Treatment Processes,* 723 pp. Humana Press, Inc., Totowa, NJ (2005).
51. L. K. Wang, Y. T. Hung, and N. K. Shammas (eds.) *Advanced Physicochemical Treatment Processes,* 690 pp. Humana Press, Inc., Totowa, NJ (2006).
52. L. K. Wang, Y. T. Hung, and N. K. Shammas (eds.) *Advanced Physicochemical Treatment Technologies*, 710 pp. Humana Press, Totowa, NJ (2007).

# 4
# Waste Chlorination and Stabilization

## Lawrence K. Wang

**CONTENTS**
  INTRODUCTION
  WASTEWATER CHLORINATION
  SLUDGE CHLORINATION AND STABILIZATION
  SEPTAGE CHLORINATION AND STABILIZATION
  SAFETY CONSIDERATIONS OF CHLORINATION PROCESSES
  RECENT ADVANCES IN WASTE DISINFECTION
  NOMENCLATURE
  ACKNOWLEDGMENTS
  REFERENCES

**Abstract** This chapter introduces three areas of waste chlorination: (a) wastewater chlorination, (b) sludge-biosolids chlorination and stabilization, and (c) septage chlorination and stabilization. Each waste chlorination area further covers the subjects of process description, glossary, design considerations, operation considerations, process equipment, process control, design examples, and application examples.

**Key Words** Chlorination · stabilization · wastes · wastewater · sludge · biosolids · septage · glossary · design · operation · equipment · control · examples · applications.

## 1. INTRODUCTION

### 1.1. Process Introduction

Chlorine is an efficient disinfectant as well as an oxidizing agent, and perhaps is the most frequently used chemical by environmental engineers and scientists since 1800s. The oldest water treatment facilities used only chlorine for water disinfection, which became the foundation of our industrial development. Today chlorine has been used in various forms for sanitary, commercial, industrial, and military applications. This chapter is a sister chapter to the following book chapters in the *Handbook of Environmental Engineering* series:

1. "Halogenation and Disinfection" chapter introduces various disinfection processes, such as chlorination, chlorine dioxide disinfection, bromination, and iodination which all involve the use of halogens (1).

2. "Potable Water Chlorination and Chloramination" chapter introduces the detailed engineering procedures for calculation of CT values for disinfection, and both conventional and innovative process equipment, including the on-site chlorine gas and hypochlorite generation facilities (2).
3. "Ozonation" chapter and "UV Disinfection" chapter introduce the disinfection/oxidation processes which use ozone and UV, respectively (1,2).
4. "Inorganic Chemical Conditioning and Stabilization" chapter introduces various chemical conditioning and stabilization processes (including chlorine stabilization) for sludge treatment (3).
5. "Pressurized Ozonation" chapter introduces a modern process involving the use of ozone and oxygen in a pressurized reactor for sludge disinfection, oxidation and stabilization (3).

This chapter deals with the same topic of chemical oxidation and disinfection, but at the advanced level. Specifically the engineering design and applications of chlorination processes for treatment of wastewater, biosolids, and septage are introduced in detail.

## 1.2. Glossary

The process involving the application of chlorine to drinking water, wastewater, industrial effluent, sludge, septage, swimming pool water, etc., to disinfect or to oxidize undesirable pathogens and compounds is termed chlorination (1–10). When chloramines (instead of chlorine) are used as the disinfectant/oxidant, the process is termed chloramination (1,8).

When wastewater is treated by chlorine, the treatment process is termed "wastewater chlorination," or simply "chlorination."

When chlorination process is used for treatment of sludge and septage, it can be called "sludge chlorination" and "septage chlorination," respectively. Either sludge chlorination or septage chlorination is a "chlorine stabilization" process (3,5–8). A device or process equipment that adds any chlorine compounds, in solid, gas, or liquid form, to water, wastewater, or sludge to kill infectious microorganisms or undesirable substances is a chlorinator (8–9). A part of treatment facility where water, wastewater, or sludge is treated by chlorine for disinfection and oxidation is a chlorine contact chamber (CCC).

Chlorine stabilization is one of a number of chemical stabilization processes, involving the use of chlorine. For instance, if lime is used in a chemical stabilization process for sludge treatment, it is a lime stabilization process.

Chlorine dosage is the amount of chlorine required to oxidize the target substance to be treated (such as water, wastewater, sludge, or septage) plus the desired chlorine residual. The target substance to be treated is termed chlorine demand. Usually the chlorine dosage is computed as mg/L concentration and the chlorine feed system set at the equivalent lb/d feed rate.

Given a desired chlorine residual of 300 mg/L and a chlorine demand of 800 mg/L, the chlorine dosage and resulting feed rate (for 12,000 gal/d throughput) are computed as follows:

$$\text{Chlorine dosage} = \text{Chlorine demand} + \text{Desired chlorine residual} \qquad (1)$$
$$\text{Chlorine dosage} = 300\,\text{mg/L} + 800\,\text{mg/L} = 1100\,\text{mg/L}$$
$$\text{Feed rate, lb/d} = (1100\,\text{mg/L}) \times (8.34\,\text{lb/MG/mg/L}) \times (0.012\,\text{MGD})$$
$$= 110\,\text{lb/d}$$

Throughput rate is the gallons of sludge fed to the unit per unit time (gpm or gpd). If the sludge is treated by chlorine (a disinfectant/oxidizing agent), the oxidized sludge is the chemical oxidation effluent. Chlorination is an oxidation process.

A total chemical process step for changing, adjusting, modifying, and improving the characteristics of sludge or septage prior to a dewatering process is termed "chemical conditioning." If only chlorine is used for the sludge or septage treatment, chlorination (or chlorine oxidation) is a "chemical conditioning" process. If the chlorine-treated sludge (or septage) needs to be further treated by another chemical (such as a neutralizing agent), then chlorination is only an intermediate process. The chlorine-treated sludge (or treated septage) is the oxidized sludge (or oxidized septage). In this case, the conditioned sludge will be the oxidized sludge that has also been conditioned by a holding tank or further chemical treatment to raise the pH and reduce the chlorine residual.

Nascent oxygen is uncombined oxygen in molecular form (O). Oxidant is an agent which oxidizes a substance by removing one or more electrons from an atom, ion, or molecule.

## 2. WASTEWATER CHLORINATION

### 2.1. Process Description

The most common use of chlorine in sewage treatment is for disinfection, which usually is the last treatment step in a secondary biological wastewater treatment plant. Where the treated secondary effluent is fed into a stream to be used for water supply or for recreational purposes, chlorination is effective in destroying the disease-producing pathogens found in treated wastewater. Other principal uses of chlorine are odor control and control of bulking in activated sludge.

Chlorine may be fed into the wastewater automatically, with the dosage depending on the degree of treatment. The wastewater then flows into a tank, where it usually is held for about 30 min to allow the chlorine to react with the pathogens (Figure 4.1). Chlorine

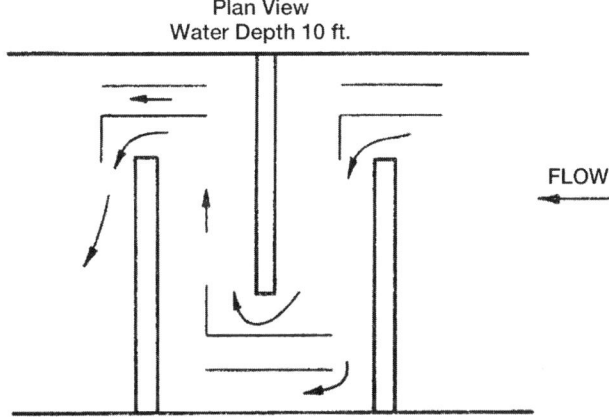

**Fig. 4.1.** Chlorine contact chamber with end-around baffles and vanes.

often is used either as a gas, or a solid or liquid compound. Liquid chlorine, or hypochlorite, has been used mostly in small systems (fewer than 5000 persons), or in large systems, where safety concerns related to handling chlorine gas outweigh economic concerns. The use of chlorine has proven to be a very effective means of disinfection.

Chlorine is also used in advanced wastewater treatment (AWT) for nitrogen removal, through a process known as "breakpoint chlorination." For nitrogen removal, enough chlorine is added to the wastewater to convert all the ammonium nitrogen to nitrogen gas. To do this, about 10 mg/L of chlorine must be added per mg/L of ammonia nitrogen in the wastewater—about 40 or 50 times more chlorine than normally used in a wastewater plant for disinfection only.

The facilities required for the chlorination process are simple. Wastewater (after secondary or tertiary treatment) flows into a mixing tank where the chlorine is added and complete mixing is provided. Because a large amount of chlorine is used and has an acidic effect on the wastewater, alkaline chemicals (such as lime) may be added to the same chamber to neutralize this acidic effect. The nitrogen gas that is formed is then released to the atmosphere. The amount of chlorine used for nitrogen control provides very effective disinfection. Because the process is just as effective in removing 1 mg/L as 20 mg/L of ammonium, breakpoint chlorination often is used as a polishing step downstream of other nitrogen removal processes.

## 2.2. Design and Operation Considerations

### 2.2.1. General Considerations

Figure 4.2 shows how residual chlorine affects coliform number. The curves show the most probable number (MPN) of coliforms remaining after 30 min of chlorine contact in a well-designed chlorine contact tank. These results should not be considered as being exact. Table 4.1 lists chlorine dosages often used for disinfection of raw and partially-treated sewage.

In breakpoint chlorination, about 10 mg/L of chlorine must be added for each mg/L of ammonia nitrogen present in the wastewater. Studies show that better pretreatment will reduce the amount of chlorine needed to reach breakpoint. Table 4.2 shows how different pretreatment processes affect the chlorine to ammonia–nitrogen ratio needed for breakpoint chlorination. The breakpoint process can result in 99+ % removal of ammonium nitrogen, reducing concentrations to less than 0.1 mg/L (as N).

To evaluate the performance of a chlorination system, an environmental engineer should check the contact time, chlorine residual, and MPN of coliform organisms after chlorination. This can be done easily in the following steps:

1. Obtain typical operating data for the chlorination system being studied. For example: (a) type of effluent = activated sludge; (b) peak plant flow = 5 MGD; (c) volume of chlorine contact chamber (CCC), v = 13,926 ft$^3$; (d) chlorine dosage = 6 mg/L; (e) chlorine residual = 1 mg/L.
2. Determine the contact time for the chlorine contact chamber based on peak flow:

$$\text{Contact time, h} = (V \text{ in ft}^3) \times (7.48 \text{ gal/ft}^3) \times (24 \text{ h/d})/(\text{Flow in gpd})$$
$$\text{Contact time, h} = (13{,}926)(7.48)(24)/(5 \times 10^6)$$
$$= 0.45 \text{ h} = 27 \text{ min} \qquad (2)$$

3. Examine the daily disinfection log sheet for chlorine feed rates and chlorine residual patterns. Compare both contact time and chlorine residual with those required by the proper regulatory agency. As a general rule, residuals between 0.2 and 1.0 mg/L after 15–30 min contact times provide good disinfection. As shown in the example, the 27 min contact time and 1.0 mg/L chlorine residual should be generally sufficient.
4. If the chlorination system does not perform as expected, the shortcomings and troubleshooting guide should be studied.

**Table 4.1**
**Chlorine Dosage Ranges**

| Waste | Chlorine dosage (mg/L) |
|---|---|
| Raw sewage | 6–12 |
| Raw sewage (septic) | 12–25 |
| Clarified sewage | 5–10 |
| Clarified sewage (septic) | 12–40 |
| Chemical precipitation effluent | 3–10 |
| Trickling filter effluent | 3–10 |
| Activated sludge effluent | 2–8 |
| Sand filter effluent | 1–5 |

*Source*: US EPA.

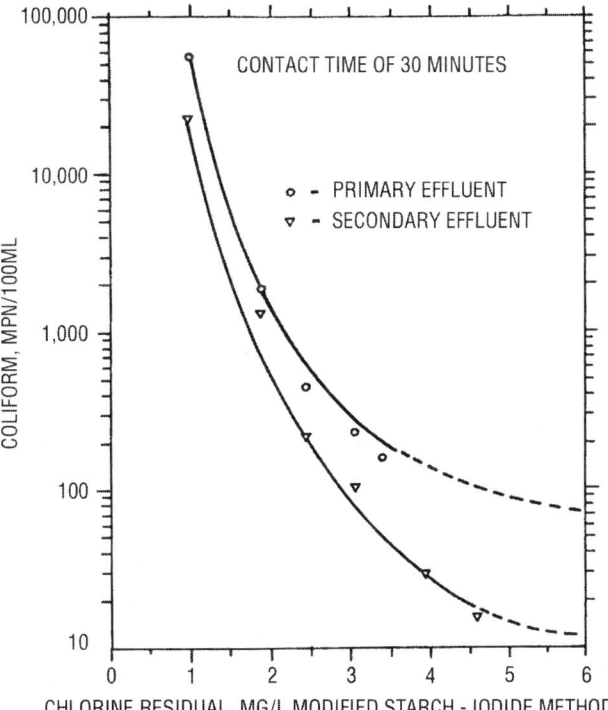

**Fig. 4.2.** MPN coliform vs. chlorine residual.

**Table 4.2**
**Effect of Pretreatment on Chlorine and Ammonia Nitrogen Breakpoint Ratio**

| Sample | Breakpoint pH | Initial $NH_4^+$-N (mg/L) | Final $NH_4^+$-N (mg/L) | Irreducible minimum residual (mg/L as $Cl_2$) | Breakpoint ratio $Cl_2$: $NH_4^+$-N (weight basis) |
|---|---|---|---|---|---|
| *Laboratory Tests* | | | | | |
| Buffered water | 6–7 | 20 | 0.1 | 0.6 | 8 : 1 |
| Raw wastewater | 6.5–7.5 | 15 | 0.2 | 7 | 9 : 1–10 : 1 |
| Lime clarified raw wastewater | 6.5–7.5 | 11.2 | 0.1 | 7 | 8 : 1–9 : 1 |
| Secondary effluent | 6.5–7.5 | 8.1 | 0.2 | 3 | 8 : 1–9 : 1 |
| Lime clarified secondary effluent | 6.5–7.5 | 9.2 | 0.1 | 4 | 8 : 1 |
| Ferric chloride clarified raw wastewater | 3.2 | 10.2 | 0.1 | 20 | 8.2 : 1 |
| *Pilot Plant Test* | | | | | |
| Filtered secondary effluent | 6–8 | 12.9–21.0 | 0.1 | 2–8.5 | 8.4 : 1–9.2 : 1 |
| Lime clarified raw wastewater—filtered | 7.0–7.3 | 9.7–12.5 | 0.4–1.2 | – | 9 : 1 |
| Alum clarified oxidation pond effluent—filtered | 6.6 | 20.6 | 0.1 | 7.6 | 9.6 : 1 |

### 2.2.2. Specific Design Procedures

For wastewater treatment, the recommended chlorine dosage for disinfection purposes should produce a chlorine residual of 0.5–1 mg/L after a specified contact time. Effective contact time of not less than 15 min at peak flow is recommended. Practical chlorine dosages recommended for wastewater disinfection and odor control are presented in below:

(a) Untreated wastewater (prechlorination) = 6–25 mg/L
(b) Primary clarification = 5–20 mg/L
(c) Chemical precipitation plant = 2–6 mg/L
(d) Trickling filter plant = 3–15 mg/L
(e) Activated sludge plant = 2–8 mg/L
(f) Multimedia filter following activated sludge plant = 1–5 mg/L

The required input data include: (a) chlorine contact tank influent flow, MGD; (b) peak flow, MGD; and (c) average flow, MGD. The design parameters include: (a) contact time at maximum flow, min; (b) length-to-width ratio; (c) number of chlorine contact tanks; (d) chlorine dosage, mg/L.

The following is recommended design procedure. The first step is to select contact time at peak flow and calculate volume of contact tank:

$$\text{VCT} = [Q_p(\text{CT})(10^6)]/(24 \times 60) \tag{3}$$

where VCT = volume of contact tank, gal, $Q_p$ = peak flow, MGD, and CT = contact time at maximum flow, min.

The second step in design is to select a side water depth and calculate surface area:

$$SA = VCT/[7.48\ SWD] \qquad (4)$$

where SA = surface area, ft$^2$, VCT = volume of contact tank, gal, and SWD = side water depth = 8 ft.

The third step in design is to select a length-to-width ratio and calculate dimensions by the following equations:

$$CTW = [SA/RLW]^{0.5} \qquad (5)$$
$$CTL = SA/CTW \qquad (6)$$

where CTW = contact tank width, ft, SA = surface area, ft$^2$, RLW = length-to-width ratio, and CTL = contact tank length, ft.

The fourth step in design is to select chlorine dosage according to the recommended chlorine dosages in this section, and then calculate chlorine requirements:

$$CR = (Q_a)(CD)(8.34) \qquad (7)$$

where CR = chlorine requirement, lb/d, $Q_a$ = average flow, MGD, and CD = chlorine dosage, mg/L.

The fifth step in design is to calculate peak chlorine requirements by the following equation:

$$PCR = (CR)(Q_p)/Q_a \qquad (8)$$

where PCR = peak chlorine requirements, lb/d, CR = chlorine requirements, lb/d, $Q_p$ = peak flow, MGD, and $Q_a$ = average flow, MGD.

The output data of wastewater disinfection design will be:

(a) Maximum flow, MGD (1 MGD = $3.785 \times 10^6$ L/d).
(b) Average flow, MGD.
(c) Contact time, min.
(d) Volume of contact tank, gal (1 gal = 3.785 L).
(e) Average chlorine requirement, lb/d (1 lb = 454 g).
(f) Peak chlorine requirement, lb/d.
(g) Tank dimensions.

## 2.3. Process Equipment and Control

The process equipment used for wastewater chlorination are similar to that for potable water chlorination, in terms of chlorine generation facilities and feeding system (1–17). The difference of the wastewater chlorination and water chlorination is their chlorine contact equipment. Many government rules and regulations specify the minimum detention time, depth, cover, mixing intensity, etc., for different types of oxidizing agents or disinfectants (17–20). The chlorine contact chamber for potable water treatment can be either indoors or outdoors, but usually is of indoors. The chlorine contact chamber for wastewater treatment is always an outdoor hydraulic structure. In general, the better the wastewater treatment plant is operated, the easier it will be to disinfect the plant effluent. Any failure to provide adequate treatment will increase the bacterial count and the chlorine requirement. High solids content and soluble organic loads increase the amount of chlorine required. Effective chlorine disinfection is dependent on the combined effect of chlorine dosage, mixing, and contact time with the wastewater. Enough disinfectant should be added to always meet the bacterial quality required by the regulatory agency. Control of the disinfection process is accomplished by measurement of the chlorine residual. General theories and process control of chlorination, disinfection, oxidation, and stabilization processes can be found from the literature (21–25).

Proper mixing is one of the most important factors in chlorine disinfection. Applying chlorine to wastewater in a well-mixed system produces a much better effluent than a system where chlorine is fed without rapid mixing, even with adequate residual and contact time. However, sufficient contact time (usually 30 min) between the chlorine and the wastewater is also needed to provide good disinfection. Usually, longer contact times are more important than higher residuals in wastewater treatment.

In breakpoint chlorination, the system must be able to meet quick changes in ammonia nitrogen concentrations, chlorine demand, pH, alkalinity, and flow. Failure to properly control chlorine dosage can result in poor nitrogen removal and chlorine overdoses. Overdoses of chlorine are a direct waste of this chemical and cause problems in adjusting the operation of the dechlorination equipment. Overdoses also can cause the direct discharge of high concentrations of chlorine residuals to the receiving water, and can result in the undesirable formation of $NCl_3$.

Usually, a base chemical is added to the breakpoint process to neutralize some of the acidity resulting from the chlorine addition. The base requirements depend on wastewater alkalinity, individual treatment processes used before breakpoint chlorination, as well as effluent pH or alkalinity restrictions by regulatory agencies.

Another consideration in breakpoint chlorination is dechlorination to remove the chlorine residual from the final effluents before it is discharged. Very often, dechlorination using sulfur dioxide or activated carbon may be needed when the breakpoint chlorination process is used. A new dechlorination technology has been introduced in another chapter of this handbook series (2). UV dechlorination is recommended by Wang (45).

In most cases, control of breakpoint chlorination requires the use of accurate and reliable automatic equipment to reduce the need for manual process control by operators. However, the operator must give special attention to this equipment and monitoring devices in order to ensure their proper operation. Table 4.3 indicates how the common process shortcomings can be compensated and improved. Table 4.4 is a wastewater chlorination process trouble-shooting guide for use by practicing environmental engineers.

### 2.4. Design Example—Design of a Wastewater Chlorine Contact Chamber

Design a chlorine contact chamber for wastewater disinfection based on the following given data and equations.

**Table 4.3**
**Common Design Shortcomings of Wastewater Chlorination and Solutions**

| Shortcomings | Solution |
|---|---|
| 1. Big changes in effluent chlorine residual when chlorine flow proportioning control device is operating properly. | 1. Install a continuous chlorine residual analyzer to control the feed rate automatically, or use a closed-loop system. |
| 2. Short-circuiting in chlorine contact tank. | 2. Make channels very narrow or Provide thorough baffling in the channel to ensure complete mixing and a sufficient contact time. |
| 3. High residual chlorine concentrations in the effluent toxic to aquatic life. | 3. Install dechlorinating systems (activated carbon, hydrogen peroxide, sulfur dioxide, sodium metabisulfite, or UV). |
| 4. Sodium hypochlorite cannot be stored for long periods of time without deteriorating. | 4. If long storage periods cannot be avoided, dilute the sodium hypochloride to slow down the rate of deterioration, or use liquid (gas) chlorine as an alternate source. |
| 5. Lack of mixing. | 5. Install mechanical mixer. |

The first step is to select contact time at peak flow and calculate volume of chlorine contact chamber (1) [use Eq. 3]:

$$\text{VCT} = [Q_p(\text{CT})(10^6)]/(24 \times 60)$$

where VCT = volume of contact tank, gal, $Q_p$ = peak flow, 2 MGD, and CT = contact time at maximum flow, 15 min; then

$$\text{VCT} = [Q_p(\text{CT})(10^6)]/(24 \times 60) = 2(15)(106)/(24 \times 60) = 20{,}833 \text{ gal}$$

The second step in design is to select a side water depth and calculate surface area [use Eq. 4]:

$$\text{SA} = \text{VCT}/[7.48 \text{ SWD}]$$

where SA = surface area, ft$^2$, VCT = volume of contact tank, 20,833 gal, and SWD = side water depth = 8 ft; then

$$\text{SA} = \text{VCT}/[7.48 \times \text{SWD}] = 20{,}833/[7.48 \times 8] = 348 \text{ ft}^2$$

The third step in design is to select a length-to-width ratio and calculate dimensions [use Eqs. 5 and 6]:

$$\text{CTW} = [\text{SA}/\text{RLW}]^{0.5}$$
$$\text{CTL} = \text{SA}/\text{CTW}$$

where CTW = contact tank width, ft, SA = surface area, 348 ft$^2$, RLW = length-to-width ratio = select 40, and CTL = contact tank length, ft; then

$$\text{CTW} = [\text{SA}/\text{RLW}]^{0.5} = [348/40]^{0.5} = 2.95 \text{ ft}$$
$$\text{CTL} = \text{SA}/\text{CTW} = 348/2.95 = 118 \text{ ft}$$

**Table 4.4**
**Troubleshooting Guide for Wastewater Chlorination Process**

| Indicators/observations | Probable cause | Check or monitor | Solutions |
|---|---|---|---|
| 1. Low chlorine gas pressure at chlorinator. | 1a. Insufficient number of cylinders connected to system. | 1a. Reduce feed rate and note if pressure rises appreciably after short period of time. If so, 1a is the cause. | 1a. Connect enough cylinders to the system so that chlorine feed rate does not exeed the withdrawal rate from the cylinders. |
| | 1b. Stoppage or flow restriction between cylinders and chlorinators. | 1b. Reduce feed rate and note if icing and cooling effect on supply lines continues. | 1b. Disassemble chlorine header system at point where cooling begins, locate stoppage and clean with solvent. |
| 2. No chlorine gas pressure at chlorinator. | 2a. Chlorine cylinders empty or not connected to system. | 2a. Visual inspection. | 2a. Connect cylinders or replace empty cylinders. |
| | 2b. Plugged or damaged pressure reducing valve. | 2b. Inspect valve. | 2b. Repair the reducing valve after shutting off cylinder valves, and decreasing gas in the header systems. |
| 3. Chlorinator will not feed any chlorine. | 3a. Pressure reducing valve in chlorinator is dirty. | 3a. Visual inspection. | 3a. 1. Disassemble chlorinator and clean valve stem and seat. 2. Precede valve with a filter-sediment trap. |
| | 3b. Chlorine cylinder hotter than chlorine control apparatus. | 3b. Cylinder area temperature | 3b. 1. Reduce temperature in cylinder area. 2. Do not connect a new cylinder which has been sitting in the sun. |
| 4. Chlorine gas escaping from chlorine pressure reducing valve (CPRV). | 4a. Main diaphragm of CPRV ruptured due to: 1. Improper assembly or fatigue. 2. Corrosion | 4a. Place ammonia bottle near termination of CPRV vent line to confirm leak. | 4a. 1. Disassemble valve and diaphragm. 2. Inspect chlorine supply systems for moisture intrusion. |

**Table 4.4** (*Continued*)

| Indicators/observations | Probable cause | Check or monitor | Solutions |
|---|---|---|---|
| 5. Inability to maintain chlorine feed rate without icing of chlorine system. | 5a. Insufficient evaporator capacity. | 5a. Reduce feed rate to about 75% of evaporator capacity. If this eliminates problem 5a is the cause. | |
| | 5b. External CPRV cartridge is clogged. | 5b. Inspect cartridge. | 5b. Flush and clean cartridge. |
| 6. Chlorination system unable to maintain water-batch temperature sufficient to keep external CPRV open. | 6a. Heating element malfunction. | 6a. Evaporator water-batch temperature. | 6a. Remove and replace heating element. |
| 7. Inability to obtain maximum feed rate from chlorinator. | 7a. Inadequate chlorine gas pressure. | 7a. Gas pressure. | 7a. Increase pressure—replace empty or low cylinders. |
| | 7b. Water pump injector clogged with deposits. | 7b. Inspect injector. | 7b. Clean injector parts using muriatic acid. Rinse with fresh water and replace in service. |
| | 7c. Leak in vacuum relief valve. | 7c. Disconnect vent line at chlorinator; place hand over vent connection to vacuum relief valve, observe if this results in more vacuum and higher chlorine feed rate. | 7c. Disassemble vacuum relief valve and replace all springs. |
| | 7d. Vacuum leak in joints, gaskets, tubing, etc. in chlorinator system. | 7d. Moisten joints with ammonia solution, or put paper containing orthotolidine at each joint in order to detect leak. | 7d. Repair all vacuum leaks by tightening joints, replacing gaskets, replacing tubing and/or compression nuts. |

(*Continued*)

**Table 4.4** (*Continued*)

| Indicators/observations | Probable cause | Check or monitor | Solutions |
|---|---|---|---|
| 8. Inability to maintain adequate chlorine feed rate. | 8a. Malfunction or deterioration of water supply pump. | 8a. Inspect pump. | 8a. Overhaul pump (if turbine pump is used, try closing down needle valve to maintain proper discharge pressure). |
| 9. Wide variation in chlorine residual produced in effluent. | 9a. Chlorine flow proportion meter capacity inadequate to meet plant flow. | 9a. Check chlorine meter capacity against plant flow meter capacity. | 9a. Replace with higher capacity chlorinator meter. |
| | 9b. Malfunctioning automatic controls. | 9b. | 9b. Call manufacturer's field service personnel. |
| | 9c. Solids settled in chlorine contact chamber. | 9c. Solids in contact chamber. | 9c. Clean chlorine contact chamber. |
| | 9d. Flow proportioning control device not zeroed or spanned correctly. | 9d. Check zero and span of control device on chlorinator. | 9d. Re-zero and span the device in accordance with manufacturer's instructions. |
| 10. Chlorine residual analyzer recorder controller does not control chlorine residual properly. | 10a. Electrodes fouled. | 10a. Visual inspection. | 10a. Clean electrodes. |
| | 10b. Loop-time too long. | 10b. Check loop-time. | 10b. Reduce loop time by doing the following: 1. Move injector closer to point of application. 2. Increase velocity in sample line to analyzer cell. 3. Move cell closer to sample point. 4. Move sample point closer to point of application. |

**Table 4.4** (*Continued*)

| Indicators/observations | Probable cause | Check or monitor | Solutions |
|---|---|---|---|
| | 10c. Insufficient potassium iodide being added for amount of residual being measured. | 10c. Potassium iodide dosage. | 10c. Adjust potassium iodide feed to correspond with residual being measured. |
| | 10d. Buffer additive system malfunctioning. | 10d. See if pH of sample going through cell is maintained. | 10d. Repair buffer additive system. |
| | 10e. Malfunctioning of analyzer cell. | 10e. Disconnect analyzer cell and apply a simulated signal to recorder mechanism. | 10e. Call authorized service personnel to repair electrical components. |
| | 10f. Poor mixing of chlorine at point of application. | 10f. Set chlorine feed rate at constant dosage and analyze a series of grab samples for consistency. | 10f. Install mixing device to cause turbulence at point of application. |
| | 10g. Rotameter tube range is improperly set. | 10g. Check tube range to see if it gives too small or too large an incremental change in feed rate. | 10g. Replace with a proper range of feed rate. |
| 11. Coliform count fails to meet required standards for disinfection. | 11a. Inadequate chlorination equipment capacity. | 11a. Check capacity of equipment. | 11a. Replace equipment as necessary to provide treatment based on maximum flow through plant. |
| | 11b. Inadequate chlorine residual control. | 11b. Continuously record residual in effluent. | 11b. Use chlorine residual analyzer to monitor and control the chlorine dosage automatically. |

(*Continued*)

**Table 4.4** (*Continued*)

| Indicators/observations | Probable cause | Check or monitor | Solutions |
|---|---|---|---|
| | 11c. Short-circuiting in contact chamber. | 11c. Contact time. | 11c. 1. Install baffling in contact chamber.<br>2. Install mixing device in contact chamber. |
| | 11d. Solids build-up in contact chamber. | 11d. Visual inspection. | 11d. Clean contact chamber to reduce solids build-up. |
| | 11e. Chlorine residual too low. | 11e. Chorine residual. | 11e. Increase contact time and/or increase chlorine feed rate. |
| 12. Chlorine residual too high in plant effluent to meet requirements. | 12a. Chlorine residual too high. | 12a. Determine toxicity level by bioassay procedures. | 12a. Install dechlorination facility (See Shortcomings). |

The fourth step in design is to select chlorine dosage according to the recommended chlorine dosages in this section, and then calculate chlorine requirements [use Eq. 7]:

$$CR = (Q_a)(CD)(8.34)$$

where CR = chlorine requirement, lb/d, $Q_a$ = average flow, 1 MGD, and CD = chlorine dosage, 8 mg/L; then

$$CR = (Q_a)(CD)(8.34) = 1 \times 8 \times 8.34 = 66.7 \text{ lb/d}$$

The fifth step in design is to calculate peak chlorine requirements [use Eq. 8]:

$$PCR = (CR)(Q_p)/Q_a$$

where PCR = peak chlorine requirements, lb/d, CR = chlorine requirements, 66.7 lb/d, $Q_p$ = peak flow, 2 MGD, and $Q_a$ = average flow, 1 MGD; then

$$PCR = (CR)(Q_p)/Q_a = (66.7 \times 2)/(1) = 133.4 \text{ lb/d}$$

Finally, the output data of wastewater disinfection design will be:

(a) Maximum flow, 2 MGD.
(b) Average flow, 1 MGD.
(c) Contact time, 15 min.
(d) Volume of contact tank, 20,833 gal.
(e) Average chlorine requirement, 66.7 lb/d.
(f) Peak chlorine requirement, 133.4 lb/d.
(g) Tank dimensions: surface area = 348 ft$^2$; side water depth = 8 ft; length–width ratio = 40; contact tank length = 118 ft.

## 2.5. Application Example—Coxsackie Sewage Treatment Plant, Coxsackie, NY, USA

This project was initiated in 1970s when the New York State Department of Health (NYSDOH) and the New York State Department of Correction (NYSDOC) undertook a program of research and development in the area of advanced biological sewage treatment at one of its correctional facilities. With a view to the impending program of sewage treatment plant construction in the entire New York State, it was decided to build an advanced sewage treatment plant and a research laboratory on the grounds of the Coxsackie Correctional Facility at West Coxsackie, New York. The plant has been performing successfully as a role model to the New York municipalities since 1973. This section only reports the operational and R&D data generated by Leo J. Hetling, Carl Beer, and Lawrence K. Wang, who were Research Director, Project Manager, and Chief Operator (NYSDEC Senior Sanitary Engineer), respectively, in 1973–1977. The plant data introduce how a typical single-sludge activated-sludge plant performs for carbonaceous oxidation, nitrification, denitrification, and phosphors removal, wastewater chlorination, and sludge chlorine stabilization (37–42).

The Coxsackie Correctional Facility is an institution for young male delinquents aged 16–21. The average age of the inmates is 18. The design inmate population is 750. During the period covered by the full analytical data of this report, the inmate population

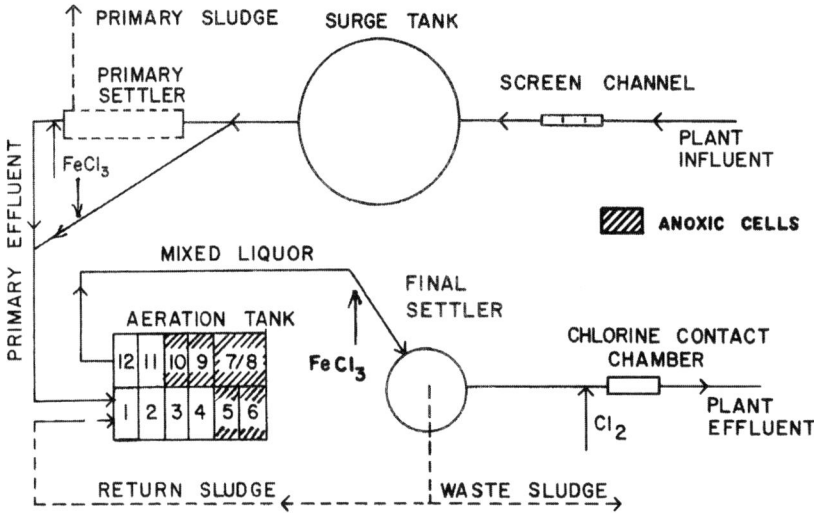

**Fig. 4.3.** Process flow diagram of Coxsackie Sewage Treatment Plant, Coxsackie, NY, USA.

was near capacity. In addition to the inmates, approx 350 prison personnel are in daytime or nighttime residence at the facility. A 306 ha (750 acres) farming operation is part of the correctional facility. Farm products are milk, vegetables, apples, and beef. The institution is located south of Albany, NY. The effluent of the single-sludge biological treatment plant is discharged to the Coxsackie Creek, a short tributary of the Hudson River. The Coxsackie Creek is classified as an intermittent stream. The New York State effluent requirements for sewage treatment plant discharging to intermittent streams are as follows: (a) 5-d BOD = 5 mg/L, maximum; (b) ammonia nitrogen, $NH_3$ = 2 mg/L maximum; and (c) DO = 7.5 mg/L minimum.

The detailed operational data can be found in another handbook by Wang, Shammas, and Hung (35). This chapter introduces the Coxsackie Sewage Treatment Plant (STP), its wastewater chlorination facility and its structural improvement. Figure 4.3 shows the flow diagram of Coxsackie STP, in which wastewater chlorine contact chamber (CCC) is the last unit process from where the plant effluent is discharged to the receiving stream, Coxsackie Creek. The hydraulic structure of CCC, illustrated by Figure 4.4, consists of mixing chamber, influent plenum, flow-through chamber, and effluent plenum.

In the CCC, chlorine solution is admixed to the final settler effluent. The chamber is equipped with a 0.25-kW (1/3 hp), 330-rpm mixer. The two plenums serve to distribute the process water over the cross section of the flow-through chamber with the help of two orifice plates mounted at the entrance and exit of that chamber. Each orifice plate carries 40 orifices of 12.7 mm (1/2 in.) diameter. The flow-through chamber is 1.52 m × 1.52 m (5 ft × 5 ft) in cross section and 4.88 m (16 ft) long.

Baffle cages were installed in the combined CCC and second-phase final settler unit (refer to Figure 4.4). The baffle cages carry Plexiglass baffles 3.2 mm (1/8 in.) thick that are inclined by 60° to the horizontal and spaced 152 mm (6 in.) on center or 305 mm (12 in.) vertically.

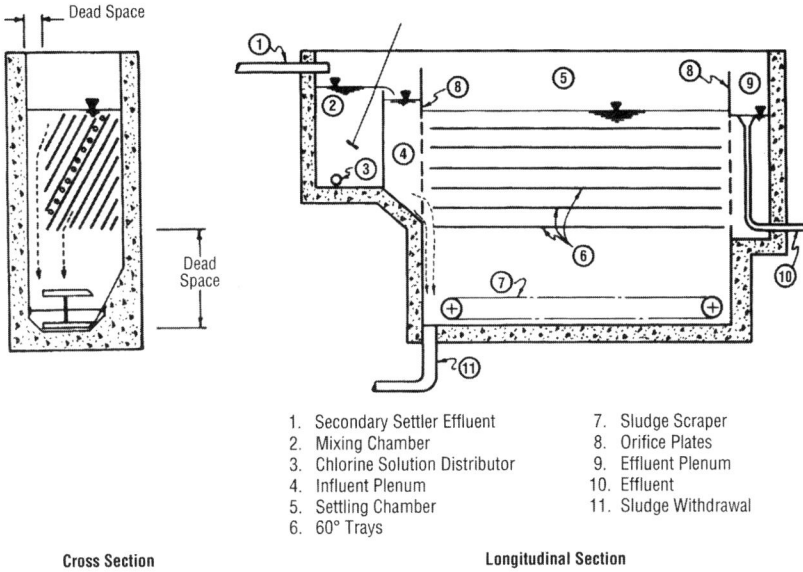

**Fig. 4.4.** Chlorine contact chamber and second-phase settler combination.

Legend:
1. Secondary Settler Effluent
2. Mixing Chamber
3. Chlorine Solution Distributor
4. Influent Plenum
5. Settling Chamber
6. 60° Trays
7. Sludge Scraper
8. Orifice Plates
9. Effluent Plenum
10. Effluent
11. Sludge Withdrawal

The data in Table 4.5 for the 5-wk period of June 28 to August 1, 1975, indicate an average 33% reduction in suspended solids (SS) from 5.4 mg/L down to 3.6 mg/L was achieved across the unit.

During the 5 wk period ending August 1, 1975, the following operating conditions prevailed: (a) average daily flow $= 629 \, \text{m}^3/\text{d}$ (166, 000 gpd); (b) average flow rate $= 7.28 \, \text{L/s}$ (0.255 cfs); (c) average flow velocity taken over the $1.525 \, \text{m} \times 1.525 \, \text{m}$ (5 ft × 5 ft) cross section $= 0.0031 \, \text{m/s}$ (0.0102 ft/s); and (d) average flow-through time for $4.88 \, \text{m}$ (16 ft) length $= 26 \, \text{min}$.

It has been proven that the operation of the subject combination unit can be improved by using corrugated sheets instead of the flat Plexiglass sheets as baffles. Using corrugated sheets drastically cut down the accumulation time of the solids on the baffles. Solids accumulate on the baffles until a certain thickness is reached. At that point, the solids slide off.

The orifice plates must be cleaned periodically of scraps of papers and bits of rags. In order to prevent such a manual operation, a combined second-phase settler and chlorine contact chamber is protected by a traveling screen.

## 3. SLUDGE CHLORINATION AND STABILIZATION

### 3.1. Process Description

Stabilization by chlorine addition has been developed and is marketed under the registered trade name Purifax. The chlorine conditioning of sludge varies greatly from the more traditional methods of biological digestion or heat conditioning. First, the reaction is almost instantaneous. Second, there is very little volatile solids reduction

Table 4.5
Effect of Chlorine Contact Chamber and Second-Phase
Settler Combination on Suspended Solids Removal

| Date 1975 | Influent SS (mg/L) | Effluent SS (mg/L) | Date 1975 | Influent SS (mg/L) | Effluent SS (mg/L) |
|---|---|---|---|---|---|
| June 28 | 2.0 | 2.6 | July 16 | 3.6 | 2.7 |
| 29 | 5.1 | 3.4 | 17 | 4.2 | 2.9 |
| 30 | 7.3 | 3.1 | 18 | 6.2 | 2.9 |
| July 1 | 5.2 | N.AV. | 19 | 3.8 | 2.7 |
| 2 | 3.8 | 2.7 | 20 | 6.3 | 4.5 |
| 3 | N.AV. | 2.7 | 21 | 5.1 | 3.8 |
| 4 | 7.9 | 1.3 | 22 | 5.5 | 3.9 |
| 5 | 5.5 | 4.7 | 23 | 6.7 | 4.6 |
| 6 | 6.3 | 5.1 | 24 | 7.6 | 5.9 |
| 7 | 4.5 | 3.9 | 25 | 7.2 | 4.5 |
| 8 | 5.6 | 4.7 | 26 | 4.9 | 4.0 |
| 9 | 6.8 | 4.5 | 27 | 5.8 | 3.7 |
| 10 | 5.0 | 4.5 | 28 | 11.4 | 3.7 |
| 11 | 5.4 | 3.7 | 29 | 5.4 | 4.1 |
| 12 | 5.1 | 3.5 | 30 | 4.3 | 3.1 |
| 13 | 3.8 | 3.3 | 31 | 4.9 | 3.8 |
| 14 | 6.0 | 4.4 | Aug. 1 | 6.9 | 3.9 |
| 15 | 4.8 | 3.5 | Ave. | 5.4 | 3.6 |

in the sludge. There is some breakdown of organic material and formation of carbon dioxide and nitrogen; however, most of the conditioning is by the substitution or addition of chlorine to the organic compound to form new compounds that are biologically inert. The sludge to be chlorinated can be either biosolids or chemical sludge.

The chemical form in which chlorine is present in wastewater is directly related to pH. The first reaction of chlorine is with ammonia (combined available chlorine). However, this is a small portion of the chlorine added for this process. Most of the chlorine (free available chlorine) ends up as either hydrochloric acid, HCl, or hypochlorous acid, HOCl. The HOCl subsequently breaks down into nascent oxygen, O, and HCl. Below pH 5, molecular chlorine, $Cl_2$, appears in solution and increases in concentration with decreasing pH. The equations for the reaction of free available chlorine in water can be summarized as follows:

$$Cl_2 + 2 H_2O \rightarrow HOCl + HCl + H_2O \tag{9}$$

Hypochlorous acid, HOCl, its subsequent by-product nascent oxygen, O, and molecular chlorine are all strong oxidants. The hydrochloric acid is not an oxidant or a disinfectant, but does lower the pH of the solution. Generally, the entire process consists of a macerator, flow meter, recirculation pump, two reaction tanks, a chlorine eductor, chlorinator, evaporator, a pressure control pump, and two holding tanks. Variations are possible with the selection of the individual units depending on the nature of the sludge.

Conventional grit removal equipment used for the plant influent will suffice for grit reduction of sludge processed through the oxidation unit. If grit removal equipment has not been provided for the plant, then it should be added to this system. The type of macerator selected depends on the type of sludge being stabilized. The resulting maximum particle size should not exceed 1/4 in. To provide optimum utilization of chlorine, the system should be preceded by a sludge holding tank, which includes some means of mixing or agitation. This is especially necessary for treating primary sludge. The primary sludge solids concentration is typically higher at the beginning of a pumping cycle and lower near the end of the cycle. Without provision of the holding tank, the chlorine requirement would be variable and over-chlorination or under-chlorination a possibility. With the holding tank in use, the chlorine requirement can be set at a constant rate. Use of sludge thickeners ahead of the conditioning unit is optional. The sludge processing rate will be reduced for thicker sludge; specifically for primary sludge or primary sludge plus trickling filter sludge greater than 4%, primary sludge plus waste-activated sludge greater than 2.6%, or waste activated sludge greater than 1%. The lower processing rates offset the reduction in volume obtained by thickening, so there is no advantage in thickening to concentrations greater than those given above for the different types of sludge.

## 3.2. Design and Operation Considerations

A schematic of the sludge chlorination process (Purifax process) is shown in Figure 4.5. The sludge is first pumped through a macerator to reduce the particle size for optimum chlorine exposure. It is then mixed with conditioned sludge ahead of the recirculation pump at a ratio of 3.8 gal of recirculated sludge for each gallon of raw sludge. The combined flow is then pumped through the first reaction tank where it is thoroughly mixed. A portion of the sludge then flows to the second reaction tank while

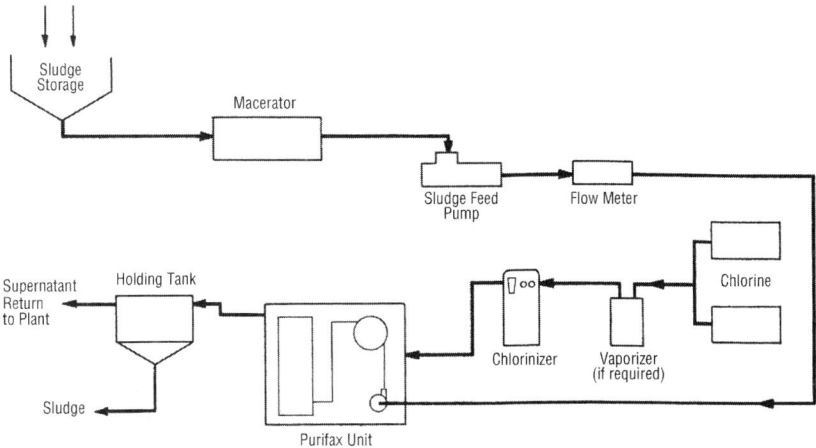

**Fig. 4.5.** Schematic of a waste sludge chlorination process system (BIF).

the remainder is recirculated. Recirculation of a portion of the sludge aids in mixing and provides better utilization of the chlorine. The recirculation rate is normally held constant. A pressure control pump at the discharge end of the second reaction tank maintains a pressure of 30–40 psi on the entire system.

Chlorine is added to the recirculated sludge line ahead of the recirculation pump. The passage of the conditioned sludge through the eductor creates a vacuum that causes chlorine gas to move from the chlorine supply into the sludge line. The recirculation of the conditioned sludge through the eductor satisfies the dilution requirements of the chlorine gas without introducing additional water into the system. The recirculation pump acts as a mixer for the raw and conditioned sludge. Almost all of the reaction between the sludge and chlorine takes place in the first reaction tank. This tank provides 3 min of detention time at design flow. The second reaction tank provides an additional 1.5 min of detention time. Operating the system under pressure forces the chlorine to penetrate into the sludge particles to insure a complete reaction.

Chlorine is supplied to the unit from a separate chlorinator located in the same room as the chlorinators for disinfecting the plant effluent. Because of the large volumes of chlorine required for the sludge chlorination (Purifax) unit, an evaporator is used ahead of the chlorinator. The sidestreams that require further treatment result from the thickening and/or dewatering processes that follow the oxidation unit. The characteristics of the supernatant vary with the type of sludge being treated and the method of thickening or dewatering. The oxidized sludge should be contained in a holding tank or reservoir for at least 48 h. This will allow time for the chlorine residual to approach zero and the pH to rise from 3.5 or 4.0 to 5.0 or 6.0. The $BOD_5$ and suspended solids (SS) concentrations obviously are quite variable but each should be less than 350 mg/L. The supernatant or filtrate sidestreams are routed to the plant headworks for treatment with the incoming sewage.

If the oxidized sludge is dewatered without provisions for the holding tank, then sodium hydroxide or lime must be added to raise the pH. The quantities of filtrate or supernatant to be treated vary with the type of process(es) used. In general, the quantity of supernatant or filtrate to be treated is minor in terms of the total treatment plant capacity. The loading rates shown in Table 4.6 apply to standard Purifax units. Chlorine dosages range from 600 to 4800 mg/L depending on the type of sludge and solids concentration. Generally, the units should be operated with the lowest concentration shown for each sludge type shown in Table 4.6. At these concentrations the chlorine dosage varies from 600 to 2000 mg/L or 0.005 to 0.017 pounds per gallon. The actual dosage used must be adjusted for each individual plant.

The stabilized sludge will have a pH of 2.5–4.5 and chlorine residual of 200–400 mg/L. The stabilized sludge will have chlorine smell and light brown color. Total solids, suspended solids, and volatile solids concentrations will be about the same as the raw sludge. When stored for 48 h, the chlorine residual will have fallen to 0 and the pH will have increased to 4.5–6.0. The organics will normally not decompose even after several days of storage.

Table 4.7 shows the expected characteristics for sidestreams from typical thickening and dewatering operations as applied to the conditioned sludge.

**Table 4.6**
**Solids Loading Rates**

| Type of sludge | Solids concentration (%) | gpm/hp |
|---|---|---|
| Primary | ≤4 | 2–3.5 |
| Primary and trickling filter | ≤4 | 2–3.5 |
| Primary and waste activated sludge | ≤2.6 | 2–3.5 |
| Primary and waste activated sludge | 4.0 | 1.5–2.5 |
| Primary and waste activated sludge | 5.0 | 1.1–2.1 |
| Waste activated sludge | 1.0 | 2.9–5.1 |
| Waste activated sludge | 2.0 | 2.2–3.9 |
| Waste activated sludge | 3.0 | 1.5–2.6 |
| Waste activated sludge | 4.0 | 0.8–1.3 |
| Anaerobic digester supernatant | 0.2 | 2.9–5.1 |
| Anaerobic digester supernatant | 0.3 | 2.5–4.5 |
| Anaerobic digester supernatant | 0.4 | 2.1–3.8 |
| Anaerobic digester supernatant | 0.5 | 1.8–3.2 |
| Septic tank sludge | 2.0 | 2.9–5.1 |
| Septic tank sludge | 3.0 | 2.5–4.5 |
| Septic tank sludge | 4.0 | 2.2–3.9 |
| Trickling filter humus | 2.0 | 2.9–5.1 |
| Trickling filter humus | 3.0 | 2.5–4.6 |
| Trickling filter humus | 4.0 | 2.2–4.0 |

## 3.3. Process Equipment and Control

### 3.3.1. Process Equipment

The process equipment for sludge chlorination or stabilization is commercially available from BIF Corporation.

### 3.3.2. Staffing Requirements

The staff requirements shown below apply to the Purifax process, macerator, pumps, and chlorination system. Dewatering or thickening operation and maintenance are not included. The labor requirements of a package chlorine treatment unit are listed below: (a) Operation = 2 h/shift/unit; and (b) maintenance = 3 h/shift/unit.

The chemical oxidation process (i.e., chlorination) is automated. The main effort is visually checking the process and operating the ancillary equipment. Most of the systems now in operation are package-type units.

### 3.3.3. Monitoring

The oxidized sludge should have a faint chlorine odor after processing, with no chlorine odor after 2 d storage.

The material should be light gray in color. If these characteristics change, the process control parameters should be checked. Each individual plant will result in processed sludge that has slightly different sensory characteristics. After the wastewater plant

**Table 4.7**
**Sidestream Characteristics**

| | |
|---|---|
| Supernatant from Conditioned Sludge Holding | |
|     $BOD_5$ | 50–150 mg/L |
|     Suspended solids | 50–200 mg/L |
|     pH | 4.5–6.0 |
|     Chlorine residual | 0 |
| Filtrate from Vacuum Filter | |
|     $BOD_5$ | 100–350 |
|     Suspended solids | 50–150 |
|     pH (with 20–30 lb/ton NaOH) | 4.5–5.5 |
|     Chlorine residual | 200–400 (unless stored) |
| Centrate from Solid Bowl Centrifuge | |
|     $BOD_5$ | 200–400 |
|     Suspended solids | 300–500 |
|     pH (with 20–30 lb/ton NaOH) | 4.5–5.5 |
|     Chlorine residual | 200–400 (unless stored) |

has operated for several weeks, then the sensory observations can be more critically reviewed. Major differences in chlorine odor may result with too little or too much chlorine. Darkening of the sludge color may result if the process is not properly operating. If either of the above occur, the chlorine residual and system pressure should be checked immediately. If these parameters are within acceptable ranges, then check for changes in the incoming sludge.

### 3.3.4. Normal Operating Procedures

#### 3.3.4.1. PRE-STARTUP AND STARTUP

The following are the pre-startup procedures: (a) Check operation of the chlorine pressure reducing valve (PRV) by turning ON the power switch on the sludge oxidation unit control panel. Turn the chlorine valve switch to the OPEN position, observe operation of valve actuator; (b) turn switch to the CLOSED position; and (c) adjust time delay relays according to the manufacturer's instruction manual.

The following are the startup procedures: (a) Adjust the chlorinator feed rate to a low setting; (b) close the chlorine pressure-reducing valve bypass valve; (c) turn on the chlorine supply to the chlorinator; (d) turn the feed pump and macerator motor starter selector switches to AUTO; (e) turn both selector switches in the PURIFAX motor starter panel to AUTO; (f) turn the power switch and alarm switch ON and the chlorine valve switch to AUTO; (g) depress the START pushbutton. When the motors start, adjust the speed of the pressure control pump to produce 30 psi at the process pressure gauge. If this is not done before the timing relay times out, the system will automatically shut down. (h) The vacuum gauge should read approx 20 in. of mercury. The vacuum gauge on the chlorinator should read the same. The pump suction gauge should read approx 5 psi. The pump discharge gauge should read approx 30 psi. (i) Check the oxidation unit for obvious sludge leaks. (j) Belt adjustment—adjust takeup on the recirculation pump belts until only a slight bow appears in the slack side. (k) Recheck the tension of new belts several times within the first 50 h of operation and adjust if necessary. (l) Thereafter check the tension periodically. (m) Install belt guard. (n) When sludge is introduced into the system, it may be necessary to readjust the speed of the pressure control pump. (o) Adjust the chlorinator feed rate to the calculated rate; and (p) check for water at each pump seal by disconnecting the seal water tubing.

#### 3.3.4.2. ROUTINE AND SHUTDOWN OPERATIONS

The system routine operation is automatic after startup. Should a problem develop causing deviation from established operating limits, the system will automatically shutdown. The system cannot be restarted until the problem causing the shutdown has been corrected. The system should be checked twice a shift.

Normal shutdown is a sequential operation initiated by depressing the STOP pushbutton. The sequence of operation is as follows:

1. Depressing the STOP pushbutton causes immediate interruption of the circuit to the chlorine pressure-reducing valve causing it to close and shut off the chlorine gas supply.
2. The chlorine pressure switch senses the loss of chlorine gas pressure and its contacts open. After a sufficient time has elapsed to evacuate chlorine gas from the piping, an OFF delay relay, 3TR, is de-energized, closing the vacuum line valve.
3. When the vacuum valve closes, its auxiliary contacts open causing interruption of the circuits to the recirculation pump and pressure control pump motor starters. Auxiliary contacts in the starters open, interrupting the circuit to the seal flushing water solenoid valve and the remotely mounted control relay. The control relay is de-energized stopping the feed pump and macerator.

### 3.3.5. Process Control Considerations

The control system is automatic, with little operator attention required. There are three variables that affect operation. They are throughput rate, chlorine feed rate, and system pressure. The throughput rate has been designed for an expected solids concentration. The process chlorine feed is set based on the expected rate of solids fed to the oxidation unit. If the actual solids concentration increases, the throughput rate should be lowered.

The chlorine feed rate is also adjusted based on throughput rate solids concentration and/or monitoring results. If the chlorine residual increases or decreases beyond the limits of the recommended range, check the chlorine demand and reset the chlorine feed.

If the above parameters are correct and the oxidized sludge characteristics are not within recommended limits, check the unit pressure. This should be between 30 and 40 psi.

### 3.3.6. Emergency Operating Procedures

The chlorine oxidation unit will not operate without power. Raw sludge must be hauled to a landfill site or temporarily stored if facilities are available. If stored, the sludge must be processed when power is restored.

Other treatment units that affect the oxidation unit include raw sludge thickening and oxidized sludge dewatering. If the raw sludge thickener is out of service, the throughput rate will not be affected unless the maximum capacity is exceeded. If this occurs, the unit hours of operation will have to be extended. The chlorine feed rate should be adjusted. The amount of adjustment is determined by the results of a chlorine demand test.

If the oxidized sludge dewatering unit is out of service, then the disposal transport system and disposal site must be expanded to handle the increased volume.

Should an emergency occur requiring immediate shutdown and over-ride of normal sequential shutdown, an EMERGENCY STOP pushbutton is provided for this purpose. This device interrupts power to all components in the oxidation unit control circuit, shuts down all motors, and closes the vacuum line valve and the chlorine pressure-reducing valve (PRV). The EMERGENCY STOP pushbutton is also used as a reset device to restore the system to normal operating status after an alarm situation has occurred.

The control system is designed to sense certain component and system failures. Pressure switches are located to sense overpressure, excessive suction, low chlorine pressure, and low eductor vacuum. A flow switch senses low flow. Motor starters sense motor overloading. Evaporator low-temperature switch senses low water temperature.

Whenever deviation from established operating limits is sensed, lights indicating the cause of the problem will be activated, an audible alarm will call attention to the problem, and the system will be automatically shutdown until the problem is corrected. The audible alarm may be switched off. The indicating lamps remain lighted until the problem is corrected and the system is reset.

A lock-out relay is included in the circuit that allows indicating alarm lamps to light, the audible alarm to sound, and prevents restarting without resetting the system when shutdown occurs in an alarm situation. It also prevents alarm devices from functioning during normal shutdown.

*3.3.7. Common Process Shortcomings and Solutions*

Table 4.8 indicates the common process shortcomings and their respective solutions. Table 4.9 is a sludge chlorination trouble shooting guide.

*3.3.8. Maintenance Considerations*

Inspect motors at regular intervals; keep motors clean and ventilation openings clear. Check valve may become clogged causing sludge to backup into chlorinator. Periodically disassemble and clean. Replace diaphragm and spring if they are deteriorated. Valve ball and seats may become scored causing sludge to back-up into diaphragm check valve. Periodically disassemble by unscrewing union nuts, with valve in CLOSED position, remove carrier and ball by pressing on flat spot on ball. Replace ball and seats if scored.

**Table 4.8**
**Common Design Shortcomings of Sludge Chlorination and Solutions**

| Shortcoming | Solution |
|---|---|
| 1. Unit improperly sized. | 1. Change operating time. |
| 2. Inadequate holding tank capacity. | 2a. Add another holding tank. |
|  | 2b. Use chemicals for pH adjustment and chlorine removal. |
| 3. Inadequate storage for new sludge during power outage or shutdown. | 3a. Store sludge in clarifiers (temporary). |
|  | 3b. Haul sludge to landfill. |

**Table 4.9**
**Troubleshooting Guide for Sludge Chlorination Process**

| Indicators/observations | Probable cause | Check or monitor | Solutions |
|---|---|---|---|
| 1. Oxidation unit shuts down (low chlorine supply pressure). | 1a. No chlorine supply pressure. | 1a. (1) Check chlorine tanks.<br>(2) Check all manual valves in supply piping.<br>(3) Check evaporator. | 1a. (1) If empty, replenish supply.<br>(2) Should be fully open.<br>(3) See manufacturer's instructions. |
| | 1b. Failure of electric chlorine pressure-reducing and shut-off valve to open. | 1b. Check chlorine valve switch on control panel (should be in AUTO position). | 1b. (1) Turn chlorine valve switch on control panel to OPEN position. If valve opens, and if chlorine pressure is indicated at chlorinator, the problem is in the control panel.<br>(2) If valve remains closed, the problem is in the valve or valve operator. |
| | 1c. Failure of chlorine pressure switch to close. | 1c. (1) Check position of relay.<br>(2) Check pressure setting and switch contacts. | 1c. (1) Should be de-energized.<br>(2) Adjust as needed. |
| | 1d. Electrical failure in control panel. | 1d. See wiring diagram. | 1d. Correct where necessary. |
| 2. Oxidation unit shutdown. | 2a. Failure of feed pump motor to operate. | 2a. (1) Check selector switch on motor starter.<br>(2) Check motor overload heaters.<br>(3) Check control relay. | 2a. (1) Should be in AUTO position.<br>(2) Correct if overloaded.<br>(3) Should be energized, if not, problem is in control panel or control relay. If relay is energized, relay has failed. |
| | 2b. Failure of feed pump to pump. | 2b. (1) Check lines for obstructions.<br>(2) Check valves.<br>(3) Pull pump. | 2b. (1) Clean lines.<br>(2) Open if necessary.<br>(3) Repair per manufacturer's instructions. |

(*Continued*)

**Table 4.9** (*Continued*)

| Indicators/observations | Probable cause | Check or monitor | Solutions |
|---|---|---|---|
| | 2c. Failure of flow meter receiver contacts to close. | 2c. (1) Check setting of auxiliary contacts. | 2c. (1) Should be set for approximately 50% of the minimum system throughput rate. |
| | | (2) Check valve in discharge piping. | (2) Should be fully open. |
| | 2d. Process pressure switch contacts opened. | 2d. (1) Check valve in discharge piping. | 2d. (1) Should be fully open. |
| | | (2) Monitor pressure control pump speed. | (2) Reduce speed. |
| | 2e. Pump suction pressure switch contacts opened. | 2e. Obstruction in inlet piping. | 2e. Remove obstruction. |
| 3. Oxidation unit shuts down (low vacuum). | 3a. Break or leak in chlorine vacuum line piping. | 3a. Locate leak. | 3a. Repair. |
| | 3b. Plugged eductor body. | 3b. Disassemble and inspect. | 3b. Remove two pipe plugs, vacuum gauge, switch assembly and vacuum line valve. Clean all openings. |
| | 3c. Recirculation pump fails to deliver required dynamic head of 30 psi. (discharge pressure minus suction pressure). | 3c. (1) Check drive belt tension. (2) Check for worn impeller. (3) Check for air in piping. | 3c. (1) Adjust. (2) Replace. (3) Bleed off air at vent plugs. |
| | 3d. Failure of pressure control pump to maintain pressure in the system. | 3d. (1) Check pump speed. (2) Check for worn impeller. | 3d. (1) Increase speed. (2) Replace. |
| | 3e. Failure of pressure control pump motor or recirculation pump motor to operate. | 3e. (1) Check selector switches on the starter panel. (2) Check motor starter overloads. | 3e. (1) Set on AUTO position. (2) Reset. |

**Table 4.9** (*Continued*)

| Indicators/observations | Probable cause | Check or monitor | Solutions |
|---|---|---|---|
| | 3f. Failure of vacuum switch contacts to close. | 3f. (1) Check switch assembly and vacuum gauge.<br>(2) Check vacuum setting and switch contact.<br>(3) Check for loss of oil. | 3f. (1) Replace.<br>(2) Clean and reset.<br>(3) Fix leak and/or refill. |
| | 3g. Electrical failure in control panel. | 3g. Check wiring diagram. | 3g. Repair. |
| 4. Macerator stopped. | 4. Jammed with debris. | 4. Inspect. | 4. Turn off power, remove obstruction. |
| 5. Indication of vacuum at oxidation unit but none at chlorinator. | 5a. Diaphragm check valve plugs. | 5a. Inspect. | 5a. Disassemble and clean. |
| | 5b. Failure of vacuum line valve ball to open. | 5b. Inspect. | 5b. (1) Replace with spare valve.<br>(2) Disassemble and replace broken ball or shaft. |
| 6. Depressing STOP button on control panel, unit continues to run longer than normal shutdown time. | 6a. Failure of electric chlorine pressure-reducing and shut-off valve to close. | 6a. Check chlorine pressure at chlorinator. | 6a. (1) If there is pressure, turn chlorine valve switch on control panel to the closed position.<br>(2) If there is still pressure, the valve is stuck open-replace.<br>(3) If there is no pressure, the problem is in the control panel (correct wiring or fuse). |
| | 6b. Failure of chlorine pressure switch contacts to open. | 6b. Check position of relay. | 6b. If energized, the pressure switch is stuck open. |
| | 6c. Failure of vacuum line valve limit switch contacts to open. | 6c. (1) Inspect.<br>(2) Check valve operator. | 6c. (1) Replace with spare valve.<br>(2) If in the OPEN position, problem is in motor, gearing, limit switches, or cams.<br>(3) If closed, problem is in the limit switch or cam. |

When reassembling valve, use caution. Only hand tighten union nuts. Check the macerator twice daily for debris.

The major concerns are contact with the low pH and the high chlorine concentration of the oxidized sludge. Human contact with the oxidized sludge should be avoided. If this occurs, shower immediately.

The macerator can be dangerous if maintenance is attempted while the unit is turned on. Be sure the power is off before doing maintenance work.

Safety practices for handling chlorine are contained in "Safety Practice for Water Utilities," AWWA Manual M3 and an AWWA/WEF/APHA handbook (24,25). The readers are referred to Section 5 for more information on safety considerations of all chlorination processes.

Generation of chlorinated hydrocarbons may be a problem (26), but the magnitude of any such problem cannot be determined at this time. The feasibility of using a chlorination step to reduce excess sludge in an activated-sludge process has been studied by Saby et al. (27). The ultimate disposal of excess waste-activated sludge (WAS) from a biological treatment plant has been a problem. Ozonation works well, but is a costly option. It was observed that the WAS production could readily be reduced by 65% once the sludge chlorination treatment was involved (27). However, the chlorination treatment also caused some problems: (a) it resulted in poor sludge settleability when part of the treated sludge was returned to the aeration tank and (b) it increased soluble COD in the plant effluent. Saby et al. (27) discovered that by integrating the immersed membrane into the activated-sludge process, these difficulties can be overcome effectively. Wang and his associates (28–35) have studied other alternative disinfectants (quaternary ammonium compounds) and oxidizing agents (ozone, chlorine dioxide) for sludge stabilization with preliminary success. Continuous research along this direction will be be needed.

### 3.4. Application Example—Coxsackie Sewage Treatment Plant, Coxsackie, NY, USA

Figure 4.3 in introduces the Coxsackie Sewage Treatment Plant (STP). The plant's special wastewater chlorination facility, a combined chlorine contact chamber (CCC) & second-phase final settler unit, is shown in Figure 4.4. This section introduces the sludge chlorination facility of Coxsackie STP and its operational data. At Coxsackie STP, data generated during 19 batches of sludge chlorination are summarized in Tables 4.10–4.12. Characteristics of the underflow from the sludge dewatering beds for nine of those batches are presented in Table 4.13. The batch size varied somewhat according to the solids concentration of the sludge treated and also according to the sludge storage space available. A typical batch consisted of $34\,m^3$ (9000 gal) applied to $111\,m^2$ ($1200\,ft^2$) of sand bed, resulting in an average sludge slurry dose of $0.31\,m$ (1 ft). However, such a depth was never attained because drainage occurred immediately upon application. At most, approx 15 cm (6 in.) of chlorinated sludge slurry were seen standing on the bed. Usually, all visible liquid had drained away by the morning following application (37–42).

Table 4.10
Chlorine Dosages Used During Sludge Chlorination Treatment

| Batch No. | Volume of sludge chlorinated | | Cl$_2$ used | | SS of sludge before treatement (mg/L) | Dry sludge treated | | Chlorine | |
|---|---|---|---|---|---|---|---|---|---|
| | (m$^3$) | (gal) | (kg) | (lb) | | (kg) | (lb) | Dosage (mg/L) | % of dry solids |
| 1 | 31.8 | (8400) | 22 | (48) | 9200 | 293 | (645) | 685 | 7 |
| 2 | 31.0 | (8200) | 27 | (60) | 7400 | 230 | (506) | 877 | 12 |
| 3 | 31.4 | (8300) | 25 | (54) | 10,100 | 318 | (699) | 780 | 8 |
| 4 | 22.4 | (6000) | 24 | (52) | 8700 | 198 | (435) | 1039 | 12 |
| 5 | 33.7 | (8900) | 36 | (78) | 10,700 | 361 | (794) | 1051 | 10 |
| 6 | 33.7 | (8900) | 33 | (72) | 12,700 | 429 | (943) | 970 | 8 |
| 7 | 37.8 | (10,000) | 21 | (46) | 8700 | 330 | (726) | 552 | 6 |
| 8 | 32.6 | (8600) | 24 | (54) | 13,400 | 437 | (961) | 753 | 6 |
| 9 | 23.5 | (6200) | 25 | (55) | 9600 | 225 | (496) | 1064 | 11 |
| 10 | 31.0 | (8200) | 36 | (80) | 9700 | 301 | (663) | 1170 | 12 |
| 11 | 32.6 | (8600) | 18 | (40) | 9800 | 320 | (703) | 558 | 6 |
| 12 | 68.6 | (18,100) | 59 | (129) | 7300 | 501 | (1102) | 855 | 12 |
| 13 | 22.7 | (6000) | 24 | (52) | 7400 | 168 | (370) | 1039 | 14 |
| 14 | 51.9 | (13,700) | 38 | (84) | 6300 | 327 | (720) | 735 | 12 |
| 15 | 37.8 | (10,000) | 35 | (76) | 7100 | 269 | (592) | 911 | 13 |
| 16 | 45.0 | (11,900) | 30 | (66) | 5400 | 244 | (536) | 665 | 12 |
| 17 | 45.0 | (11,900) | 24 | (54) | 5600 | 253 | (556) | 544 | 10 |
| 18 | 22.7 | (6000) | 24 | (54) | 7200 | 164 | (360) | 1079 | 15 |
| 19 | 42.4 | (11,200) | 39 | (87) | 4200 | 178 | (392) | 931 | 22 |
| Total | 678.0 | (179,100) | 564 | (1241) | | 5546 | (12,199) | 831* | 10* |

Batch 18: Primary sludge; all other batches were a mixture of primary sludge and WAS.
*Weighted mean.

## Table 4.11
### Effect of Sludge Chlorination on Solids Concentration, Filterability, and pH

| | SS | | | VSS | | | Filterability | | | pH | |
|---|---|---|---|---|---|---|---|---|---|---|---|
| Batch no. | Before (mg/L) | After (mg/L) | Change (%) | Before (mg/L) | After (mg/L) | Change (%) | Before (mL/30 sec) | After (mL/30 sec) | Change (%) | Before | After |
| 1 | 9190 | 8680 | −6 | 4950 | 5020 | +1 | 122 | 182 | +49 | 6.45 | 2.6 |
| 2 | 7400 | 7240 | −2 | 3780 | 3680 | −3 | 124 | 172 | +40 | 6.3 | 2.65 |
| 3 | 10,100 | 9540 | −6 | 5680 | 5580 | −2 | 20 | 97 | +385 | 7.0 | 2.5 |
| 4 | 8730 | 8070 | −8 | 5100 | 5030 | −1 | 158 | 230 | +46 | 6.7 | 2.5 |
| 5 | 10,690 | 10,420 | −3 | 6190 | 6470 | +5 | 16 | 69 | +331 | 6.75 | 2.7 |
| 6 | 12,705 | 11,765 | −7 | 7480 | 7390 | −1 | 12 | 67 | +458 | 7.1 | 2.4 |
| 7 | 8680 | 8020 | −8 | 4100 | 4440 | +8 | 120 | 162 | +35 | 6.0 | 2.5 |
| 8 | 13,440 | 12,620 | −6 | 7040 | 7420 | +5 | 14 | 63 | +350 | 6.4 | 3.0 |
| 9 | 9590 | 9220 | −4 | 5910 | 6010 | +2 | 10 | 64 | +540 | 6.6 | 3.1 |
| 10 | 9680 | 9720 | — | 5640 | 6080 | +8 | 34 | 76 | +124 | 6.7 | 2.75 |
| 11 | 9780 | 9460 | −3 | 5550 | 5600 | +9 | 68 | 99 | +46 | 6.0 | 3.0 |
| 12 | 7300 | 6920 | −5 | 4860 | 4820 | −1 | 16 | 67 | +319 | 6.5 | 2.6 |
| 13 | 7360 | 5920 | −20 | 4720 | 4130 | −13 | 88 | 144 | +64 | 5.85 | 2.4 |
| 14 | 6260 | 6710 | −7 | 4100 | 4890 | +19 | 84 | 152 | +80 | 6.9 | 3.2 |
| 15 | 7050 | 6330 | −10 | 4220 | 4030 | −5 | 50 | 84 | +68 | 6.8 | 2.4 |
| 16 | 5440 | 4950 | −9 | 3320 | 3340 | — | 51 | 82 | +61 | 6.9 | 2.3 |
| 17 | 5600 | 5430 | −3 | 3460 | 3560 | +3 | 30 | 77 | +157 | 6.9 | 2.5 |
| 18 | 7220 | 6940 | −4 | 4580 | 4820 | +5 | 10 | 83 | +730 | 7.2 | 2.25 |
| 19 | 4220 | 4020 | −5 | 2780 | 2920 | −5 | 22 | 24 | +10 | 7.4 | 2.7 |
| Weighted mean | 8180 | 7760 | −5 | 4790 | 4900 | +2 | 52 | 101 | +49 | | |

**Table 4.12**
**Effect of Sludge Chlorination on Slurry Filtrate Quality**

| Batch no. | TOC Before (mg/L) | TOC After (mg/L) | TOC Change (mg/L) | $NH_4^+ - N$ Before (%) | $NH_4^+ - N$ After (mg/L) | $NH_4^+ - N$ Change (mg/L) | $NO_2^- + NO_3^- N$ Before (mg/L) | $NO_2^- + NO_3^- N$ After (mg/L) | $NO_2^- + NO_3^- N$ Change (mg/L) | Dissolved orthophosphorus Before (mg/L) | After (mg/L) | Change (mg/L) | Chlorine residual After (mg/L) |
|---|---|---|---|---|---|---|---|---|---|---|---|---|---|
| 1 | 17.5 | 63.5 | +363 | 1.3 | 2.3 | +1.0 | 42.0 | 42.0 | 0 | 0.5 | 2.4 | +1.9 | 133 |
| 2 | | | | 0.6 | 1.2 | +0.6 | 74.0 | 71.5 | −2.5 | 1.8 | 8.8 | +7.0 | 102 |
| 3 | | | | 8.9 | 9.2 | +0.3 | 0.2 | 1.0 | +0.8 | 2.1 | 1.6 | −0.5 | 111 |
| 4 | | | | 1.0 | 1.7 | +0.7 | 35.5 | 34.5 | −1.0 | 1.9 | 6.1 | +4.2 | 233 |
| 5 | | | | 9.5 | 13.6 | +4.1 | 37.5 | 31.2 | −6.3 | 2.2 | 2.2 | — | 142 |
| 6 | 115 | 184 | +183 | 20.5 | 19.6 | −0.9 | 5.7 | 0.4 | −4.3 | 1.6 | 2.0 | +0.4 | 164 |
| 7 | 11 | 71 | +545 | 0.5 | 2.2 | +1.7 | 59.0 | 56.0 | −3.0 | 0.6 | 2.6 | +2.0 | 191 |
| 8 | 150 | 188 | +253 | 22.0 | 22.0 | 0 | 62.5 | 83.5 | +21.0 | 5.1 | 3.0 | −2.1 | 121 |
| 9 | | | | 72.6 | 61.2 | −11.1 | 0.3 | 0.3 | — | 4.7 | 1.3 | −3.4 | 22 |
| 10 | | | | 1.3 | 3.6 | −2.3 | 5.1 | 19.9 | +14.8 | 0.8 | 0.8 | — | 89 |
| 11 | | | | 2.8 | 28.5 | +25.7 | 3.8 | 15.0 | +11.2 | 0.4 | 1.6 | +1.2 | 150 |
| 12 | | | | 1.7 | 0.2 | −1.5 | 7.6 | 4.0 | −3.6 | 0.92 | 1.5 | +0.6 | 126 |
| 13 | 20.5 | 46 | +124 | 1.6 | 2.4 | +0.8 | 2.7 | 0.8 | −1.9 | 0.20 | 3.15 | +3.0 | 244 |
| 14 | 24 | 61.5 | +156 | 8.8 | 11.7 | +2.9 | 0.8 | 6.9 | +6.1 | 0.31 | 1.40 | +1.1 | 121 |
| 15 | 66 | 74.5 | +13 | 0.1 | 9.1 | +9.0 | 6.5 | 4.3 | −2.2 | 4.53 | 5.65 | +1.1 | 171 |
| 16 | 20 | 49 | +145 | 0.4 | 0.5 | +0.1 | 0.5 | 1.5 | +1.0 | 0.33 | 4.70 | +4.4 | 197 |
| 17 | 39 | 50 | +28 | 0.2 | 0.1 | −0.1 | 3.8 | 3.7 | −0.1 | 0.61 | 1.95 | +1.3 | 74 |
| 18 | | | | 0.1 | 0.3 | +0.2 | 1.7 | 0.9 | −0.8 | 3.32 | 5.50 | +2.2 | 315 |
| 19 | | | | 0.2 | 0.2 | — | 8.7 | 0.6 | −8.1 | 0.92 | 3.96 | +3.0 | 168 |
| Weighted mean | | | | 6.7 | 8.7 | +2.0 | 17.6 | 18.4 | +0.8 | 1.6 | 3.0 | +1.4 | 146 |

Table 4.13
Quality of Sludge Dewatering Bed Underflow During Sludge Chlorination Treatment

| Batch no. | Type of bed* | Chlorine dosage (mg/L) | Constituents of bed filtrate** | | | | | | | | |
|---|---|---|---|---|---|---|---|---|---|---|---|
| | | | SS (mg/L) | VSS (mg/L) | pH | Chlorine residual (mg/L) | TOC (mg/L) | COD (mg/L) | Total alk. (mg/L as $CaCO_3$) | $NH_4^+ - N$ (mg/L) | $NO_2^- +$ $NO_3^-N$ (mg/L) | Dissolved orthophosphorus (mg/L) |
| 11 | CTF | 558 | 21 | 13 | 6.6 | 2.7 | N. AV. | 193 | 78 | 8 | 5 | 1.4 |
| 12 | CTF | 855 | 76 | 24 | 6.2 | 0.7 | N. AV. | 208 | 115 | 12 | 15 | 1.2 |
| 13 | SSB | 1039 | 29 | 12 | 5.2 | 0.1 | 43 | 147 | 42 | 36 | 11 | 1.8 |
| 14 | CTF | 735 | 14 | N. AV. | 6.3 | 2.7 | 39 | 110 | 130 | 11 | 1 | 0.4 |
| 15 | CTF | 911 | 40 | 19 | 6.7 | 0.1 | 44 | 241 | 90 | 24 | 17 | 1.3 |
| 16 | SSB | 665 | 9 | 2 | 5.8 | 5.3 | 51 | 142 | 87 | 12 | 14 | 0.5 |
| 17 | CTF | 544 | 64 | 43 | 6.7 | 0.1 | 60 | 178 | 73 | 15 | 12 | 1.0 |
| 18 | CTF | 1079 | 25 | 22 | 6.3 | 0.1 | N. AV. | 199 | 52 | 21 | 8 | 1.0 |
| 19 | CTF | 931 | 55 | 27 | 6.9 | 2.7 | N. AV. | 211 | 127 | 6 | 15 | 1.3 |

*CTF = Converted trickling filter.
SSB = Standard sand drying bed.
**Grab samples.

The slurry was treated at a rate of approx 2.5 L/s (40 gpm). The weighted average of the chlorine dosage was 830 mg/L or 10% of the dry weight of the sludge chlorinated. The chlorine dosage was adjusted to produce a pH of 2.3–2.8 in the chlorinated slurry. The pH of the unchlorinated slurry indicates that considerable nitrification had occurred on some batches before treatment.

During chlorination, samples of unchlorinated and chlorinated sludge slurry were withdrawn and subjected to the tests reflected in Tables 4.10–4.12. Glass fiber filters were used to determine some of the parameters in Table 4.13.

There was only a slight reduction in SS due to chlorination, approx 5%. It should be noted in this connection that solids reduction due to acidification with $H_2SO_4$ to pH 2 was found to be between 50 and 60% in unrelated laboratory tests on the waste sludge of this plant.

The increase in filterability, 49% on the average, was of course the most important change in sludge characteristics brought about by sludge chlorination. The increase of filterability was greater for sludge of a low initial filterability. The weighted average filterability was increased from 52 to 101 mL/30 s. The mechanisms of this increase remain to be explained.

The changes in nutrient concentration were somewhat erratic but on the average negligible, i.e., under 3 mg/L.

The chlorine residual in the chlorinated sludge was approx 150 mg/L. A significant increase in the TOC of the sludge supernatant was observed from a range of 11–150 mg/L to a range of 46–188 mg/L. The sludge dewatering beds underflow did not impose any significant load on the Coxsackie activated sludge system. The following constituent ranges were determined: (a) SS = 9–76 mg/L; (b) VSS = 2–43 mg/L; (c) pH = 5.2–6.9; (d) chlorine residual = 0.1–5.3 mg/L; (e) TOC = 39–60 mg/L; (f) COD = 142–241 mg/L; (g) alkalinity = 42–130 mg/L as $CaCO_3$; (h) ammonia–N = 6–36 mg/L; (i) nitrate–nitrite–N = 1–17 mg/L; and (h) dissolved orthophosphorus = 0.4–1.8 mg/L.

## 4. SEPTAGE CHLORINATION AND STABILIZATION

### 4.1. Process Description

The septage chlorination process (Purifax) utilizes chlorine gas in solution to oxidize various types of waste sludge (biosolids), including septage. Chlorination is one type of chemical oxidation process that stabilizes sludge and septage both by reducing the number of organisms present and by making organic substrates less suitable for bacterial metabolism and growth.

The septage chlorination process involves disinfection and oxidation of several septage constituents with high dosages of chlorine gas, which is applied directly to the septage in an enclosed reactor for a short time. Because of the reaction of chlorine gas with the septage, significant quantities of hydrochloric acid are formed, and the stablilized septage has low pH (about 2). The reactor vessel is moderately pressurized (30–40 psi) to ensure more complete absorption of the chlorine gas as well as adequate chlorination penetration into the larger particles in the sludge. At these pressures, the gases formed are supersaturated in the treated septage. When discharged from the reactor vessel at atmospheric pressure, these gases come out of solution as fine bubbles that float the septage solids. The process is followed by dewatering, generally on sand beds. Septage chlorination is also termed chlorine stabilization, which, like lime stabilization, does not completely destroy organic matter or solids during septage treatment. It can, however, produce a relatively biologically stable end-product, which is dewaterable and which does not have an offensive odor. Because chlorine reactions with sludge and septage are very rapid, reactor volumes are relatively small; therefore, compared with biological digestion processes, septage chlorination system sizes are generally smaller, and capital costs may be lower, depending on the site-specific circumstances. In addition, septage chlorination systems can be run intermittently (unlike biological processes) so long as sufficient storage volume is available both upstream and downstream of the reactor. As a result, operating costs are more directly dependent on septage production rates. Septage treatment facilities utilizing chlorination include Babylon, NY; Ventura, CA; Putnam, CT; and Bridgeport, CT, all in USA.

## 4.2. Design and Operation Considerations

An investigation, of the septage chlorination process were conducted at the Lebanon, Ohio, USA treatment plant which addressed chlorine requirements, dewatering rate, and sand bed underdrainage quality (43). The study concluded the following:

1. The chlorination process, in conjunction with sand bed dewatering, was an effective septage treatment method.
2. The sand bed underdrainage quality, compared with untreated septage, indicated the following removals: COD, 98%; $BOD_5$, 95–97%; total phosphorus, 99%; ammonia, 55–75%.
3. Mass balance calculations indicate that the sand dewatering beds following the chlorination process were the site of the majority of the organic and nutrient removal. It is possible that after repeated application, the removal capacity of the sand would be exhausted.
4. Large dosages of chlorine (1000–3000 mg/L) were required for the process to operate satisfactorily.
5. Chlorinated organics formed during processing appeared to be tied up in the sludge solids. The ultimate fate of these organics and their effects on the environment are not well documented.

The study (43) also showed that chlorine treatment of septage produced a solids fraction with greatly reduced total and fecal coliform concentrations, although coliform concentrations in the sand bed underdrainage were quite high, as summarized in Table 4.14. The average chlorine dose used during the study was 0.0021 kg chlorine per liter septage or 0.115 kg chlorine per kg dry solids.

A schematic diagram of a chlorine oxidation system is shown in Figure 4.6. The heart of the septage chlorination (Purifax) system consists of a disintegrator, a recirculation pump, two reaction tanks, a chlorine eductor, and a pressure-control pump. The chlorine can be fed to the system through a chlorinator and/or evaporator. An influent feed pump and flow meter should also be provided.

Raw septage is pumped through the disintegrator to reduce particle size and increase particle surface area for contact with the chlorine. Chlorinated septage from the first reactor is mixed with raw septage just prior to reaching the recirculation pump. The combined flow then passes through the first reaction tank. Chlorine is added to the system

### Table 4.14
### Bacteriological Data Generated From Chlorination Treatment of Septage[a]

|  | Total coliform (counts/100 mL) | Fecal coliform (counts/100 mL) |
|---|---|---|
| Raw septage | $4.4 \times 10^7$ | $5.3 \times 10^6$ |
| Purifax™ treated septage[b] | 200 | 200 |
| Sand bed underdrainage | $6.9 \times 10^6$ | $3.2 \times 10^4$ |

[a] Values are averages of four runs.
[b] Dewatered solids.
*Source*: US EPA.

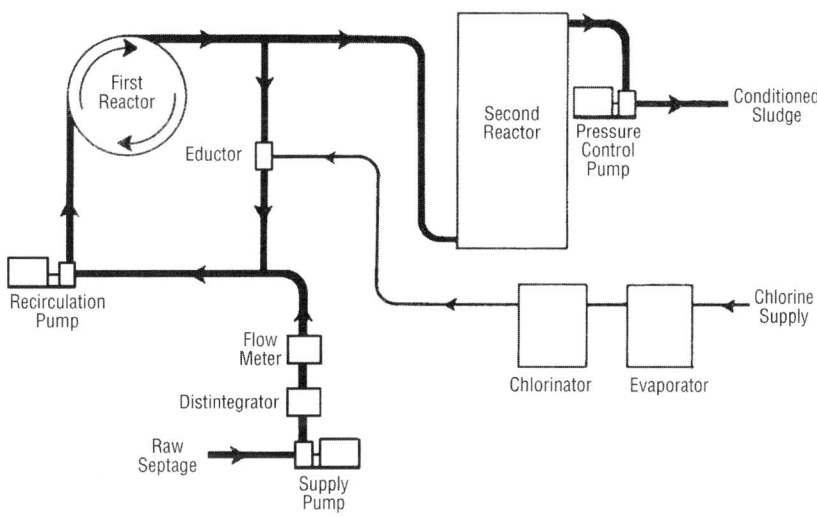

**Fig. 4.6.** Schematic of a septage chlorination system (US EPA).

by means of an eductor in the recirculation loop. Recirculation aids mixing and efficient chlorine use. The ratio of recirculated reacted product to raw septage is normally about 7 to 1. System pressure (30–40 psi) is maintained by a pressure-control pump located at the discharge of the second reactor tank, which has been provided to increase system detention time to allow for a more complete reaction between septage and chlorine.

A holding/equalization tank should be provided upstream of the oxidation system. Mechanical mixing can be used, although air mixing is preferable because it enhances aerobic conditions and reduces odors. A particular benefit of septage chlorination treatment is that odor can be controlled in the holding tank by returning a portion of the filtrate or supernatant from the dewatering process. Ventilation of such tanks must be provided. A downstream holding tank is beneficial in that it ensures optimum functioning of subsequent processes, and it allows the chlorine residual to drop from approx 200 mg/L to about 0, and the pH to rise to between 4.5 and 6.5. This process takes approx 48 h.

### 4.3. Process Equipment and Control

The septage chlorination process equipment (Purifax) is commercially available over a wide range of flow capacities from 55 m$^3$/d (10 gpm) to a theoretically unlimited maximum. Sizing information is available from the manufacturer of Purifax. The system can be dimensioned such that the daily volume of septage is treated in 4–6 h. Most septage chlorination units are of a prefabricated, modular design, completely self-contained and skid-mounted.

Chlorine dosages vary from 700 to 3000 mg/L, depending on the solids content of the septage and the amount of chlorine-demanding substances present. (9). These substances include ammonia, amino acids, proteins, carbonaceous material, and hydrogen sulfide. The Babylon, New York septage treatment facility uses about 0.6 kg Cl$_2$/L influent (5 lb per 1000 gal).

The chlorine dosage is approx 0.7 kg Cl$_2$/L influent (6 lb/1000 gal) for septage with a suspended solids concentration of 1.2%. The chlorine demand varies in proportion to the solids concentration. For example, if the solids concentration were to double, the chlorine concentration would double as well.

### 4.4. Design Criteria

A summary of the typical design criteria for chlorine stabilization of septage is presented in below:

Chlorination system size = to treat daily septage volume within 4–6 h
Chlorine dosage = 0.7 kg/L for 1.2% TS, chlorine demand varies directly with TS

Limitations of the chlorine stabilization process center on chemical, operational, and environmental factors. From a chemical standpoint, the low pH of chlorine-stabilized septage may require neutralization prior to mechanical dewatering or before being applied to acid soils. Costs of neutralization are in addition to chlorine costs. Chlorine stabilization does not reduce sludge mass nor produce methane gas as a by-product for energy generation. The process consumes relatively large amounts of chlorine. Special safety and handling precautions must be used when employing this system. If high alkalinity wastes are processed, $CO_2$ generated during chlorination may promote cavitation in downstream pumps. The potential for production of carcinogenic compounds by the chlorination process has been a major concern, because these compounds may leach into the ground or contaminate surface waters as a result of sludge or liquid effluent disposal.

The effluent (filtrate, supernatant) from the dewatering step is not suitable for direct discharge into surface waters. Infiltration/percolation beds have been used for effluent disposal. Alternative disposal methods have included direct recycle to a treatment plant or direct discharge following activated carbon adsorption.

The major parameters used to control the septage chlorination process are treated septage color, effluent pH, and effluent chlorine residual. The chlorine dose can be adjusted until the effluent stream is a light buff color with a pH of 2–2.5, and a chlorine residual of 150–200 mg/L (43).

# 5. SAFETY CONSIDERATIONS OF CHLORINATION PROCESSES

Each safety program should include a list of safety rules for all employees to learn and use. An inspector cannot enforce safety rules but a written set of safety rules should be available to all employees. The safety rules should include the following areas:

1. Personal hygiene
2. Fire protection
3. Protection around openings
4. Tools and safety equipment
5. Accidents and first aid
6. Safety drill procedures
7. Gases
8. Created hazards
9. Accident prevention
10. Driving safety
11. In-plant traffic
12. Machinery guards
13. Ventilation
14. Gas utilization
15. Structures
16. Housekeeping
17. Safe operation
18. Electrical equipment
19. Procedures for one-man shifts
20. Laboratory

A detailed, well-written safety manual should include the above plus specific rules for individual unit processes.

Chlorine is a highly toxic gas, which may be fatal if inhaled in sufficient quantity. The presence of chlorine is easily detected however. A concentration of 3.5 parts per million of chlorine by volume is detectable. At concentrations between 15 and 30 parts per million, significant irritation of the mucus membranes and nasal linings will occur. Exposure to chlorine at a concentration of 1000 parts per million will result in fatalities in a very short exposure time. Most chlorination facilities using gaseous or liquid chlorine are designed to rigorous safety standards presenting minimum hazards to operating personnel. An adequately designed facility will have continuously monitoring chlorine leak detectors, which sound an alarm in the event of a leak. The following safety requirements should be met for any chlorine application facility.

1. Chlorination equipment should always be placed in an adequately ventilated room and isolated from other working areas. Ventilation should be provided with fan at floor level since chlorine is heavier than air. Access should be from an outside door.
2. Provisions should be made to continuously ventilate the area surrounding the chlorine cylinders and the chlorination equipment.
3. Equipment should be provided for continuous monitoring of the air in chlorine storage and application area.
4. Proper instructions and supervision to workers charged with responsibility of chlorination equipment should be provided.

5. Proper and approved self-contained breathing apparatus for persons working where there is a possibility of exposure to chlorine gas fumes should be provided, should be stored outside the area of danger, and should be quickly accessible.
6. Combustible or inflammable materials should never be stored near chlorine containers or application equipment.
7. Heat should never be applied to any chlorine container.
8. A water supply to keep chlorine containers cool in case of fire should be provided.
9. Several appropriate emergency container leak repair kits should be stored near the chlorine application facility.
10. Plant safety rules should require that any leak in storage cylinders or application equipment be attended by at least two persons wearing self-contained breathing apparatus.
11. Emergency shower and eye wash facilities should be provided adjacent to entry doors into chlorine storage or application facility.
12. First aid procedures should be developed and all personnel handling chlorine should be familiar with their application. These procedures should be posted in the chlorine area.
13. Fire protection should be provided by Class C fire extinguishers (for energized electrical equipment) and located in the area immediately adjacent to the chlorination room.
14. Procedures should be developed to handle chlorine leaks from storage cylinders or application equipment. Periodically operating personnel should review these procedures in a hypothetical emergency situation.

## 6. RECENT ADVANCES IN WASTE DISINFECTION

So far chlorination is still the most popular disinfection process for treating either wastewater or biosolids. Since the reactions between chlorine and organic matters (precursors) produce trihalomethane (THM) and other cancer-causing disinfection byproducts (DBPs), many new technologies have been developed as possible alternatives for eliminating or at least reducing the DBPs (45).

For wastewater disinfection, the first two (#a and #b) of the following are of the conventional technologies, and all the remaining are alternative technologies (45, 46):

(a) At the end of a wastewater treatment train (such as preliminary treatment, primary clarification, biological treatment, secondary or final clarification, optional tertiary treatment and disinfection), chlorine is used as a disinfectant for destruction-inactivation of pathogenic microorganisms in the plant effluent before the effluent is discharged into a receiving water.
(b) An inorganic chemical reducing agent, such as sulfur dioxide, sodium thiosulfate, sodium sulfite, sodium bisulfite, sodium metabisulfite, or calcium thiosulfate, is used for dechlorination after the above chlorine disinfection in #a, so less DBPs will be produced, and there will be no harmful extra residual chlorine to be released to the environment.
(c) An organic chemical, such as ascorbic acid (i.e. Vitamin C), or sodium ascorbate, is used for dechlorination after the above chlorine disinfection in #a, so less DBPs will be produced, and there will be no harmful extra residual chlorine to be released to the environment.
(d) An alternative disinfectant, such as chlorine dioxide, or ozone, replaces chlorine for disinfection in #a, so less DBPs will be produced.
(e) Hydrogen peroxide is used for neutralization of residual chlorine after the above chlorine disinfection in #a, so less DBPs will be produced, and there will be no harmful extra residual chlorine to be released to the environment.
(f) A physical process, such as UV, is used for dechlorination after the above chlorine disinfection in #a, so less DBPs will be produced, and there will be no harmful extra residual

chlorine to be released to the environment. The combined chlorination-UV process is also an Advanced Oxidation Process (AOP), which is highly effective for destruction-inactivation of pathogenic microorganisms, but does produce some DBPs.

(g) A newly developed AOP process involving the use of either combined UV-$H_2O_2$, or combined UV-ozone replaces chlorine disinfection in #a, so there will be no DBPs, nor harmful extra residual chlorine, except that the cost of this AOP process is still high at present.

(h) A newly developed UV-photocatalysis process (i.e. combined UV-$TiO_2$ photocatalytic oxidation) replaces chlorine disinfection in #a, so there will be no DBPs, nor harmful extra residual chlorine, except that the cost of this AOP process is very high at present.

(i) An adsorbent, such as granular activated carbon, or polymeric absorbent, is used for dechlorination after the above chlorine disinfection in #a, so less DBPs will be produced, and there will be no harmful extra residual chlorine to be released to the environment. This adsorption technology is technically feasible, but economically unfeasible.

(j) UV process replaces chlorine for disinfection in #a, when tertiary treatment in #a is either sand filtration or membrane filtration.

(k) UV process replaces chlorine for disinfection in #a, when the combined biological treatment and final clarification in #a is a membrane bioreactor (MBR).

For disinfection of biosolids and septage, DBPs production shall be significantly reduced or totally eliminated. While chlorination of biosolids and septage is still practiced, it is gradually being phased out in favor of other alternatives (1–3, 45), such as (a) chlorine dioxide disinfection; (b) ozonation; (c) pressurized ozonation; (d) irradiation; (e) lime stabilization; (f) thermal processes; (g) autothermal thermophilic aerobic digestion; and (h) high temperature composting.

## NOMENCLATURE

| | |
|---|---|
| CD | chlorine dosage, mg/L |
| CR | chlorine requirement, lb/d |
| CT | contact time at maximum flow, min |
| CTL | contact tank length, ft |
| CTW | contact tank width, ft |
| PCR | peak chlorine requirements, lb/d |
| $Q_a$ | average flow, MGD |
| $Q_p$ | peak flow, MGD |
| RLW | length-to-width ratio |
| SA | surface area, $ft^2$ |
| SWD | side water depth, ft |
| $V$ | volume, $ft^3$ |
| VCT | volume of contact tank, gal |

## ACKNOWLEDGMENTS

Coxsackie Sewage Treatment Plant (STP) in New York, USA, is a model single sludge biological treatment plant that was built and operated under the co-sponsorship of the New York State Department of Health (NYSDOH), the New York State Department of Environmental Conservation (NYSDEC), and the New York State Department of

Correction (NYSDOC). The single sludge plant (including secondary biological oxidation, tertiary nutrients removal, wastewater chlorination, and sludge chlorine stabilization) has been performing successfully and has been a role model to the New York municipalities since 1974. Leo J. Hetling (PE, PhD), Carl Beer (PE), and Lawrence K. Wang (PhD, PE) were Research Director, Project Manager, and Operator (NYSDEC Senior Sanitary Engineer), respectively, in 1973–1977. This chapter was written in memory of late Carl Beer, who was a pioneer in the field of single sludge biological systems. Both Carl Beer and Lawrence K. Wang (author of this chapter) received a Kenneth Research Award from the New York Water Environment Association for this environmental engineering contribution (37–42).

## REFERENCES

1. L. K. Wang, Y. T. Hung, and N. K. Shammas (eds.), *Physicochemical Treatment Processes*. Humana Press, Totowa, NJ 723 pp. (2005).
2. L. K. Wang, Y. T. Hung, and N. K. Shammas (eds.), *Advanced Physicochemical Treatment Processes*. Humana Press, Totowa, NJ 690 p. (2006).
3. L. K. Wang, N. K. Shammas, and Y. T. Hung (eds.), *Biosolids Treatment Processes*. Humana Press, Totowa, NJ 820 p. (2007).
4. U.S. Environmental Protection Agency, *Technologies for Upgrading Existing or Designing New Drinking Water Treatment Facilities*. US EPA/625-4-89-023, Washington, DC (1990).
5. U.S. Environmental Protection Agency, *Septage Treatment and Disposal*. US EPA/625-6-84-009, Washington, DC (1984).
6. U.S. Environmental Protection Agency, *Sludge Handling and Conditioning*. US EPA/430-9-78-002, Washington, DC (1978).
7. U.S. Environmental Protection Agency, *Performance Evaluation and Troubleshooting at Municipal Wastewater Treatment Facilities*. US EPA/430-9-78-001, Washington, DC (1978).
8. L. K. Wang, *Environmental Engineering Glossary*. Calspan Corporation, Buffalo, NY 424 p. (1974).
9. U.S. Environmental Protection Agency, *Design of Wastewater Treatment Facilities Major Systems*. US EPA/430-9-79-008, Washington, DC (1978).
10. L. K. Wang, Y. T. Hung, H. H. Lo, and C. Yapijakis (eds.), *Handbook of Industrial and Hazardous Wastes Treatment*. CRC Press/Marcel Dekker, New York 1345 p. (2004).
11. P. S. Singer, THM Control using alternate oxidant and disinfectant strategies. *Proceedings 1986 AWWA Conference*, American Water Works Association, Denver, pp. 999–1017 (1986).
12. W. R. McKeon, J. J. Muldowney and B. S. Aptowicz, The evolution of a modified strategy to reduce trihalomethane formation. *Proceedings 1986 AWWA Conference*, American Water Works Association, Denver pp. 967–997 (1986).
13. L. W. Casson, J. W. Bess, and T. J. Navin, *Ultra Pure Hypochlorite*. Department of Civil and Environmental Engineering, University of Pittsburgh, Pittsburgh (2005).
14. L. K. Wang, *The State-of-the-art Technologies for Water Treatment and Management*. United Nations Industrial Development Organization (UNIDO), Vienna, Austria, UNIDO Training Manual No. 8-8-95, August (1995).
15. M. Krofta and L. K. Wang, *Removal of Trihalomethane Precursors and Coliform Bacteria by Lenox Flotation-Filtration Plant*, Water Quality and Public Health Conference. PB83-244053. National Technical Information Service, Springfield, VA, pp. 17–29 (1983).

16. M. Krofta and L. K. Wang, *Development of a New Water Treatment Process for Decreasing the Potential for THM Formation*, PB81-202541, National Technical Information Service, Springfield, VA (1987).
17. L. K. Wang, *Standards and Guides of Water Treatment and Water Distribution Systems.* PB88-177902/AS. National Technical Information Service, Springfield, VA (1987).
18. L. K. Wang, *Drinking Water Standards and Regulations*, PB88-178058/AS, National Technical Information Service, Springfield, VA (1987).
19. U.S. Environmental Protection Agency, *Guidance Manual for Compliance with the Filtration and Disinfection Requirements for Public Water Systems Using Surface Water Sources.* 684-003/41811. U.S. Government Printing Office, Washington, DC (1992).
20. State of Missouri. *Missouri State Operating Permit.* Department of Natural Resources, Missouri Clean Water Commission, June 25 (2004).
21. L. K. Wang, M. H. S. Wang, G. G. Peery, and R. C. M. Cheugn, General theories of chemical disinfection and sterilization of sludge, part I, *Water and Sewage Works*, 125, 7, 30–32 (1978).
22. L. K. Wang, M. H. S. Wang, G. G. Peery, and R. C. M. Cheugn, General theories of chemical disinfection and sterilization of sludge, part II, *Water and Sewage Works*, 125, 8, 58–62 (1978).
23. L. K. Wang, M. H. S. Wang, G. G. Peery, and R. C. M. Cheugn, General theories of chemical disinfection and sterilization of sludge, part III, *Water and Sewage Works*, 125, 9, 99–104 (1978).
24. American Water Works Association, *Safety Practice for Water Utilities*, Manual No. M3, Denver (2004).
25. AWWA/WEF/APHA, *Standard Methods for the Examination of Water and Wastewater.* American Public Health Association, Washington, DC (2004).
26. A. A. Stevens, L. A. Moore, C. J. Slocum, B. L. Smith, D. R. Seeger, and J. C. Ireland, By-products of chlorination at ten operating utilities. *Proceedings of the 6$^{th}$ Conference on Water Chlorination: Environmental Impact and Health Effects.* Oak Ridge Associated Universities, Oak Ridge, TN, May 3–8 (1987).
27. S. Saby, M. Djafer, and G. H. Chen, Feasibility of using a chlorination step to reduce excess sludge in activated sludge process. *Water Research*, 36, 3, 656–666 (2002).
28. L. K. Wang, A potential organic disinfectant for water purification, *Journal of the New England Water Works Association*, 89, 3, 250–270 (1975).
29. L. K. Wang, Disinfection with quaternary ammonium compounds, *Water Resources Bulletin, Journal of American Water Resources Association*, 11, 5, 919–933 (1975).
30. L. K. Wang, Thickening of sewage sludge with quaternary ammonium compounds and magnetic fields, *Proceedings of the Third National Conference on Complete Water Reuse*, pp. 252–258, June (1976).
31. L. K. Wang. Cationic surface active agent as bactericide, *Industrial and Engineering Chemistry, Product Research and Development*, 14, 4, 308–312 (1975).
32. L. K. Wang, *Pretreatment and Ozonation of Cooling Tower Water, Part I*, PB84-192053. National Technical Information Service, Springfield, VA, April (1983).
33. L. K. Wang, *Pretreatment and Ozonation of Cooling Tower Water, Part II*, PB84-192046. National Technical Information Service, Springfield, VA. August (1983).
34. L. K. Wang, J. V. Krougzek, and U. Kounitson, *Case Studies of Cleaner Production and Site Rededication*, United Nations Industrial Development Organization (UNIDO) Vienna, Austria, UNIDO-Registry No. DTT-5-4-95, April (1995).
35. L. K. Wang, N. K. Shammas, and Y. T. Hung (eds.), *Advanced Biological Treatment Processes.* Humana Press, Totowa, NJ (2008).

36. M. K. Stenstrom, *Optimization of Chlorination of Activated Sludge Plant Effluent by Ammonia Control*. Technical paper presented at the 1980 Joint automatic Control Conference, San Francisco, August 13–15 (1980).
37. C. Beer and L. K. Wang, Full scale operations of plug flow activated sludge systems, *Journal of New England Water Pollution Control Association*, 9, 2, 145–173 (1975).
38. C. Beer and L. K. Wang, Activated sludge systems using nitrate respiration—design considerations, *Journal of Water Pollution Control Federation*, 50, 9, 2120–2131 (1978).
39. C. Beer, L. J. Hetling, and L. K. Wang, *Full-Scale Operation of Plug flow Activated Sludge Systems*. NYSDEC Technical Report No. 42. New York State Department of Environmental Conservation, Albany, NY (1975).
40. C. Beer, J. F. Bergenthal, and L. K. Wang, A study of endogenous nitrate respiration of activated sludge. *Proceedings of the $9^{th}$ Mid-Atlantic Industrial Waste Conference*, Bucknell University, Lewisburg, PA (1977).
41. U.S. Environmental Protection Agency, *A Study of Nitrate Respiration in the Activated Sludge Process*. US EPA/600-2-80-154, Washington, DC (1978).
42. C. Beer and L. K. Wang, *Process Design of Single Sludge Activated Sludge Systems Using Nitrate Respiration*. Report No. 50. N.Y. State Department of Environmental Conservation, Albany, NY (1977). (Kenneth Research Award from NY Water Environment Association.)
43. U.S. Environmental Protection Agency, *Evaluation of the Purifax Process for the Treatment of Septic Tank Sludges*, Cincinnati, OH (1975).
44. U.S. Environmental Protection Agency, *Inspectors Guide for Evaluation of Municipal Wastewater Treatment Plants*. US EPA-430-9-79-010. Cincinnati, OH (1979).
45. L. K. Wang, *New Technologies for Water and Wastewater Treatment*. NYSAWWA & NYWEA Joint Tiff Symposium. Liverpool, NY. Nov. 15–17 (2005).
46. L. K. Wang, *Recent Advances in UV Technologies*. 2008 National Engineers Week Meeting, Albany, NY. Feb. 14–15 (2008).

# 5
# Storage of Sewage Sludge and Biosolids

## Nazih K. Shammas and Lawrence K. Wang

**CONTENTS**
  INTRODUCTION
  WASTEWATER TREATMENT STORAGE
  FACILITIES DEDICATED TO STORAGE OF LIQUID SLUDGE
  FACILITIES DEDICATED TO STORAGE OF DEWATERED SLUDGE
  FIELD STORAGE OF BIOSOLIDS
  DESIGN EXAMPLES
  NOMENCLATURE
  REFERENCES
  APPENDIX

**Abstract** Storage is an integral part of every wastewater solids treatment and disposal/reuse system, since it is necessary to ensuring that the system will be used to full capacity. Storage allows different processes to operate on schedules that best fit the overall system objectives and precludes the need to force all processes to operate on the same schedule. Recent emphasis on the control of wastewater solids treatment and disposal/reuse mandates that effective storage be provided. Storage that is compatible with the objectives of a system must be incorporated into its design to enhance both the system's reliability and its efficiency.

This chapter discusses the following topics: need, risk, and benefits of storage; wastewater treatment plant storage; facilities for the storage of liquid sludges (holding tanks, facultative sludge lagoons anaerobic liquid sludge lagoons, aerated storage basins); facilities for the storage of dewatered sludges (drying sludge lagoons, confined hoppers or bins, unconfined stockpiles); field storage of biosolids; and design examples.

**Key Words** Management of biosolids storage • sludge storage • facilities for storage • field storage • tanks • lagoons • basins • bins • stockpiles.

## 1. INTRODUCTION

Storage is an integral part of every wastewater solids treatment and disposal/reuse system, since it is necessary to ensuring that the system will be used to full capacity. The purposes of sludge/biosolids storage are as follows (1, 2):

From: *Handbook of Environmental Engineering, Volume 7: Biosolids Engineering and Management*
Edited by: L. K. Wang, N. K. Shammas and Y. T. Hung © The Humana Press, Totowa, NJ

1. Reduce the pathogen population by aeration and mixing
2. Further stabilize the sludge
3. Equalize short-term peak loads
4. Prepare the sludge for further processing
5. Further thicken the sludge/biosolids as a result of the detention afforded by storage
6. Provide the means for loading the processed sludge (biosolids) into the reuse system

Recent emphasis on the control of wastewater solids treatment and disposal/reuse mandates that effective storage be provided. Storage that is compatible with the objectives of a system must be incorporated into its design to enhance both the system's reliability and its efficiency.

## 1.1. Need for Storage

Storage allows different processes to operate on schedules that best fit the overall system objectives and precludes the need to force all processes to operate on the same schedule. For example, solids are generated from the wastewater treatment system 24 hours a day, but it may be most convenient to operate the solids processing system only on the day shift. Therefore, solids must be stored during off-hours. Storage must also be provided between adjacent treatment processes or disposal systems, which operate at different rates, for example, between centrifuges that discharge solids at 100 t/h (91 T/h) (where t is the English ton and T is the metric ton) and incinerators that must be fed at 50 t/h (45 T/h). In addition, it must be provided upstream from virtually any land disposal or reuse system, since sludge/biosolids can usually be applied to land only part of the year, whereas the waste treatment plant generates solids all year around (3).

## 1.2. Risks and Benefits of Solids Storage Within Wastewater Treatment Systems

Stored solids can be washed from the wastewater treatment system, thereby degrading effluent quality. They may also become septic, with the same effect. As a general rule, solids should not be stored in wastewater treatment systems unless storage provides benefits that clearly outweigh the risks involved. For many small plants, if sludge processing units are operated only on the day shift, the benefits do outweigh the risks. These plants frequently store solids within the wastewater treatment process for periods as long as 24 hours. Large plants, which typically process sludge around the clock, make less frequent use of storage within the wastewater treatment system. The main exception to this rule is the storage of solids within wastewater stabilization ponds, where solids and dead algae settle to the bottom of the ponds and anaerobically decompose. These solids are seldom removed and often accumulate for many years with no deleterious effect.

## 1.3. Storage Within Wastewater Sludge Treatment Processes

Solids can be stored within sludge treatment processes with fewer adverse effects than if they were stored within the wastewater treatment system. Whereas the processes of disinfection, conditioning, mechanical dewatering, high-temperature conversion, and heat drying do not provide storage, those of gravity thickening, anaerobic and aerobic digestion, air drying, and composting do. Used judiciously, these processes can store enough solids to enable necessary adjustments to be made in rates of flow between processes. One or two of these processes can provide cost-effective storage for periods exceeding 1 month. However, because of process limitations, some cannot provide storage for minimum periods of 3 to 4 days even though they can store for periods of 3 to 4 weeks and longer.

## 1.4. Field Storage of Biosolids

Field storage of biosolids is necessary during inclement weather when land application sites are not accessible and during winter months when land application to snow covered and frozen soil is prohibited or restricted. Storage also may be needed to accommodate seasonal restrictions on land availability due to crop rotations or equipment availability. For small generators, storage allows accumulation of enough material to efficiently complete land application in a single spreading operation. Well-planned and well-managed storage options not only provide operational flexibility at the treatment facility, but also improve the agronomic, environmental, and public acceptance aspects of biosolids use.

A United States Environmental Protection Agency (US EPA) and U.S. Department of Agriculture (USDA) guide to field storage of biosolids (4) provides a set of consistent recommended management practices for the field storage of biosolids. It identifies three critical control points for managing the system:

1. The wastewater treatment plant (WWTP)
2. The transportation process
3. The field storage site

The US EPA and the USDA stress the continuing need for partnership and good communication between the biosolids generators and managers responsible for storage and land application to ensure community-friendly operations. The management practices include three critical issues that have potential environmental, public health, and community relations impacts (4):

1. Odors
2. Water quality
3. Pathogens

## 1.5. Effects of Storage on Wastewater Solids

If wastewater solids are to be stored for any extended period of time, they must be stable. Stable liquid sludge with less than 10% solids can be stored in facultative sludge lagoons, anaerobic sludge lagoons, or aerated basins. When it is air dried to greater than 30% to 40% solids, stable sludge (biosolids) can be stored safely and without odors in relatively small, confined structures or in unconfined stockpiles. It is impractical to store unstabilized dewatered or partially dried sludge (sludge containing more than 10% and less than 30% solids) for much longer than 3 to 4 days because septic conditions and problems associated with septicity (odors, poor solids transport properties) can develop.

Wastewater solids are usually stored in concentrated form. If these solids are biodegradable, indigenous oxygen supplies can rapidly be depleted and anaerobic decomposition begins. Anaerobic decomposition is often, but not always, accompanied by the production of undesirable odors (5). However, anaerobic decomposition will not occur in the following situations (3):

1. The biodegradable materials present are insufficient to support biological activity. For example, screenings and grit are relatively nonodorous, provided they have been processed and transported hydraulically prior to final dewatering. The washing action that occurs during these operations reduces the concentration of putrescible organic material. Conversely, if processed and transported mechanically (that is, without washing), they may be the source of strong odors when subsequently stored.
2. Oxidizing conditions can be maintained. Agents such as oxygen, chlorine, and hydrogen peroxide can be used to this end if the sludge is in liquid form. Forced aeration or physical manipulation can be used to maintain the aerobic condition if solids are dewatered and managed, as is done in composting.
3. Moisture is reduced to discourage biological activity. For example, air-dried stabilized sludge with solids content greater than 40% to 50% and unstabilized heat-dried sludges can be stored indefinitely without causing a nuisance, provided rewetting does not occur.
4. The pH is adjusted to values above approximately 12 and below approximately 4 by adding chemicals like lime or chlorine. Note that pH extremes must be maintained. These treatments do not destroy putrescible materials, and the biocidal effects caused by extreme pH are lost as the pH drifts toward neutral values as the result of interaction with atmospheric carbon dioxide.

The fact that anaerobic digesters and facultative sludge lagoons have operated without nuisance odors clearly indicates that storage can be accomplished under anaerobic conditions without adverse effects. Work on facultative sludge lagoons in Sacramento, California, documents these conclusions (6).

Nuisance odors will not develop in anaerobic storage when sufficient methane bacteria are present. If the methane bacteria are destroyed, however, serious odor problems may result. As an example, consider anaerobically digested sludge that is placed on a drying bed or in a drying lagoon. The top layer of sludge is dewatered, and methane bacteria die as the sludge aerates and dries. Odor levels are extremely low, since the sludge is too dry to support anaerobic biological activity. Should the surface of the sludge be rewetted (for example, by rainfall or surface flooding), however, anaerobic activity would resume, the organic acid concentration would rapidly increase, and odors would increase to nuisance levels. Odor problems that were experienced with approximately 580 acres (235 hectares [ha]) of drying lagoons in San Jose, California, immediately following a rainstorm, is an example of this type of problem (7). Not all the effects of solids storage are negative. Storage of anaerobically digested sludge in the liquid state can be beneficial for its ultimate disposal. If such sludge is stored for 1 or 2 years without being contaminated by freshly digested sludge, its organics content (40% to 50%) and its content of pathogenic bacteria, viruses, and parasites will be greatly reduced (1, 8).

## 1.6. Types of Storage

Wastewater solids may be stored in facilities within the treatment system, within the sludge treatment and disposal system, and within tanks, lagoons, bins, or stockpiles provided primarily for storage. This latter group is divided into two divisions—those provided for either liquid or dewatered sludge. The use of wastewater and sludge treatment facilities for solids storage must not adversely affect their treatment capability. If this potential exists, then facilities dedicated primarily to storage must be provided.

There are three methods of storage (3):

1. *Single-phase concentration*: Solids accumulate in a completely mixed vessel as a result of increasing concentration. The solids concentration is uniform throughout and vessel volume is constant. For example, solids build up within the aeration reactor of an activated sludge system if solids are not wasted.
2. *Two-phase concentration*: Storage is within the solids layer of a liquid/solids separation device. The volume of the solids layer increases; however, total system volume remains constant. For example, solids are accumulated in a gravity thickener by terminating sludge withdrawal from the thickener and allowing the sludge blanket to build up.
3. *Displacement*: Solids are stored as a result of changing total system volume. For example, solids can accumulate within digesters with floating covers by displacement storage, since the covers can rise to accommodate greater volumes of sludge.

Storage may be accomplished by two or three methods operating in concert. For example, solids can accumulate in a floating cover equipped secondary digester by simultaneous two-phase concentration and displacement. Storage may be further categorized as follows by detention time:

1. Equalization storage solids detention time should not exceed 3 to 4 days.
2. Short-term storage solids detention time should not exceed 3 to 4 weeks.
3. Long-term storage solids detention time is greater than 1 month.

Table 5.1 lists wastewater solids storage by type, facility, method, and detention time category. The table also provides useful information, limitations, and other comments for each type of storage.

## 2. WASTEWATER TREATMENT STORAGE

Influent variability and fixed effluent requirements make operational flexibility a necessity for every wastewater treatment plant. One of the most cost-effective means of providing flexibility for small plants is to ensure that treatment processes contain storage within themselves.

### 2.1. Storage Within Wastewater Treatment Processes

Listed in Table 5.1 are several wastewater treatment processes that can provide solids storage. The following subsections describe ways in which this storage can be used effectively.

#### 2.1.1. Grit Removal

Grit removal basins and channels (9) may be used to store unusually heavy grit loadings, which, when combined sewer systems are involved, generally arrive at the treatment plant after a dry spell and during the first flush of a storm. Storage must be provided to contain all of the grit that could accumulate during the storm. The required storage volume is a function of grit loading and the rate at which the grit can be transferred out of the basin or channel. Where grit is transferred manually (for example, in small plants with duplicate channels), the designer may wish to provide

Table 5.1
Wastewater solids storage applicability

| Type | Method | Detention time | | | Comments |
|---|---|---|---|---|---|
| | | Equalizing 3 to 4 days | Short term 3 to 4 weeks | Long term >1 month | |
| Storage within wastewater treatment processes | | | | | Use of wastewater treatment processes for storage must not adversely affect treatment efficiency |
| Grit removal | Two-phase concentration | x | x | — | Storage time depends on sewer system grit loading to plant |
| Primary sedimentation | Two-phase concentration | x | — | — | Temperature sensitive; storage for over 24 hours |
| Aeration reactors | Single-phase concentration | x | x | — | Storage within extended aeration systems, for example, oxidation ditches, can exceed 3 weeks if accomplished in conjunction with secondary sedimentation concentration |
| Secondary sedimentation | Two-phase concentration | x | — | — | Highly temperature sensitive; storage for over 8 hours requires chemicals |
| Imhoff tanks | Two-phase concentration | — | x | x | Lightly loaded systems can store for over 6 months; most systems will require solids removal every 4 to 6 weeks |
| Community septic tanks | Two-phase concentration | — | — | x | Sludge from many septic tanks is removed only once in several years |
| Wastewater stabilization ponds | Single- and two-phase concentration | — | — | x | Aerated ponds operate like aeration reactors; other ponds use two-phase concentration and can store solids for many years |
| Storage within sludge treatment processes | | | | | Use of sludge treatment processes for storage must not adversely affect sludge treatment efficiency |
| Gravity thickeners | Two-phase concentration | x | — | — | Temperature sensitive; usually not used with WAS; storage for over 24 hours requires chemicals |

Table 5.1 (*Continued*)

|  |  | Detention time | | | |
|---|---|---|---|---|---|
| Type | Method | Equalizing 3 to 4 days | Short term 3 to 4 weeks | Long term >1 month | Comments |
| Anaerobic digesters | Single- and two-phase concentration and displacement | x | x | — | Floating covers allow for displacement storage; two-phase concentration storage impracticable if WAS present; single-phase concentration storage possible if digesters operated in conjunction with primary sedimentation concentration changes |
| Aerobic digesters | Single- and two-phase concentration and displacement | x | x | — | Decanting can be limiting; short-term storage possible if digesters operated in conjunction with sedimentation concentration; displacement storage requires aeration systems that will operate with variable level |
| Composting | Two-phase concentration and displacement | — | x | x | Evaporation with process accomplishes two-phase concentration; processed solids not removable for 3 to 4 weeks |
| Drying beds | Two-phase concentration and displacement | — | x | x | Initial settling accomplishes two-phase concentration; processed solids not normally removable for 3 to 4 weeks |
| Facilities provided primarily for storage of liquid sludge | | | | | |
| Holding tanks | Single- and two-phase concentration and displacement | x | — | — | Storage limited to equalizing by high costs of detention and continuous mixing |

(*Continued*)

Table 5.1 (Continued)

| Type | Method | Detention time | | | Comments |
|---|---|---|---|---|---|
| | | Equalizing 3 to 4 days | Short term 3 to 4 weeks | Long term >1 month | |
| Facultative sludge lagoons | Two-phase concentration | — | x | x | Time required for initial settling limits storage to short or long term; mechanics of sludge removal makes short-term storage very expensive; odor-free operation requires anaerobically digested solids; organic loadings must be restricted and surface agitation provided; odor mitigation required when surface area exceeds 30 to 40 acres |
| Anaerobic liquid sludge lagoons | Two-phase concentration | — | x | x | Time required for initial settling limits storage to short or long term; mechanics of sludge removal makes short-term storage very expensive; odor minimization requires anaerobic digested solids; usually operated without organic loading restriction; no surface agitation provided; potential odor risk high, although no quantifying data available |
| Aerated storage | Single- and two-phase concentration and displacement | x | x | — | High energy demand usually restricts detention time; same limits as aerobic digesters |

200

Table 5.1 (Continued)

| Type | Method | Detention time | | | Comments |
| --- | --- | --- | --- | --- | --- |
| | | Equalizing 3 to 4 days | Short term 3 to 4 weeks | Long term >1 month | |
| Facilities provided primarily for storage of dewatered sludge | | | | | |
| Sludge drying lagoons | Two-phase concentration and displacement | — | — | x | Initial settling accomplishes two-phase concentration; process solids not normally removable for 1 to 2 months; odor minimization requires anaerobically digested solids; can be odorous if aerobically stabilized surface layers begin to decompose anaerobically when rewetted |
| Confined hoppers or bins | Displacement | x | x | — | Noist (15% to 30% solids) dewatered sludge can present major material management and odor production problems if storage time exceeds 3 to 4 days; structures usually too expensive for long-term storage; short-term storage can be successful with dry (greater than 30% to 40% solids) stabilized sludges |
| Unconfined stockpiles | Displacement | x | x | x | Requires stabilized dry (greater than 30% to 40% solids) sludge; stockpiles are usually covered in very wet climates; natural freeze drying is possible |

WAS, waste-activated sludge.
*Source*: US EPA (3).

storage sufficient to hold grit during periods when the plant may be unattended (long weekends).

Special techniques or equipment may be needed to transfer heavy grit accumulations. If grit is transferred mechanically (by flight, bucket, and screw conveyors), the equipment must be able to start while the entire basin or channel is filled with grit. If grit is transferred hydraulically, air agitation should be used to loosen up the accumulated solids during the removal operation. Hydraulic removal can be accomplished by eductors, air-lift pumps, or special centrifugal pumps. When special centrifugal (torque-flow or vortex) pumps are used, the grit should be loosened up in the immediate vicinity of the pump suction by a high-velocity water jet. More design information on grit removal facilities is available elsewhere (1, 9, 10).

### 2.1.2. Primary Sedimentation

If storage is provided in primary sedimentation (9), solids processing systems can operate at rates independent of the rate at which solids are removed from the wastewater. This is especially useful for small plants that are not manned continuously and for any size plant that experiences large diurnal or seasonal fluctuations in settleable solids.

Concentration of sludge removed from the primary sedimentation tank may be controlled to some degree if the depth of the sludge layer in the sludge removal hoppers is controlled. Hopper sides should be sloped at least 60° off the horizontal so that solids can flow by gravity to the pump suction. Primary sedimentation tank storage capacity should be sufficient to allow suitably sized primary sludge pumps to remove the peak sludge loadings. Otherwise the solids may interfere with the gathering function of the longitudinal sludge collectors in rectangular tanks or the main collector in circular clarifiers.

Efficient primary sedimentation storage requires the use of a control timer, density, and blanket-level instrumentation. Ideally, all three factors can control primary sludge pump operations. Blanket level sets the time when the pump starts; control timers set the cyclical period when the pumps can share the discharge piping (if necessary) and the minimum pump operating period if the density of the pumped sludge is below the required concentration; and density sets the time when the pump shuts down. More design information on primary sedimentation tank design is available elsewhere (1, 9, 10).

### 2.1.3. Aeration Reactors and Secondary Sedimentation

Solids are stored in aeration reactors and secondary sedimentation tanks whenever there is an increase in the solids concentration of the mixed liquor. This solids increase requires the two processes to be operated as one, with the sedimentation tank providing the two-phase concentration necessary to fully utilize the single-phase concentration storage capabilities of the reactors. Reactors should be designed with the flexibility to operate either in the plug flow, step feed, re-aeration, or contact stabilization modes or any combination of these. Given a fixed reactor size, maximum solids storage capability is provided when the process operates in a combination of the re-aeration and contact stabilization modes. Often the ability to switch between complete plug flow and partial re-aeration modes allows the solids to be removed from the hydraulic flow stream and prevents their loss when that stream receives a shock loading. Operation in the step feed mode also minimizes the solid loading rates to the secondary sedimentation tanks. This solids storage flexibility should be provided regardless of whether the source of aeration comes from dissolved air or pure oxygen. Plug flow nitrifying aeration systems, which are often required to retain solids for 2 to 3 weeks, operate at maximum efficiency when the hydraulic and organic loadings have the least diurnal fluctuation. This uniformity is often achieved in smaller plants through upstream flow equalization. Oxidation ditches are a simple type of aeration reactor found in many small treatment plants. More design information on aeration reactors and flow equalization is available elsewhere (1, 11–16).

Secondary sedimentation tank two-phase concentration storage is vital to the successful operation of an aeration system. Design of secondary sedimentation facilities usually involves the use of the solids flux theory, which is discussed in detail elsewhere (9, 17, 18). To take maximum advantage of the concentration capabilities, secondary sedimentation tanks are usually from 150% to 200% deeper than primary sedimentation tanks (14 to 20 ft [4.3 to 6.1 m]). Blanket-level instrumentation is commonly used to keep track of sludge storage levels within the secondary sedimentation tanks. More design information on secondary sedimentation tanks is available elsewhere (1, 10, 12).

## 2.1.4. Imhoff and Community Septic Tanks

Both the Imhoff and the community septic tanks were in use long before most of sludge treatment processes. For this reason, it is not surprising that their design includes significant sludge storage capabilities. Imhoff tanks are still in use in many of the older treatment plants, and they still provide those plants with extensive solids storage capacity in what are essentially unheated low rate anaerobic digesters. The storage capacity of Imhoff and septic tanks is part of the empirical design criteria for these facilities. While their future use may be limited because of today's secondary treatment mandate, both processes offer low-cost primary treatment for upgrading small community wastewater stabilization pond facilities. In Newman, California, community septic tanks were upgraded to provide primary treatment for a 0.76-MGD (million gallons per day) (33.3-L/s) complete treatment plant with pond stabilization for secondary treatment and overland flow for tertiary treatment (19). More information on Imhoff and community septic tank design is available elsewhere (1, 20, 21).

## 2.1.5. Wastewater Stabilization Ponds

Wastewater stabilization ponds are cost-effective because of their ability to store solids. Pure aerobic wastewater stabilization ponds provide only single-phase concentration type storage, whereas the more commonly used anaerobic and facultative ponds can provide for long-term, two-phase, concentration-type storage of removed settleable and created biological solids. When debris is thoroughly removed from their influent, secondary facultative ponds can store most of the wastewater solids from a large secondary treatment plant for many years. In Sunnyvale, California, secondary treatment facultative stabilization ponds covering 425 acres (172 ha) have been receiving the majority of sewage solids from a 15-MGD (657-L/s) plant for the past 10 years with no ill effects. Sunnyvale's tertiary treatment facilities for algae and nitrogen removal return all solids removed by dissolved air flotation and gravity filtration to the ponds (21). Primary sludge is removed from the plant before the primary effluent is discharged into the pond and anaerobically stabilized in complete-mix digesters. Supernatant from these digesters is discharged daily into the plant's influent. Most of the solids eventually find their way to the facultative pond. Bottom sludge is withdrawn every week or 10 days from the complete-mix digesters and discharged to anaerobic sludge lagoons. The primary sedimentation effluent and the uncaptured and unrecycled contents of the supernatant merge with the anaerobic bottom layers in the secondary treatment facultative stabilization ponds.

Primary wastewater (usually anaerobic stabilization) ponds that receive raw sewage must be drained and cleaned approximately every 5 to 10 years, depending on loadings. Secondary wastewater (usually facultative stabilization) ponds that are sufficiently deep (6 to 8 ft [1.8 to 2.4 m]) and that receive only those solids generated by biological activity probably never require cleaning. More design information on wastewater stabilization ponds is available elsewhere (22, 23).

*2.1.6. Evaporation Lagoons*

The evaporation lagoon may be described as an open holding facility that depends solely on climatic conditions such as evaporation, precipitation, temperature, humidity, and wind velocity to effect dissipation (evaporation) of on-site wastewater. Individual lagoons may be considered as an alternate means of wastewater disposal on individual pieces of property. The basic impetus to consider this system is to allow building and other land uses on properties that have soil conditions not conducive to the workability and acceptability of the conventional on-site drain-field or leach-bed disposal systems (2). The lagoon is often preceded by septic tanks or aerobic units in order to provide a more acceptable influent to and minimize sludge removal from the lagoon.

Generally, if the annual evaporation rate exceeds the annual precipitation, this method of disposal may at least be considered. The deciding factor then becomes the required land area and holding volume. It should be noted that for unlined on-site installations such as homes and small industrial applications, there may also be a certain amount of infiltration or percolation in the initial period of operation. However, after a time, it may be expected that solids deposition will eventually clog the surface to the point where infiltration is eliminated. The potential impact of wastewater infiltration to the groundwater, and particularly on-site water supplies, should be evaluated in any event and, if necessary, lagoon lining may be utilized to alleviate the problem. Moreover, the following environmental impact and limitations to the lagoon's use should be taken into consideration:

1. Potential for odors and health hazard when not properly designed
2. Public access restrictions are necessary
3. May adversely affect surrounding property values
4. Land area requirements
5. Dependence on meteorological and climatological conditions
6. May require provision to add makeup water to maintain a minimum depth during dry and hot seasons

# Storage of Sewage Sludge and Biosolids

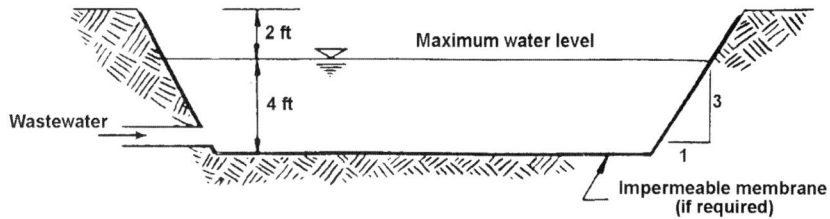

**Fig. 5.1.** Cross section of evaporation lagoon. *Source*: US EPA (2).

### 2.1.6.1. PERFORMANCE

The performance of evaporation lagoons is necessarily site-specific; therefore, the following data are presented on the basis of the net annual evaporation rate that may exist in a certain area (2):

| Net annual evaporation (inches) (True annual evaporation – Annual precipitation) | Lagoon performance (gallons of water evaporated/ft$^2$/yr) |
|---|---|
| 5 | 3.1 |
| 10 | 6.2 |
| 15 | 9.4 |
| 20 | 12.5 |
| 40 | 24.9 |
| 60 | 37.4 |

### 2.1.6.2. DESIGN CRITERIA

According to the US EPA (2) and E.M. Pickett (24), the recommended design criteria for evaporation lagoons are as follows:

1. Hydraulic loading
2. Anticipated flow of wastewater
3. Evaporation rates (10-yr minimum of monthly data)
4. Precipitation rates
5. Depth 2 to 4 ft
6. Banks more than 2 ft higher than maximum water level (Figure 5.1)
7. Level bottom

The rate of wastewater flow may be anticipated to be in the range of 50 gallons (gal)/person/d, depending on the individual site location. Precipitation and evaporation data for most areas can be readily found in weather bureau records. A 12-month mass balance should be utilized to properly determine design sizing.

### 2.1.6.3. ENERGY AND COST

Lagoons are gravity fed from source. Where pumping is required, energy requirements may be approximated by using the following equation, assuming a wire to water efficiency of 60%:

$$hp = \frac{62.4QH}{550} \tag{1a}$$

where

$hp$ = horsepower
$Q$ = flow, ft³/s
$H$ = discharge head, ft

$$E = \frac{62.4QH}{550} \frac{(24)(365)}{(7.48)(24)(60)(60)} \frac{0.7457}{0.60}$$
$$E = 0.0019 \, QH \tag{1b}$$

where

$E$ = energy, kwh/yr
$Q$ = flow, gal/d
$H$ = discharge head, ft

Land costs associated with the individual total retention lagoon are site specific and are not included. Typical excavation and liner (plastic) costs associated with a two-bedroom residence may be estimated as follows (2):

|  | Unit price 2007 (USD) | Cost 2007 (USD) |
|---|---|---|
| Excavation and hauling (750 yd³) | 1.73/yd³ | 1300 |
| Liner (10 mil PVC) (21,000 ft²) | 0.25/ft² | 5250 |
| Supervision and hand labor | Lump sum | 1850 |
| Subtotal |  | 8400 |

(To the above must be added fencing, septic tank and ancillary costs)

## 2.2. Storage Within Wastewater Sludge Treatment Processes

Table 5.1 lists wastewater sludge treatment processes that provide some degree of solids storage. The following subsections discuss how much of this storage capability can be used and how its use can be made as effective as possible.

### 2.2.1. Gravity Thickeners

Gravity thickeners separate liquid from primary and fixed-growth biological secondary solids (25). In this sense, they function like primary and secondary sedimentation facilities. Cool temperatures and chemicals that retard septicity enable gravity thickeners to store sludge for several days. Equipment precautions recommended for primary sedimentation facilities apply to gravity thickeners. Using the same type of calculation indicated in the primary sedimentation design examples, storage capacity can be increased by providing extra depth.

### 2.2.2. Anaerobic Digesters

Anaerobic digesters (26) provide all three types of storage. Those with floating covers have the flexibility to store about 20% to 25% of the digester's volume. The cover movement is used to provide displacement storage. Fixed cover digesters must be protected from excessive vacuum or pressure conditions whenever an attempt is made to achieve displacement storage.

Secondary digesters can be used for two-phase concentration storage by means of liquid-solids separation as long as they are not treating stabilized biological suspended growth secondary sludge (activated sludge). Biological fixed growth secondary sludge normally does not use secondary digester, two-phase concentration storage. More and more treatment plants are finding that the stabilization of activated sludge has a major impact on digester operation. Without waste-activated sludge, the liquid-solids separation process in secondary digesters can concentrate and store solids for considerable periods of time. These time periods usually equal the time required to fill the secondary digester at design flow rates and, depending on the quality of acceptable supernatant, can often be extended.

All digesters can be used to provide equalization storage. Digesters may be used to equalize peak loadings and thereby make downstream dewatering more cost-effective as the example in the last section of this chapter illustrates.

### 2.2.3. Aerobic Digesters

To use an aerobic digester for two-phase concentration type storage, the normally highly agitated contents must be made quiescent and the solids made to settle from the liquor before the whole mass becomes anaerobic and starts to decompose and create nuisance odors. Chemical treatment can facilitate solids settling. Without successful decanting, only single-phase concentration-type storage and displacement-type storage can be used by aerobic digesters. When displacement type storage is used with a fixed surface area, the liquid surface must rise or fall. Under such conditions, the aeration and mixing source must automatically adjust to such changes. Floating mechanical units and fixed-bottom mounted diffusers are both adaptable to these requirements; fixed mechanical aerators are not. Long-term storage in aerobic digesters will have a relatively low capital cost and a very high operating (energy) cost. Often, evaporation can account for significant concentration of the stored solids. As long as the solids remain aerobic throughout the digester, the odor impact of such storage is minimal. More information on aerobic digesters is available elsewhere (27).

### 2.2.4. Composting

Composting is one of the two wastewater solids processes with storage capabilities that are effective for long-term storage. Once the wastewater solids have been stabilized by composting, the curing step can be extended as long as storage is required. This curing step usually involves nothing more than the placing of the composted material in unconfined stockpiles exposed to the atmosphere. As long as there are no site restrictions, this method of storage can be very economical, for it is actually just another use of the time needed for curing and removing the material to its point of final disposal. More design information on composting is available elsewhere (20, 28).

### 2.2.5. Drying Beds

Drying beds are used extensively by many smaller plants in conjunction with anaerobic and aerobic digestion. They are operated on a fill and draw basis and are often used to provide two-phase concentration-type and displacement-type storage between production and disposal. To ensure adequate storage capability, the designer should allow for up to 50% excess drying bed area. More design information on sludge drying beds is available elsewhere (29).

## 3. FACILITIES DEDICATED TO STORAGE OF LIQUID SLUDGE

When solids storage within wastewater treatment processes and sludge solids treatment processes cannot provide the operational flexibility necessary to maintain cost-effective solids treatment and disposal, these within-process storage capabilities must be augmented with special dedicated storage facilities. These dedicated storage facilities can provide storage for sludge in either the liquid or dewatered state, and, depending on design considerations and upstream treatment, may be utilized for any of the three detention times listed in Table 5.1.

Usually, dedicated liquid storage facilities consist of one of the three types listed in Table 5.1. Although listed as primarily for storage of liquid sludge, any of these facilities that are used for anything other than equalizing storage (for 3 to 4 days) will also provide some degree of solids treatment. Holding tanks, without air agitation, and facultative sludge lagoons usually continue anaerobic stabilization. Holding tanks, with air agitation, and aeration basins continue aerobic stabilization. As these are side benefits to the main design functions of these facilities, they have been ignored for the purpose of these classifications. However, if the storage is for the long term, then the additional treatment afforded certainly must be taken into account in setting final disposal/reuse criteria.

### 3.1. Holding Tanks

Holding tanks are commonly provided as an integral part of most conditioning processes and many stabilization processes. Holding tanks may be used for blending different materials as well as for equalizing storage, thereby ensuring that the downstream sludge treatment process is uniformly loaded, both in quality and quantity. Holding tanks also often provide the decanting facilities for sludge treatment processes, such as thermal conditioning, which create products that support two-phase concentration.

Holding tanks that are to be used for blending must be maintained in a homogeneous condition. Such tanks can thus provide only single-phase concentration type storage or displacement type storage. Usually such tanks are relatively small, with detention times measured in hours instead of days. Most of the storage, therefore, is provided by volume adjustments. Holding tanks that involve blending and provide equalizing storage are usually limited to a batch, or a near-batch, type of operation or continuous level adjustments. Tank contents can be mixed by mechanical impellers, hydraulic recirculation, or gas agitation. Each method's applicability may be restricted by the type of material requiring the blending. For example, mechanical impellers are not applicable when unground sludge is being stored. The use of gas agitation and recirculation mixing is normally only limited by the volume that must be blended.

If the holding tank is located downstream from a sludge treatment process, special precautions may be required. For example, if downstream from anaerobic digestion and planned for more than a few hours of storage, the blending tank should be designed with a cover and be equipped to collect and remove combustible digester gas. If downstream from chlorine stabilization, it should be designed to function in a very low pH (acid) environment. Whatever its function, however, the holding tank must be designed to eliminate the production of malodorous gaseous discharges. This elimination is made especially difficult when the holding tank must provide equalizing storage and operate on a batch basis. Unless the solids supplied to the holding tank are completely stabilized biosolids, the tank's use for extended periods of storage will result in the creation of nuisance odors.

Even short periods of storage of unstabilized primary and secondary sludges in a holding tank can produce nuisance odors if no form of temporary inhibiting treatment has been applied. Decant tanks following thermal conditioning often create major odor problems. There are many ways of dealing with the odorous gases created by these holding tanks, for example, by passing the gases back through the aeration system, activated carbon filters, chemical scrubbers, and incinerators. The best design solution, however, is to minimize their creation.

*3.1.1. Design Criteria*

Rectangular or cylindrical tanks can be used for storage, and agitators can be used for mixing. Sludge and biosolids can also be mixed by the use of recycle. Air or pure oxygen can be used for aeration and mixing. Sludge storage can also be used for chemical, tertiary, as well as other sludges. Sludge scraper mechanisms with picket arms would then be required (30, 31). Design criteria for holding tanks includes the following (13, 30):

1. Tank floor slope is generally 1:12. The increased depth near the center of the tank can serve to compact the sludge. Sludge concentration increases as a function of the depth of the sludge blanket.
2. The effluent line should draw compressed solids out of the bottom of the tank.
3. Mixing by air diffusion requires at least $25\,\text{ft/min}/1000\,\text{ft}^3$.
4. Mixing by agitators (mixers) requires approximately $1.0\,\text{hp}/1000\,\text{ft}^3$.

As an example, the Sacramento, California, Regional Wastewater Treatment Plant was provided with a holding or blending tank that is capable of receiving the daily anaerobically digested sludge discharged from nine complete-mix digesters (32). This digested sludge discharge was expected to vary from 0.56 to 0.94 MGD (24.5 to 41.2 L/s) over the following 20 years. The blending tank was 110 ft (33.5 m) in diameter and had a 38.5-ft (11.7-m) sidewater depth. It was provided with a Downes-type floating cover that has a vertical movement of at least 14 ft (4.3 m). This floating cover movement allowed the blending tank to mix the entire daily discharge from all the nine digesters prior to discharging its daily accumulation to downstream facultative sludge lagoons. This blending tank provided a complete separation between the operational control of the complete-mix tank and the controlled feeding of the 20 lagoons. Total solids retention time of the blended sludges was at least 3 days, and approximately one third of the liquid contents of the blending tank were displaced each day. Except for the provision for the extra floating cover travel and the use of bottom mounted gas diffusers, the blending tank had the same design as the four complete-mix tanks. This method of blending and containment minimized the release of odorous gases and maintained a safe control on the production and use of the digester gas during the blending operation. Figure 5.2 is a sectional sketch of the blending digester.

In Aliso, California, two 26,000-gal blending tanks were proposed to blend and equalize unstabilized sludge flows from several sources at the Aliso Regional Solids Stabilization Facility (33). These tanks were provided with hydraulic mixing and fixed covers. The gas cap above the varying liquid level within the fixed covered tanks was maintained at a constant pressure by an inter-tie with the low-pressure digester gas system. This inter-tie eliminated the need for special odor control equipment, minimized the danger from the possible production of an explosive gas-air mixture, and negated the need for some highly complicated pressure control system to protect against a rapid drawdown that might pull a vacuum or air into the blending tank. Figure 5.3 shows a sectional elevation of this raw sludge blending tank. More examples on the use of storage tanks are available elsewhere (34–37).

While very little specific design guidance is provided in the literature for sludge holding tanks, the major issue that must be dealt with is the same as for most sludge treatment processes material management. Wastewater sludge can contain almost anything. If a holding tank design is to incorporate mechanical mixing, which can be incapacitated by stringy material, the designer must make sure that material removed or cut up before reaching the blending tank. Hydraulic mixing pumps must be of the nonclog type or the material reduced in particle size by grinding so that it can pass through the minimum clearances of the type of pump used.

The other major design problem involves the control of odors that are so often an integral part of any type of sludge holding tank. The Sacramento and Aliso holding tank design examples indicate two very successful means of dealing with such odors (that is, containing and incorporating them with the low pressure digester gas system). In many locations stabilized material is held within the holding tank only a few hours. Under these circumstances, their design depends on minimum odor generation, a reasonable assumption given the short retention period. Often decant tanks and conditioning blending tanks cannot depend on either of these methods of odor control. The designer should be very aware that when such a situation exists it will be expected that odors will be confined and treated to the point where their discharge ceases to create a nuisance.

### 3.1.2. Costs of Holding Tanks

The following points were considered in evaluating the costs for sludge and biosolids holding tanks (2, 38):

1. Construction cost includes the storage tank and air supply system.
2. Operation cost includes storage of thickened primary and secondary sludge (1900 lb/MG (million gallons); at 4% solids);
3. Mixing by diffused air (25 CFM (Cubic feet per minute)/1000 ft$^3$) or approximately 130 hp/MG of sludge.
4. Energy, kwh/yr = 4242 × MGD of plant throughput assuming that all conventional activated sludge plant sludges pass through the storage tank.
5. To adjust costs for other sludge quantities and concentrations, enter the curves at effective flow:

$$Q_{Effective} = Q_{Design} \frac{(NewDesignSludgeMass)(4\%)}{(1900 lb/MG)(NewDesignSludgeConcentraation)} \quad (2)$$

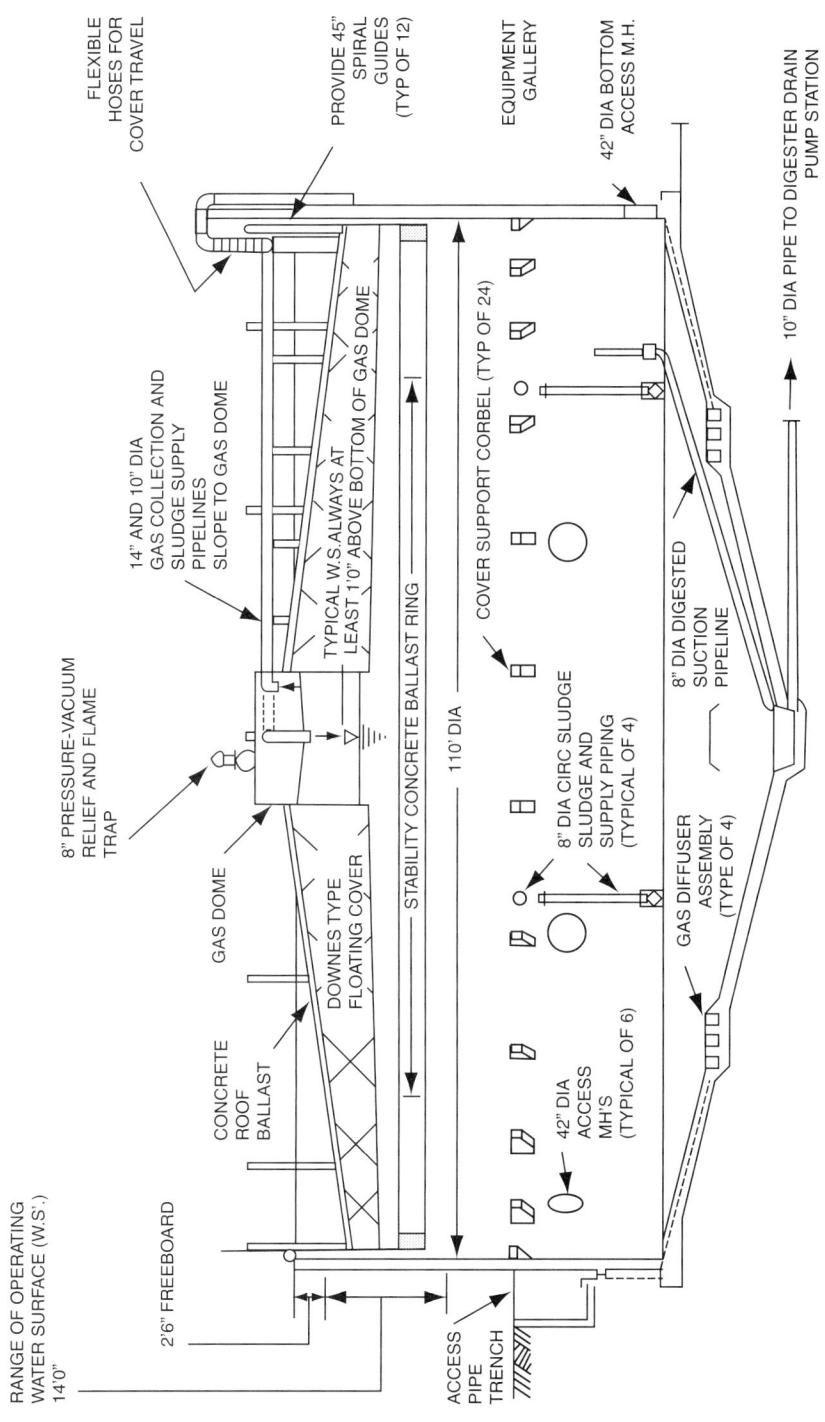

**Fig. 5.2.** Blending digester. MH, Manhole. *Source:* US EPA (3).

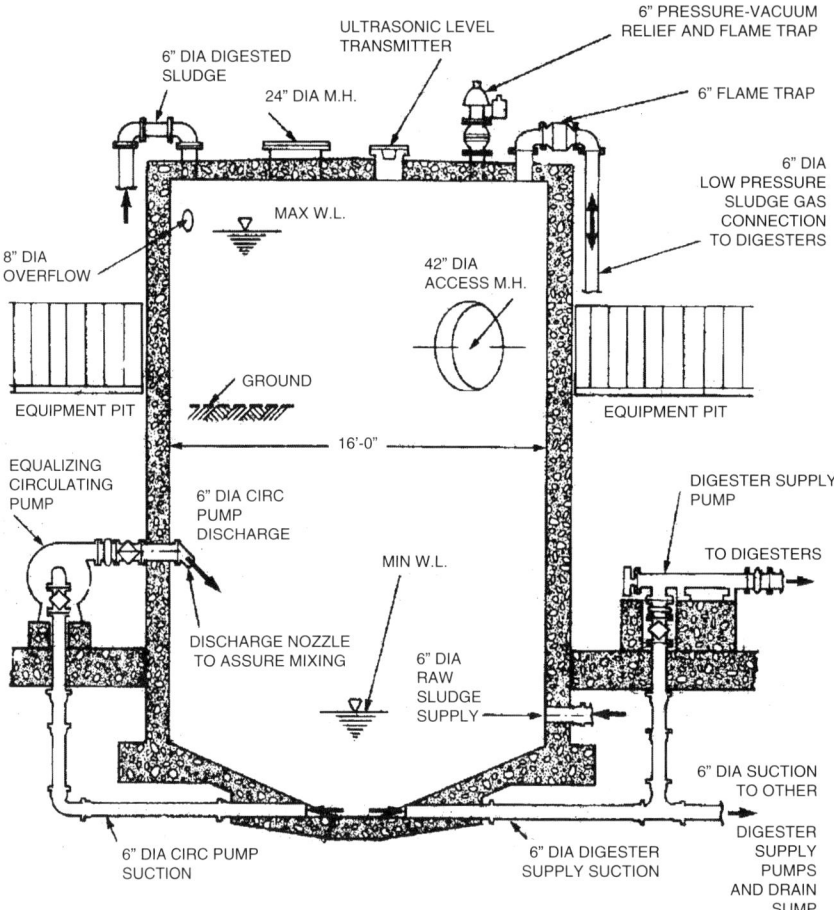

**Fig. 5.3.** Sludge equalization tank. WL, water level. *Source*: US EPA (3).

where

$Q_{\text{Effective}}$ = effective flow
$Q_{\text{Design}}$ = design flow

The construction costs for pipelines and pumping stations are shown in Figure 5.4 and operation and maintenance costs are shown in Figure 5.5 (2, 38). The costs (2007 basis) can be calculated by using the U.S. Army Corps of Engineers Civil Works Construction Yearly Average Cost Index for Utilities (39) as shown in the appendix to this chapter.

$$\text{Cost (2007 basis)} = \frac{2007\, Index}{1980\, Index} \times \text{Cost from Figs. 5.4 and 5.5}$$

$$= \frac{539.74}{277.60} \times \text{Cost from Figs. 5.4 and 5.5}$$

$$= 1.94 \times \text{Cost from Figs. 5.4 and 5.5} \qquad (3)$$

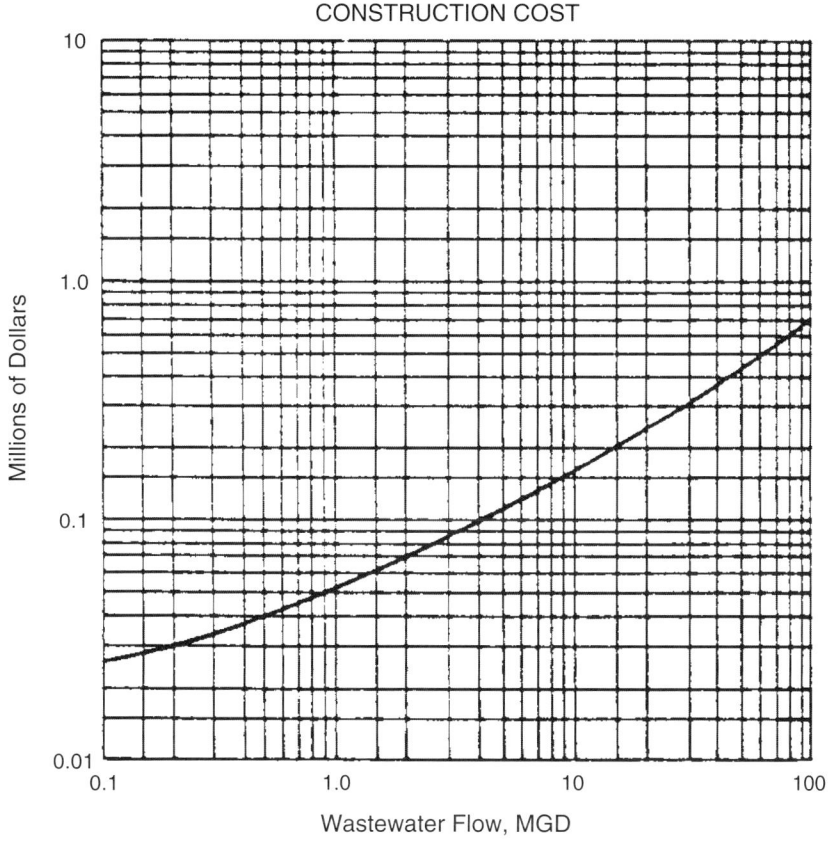

**Fig. 5.4.** Construction cost for sewage-sludge and biosolids holding tank. *Note*: Cost (2007 basis, in U.S. dollars) = 1.94 × Cost from Figure. *Source*: US EPA (2).

## 3.2. Facultative Sludge Lagoons

Sludge lagoons have been used for years to store wastewater solids. Unfortunately, most of this use has been with complete disregard to the aesthetic impact on the surrounding environment. Such misuse has become so widespread that just the use of the term *sludge lagoon* is often enough to eliminate their consideration in present-day alternatives analyses.

Studies in Sacramento, California, based on the successful operation of facultative sludge lagoons in Auckland, New Zealand, indicate that sludge lagoons can be designed to be environmentally acceptable and still remain extremely cost-effective (40, 41). The facility studied in Sacramento provides storage for at least 5 years of sludge production. The sludge stored in the facultative sludge lagoon continued to stabilize without creating an odor level unacceptable to the surrounding neighborhood. Table 5.2 lists the advantages and limitations of using facultative sludge lagoons for long-term storage.

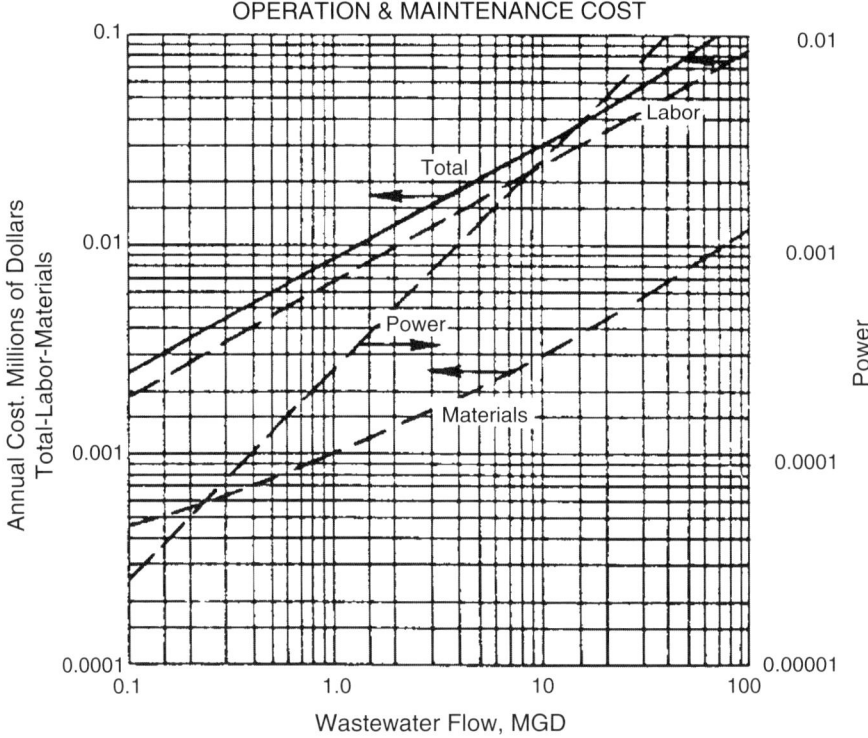

**Fig. 5.5.** Operation and maintenance cost for sewage-sludge and biosolids holding tank. *Note*: Cost (2007 basis, in U.S. dollars) = 1.94 × Cost from Figure. *Source*: US EPA (2).

*3.2.1. Theory*

Facultative sludge lagoons (FSLs) are designed to maintain an aerobic surface layer free of scum or membrane-type film buildup. The aerobic layer is maintained by keeping the annual organic loading to the lagoon at or below a critical area loading rate and by using surface mixers to provide agitation and mixing of the aerobic surface layer. The aerobic surface layer of the FSLs is usually from 1 to 3 ft (0.30 to 0.91 m) in depth and supports a very dense population of between $50 \times 10^3$ and $6 \times 10^6$ organisms/mL of algae (usually *Chorella*). Dissolved oxygen is supplied to this layer by algal photosynthesis, by direct surface transfer from the atmosphere, and by the surface mixers. The oxygen is used by the bacteria in the aerobic degradation of colloidal and soluble organic matter in the digested sludge liquor, while the digested sludge solids settle to the bottom of the basins and continue their anaerobic decomposition. Sludge liquor or supernatant is periodically returned to the plant's liquid process stream.

The nutrient and carbon dioxide released in both the aerobic and anaerobic degradation of the remaining organic matter within the digested sludge are, in turn, used by the algae in the cyclic-symbiotic relationship. This vigorous relationship maintains the pH of the FSL surface layer between 7.5 and 8.5, which effectively minimizes any hydrogen

**Table 5.2**
**Advantages and limitations of using facultative sludge lagoons for long-term storage**

| Advantages | Limitations |
|---|---|
| Provides long-term storage with acceptable environmental impacts (odor and groundwater contamination risks are minimized) | Can only be used following anaerobic stabilization; if acid phase of digestion takes place in lagoons, they will stink |
| Continues anaerobic stabilization, with up to 45% vs. reduction in first year | Large acreages require special odor mitigation measures |
| Decanting ability ensures minimum solids recycle with supernatant (usually less than 500 mg/L) and maximum concentration for storage and efficient harvesting (>6% solids) starting with digested sludge of <2% solids | Requires large areas of land, for example, 15 to 20 gross acres (6 to 8 ha) for 10 MGD, (438 L/s) 200 gross acres (80 ha) for 136 MGD (6000 L/s) carbonaceous activated sludge plants |
| Long-term liquid storage is one of few natural (no external energy input) means of reducing pathogen content of sludges | Must be protected from flooding |
| | Supernatant will contain 300–600 mg/L of TKN mostly ammonia |
| Energy and operational effort requirements are very minimal | Magnesium ammonia phosphate (struvite) deposition requires special supernatant design |
| Once established, buffering capacity is almost impossible to upset | |
| Allows for all tributary digesters to operate as primary complete-mix units (one blending unit may be required for large installations) | |
| Provides environmentally acceptable place for disposal of digester contents during periodic cleaning operations | |
| Sludge harvesting is completely independent from sludge production | |

TKN, total Kjeldahl nitrogen.

sulfide ($H_2S$) release and is believed to be one of the major keys to the successful operation of this sludge storage process.

Facultative sludge lagoons must operate in conjunction with anaerobic digesters. They cannot function properly (without major environmental impacts) when supplied with either unstabilized or aerobically digested sludge. If the acid phase of anaerobic stabilization becomes predominant, the lagoon will stink. Figure 5.6 is a schematic representation of the reactions in a typical FSL.

### 3.2.2. Usage Status

Facultative sludge lagoons were installed initially in 1960 in the Auckland, New Zealand, Manukau sewage treatment plant to provide for the storage and disposal of that plant's anaerobically digested primary sludge. Although lagoons were installed at

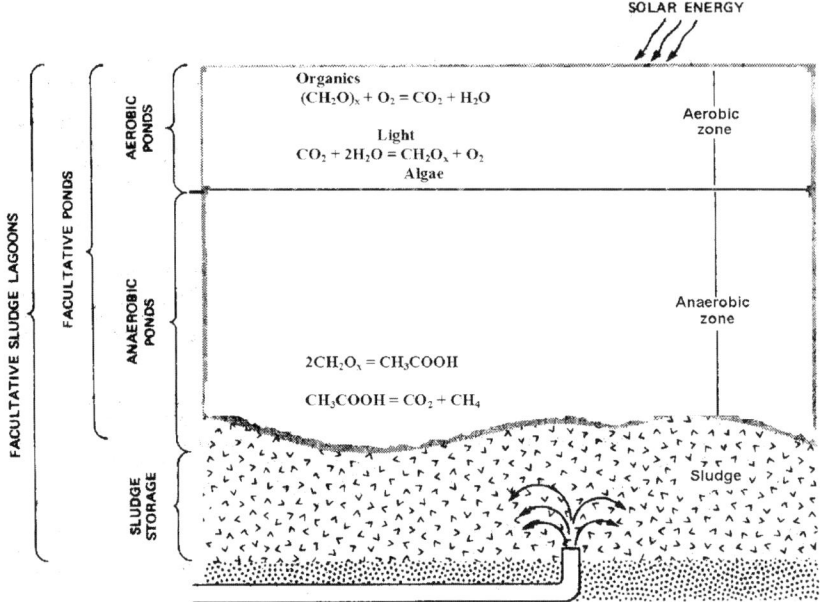

**Fig. 5.6.** Schematic representation of a facultative sludge lagoon (FSL). *Source*: US EPA (3).

Dublin–San Ramon, California, in 1965, Medford, Oregon, in 1971, and in other sites in the United States since 1960 in an attempt to duplicate the successful Auckland installation, it was not until 1974 that the area loading became the critical criterion for their success. Studies in Sacramento since 1974, with approximately 40 acres (16.2 ha) of FSLs, have determined that the standard annual loading rate can be doubled during the warm, long, sunny days of July, August, and September (3). Reduced algae activity during the colder winter months indicates that the standard loading rate should not be exceeded.

Since 1974, additional FSLs have been placed in service at Corvallis, Oregon, entailing 4.5 acres (1.82 ha) and Salinas, California, entailing 6.0 acres (2.43 ha). Other FSLs were built: Eugene-Springfield, Oregon, 25 acres (10.1 ha); Red Bluff, California, 0.93 acres (0.38 ha); Sacramento, California, 84 acres (34 ha); Flagstaff, Arizona, 7.3 acres (2.95 ha); and Colorado Springs, Colorado, 60 acres (24.3 ha). Successful operation was experienced in winter under freezing conditions at Corvallis, Oregon. Experience to date indicates the design criteria established at Sacramento are applicable under other climatic conditions (3).

### 3.2.3. Design Criteria

Design considerations for the FSLs include the following (3):

1. Area loading rate
2. Surface agitation requirements
3. Dimensional and layout limitations
4. Physical factors

## 3.2.3.1. AREA LOADING RATE

To maintain an aerobic top layer, the annual organic loading rate to that FSL must be at or below 20 lb of volatile solids (VS)/1000 ft$^2$/d (1.0 Ton VS/ha/d). Lagoons have been found to be capable of receiving the equivalent of the daily organic loading rate every second, third, or fourth day without experiencing any upset. That is, lagoons have assimilated up to four times the normal daily loadings as long as they have had 3 days of rest between loadings. Loadings as high as 40 lb VS/1,000 ft$^2$/d (2.0 T VS/ha/d) have been successfully assimilated for several months during the warm summer and fall. Experiments on small basins loaded to failure indicate that peak loadings up to 90 lb VS/1,000 ft$^2$/d (4.5 T VS/ha/d) can be tolerated during the summer and fall as long as they do not occur for more than 1 week (3).

## 3.2.3.2. SURFACE AGITATION REQUIREMENTS

Experiments on FSLs that were continuously loaded at the standard rate indicate that they cannot function in an environmentally acceptable manner without daily operation of surface agitation equipment. Observations indicate the brush-type mixer is required to break up the surface film that forms during the feeding of the lagoon. If this film is not dissipated, a major source of oxygen transfer to the surface layer is eliminated. The FSLs with surface areas of from 4 to 7 acres (1.6 to 2.8 ha) require the operation of two surface mixers from 6 to 12 hours a day to successfully maintain scum-free surface conditions. All successful installations have used brush-type floating surface mixers to achieve the necessary surface agitation. Experiments indicate that two brush-type mixers with 8-ft-long (2.4-m) rotors turning at approximately 70 rotations per minute (rpm) and driven by 15 hp (11.2 kW) motors are required for a 4 to 7 acre (1.6 to 2.8 ha) lagoon. The mixers need to operate 12 hours a day. Lagoons of much less than 4 acres (1.62 ha) should be able to achieve the same results with two mixers with 6-ft-long (1.8-m) rotors and 5-hp (3.7-kW) motors. Operation time is expected to be about the same number of hours per day. The FSLs of larger than 7 acres (2.8 ha) have not been found to be cost-effective because of the need to take the lagoons out of service during sludge removal operations (3).

Brush-type mixers have been used to limit the agitation to the surface layer of the FSLs. So far this has been an acceptable application; however, there is some question as to their applicability for very cold climates. Several submerged pump-type floating aerators have been reviewed, and they could be adapted to provide the necessary surface agitation if the brush type could not function under severe freezing conditions. Two mixers are used per FSL to ensure maximum scum breakup in those areas of the lagoon where the prevailing wind deposits the daily loading of scum. The agitation and mixing action of the two mixers located at opposite ends or sides of the lagoon also act to maintain equal distribution of the anaerobic solids.

## 3.2.3.3. DIMENSIONAL AND LAYOUT LIMITATIONS

The FSL size is usually determined by the number of lagoons required to ensure adequate surface area while sludge is removed from a lagoon. If the removed sludge is to be reused, several spare lagoons are required to keep full lagoons out of service for

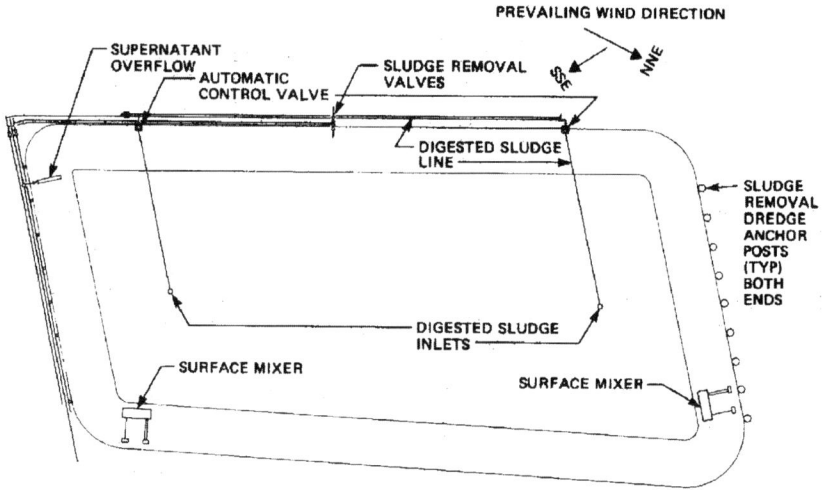

**Fig. 5.7.** Typical FSL layout. *Source*: US EPA (3).

the 2- to 3-year pathogen die-off period (3). The maximum area for a single lagoon is somewhat arbitrary but is based on the most practical size for loading, surface agitation, mixing, and removal requirements. Large, 4 to 7 acre (1.6 to 2.8 ha) individual lagoons would be applicable only to plants with over 70 acres (28 ha) of FSLs; FSLs as small as 150 ft (45.7 m) on a side have been operated successfully.

Lagoon depth was established by the practical limitation of commercially available dredges with a proven capability of removing wastewater solids from beneath liquid surfaces. Equipment that meets this requirement is available to extract sludge from FSLs up to 11.5 and 15 ft (3.5 and 4.7 m) of depth. For plants of up to 10 MGD (440 L/s), the 11.5-ft (3.5-m) depth dredge should be adequate. For plants >10 MGD (440 L/s) the 15-foot (4.7-m) depth should be used to provide additional storage flexibility. If surface agitation must be maintained by submerged pump type aerators, it may be necessary to employ the deepest lagoon possible to ensure adequate separation between the aerobic zone and the anaerobic settling zone of the FSL. Contractors can supply dredge equipment for a lagoon, either with or without the manpower to operate it.

The FSLs are usually best designed to have a long and a short dimension with the shortest dimension oriented parallel to the direction of the maximum prevailing wind. The longer side is made conducive to efficient dredge operation, while the short side's parallel orientation to the prevailing wind direction helps to minimize wave erosion on the surrounding levees. Figure 5.7 is a typical FSL layout, and Figure 5.8 is a typical FSL cross section.

When the area of FSLs exceeds 40 acres (16.2 ha), the potential cumulative effect of large odor emission areas to the vicinity must be considered. Figure 5.9 shows the layout for the 124 acres (50.2 ha) of Sacramento FSLs that were sited on the basis of the least odor risk to surrounding areas.

Work at Sacramento has also determined that batteries of FSLs totaling 50 to 60 acres (20 to 24 ha) are about the maximum size for most effectively reducing the transport of odors.

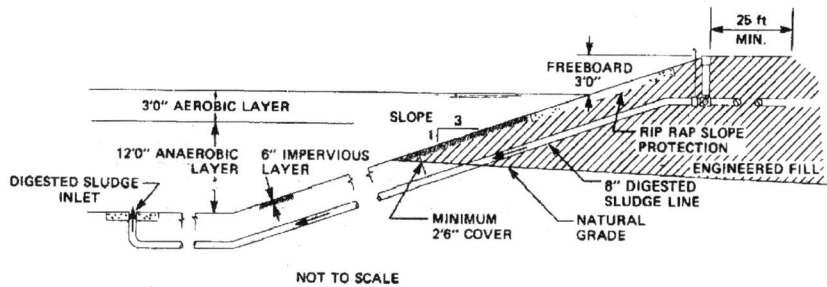

**Fig. 5.8.** Typical FSL cross section. *Source*: US EPA (3).

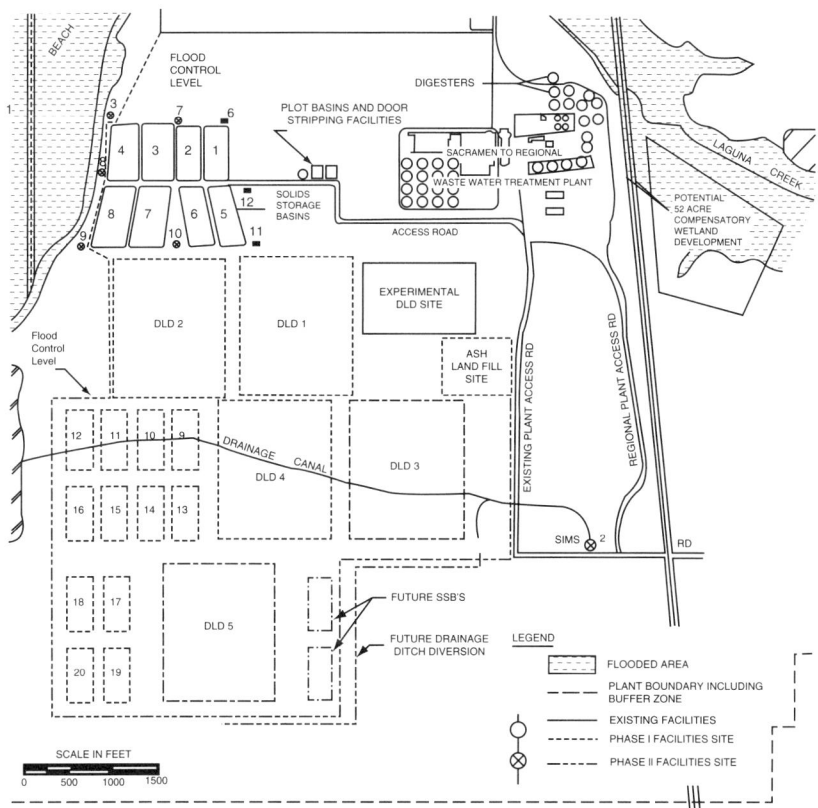

**Fig. 5.9.** Layout for 124 acres of FSL, Sacramento Regional wastewater treatment plant (WWTP). SSB = Solids storage basin. *Source*: US EPA (3).

3.2.3.4. Physical Considerations

Many of the detailed physical considerations applied to the final design of the Sacramento FSLs are shown on Figures 5.8 and 5.9. Supernatant withdrawal is located upstream from the prevailing winds to minimize scum buildup in its vicinity. The FSL supernatant will precipitate magnesium ammonia phosphate (struvite) on any rough surface that is not completely submerged. It has also been found to precipitate inside cavitating pumps. This crystalline material can completely clog cast iron fittings and pump valves when the surface goes through a fill-and-draw cycle or when its operation results in the presence of diffused air. The most practical approach to successful elimination of this problem has been to use polyvinyl chloride (PVC) piping throughout the FSL supernatant process and to design the process for gravity return to the plant influent, with a minimum of critical depth conditions. If pumping is required, submerged slow-speed nonclog centrifugal pumps with low suction and discharge velocities (to minimize cavitation) will be the most trouble-free. All equipment that cannot be PVC or other smooth nonmetallic material should be coated with a smooth, impervious surface.

Two digested sludge feed lines are provided, each with its own automatic valve, to ensure adequate distribution of solids over the whole volume of the FSL. Surface mixers are downstream of the prevailing winds. The harvested sludge dredge hookup is centrally located. Lagoon dike slopes are conservative—three horizontal to one vertical—with adequate riprap provided in the working zone of the surface level. Sufficient freeboard is provided to protect against any conceivable overtopping of the dikes. Digested sludge feed pipelines are located directly below the bottom of the lagoons, with the inlet surrounded by a protective concrete surface. All piping within the basin is out of the way of future dredging operations.

Many of the physical considerations for the basins have been required by the State Dam Safety Agency. Larger FSLs most probably will come under some regulatory agency whose responsibility involves seeing that earthen structures used to confine large quantities of liquid a significant height above the existing ground surface are safe. It is wise to check early to ascertain what, where, and how these agencies will be involved in FSL design.

### 3.2.4. Operational Considerations

Operational considerations can be divided into three categories: the loading or placement of sludge into the FSLs, their routine operation, and the removal of their solids. The considerations listed below were developed during the 5 years of the study on the Sacramento lagoons (3).

#### 3.2.4.1. STARTUP AND LOADING

The FSLs should be initially filled with effluent. Ideally, that effluent should then have about 3 to 6 weeks to develop an aerobic surface layer prior to the introduction of digested sludge. All FSLs should be loaded daily, with the loading distributed equally between FSLs. Loadings should be held below 20 lb VS/1000 ft$^2$/d (1.0 T VS/ha/d) on an average annual basis. As indicated earlier, considerable flexibility does exist. Loads can vary from day to day, and batch or intermittent loading of once every 4 days or less is acceptable. Shock loadings, such as with digester cleanings, should be distributed to all, operating FSLs in proportion to the quantity of sludge inventory they possess. The FSLs should be loaded during periods of favorable atmospheric conditions, particularly just above ground surface, to maximize odor dispersion. The fixed and volatile sludge solids loadings to the FSLs and their volatile contents should be monitored quarterly.

### 3.2.4.2. DAILY ROUTINE

Surface mixers should operate for a period of between 6 and 12 hours. Operation should not coincide with FSL loading and should always be during the hours of minimum human exposure (usually midnight to 5 a.m.) and during periods of favorable atmospheric conditions. The FSL supernatant return to the wastewater treatment process should be regulated to minimize shock loadings of high ammonia. Supernatant return flows should be monitored so that their potential impact on the liquid treatment process can be discerned. The sludge blanket in a lagoon should not be allowed to rise higher than 2 ft below the operating water surface.

### 3.2.4.3. SLUDGE REMOVAL

The FSLs that are to be emptied of accumulated solids should be removed from routine operation at least 30 days prior to the removal of any solids. Pathogen safe reuse requires removal from operations for 2 to 3 years (3). Sludge removal should be limited to those FSLs that are concentrating the sludge solids to 6% to 8%. During FSL sludge removal operations, the water surface level should not be allowed to drop more than 12 to 18 inches (in) (30 to 46 cm) below its normal operating level.

### 3.2.5. Energy Impacts

Energy requirements of FSLs are relatively small because FSLs use solar energy. The sun supplies the needed energy for the algal photosynthesis. In turn, the algal cells supply the dissolved oxygen to support the aerobic bacterial action in the surface layer. The only outside power used in normal FSL operation is for surface agitation, supernatant pumping and treatment, and the supply and removal of the sludge. For the 124-acre (50.2-ha) Sacramento installation, it was calculated that these energy requirements could equal $31,700 \times 10^6$ British thermal units (BTU)/yr (33,400 GJ (Giga Joule = $10^9$ Joule)/yr) when the FSLs became fully loaded (42). As loading is based on area, the energy impact of FSLs will be $255 \times 10^6$ BTU/yr/acre (670 GJ/yr/ha). With maximum odor source control and transport reduction measures, this energy use will increase to $294 \times 10^6$ BTU/yr/acre (765 GJ/yr/ha). As no chemicals or major structures are involved, all FSL energy impacts are direct. There are no secondary impacts.

### 3.2.6. Actual Performance Data

The figures and tables discussed in this section report the actual performance of the eight FSLs in operation at the Sacramento Central Wastewater Treatment Plant. Although the plant is designed as a 24-MGD (1.1-m$^3$/s) carbonaceous activated sludge secondary wastewater treatment plant with anaerobic digestion for solids stabilization, it treats the total solids from three upstream secondary treatment plants, the total annual flow of which is considerably greater than its own. Solids from those upstream plants are transported to the central plant by its tributary sewer collection system. The central plant also receives a substantial solids loading (up to 35% daily surcharge) from seasonal canning operations. Table 5.3 lists the FSL loadings for a period of 4 years. Figure 5.10 summarizes typical surface layer data for four of the FSLs, and Table 5.4 summarizes the FSLs design data. The overall volatile solids (VS) reduction in the lagoons is 42%.

Recycled FSL supernatant quality is given in Table 5.5, and complete mineral, heavy metals, and chlorinated hydrocarbon data for digested, FSL, and harvested solids is provided in Table 5.6. While the specific conductance in the supernatant remains high

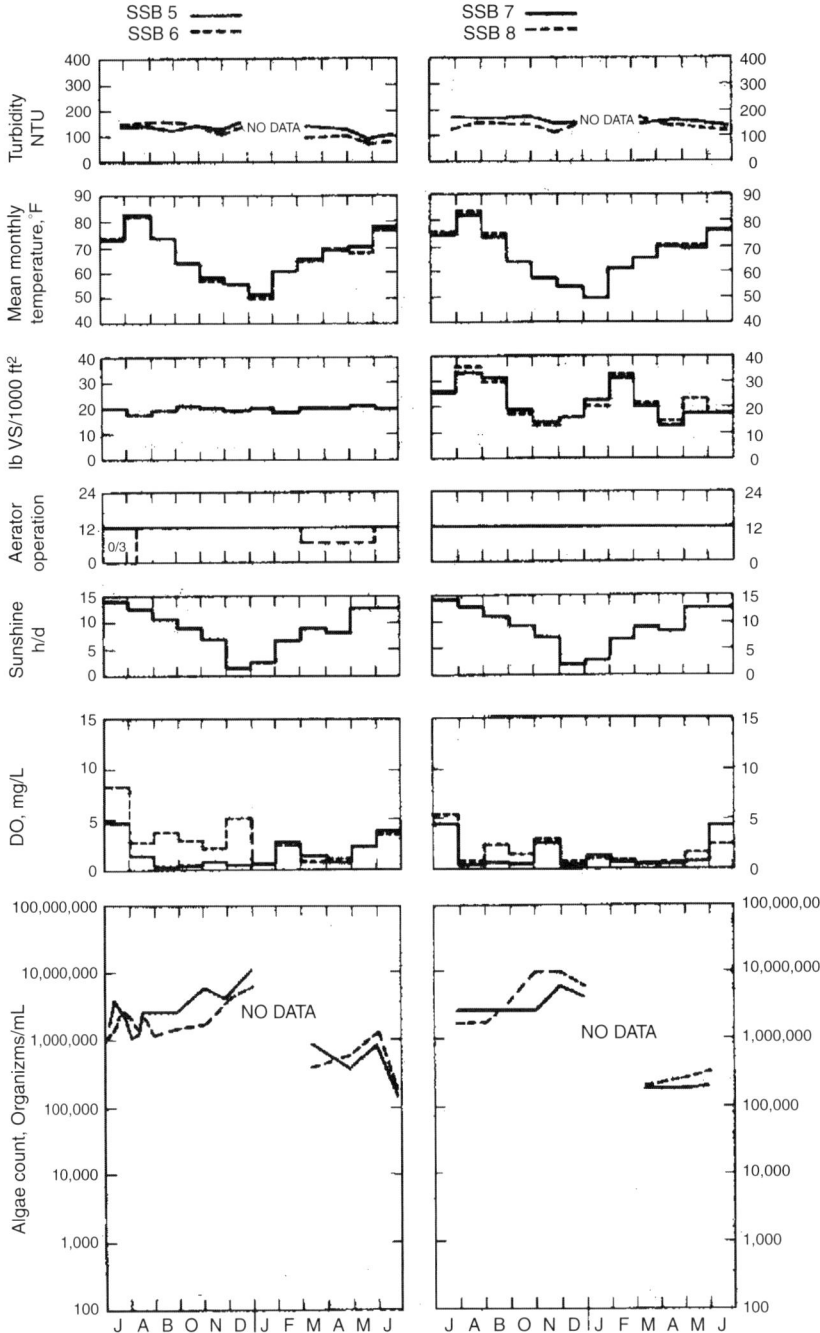

**Fig. 5.10.** Sacramento Central WWTP surface layer monitoring data for FSLs 5 to 8. *Source*: US EPA (3).

### Table 5.3
**Sacramento Central Wastewater Treatment Plant volatile reductions, digested sludge quantities and facultative sludge lagoon (FSL) area loadings**

| Year | Digester volatile reduction (%) | Digested solids to FSLs | | | FSL loading |
|---|---|---|---|---|---|
| | | Annual average total solids ($10^3$ lb/d$^a$) | Volatile (%) | Solids (%) | Annual average (lb VS/$10^3$ ft$^2$/d$^a$) |
| 1 | 52 | 44.1 | 63 | 1.7 | 22.5 |
| 2 | 50 | 35.9 | 67 | 1.6 | 15.9 |
| 3 | 51 | 44.0 | 68 | 1.6 | 17.1 |
| 4 | 45 | 52.7 | 66 | 1.6 | 20.7 |

$^a$ Dry solids.
VS, volatile solids.
*Source*: US EPA (3).

### Table 5.4
**Sacramento Central Wastewater Treatment Plant FSL design data**

| FSL | Depth from water surface to bottom, ft | Area at water surface 1000 ft$^2$ (acres) | Volume below sludge blanket, 1000 ft$^3$ | Loading capacity of basin,$^a$ 1000 lb VS/d |
|---|---|---|---|---|
| 1 | 11 | 164.0 (3.8) | 1030.4 | 3.28 |
| 2 | 11 | 164.0 (3.8) | 1030.4 | 3.28 |
| 3 | 14 | 244.2 (5.6) | 2137.0 | 4.88 |
| 4 | 14 | 229.0 (5.3) | 1983.0 | 4.58 |
| 5 | 15 | 204.2 (4.7) | 1851.0 | 4.08 |
| 6 | 15 | 204.2 (4.7) | 1850.0 | 4.08 |
| 7 | 15 | 270.0 (6.2) | 2689.0 | 5.40 |
| 8 | 15 | 270.0 (6.2) | 2689.0 | 5.40 |
| Total | | 1749.6 (40.1) | 15,259.8 | 31.80 |

$^a$Capacity of lagoon based on a design volatile solids (VS) loading of 20 lb/1000 ft$^2$ of water surface area per day.
*Source*: US EPA (3).

(2500 to 4300 mhos/cm), the supernatant contains very little of the heavy metals. Rainfall increases the quantity of supernatant and decreases its strength. Winter-specific conductivity always dropped in Sacramento following significant rainfall. The only solution to this problem would seem to be to reduce the heavy metals concentrations in the unstabilized sludge.

#### 3.2.7. Public Health and Environmental Impact

The FSLs have been found to have insignificant environmental impact in Sacramento during 5 years of study (3):

1. No vector impact
2. No groundwater impact
3. Controlled pathogen impact
4. Acceptable odor impact

**Table 5.5**
**Sacramento Central Wastewater Treatment Plant recycled FSL supernatant quality**

| Constituent[a] | Month/day | | | | | | Average |
|---|---|---|---|---|---|---|---|
| | 10/5 | 10/6 | 10/7 | 10/11 | 10/30 | 12/20 | |
| BOD | 140 | 140 | 140 | 96 | 200 | 110 | 143 |
| TPO$_4$ | 51 | 50 | 66 | 120 | 110 | 80 | 79 |
| Sulfides | 0 | 0 | 0 | 0 | 0 | 0 | – |
| COD | – | – | – | 910 | 960 | 874 | 935 |
| TKN | – | – | – | 220 | 360 | 394 | 290 |
| pH | – | – | – | 7.7 | 7.7 | 7.8 | 7.7 |
| SS | – | – | – | 470 | 420 | 728 | 445 |
| NH$_3$–N | – | – | – | – | 300 | 335 | 300 |

[a]In mg/L except for pH.
BOD, biochemical oxygen demand; COD = chemical oxygen demand; TKN, total Kjeldahl nitrogen.
*Source:* US EPA (3).

#### 3.2.7.1. VECTOR IMPACTS

Rodents and flies have apparently not bred around the FSLs for the 5-year operation period. Scum control is obviously the key to elimination of this problem.

#### 3.2.7.2. GROUNDWATER IMPACTS

Groundwater contamination is nonexistent. Monitoring wells surrounding the 40 acres (16.2 ha) of existing FSLs in Sacramento have been sampled monthly and have never shown any indication of groundwater contamination traceable to the lagoons. Tests show that sludge that settles to the bottom quickly and effectively seals off the lagoon contents from the surrounding soils. Undisturbed soil samples taken directly from the bottom of a lagoon with a limited history (1 to 2 years) and a lagoon with a long history (4 to 5 years) confirm that the FSL contents have a limited penetration into the surrounding soils. These studies indicate that the sealing of FSLs is a combination of soil pore plugging by suspended and colloidal materials within the sludge and the formation of mucus-like materials that create an impermeable membrane between the stored sludge and the underlying soil. Sandy soils take longer to seal than silty clay soils, but both achieve complete sealing in 2 to 3 months.

The 2- to 6-inch (5.08 to 15.24 cm) engineered fill seal provided over the natural bottom and side slopes of the FSL cross-section in Figure 5.8 ensures that none of the FSL startup sewage or diluted sludge content escapes during the natural sealing process.

#### 3.2.7.3. PATHOGEN IMPACTS

It has been recognized for many years that long-term liquid storage significantly reduces the pathogenic microorganism content in sludge (3). Studies in Sacramento confirm this for the most common bacteria. Figure 5.11 shows that the fecal coliform population decreases as the sludge passes through the sludge management system. Studies of parasitic protozoans and their cysts, helminths, and eggs (ova), and studies of viruses were inconclusive either because insufficient numbers were found or the

## Table 5.6
Sacramento Central Wastewater Treatment Plant comparison of digested FSL and removed sludge analytical data

| Constituent | Digested sludge[a] | Stored sludge[a] | | | | | | | | Removed sludge[a] |
|---|---|---|---|---|---|---|---|---|---|---|
| | | FSL 1 | FSL 2 | FSL 3 | FSL 4 | FSL 5 | FSL 6 | FSL 7 | FSL 8 | |
| mg/L | | | | | | | | | | |
| Alkalinity[b] | 2556 | 2653 | 2676 | 2638 | 2348 | 1940 | 1687 | 2239 | 2175 | 2069 |
| Chloride[c] | 143 | 178 | 225 | 204 | 209 | 169 | 166 | 171 | 186 | 171 |
| Ammonia[c] | 444 | 685 | 765 | 751 | 649 | 502 | 452 | 613 | 600 | 573 |
| Soluble phosphorus (P)[c] | 65 | 44 | 38 | 49 | 33 | 28 | 50 | 51 | 49 | 45 |
| Sulfate[c] | 38 | 87 | 97 | 91 | 113 | 73 | 77 | 68 | 49 | 151 |
| % dry weight | | | | | | | | | | |
| Total phosphorus (P) | 1.6 | 2.0 | 1.9 | 1.7 | 1.8 | 1.4 | 1.6 | 1.6 | 1.4 | 1.9 |
| Total nitrogen (N) | 8.7 | 5.1 | 5.2 | 5.2 | 4.1 | 5.4 | 6.2 | 5.8 | 5.1 | 5.9 |
| mg/kg dry weight | | | | | | | | | | |
| Calcium | 21,000 | 27,000 | 25,000 | 21,000 | 28,000 | 28,000 | 24,000 | 26,000 | 21,000 | 24,000 |
| Magnesium | 5800 | 8200 | 7900 | 7900 | 6300 | 5500 | 5300 | 6300 | 3500 | 8600 |
| Potassium | 5500 | 3200 | 3900 | 3800 | 2900 | 2600 | 3000 | 3100 | 3200 | 4500 |
| Sodium | 9200 | 3100 | 3450 | 3500 | 3300 | 4100 | 5600 | 4600 | 4200 | 5400 |
| Arsenic | 47 | 75 | 72 | 89 | 101 | 22 | 28 | 82 | 62 | 15.4 |
| Beryllium | <2.2 | <1.1 | <1.1 | <1.0 | <1.1 | <1.4 | <1.5 | <1.0 | <1.2 | <1.3 |
| Cadmium | 12 | 24 | 26 | 19 | 16 | 14 | 18 | 21 | 17 | 19 |
| Chromium | 165 | 218 | 245 | 224 | 243 | 173 | 220 | 278 | 188 | 181 |
| Copper | 340 | 410 | 398 | 385 | 721 | 400 | 477 | 456 | 353 | 384 |
| Lead | 185 | 134 | 123 | 96 | 134 | 116 | 183 | 153 | 121 | 159 |
| Mercury | 3.7 | 5.3 | 5.1 | 5.3 | 5.2 | 5.0 | 5.8 | 5.8 | 4.2 | 5.6 |
| Molybdenum | <22 | <13.4 | <16 | <14 | <12.5 | <13.7 | <15.4 | <12.2 | <11.8 | <13 |
| Nickel | 63 | 58 | 72 | 70 | 115 | 46 | 48 | 60 | 53 | 77 |
| Selenium | 1.6 | 1.7 | 1.4 | 1.6 | 1.4 | 4.1 | 3.2 | 2.6 | 1.4 | 5.6 |
| Silver | 28 | 26 | 26 | 26 | 23 | 34 | 38 | 35 | 27 | 28 |

(*Continued*)

Table 5.6 (*Continued*)

| Constituent | Digested sludge[a] | Stored sludge[a] | | | | | | | | Removed sludge[a] |
|---|---|---|---|---|---|---|---|---|---|---|
| | | FSL 1 | FSL 2 | FSL 3 | FSL 4 | FSL 5 | FSL 6 | FSL 7 | FSL 8 | |
| Zinc | 930 | 1700 | 1500 | 1300 | 1325 | 1207 | 1400 | 1400 | 1090 | 1200 |
| PCB 1242 | e | <2.8 | <3.1 | <2.9 | <2.6 | <2.3 | <2.6 | <3.0 | <3.0 | <2.1 |
| PCB 1254 | e | 5.5 | 5.3 | 4.0 | 4.8 | 4.7 | 3.8 | 6.6 | 3.3 | 4.6 |
| Tech chlordane | e | 3.8 | 4.0 | 3.6 | 4.0 | 3.9 | 4.2 | 5.9 | 3.8 | 5.0 |
| Other pesticidals[d] | e | 0.30 | 0.27 | 0.25 | 0.22 | 0.25 | 0.25 | 0.27 | 0.23 | <0.7 |
| Units as noted | | | | | | | | | | |
| Cd/Zn ratio, % | 1.3 | 1.4 | 1.7 | 1.5 | 1.0 | 1.1 | 1.3 | 1.5 | 1.5 | 1.5 |
| Total solids, % | 1.7 | 7.0 | 6.3 | 6.1 | 7.6 | 4.7 | 3.4 | 4.8 | 5.7 | 4.1 |
| Volatile solids, % of total | 68 | 55 | 55 | 53 | 52 | 60 | 62 | 61 | 52 | 54 |
| pH | 7.5 | 7.3 | 7.3 | 7.3 | 7.3 | 7.2 | 7.4 | 7.3 | 7.3 | 7.4 |
| Specific conductance[c], μmhos/cm | 4742 | 5109 | 5847 | 5743 | 4914 | 4434 | 4093 | 5061 | 4760 | 4731 |

[a]Values are averages from samples collected during 1977.
[b]As $CaCO_3$, determined by potentiometric titration of supernatant.
[c]Determined on supernatant; other determinations run on solution resulting from acid digestion of whole sample.
[d]Other pesticidals include residues such as DDT, DDE, dieldrin, etc.
[e]Analysis not performed.
μmhos, The SI derived unit of electrical conductance, equal to one ampere per volt. It is equivalent to the reciprocal of the ohm unit. Also called siemens.
*Source*: US EPA (3).

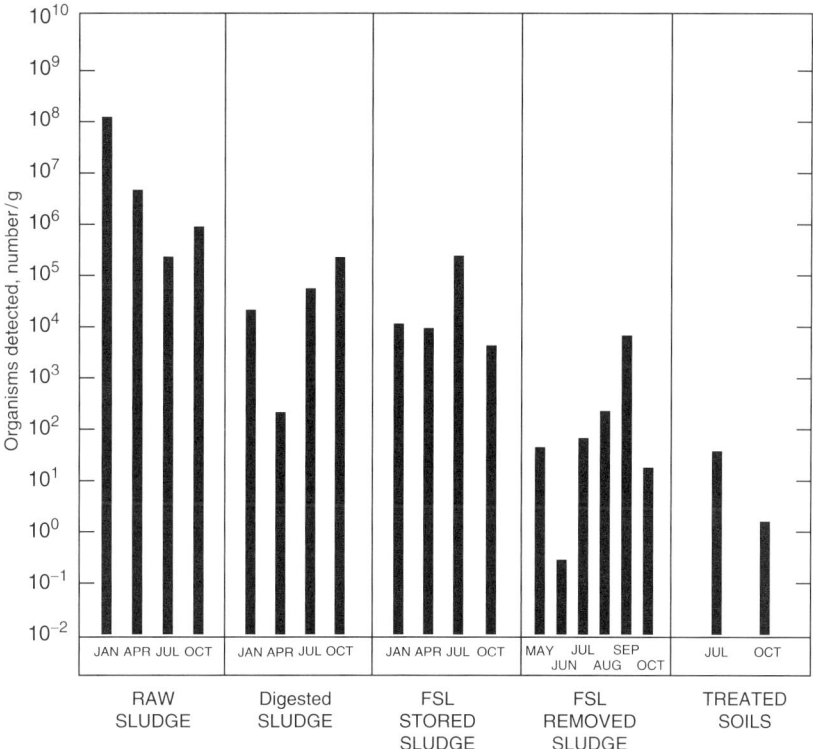

**Fig. 5.11.** Sacramento Central WWTP fecal coliform populations for various locations in the solids treatment/disposal process. *Source*: US EPA (3).

techniques required for reasonable reproducibility were unavailable to the project. The system of disposal selected, that of dedicated land disposal, made further investigatory work unnecessary.

3.2.7.4. ODOR IMPACTS

Odor impacts change in direct proportion to the FSLs surface area. In most small plants (those requiring <40 acres [16.2 ha] of FSLs), controlling the loading rate, using adequate surface agitation, providing sufficient buffering area, and carefully selecting the best time periods for feeding and surface agitation operation are sufficient to achieve acceptable levels of odor risk. Table 5.7 shows the annual odor risk analysis developed for the existing 40 acres (16.2 ha) of FSLs at the Sacramento site before the installation of the barriers and wind machines (6). No high-technology mitigation has been required to maintain this acceptable risk level. For larger areas of FSLs, additional odor control measures would probably be required. These might include the installation of a blender digester to keep raw sludge from short circuiting to the FSLs, vacuum vaporization to remove entrained odors from the digested sludge prior to its discharge into the FSLs, separation of batteries of FSLs, construction of special 12-ft (3.7-m) high barriers around the FSLs to ensure maximum odor dispersion at low wind speeds, and the use of wind machines to aid odor dispersion when the atmosphere is calm.

Table 5.7
Sacramento Central Wastewater Treatment Plant odor risk for 40 acres of FSL[a]

| Downwind odor concentration, C | Annual events, d | | | | | | | | |
|---|---|---|---|---|---|---|---|---|---|
| | Direction toward which wind is blowing | | | | | | | | |
| | N | NE | E | SE | S | SW | W | NW | Total |
| 2[b] | 2.8 | 2.1 | 3.2 | 7.3 | 11.5 | 6.7 | 4.1 | 3.1 | 38.9 |
| 5[c] | 0.3 | 0.3 | 0.5 | 1.2 | 1.6 | 0.8 | 0.4 | 0.3 | 5.4 |
| 10[d] | 0.08 | 0.06 | 0.10 | 0.20 | 0.3 | 0.2 | 0.1 | 0.1 | 1.1 |

[a]Includes source control mitigation—controlled organic surface loading rate, adequate surface mixers, and controlled feeding and mixer operating times and odor transport mitigation—2000 to 5000 ft of buffer.
[b]2 ou/cf barely detectable ambient odor criteria.
[c]5 ou/cf threshold complaint conditions.
[d]10 ou/cf consistent complaint conditions.
ou/cf = odor units/cubic feet.
*Source*: US EPA (3).

Table 5.8
Sacramento Regional Wastewater Treatment Plant ultimate odor risk for 124 acres of FSL[a]

| Downwind odor concentration, C | Annual events, d | | | | | | | | |
|---|---|---|---|---|---|---|---|---|---|
| | Direction toward which wind is blowing | | | | | | | | |
| | N | NE | E | SE | S | SW | W | NW | Total |
| 2[b] | 0.44 | 0.15 | 0.18 | 0.41 | 0.85 | 0.31 | 0.22 | 0.33 | 2.9 |
| 5[c] | 0.08 | 0.02 | 0.03 | 0.06 | 0.13 | 0.04 | 0.03 | 0.05 | 0.44 |
| 10[d] | 0.02 | <0.01 | <0.01 | 0.01 | 0.02 | 0.01 | 0.00 | 0.01 | <0.09 |

[a]Includes source control mitigation—controlled organic surface loading rate, adequate surface mixers, blending digester, vacuum vaporization and controlled feeding and mixer operation times, and odor transport mitigation—2000 to 5000 ft of buffer and, separation of groups of FSLs, barriers and wind machines.
[b]2 ou/cf barely detectable ambient odor criteria.
[c]5 ou/cf threshold complaint conditions.
[d]10 ou/cf consistent complaint conditions.
*Source*: US EPA (3).

The odors from 40 acres (16.2 ha) of FSLs in Sacramento have proven to be completely acceptable. An analysis of the annual odor risks for the 124 acres (50.2 ha) of FSLs is shown in Table 5.8. This analysis shows that with the installation of complete control measures, the incidence of threshold complaint odor levels at the plant boundary (2000 to 5000 ft [610 to 1520 m] downwind) will be less than once every 2 years, regardless of wind direction, and once every 7 years for the worst specific wind direction. This level of odor risk was found to be acceptable in the public environmental impact hearings.

## 3.2.8. Cost Information

The major elements involved in determining FSL costs are land and earth moving. Both are usually quite site specific. Normally, land costs vary less predictably than construction costs. A typical FSL storage facility for a l0-MGD (438-L/s) secondary carbonaceous activated sludge treatment plant with primary sedimentation, anaerobic digestion, and normal strength domestic and industrial wastewater will cost about USD 3,440,000 to construct and USD 57,000/year to operate. Construction costs are reported on the basis of 2007 and do not include the cost of land. Operation costs are calculated on 2007 wage rates and do not include dredge operators or any other removal costs (3).

Construction costs include the installation of three complete 4-acre (1.62 ha) FSLs. This is assumed to be the capacity needed to meet the annual digested sludge loading rate criterion of 20 lb VS/1000 ft$^2$/d (1.0 T VS/ha/d). It is based on a conservative unstabilized sludge production rate and a nominal 50% volatile solids reduction in the anaerobic digesters. The three lagoons will provide capacity for daily loading, digester cleaning, and maintenance and storage for intermittent removal to dedicated land disposal. The FSLs are assumed to be 15 ft (4.6 m) in depth and have 3:1 dike side slopes. If they are required, purchase of the dredge and booster pump would add another USD 344,000 to USD 412,000 to the construction costs.

Odor control costs, including blending digester, vacuum vaporizer, barriers, and wind machine could increase the construction costs another USD 572,000 and the operation costs another USD 57,000/yr. As indicated by the odor impact assessment, sufficient area to ensure maintenance of loading criteria, together with surface agitators and proper buffer, would make it possible to avoid the cost of the aforementioned more extensive odor mitigation facilities.

Construction costs for the 124 acres (50.2 ha) of FSLs with complete odor mitigation facilities for the Sacramento Regional Wastewater Treatment Plant are estimated to be USD 65,720,000. This includes almost USD 7,560,000 for the existing 40 acres (16.2 ha) of FSLs with barrier wall and wind machines. This acreage will store the solids from a l36-MGD (5960-L/s) secondary carbonaceous activated sludge treatment plant. Operation costs are estimated at USD 1,488,000/yr.

## 3.3. Anaerobic Liquid Sludge Lagoons

Many such lagoons are being operated throughout the United States. One system that has collected some meaningful data is the 220.2 acres (89.1 ha) in operation at the Metropolitan Sanitary District of Greater Chicago (MSDGC) Prairie Plan land reclamation project in Fulton County, Illinois. R.R. Rimkus (3, 31), chief of maintenance and operations for MSDGC, provided the layout shown in Figure 5.12 of the four lagoons at this site. He reports that lagoons 1 and 2 have been in service for 8 and 7 years, respectively, and lagoons 3a and 3b for 6 years. Lagoons 1 and 2 have an average depth of 35 ft (10.7 m) plus or minus 1 ft (0.3 m), and lagoons 3a and 3b are about 18 ft (5.5 m) deep. Lagoons 3a and 3b are utilized more for supernatant treatment and storage.

Rimkus further reports that barged anaerobically digested waste activated sludge from Chicago is discharged into Fulton County lagoons 1 and 2 throughout the year, when river shipment conditions permit, at a frequency of about 20 days per month. Solids

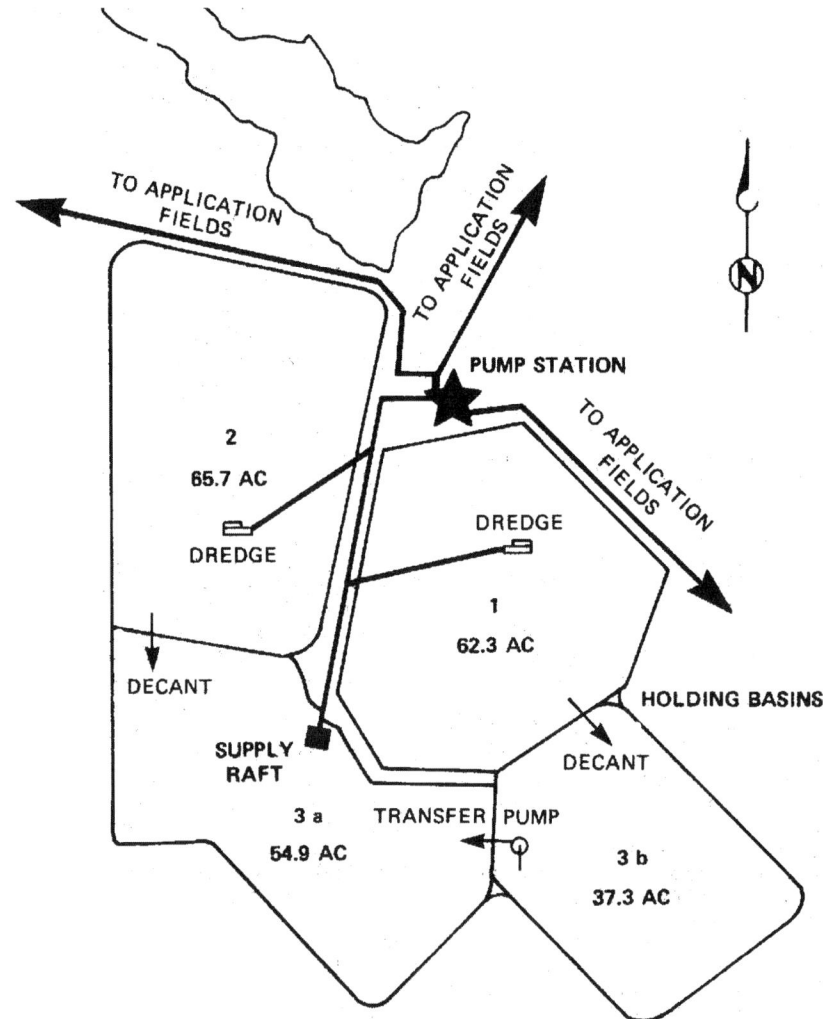

**Fig. 5.12.** Anaerobic liquid sludge lagoons, Prairie Plan Land Reclamation Project, Greater Chicago, Illinois. AC, acre. *Source*: US EPA (3).

loading varies between 65,000 and 95,000 dry t (59,000 to 86,200 T) per year. Based on the total loading received by lagoons 1 and 2 and the volatile solids content of the digested sludge equaling 57%, the organic loading rate to the Fulton County lagoons varies between 36 and 50 lb VS/1000 ft$^2$/d (1.7 to 2.4 T/ha/d). This is considerably above the 20 lb VS/1000 ft$^2$/d (1.0 T/ha/d) established in Sacramento to maintain facultative conditions within the lagoons. If the area of all four lagoons is considered, this organic loading rate drops to 21 to 29 lb VS/1000 ft$^2$/d (1.0 to 1.4 T/ha/d), which is close to the facultative sludge lagoon concept.

Rimkus (31) reports that the solids concentration of sludge pumped from the barge to the lagoons varies from 4% to 6% by weight. Further, the sludge pumped from lagoons to fields varied from 3.57% to 5.93% by weight. The average annual quantity

Table 5.9
Removed sludge—Prairie Plan Land Reclamation Project, Greater Chicago

| Constituent | Minimum mg/L | Maximum mg/L | Mean mg/L | Mean content lb/dry ton |
|---|---|---|---|---|
| Total phosphorus, P | 900 | 2960 | 1416 | 59.6 |
| Kjeldahl nitrogen, N | 1276 | 2905 | 2329 | 98.1 |
| Nitrogen as ammonia, N–$NH_3$ | 772 | 1338 | 1046 | 44.0 |
| Alkalinity as $CaCO_3$ | 1640 | 5750 | 3630 | 153 |
| Chloride, Cl | 228 | 752 | 388 | 16.3 |
| Iron, Fe | 1000 | 2900 | 1938 | 81.6 |
| Zinc, Zn | 87 | 231 | 171 | 7.2 |
| Copper, Cu | 44.8 | 124 | 81.6 | 3.44 |
| Nickel, Ni | 9 | 28 | 18 | 0.76 |
| Magnesium, Mg | 8.5 | 28.3 | 18.0 | 0.758 |
| Potassium, K | 80 | 200 | 166 | 6.99 |
| Sodium, Na | 30 | 120 | 88 | 3.7 |
| Manganese, Mn | 80 | 810 | 450 | 18.9 |
| Calcium, Ca | 710 | 1800 | 1185 | 49.9 |
| Lead, Pb | 25.9 | 54.5 | 42.1 | 1.77 |
| Chromium, Cr | 90.6 | 513 | 175 | 7.37 |
| Cadmium, Cd | 7.5 | 20.2 | 13.2 | 0.556 |
| Aluminum, Al | 340 | 900 | 679 | 28.6 |
| Mercury, Hg | 0.132 | 1.920 | 0.417 | 0.018 |
| Total solids, % | 3.57 | 5.93 | 4.75 | 2000 |
| Total volatile solids, % | 43.5 | 50.0 | 47.5 | 950 |
| pH, units | 7.2 | 7.9 | — | — |
| EC, µmhos/cm | 2500 | 6800 | 4675 | — |

*Source*: US EPA (3).

of removed sludge is 60,000 dry t (54,400 T). Mean value for volatile solids content of removed sludge was 47.5%. If the barged sludge volatile content is 57%, then the lagoons are reducing the volatile solids by 17%. Data for sludge removed are given in Table 5.9. Sludge removal is usually accomplished in about 115 days, between May 1 and November 15.

According to Rimkus (31), Fulton County supernatant is disposed of on 1320 acres (534.2 ha) of alfalfa-brome hay fields. Average annual quantity to dispose equals 700,000 wet t (634,900 T) with average ammonia content of 109.9 mg/L and an average total Kjeldahl nitrogen (TKN) content of 156.4 mg/L. Table 5.10 lists other data on lagoon supernatant. Dissolved oxygen (DO) measurements taken in the summer and fall in lagoons 3a and 3b indicate the surface DO ranged between 0.9 and 8.5 mg/L, while the bottom DO ranged between 0.4 and 2.6 mg/L. The lowest lagoon temperature during this period was 40.6°F (4.8°C). The lagoon surface is frozen between 45 and 60 days a year, with scum buildup experienced only during periods of new sludge input. No surface agitation equipment is used on any of the lagoons. The nearest residence to the lagoon is approximately 6000 ft (1800 m) from

**Table 5.10**
**Supernatant—Prairie Plan Reclamation Project, Greater Chicago**

| Constituent | Mean value mg/L | Range mg/L |
|---|---|---|
| BOD, total | 170 | 28–466 |
| BOD, soluble | 62 | 20–114 |
| COD, total | 951 | 325–2120 |
| COD, soluble | 695 | 328–1026 |
| TSS | 276 | 52–1041 |

*Source*: US EPA (3).

the perimeter of the installation. No information is available regarding odors or odor complaints.

### 3.4. Aerated Storage Basins

To use aerated storage basins successfully for wastewater solids, a design must meet the following criteria (3):

1. Basin contents must be sufficiently mixed to ensure uniformity of solids concentration and complete dissemination of oxygen.
2. Sufficient oxygen must be available to maintain aerobic conditions throughout the basin at maximum attainable solids concentration.
3. Liquid level variation must be sufficient to accommodate maximum storage needs under anticipated rainfall.

#### 3.4.1. Mixing Requirements

Equipment required for aerated storage basins is similar to that for aerobic digestion. Unfortunately for the designer, mixing capability for various types of static or mechanical aeration devices varies greatly. Fixed or floating turbine or propeller-type aerators are often affected by very limited side boundaries, while brush-type aerators and aspirating pumps often have almost unlimited side boundaries but rather restricted vertical mixing capabilities. Submerged static aeration devices are excellent for vertical mixing but are always limited by very confined side boundaries. The designer should rely on a performance-type specification to achieve the desired results. The equipment supplier should be given information about the configuration of the basin, its liquid level operating range, the maximum solids concentration expected, and the level of dissolved oxygen to be maintained. The designer is expected to have established the most cost-effective basin configuration based on loading, site-specific conditions, and available aeration equipment requirements. A maximum horsepower limit should be established, and the specifications should include a bonus to be added to the bid price and a penalty to be subtracted from the bid price based on the energy costs involved when the equipment meets the required performance. A guarantee should be used to ensure that the final installation will meet the performance requirement.

## 3.4.2. Oxygen Requirements

Oxygen requirements to maintain aerobic conditions within an aerobic storage basin will be considerably less than that required for aerobic digesters if the material being stored has been stabilized prior to its introduction to the basin. Minimum measurable dissolved oxygen levels of about 0.5 mg/L are quite adequate to maintain a basin free from anaerobic activity, as long as it is provided with adequate mixing. If the basin influent is not sufficiently stabilized to minimize oxygen requirements, then the aerobic storage basin must be designed for oxygen requirements similar to aerobic digesters. Oxygen transfer capabilities are similar to mixing capabilities for the various types of applicable equipment. The design should therefore include oxygen transfer requirements as part of the performance requirement indicated in the preceding section on mixing specifications.

## 3.4.3. Level Variability

Often, aerated storage basins cannot be decanted, because solids settle when the aerator is turned off, and anaerobic decomposition may also occur, resulting in odor production. Attempts at in-basin decanting without aeration and mixer shutdown will usually result in the recycling of the concentrated solids back to the liquid process. Separate continuous decanting is usually possible either by sedimentation or dissolved air flotation. Evaporation will also quite often result in significant liquid removal. Aerobic storage basins that do not have separate decanting facilities must be operated on single-phase concentration or displacement storage concepts.

The single-phase concentration concept will function as described for aerobic digesters. The displacement concept, however, will require liquid level variability and make aerated storage basin equipment installation quite complicated. Under such conditions, this equipment must be capable of maintaining adequate mixing and oxygen transfer over the complete range of liquid level variation. This requirement may cause this equipment to have varying mixing and aeration capabilities, depending on the basin depth. Variable speed drives, multispeed drives, or variation in the quantity of diffused air should be investigated. At no time should the equipment be operated under conditions that will waste energy.

## 4. FACILITIES DEDICATED TO STORAGE OF DEWATERED SLUDGE

Dedicated dewatered sludge storage of wastewater solids can include the storage of easily managed dry solids (>60% solids) or hard-to-manage wet solids (15% to 60% solids). Dry solids are usually the product of heat-drying, high temperature conversion, or air-drying processes, and can be stored by standard dry material storage techniques. Descriptions of these techniques are readily available in materials processing textbooks, and, if desired, more detailed data are available elsewhere (43, 44). The storage of wet solids is another matter, however. The successful application of common storage techniques to this normally unstable organic material is practically impossible. The most commonly accepted methods of providing dedicated storage for wet organic material involves the use of drying sludge lagoons, placing the material in some type of confined structure or placing it in unconfined stockpiles. All three methods can involve special design considerations.

## 4.1. Drying Sludge Lagoons

Drying sludge lagoons are probably the most universally practiced method of storing of wet organic sludge. Actually, the material arrives at the lagoons in a liquid form, but as described under Chicago's actual performance data, most of the storage capability is derived while the material is in a partially dewatered state. Unfortunately, many existing applications of this method of storage are being operated with sludge that has not been anaerobically stabilized prior to its discharge to the lagoons. In some cases, drying sludge lagoons are used after aerobic digestion, and in other cases they have been used as digesters with no upstream stabilization. In these instances, odors that are quite unacceptable to the surrounding community are produced. When such lagoons are considered a means of ultimate disposal, they are called "permanent lagoons." Because permanent sludge lagoons have sometimes been the source of strong odors, they are often rejected as a means to store sludge, either in the liquid or semisolid state (30).

### 4.1.1. Performance Data

Several reasonably successful drying sludge lagoon operations exist. An investigation of their actual performance, however, indicates that these lagoons are acceptable because they receive adequately stabilized anaerobically digested sludge (biosolids) and do not normally generate the odors associated with the acid phase of anaerobic stabilization.

#### 4.1.1.1. SAN JOSE, CALIFORNIA

The San Jose/Santa Clara Water Pollution Control Plant in San Jose, California, is a secondary treatment plant that operates on the Kraus modification of the activated sludge process during its seasonal canning loading period. The plant stores its anaerobically digested primary and waste activated sludge in 73 sludge lagoons on 580 acres (235 ha) of land immediately adjacent to the plant (7). The plant operated both anaerobic liquid sludge lagoons and drying sludge lagoons, with 35 either filled or more than half filled with liquid sludge and 32 containing 2 ft (0.60 m) or less of dried sludge. Three lagoons have never been used, and three have been dredged and are empty. The drying sludge lagoons were filled in layers of approximately 1 ft (0.30 m), and each layer was allowed to dry by evaporation prior to the addition of the next layer. The drying lagoon operation took place over a 2-year period, when operational limitations and odor production resulted in the return to anaerobic liquid sludge lagoon storage. Liquid sludge lagoon storage had been practiced in the prior period.

Residents living in areas near the sludge lagoons have become increasingly concerned about odors produced by the lagoons. Several complaints were registered with the Air Pollution Control Board. The area most affected is a residential community just southeast of the plant. Correlation of complaints with atmospheric conditions indicates that the greatest odor risk occurs with a northwest wind and when dry weather is followed by heavy rain. This points to the danger of rewetting the dried surface

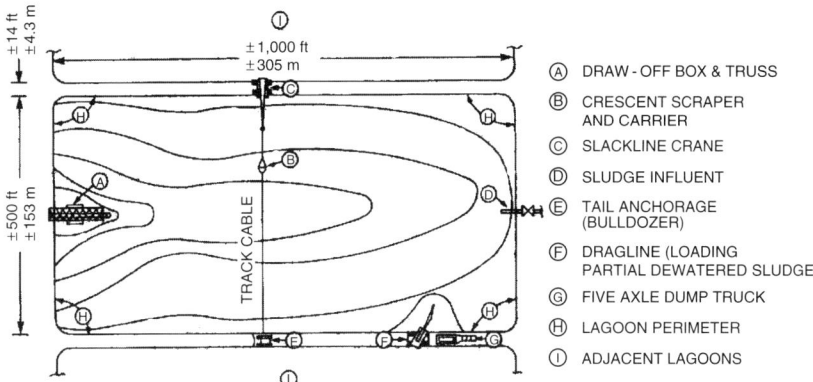

**Fig. 5.13.** Plan view of drying sludge lagoon near West-Southwest WWTP, Chicago, Illinois. *Source*: US EPA (3).

layers and anaerobically stabilized material and confirms that this can create strong odors.

#### 4.1.1.2. CHICAGO

The MSDGC operates 30 drying sludge lagoons, each with an average storage capacity of 200,000 yd$^3$ (153,000 m$^3$) and a storage depth of 16 ft (4.9 m) (31). Figure 5.13 shows a plan view of a typical lagoon. Anaerobically digested sludge is pumped to the MSDGC lagoons at a solids content of about 4%. Volatile content of this material is approximately 57%. Sludge is usually applied to each available lagoon in 6-in (152-mm) layers in rotation. Rotations are repeated.

Supernatant appears on the lagoon surface approximately 5 to 7 days after each fresh sludge application. It is then drained from the surface and returned to the West-Southwest Sewage Treatment Works by removing one or more stop logs from the draw-off box. Once the supernatant is decanted, the 8% to 10% solids sludge is further concentrated by evaporation. Evaporation tapers off, however, as an aerobic sludge crust develops. Supply sludge concentration (4% solids) is beneficial, as it covers the entire lagoon surface with only a slight gradient from the point of application. Any higher concentration would inhibit this coverage and reduce the evaporative surface area per unit volume. Lagoons that have been filled to capacity by this method have an average solids content of 18% to 22% by weight. Volatile solids content of this material is in the range of 35% to 40%, indicating that the lagoons are producing about a 34% volatile solids reduction.

Once the drying sludge lagoons are filled, they are taken out of service and preconditioned to provide an improved drainage gradient. For this purpose, the sludge is excavated from the area adjacent to the draw-off box and the slope within the lagoon is allowed to stabilize to the point at which the area remains reasonably free of solids. Excavation is by pump with nearby mixers and additional water, if necessary, to ensure sludge fluidity. Figure 5.14 illustrates a cross section of this area after preconditioning is complete. When the sludge has stabilized, the lagoon is left dormant through the

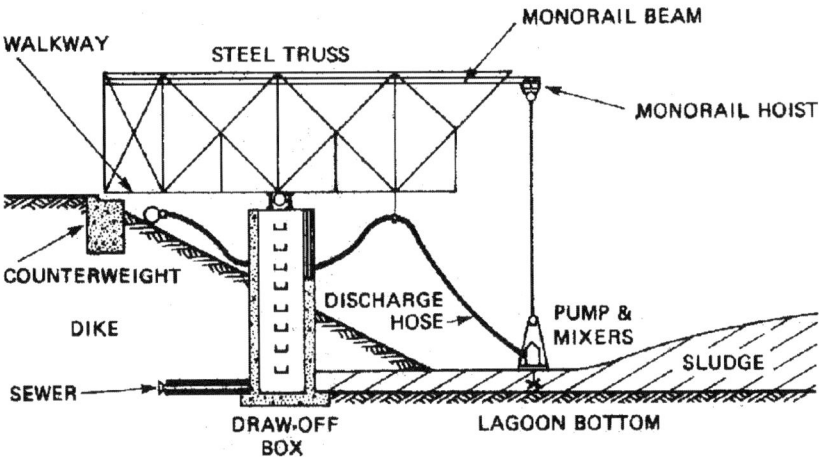

**Fig. 5.14.** Cross section of draw-off box area drying sludge lagoon near West-Southwest WWTP, Chicago, Illinois. *Source*: US EPA (3).

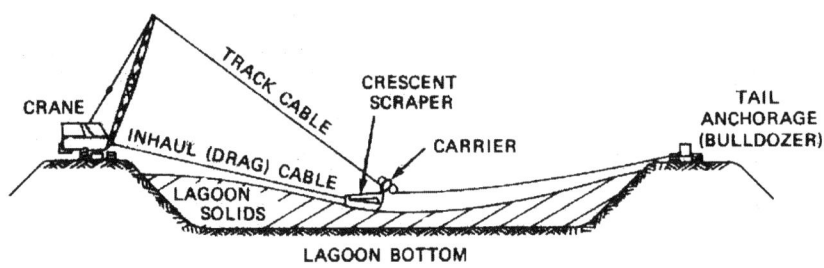

**Fig. 5.15.** Cross section of drying sludge lagoon with slackline cable near West-Southwest WWTP, Chicago, Illinois. *Source*: US EPA (3).

following winter and early spring. Trapped water and rainfall runoff are drained by gravity to the draw-off structure.

Once relatively dry weather returns, a slack-line cable system is utilized with a dragline crane to further condition the sludge. The slack-line system, which is shown in Figure 5.15, is used to improve the lagoon surface drainage and to scrape as much of the dried crust as possible to the side of the lagoon. This system provides the following four operational benefits (3):

1. Drier sludge is scraped to the side, where it can be reached by portable drag-line or clamshell and loaded onto dump trucks.
2. Piling sludge along sides improves lagoon drainage pattern and profile.
3. Removal of crust exposes wetter sludge to atmosphere for optimum evaporation.
4. Some of dried crust mixes with wetter material during removal and increases the wet sludge solids content.

Once the sludge crust is scraped to the side of the lagoon, it is removed by portable drag-line or clamshell, loaded onto watertight five-axle dump trucks, and delivered to the general public for reuse. This lagoon sludge, at its time of delivery, usually has an average solids content of 30% to 35% by weight. Tree nurseries, sod farms, landfills, and stripped land are among the major users of this material. The MSDGC disposes of the sludge production that exceeds 100,000 dry t (90,700 T) at a 2007 cost of USD 40.67/dry t (USD 44.84/T). Preconditioning costs are approximately USD 7.00/dry t (USD 7.58/T), which makes the cost for the whole operation USD 47.67 dry t (USD 52.42/T). Preconditioning is accomplished by MSDGC manpower and equipment, and the services of the slack-line, drag-line, and trucks are contracted out. The overall operation requires little capital investment, minimal lead time, and limited effort. Natural processes are optimized and odors minimized. The level of odor involved has not been qualified.

## 4.2. Confined Hoppers or Bins

A designer is often tempted to take advantage of the volumetric reduction in material provided by the dewatering process and lay out the sludge disposal system based on short- and long-term storage (3 weeks to >6 months) of the dewatered product. If the product is too wet (<30% solids), several problems may arise with this type of storage. These problems include continuing decomposition, liquefaction, and concentration and consolidation. Although each may have its own result, all three problems are interrelated and combine to limit the use of this type of storage to equalization storage and then only if special attention is given to controlling the difficulties. A brief description of some of these difficulties is given in the following subsections.

### 4.2.1. Continuing Decomposition

Unless it is stabilized to nonreactive levels (<50% by weight), the biodegradable volatile organic material of wastewater solids will continue to decompose if the moisture content is too high (solids content <30%). This decomposition will reduce organic material and generate gaseous by-products. Depending on the stage and sometimes the type of stabilization employed prior to dewatering, the method of conditioning for dewatering, and the dewatering method itself, gaseous by-products may or may not be odorous. For example, a biodegradable volatile content of <50% would result in strong odors; aerobically stabilized dewatered sludge would be more subject to strong odors than anaerobically stabilized dewatered sludge; polymer-conditioned dewatered sludge would be more subject to strong odors than lime and ferric conditioned dewatered sludge; and centrifuged dewatered sludge would be more subject to strong odors than vacuum filtered dewatered sludge.

Enclosed structures are often used in this type of storage to ensure odor-free operation. Such structures may be extremely hazardous if the designer fails to recognize the potentially explosive nature of some of these gaseous by-products and ensure that they are never mixed with air within the combustible range. If such protection involves the replacement of the displaced volume, it may become the limiting feature of the storage structure's ability to manage the sludge.

One solution to this problem is to treat the volume above the solids as part of the digester gas storage system. However, this is practical only if the overall solids treatment system uses anaerobic digestion for stabilization and the gas collection system has sufficient capacity to fill the void created by storage discharge within the required period of time. Major problems of such a system are the sealing of sludge supply and discharge and ensuring accessibility for maintenance.

To eliminate the discharge and supply problems and ensure convenient access to the storage loading equipment, the enclosed area of the storage structures should be sufficiently ventilated. The area must be ventilated with about 20 to 30 air changes an hour. Air movement should be felt by the operators who work in the area. To ensure ventilation of all areas, regardless of any continuously or intermittently operating openings, both supply and exhaust air should be managed by powered fans. All exhaust air should pass through an odor removal system. The quantity of exhaust ventilation air should be slightly greater than the quantity of supply ventilation air to ensure a negative pressure within the area and minimize leakage that might bypass the odor removal system. The atmosphere of enclosed areas should be monitored with hydrocarbon detectors to provide ample warning if the gas release begins to develop dangerous mixtures of methane and air.

*4.2.2. Liquefaction*

When the reduction of putrescible organic material is carried out within a confined structure used for short- or long-range storage (3 to 4 weeks or longer), the liquefaction of dewatered solids occurs. Liquefaction is negligible when the storage is limited to equalization (3 to 4 days). The designer must be aware of the effects of this liquefaction and realize that as the liquid or moisture content of the sludge increases, the difficulties of transport also increase. An example of this liquefaction can be seen in Example 3 at the end of this chapter.

*4.2.3. Concentration and Consolidation*

The material handling properties of the dewatered sludge entering the storage facilities often do not resemble those of the material discharged from the same facility. The method of controlling the discharge must be flexible enough to adapt to these changes in properties at any time. A live bottom discharge for variable positive control and backup isolating valves for positive shutoff if the live bottom equipment fails or the material starts to run like water are mandatory when the volume of storage greatly exceeds the volumetric capacity of the transport system receiving the discharge. As long as the storage structure's volume does not exceed the capacity of the transport system receiving the discharge, and that transport system is of the bulk handling type (for example, truck, rail car, or barge), the discharge control can be a simple open-close valve. Water-collecting, tracked, hopper valves with remote motor or air cylinder operation can be used for this control. Facilities whose storage volume exceeds the discharge transport system capacity or whose transport system is of the continuous rate type (for example, conveyor belts, screw conveyors, and pipelines) must be provided with a discharge system capable of infinite variability under all degrees of moisture content or concentration. Such systems must be provided with remote controls that are capable of detecting overloads prior to their overwhelming the transport system. The controls must be capable of automatically closing the discharge control system's backup and open-close isolating valve. Sonic level detectors and capacitance probes can be used for this function.

The use of polymers to condition the sludge prior to dewatering can have a major effect on its ability to be stored conveniently in the dewatered state. Hansen et al. (45) report that high polymer doses used experimentally (testing a belt filter press) at the Los

Angeles County plant created a dewatered sludge that was quite viscous. This material tended to act like glue and was extremely difficult to remove from conveyors, especially at transfer points and the head point above the hoppers. The material could be stored, but required a positive type of unloading system at the storage discharge to ensure that the lumps were pushed onto the discharge conveyor.

When exceptionally dry dewatered sludge (greater than 30% solids) is stored, bridging can be a very difficult problem. None of the facilities investigated had successfully solved this problem. It is suggested that any large system that anticipates storing dewatered sludge much dryer than 30% solids should set up a test facility to develop a reliable system for overcoming this difficulty.

*4.2.4. Performance Data*

Probably one of the most successful confined bin dewatered sludge storage facilities is located at the County Sanitation Districts of Los Angeles County Joint Water Pollution Control Plant (JWPCP) in Carson, California. It provides advanced primary wastewater treatment for about 350 MGD ($15.3\,\text{m}^3/\text{s}$) of wastewater. It also receives the sludge from five tertiary treatment plants that employ activated sludge followed by multimedia filtration and have a combined capacity of 120 MGD ($5.2\,\text{m}^3/\text{s}$). Sludge from all six plants is treated at the JWPCP using the anaerobic stabilization (digestion), dewatering (centrifugation), and composting (windrow) processes.

The US EPA (3) reported the centrifuges were producing about 400 to 600 wet t (360 to 540 T) of dewatered digested sludge each day with a 25% average solids content. Twelve storage bins, each capable of holding 550 wet t (500 T) of dewatered sludge, are provided to equalize 24-hour-a-day centrifuge production with 10-hour-a-day windrow construction. The storage bins also provide the 5-day storage needed to ensure continuous dewatering when both the composting and backup sanitary fill disposal options are unavailable due to excessive rainfall. The facilities have been in service for many years, and according to Hansen et al. (45), the maximum period of disposal unavailability has not exceeded 2 days to date, although there have been times when all 12 of the bins have been filled with dewatered sludge. An isometric sketch of the JWPCP storage and truck loading station is shown in Figure 5.16.

The upper and end areas of the JWPCP storage facilities are completely enclosed with metal cladding and equipped with positive supply and exhaust ventilation. The potentially most odorous ventilation air is passed through caustic wet scrubbers prior to discharge to the atmosphere. There is some indication that additional activated carbon scrubbing may be required to ensure complete removal of all odors. Although the supply and discharge areas of the storage buildings are continuously monitored for explosive conditions, Hansen et al. (45) report that little methane gas seems to be generated as long as the dewatered sludge solids content is greater than 18% and the sludge is not left in storage more than a few days. Bubbles that can be observed in the standing water on top of the stored sludge attest to the fact that decomposition is continuing in the bins.

Each storage bin is fabricated of steel, is 30 ft in diameter, and tapers at the bottom to a 5-ft-square discharge. The taper is at 30° off the vertical. Hansen et al. (45) report that this taper seems to eliminate bridging, except during the storage of extremely dry

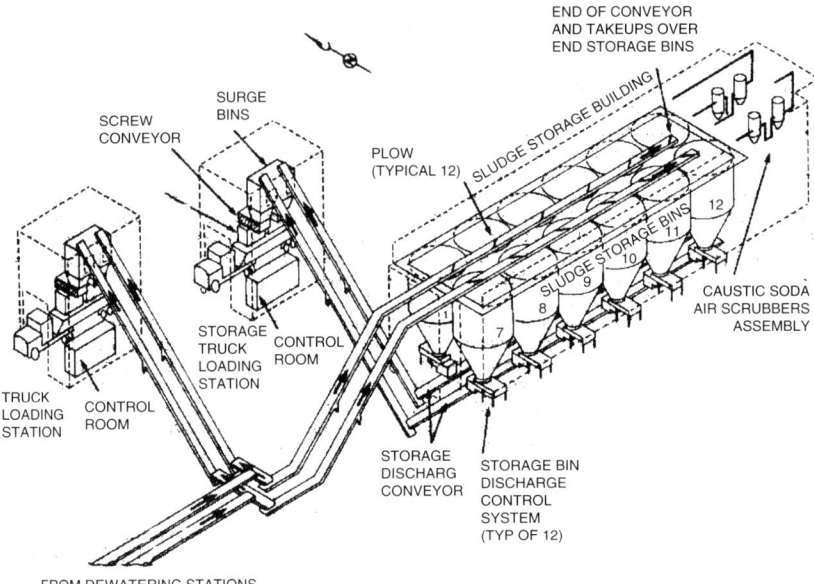

**Fig. 5.16.** Isometric of sludge storage and truck loading station, Joint Water Pollution Control Plant, Los Angeles County, California. *Source*: US EPA (3).

(greater than 30% solids) sludge. The 5-ft-square (1.5-m-square) discharge is equipped with five 12-in-diameter (30.5-cm) continuous screw conveyors (live-bottom system) that can be operated in any combination or number to positively control the stored sludge discharge to the discharge conveyor belt. Normal operation requires only the three middle screw conveyors to be in service. A cylinder-type plug valve with five 10-in-long (25.4-cm) by 8-in-wide (20.3-cm) openings has been provided to ensure positive isolation between the live-bottom system and the discharge conveyor. The plug valve is fabricated of 0.406-in (10.3-mm) steel wall, 12-in (30.5-cm) outside diameter steel pipe, approximately 5 ft (1.5 m) long, and is actuated by a pneumatic cylinder that positively rotates the valve 90° from a full open to a tight shutoff position. An isolating bull gate that can be hydraulically forced between the bottom of the storage bin and the top of the live-bottom assembly is also provided. It can be used to cut off sludge discharge should the live-bottom assembly fail with a load in the hopper. It has been suggested that a hydraulically operated gate valve or knife-gate valve could also be used to provide this isolation. An isometric view of this discharge control system is shown on Figure 5.17.

Hansen et al. (45) report that the storage facilities were built at a cost of USD 11 million (2007 basis). Sludge variability during startup created several problems that have been successfully solved by simplifying the supply to the storage tanks by equipping each with a plow and moving the end of the supply belts over the end hoppers, by providing the live-bottom discharge system with a positive discharge isolation valve, and by increasing the ventilation level in the supply and storage areas to achieve the "breeze" atmosphere necessary to satisfy operator safety concerns.

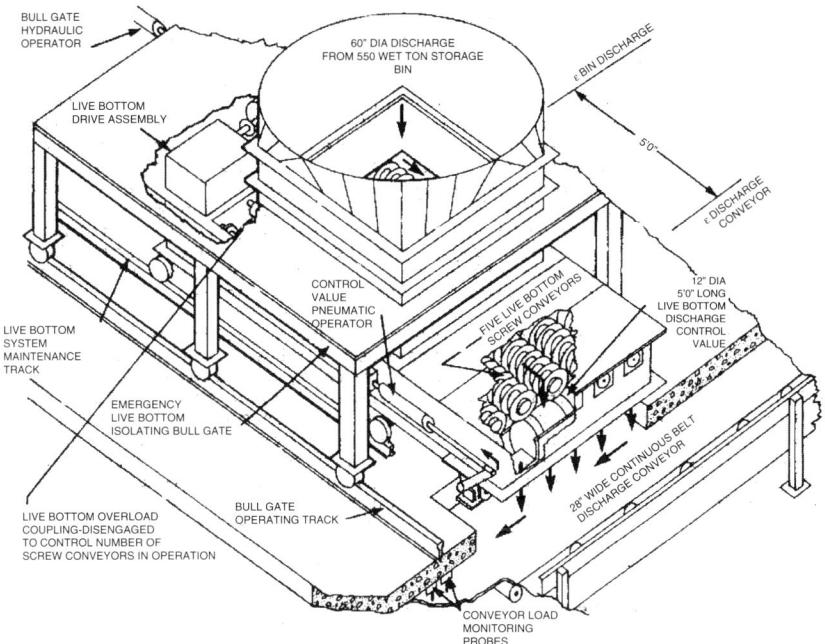

**Fig. 5.17.** Storage bin discharge control system, Joint Water Pollution Control Plant, Los Angeles County, California. *Source*: US EPA (3).

### 4.3. Unconfined Stockpiles

Unconfined stockpiles are a major method of providing long-term storage for dewatered sludge (46). This method is used primarily for the storage of air-dried, anaerobic or aerobic stabilized sludge at thousands of small plants across the country. Probably the largest storage and weathering installation is operated by the MSDGC at their West-Southwest Sewage Treatment Works (WSW-STW). All of the air-dried Imhoff sludge at WSW-STW is stored and aged up to several years on between 50 and 100 acres (20 and 40 ha) of land, and then the biosolids product is made available for delivery to the public as "Nu-Earth" (31). The air-dried material weathers to less than 50% moisture after 1 to 2 years of aging.

Unconfined stockpiles of mechanically dewatered stabilized sludge, which has less than 25% solids, usually are destroyed (lose all semblance of stability) when exposed to extensive rainfall. While it is possible to maintain such a stockpile for equalizing or short-term storage, especially in very dry climates like the southwest, long-term storage is usually quite impossible. Stabilized sludges with a high chemical content (greater than 40% lime plus some ferric) or a very low organic content (less than 50% volatile solids) sometimes prove to be exceptions. Highly stabilized lagooned sludges can also be one of these exceptions. Such open stockpiles usually quickly absorb atmospheric moisture and rapidly deteriorate in climates with intense or frequent rainfall.

Covered stockpiles are often used in those areas where rainfall is intense or frequent to ensure the dewatered biosolids integrity during periods of equalizing storage. Such stockpiling is usually limited because of the expense of developing covered areas of sufficient size to provide adequate storage area and equipment accessibility. The North Shore Sanitary District (NSSD) (47), north of Chicago, disposes of its anaerobically stabilized (digested) dewatered sludge in deep trenches on a 300-acre (121-ha) site. On 10 to 20 days a year, the NSSD disposal operation is abandoned due to wet conditions, and the dewatered sludge is stored in a covered and enclosed building for disposal within a few days. The building is enclosed to maintain odor control. The district also frequently liberally sprinkles the dewatered sludge with lime during transport and storage to maintain odor control.

Unfortunately, no quantitative work has been published regarding the odor risk of stockpiling dewatered sludge. Drying lagoons, like those operated at San Jose, California, do create malodorous conditions in surrounding urban areas during or immediately after being wetted by rainfall. Work in Sacramento, however, indicates that odors are generated cumulatively in direct relationship to the area covered by the odor producing sludge (6). Good housekeeping around such stockpiles is mandatory to ensure proper rodent control.

## 5. FIELD STORAGE OF BIOSOLIDS

Successful biosolids land application programs have provisions to deal with daily biosolids production in the event that biosolids cannot be land applied immediately. This contingency planning generally includes storage as well as other backup options, such as landfill disposal, incineration, or alternative treatment and use, including composting, heat drying, and advanced alkaline stabilization.

The focus of this section is on management practices for field storage of biosolids prior to land application, as distinguished from land application and spreading. It stresses recommended management practices for three critical control points: the wastewater treatment facility, the transportation system, and the field storage site. The term *critical control point* means a location, event, or process point at which specific monitoring and responsive management practices should be applied. If these points are controlled, the objectives and goals of a responsible and community-friendly practice can be achieved. Equally important is the continuing need for partnership and good communication among biosolids generators, storage site managers, and land appliers.

The term *field storage* refers to temporary or seasonal storage. Storage operations involve an area of land or facilities constructed to hold biosolids until material is land applied on designated and approved sites. More permanently constructed storage facilities can involve state or locally permitted areas of land or facilities used to store biosolids. The permissible time limits for field storage vary by state and local jurisdiction. They are usually located at or near the land application site, and are managed so that biosolids come and go on a relatively short cycle, based on weather conditions, crop rotations, and land or equipment availability. Alternatively, storage sites are used to accumulate enough material to conduct an efficient spreading operation.

It is very clear to all biosolids generators, transporters, storers, land appliers, and local officials that malodors are the greatest reason for public concern about storage sites. Much of the US EPA-USDA guidance document (4) seeks to provide information and strategies useful in minimizing odor problems.

## 5.1. Management of Storage

This subsection explains the general principles underlying the management of biosolids in storage situations. Biosolids managers should keep these concepts in mind as they assess their storage needs and options and develop a management plan suited to their unique situation.

### 5.1.1. Critical Control Points (Key Management Areas)

Even with a wide variety of biosolids and the numerous types of field situations that are encountered throughout the U.S., all field storage operations can be broken down into three areas that are critical for good management (4):

- Critical control point 1: the biosolids generating facility (WWTP)
- Critical control point 2: transportation process
- Critical control point 3: field storage site

Activities in each of these areas are critical to the overall success of biosolids storage operations. For instance, the level of treatment and posttreatment handling at the generating facility may affect the odor characteristics once biosolids reach the field site.

This subsection provides detailed descriptions of the practices recommended for management of these areas as well as explanations of their importance relative to odors, water quality, pathogens, and community acceptance. Biosolids managers are encouraged to carefully analyze their own particular situations and select the most feasible combination of practices for their unique situation (48).

Table 5.11 highlights the main issues and some of the self-monitoring activities and control options involved in each of these management areas. Complete descriptions of these practices are provided in subsequent subsections.

### 5.1.2. Variables Related to Intensity of Management

There are five variables that affect the level or intensity of management required for successful field storage of biosolids (4):

1. *Stability of biosolids*: Material that is less well stabilized generally has a greater potential to generate unacceptable levels of odorous compounds.
2. *Water content of biosolids*: Liquid and some semisolid material require pumping equipment and constructed storage facilities.
3. *Length of storage period*: Longer storage periods increase the potential for exposure to wet or hot weather and a resumption of microbial decomposition leading to the generation of odorous compounds.
4. *Volume of stored material*: Management requirements in terms of site design, operation, and the potential scale of odor or water quality impacts may increase with the volume of material stored.
5. *Climate and weather conditions*: Warm humid weather or wet conditions generally increase management requirements as compared to storage during dry or cold conditions.

These variables are interrelated, and therefore exceptions to particular points may occur when mitigated by other variables. For instance, a large volume of a well-stabilized

Table 5.11
**Overview of management control points for field stored materials**

| Issues | Self-monitoring checklist | Control options |
|---|---|---|
| 1. Biosolids generating facility<br>• Odors and aesthetics<br>• Consistency of biosolids<br>• Biosolids treatment | • Assess biosolids to determine:<br>  503 treatment criteria for pathogens and VAR<br>  Degree of stability and odor potential includes factors such as volatile solids content; lime, polymer and iron usage, and pH<br>  Physical consistency<br>  Ratio of primary to secondary<br>  Cleanliness of equipment<br>  Time of retention after treatment | • Generator, storer, and land applier communicate about status of biosolids treatment or problems<br>• Reduce posttreatment retention<br>• Have options to divert unacceptable loads<br>• Reevaluate treatment and handling practices to address chronic issues<br>• Provide further treatment<br>• Provide vehicle cleaning station |
| 2. Transportation<br>• Odors and aesthetics<br>• Traffic and safety | • Proper equipment in compliance with state and federal transportation regulations<br>• Regular inspections and maintenance of vehicles and equipment<br>• Suitable haul routes<br>• Vehicles and equipment kept clean | • Train drivers<br>• Plan/inspect haul routes, minimize time in transport<br>• Emergency spill plan and supplies in place<br>• Maintain and clean trucks and equipment regularly |
| 3. Field storage site<br>• Odors and aesthetics<br>• Water quality and environmental protection<br>• Safety and health protection | • Proper site location and suitability<br>• Proper design of field storage or constructed facility<br>  ○ Run-on and run-off controls<br>  ○ Accumulated water control<br>  ○ Buffers<br>• Biosolids quality vs. length and amount in storage<br>• Operations and maintenance plan<br>• Odor prevention and mitigation plan<br>• Spill control and response plan<br>• Safety plan | • Regular self-inspections of site and operations<br>• Consistent implementation of management plans<br>• Self-monitoring of biosolids quality and condition<br>• Revision of management plans when necessary<br>  ○ Change amount or length of storage<br>  ○ Implement odor control and mitigation measures<br>  ○ Implement additional structural or site management practices<br>• Remove stored biosolids when atmospheric conditions are conducive to low odor impacts on neighbors |

*Source:* US EPA (4).

biosolids may be less management intensive in terms of preventing nuisance odors than the storage of a small volume of a less well-stabilized material.

*5.1.3. Need for Partnerships*

It is recognized by experienced biosolids management teams that partnership and good communication between the biosolids generator and the biosolids manager responsible for storage and land application is essential to optimizing the management of biosolids destined for storage. Successful storage programs require coordination of management activities at the generating facility, in transit, and at the storage site.

Likewise, good communication links are necessary between the biosolids manager and the biosolids users, local governments, and citizens of communities where biosolids storage activities are located.

The absence of such partnerships has often resulted in odors or other problems with subsequent unfavorable community acceptance, political, or economic consequences. Land appliers overwhelmed with community-relations problems may be forced to cease land application and seek alternative management options. These are typically more costly to consumers than field storage and land application, or result in lost economic and environmental benefits to farmers, landowners, and diversion of biosolids to non-beneficial uses, such as land filling or incineration.

*5.2. Odors*

Malodors are the single most important cause of public dissatisfaction with biosolids or other organics recycling and utilization projects. Thus, odor management is a high priority. Experience and practice have demonstrated that biosolids and other organic by-products, such as animal manure, landscape trimmings, and food processing residuals, can be handled and processed without release of excessive malodorous compounds. However, if any of these materials, including biosolids, are poorly managed, then objectionable odors may develop during storage, and public acceptance of such a project will erode (49–51).

The malodorous compounds (odorants) associated with biosolids, manures, and other organic materials are the volatile emissions generated from the chemical and microbial decomposition of organic nutrients. When inhaled, these odorants interact with the odor sensing apparatus (olfactory system) and the person perceives odor.

Individual sensitivity to the quality and intensity of an odorant can vary significantly, and this variability accounts for the difference in sensory and physical responses experienced by individuals who inhale the same amounts and types of compounds (52). This distinction between *odor*, which is a sensation, and *odorant*, which is a volatile chemical compound, is important for everyone who deals with the odor issue to recognize. When odorants are emitted into the air, individuals may or may not perceive an odor. With biosolids, three conditions are necessary to create malodorous conditions (4):

1. Emission: Presence of an odorous volatile chemical (odorant)
2. Transport: Topographic and atmospheric conditions conducive to transport of the odorant with minimal dilution
3. Perception: People are present and they perceive odor

When people perceive what they regard as unacceptable amounts or types of odor, odorous emissions can become a problem.

*5.2.1. Primary Biosolids Odorants*

The odorous compounds generated, and most often detectable, at significant levels during biosolids treatment, storage, and use are ammonia, amines, and reduced sulfur-containing compounds. Amines can be produced in easily detectable quantities during high-temperature processes. Amines include methylamine, ethylamine, trimethylamine, and diethylamine. Amines often accompany ammonia emissions, and if chlorine is used chloramines may be released. The sulfur compounds include compounds such as hydrogen sulfide, dimethyl sulfide, dimethyl disulfide, and methyl mercaptan. The potential for these compounds to be annoying is based in part on their individual and combined quantity, intensity, pervasiveness, and character (53).

Amines and reduced-sulfur compounds may be detectable and perceived at greater distances from a storage facility than ammonia because they are more persistent (pervasive), intense, and have very low odor detection thresholds. Although ammonia is usually the primary odor associated with limed or alkaline stabilized biosolids, it has an intense odor that can often mask other odors, such as reduced-sulfur compounds. However, because the detection threshold for ammonia is much greater than that of many of the reduced sulfur compounds, the odors of reduced-sulfur or amine compounds are more likely than that of ammonia to be detected at distances from the site where ammonia is no longer above its odor threshold concentration.

*5.2.2. Odor Management: A Partnership Effort*

There is no doubt that untreated wastewater solids have inherently undesirable odor qualities. However, many current treatment processes have the capacity to produce biosolids that are minimally odorous. Despite this, occasional malodorous batches can occur, and thus biosolids generators, storers, and land appliers should make provisions to handle these appropriately. These provisions rely on close communication and working linkages among the biosolids management partners (i.e., generator, transporter, storer, and applier). Good management of each process technology and a cooperative effort among the biosolids management partners to ensure proper transportation, handling, and storage of the materials can minimize the potential for unacceptable odor concentrations at storage sites. Minimizing odor during storage entails the following steps (4):

1. Stabilize biosolids at the WWTP as much as possible
2. Avoid use of polymers that lead to malodor
3. Maintain proper pH during treatment
4. Meet Part 503 vector attraction reduction
5. Locate storage at remote sites
6. Minimize duration of storage
7. Assess meteorological conditions before loading and unloading
8. Ensure good housekeeping

## 5.2.3. Factors Affecting Ultimate Odor Potential at Critical Control Point 1: The Wastewater Treatment Plant

When an odor situation cannot be averted, management of the emissions and quick response through mitigation practices are required to avoid creating nuisance odor situations. At the WWTP, which is critical control point 1, this coordination includes the following (4, 54, 55):

1. Assessing the stability of the biosolids before they leave the WWTP.
2. Having contingency plans to provide remedial treatment, or diversion of unacceptably odorous material to suitable land application or disposal sites.
3. Notifying the storer and land applier of any changes in mixing (primary or secondary solids), polymer or other additives, pH, moisture content, or stability.

Decisions relative to odor control are a series of trade-offs involving higher degrees of treatment at the WWTP versus the intensity of management at the off-site storage locations. Ensuring that the odor of biosolids leaving the WWTP is minimized is a key consideration, since it is more difficult to treat an odor problem that originated at the WWTP once the biosolids are placed at the storage site. In all cases, the temporary measures invoked to deal with unexpected and unanticipated events that lead to odors must be considered only as such. Persistent problems require an examination of the treatment and handling processes to develop a better management approach.

### 5.2.3.1. STABILITY

The success of the various solids treatment technologies with regard to malodor reduction depends largely on the degree of stabilization achieved in the biosolids before it leaves the treatment facility and the preservation of stability until used. Wastewater treatment technologies differ in their capacity to stabilize biosolids.

The potential for odorous emissions depends partly on the extent to which organic matter and nutrients are present in forms that microbes readily use. Stabilization processes may either (1) decrease the level of volatile organic compounds and the availability of nutrients to reduce the potential for microbial generation of odors, or (2) change the physical or chemical characteristics of the biosolids in a way that inhibits microbial growth (56). Table 5.12 lists seven commonly used stabilization or processing methods. Odor issues associated with each method or process are shown along with appropriate corresponding prevention or remediation approaches.

The types of treatment and stabilization processes used at a WWTP are primary factors influencing the type and level of odors that may be potentially generated by a particular biosolids. Other factors at the wastewater treatment plant that may affect the odor potential of biosolids include the following (4):

1. Periodic changes in influent characteristics (e.g., fish wastes, textile wastes, and other wastewaters with high odor characteristics)
2. Type of polymer used and its susceptibility to decomposition and release of intense and pervasive odorants such as amines when biosolids are heated or treated with strong alkaline materials
3. Blending of primary and secondary biosolids, which may create anaerobic conditions or stimulate a resumption of microbial decomposition

**Table 5.12**
**Prevention and management of odorous emissions associated with biosolids stabilization or processing methods**

| Stabilization and processing methods | Potential causes of odorous emissions | Long-term potential solutions | Short-term temporary solutions |
|---|---|---|---|
| Anaerobic digestion | "Sour", overloaded or thermophilic digester; volatilization of fatty acids and sulfur-compounds | Optimize digester; don't overload | Apply topical lime to stored biosolids |
| Aerobic digestion | Low solids retention time; high organic loading, poor aeration | Increase retention time and aeration; lower organic load | |
| Drying beds | Incomplete digestion of biosolids being dried | Optimize digestion | |
| Compost | Poor mixing of bulking agent; poor aeration; improperly operating biofilters | Mix better; adjust mix ratio and aeration rate; improve biofilter function | Aerate more effectively; remix; re-compost |
| Alkaline stabilization | Addition of insufficient alkaline material so pH drops below 9, microbial decomposition may occur with generation of odorous compounds; check compatibility of polymer with high pH and other additives, e.g., $FeCl_3$ | Increase pH; provide finer mesh grade of alkaline material and mix better to avoid inadequate contact with biosolids | Check pH; apply topical lime |
| Thermal conditioning and drying | High temperature volatilization of fatty acids and sulfur compounds | Use secondary treatment biosolids; primary solids are less stable and more odorous when heated | Apply topical lime if biosolids are still liquid |

*Source*: US EPA (4).

4. Completeness of blending and mixing, and quality of products used for stabilization (i.e., type of lime and granule size)
5. Effectiveness and consistency of vector attraction reduction method
6. Handling, storage time, and storage method when stabilized biosolids are held at the WWTP prior to transport (e.g., anaerobic conditions developing in enclosed holding tanks when material is held for several days during hot weather)

#### 5.2.3.2. Vector Attraction Reduction

Stabilization treatment may include processes at the WWTP to reduce the attraction of vectors to biosolids as outlined in the Part 503 rule (see Chapter 6). The effectiveness and consistency of these treatments may also help to minimize odor potential. Odor is typically less of a problem for biosolids that fully meet one of the first eight Part 503 vector attraction reduction (VAR) options (57). However, sometimes it is necessary to store materials that will meet VAR by options 9 or 10 (injection or soil incorporation). In such cases, increased management intensity (e.g., storage for short periods of time, storage during cold weather, storage at remote locations, etc.) and self-monitoring for unacceptable odor levels may be needed to prevent nuisance odor conditions.

### 5.2.4. Factors Affecting Ultimate Odor Potential at Critical Control Point 2: The Transportation Process

The process of transporting biosolids from the generating facility to the field storage site may impede traffic, be unsightly, and can potentially emit nuisance odors into the community. The transportation process must be properly managed as to minimize these problems, including the public's exposure to biosolids odors. One way to reduce public exposure to odors is to choose a hauling route that avoids densely populated residential areas. The fewer residences located along a hauling route, the less likely the general public will be annoyed by the traffic and biosolids odors. Making sure that the trucks used to haul biosolids are clean and well maintained is another effective way to keep road surfaces clean and control odors during biosolids transport. Trucks should be cleaned before leaving the generating facility and after the biosolids have been deposited on the field storage site. These steps are important because odor concerns are exacerbated by increased road congestion, and by biosolids adhering to trucks and roadways.

### 5.2.5. Factors Affecting Ultimate Odor Potential at Critical Control Point 3: The Field Storage Site

In most cases, biosolids produced at WWTPs with well-operated stabilization processes can be stored off-site without creating odor nuisances. However, if certain conditions occur while material is in storage, the potential for odorous emissions (sulfur- or amine-containing compounds or ammonia) will increase. Specific storage site conditions that contribute to generation of odorants include the following (4, 58):

1. Meteorological conditions
2. Distance to sensitive receptors (i.e., housing developments)
3. pH drops below 9 in lime-stabilized biosolids
4. Anaerobic or deficient oxygen conditions within the biosolids
5. Storage of primary biosolids with waste-activated (digested) biosolids
6. Rewetting of dried material
7. Ponded water in contact with stored biosolids

8. Prolonged storage of inadequately stabilized biosolids
9. Inadequate handling methods
10. Deficient housekeeping and spill control

Meteorological conditions such as wind speed and direction, cloud conditions, relative humidity, and temperature can change with the season, day to day, and even with the time of day. Warm temperatures and high humidity increase the potential for odor nuisances, while cold, dry conditions reduce the potential for nuisance complaints.

Most odors from a biosolids storage site are area-source, rather than point-source, ground level emissions. Under moderate atmospheric stability (e.g., partly sunny, wind speeds 8 to 12 mph, moderate turbulence) and on a flat terrain area, source odorants undergo fairly rapid dilution as the distance from the source increases. As such, concentrations of odorants will likely not be objectionable to neighbors if the biosolids are reasonably well stabilized. Conversely, pervasive odorants from poorly stabilized batches can be detected at considerable distances from the source. Rough terrain, valleys, and other topographical features can increase the complexity of airflow patterns. Odor dispersion analysis can help site managers schedule operations to avoid high odor concentrations from developing at sensitive downwind locations.

Odorants emitted from ground-level sources remain most concentrated during periods of high atmospheric stability, such as occur with air temperature inversions and low wind speeds at night and very early morning. This means that odor complaints may be higher during nonbusiness hours. Dispersion is enhanced once the sun has warmed the soil surface. For permanent constructed facilities, a basic wind dispersion analysis of the site, including seasonal and annual prevailing wind direction, and typical meteorological conditions for the area will help site operators plan activities so as to minimize odorous emission impacts downwind (4, 59).

## 5.3. Water Quality

This subsection describes the types of water quality impacts potentially attributable to specific nutrients and pollutants in stored biosolids and other organic materials. In addition, key concepts in construction and management of storage systems that are known to work well in preventing water quality impacts from biosolids storage are discussed and related specifically to recommended storage management practices.

Measurements of the following constituents of organic and inorganic materials stored on or applied to soil are customarily made to assess their potential impact on water quality. Table 5.13 summarizes these components relative to biosolids storage and their potential impacts on water quality:

1. Nutrients
2. Organic matter
3. Pathogens
4. Metals

Assessment of the presence of constituents such as nutrients, organic matter, pathogens, or metals is the first step in developing effective water quality protection practices for stored materials. The second step is to examine the possible ways of transport. Constituents can have an impact on water quality only if significant amounts of

**Table 5.13**
**Potential ground and surface water quality impacts resulting from improper management of water at storage sites**

| Biosolids constituent | Potential water quality impacts | Behavior, transport mechanism, and mitigating factors |
|---|---|---|
| Nitrogen | Eutrophication; human/livestock/poultry health effects | Nitrate-N, nitrite-N, and ammonium-N are water soluble and can move in runoff or leachate |
| Phosphorus | Eutrophication | Predominately particulate-bound transported by erosive surface runoff |
| Organic matter | Depletes oxygen levels in water | Soluble and particulate-bound movement of organic matter in surface runoff |
| Particulates | Siltation or turbidity; carrier for other pollutants | Mass transport in surface runoff |
| Pathogens | Transmission of viable disease-causing bacteria, viruses or parasites | Insignificant levels in Class A biosolids, potentially present in Class B materials; physical transport in sediment, runoff, and leachate from Class B biosolids is possible |
| Regulated metals | Toxic effects | Not very water soluble; reduced through pretreatment programs and Part 503 limits |
| Toxic organic chemicals | Toxic effects | Reduced through industrial pretreatment programs and WWTP processes |

WWTP, wastewater treatment plant.
*Source*: US EPA (4).

the material reach surface or ground water. Good storage design and use of appropriate management practices effectively block potential transport pathways. Movement of constituents is driven primarily by the following (4):

1. Precipitation events
2. Runoff and erosion of soluble and particulate components (including nutrients, organic matter, and pathogens to surface waters)
3. Leaching to ground water of soluble nutrients and compounds

In addition, wind erosion can contribute to loss of dry or composted material under arid, windy conditions that may also impact water quality.

*5.3.1. Nutrients, Organic Matter, and Impacts on Surface Water*

The content and form of nitrogen (N) and phosphorus (P), which must be taken into consideration in specific biosolids, vary depending on wastewater sources and treatment processes. Like all organic residuals, biosolids contain significant amounts of N and P.

Proper storage conserves these nutrients until crops can use them during the growing season. Good management of stored organic residuals is needed to prevent excess amounts of organic or inorganic N from entering surface or ground water.

Runoff of nutrients can contribute to eutrophication of surface water, which may impair its use for fisheries, recreation, industry, and drinking water source. Nitrogen is the primary contributor to eutrophication in brackish and saline waters (e.g., estuaries) and to some extent in freshwater systems. However, P concentration is usually the controlling eutrophication factor in freshwater. Both nitrate and ammonia are water soluble, and thus are transported in leachate and runoff. Organically bound N does not interact in the environment until it is mineralized into water-soluble nitrate. Ammonia can be toxic to fish.

Excess nutrients and organic matter in surface water can increase the growth of undesirable algae and aquatic weeds. The carbon and nutrients in organic matter serve as food for bacteria, thus adding organic matter and nutrients to water can directly increase the biochemical oxygen demand (BOD), deplete dissolved oxygen levels in water, and accelerate eutrophication. Low oxygen levels result from high BOD stress fish, shellfish, and other aquatic invertebrates. In a worst-case scenario, such as a direct spill of material from a storage facility into a waterway, heavy organic (BOD) and ammonia-N loadings could deplete oxygen levels rapidly and lead to septic conditions and fish kills.

The majority of P binds to mineral and organic particles, and therefore is subject to runoff and erosion rather than leaching, except under conditions of very sandy soils with low P binding capacity. Eroded particulates also serve as a physical substrate to convey adsorbed P, metals, and other potential pollutants in runoff.

### 5.3.2. Nutrients and Groundwater

The main concern with groundwater impacts of longer-term stockpiles (organic or inorganic) is the potential for leaching of soluble nitrate–N, which can impact local wells or eventually discharge to surface waters and contribute to eutrophication. Such situations have occurred in agricultural regions of the U.S. where excessive amounts of inorganic fertilizer or animal manures have been applied over several years. The high nitrate levels in wells have resulted in some cases of methemoglobinemia in susceptible infants. This rare condition reduces the blood's ability to carry oxygen efficiently, hence the condition's other name—blue-baby syndrome. Elevated nitrate in water can have the same effect on immature horses and pigs and can cause abortions in cattle. Water management practices at storage sites must be adequate to protect against such impacts. Phosphorus is not a drinking water concern, because it is not a health concern for humans or animals as nitrate is, and it binds to iron and soil minerals and has low water solubility.

### 5.3.3. Pathogenic Organisms

In the U.S., biosolids that are stored prior to land application must have been treated to meet US EPA Part 503 Class A or Class B pathogen density limits. The requirements for these types of biosolids can also include restricted access to field sites (Class B) to protect humans and animals from infection that might potentially result from direct contact with biosolids. Protection of water sources from contamination by residual pathogens or parasites in Class B biosolids can be accomplished through proper site selection, buffers, and management practices (60–62).

In general, soil is an effective barrier to the movement of pathogens via leachate into groundwater. Both organic matter and clay minerals in soil physically filter, adsorb, and immobilize microorganisms, including protozoan cysts and parasitic worm ova. However, sandy soils are typically very porous and cannot adsorb or immobilize microbes as well as clay and loam soils containing organic matter, thus they are not as effective retardants to the movement of pathogens. Soils in general are subject to a range of physical, chemical, and biological conditions that destroy pathogens such as extremes of wetness and dryness, temperature variations, and attack by natural soil microbes.

*5.3.4. Metals and Synthetic Organic Chemicals*

Like other residuals, biosolids may also contain measurable levels of metals and synthetic organic chemicals. In terms of organic and inorganic residuals, the same management practices that effectively isolate nutrients from surface and groundwater resources during storage are equally effective in containing any metals or synthetic organic chemicals. The potential for water quality impact from metals or synthetic organic chemicals present in biosolids is minimal from the outset because of their inherently low levels. Biosolids suitable for land application must meet stringent quality standards for metal concentrations under Part 503 regulations (see Chapter 6).

With the widespread implementation of industrial pretreatment programs, biosolids used in land application increasingly comply with the most conservative of Part 503 metal standards (63). In addition, metals in biosolids are bound strongly with other biosolids constituents and are not highly water soluble, hence they cannot leach into ground water. According to a review by the National Research Council (NRC), toxic organic compounds typically are not found in biosolids in significant levels (64). This is primarily attributable to effective industrial pretreatment programs and to the destruction or volatilization of organics during the treatment process. The NRC report also noted that "PCBs [polychlorinated biphenyls] and detergents are the only classes of synthetic organic compounds that occur in biosolids at concentrations above levels found in conventional irrigation water or soil additives." The PCBs bind to particulates and are relatively water insoluble and so are not susceptible to leaching. In addition, the low PCB levels in biosolids continue to decline due to enactment of a ban on production and use of PCBs in the U.S. Detergent compounds including surfactants and binders have been found in biosolids in relatively high concentrations (0.5 to 4.0 g/kg dry weight); however, they bind to biosolids organic matter, rapidly biodegrade, and do not readily leach.

*5.3.5. Management Approaches*

Water quality protection practices are based on three key concepts:

1. Keep clean runoff clean by minimizing contact with stored biosolids.
2. Properly manage water that comes into contact with stored biosolids.
3. Prevent movement of the biosolids into water resources.

Minimizing the amount of water that comes into contact with stored biosolids is the first step in keeping nutrients and pollutants out of water resources. Practices used under various storage scenarios to achieve this include the following (4):

1. Proper site selection to avoid run-on, flooding, or high water tables that intercept stored biosolids
2. Installation of up-slope diversions to channel runoff away from a field stockpile or constructed storage facility
3. Containment of biosolids in enclosed structures or tanks

Any significant precipitation or up-slope runoff that comes in contact with stored biosolids may contribute nutrients, pathogens, or pollutants. Whether this water accumulates on or near the biosolids, runs off, or leaches through the soil, it has the potential to transport contaminants to water resources. Practices to address this issue include the following:

1. Proper shaping of field stockpiles to shed water and avoid puddles of water, or infiltration of water through a stockpile and subsequent loss through runoff or leaching
2. Construction of enclosed storage facilities or tanks
3. Construct lagoons/pads with impervious earthen, concrete, or geotextile liners
4. Removal of accumulated water to sites where liquid may be applied
5. Providing buffers between storage areas and waterways

For permanent, long-term storage facilities, an impermeable liner (i.e., earthen, geotextile, or concrete) is recommended to ensure against leaching. For all constructed storage facilities, site soils and water table investigations are essential to ensure stable foundations. Soil settling and shifting can result in leakage through cracks. High water tables may float concrete pads or rupture the watertight seals of lagoons.

For short-term field storage, liners are generally unnecessary. Proper shaping of stockpiles encourages shedding of precipitation to prevent infiltration of water and subsequent leaching. Stockpiles should not be located on soils in environmentally sensitive areas with extremely high hydraulic conductivities with excessive infiltration rates, areas with very shallow seasonal high water tables or depths to bedrock, or areas adjacent to or on limestone features such as sinkholes or rock outcrops.

Accumulated water (i.e., precipitation) forms a separate layer on top of liquid or semisolid biosolids or collects in puddles after contact with the material. Overflow or runoff of this water to surface or ground water resources can be prevented or minimized by the following (4):

1. For open storage facilities:
   a. Use sumps or gravity flow to direct accumulated water to on-site filter strips or treatment ponds.
   b. Mix accumulated water with biosolids for removal to land application site.
   c. Decant and transport water accumulations off-site to treatment facilities.
   d. Apply to the land through irrigation systems (taking care not to exceed hydraulic loading rates to prevent ponding or runoff).
2. For constructed facilities:
   a. Build a roof to keep precipitation off the material.
   b. Pads should have adequate slope to prevent ponding and appropriate flow management.

Direct deposition of biosolids in waterways has the greatest potential for significantly impacting water quality through additions of nutrients, organic matter, pathogens or pollutants. Management practices to prevent this occurrence include the following:

- Adequate buffers between storage area and water resources
- Proper storage methods for the physical consistency of the biosolids
- Proper design and maintenance of constructed storage facilities
- A spill response and control plan supported by staff training and the availability of the necessary supplies and equipment

Proper materials management is an essential measure in water quality protection for all biosolids storage facilities and field stockpiling sites. Well-designed storage operations optimize water quality protection measures by including the following (4):

- Structural elements to minimize the potential for accidental spills
- Operational procedures to reduce potential accidents
- Contingency plans to promptly mitigate spills if they do occur

Preventative measures for field stockpiles include the following:

- Proper site selection including buffer distances and slopes
- Proper vehicle and equipment safety features (e.g., waterproof seals on trailer tailgates), maintenance, and operator training
- Adequate staff training and proper operation of site to prevent accidental spills or losses of material to water resources (e.g., truck rollovers, excess residuals left in loading areas)
- Written spill cleanup and contingency plans and advanced preparation (e.g., equipping storage sites and vehicles with appropriate cleanup tools, and staff drills to ensure rapid and effective response to spills

Preventative measures for constructed facilities include the following:

- Soil strength and suitability assessments prior to construction to avoid uneven settling and other problems that lead to cracks or leaks
- Adequate design volumes, including space for precipitation accumulations
- Use of good engineering construction practices to prevent structural failures and malfunctions (e.g., impermeable liners or backflow regulators on gravity systems, paving and curbing of off-loading pads for permanent facilities)
- Proper vehicle and equipment safety features (e.g., waterproof seals on trailer tailgates), maintenance, and operator training
- Adequate staff training and proper operation of site to prevent accidental spills or losses of material to water resources (e.g., truck rollovers, overtopping of freeboard)
- Written spill cleanup and contingency plans and advanced preparation (e.g., equip sites and vehicles with cleanup tools; conduct staff drills to prepare for effective spill response)

Managers of stored biosolids need to assess the nature of their biosolids, the operational requirements, the limitations of their land application program, and the storage option most suitable for their operation to select the best combination of design and management practices for their specific situation.

## 5.4. Pathogens

Untreated wastewater contains pathogens, such as viruses, bacteria, and animal and human parasites (protozoa and helminths), that may cause various human diseases and illnesses. Often these pathogens are or become attached to the separated wastewater solids. It is precisely because of the potential presence of pathogens in untreated wastewater that treatment processes are used to clean wastewater prior to discharge to streams. This is also the reason that wastewater residuals must be subjected to additional pathogen reduction treatment prior to land application of the biosolids (65–68).

These treatment processes in the U.S. are carefully regulated and monitored to ensure a consistent level of treatment and pathogen destruction (63, 69). The combination of treatment and appropriate biosolids management at land application sites has proven to be effective in preventing the transmission of pathogens that can cause disease. Incidents of infectious disease, through either direct exposure or food or water pathways, have not been documented from land application of biosolids in the U.S. since this combination of regulated practices has been implemented.

The potential exposure to pathogens during proper biosolids storage is no greater than that associated with direct land application. This section describes prudent management practices recommended to safely store biosolids in a manner that limits the potential for transmission of disease agents.

### 5.4.1. Biosolids Products Characteristics

Biosolids destined for beneficial use in land application must meet pathogen reduction criteria for either Class A or Class B according to Part 503 rules (see Chapter 6). Only biosolids intended for and that meet Part 503 criteria for safe land application should be placed in a field stockpile or a constructed storage facility (63). The two classes of biosolids have different characteristics that influence storage management considerations. Documentation of Class A or B treatment may be achieved either through testing of the final product for specific pathogens or indicator organisms or by use of approved treatment processes.

In addition to the pathogen reduction requirement, biosolids must also be treated to reduce their attractiveness to vectors, such as rodents, flies, mosquitoes, and others, that are capable of transmitting pathogens. Part 503 rules specify analytical standards and treatment processes to achieve VAR requirements (63).

*Class A*: Class A biosolids typically are treated by processes to further reduce pathogens (PFRP) such as composting, pasteurization, drying or heat treatment, advanced alkaline treatment (70), or by testing and meeting the pathogen density limits in Part 503. Class A pathogen reduction reduces the level of pathogenic organisms in the biosolids to a level that does not pose a risk of infectious disease transmission through casual contact or ingestion.

*EQ Class A*: Class A biosolids that also meet one of Part 503 VAR options 1 to 8 (71, 72) and meet the metals limits (Part 503 Table 5.3) are designated as exceptional quality (EQ). These products are exempted from the Part 503 general requirements, management practices, and site restrictions, and may be generally marketed and distributed.

*Class B*: Class B biosolids typically are treated by processes to significantly reduce pathogens (PSRP) such as aerobic digestion, anaerobic digestion, air drying, and lime stabilization (70). As an alternative, producers may document compliance by analyzing the material for fecal coliform levels. When Class B requirements are met, the level of pathogenic organisms is significantly reduced, but pathogens are still present. In this case, other precautionary measures required by the Part 503 rule, such as site and crop harvesting restrictions, are implemented to protection of public health.

*5.4.2. Biosolids Storage Considerations*

#### 5.4.2.1. PATHOGENS IN STORED CLASS A BIOSOLIDS

In general, storage of Class A biosolids presents few pathogen concerns due to the level of pathogen reduction achieved by the treatment processes. The potential for exposure to viruses or parasites (helminth ova) in a Class A product is insignificant as a result of treatment and because these organisms cannot grow outside a suitable host organism. This potential does not increase during storage. Treatment also reduces bacterial pathogens to safe levels. However, bacteria depend on readily available sources of nutrients, adequate water, and favorable environmental conditions, and can grow without a host organism. In specific and very limited situations, the necessary combinations of these factors have been found to occur in stored Class A biosolids. Here are three examples of these circumstances (4):

1. If Class A biosolids compost that is no longer self-heating is blended with green or unstabilized organic materials, such as fresh yard trimmings, fresh hay, or green woodchips, the bacterial population can grow rapidly. This is because these fresh materials contain readily available carbon that bacteria need and the compost lacks. If these types of mixtures are managed as self-heating compost piles, in which time/temperature conditions adequate to destroy bacterial pathogens are achieved, then the final products will also contain undetectable levels of pathogens, as do Class A biosolids. At such low concentrations, disease will not be transmitted even with direct contact with biosolids. If Class A biosolids are mixed with products that contain unavailable carbon sources, such as cellulose and lignin in paper and wood processing residuals, pathogen concentrations will remain undetectable because these nutrients cannot be used by pathogens.
2. If a Class A product is inadequately composted, or its nutrients are not well stabilized, bacterial pathogen growth will not occur as long as the material is kept very dry, that is, a total solids content of 80% or greater. However, if such dry materials take on moisture during storage, and nutrients, pH, temperature, and other environmental conditions are favorable, pathogen and microbial regrowth could occur. Thus, preparers should be aware that if they conduct various types of blending or permit water content to increase in heat-dried Class A products, the potential for temporary increases in bacterial growth exists.

    It is important to recognize that growth during storage is usually a temporary condition in which bacterial populations increase in response to the sudden availability of a food source, but decline to previous low levels once it is consumed. The growth and presence of nonpathogenic microorganisms in biosolids act to counterbalance the stimulating effect of nutrients on bacterial growth through the natural competition for nutrients.

    If pathogen regrowth occurs, the material should be held in storage until populations decline to acceptable levels or it should be re-treated to meet standard pathogen limits. The potential for pathogen growth should be considered in establishing appropriate storage conditions and in blending or augmenting Class A biosolids with other organic materials.
3. If the pH of Class A alkaline-stabilized material drops significantly during extended storage and the color, consistency, or odor of the product has deteriorated, then retesting for pathogens may be advisable. Significant decreases in pH on occasion have been associated with increases in the level of fecal coliform above the 1000 MPN (most probable number)/g regulatory limit.

#### 5.4.2.2. PATHOGENS IN STORED CLASS B BIOSOLIDS

The probable presence of pathogenic organisms is assumed for biosolids treated to Class B pathogen reduction standards. Likewise, Class B biosolids blended with any other organic materials, such as leaves, sawdust, and woodchips, for whatever reason, is not considered to alter the pathogen status. For this reason, storage practices should provide a level of protection equivalent to Class B site restrictions for use to minimize human, animal, or environmental exposure to disease-causing organisms either through direct contact or via the food chain (73, 74).

#### 5.4.2.3. ACCUMULATED WATER

Ponded water that has contacted stored biosolids may contain nutrients and have a moderate enough pH to provide a favorable medium for growth of bacteria, including pathogens. This may occur even when the bulk of the stored product is dry. In addition, according to the preliminary risk assessments for land application of biosolids, the highest risk pathways for viruses, bacteria, and parasites involve direct human contact with biosolids or with surface waters that have been contaminated by runoff and sediment, particularly immediately after a rainfall. Therefore, management of stormwater to minimize contact with biosolids and properly dealing with any water that accumulates in contact with stored biosolids is essential.

#### 5.4.2.4. REQUIRED RETESTING

*Class A and EQ*: For EQ biosolids the Part 503 requirements to test stored materials prior to use depends on who has control of the stored material. If the material remains in the control of the original preparer (directly or indirectly through a contracted processor or applier), the material must be retested prior to final use. If a preparer gives or sells EQ biosolids to a second party, for instance a landscaper, who then stores the material before land application, testing for pathogens, is not required under Part 503 (Figure 5.18 and Table 5.14).

*Class B material*: For Class B biosolids, any mixture of a Class B biosolids and a nonhazardous material is considered as a product derived from biosolids, and hence, by definition, biosolids. Thus, if either a preparer or a land applier blends ground green waste with Class B biosolids, and then plans to till that mixture into the soil the mixture would still need to meet the Part 503 Class B standard and site restrictions (i.e., pollutants, pathogen, and vector attraction reduction requirements). The party who mixes the biosolids with another material is the preparer, as defined in Part 503 (Figure 5.18 and Table 5.14).

### 5.4.3. Storage Site Management

Three conditions are necessary to produce infectious disease (4):

1. The disease agent must be present in sufficient concentrations to be infectious.
2. Susceptible individuals must come in contact with the agent in a manner that causes infection.
3. The agent must be able to overcome the physical and immunological barriers of the individual.

Proper management practices break the chain of transmission either by keeping susceptible individuals or animals from direct contact with stored materials or by preventing the movement of any residual pathogens or parasites in stored materials into

**Table 5.14**
**Part 503 pathogen density limits: biosolids pathogen standards can be satisfied by determining the geometric mean of seven samples of biosolids after treatment for the following:**

| Pathogen or Indicator | Standard density limits (dry wt) | |
|---|---|---|
| Class A | | |
| • Salmonella | <3 MPN/4 g total solids | or |
| • Fecal coliforms | <1000 MPN/g | and |
| • Enteric viruses | <1 PFU/4 g total solids | and |
| • Viable helminth ova | <1/4 g total solids | |
| Class B | | |
| Fecal coliform density | <2,000,000 MPN/g total solids (dry wt. basis) | |

MPN = most probable number; PFU = plaque forming unit.
*Source*: US EPA (4).

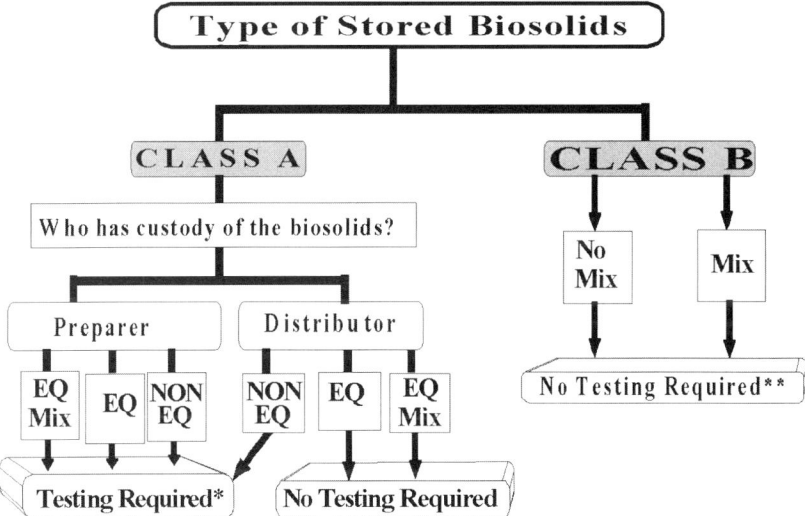

\* Before custody of the biosolids is transferred to the distributor, OR when something other than EQ biosolids is mixed with NON-EQ biosolids after the preparer has released control of it.

If anything is mixed with NON-EQ biosolids, the mixture is subject to the land application general requirements and management practices when it is land-applied.

\*\* When used according to Class B site restrictions

**Fig. 5.18.** Decision tree diagram showing the interrelationship between biosolids pathogen reduction status (Classes A, B, and EQ), current custodian, and mixing with non-biosolids material relative to testing and retesting requirements. *Source*: US EPA (4).

the environment in a way that would be harmful. Biosolids regulations are designed to address the first two of the three conditions that produce infections disease:

1. Biosolids that meet rigorous Class A pathogen reduction standards do not have detectable levels of pathogens and are exempt from site restrictions.
2. For Class B biosolids, the risk of transmission of infectious disease agents is reduced to acceptable levels by a combination of treatment to reduce pathogen levels and management practices to minimize the potential for exposure of susceptible individuals to pathogens or parasites.

Management options to restrict potential movement of pathogens are as follows (4, 75, 76):

1. Use of appropriate buffers or filter strips to control runoff from field stockpiles.
2. Diverting stormwater runoff away from the stored biosolids.
3. Practices such as stormwater containment ponds or collection and irrigation systems for uncovered constructed storage pads or lagoons.
4. Enclosure of long-term storage of biosolids in facilities with roofed structures to prevent contact with precipitation or runoff where feasible.
5. Restriction of public access to field storage sites. Constructed facilities may warrant fencing, but fencing of field storage stockpiles is needed only if storage will occur in areas that are accessible to livestock.
6. Any runoff that has been in contact with the biosolids should be kept isolated from any adjacent fruit or vegetable crops that would be harvested, sold in the fresh market, and potentially consumed raw.

### 5.4.4. Worker Safety

Worker safety is always a primary consideration, and basic hygiene training similar to that of workers at a wastewater treatment plant should be provided to biosolids haulers and storage site staff. The use of good personal hygiene and work habits form the basis of a worker protection program for those handling biosolids. Some specific recommendations include the following (4):

1. Wash hands thoroughly with soap and water after contact with biosolids.
2. Avoid touching face, mouth, eyes, nose, genitalia, or open sores and cuts.
3. Wash hands before eating, drinking, smoking, or using the restroom.
4. Eat in designated areas away from biosolids handling activities.
5. Do not smoke or chew tobacco or gum while working with biosolids.
6. Use gloves to protect against skin abrasions or contact between abrasions and biosolids, or surfaces exposed to biosolids, when they occur unexpectedly.
7. Remove excess biosolids from shoes prior to entering vehicle.
8. Keep wounds covered with clean, dry bandages.
9. Flush eyes thoroughly, but gently, if biosolids contact eyes.
10. Change into clean work clothing on a daily basis and, if possible, before going home; reserve work boots for use at storage sites or during biosolids transport.

The Centers for Disease Control and Prevention recommends that immunizations for diphtheria and tetanus be current for the general public, including all wastewater workers. Boosters are recommended every 10 years. The tetanus booster should be repeated in the case of a wound that becomes dirty, if the previous booster is more than 5 years old. Consult a doctor regarding direct exposure through an open

*Storage of Sewage Sludge and Biosolids* 261

wound, eyes, nose, or mouth. It should be noted that a hepatitis A vaccine has been developed and is available to the general public. Consequently, it is recommended that those working with biosolids receive this vaccination as an additional protection.

## 6. DESIGN EXAMPLES

### Example 1

The designer of a 7.5-MGD (0.33-m$^3$/s) average design flow wastewater treatment plant wishes to determine the available sludge storage volume in three rectangular primary sedimentation tanks. The tanks are designed to treat a peak wet weather flow of 20 MGD (0.88 m$^3$/s). Available storage will determine the maximum time allowed between sludge pumping cycles and the maximum capacity of the sludge pumps.

Tank design is based on conservative experience involving overflow rates and mean velocities at average design flows. Each tank is 110 ft long (33.5 m) and 19 ft wide (5.8 m), with an average sidewater depth of 10 ft (3.05 m). Longitudinal collectors operating continuously bring the settled sludge to the head end of the tank, where it is conveyed to the sludge removal hopper by cross-collectors. The sludge is then pumped from the removal hopper on a timed cycle with density and blanket-level instrumentation. Cross-collection channels and sludge removal hoppers have been laid out to aid in the concentration, storage, and removal of the collected sludge by providing steep side slopes, ample depths, and short suction pipelines. Combined storage volume of the cross-collector channel and removal hopper of the selected tank design is approximately 350 ft$^3$ (9.9 m$^3$) for each tank.

It is assumed that peak and wet weather flows will be of at least 8 hours' duration and will have an average suspended solids content of 200 mg/L. Primary sedimentation tank removal efficiency is assumed to be only 50% at peak wet weather flow, down from 60% at average design flow, because of higher overflow rate and higher mean velocity.

### Solution

The solids collected in each of the three primary sedimentation tanks during the storm can be calculated as follows:

Peak sewage flow = 20 MGD = 20 MG = million gallons/d
Total suspended solids (TSS) of sewage flow = 200 mg/L = 200 (8.34) lb/MG
Number of sedimentation tanks = 3
TSS removal efficiency = 50 % = 0.50

$$\frac{(20 MG/d)(0.50)(200)(8.34 lb/MG)}{(3)(24 h/d)} = 231 \text{ lb/h } (105 \text{ kg/h}) \qquad (4)$$

Primary sludge solids concentration and wet bulk specific gravity are assumed to be 6% and 1.07, respectively. Using these assumptions, the volume produced in each tank can be calculated as follows:

$$\frac{231\,lb/h}{(0.06)(1.07)(8.34\,lb/gal)(7.48\,gal/ft^3)} = 58\,\text{ft}^3/\text{h}\ (1.6\,\text{m}^3/\text{h}) \qquad (5)$$

By dividing this production into the storage volume available, one can find the maximum period of time between pump cycles to be slightly greater than 6 h:

$$\frac{350\,ft^3}{58\,ft^3/h} = 6.02\,\text{h} \qquad (6)$$

The primary sludge piping to the digester is arranged so that only one primary sludge pump can operate at a time. To ensure sufficient pumping capacity to handle the peak wet weather sludge, it is necessary that each pump operate only one third of the time. Each pump, therefore, must have the capacity to remove all of the sludge stored during the 6-hour cycle in 2 hours. This capacity is calculated as follows:

$$\frac{(231\,lb/h)(6h/2h)}{(0.06)(8.34 \times 1.07)\,lb/gal\,(60\,\min/h)} = 21.6\,\text{gpm}\ (1.36\,\text{L/s}) \qquad (7)$$

As an additional safety factor, to ensure maximum reliability and operational flexibility, this pumping rate is doubled and rounded off to 50 gallons per minute (gpm) (3.2 L/s). The pump selected (a diaphragm pump, see Chapter 1) can be adjusted down to 25 gpm (1.6 L/s) if higher flow rates are found to pull liquid instead of concentrating solids.

*Example 2*

This example illustrates how the digester storage can be used to "damp-out" solids surges and thus prevent overloading of downstream dewatering units.

Consider a primary wastewater treatment plant with the flow scheme and average loads shown in Figure 5.19. Average loading to the dewatering units is 103,313 lb/d (46,904 kg/d). Dewatering unit capacity is 200,000 lb/d (90,800 kg/d); under average loading conditions, the dewatering units are clearly not stressed. The treatment plant, however, receives flow from a combined sanitary/storm sewer network. During storms, hydraulic loadings increase dramatically as a result of infiltration and inflow to the sewer system. Plant solids loadings also increase sharply as the result of solids being carried into the sewer by runoff and the scouring of previously accumulated materials from the sewer system.

From historical records, the peak 5-day solids loading (average load for the most heavily loaded 5 consecutive days) is 433,000 lb/d (196,582 kg/d). This is 2.57 times greater than the average digester load. If the storage available upstream of the dewatering units is not utilized, dewatering unit loading would also be 2.57 times the average value or 265,000 lb/d (120,310 kg/d). The dewatering units would therefore be overloaded. Overload can be prevented, however, if digester storage is properly utilized. Solids can be stored within the digester so that, during peak loading periods, dewatering capacity is not exceeded. The accumulated solids can be released after the storm has passed and the dewatering units are no longer stressed.

Solids may be stored in the digesters by either of two mechanisms, acting either singly or in concert:

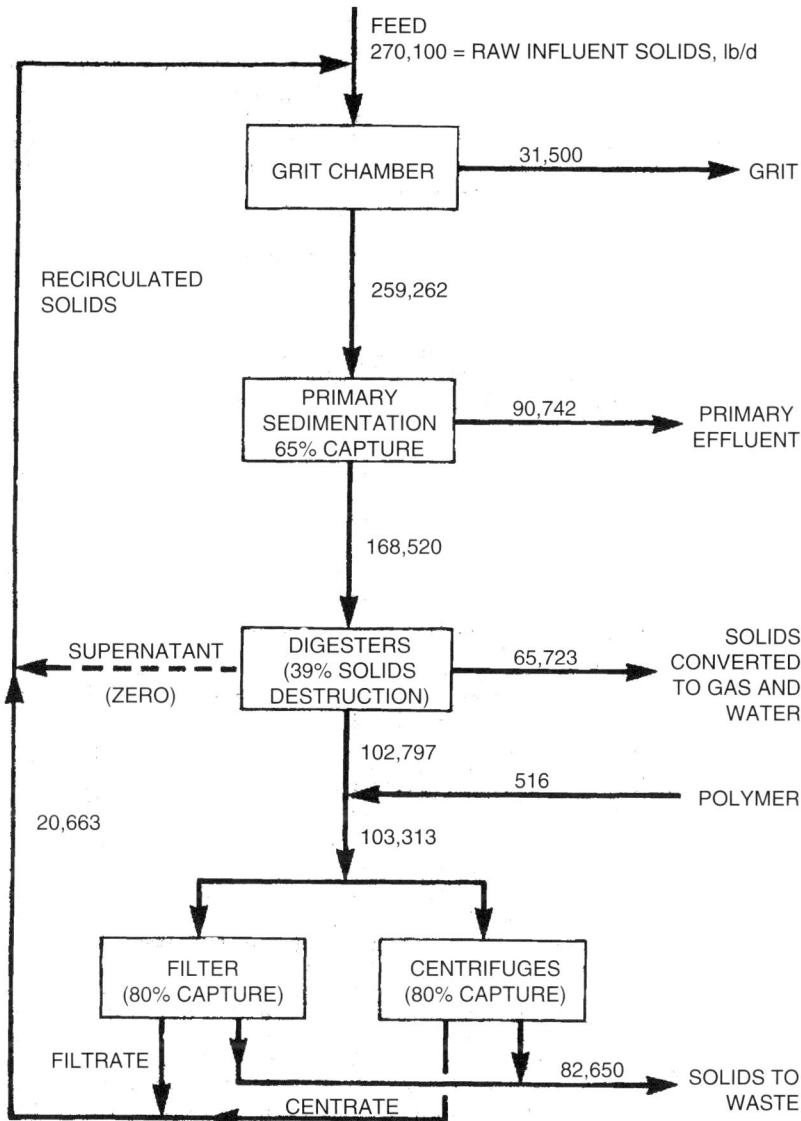

**Fig. 5.19.** Solids balance and flow diagram; design example 2, single-phase concentration and displacement storage. *Source*: US EPA (3).

1. The digester working volume is increased by allowing the floating covers to rise (displacement storage).
2. The digester feed is thickened to a greater degree than previously. As a result, the solids concentration of the digested material increases (single-phase concentration).

*Solution*

The following analysis examines how one of several possible operating strategies can be implemented. It is assumed that the system is at average conditions (Figure 5.19) when a large storm occurs and for the next 5 days average digester loadings increase to

433,000 lb/d (196,582 kg/d). At the onset of the storm, the plant operator decides to ease a potentially serious dewatering overload situation by (1) allowing the floating covers to rise at the rate of 1 ft/d (0.305 m/d) and (2) by thickening the raw sludge withdrawn from the primary sedimentation basin from the normal 5% concentration to 7% concentration. The additional thickening is accomplished by allowing sludge to accumulate to greater depths in the primary sedimentation tanks cross-collection trough and sludge hoppers. The intent of these two operations is to maintain digested solids mass flow rates below 200,000 lb/d (90,800 kg/d) to prevent dewatering unit overload.

The effects of these operations can be estimated from an unsteady state analysis of digester operations. The basic predictive equation is derived by an unsteady state mass balance:

1. The solids mass balance equation:

$$\text{Solids in} - \text{Solids out} - \text{Solids destroyed} = \text{Solids accumulated} \tag{8}$$

   a. Solids in $= QC_i$
   b. Solids out $= (Q - k)C$
   c. Solids destroyed $= QC_i X$
   d. Solids accumulated $= d(VC)/dt$
   where
   $Q =$ digester feed rate, volume per time
   $C_i =$ digester feed solids concentration, mass per volume
   $C =$ digester sludge solids concentration, mass per volume
   $k =$ rate of liquid accumulation in the digesters due to rise of floating covers, volume per time
   $X =$ fraction of digester feed destroyed by digestion, dimensionless
   $V =$ digester liquid volume
   $t =$ time

2. Summing the terms:

$$QC_i - (Q - k)C - QC_i X = \frac{d(VC)}{dt} \tag{9}$$

3. The right-hand side of Eq. (9) can be further developed:

$$\frac{d(VC)}{dt} = V\frac{dC}{dt} + C\frac{dV}{dt} = V\frac{dC}{dt} + Ck \tag{10}$$

4. Simplifying:

$$QC_i(1 - X) - QC = V\frac{dC}{dt} \tag{11}$$

5. Make the simplifying assumptions that digester feed flow, feed concentration, and liquid volume are constant for the period $t$. Equation 11 can then be integrated and solved for $C$:

$$C = C_i(1 - X) - [C_i(1 - X) - C_o]e^{-\frac{Q}{V}t} \tag{12}$$

Equation 12 predicts digested sludge solids concentration at any time beyond initiation of the operating strategy. $C_o$ is defined as digested sludge concentration at the time the operating strategy is put into operation. Note that the product of digested

Table 5.15
Calculations for digester effluent mass flow rate from equation 12

| Operating strategy | Time after start of storm, d | Digester loading, lb/d | Digester feed solids concentration, % | Digester feed rate, gpd | Digester volume increase/day due to rise of floating covers, gpd | Digester volume, gal | Fractional solids destruction | Digester effluent flow, gpd | Digested sludge solids concentration, % | Dewatering unit feed rate, lb/d |
|---|---|---|---|---|---|---|---|---|---|---|
| A. Floating cover rise = 1 ft/d; digester feed thickened to 7% | $0^-$ | 168,520 | 5 | 396,051 | 0 | $5.97 \times 10^6$ | 0.39 | 396,051 | 3.05 | 102,797 |
| | $0^+$ | 433,000 | 7 | 726,875 | 0 | $5.97 \times 10^6$ | 0.20 | 550,632 | 3.05 | 142,919 |
| | 1 | 433,000 | 7 | 726,875 | 176,243 | $6.05 \times 10^6$ | 0.20 | 550,632 | 3.38 | 156,429 |
| | 2 | 433,000 | 7 | 726,875 | 176,243 | $6.23 \times 10^6$ | 0.20 | 550,632 | 3.58 | 167,772 |
| | 3 | 433,000 | 7 | 726,875 | 176,243 | $6.41 \times 10^6$ | 0.20 | 590,632 | 3.78 | 177,373 |
| | 4 | 433,000 | 7 | 726,875 | 176,243 | $6.58 \times 10^6$ | 0.20 | 590,632 | 3.96 | 185,561 |
| | 5 | 433,000 | 7 | 726,875 | 176,243 | $6.76 \times 10^6$ | 0.20 | 550,632 | 4.11 | 192,594 |
| B. Floating cover rise = 1 ft/d; digester feed remains at 5% | $0^-$ | 168,520 | 5 | 396,051 | 0 | $5.97 \times 10^6$ | 0.39 | 396,051 | 3.05 | 102,797 |
| | $0^+$ | 433,000 | 5 | 1,017,626 | 0 | $5.97 \times 10^6$ | 0.20 | 841,383 | 3.05 | 218,385 |
| | 1 | 433,000 | 5 | 1,017,626 | 176,243 | $6.05 \times 10^6$ | 0.20 | 841,383 | 3.20 | 228,903 |
| | 2 | 433,000 | 5 | 1,017,626 | 176,243 | $6.23 \times 10^6$ | 0.20 | 841,383 | 3.31 | 237,330 |
| | 3 | 433,000 | 5 | 1,017,626 | 176,243 | $6.41 \times 10^6$ | 0.20 | 841,383 | 3.41 | 244,156 |
| | 4 | 433,000 | 5 | 1,017,626 | 176,243 | $6.58 \times 10^6$ | 0.20 | 841,383 | 3.49 | 249,740 |
| | 5 | 433,000 | 5 | 1,017,626 | 176,243 | $6.76 \times 10^6$ | 0.20 | 841,383 | 3.56 | 254,989 |
| C. Floating covers are not allowed to rise; digester feed thickened to 7% | $0^-$ | 168,520 | 5 | 396,051 | 0 | $5.97 \times 10^6$ | 0.39 | 396,051 | 3.05 | 102,797 |
| | $0^+$ | 433,000 | 7 | 726,875 | 0 | $5.97 \times 10^6$ | 0.20 | 726,875 | 3.05 | 188,664 |
| | 1 | 433,000 | 7 | 726,875 | 0 | $5.97 \times 10^6$ | 0.20 | 726,875 | 3.34 | 206,745 |
| | 2 | 433,000 | 7 | 726,875 | 0 | $5.97 \times 10^6$ | 0.20 | 726,875 | 3.60 | 222,755 |
| | 3 | 433,000 | 7 | 726,875 | 0 | $5.97 \times 10^6$ | 0.20 | 726,875 | 3.83 | 236,928 |
| | 4 | 433,000 | 7 | 726,875 | 0 | $5.97 \times 10^6$ | 0.20 | 726,875 | 4.03 | 249,478 |
| | 5 | 433,000 | 7 | 726,875 | 0 | $5.97 \times 10^6$ | 0.20 | 726,875 | 4.21 | 260,588 |

gpd, gallons per day.
*Source*: US EPA (3).

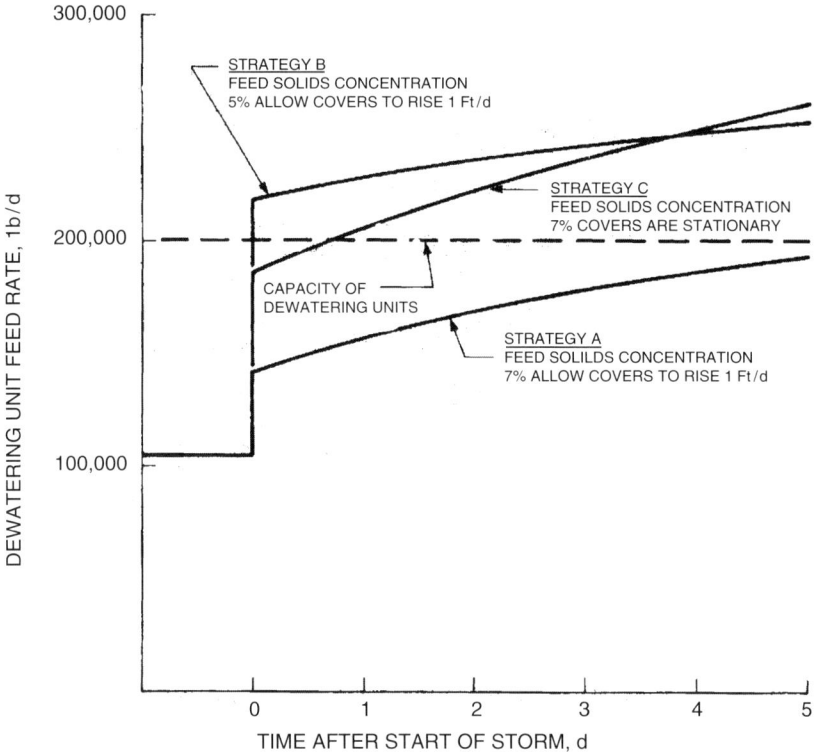

**Fig. 5.20.** Effect of various operating strategies on dewatering unit feed rates. *Source*: US EPA (3).

sludge concentration ($C$) and digester effluent liquid flow ($Q - k$) is the load that the dewatering units must process.

Calculations related to the operating strategy just described are summarized in Table 5.15 part A. The digested solids mass flow rates are calculated just before the storm ($t = 0^-$), immediately after the storm begins ($t = 0^+$) and for each of the next 5 consecutive days. It is assumed that digester loading increases in one step from 168,520 lb/d (76,508 kg/d) to 433,000 lb/d (196,582 kg/d) at $t = 0$. Digested sludge liquid volume at the beginning of the storm is $5.97 \times 10^6$ gal ($22,600 \, m^3$). Each 1 ft (0.305 m) of cover rise increases tank volume by 176,243 gal ($667 \, m^3$). Solids destruction within the digesters is assumed to be 39% ($X = 0.39$) during average conditions, dropping to 20% ($X = 0.20$) during the storm due to decreased digester retention time. The calculation shows that dewatering capacity (200,000 lb/d [90,800 kg/d]) is not exceeded during the storm, thus the operating strategy has been successful. Calculations for two other strategies (Table 5.15 parts B and C), which were not successful are also included. The results are shown graphically in Figure 5.20.

## Example 3

An example of liquefaction in which no evaporation or additional moisture is added during storage, can be seen in the following calculations:

| | |
|---|---|
| Dewatered digested biosolids (polymer conditioner used) | |
| Solids to be stored | 1000 dry t (907 T) |
| Total solids (TS) content | 20% |
| Volatile solids | 65% |
| Assumed reduction of VS during 6 months storage | 20% |
| Water content of dewatered biosolids | 5000 t (4,535 T) |
| VS, at start of storage | 650 dry t (590 T) |
| VS, at end of storage | 520 dry t (472 T) |
| Fixed solids, unchanged | 350 dry t (317 T) |
| TS, at end of storage | 870 dry t (789 T) |
| Total solids content at end of storage | 14.8% |

The example indicates how a reasonably dry, dewatered digested sludge (20% solids) can be liquefied to a fairly wet, digested sludge (14.8%) if the putrescible organic material continues to be reduced. The speed of this reduction is greatly affected by temperature and organic content in the dewatered sludge. Thus, liquefaction will be a greater problem in warm climates or during the hot summer seasons. If lime and ferric chemicals are used to condition the digested sludge for dewatering, liquefaction will be greatly reduced, both because of the lower overall organic content of the material and the inhibiting effects of the chemicals on the bacterial reduction of the putrescible organic matter.

## NOMENCLATURE

$C$ = digester sludge solids concentration, mass per volume
$C_i$ = digester feed solids concentration, mass per volume
$C_o$ = digested sludge concentration at time $t$ is zero
$E$ = energy, kWh/yr
$H$ = discharge head, ft
$k$ = rate of liquid accumulation in the digesters due to rise of floating covers, volume per time
$Q$ = flow, volume per time
$Q_{Effective}$ = effective flow, volume per time
$Q_{Design}$ = design flow, volume per time
$t$ = time
$V$ = digester liquid volume
$X$ = fraction of digester feed destroyed by digestion, dimensionless

## REFERENCES

1. Metcalf and Eddy, Inc., *Wastewater Engineering Treatment and Reuse*, 4th ed., McGraw-Hill, New York (2003).
2. U.S. Environmental Protection Agency, *Innovative and Alternative Technology Assessment Manual*, US EPA 430978009, NTIS # PB81-103277, Washington, DC, 117 pages, February (1980).
3. U.S. Environmental Protection Agency, *Process Design Manual, Sludge Treatment and Disposal*, US EPA 625/1-79-011, Cincinnati, OH, September (1979).

4. U.S. Environmental Protection Agency and U.S. Department of Agriculture, *Guide to Field Storage of Biosolids*, US EPA/832-B-00-007, US EPA, Washington, DC, and USDA, Beltsville, MD, July (2000).
5. D. Glindemann, J. T. Novak, S. N. Murthy, S. Gerwin, R. Forbes, and M. Higgins, Standardized biosolids-incubation; headspace odor measurement and odor production consumption cycles. In: *Proceedings Water Environment Federation and AWWA Odors and Air Emissions Conference*, Bellevue, Washington (2004).
6. Sacramento Area Consultants, *Sewage Sludge Management Program Final Report*, Volume 4, Sacramento Regional County Sanitation District. Sacramento, CA, September (1979).
7. San Francisco Bay Region Wastewater Solids Study, *San Francisco Bay Region Sludge Management Plan, Volume VI San Jose/Santa Clara Project and Environmental Impact Report*, Oakland, CA, December (1978).
8. J. B. Farrell, *Best Management Technology Definitions for (a) Sludge Stabilization and (b) Additional Pathogen Reducing Processes*, U.S. Environmental Protection Agency, Cincinnati, OH, November (1978).
9. N. K. Shammas, I. J. Kumar, S. Y. Chang, and Y. T. Hung, Sedimentation. In: *Physicochemical Treatment Processes*, L. K. Wang, Y. T. Hung and N. K. Shammas (eds.), Humana Press, Totowa, NJ (2005).
10. Water Environment Federation, *Design of Municipal Wastewater Treatment Plants*, 4$^{th}$ ed., *MOP 8*, Alexandria, VA (1998).
11. U.S. Environmental Protection Agency, *Evaluation of Flow Equalization in Municipal Wastewater Treatment*, US EPA-600/2-79-096, Cincinnati, OH, May (1979).
12. U.S. Environmental Protection Agency, *Upgrading Existing Wastewater Treatment Plants*, Cincinnati, OH, October (1974).
13. W. L. Berk, *The Design, Construction and Operation of the Oxidation Ditch*, RAD-211, Lakeside Equipment Corporation, Bartlett, IL (1978).
14. R. K. Goel, J. R. V. Flora, and J. P. Chen, Flow equalization and neutralization. In: *Physicochemical Treatment Processes*, L. K. Wang, Y. T. Hung and N. K. Shammas, (eds.), Humana Press, Totowa, NJ (2005).
15. L. K. Wang, Z. Wu, and N. K. Shammas, Activated sludge processes. In: *Biological Treatment Processes*, L. K. Wang, N. C. Pereira, and Y. T. Hung (eds.), Humana Press, Totowa, NJ (2008).
16. N. K. Shammas and L. K. Wang, Pure oxygen activated sludge processes. In: *Biological Treatment Processes*, L. K. Wang, N. C. Pereira and Y. T. Hung (eds.), Humana Press, Totowa, NJ (2008).
17. R. I. Dick, E. L. Thakston and W. W. Eckenfelder, Jr., (Eds.), *Water Quality Engineering New Concepts and Developments*, Jenkins Publishing Co., Austin, TX and New York (1972).
18. T. M Keinath, M. D. Ryckman, C. H. Dana, Jr., and D. A. Hofer, *Design and Operational Criteria for Thickening of Biological Sludges, Parts I, II, III, IV*. Water Resources Research Institute, Clemson University, September (1976).
19. D. L. Tucker, and N. D. Vivado, Design of an overland flow system at Newman, California. In: *Proceedings of the 51st Annual Water Pollution Control Federation Conference*, Anaheim, CA, October (1978).
20. B. G. Liptak (ed.), *Environmental Engineers Handbook, Volume I Water Pollution*, Chilton Book Company, Radnor, PA (1974).
21. Water Resources Control Board, *Final Report—Phase I, Rural Wastewater Disposal Alternative*, State of California, Sacramento, CA, September (1977).
22. U.S. Environmental Protection Agency, *Upgrading Lagoons*, Technology Transfer Seminar Publication, US EPA-626/4-73-0016, Cincinnati, OH, revised, June (1977).

23. N. K. Shammas, L. K. Wang, and Z. Wu, Waste stabilization ponds and lagoons. In: *Biological Treatment Processes*, L. K. Wang, N. C. Pereira, and Y. T. Hung (eds.), Humana Press, Totowa, NJ (2008).
24. E. M. Pickett, Evapotranspiration and individual lagoons. In: *Proceedings of the Northwest On-Site Waste Water Disposal Short Course*, University of Washington, December 8–9 (1976).
25. N. K. Shammas and L. K. Wang, Gravity thickeners. In: *Biosolids Treatment Processes*, L. K. Wang, N. K. Shammas, and Y. T. Hung (eds.), Humana Press, Totowa, NJ (2007).
26. J. R. Taricska, D. A. Long, J. P. Chen, Y. T. Hung, and S. W. Zou, Anaerobic digestion. In: *Biological Treatment Processes*, L. K. Wang, N. C. Pereira, and Y. T. Hung (eds.), Humana Press, Totowa, NJ (2007).
27. N. K. Shammas and L. K. Wang, Aerobic digestion. In: *Biological Treatment Processes*, L. K. Wang, N. C. Pereira, and Y. T. Hung (eds.), Humana Press, Totowa, NJ (2008).
28. N. K. Shammas and L. K. Wang, Composting. In: *Biological Treatment Processes*, L. K. Wang, N. C. Pereira, and Y. T. Hung (eds.), Humana Press, Totowa, NJ (2008).
29. L. K. Wang, Y. Li, N. K. Shammas and G. P. Sakellaropoulos, Drying beds. In: *Biosolids Treatment Processes*, L. K. Wang, N. K. Shammas, and Y. T. Hung (eds.), Humana Press, Totowa, NJ (2007).
30. P. A. Vesilind, *Treatment and Disposal of Wastewater Sludges*, Ann Arbor Science Publishers, Ann Arbor, MI (1974).
31. R. R. Rimkus, J. M. Ryan, and R. W. Dring, A New approach to dewatering and disposal of lagooned digested Sludge. In: *Proceedings of the Annual Convention, ASCE*, Chicago, October (1978).
32. Sacramento Area Consultants, *Sewage Sludge Management Program Final Report, Volume 7, Environmental Impact Report and Advanced Site Planning*, Sacramento Regional County Sanitation District, Sacramento, CA, September (1979).
33. Brown and Caldwell, *Joint Regional Wastewater and Solids Treatment Facility Project Design Report*, Moulton-Miguel Water District, Aliso Water Management Agency, March (1978).
34. Georgia State Environmental Protection Division, *Guidelines for Land Application of Sewage Sludge (Biosolids) at Agronomic Rates*, June (2006).
35. M. Lang, National manual of good practice for biosolids. In: *Water Environment Federation Biosolids Specialty Conference*, San Diego, CA, February 21 (2001).
36. City Of Albany, *Biosolids Management Program*, DEQ/US EPA 2005 Annual Report, Albany, NY, January 25 (2006).
37. G. J. Humpal and J. S. Cannestra, *Offsite Biosolids Storage and Pumped Injection System at WalCoMet*, Walworth County Metropolitan Sewage District (WalCoMet), www.strand.com/conf_papers/walco-abst.pdf, Delavan, WI (2007).
38. U.S. Environmental Protection Agency, *Areawide Assessment Procedures Manual*, Vol. III, Report No. 600/9-76-014, Washington, DC, July (1976).
39. U.S. Army Corps of Engineers, *Civil Works Construction Cost Index System Manual*, 110-2-1304, Washington, DC, pp. 44, http://www.nww.usace.army.mil/ cost (2007).
40. Sacramento Area Consultants, *Sewage Sludge Management Program Final Report, Volume 2, SSB Operation and Performance*, Sacramento Regional County Sanitation District, Sacramento, CA, September (1979).
41. Sacramento Area Consultants, *Study of Wastewater Solids Processing and Disposal, Appendix C*, Sacramento Regional County Sanitation District, Sacramento, CA, June (1975).

42. Sacramento Area Consultants, *Innovative and Alternative Technology Documentation Sacramento Regional Wastewater Treatment Plant—Solids Project*, Sacramento Regional County Sanitation District. Sacramento, CA, April (1979).
43. H. C. Hawk (ed.) *Bulk Materials Handling*, University of Pittsburgh, School of Engineering, Pittsburgh, PA (1971).
44. National Lime Association, *Lime Handling Application and Storage in Treatment Processes, Bulletin 213*, Washington, DC, 2$^{nd}$ ed. (1971).
45. B. E. Hansen, D. L. Smith, and W. E. Garrison, Start-up problems of sludge dewatering facility. In: *Proceedings of the 51st Annual Water Pollution Control Federation Conference*, Anaheim, CA, October (1978).
46. Nova Scotia Environment and Labor, *Guidelines for Land Application and Storage of Biosolids in Nova Scotia*, Environmental Monitoring and Compliance, Nova Scotia, Canada, www.gov.ns.ca/enla/water/docs/BiosolidGuidelines.pdf, May (2004).
47. G. D. Lukasik and J. W. Cormack, *Development and Operation of a Sanitary Landfill for Sludge Disposal—North Shore Sanitary District*, North Shore Sanitary District (1976).
48. Carollo, *Biosolids Management*, Planners & Designers, http://www.carollo.com/231/section.aspx/download/44 (2007).
49. D. Borgatti, G. A. Romano, T. J. Rabbitt, and T. J. Acquaro, 1996 odor control program for the Springfield Regional WWTP. In: *New England WEA Annual Conference*, Boston, 26–29 January (1997).
50. R. T. Haug, An essay on the elements of odor management, *Biocycle*, 30, 10, 60–67 (1990).
51. J. M. Walker, Fundamentals of odor control, *Biocycle* 30, 9, 50–55 (1991).
52. P. Rosenfeld, *Characterization, Quantification, and Control of Odor Emissions from Biosolids Application to Forest Soil*, Ph.D. Dissertation, University of Washington, Seattle, WA (1999).
53. M. J. Higgins, G. Adams, T. Card, Y. Chen, Z. Erdal, R. H. Forbes, Jr., D. Glindemann, J. R. Hargreaves, L. Hentz, D. McEwen, S. N. Murthy, J. T. Novak, and J. Witherspoon, Relationship between biochemical constituents and production of odor causing compounds from anaerobically digested biosolids. In: *Odors and Air Emissions*, Water Environment Federation, Alexandria, VA (2004).
54. C. Wilber and C. Murray, Odor source evaluation, *BioCycle* 31, 3, 68–72 (1990).
55. C. Wilber (ed.), *Operations and Design at the Wastewater Treatment Plant to Control Ultimate Recycling and Disposal Odors of Biosolids*, U.S. Environmental Protection Agency, Washington, DC (2000).
56. M. S. Switzenbaum, L. H. Moss, E. Epstein and A. B. Pincince, *Defining Biosolids Stability: A Basis for Public and Regulatory Acceptance*, WERF Project 94-REM-1 Final Report, Water Environment Research Foundation, Alexandria, VA (1997).
57. J. E Smith and J. B. Farrell, *Vector Attraction Reduction Issues Associated with the Part 503 Regulations and Supplemental Guidance*, U.S. Environmental Protection Agency, Cincinnati, OH (1992).
58. C. M. McGinley, D. L. McGinley, and K. J. McGinley, Odor school—curriculum development for training odor investigators. In: *Air Water Management Association International Specialty Conference Proceedings Odors and Environmental Air*, pp. 121–127, Bloomington, MN, 13–15 September (1995).
59. F.V. Wilby, Variation in recognition odor threshold of a panel. *Journal of Air Pollution Control Association*, 19, 2, 96–100 (1969).
60. C. Lue-Hing, D. R. Zenz, and R. Kuchenrither, *Municipal Sludge Management-Processing, Utilization, Disposal, Water Quality Management Library*, Vol. 4, Technomic Publishing Co., Lancaster, PA (1992).

61. C. P. Gerba, Pathogens. In: *Proceedings: Workshop on Utilization of Municipal Wastewater and Sludge on Land*. A. L. Page, T. L. Gleason, III, J. S., Jr., I. K. Iskandar and L. E Sommers (eds.), University of California, Riverside, CA (1983).
62. U.S. Environmental Protection Agency, Preliminary Risk Assessment for Parasites in Municipal Sewage Sludge Applied to Land. US EPA Pub. No. 600/6-91/001, Cincinnati, OH (1991).
63. Code of Federal Regulations (CFR), *Standards for the Use and Disposal of Sewage Sludge, Title 40*, Volume 3, Parts 425 to 699, Federal Register February 19, 1993 (58 FR 9248), U.S. Government Printing Office, Washington, DC (1993).
64. National Research Council, *Use of Reclaimed Water and Sludge in Food Crop Production*, National Academy Press, Washington, DC, 178 pp. (1996).
65. U.S. Environmental Protection Agency, *Environmental Regulations and Technology—Control of Pathogens in Municipal Wastewater Sludge*, US EPA Pub. No. 625/10-89/006, Cincinnati, OH (1989).
66. U.S. Environmental Protection Agency, *Environmental Regulations and Technology—Control of Pathogens and Vector Attraction in Sewage Sludge*, US EPA Pub. No. 625/R-92/013, Cincinnati, OH (1992).
67. U.S. Environmental Protection Agency, *Preliminary Risk Assessment for Bacteria in Municipal Sewage Sludge Applied to Land*, US EPA Pub. No. 600/6-91/006, Columbus, OH (1991).
68. U.S. Environmental Protection Agency, *Preliminary Risk Assessment for Viruses in Municipal Sewage Sludge Applied to Land*, US EPA Pub. No. 600/R-92/064, Columbus, OH (1992).
69. U.S. Environmental Protection Agency, *Technical Support Document for Pathogen Reduction in Sewage Sludge*, Publication no. PB 89-136618, National Technical Information Service, Springfield, VA (1989).
70. L. K. Wang, N. K Shammas, and Y. T. Hung (eds.), *Biosolids Treatment Processes*, Humana Press, Totowa, NJ, 820 p. (2007).
71. J. B. Farrell, V. Bhide, and J. E. Smith, Jr., Development of US EPA's new methods to quantify vector attraction of wastewater sludges, *Water Environmental Research* 68, 3, 286–294 (1996).
72. J. E. Smith, Jr., and J. B. Farrell, Vector attraction reduction issues associated with the part 503 regulations and supplemental guidance. In: *Proceedings of the Water Environment Federation's Conference, International Management of Water and Wastewater Solids for the 21st Century: A Global Perspective*, Washington, DC, 1311–1330, June 19–22 (1994).
73. U.S. Environmental Protection Agency, *Health Effects of Land Application of Municipal Sludge*, US EPA 600/1-85/015, Research Triangle Park, NC (1985).
74. W. A. Yanko, A. S. Walker, J. L. Jackson, L. L. Libao, and A. L. Gracia, Enumerating Salmonella in biosolids for compliance with pathogen regulations, *Water Environmental Research* 67, 3, 364–370 (1995).
75. C. Fulhage, *Collection and Storage of Biosolids*, Department of Agricultural Engineering, University of Missouri, http://muextension.missouri.edu/explore/ envqual/wq0431.htm, June (2006).
76. Metro Wastewater Reclamation District, *Environmental Management System for Biosolids, EMS Program Performance Report*, http://www.metrowastewater.com/documents/EMS_2005_Performance_Report.pdf, Denver, CO (2005).

# APPENDIX

**U.S. Army Corps of Engineers Civil Works Construction Yearly Average Cost Index for Utilities (39)**

| Year | Index | Year | Index |
|------|-------|------|-------|
| 1967 | 100 | 1988 | 369.45 |
| 1968 | 104.83 | 1989 | 383.14 |
| 1969 | 112.17 | 1990 | 386.75 |
| 1970 | 119.75 | 1991 | 392.35 |
| 1971 | 131.73 | 1992 | 399.07 |
| 1972 | 141.94 | 1993 | 410.63 |
| 1973 | 149.36 | 1994 | 424.91 |
| 1974 | 170.45 | 1995 | 439.72 |
| 1975 | 190.49 | 1996 | 445.58 |
| 1976 | 202.61 | 1997 | 454.99 |
| 1977 | 215.84 | 1998 | 459.40 |
| 1978 | 235.78 | 1999 | 460.16 |
| 1979 | 257.20 | 2000 | 468.05 |
| 1980 | 277.60 | 2001 | 472.18 |
| 1981 | 302.25 | 2002 | 484.41 |
| 1982 | 320.13 | 2003 | 495.72 |
| 1983 | 330.82 | 2004 | 506.13 |
| 1984 | 341.06 | 2005 | 516.75 |
| 1985 | 346.12 | 2006 | 528.12 |
| 1986 | 347.33 | 2007 | 539.74 |
| 1987 | 353.35 |      |        |

# 6
# Regulations and Costs of Biosolids Disposal and Reuse

## Nazih K. Shammas and Lawrence K. Wang

**CONTENTS**
  INTRODUCTION
  LAND APPLICATION OF BIOSOLIDS
  SURFACE DISPOSAL OF BIOSOLIDS
  INCINERATION OF BIOSOLIDS
  PATHOGEN AND VECTOR ATTRACTION REDUCTION REQUIREMENTS
  COSTS
  ACRONYMS
  NOMENCLATURE
  APPENDIX

**Abstract** In February 1993, United States federal standards for the use or disposal of biosolids (Title 40 of the Code of Federal Regulations [CFR], Part 503) were enacted. As required by the Clean Water Act Amendments, the United States Environmental Protection Agency developed a new regulation to protect public health and the environment from any reasonably anticipated adverse effects of certain pollutants that might be present in sewage sludge biosolids. This Part 503 rule established the requirements for the final use or disposal of sewage sludge (biosolids) when biosolids are applied to land to condition the soil or fertilize crops or other vegetation grown in the soil, placed on a surface disposal site for final disposal (landfill) or fired in a biosolids incinerator.

This chapter addresses the Part 503 rule and discusses its five subparts: general provisions, requirements for land application, surface disposal, pathogen and vector attraction reduction, and incineration. The chapter also covers the pollutant limits; total hydrocarbons, pathogen, and vector attraction reduction practices; general requirements and management; frequency of monitoring; record keeping and reporting requirements; and costs.

**Key Words** Sewage sludge • biosolids • US CFR Part 503 Rule • land application • surface disposal • landfill • incineration • pollutants • hydrocarbons • pathogens • vector attraction • costs • Cost Index.

From: *Handbook of Environmental Engineering, Volume 7: Biosolids Engineering and Management*
Edited by: L. K. Wang, N. K. Shammas and Y. T. Hung © The Humana Press, Totowa, NJ

# 1. INTRODUCTION

## 1.1. Historical Background

The United States Clean Water Act (CWA) of 1977 (1) contained two major provisions for wastewater solids utilization and disposal. Section 405 required the U.S. Environmental Protection Agency (US EPA) to issue guidelines and regulations for the disposal and reuse of wastewater solids. The other major provision was intended to encourage sludge utilization. This provision, Section 307, required pretreatment of industrial wastes if such wastes inhibit wastewater treatment or sludge utilization. This should increase the potential for sludge reuse.

The Resource Conservation and Recovery Act (RCRA) of 1976 (2) required that solid wastes be utilized or disposed of in a safe and environmentally acceptable manner. Wastewater solids were included by definition in provisions relating to solid waste management. The US EPA developed guidelines and criteria to implement the provisions of this act. These guidelines and criteria fall into three general categories: (a) treatment and disposal of potentially hazardous solid wastes (wastewater solids were excluded from this category); (b) criteria and standards for solid waste disposal facilities; and (c) criteria defining the limits for solid waste application to agricultural lands.

The Toxic Substances Control Act (TSCA) of 1976 (3) authorized the US EPA to obtain production and test data from industry on selected chemical substances and to regulate them where they pose an unreasonable risk to the environment. This act, in combination with other federal legislation, should help reduce the amount of pollutants discharged to the municipal system from manufacturing processes. Of particular significance to wastewater solids utilization is the fact that the act prohibited the production of polychlorinated biphenyls (PCBs) after January 1979 and the commercial distribution of PCBs after July 1979. The PCBs can be concentrated in wastewater sludges and is a chemical constituent of concern in meeting proposed utilization criteria. Sludge PCB levels started to decrease once PCBs no longer entered the waste treatment system.

Several large cities, including New York and Philadelphia, as well as some smaller cities in the New York–New Jersey area were disposing of their wastewater solids by barging them to the ocean. The 1977 amendments to the Marine Protection, Research, and Sanctuaries Act (MPRSA) of 1972 (4), as well as other laws and regulations, prohibited disposal of "sewage sludge" by barging since December 31, 1981. In addition, federal construction funds were no longer available for wastewater solids treatment and disposal systems that included any type of ocean disposal, either by barge or pipeline.

The National Environmental Policy Act (NEPA) of 1969 (5) required that the federal government consider the environmental effects of many actions. Municipal wastewater treatment systems, including solids treatment, utilization, and disposal systems were covered by this act because of their potential effect on the environment and because they were funded by federal construction grants. Most states have similar policy acts. The acts, which require reports and hearings, ensured that the environmental consequences of proposed operations are considered, and also provided the designer with a useful forum to develop public response. They did, however, usually lengthen the facility planning and design process.

While most states and municipalities followed federal guidelines, many have formulated more restrictive measures. For example, localities that apply sludge to land on which food crops are grown may wish to analyze their sludges more frequently than required by federal guidelines or limit sludge application rates more severely.

As indicated, Section 405 of the Clean Water Act of 1977 required the US EPA to promulgate regulations governing the issuance of permits for the disposal of sewage sludge relative to Section 402 National Pollutant Discharge Elimination System (NPDES) permits and to develop and publish from time to time regulations providing guidelines for the disposal and utilization of sludge. These regulations are to identify uses for sludge, specify factors to be taken into account in determining the measures and practices applicable to each such use or disposal (including publication of information on costs), and identify concentrations of pollutants that interfere with each such use or disposal.

This broad authority to issue regulations covering different sludge management practices has been viewed as a mechanism to allow the US EPA to bring together all of the regulations that have been or will be issued under various legislative authorities for controlling municipal sludge management at a single location in the Code of Federal Regulations (CFR), under the joint authority of Section 405. Therefore, regulations on air emission controls were to be issued under the joint authority of Section 405 of the Clean Water Act and various sections of the Clean Air Act (CAA) of 1970 (6); regulations on land disposal and land application under joint authority with the RCRA; regulations on ocean disposal under joint authority with the MPRSA, and so forth. Regulations covering practices not influenced by other authorities (for example, home use, give-away or sale of sludge derived products) were to be issued solely under the broad authority of Section 405.

In February 1993, federal standards for the use or disposal of biosolids (40 CFR Part 503) were enacted (7). As required by the Clean Water Act Amendments of 1987, the US EPA developed a new regulation to protect public health and the environment from any reasonably anticipated adverse effects of certain pollutants that might be present in sewage sludge biosolids. This Part 503 rule, or simply the 503 rule, addresses land application and beneficial use of biosolids.

## 1.2. Background of the Part 503 Rule

This chapter refers exclusively to sewage sludge as biosolids. Biosolids are a primarily organic solid product produced by wastewater treatment processes that can be beneficially recycled. The fact that the biosolids can be recycled does not preclude their being disposed. The Part 503 rule establishes requirements for the final use or disposal of sewage sludge (biosolids) when biosolids are (8)

1. applied to land to condition the soil or fertilize crops or other vegetation grown in the soil,
2. placed on a surface disposal site for final disposal, or
3. fired in a biosolids incinerator.

The rule also indicates that if biosolids are placed in a municipal solid waste landfill, the biosolids must meet the provisions of 40 CFR Part 258.

The Part 503 rule is designed to protect public health and the environment from any reasonably anticipated adverse effects of certain pollutants and contaminants that may be present in biosolids. The provisions of the Part 503 rule are consistent with the US EPA's policy of promoting beneficial uses of biosolids (9). Land application takes advantage of the soil conditioning and fertilizing properties of biosolids.

It is important to note that persons using or disposing of biosolids are subject to state and possibly local regulations as well. Furthermore, these state and other regulations may be more stringent generally than the federal Part 503 rule, may define biosolids differently, or may regulate certain types of biosolids more stringently than does the Part 503 rule.

### 1.3. Risk Assessment Basis of the Part 503 Rule

Many of the requirements of the Part 503 rule are based on the results of an extensive multimedia risk assessment. This risk assessment was more comprehensive than for any previous federal biosolids rule-making effort, the earliest of which began in the mid-1970s. Research results and operating experience over the past three decades have greatly expanded the understanding of the risks and benefits of using or disposing of biosolids.

Development of the Part 503 rule began in 1984. During this extensive effort, the US EPA addressed 25 pollutants using 14 exposure pathways in the risk assessment. In this assessment, the US EPA also developed a new methodology that provided for the protection of the environment and public health. The new method for conducting the multimedia risk assessment was reviewed and approved by the US EPA's Science Advisory Board.

The US EPA proposed the Part 503 rule in February 1989. During the 4 years between the publication of the proposed and final rule, the data, models, and assumptions used in the risk assessment process were reviewed and revised in an effort involving internationally recognized experts working closely with the US EPA. This process has resulted in the establishment of state-of-the-art risk-based standards for controlling the use or disposal of biosolids. Detailed information describing the risk assessment and technical basis of the Part 503 standards is contained in the Preamble to the Part 503 rule and in several technical support documents (10).

### 1.4. Overview of the Rule

The Part 503 rule includes five subparts (8):

1. General provisions
2. Requirements for land application
3. Requirements for surface disposal
4. Requirements for pathogen and vector attraction reduction
5. Requirements for incineration

For each of the regulated use or disposal practices, a Part 503 standard includes general requirements, pollutant limits, management practices, operational standards, and requirements for the frequency of monitoring, record keeping, and reporting, as shown in Figure 6.1. For the most part, the requirements of the Part 503 rule are self-implementing and must be followed even without the issuance of a permit (8, 11).

# Regulations and Costs of Biosolids Disposal and Reuse

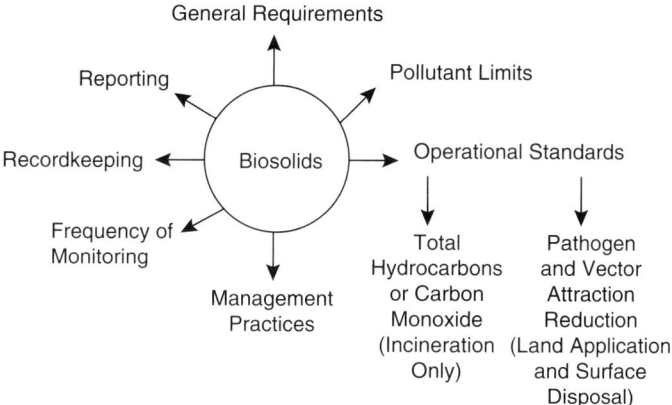

**Fig. 6.1.** Part 503 standards for regulated use and disposal practices. *Source*: US EPA (8).

Part 503 specifies certain exclusions from the rule. These exclusions are listed in Table 6.1. Also listed in Table 6.1 are the federal regulations that apply to biosolids-related activities not covered by the Part 503 rule.

## 2. LAND APPLICATION OF BIOSOLIDS

Land application is the application of biosolids to land to either condition the soil or to fertilize crops or other vegetation grown in the soil. More than half of the biosolids production in the U.S. is currently being used beneficially to improve soils. The categories or types of land that benefit from the application of biosolids are as follows (12):

1. Agricultural land, forests, and reclamation sites, collectively called "nonpublic contact sites" (areas not frequently visited by the public).
2. Public parks, plant nurseries, roadsides, golf courses, lawns, and home gardens, collectively called "public contact sites" (areas where people are likely to come into contact with biosolids applied to land). The Part 503 rule, however, does not regard lawns and home gardens as public contact sites, and fewer types of biosolids may be land applied to these sites (i.e., cumulative pollutant loading rate [CPLR] biosolids are not permitted on lawns and home gardens given the considerable difficulty of tracking cumulative levels of metals in biosolids applied to such sites).

Biosolids can be either applied to land in bulk or sold or given away in bags or other containers for land application. The term *biosolids in bulk* refers to biosolids that are marketed or given to manufacturers of products that contain biosolids. The term *biosolids in bags* generally refers to biosolids in amounts that are bagged and generally marketed for use on smaller units of land such as lawns and home gardens. The term *other containers* is defined in the Part 503 rule as open or closed receptacles (e.g., buckets, boxes, or cartons) or vehicles with a load capacity of 1 metric ton or less. (Most pickup trucks as well as trailers pulled by an automobile would meet the regulatory definition of other containers.)

Biosolids are generally land applied using one of several techniques. The biosolids may be sprayed or spread on the soil surface and left on the surface (e.g., on pastures, range and forest land, or lawn). They also may be tilled (incorporated) into the soil after

Table 6.1
Exclusions from Part 503 rule

| Subjects for which no requirements are included in Part 503 rule | Applicable U.S. federal regulation |
|---|---|
| *Treatment of biosolids* Processes used to treat sewage sludge prior to final use or disposal (e.g., thickening, dewatering, storage, heat drying) | None (except for operational parameters used to meet the Part 503 pathogen and vector attraction reduction requirements) |
| *Selection of use or disposal Practice* The selection of a biosolids use or disposal practice | None (the determination of the biosolids use or the disposal practice is a local decision) |
| *Incineration of biosolids with other wastes* Biosolids co-fired in an incinerator with other wastes (other than as an auxiliary fuel) | 40 CFR Parts 60, 61 |
| *Storage of biosolids* Placement of biosolids on land for 2 years or less (or longer when demonstrated not to be a surface disposal site but rather, based on practices, constitutes treatment or temporary storage) | None |
| *Industrial sludge* Sludge generated at an industrial facility during the treatment of industrial wastewater with or without combined domestic sewage | 40 CFR Part 257 if land applied 40 CFR Part 258 if placed in a municipal solid waste landfill |
| *Hazardous sewage sludge* Sewage sludge determined to be hazardous in accordance with 40 CFR Part 261, *Identification and Listing of Hazardous Waste* | 40 CFR Parts 261–268 |

| | |
|---|---|
| *Sewage sludge containing PCBs 250 mg/kg* | |
| Sewage sludge with a concentration of polychlorinated biphenyls (PCBs) equal to or greater than 50 milligrams per kilogram of total solids (dry-weight basis) | 40 CFR Part 761 |
| *Incinerator ash* | |
| Ash generated during the firing of biosolids in a biosolid incinerator | 40 CFR Part 257 if land applied |
| | 40 CFR Part 258 if placed in a municipal solid waste landfill |
| | 40 CFR Parts 261–268 if hazardous |
| *Grit and screenings* | |
| Grit (e.g., sand, gravel, cinders) or screenings (e.g., relatively large materials such as rags) generated during preliminary treatment of domestic sewage in a treatment works | 40 CFR Part 257 if land applied |
| | 40 CFR Part 258 if placed in a municipal solid waste landfill |
| *Drinking water sludge* | |
| Sludge generated during the treatment of either surface water or ground water used for drinking water | 40 CFR Part 257 if land applied |
| | 40 CFR Part 258 if placed in a municipal solid waste landfill |
| *Certain nondomestic septage* | |
| Septage that contains industrial or commercial septage, including grease-trap pumpings | 40 CFR Part 257 if land applied |
| | 40 CFR Part 258 if placed in a municipal solid waste landfill |

being surface applied or injected directly below the surface for producing row crops or other vegetation and for establishing lawns (13, 14).

Biosolids in a liquid state can be applied using tractors, tank wagons, irrigation systems, or special application vehicles. Dewatered biosolids are typically applied to land using equipment similar to that used for applying limestone, animal manures, or commercial fertilizers. Both liquid and dewatered biosolids are applied to land with or without subsequent incorporation into the soil. Because biosolids are typically treated before being land applied, their use poses a low degree of risk. This section discusses approaches for meeting the requirements of the Part 503 rule for the land application of biosolids.

Biosolids applied to the land must meet risk-based pollutant limits specified in Part 503. Operational standards to control disease-causing organisms called pathogens and to reduce the attraction of vectors (e.g., flies, mosquitoes, and other potential disease-carrying organisms) to the biosolids must also be met. In addition, there are general requirements, management practices, and frequency of monitoring, record keeping, and reporting requirements that must be met. Each of these land application requirements is discussed below (8).

### 2.1. Pollutant Limits, and Pathogen and Vector Attraction Reduction Requirements

First, all biosolids applied to the land must meet the ceiling concentrations for pollutants, listed in the first column of Table 6.2. The ceiling concentrations are the maximum concentration limits for 10 heavy metal pollutants in biosolids; specifically, arsenic, cadmium, chromium, copper, lead, mercury, molybdenum, nickel, selenium, and zinc. If a limit for any one of the pollutants is exceeded, the biosolids cannot be applied to the land until such time that the ceiling concentration limits are no longer exceeded. The ceiling concentrations for pollutants are included in Part 503 to prevent the land application of biosolids with the highest levels of pollutants and to encourage pretreatment efforts that will result in lower levels of pollutants.

Second, biosolids applied to the land must also meet either pollutant concentration limits, cumulative pollutant loading rate limits, or annual pollutant loading rate limits for these same heavy metals.

Third, *e*ither Class A or Class B pathogen requirements (summarized in Table 6.3) and site restrictions (Table 6.4) must be met before the biosolids can be land applied; the two classes differ depending on the level of pathogen reduction that has been obtained.

Fourth, one of 10 options specified in Part 503 and summarized in Table 6.5 to achieve vector attraction reduction (VAR) must be met when biosolids are applied to the land.

### 2.2. Options for Meeting Land Application Requirements

The Part 503 requirements can be grouped into four options for meeting pollutant limits and pathogen and vector attraction reduction operational standards when biosolids are applied to the land. The options include the following:

1. The exceptional quality (EQ) option
2. The pollutant concentration (PC) option

**Table 6.2**
**Pollutant limits**

| Pollutant | Ceiling concentration limits for all biosolids applied to land mg/kg[a] | Pollutant concentration limits for EQ and PC biosolids mg/kg[a] | Cumulative pollutant loading rate limits for CPLR biosolids kg/ha | Annual pollutant loading rate limits for APLR biosolids kg/ha/yr |
|---|---|---|---|---|
| Arsenic | 75 | 41 | 41 | 2.0 |
| Cadmium | 85 | 39 | 39 | 1.9 |
| Chromium | 3000 | 1200 | 3000 | 150 |
| Copper | 4300 | 1500 | 1500 | 75 |
| Lead | 840 | 300 | 300 | 15 |
| Mercury | 57 | 17 | 17 | 0.85 |
| Molybdenum[b] | 75 | — | — | — |
| Nickel | 420 | 420 | 420 | 21 |
| Selenium | 100 | 36 | 100 | 5.0 |
| Zinc | 7500 | 2800 | 2800 | 140 |
| Applies to: | All biosolids that are land applied | Bulk biosolids and bagged biosolids | Bulk biosolids | Bagged biosolids[c] |

[a]Dry-weight basis.
[b]As a result of the amendment to the rule, the limits for molybdenum were deleted from the Part 503 rule.
[c]Bagged biosolids are sold or given away in a bag or other container.
APLR, annual pollutant loading rate; CPLR, cumulative pollutant loading rate; EQ, exceptional quality; PC, pollutant concentration.

**Table 6.3**
**Summary of Class A and Class B Pathogen Reduction Requirements**

**Class A**

In addition to meeting the requirements in one of the six alternatives listed below, fecal coliform or *Salmonella* sp. bacteria levels must meet specific density requirements at the time of biosolids use or disposal or when prepared for sale or give-away

*Alternative 1: thermally treated biosolids*
Use one of four time-temperature regimens
*Alternative 2: biosolids treated in a high pH–high temperature process*
Specifies pH, temperature, and air-drying requirements
*Alternative 3: for biosolids treated in other processes*
Demonstrate that the process can reduce enteric viruses and viable helminth ova; maintain operating conditions used in the demonstration
*Alternative 4: biosolids treated in unknown processes*
Demonstration of the process is unnecessary; instead, test for pathogens—*Salmonella* sp. or fecal coliform bacteria, enteric viruses, and viable helminth ova—at the time the biosolids are used or disposed of or are prepared for sale or give away
*Alternative 5: use of PFRP*
Biosolids are treated in one of the processes to further reduce pathogens (PFRP)
*Alternative 6: use of a process equivalent to PFRP*
Biosolids are treated in a process equivalent to one of the PFRPs, as determined by the permitting authority

**Class B**

The requirements in one of the three alternatives below must be met

*Alternative 1: monitoring of indicator organisms*
Test for fecal coliform density as an indicator for all pathogens at the time of biosolids use or disposal
*Alternative 2: use of PSRP*
Biosolids are treated in one of the processes to significantly reduce pathogens (PSRP)
*Alternative 3: use of processes equivalent to PSRP*
Biosolids are treated in a process equivalent to one of the PSRPs, as determined by the permitting authority

3. The cumulative pollutant loading rate (CPLR) option
4. The annual pollutant loading rate (APLR) option

It is very important to realize that each option is equally protective of public health and the environment; that is, EQ, PC, CPLR, and APLR biosolids used in accordance with the Part 503 rule are equally safe. This safety is ensured by the combination of pollutant limits and management practices imposed by each option.

Whichever option is chosen, at a minimum, the ceiling concentrations for pollutants (Table 6.2) and the frequency of monitoring (see Table 6.11), reporting, and recordkeeping requirements must be met. The four options are summarized in Table 6.6, illustrated in Figure 6.2, and discussed in detail below.

**Table 6.4**
**Restrictions for the harvesting of crops and turf, grazing of animals, and public access on sites where Class B biosolids are applied**

Restrictions for the harvesting of crops* and turf

1. Food crops, feed crops, and fiber crops, whose edible parts do not touch the surface of the soil, shall not be harvested until *30 days* after biosolids application
2. Food crops with harvested parts that touch the biosolids/soil mixture and are totally above ground shall not be harvested until *14 months* after application of biosolids
3. Food crops with harvested parts below the land surface where biosolids remain on the land surface for 4 months or longer prior to incorporation into the soil shall not be harvested until *20 months* after biosolids application
4. Food crops with harvested parts below the land surface where biosolids remain on the land surface for less than 4 months prior to incorporation shall not be harvested until *38 months* after biosolids application
5. Turf grown on land where biosolids are applied shall not be harvested until *1 year* after application of the biosolids when the harvested turf is placed on either land with a high potential for public exposure or a lawn, unless otherwise specified by the permitting authority

Restriction for the grazing of animals

1. Animals shall not be grazed on land until *30 days* after application of biosolids to the land

Restrictions for public contact

1. Access to land with a high potential for public exposure, such as a park or ball field, is restricted for *1 year* after biosolids application; examples of restricted access include posting with no trespassing signs, and fencing
2. Access to land with a low potential for public exposure (e.g., private farmland) is restricted for *30 days* after biosolids application; an example of restricted access is remoteness

*Examples of crops impacted by Class B pathogen requirements are listed in Table 6.9.

Depending on the land application option under consideration, site restrictions (Table 6.4), general requirements, and management practices also apply. These additional restrictions, requirements, and practices are summarized in Tables 6.7 and 6.8 and discussed in greater detail at a later point of this chapter. Table 6.7 graphically displays the level of required regulatory control for each option. The types of land onto which these different biosolids may be applied are listed in Table 6.8.

*2.2.1. Option 1: Exceptional Quality Biosolids*

For biosolids to qualify under the EQ option, the following requirements must be met:

1. The ceiling concentrations for pollutants in Table 6.2 may not be exceeded.
2. The pollutant concentration limits in Table 6.2 may not be exceeded.
3. One of the Class A pathogen requirements in Table 6.3 must be met.
4. One of the first eight vector attraction reduction options in Table 6.5 must be achieved.

Methods that typically achieve the pathogen and vector attraction reduction requirements and allow biosolids to meet EQ requirements include alkaline stabilization, composting, and heat drying (15). The Part 503 frequency of monitoring, record keeping, and reporting requirements also must be met for EQ biosolids.

**Table 6.5**
**Summary of vector attraction reduction options***

| Option | Requirements that must be met |
|---|---|
| *Option 1* | Reduce the mass of volatile solids by a minimum of 38% |
| *Option 2* | Demonstrate vector attraction reduction with additional anaerobic digestion in a bench-scale unit |
| *Option 3* | Demonstrate vector attraction reduction with additional aerobic digestion in a bench-scale unit |
| *Option 4* | Meet a specific oxygen uptake rate for aerobically treated biosolids |
| *Option 5* | Use aerobic processes at greater than 40°C (average temperatures 45°C) for 14 days or longer (e.g., during biosolids composting) |
| *Option 6* | Add alkaline materials to raise the pH under specified conditions |
| *Option 7* | Reduce moisture content of biosolids that do not contain unstabilized solids from other than primary treatment to at least 75% solids |
| *Option 8* | Reduce moisture content of biosolids with unstabilized solids to at least 90% |
| *Option 9* | Inject biosolids beneath the soil surface within a specified time, depending on the level of pathogen treatment |
| *Option 10* | Incorporate biosolids applied to or placed on the land surface within specified time periods after application to or placement on the land surface |

*Note: Details of each vector attraction reduction option are provided later in the chapter.

Once biosolids meet EQ requirements, they are not subject to the land application general requirements and management practices in Part 503, with one possible exception: if the regional administrator or the state director determines, on a case-by-case basis, that such requirements are necessary to protect public health and the environment (this exception applies only to bulk biosolids). Once biosolids have been established as meeting EQ requirements, whether in bulk form or in bags or other containers, they can generally be applied as freely as any other fertilizer or soil amendment to any type of land. While not required by the Part 503 rule, EQ biosolids should be applied at a rate that does not exceed the agronomic rate that supplies the nitrogen needs of the plants being grown, just as for any other commercial fertilizer or soil amending material that contains nitrogen.

### 2.2.2. Option 2: Pollutant Concentration Biosolids

To qualify under the PC option, biosolids must meet several requirements, including the following:

1. The ceiling concentration for pollutants in Table 6.2 may not be exceeded.
2. The pollutant concentration limits in Table 6.2 may not be exceeded.
3. One of three Class B pathogen requirements must be met (Table 6.3), as well as Class B site restrictions (Tables 6.4 and 6.9).
4. One of 10 vector attraction reduction options must be achieved (Table 6.5).
5. Applicable site restrictions (Tables 6.4, 6.7, and 6.8), general requirements, and management practices must be met.
6. Frequency of monitoring (see Table 6.11) as well as record-keeping and reporting requirements must be met.

**Table 6.6**
**Options for meeting pollutant limits and pathogen and vector attraction reduction requirements for land application**

| Option* | Pollutant limits | Pathogen requirements | Vector attraction reduction requirements |
|---|---|---|---|
| Exceptional quality (EQ) biosolids | Bulk or bagged biosolids meet pollutant concentration limits in Table 6.2 | Any one of the Class A requirements in Table 6.3 | Any one of the requirements in options 1 through 8 in Table 6.5 |
| Pollutant concentration (PC) biosolids | Bulk biosolids meet pollutant concentration limits in Table 6.2 | Any one of the Class B requirements in Tables 6.3 and 6.4 | Any one of the 10 requirements in Table 6.5 |
|  |  | Any one of the Class A requirements in Table 6.3 | Requirements 9 or 10 in Table 6.5 |
| Cumulative pollutant loading rate (CPLR) biosolids | Bulk biosolids applied subject to cumulative pollutant loading rate (CPLR) limits in Table 6.2 | Any one of the Class A or Class B requirements in Tables 6.3 and 6.4 | Any one of the 10 requirements in Table 6.5 |
| Annual pollutant loading rate (APLR) biosolids | Bagged biosolids applied subject to annual pollutant loading rate (APLR) limits in Table 6.2 | Any one of the Class A requirements in Table 6.3 | Any one of the first 8 requirements in Table 6.5 |

*Each of these options also requires that the biosolids meet the ceiling concentrations for pollutants listed in Table 6.2, and that the frequency of monitoring requirements and record keeping and reporting requirements are met. In addition, the general requirements and the management practices have to be met when biosolids are land applied (except for EQ biosolids).

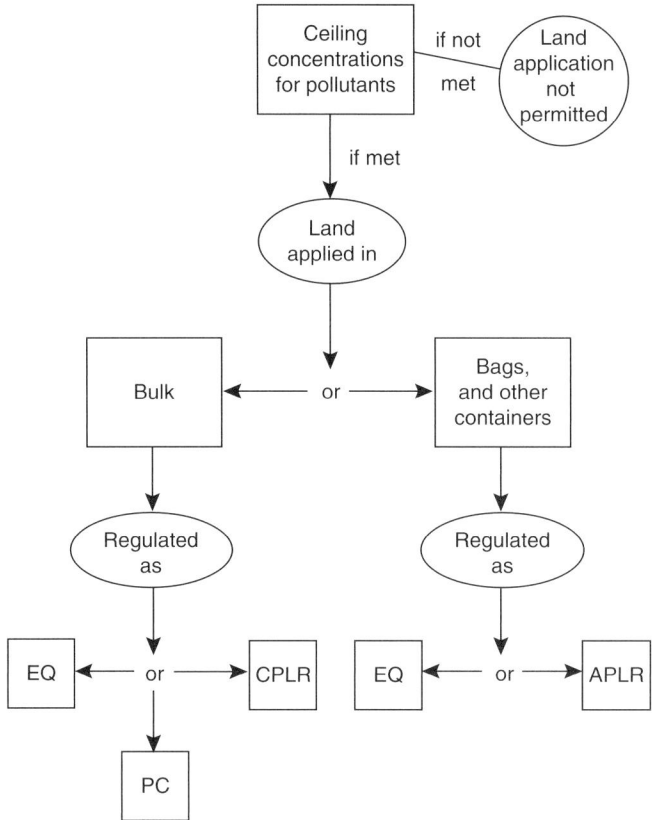

**Fig. 6.2.** Options for meeting certain Part 503 land application requirements. APLR, annual pollutant loading rate; CPLR, cumulative pollutant loading rate; EQ, exceptional quality; PC, pollutant concentration. *Source*: US EPA (8).

Class A biosolids meeting VAR requirements 9 and 10 in Table 6.5 are another type of biosolids material that would fit in the PC category. Thus, PC biosolids must meet more requirements than EQ biosolids, but are subject to fewer requirements than CPLR biosolids. The majority of biosolids in the U.S. could be characterized as PC biosolids (8, 16, 17).

*2.2.3. Option 3: Cumulative Pollutant Loading Rate Biosolids*

The third option for meeting land application requirements allows bulk biosolids that do not meet the pollutant concentration limits in Table 6.2 to be land applied as safely as EQ and PC biosolids. To qualify as CPLR biosolids, the following requirements must be met:

1. The ceiling concentrations for pollutants in Table 6.2 may not be exceeded.
2. The CPLRs listed in Table 6.2 may be not be exceeded.
3. Either the Class A or Class B pathogen requirements in Table 6.3 must be met.
4. One of the 10 vector attraction reduction options in Table 6.5 must be met.

**Table 6.7**
**Summary of regulatory requirements for different types of biosolids**

| Type of biosolids and class of pathogens | Meet ceiling concentration for pollutants | Meet pollutant concentration limits | Site restrictions | General requirements and management practices | Track added pollutants |
|---|---|---|---|---|---|
| EQ Bag or bulk Class A | Yes | Yes | No | No | No |
| PC Bulk only Class A[a] | Yes | Yes | No | Yes | No |
| PC Bulk only Class B | Yes | Yes | Yes | Yes | No |
| CPLR Bulk only Class A | Yes | No | No | Yes | Yes |
| CPLR Bulk only Class B | Yes | No | Yes | Yes | Yes |
| APLR Bag only Class A | Yes | No | No | Yes[b] | Yes[c] |

[a]Biosolids meeting Class A pathogen reduction requirements but following options 9 or 10 vector attraction reduction requirements are also considered PC biosolids.
[b]The only general and management practice requirement that must be met is a labeling requirement.
[c]The amount of biosolids that can be applied to a site during the year must be consistent with the annual whole sludge application rate (AWSAR) for the biosolids that does not cause any of the ALPRs to be exceeded.

Table 6.8
Types of land onto which different types of biosolids may be applied

| Biosolids option | Pathogen class | VAR[a] options | Type of land | Other restrictions |
|---|---|---|---|---|
| EQ | A | 1–8 | All[b] | None |
| PC | A | 9 or 10 | All except lawn and home gardens[c] | Management practices |
|  | B | 1–10 | All except lawn and home gardens[c] | Management practices and site restrictions |
| CPLR | A | 1–10 | All except lawn and home garden[d] | Management practices |
|  | B | 1–10 | All except lawn and home garden[c,d] | Management practices and site restrictions |
| APLR | A | 1–8 | All, but most likely lawns and home gardens | Labeling management practice |

[a]Vector attraction reduction.
[b]Agricultural land, forest, reclamation sites, and lawns and home gardens.
[c]It is not possible to impose site restrictions on lawns and home gardens.
[d]It is not possible to track cumulative additions of pollutants on lawns and home gardens.

Table 6.9
Examples of crops impacted by site restrictions for Class B biosolids

| | Harvested parts that: | |
|---|---|---|
| Usually do not touch the soil/biosolids mixture | Usually touch the soil/biosolids mixture | Are below the soil/biosolids mixture |
| Peaches | Melons | Potatoes |
| Apples | Strawberries | Yams |
| Oranges | Eggplant | Sweet potatoes |
| Grapefruit | Squash | Rutabaga |
| Corn | Tomatoes | Peanuts |
| Wheat | Cucumbers | Onions |
| Oats | Celery | Leeks |
| Barley | Cabbage | Radishes |
| Cotton | Lettuce | Turnips |
| Soybeans |  | Beets |

5. Applicable site restrictions (Tables 6.4, 6.7, and 6.8), general requirements, and management practices must be met.
6. Frequency of monitoring (see Table 6.11) as well as record-keeping and reporting requirements must be met.

The CPLR is the maximum amount of regulated pollutants in biosolids that can be applied to a site. When the CPLR for any one of the 10 heavy metals listed in Table 6.2

is reached at a site, no additional bulk biosolids, subject to the CPLR limits, may be applied to the site.

### 2.2.4. Option 4: Annual Pollutant Loading Rate (APLR) Biosolids

The fourth option only applies to biosolids that are sold or given away in a bag or other container for application to land. Under this option, the following requirements must be met:

1. The ceiling concentrations for pollutants in Table 6.2 may not be exceeded.
2. The APLRs listed in Table 6.2 may not be exceeded.
3. The Class A pathogen requirements in Table 6.3 must be met.
4. One of the first eight vector attraction reduction options in Table 6.5 must be met.
5. Applicable site restrictions (Tables 6.4, 6.7, and 6.8), general requirements, and management practices must be met.
6. The frequency of monitoring (see Table 6.11) as well as record-keeping and reporting requirements must be met.

An APLR is the maximum amount of regulated pollutants in biosolids that can be applied to a site in any 1 year. The APLRs rather than the CPLRs are used for biosolids sold or given away in a bag or other container because tracking the amount of pollutants applied in biosolids is not feasible in this situation.

Labeling for bagged or containerized APLR biosolids is required. To meet the labeling requirement, the preparer of biosolids must calculate the amount of biosolids that can be applied to a site during the year so that none of the APLRs are exceeded. This amount of biosolids is referred to as the annual whole sludge application rate (AWSAR). The AWSAR can be determined once the pollutant concentrations in the biosolids are known. The AWSAR must be calculated for each of the 10 metals listed in Table 6.2, and the lowest AWSAR for the 10 metals is the allowable AWSAR for the biosolids. The AWSAR on the required label or information sheet has to be equal to or less than the AWSAR calculated using the procedure shown below.

The procedure to determine the annual whole sludge (biosolids) application rate for biosolids sold or given away in a bag or other container is as follows:

1. Analyze a sample of the biosolids to determine the concentration of each of the 10 regulated metals in the biosolids.
2. Using the pollutant concentrations from step 1 and the APLRs from Table 6.2, calculate an AWSAR for each pollutant using Eq. 1:

$$AWSAR = \frac{APLR}{(C)(0.001)} \quad (1)$$

where
$AWSAR$ = annual whole sludge (biosolids) application rate (dry T of biosolids/ha/yr)
$APLR$ = annual pollutant loading rate (in Table 6.2 kg of pollutant/ha/yr)
$C$ = Pollutant concentration (mg of pollutant/kg of biosolids, dry weight)
$0.001$ = a conversion factor

3. The AWSAR for the biosolids is the lowest AWSAR calculated for each pollutant in step 2.

**Table 6.10**
**Calculation of the annual whole sludge (biosolids) application rate for Example 1**

| Metal | Biosolids concentrations (mg/kg) | APLR (kg/ha/yr) | AWSAR = APLR/C (0.001) = T/ha |
|---|---|---|---|
| Arsenic | 10 | 2.0 | $2/(10 \times 0.001) = 200$ |
| Cadmium | 10 | 1.9 | $1.9/(10 \times 0.001) = 190$ |
| Chromium | 1000 | 150 | $150/(1000 \times 0.001) = 150$ |
| Copper | 3750 | 75 | $75/(3750 \times 0.001) = 20$ |
| Lead | 150 | 15 | $15/(150 \times 0.001) = 100$ |
| Mercury | 2 | 0.85 | $0.85/(2 \times 0.001) = 425$ |
| Nickel | 100 | 21 | $21/(100 \times 0.001) = 210$ |
| Selenium | 15 | 5.0 | $5/(15 \times 0.001) = 333$ |
| Zinc | 2000 | 140 | $140/(2000 \times 0.001) = 70$ |

While not required by the Part 503 rule, it would also be good practice to provide information about the nitrogen content of the biosolids as well as the AWSAR on the label or information sheet that accompanies the biosolids. Calculations that can be useful for determining how much nitrogen is being applied to land relative to the AWSAR and the nitrogen requirements of the plants being grown are given in later following section.

EXAMPLE 1

Determine the annual whole sludge (biosolids) application rate for biosolids sold or given away in a bag or other container.

SOLUTION

1. Biosolids to be applied to land are analyzed for each of the 10 metals regulated in Part 503. Analysis of the biosolids indicates the pollutant concentration in the second column of Table 6.10.
2. Using these test results and the APLR for each pollutant from Table 6.2, the AWSAR for all the pollutants are calculated as shown in the fourth column of Table 6.10.
3. The AWSAR for the biosolids is the lowest AWSAR calculated for all 10 metals. In our example, the lowest AWSAR is for copper at 20 T of biosolids/ha/yr. Therefore, the controlling AWSAR to be used for the biosolids is 20 T/ha/yr. The 20 T of biosolids/ha is the same as 410 lb of biosolids/1,000 ft$^2$ (since 1 T = 2205 lb; 1 ha = 107,600 ft$^2$).
4. The AWSAR on the label or information sheet would have to be equal to or less than 410 lb/1000 ft$^2$.

## 2.3. General Requirements and Management Practices

The Part 503 general requirements and management practices must be met for all but EQ biosolids. The specific general requirements and kinds of management practices that apply to each type of biosolids are discussed below.

### 2.3.1. Endangered Species

The Part 503 rule prohibits the application of bulk biosolids to land if it is likely to adversely affect endangered or threatened species or their designated critical habitat. Any direct or indirect action that reduces the likelihood of survival and recovery of an endangered or threatened species is considered an adverse effect. Critical habitat is any place where an endangered or threatened species lives and grows during its life cycle. The U.S. Fish and Wildlife Service (FWS) has a Threatened and Endangered Species System (TEES) on its Web site that lists all endangered species (18).

Practices that involve applying biosolids to lands (subjected to normal tillage, cropping, and grazing practices, or mining, forestry, and other activities that by their nature are associated with turning the soil and affecting vegetation) are not likely to result in any increase in negative impacts on endangered species and in fact may be beneficial given the nutritive and soil-building properties of biosolids. It is the responsibility of the land applier, however, to determine if the application of biosolids might cause an adverse effect on an endangered species or its critical habitat. Moreover, the Part 503 rule requires the land applier to certify that the applicable management practices have been met, including the requirement concerning endangered species, and those records are kept indicating how the applicable management practices have been met.

*2.3.2. Flooded, Frozen, or Snow-Covered Land*

Application of biosolids to flooded, frozen, or snow-covered land is not prohibited by the Part 503 rule. Appliers must ensure, however, that biosolids applied to such land does not enter surface waters or wetlands unless specifically authorized by a permit issued under the Clean Water Act (CWA) (1). Some common runoff controls include slope restrictions, buffer zones/filter strips, tillage to create a roughened soil surface, crop residue or vegetation, berms, dikes, silt fences, diversions, siltation basins, and terraces.

*2.3.3. Distance to U.S. Waters*

Bulk biosolids may not be applied within 10 m (33 ft) of any waters of the U.S. (e.g., intermittent following streams, creeks, rivers, wetlands, or lakes) unless otherwise specified by the permitting authority. Permitting authorities can allow exceptions to this requirement if the application of biosolids is expected to enhance the local environment. For example, biosolids application may help revegetate a stream bank and otherwise minimize erosion.

*2.3.4. Agronomic Rate*

The agronomic rate for biosolids application is a rate that is designed to provide the amount of nitrogen needed by a crop or vegetation to attain a desired yield while minimizing the amount of nitrogen that will pass below the root zone of the crop or vegetation to the ground water. Crop-available nitrogen in biosolids that is applied in excess of the agronomic rate could result in nitrate contamination of the ground water. The Part 503 rule requires that the rate of land application for bulk biosolids be equal to or less than the agronomic rate, except in the case of a reclamation site where a different rate of application is allowed by the permitting authority.

Although the preparer is required to supply the land applier with information on the nitrogen content of the biosolids, the land applier is responsible for determining that the biosolids are applied at a rate that does not exceed the agronomic rate for that site. Procedures for the design of the agronomic rate differ depending on such factors as the total and available nitrogen content of the biosolids, nitrogen losses, nitrogen from sources other than biosolids (including estimates or measurements of available nitrogen already present in the soil), and the requirements for the expected yield of crop or vegetation. An example for calculation of the nitrogen supplied by biosolids based on the AWSAR is provided below.

Example 2

Methodology for the biosolids applier to determine the amount of nitrogen provided by the AWSAR relative to the agronomic rate: In Example 1, the AWSAR for the biosolids in the example calculation was determined to be 410 lb of biosolids/1000 ft$^2$ of land. If biosolids were to be placed on a lawn that has a nitrogen requirement of about 200 lb of available nitrogen/acre/yr, determine the amount of nitrogen provided by the AWSAR relative to the agronomic rate if the AWSAR was used.

Solution

1. The nitrogen content of the biosolids indicated on the label is 1% total nitrogen and 0.4% available nitrogen the first year.
2. The AWSAR is 410 lb of biosolids/1000 ft$^2$, which is 17,860 lb of biosolids/acre:

$$(410\,\text{lb biosolids}/1000\,\text{ft}^2)\,(43{,}560\,\text{ft}^2/\text{acre})\,(0.001) = 17{,}860\,\text{lb biosolids}/\text{acre} \quad (2)$$

3. The available nitrogen from the biosolids is 71 lb/acre:

$$(17{,}860\,\text{lb biosolids}/\text{acre})\,(0.004) = 71\,\text{lb}/\text{acre} \quad (3)$$

4. Since the biosolids application will only provide 71 lb of the total 200 lb of nitrogen required, in this case the AWSAR for the biosolids will not cause the agronomic rate for nitrogen to be exceeded and an additional 129 lb/acre of nitrogen would be needed from some other source to supply the total nitrogen requirement of the lawn.

### 2.4. Frequency of Monitoring Requirements

Pollutants, pathogen densities, and vector attraction reduction must be monitored when biosolids are applied to the land. This monitoring ensures that pollutant limits and pathogen and vector attraction reduction requirements are being met (19). The required frequency of monitoring is 1, 4, 6, or 12 times per year, depending on the number of metric tons (T) (dry-weight basis) of biosolids used or disposed in that year. This frequency is presented in Table 6.11. Frequency of monitoring requirements must be met regardless of which option is chosen for meeting pollutant limits and pathogen and vector attraction reduction requirements, with the exception of Class B pathogen alternative 2.

### 2.5. Record-Keeping and Reporting Requirements

Part 503 requires that certain records be kept by the person who prepares biosolids for application to the land and the person who applies biosolids to the land. Some of the records that must be kept when biosolids are applied to the land include statements certifying whether certain land application requirements are met (19, 20).

The certifier should periodically check the performance of his or her employees to verify that the Part 503 requirements are being met. Then, when a federal or state inspector checks the employee's logs, office records, and performance in the field, the inspector should find that the required management practices are being followed and that any applicable pathogen and vector attraction reduction requirements, including

**Table 6.11**
**Frequency of monitoring for pollutants, pathogen densities, and vector attraction reduction**

| Amounts of biosolids** T*/yr | Amount of biosolids | | Frequency |
|---|---|---|---|
| | Average t*/d | t*/yr | |
| >0 to <290 | >0 to <0.85 | >0 to <320 | Once per year |
| ≥290 to <1500 | 0.85 to <4.5 | 320 to <1650 | Once per quarter (4 times/yr) |
| ≥1500 to <15,000 | 4.5 to <45 | 1650 to <16,500 | Once per 60 days (6 times/yr) |
| ≥15,000 | ≥45 | ≥16,500 | Once per month (12 times/yr) |

*T = Metric ton = 1000 kg; and t = English ton = 2000 lb
**Either the amount of bulk biosolids applied to the land or the amount of biosolids received by a person who prepares biosolids for sale or giveaway in a bag or other container for application to the land (dry-weight basis).

associated crop harvesting, animal grazing, and site access restrictions, are being met. Even if the preparer/applier is not required to report this information, he or she must keep these records for 5 years, or indefinitely for cumulative amounts of pollutants added to any site by CPLR biosolids. These required records may be requested for review at any time by the permitting or enforcement authority.

Some facilities are not subject to any Part 503 reporting requirements. However, all Class I treatment works, treatment works serving a population of 10,000 or more, and treatment works with a 1 million gallons per day (MGD) or greater design flow have reporting responsibilities. Each year, facilities with reporting requirements must submit the information contained in their records.

### 2.6. Domestic Septage

Part 503 imposes separate requirements for domestic septage applied to agricultural land, forest, or a reclamation site (i.e., nonpublic-contact sites). The simplified rule for application of domestic septage to such sites is explained in the US EPA guide and manual (21, 22). If domestic septage is applied to public contact sites or home lawns and gardens, the same requirements must be met as for bulk biosolids applied to the land including general requirements, pollutant limits, pathogen and vector attraction reduction requirements, management practices, frequency of monitoring requirements, and record-keeping and reporting requirements.

### 2.7. Liability Issues and Enforcement Oversight

The Part 503 rule is self-implementing and its provisions must be followed whether or not a permit is issued. The state rules, which may be different from and more stringent than the Part 503 rule, may also apply.

The US EPA's Part 503 rule concerning the use or disposal of biosolids includes enforcement measures regarding the proper testing and application of biosolids. Landowners (including their lenders) and leaseholders who use biosolids beneficially as a fertilizer substitute or soil conditioner in accordance with the US EPA's Part 503 rule are protected from liability under the Superfund legislation (23) as well as any enforcement action from the US EPA under the Part 503 rule. Where the federal requirements are not followed, appliers of biosolids are vulnerable to the US EPA enforcement actions or citizen-initiated suits and can be required to remediate any problems for which they are found liable.

The US EPA oversight of land application practices includes a program for administering permits and for monitoring, reporting, and inspecting. As with wastewater discharge standards and requirements, preparers and land appliers are required to keep detailed records and class I biosolids management facilities must self-report on their activities during the preceding calendar year. The reports must include information on biosolids quality. In the case of CPLR biosolids, a field-by-field analysis of the site activity must also be reported, including information on management practices and on the cumulative application of metals. Hence, the US EPA will know the quality of the biosolids and where they are going, in accordance with US EPA Part 503 requirements.

The US EPA does not rely solely on the word of the regulated community. It conducts routine sampling and inspections of these facilities. If discrepancies are identified, enforcement actions will be taken. Enforcement actions can include fines of up to $25,000 per day per violation, injunctive relief, or criminal imprisonment.

## 3. SURFACE DISPOSAL OF BIOSOLIDS

The Part 503 rule defines an activity as surface disposal if biosolids are placed on an area of land for final disposal. Some surface disposal sites may be used for beneficial purposes as well as for final disposal. Owners and operators of surface disposal sites and anyone who prepares biosolids for final disposal of only biosolids on a surface disposal site must meet the requirements in subpart C of the Part 503 rule. These requirements are described in this chapter (8, 24).

Surface disposal sites include monofills, surface impoundments and lagoons, waste piles, dedicated disposal sites, and dedicated beneficial use sites.

1. *Monofills* are landfills where only biosolids are disposed. Monofills include trenches and area fills. In trenches, biosolids are placed in an excavated area that can be a wide, shallow trench or a narrow, deep trench. In area fills, biosolids are placed on the original ground surface in mounds, layers, or diked containments. With area fills, excavation is not required (as it is with trenches) because biosolids are not placed below the ground surface. Area fills often are used when shallow bedrock or ground water is present.
2. *Surface impoundments and lagoons* are disposal sites where biosolids with high water content are placed in an open, excavated area. If lagoons are used for treatment, they are not considered surface disposal sites.
3. *Waste piles* are mounds of dewatered biosolids placed on the soil surface for final disposal.
4. *Dedicated disposal sites* receive repeated applications of biosolids for the sole purpose of final disposal. Such sites often are located at publicly owned treatment works (POTW) sites.
5. *Dedicated beneficial use sites* are surface disposal sites where biosolids are placed on the land at higher rates or with higher pollutant concentrations than are allowed when biosolids are land applied for farming or reclamation. Such sites might receive repeated applications of biosolids. In contrast to dedicated disposal sites, dedicated beneficial use sites are used to grow crops for beneficial purposes. For such sites, the permitting authority will issue a permit that specifies appropriate management practices that ensure the protection of public health and the environment from any reasonably anticipated adverse effects of certain pollutants that may be present in biosolids if crops are grown or animals are grazed.

It is important to be able to differentiate among surface disposal, storage, land application, and treatment. An activity is considered storage if biosolids are placed and remain on land for 2 years or less. If biosolids remain on land for longer than 2 years, this land is considered an active biosolids unit and the surface disposal requirements in Part 503 have to be met. An active biosolids unit is the area, trench, waste pile, or lagoon where biosolids are currently being placed. However, biosolids can remain on the land for longer than 2 years, but the person who prepares the biosolids must demonstrate that the site is not an active biosolids unit.

Any practice in which biosolids that meet pollutant concentrations, CPLRs, or APLRs, as well as ceiling limits are applied to land at agronomic rates to condition soil or to fertilize crops or vegetation is considered land application, not surface disposal.

The surface disposal provisions of the Part 503 rule do not apply when biosolids are treated on the land, such as in a treatment lagoon or stabilization pond, and treatment could be for an indefinite period. Placement of biosolids on the land in a municipal solid waste landfill also is not considered surface disposal under Part 503, but would be covered under 40 CFR Part 258 instead.

Part 503 standard and regulatory requirements for surface disposal of biosolids include the following (8):

1. General requirements
2. Pollutant limits
3. Management practices
4. Operational standards for pathogen and vector attraction reduction
5. Frequency of monitoring requirements
6. Record-keeping requirements
7. Reporting requirements

### 3.1. General Requirements for Surface Disposal Sites

No biosolids can be placed on an active biosolids unit unless the requirements described in this section are met.

If an active biosolids unit is located within 60 m of a geologic fault with displacement in Holocene time (i.e., relatively recently), located in an unstable area, or located in a wetland, the unit must have been closed in 1994. There are two exceptions to this requirement: (a) if the permitting authority has indicated that the location of a specific unit within 60 m of a fault with displacement in Holocene time is acceptable, or (b) if a permit was issued under the CWA that allows the unit to be located in a wetland.

If an active biosolids unit is about to be closed, the owner/operator of the unit must provide the permitting authority with a written plan at least 180 days prior to the closing. The plan must describe closure and postclosure activities and systems including (a) the operation and maintenance of the leachate collection system for 3 years after closure (if the unit has such a system); (b) the system used to monitor the air for methane gas for 3 years after closure (if biosolids units are covered); and (c) measures to restrict public access for 3 years after closure.

The permitting authority may determine that the closure plan must include provisions for monitoring the air for methane gas or leachate collection for more than 3 years. For example, if the biosolids placed on the surface disposal site were not stabilized, it may be necessary to monitor the air for methane gas and restrict access for a longer period to protect public health and the environment. Also, in areas of high rainfall, the permitting authority may deem it necessary to collect leachate for a longer period to ensure that the integrity of the liner is maintained.

Should ownership of a surface disposal site change hands, the owner must provide the subsequent owner with written notification that biosolids were placed on the land.

The notification required for the subsequent owner of a surface disposal site varies depending on when the land was sold and the provisions of the closure plan. For instance, if a surface disposal site was covered, had a liner, and was sold 1 year after closure, the notification should inform the next owner that the property was used to dispose of biosolids and that the new owner must operate the leachate collection system, monitor the air for methane gas, and restrict public access for an additional 2 years.

### 3.2. Pollutant Limits for Biosolids Placed on Surface Disposal Sites

For surface disposal, a pollutant limit is the amount of pollutant allowed per unit amount of biosolids. Subpart C of Part 503 sets pollutant limits for arsenic, chromium, and nickel in biosolids. These limits apply only to active biosolids units without liners and leachate collection systems.

A liner is a layer of relatively impervious soil, such as clay, or a layer of synthetic material that covers the bottom of an active biosolids unit and has a hydraulic conductivity of $1 \times 10^{-7}$ cm/s or less. The liner slows the seeping of liquid on the surface disposal site into the ground water below. A leachate collection system is a system or device installed immediately above a liner that collects and removes leachate as it seeps through the disposal site. Biosolids placed on an active biosolids unit with a liner and leachate collection system do not have to meet pollutant limits, based on the assumption that these systems prevent pollutants from migrating to ground water.

There are two options for meeting the pollutant limits for arsenic, chromium, and nickel in active biosolids units without using liners and leachate collection systems:

*Option 1*: Make sure that the levels of arsenic, chromium, and nickel are not above the levels listed in Table 6.12, which are based on the distance between the active biosolids unit's boundary and the property line of the surface disposal site.
*Option 2*: Meet the site-specific pollutant limits for arsenic, chromium, and nickel, if site-specific limits have been set by the permitting authority.

The limits for the first option are based on how far the boundary of each active biosolids unit is from the surface disposal site property line. For example, the limits are 73 mg/kg for arsenic, 600 mg/kg for chromium, and 420 mg/kg for nickel if the boundary of the active biosolids unit closest to the site's property line is greater than 150 m away.

There may be more than one active biosolids unit at a surface disposal site. If the boundary of a second active biosolids unit on the same site is 75 m from the property line, then the arsenic limit for that second unit would be 46 mg/kg. Thus different active biosolids units on the same site can have different pollutant limits, based on the closest distance between the active biosolids unit boundaries and the property line of the surface disposal site.

**Table 6.12**
**Option 1: pollutant limits**

| Distance from the boundary of active biosolids unit to surface disposal site property line (m) | Pollutant concentration* | | |
|---|---|---|---|
| | Arsenic (mg/kg) | Chromium (mg/kg) | Nickel (mg/kg) |
| 0 to less than 25 | 30 | 200 | 210 |
| 25 to less than 50 | 34 | 220 | 240 |
| 50 to less than 75 | 39 | 260 | 270 |
| 75 to less than 100 | 46 | 300 | 320 |
| 100 to less than 125 | 53 | 360 | 390 |
| 125 to less than 150 | 62 | 450 | 420 |
| Equal to or greater than 150 | 73 | 600 | 420 |

*Dry-weight basis (basically, 100% solids content).

The second option for meeting pollutant limits is to meet site-specific limits set by the permitting authority, who would determine the limits after evaluating site data. The owner/operator of the surface disposal site must request site-specific limits when applying for a permit. The permitting authority then must determine whether site-specific pollutant limits are appropriate for the particular site. Site-specific limits may be justified if the site conditions vary significantly from those assumed in the risk assessment used to derive the Part 503 pollutant limits. In general, if the depth to ground water is considerable or a natural clay layer underlies the site, the permittee may consider requesting site-specific pollutant limits.

### 3.3. Management Practices for Surface Disposal of Biosolids

The Part 503 rule includes management practices that must be followed when biosolids are placed on a surface disposal site. Most of these management practices apply to all surface disposal sites. A few, however, apply only to sites with liners and leachate collection systems or to sites with covers. (A cover can be soil or other material placed over the biosolids.) The required management practices for surface disposal sites are discussed below (25).

#### 3.3.1. Protection of Threatened or Endangered Species

This requirement applies to persons who place biosolids on land where there is potential for harming certain species of plants, fish, or wildlife or their habitat. Biosolids cannot be placed on an active biosolids unit in a surface disposal site where such disposal is likely to have an adverse affect on a threatened or endangered animal or plant species or its critical habitat. Threatened or endangered species and their critical habitats are listed on the FWS Web site (18). "Critical habitat" is defined as any place where a threatened or endangered species lives and grows during any stage of its life cycle. Any direct or indirect action in a critical habitat that diminishes the likelihood of survival and recovery of a listed species is considered destruction or adverse modification of a critical habitat.

### 3.3.2. Restriction of Base Flood Flow

An active biosolids unit in a surface disposal site must not restrict the flow of a base flood. A base flood is a flood that has a 1% chance of occurring in any given year (or a flood that is likely to occur once in 100 years). This management practice reduces the possibility that an active biosolids unit might negatively affect the ability of an area to absorb the flows of a base flood. The practice also helps to prevent surface water contamination and to protect the public from the possibility of a base flood releasing biosolids to the environment.

To determine whether a surface disposal site is in a 100-year flood plain, one can consult the flood insurance rate maps (FIRMs) and flood boundary and floodway maps published by the Federal Emergency Management Agency (FEMA). States, counties, and towns usually have maps delineating flood plains as well. Other agencies that maintain flood zone maps are the U.S. Army Corps of Engineers (ACE), the U.S. Geological Survey (USGS), the U.S. Soil Conservation Service (SCS), and the Bureau of Land Management (BLM).

If the owner/operator of a surface disposal site determines that the site is within a 100-year flood zone, the permitting authority has ultimate responsibility for determining whether the active biosolids unit will restrict the flow of a base flood. This assessment considers the flood plain storage capacity and the floodwater velocities that would exist with and without the presence of the biosolids unit. If the presence of the unit would cause the base flood level to rise one additional foot, then the unit restricts the flow of a base flood, potentially causing more flood damage than would otherwise occur.

If the permitting authority determines that the active biosolids unit will restrict the flow of the base flood, it may require the unit to close or it may require remedial action to avoid the restriction of the base flood flow. Such actions might include constructing embankments or implementing an alternative unit design intended to prevent the unit from being damaged by floodwaters.

### 3.3.3. Geological Stability

Three of the management practices concern the distance of active biosolids units from certain types of geologic formations. These management practices help ensure that biosolids units are located in geologically stable areas or that the units can withstand certain ground movements. The geologic formations covered by these management practices are fault areas with displacement in Holocene time, unstable areas, and seismic impact zones:

*Fault*: A crack in the earth along which the ground on either side may move. Such ground movement is called displacement. An active biosolids unit must be located at least 60 m from a fault that has displacement measured in Holocene time (recent geological time of approximately the last 11,000 years). Requiring this distance from a fault helps ensure both that the structures of a biosolids unit will not be damaged if ground movement occurs in a fault area and that leachate will not spread through faults into the environment.

*Unstable area*: Land where natural or human activities might occur that could damage the structures of an active biosolids unit and allow the release of pollutants into the environment. Unstable areas include land where large amounts of soil are moved, such as by landslides, where the surface lowers or collapses when underlying limestone or other materials dissolve. An active biosolids unit cannot be located in an unstable area. This restriction protects the structures of a biosolids unit from damage by natural or human forces.

*Seismic impact zone*: An area in which certain types of ground movements (horizontal ground level acceleration) have a 10% or greater chance of occurring at a certain level (measured as 0.10 gravity) once in 250 years. The USGS keeps records of the location of these areas. When a surface disposal site is located in a seismic impact zone, each active biosolids unit must be designed to withstand the maximum recorded horizontal ground level acceleration. This management practice helps ensure that the containment structures, such as the liner and leachate collection system, of a biosolids unit will not crack or collapse because of ground movement and that leachate will not be released due to seismic activity. Various seismic design methods have been developed for biosolids units located in seismic impact zones. Appropriate design modifications may include shallower unit side slopes and a more conservative design for dikes and runoff controls. Also, contingencies for the leachate collection system should be considered, in case the primary system becomes ineffective.

Individuals can determine whether a property is within a geologically unstable area using maps that are available through the USGS, Earth Science Information Center, Reston, Virginia.

### 3.3.4. Protection of Wetlands

Wetlands are areas in which the soils are filled with water (saturated) during part of the year and that support vegetation typically found in saturated soils. Examples of wetlands include swamps, marshes, and bogs. Wetlands perform important ecological functions, such as holding floodwaters, serving as habitat, and providing sources of food for numerous species including 60% of the endangered species, and reducing soil erosion. Wetlands also hold pollutants, preventing them from contaminating other areas.

An active biosolids unit cannot be located in a wetland unless the owner/operator holds a valid permit (NPDES permit). Controlling where active biosolids units are located protects wetlands from possible contamination when biosolids are placed in an active biosolids unit. Criteria for identifying wetlands have been developed by a federal task force and appear in an ACE manual (26).

### 3.3.5. Collection of Runoff

Runoff is rainwater or other liquid that drains over the land and runs off the land surface. Runoff from an active biosolids unit might be contaminated with biosolids. Runoff from an active biosolids unit must be collected and disposed according to the permit requirements of the NPDES and any other applicable requirements. The runoff collection system must have the capacity to handle runoff from a 25-year, 24-hour storm event (a storm that is likely to occur once in 25 years for a 24-hour period). This requirement helps ensure that runoff (which may contain pollutants) from an active biosolids unit is not released into the environment. The peak flow of water and the total runoff volume of water during the 25-year, 24-hour storm must be calculated to ensure that the extent of stormwater controls is adequate to collect runoff from such a storm.

### 3.3.6. Collection of Leachate

Leachate is fluid from excess moisture in biosolids or from rainwater percolating down through the active biosolids unit from the land surface. If an active biosolids unit has a liner and a leachate collection system, two additional management practices in the Part 503 rule apply.

The first management practice requires that the leachate collection system be operated and maintained according to design requirements and engineering recommendations. The owner/operator of the surface disposal site is responsible for ensuring that the system is always operating according to design specifications and is properly and routinely maintained (e.g., pumps are periodically cleaned and serviced, and the system is periodically inspected to detect clogs and flushed to remove deposited solids).

The second management practice requires that leachate be collected and disposed in accordance with applicable requirements. Leachate should be collected and pumped out by a system placed immediately above a liner. If leachate is discharged to surface water as a point source, then an NPDES permit is required. Otherwise, leachate may be used to irrigate adjacent land or discharged to a POTW. It is recommended that the leachate be tested to determine whether some treatment is appropriate before irrigating or discharging it to a POTW.

Both management practices must be followed while the unit is active and then for 3 years after the unit is closed, or for a longer period if required by the permitting authority.

The Part 503 rule regulates active biosolids units without liners and leachate collection systems through the pollutant limits and through other management practices in the regulation.

### 3.3.7. Limitations on Methane Gas Concentrations

The Part 503 rule includes a management practice that limits concentrations of methane gas in air because of its explosive potential. Methane, an odorless and highly combustible gas, is generated at surface disposal sites when biosolids are covered by soil or other material (e.g., geomembranes), either daily or at closure. The gas can migrate and be released into the environment. To protect site personnel and the public from risks of explosions, air must be monitored for methane gas continuously within any structure on the site and at the property line of the surface disposal site. Only surface disposal sites that cover biosolids units (either daily or at closure) must meet this management practice. When biosolids units are not covered, the air does not have to be monitored for methane gas.

This management practice limits the amount of methane gas in air in both active and closed biosolids units. When a cover is placed on an active biosolids unit, the methane gas concentration in air in any structure within the property line of a surface disposal site must be less than 25% of the lower explosive limit (LEL) (i.e., 1.25%). The LEL is the lowest percentage (by volume) of methane gas in air that supports a flame under certain conditions (at 25°C and atmospheric pressure). For methane, the LEL is 5%. Therefore, if 5% of the LEL is 50,000 ppmv (part per million by volume) methane, then air in any structure within the property line must not exceed 12,500 ppmv methane.

A methane gas–monitoring device must be installed so that methane concentrations in the air inside all structures on the property are continuously measured and the measurement can be read by any individual before entering the structure. (The act of entering the building could create enough of a spark to ignite explosive levels of methane gas.)

For air at the property line of a surface disposal site with a covered biosolids unit, the limit for methane gas concentration is the LEL (i.e., 5%). In some cases, the permitting authority may determine that a methane monitoring device at one downwind location on the property line is adequate to meet this requirement because the wind patterns are consistent. In other cases, where wind conditions at the site are highly variable, more than one device may be necessary to provide adequate protection.

Methane gas concentrations must be monitored at all times when the biosolids units are active and for 3 years after the last active biosolids unit on the site is closed. If unstabilized biosolids are disposed at a site, the permitting authority may require methane gas to be monitored for longer than 3 years after closure because of the higher potential for methane generation with unstabilized biosolids.

### 3.3.8. Restrictions on Crop Production

Food, feed, or fiber crops may not be grown on an active biosolids unit unless the owner or operator of the surface disposal site can demonstrate to the permitting authority that, through management practices, public health and the environment are protected from any reasonably anticipated adverse effects of certain pollutants that may be present in biosolids. If the owner/operator wishes to grow crops on the site, he or she must obtain a permit that requires the implementation of certain management practices to ensure that the levels of pollutants taken up by crops do not negatively affect the food chain in regard to animals or humans.

These special management practices might include testing crops and animal tissue for the presence of pollutants if animal feed is produced on the site, or setting a monitoring schedule for the crops and any animal feed products derived from crops grown on the site.

### 3.3.9. Restrictions on Grazing

Animals must not be grazed on an active biosolids unit unless the owner/operator of a surface disposal site can demonstrate to the permitting authority that public health and the environment are protected from any reasonably anticipated adverse effects of certain pollutants that may be present in biosolids. If the owner/operator wishes to graze animals on the site, he or she must obtain a permit. The permit could require specified management practices, such as monitoring the concentration of pollutants in any animal product (dairy or meat). This restriction on grazing helps ensure that unsafe levels of pollutants do not find their way into animals from which people obtain food. A site where a special permit allows the production of crops or grazing is considered *a* dedicated beneficial use site.

### 3.3.10. Restrictions on Public Access

Public access to a surface disposal site must be restricted while an active biosolids unit is on the site and then for 3 years after the last active biosolids unit has been closed. This management practice helps to minimize public contact with any pollutants, including pathogens that may be present in biosolids placed on an active biosolids unit. It also keeps people away from areas where there is a potential for a methane gas explosion, as discussed above.

Fencing off an area and installing gates that lock might be necessary to restrict access in densely populated areas. Natural barriers, such as hedges, trees, embankments, or ditches, along with warning signs, might be adequate in less-populated areas. In remote areas, it might be sufficient to post warning signs that say, "Do not enter," "No trespassing," or "Access restricted to authorized personnel only."

### 3.3.11. Protection of Ground Water

This management practice states that biosolids placed on an active biosolids unit must not contaminate an aquifer. An aquifer is an area below the ground that can yield water in large enough quantities to supply wells or springs. Contaminating an aquifer in this instance means introducing a substance that can cause the level of nitrate in ground water to increase above regulated limits. This management practice also requires that the owner/operator obtain proof that ground water is not contaminated. This proof must be either by way of (a) a ground-water monitoring program developed by a qualified ground-water scientist, or (b) certification by a ground-water scientist that ground water will not be contaminated by the disposal of biosolids at the site.

Usually, certification is an option only if the site has a liner and a leachate collection system. It is generally infeasible for a ground-water scientist to certify that ground water will not be contaminated in the absence of a liner, unless the depth to ground water is considerable and there is a natural clay layer under the soil or unless the amount of biosolids placed on the site is quite small (e.g., at the agronomic or reclamation rate).

Only nitrate-nitrogen levels in ground water are addressed by this management practice. Nitrate-nitrogen levels in ground water must not exceed the maximum contaminant level (MCL) of 10 mg/L or must not increase an existing exceedance of the ground water MCL for nitrate-nitrogen. Potential pollutants other than nitrate are addressed by the pollutant limits.

### 3.4. Pathogen and Vector Attraction Reduction Requirements for Surface Disposal Sites

Pathogens are disease-causing organisms, such as certain bacteria and viruses that might be present in biosolids. Vectors are animals, such as rats or insects, that might be attracted to biosolids and can spread disease after coming into contact with the biosolids. The Part 503 rule includes requirements concerning the control of pathogens and the reduction of vector attraction for biosolids placed on a surface disposal site. Biosolids can be placed on an active biosolids unit only if the pathogen and vector attraction reduction requirements are met (Table 6.13).

**Table 6.13**
**Pathogen and vector attraction reduction requirements and options for surface disposal sites**

A. Pathogen reduction requirements
*Options (must meet one of these)*:
1. Place a daily cover on the active biosolids unit
2. Meet one of six Class A pathogen reduction requirements
3. Meet one of three Class B pathogen reduction requirements, except site restrictions

B. Vector attraction reduction requirements
*Options (must meet one of these)*:
1. Place a daily cover on the active biosolids unit
2. Reduce volatile solids content by a minimum of 38% or less under specific laboratory test conditions with anaerobically and aerobically digested biosolids
3. Meet a specific oxygen uptake rate (SOUR)
4. Treat the biosolids in an aerobic process for a specified number of days at a specified temperature
5. Raise the pH of the biosolids with an alkaline material to a specified level for a specified time
6. Meet a minimum percent solids content inject or incorporate the biosolids into soil

For pathogen reduction, the biosolids placed on an active biosolids unit must meet either Class A or Class B pathogen requirements, or a cover (soil or other material) must be placed on the active biosolids unit at the end of each day. If a daily cover is placed on the active biosolids unit, no other pathogen reduction requirements apply. If the biosolids meet Class B requirements, the site restrictions that apply to Class B do not have to be followed because the management practices for surface disposal already include these site restrictions.

For vector attraction reduction, one of several options listed in Table 6.13 must be met. Representative samples of biosolids must be collected and analyzed to demonstrate that the pathogen and vector attraction reduction requirements have been met.

Pathogen and vector attraction reduction requirements, including Class A and Class B pathogen requirements, are discussed in more detail in a following section. In most cases, owners or operators of surface disposal sites will place a daily cover on the biosolids unit to meet pathogen and vector attraction reduction requirements.

### 3.5. Frequency of Monitoring Requirements for Surface Disposal Sites

The monitoring of several different parameters is required at surface disposal sites, as shown in Table 6.14. Monitoring is required for surface disposal sites without liners to determine levels of arsenic, chromium, and nickel in biosolids. Monitoring is required in both lined and unlined sites to show that the chosen pathogen and vector attraction reduction requirement is being met and to measure the amount of methane gas in air at a covered surface disposal site. How frequently biosolids must be monitored is determined according to the amount of biosolids placed on an active biosolids unit, as shown in

**Table 6.14**
**Monitoring required at surface disposal sites**

| Parameter | Biosolids or air |
|---|---|
| Arsenic | Biosolids |
| Chromium | Biosolids |
| Nickel | Biosolids |
| Pathogens | Biosolids for several options |
| Vector attraction reduction | Biosolids for several options |
| Methane gas | Air in each structure on site |
| Methane gas | Air at surface disposal site property line |

**Table 6.15**
**Frequency of monitoring for surface disposal of biosolids**

| Amount of biosolids** T/yr* | Amount of biosolids | | Minimum frequency |
|---|---|---|---|
| | Average t/d* | t/yr* | |
| Greater than zero but less than 290 | >0 to <0.85 | >0 to <320 | Once per year |
| Equal to or greater than 290 but less than 1500 | 0.85 to <4.5 | 320 to <1650 | Once per quarter (4 times/yr) |
| Equal to or greater than 1500 but less than 15,000 | 4.5 to <45 | 1650 to 16,500 | Once per 60 days (6 time/yr) |
| Equal to or greater than 15,000 | ≥45 | ≥16,500 | Once per month (12 times/yr) |
| Methane gas in air | | | Continuously with methane monitoring device if biosolids unit is covered |

*T = Metric ton = 1000 kg; and t = English ton = 2000 lb.
**Amount of biosolids (other than domestic septage) placed on active biosolids units-dry-weight basis.

Table 6.15. The permitting authority may require more frequent monitoring, for example, if the pollutant and pathogen levels in the biosolids are highly variable.

After biosolids have been monitored for 2 years at the frequency specified in Table 6.15, the permitting authority may reduce the frequency of monitoring for arsenic, chromium, nickel, and, under limited circumstances, pathogens in biosolids placed on an active biosolids unit. The frequency may be reduced, for example, if the pollutant levels in biosolids do not vary greatly or if pathogens are never detected when using Class A, alternative 3, to meet pathogen reduction requirements. At the least, monitoring must be performed once a year.

Methane gas in air must be monitored continuously, both at the property line of the surface disposal site and within each structure at the site, if an active biosolids unit is covered. Methane gas monitors can be installed permanently to continuously test the air and provide readings of methane levels as a percent of the LEL. Monitoring must be continued as long as any covered biosolids unit on the site is active and then for 3 years after the last biosolids unit has been closed, if covered at closure.

### 3.6. Record-Keeping and Reporting Requirements for Surface Disposal Sites

Certain information must be recorded and kept for 5 years from the time the biosolids are placed on a surface disposal site. A separate set of records must be kept by a person who prepares biosolids for placement on a surface disposal site and by the owner/operator of a surface disposal site.

The preparer of biosolids to be placed on an active biosolids unit must develop and keep the following information for 5 years:

1. The concentrations of arsenic, chromium, and nickel in biosolids for active biosolids disposal units without a liner and leachate collection system with boundaries that are 150 m or more from the surface disposal site's property line
2. A certification statement that the requirements are met
3. A description of how certain pathogen and vector attraction reduction requirements are met

An owner/operator of a surface disposal site on which biosolids are placed must develop and keep the following information for 5 years:

1. The concentrations of arsenic, chromium, and nickel in biosolids for active biosolids units with boundaries less than 150 m from the property line or for active biosolids units with site-specific limits
2. A certification statement that the requirements are met
3. A description of how the management practices for surface disposal sites are being met
4. A description of how certain vector attraction reduction requirements are being met

The Part 503 regulation includes reporting requirements only for those class 1 and 1-MGD-or-greater facilities. These facilities must present the information developed for record-keeping purposes to the permitting authority.

### 3.7. Regulatory Requirements for Surface Disposal of Domestic Septage

The regulatory requirements for the surface disposal of septage are not as extensive as those for biosolids. The requirements for surface disposal of domestic septage include meeting the same management practices that are required for the surface disposal of biosolids and one of the VAR alternatives 9 to 12 (listed in a later section). Note that VAR alternative 12 would require a determination that the pH of the septage had been raised to 12 for a period of 30 minutes. The person who places the domestic septage on the surface disposal site must certify that VAR has been achieved and write a description of how it was achieved. The certification and description must be kept on file for 5 years. There are no pathogen requirements for the surface disposal of domestic septage (24).

## 4. INCINERATION OF BIOSOLIDS

Biosolids incineration is the firing of biosolids at high temperatures in an enclosed device. Anyone who fires biosolids in an incinerator, except as described below, must meet the requirements in subpart E of the Part 503 rule (8).

Incineration systems generally consist of an incinerator (furnace) and one or more air pollution control devices (APCDs). The most commonly used incinerators are multiple-hearth, fluidized-bed, and electric infrared furnaces (27; also Chapters 10 and 11). Most APCDs are used either to remove small particles and their adhering metals in the exhaust gas or to further decompose organics. Examples of metal-removing APCDs are wet scrubbers (28), dry and wet electrostatic precipitators (29), and fabric filters (30). Afterburners (31), another type of APCD, are used to burn organics in exhaust gases more completely.

Auxiliary fuel is often used to enhance the burning of biosolids. Any additives to biosolids that are fired in a biosolids incinerator, such as natural gas, fuel oil, grit, screenings, scum, wood chips, coal, dewatering chemicals, and municipal solid waste, is considered auxiliary fuel. If municipal solid waste accounts for more than 30% (by dry weight) of the mixture of biosolids and auxiliary fuel, however, the municipal solid waste is not considered auxiliary fuel under Part 503. Instead, the process would be covered by 40 CFR Parts 60 and 61.

Nonhazardous incinerator ash generated during the firing of biosolids is not covered by the Part 503 rule when it is used or disposed. Instead, it must be disposed according to the solid waste disposal regulations in 40 CFR Part 258 (32, 33); however, if the ash is applied to the land or placed on other than a municipal solid waste landfill, the regulations in 40 CFR Part 257 must be followed. Hazardous wastes are not considered auxiliary fuels under Part 503. Thus, an incinerator that burns hazardous wastes with biosolids is considered a hazardous waste incinerator, not a biosolids incinerator, and is covered by 40 CFR Parts 261 through 268.

The requirements for biosolids incineration, including pollutant limits for seven metals, limits for total hydrocarbons, general requirements, management practices, frequency of monitoring requirements, and record-keeping and reporting requirements are discussed in this section.

### 4.1. Pollutant Limits for Biosolids Fired in a Biosolids Incinerator

A pollutant limit is the amount of pollutant allowed per unit amount of biosolids before incineration. Part 503 rule regulates seven metals: arsenic, beryllium, cadmium, chromium, lead, mercury, and nickel. The limits protect human health from the reasonably anticipated harmful effects of these pollutants when biosolids are incinerated. The approaches for determining the limit for each pollutant and for total hydrocarbons (THCs) are summarized in Table 6.16 and discussed below.

#### 4.1.1. Beryllium and Mercury Pollutant Limits

Levels of beryllium and mercury emitted from a biosolids incinerator must meet the National Emission Standards for Hazardous Air Pollutants (NESHAPs).

The NESHAP for beryllium requires that the total quantity of beryllium emitted from each incinerator not exceed 10 g during any 24-hour period. The NESHAP for beryllium does not apply if written approval has been obtained from the regional administrator (1) when the ambient concentration of beryllium in the proximity of the biosolids incinerator does not exceed 0.01 µg/m$^3$ when averaged over a 30-day period, or (2)

## Table 6.16
### Summary of pollutant limits for biosolids incineration

| Pollutant | How to figure out pollutant limits | Determine dispersion factor (DF) | Determine control efficiency (CE) | Use National Ambient Air Quality Standard (NAAQS) | Use risk specific concentration (RSC) | Use correction factor for oxygen | Use correction factor for moisture |
|---|---|---|---|---|---|---|---|
| **Pollutant limits** | | | | | | | |
| Arsenic | Use equation for arsenic | Yes | Yes | No | Yes | No | No |
| Beryllium | Use NESHAPs[a] | No | No | No | No | No | No |
| Cadmium | Use equation for cadmium | Yes | Yes | No | Yes | No | No |
| Chromium | Use equation for chromium | Yes | Yes | No | Yes | No | No |
| Lead | Use equation for lead | Yes | Yes | Yes | No | No | No |
| Mercury | Use NESHAPs[a] | No | No | No | No | No | No |
| Nickel | Use equation for nickel | Yes | Yes | No | Yes | No | No |
| **Operational standard** | | | | | | | |
| THC or CO[b] | Limit is 100 ppmv | NO | NO | NO | NO | Yes | Yes |

[a] National Emissions Standards for Hazardous Air Pollutant requirements are summarized in the text.
[b] THC or CO determinations are technology-based standards that in the judgment of the US EPA protect public health and the environment from the reasonably anticipated adverse effects of organic pollutants in the exit gas of biosolids incinerator.

if the biosolids incinerator operator can prove (with historical data) that the biosolids fired in the incinerator do not contain beryllium (8).

The NESHAP for mercury requires that the total quantity of mercury emitted into the atmosphere from all incinerators at a given site does not exceed 3200 g during any 24-hour period (8).

### 4.1.2. Control Efficiency, Dispersion Factor, Feed Rate, and Pollutant Limit Calculations for Lead

A person firing biosolids (e.g., the manager of an incineration operation) must determine the pollutant limit for lead in biosolids by using Eq. 4:

$$C_{lead} = \frac{(0.1)(NAAQS)(86{,}400)}{(DF)(1-CE)(SF)} \qquad (4)$$

where

$C_{lead}$ = the pollutant limit (allowable daily concentration of lead in biosolids, dry-weight basis), mg/kg

0.1 = the allowable ground level concentration of lead from biosolids is 10% of the NAAQS for lead

$NAAQS$ = National Ambient Air Quality Standard for lead, µg/m$^3$ (currently this standard is 1.5 µg/m$^3$)

$DF$ = dispersion factor based on an air dispersion model, µg/m$^3$/g/s

$CE$ = biosolids incinerator control efficiency for lead based on a performance test, fraction

$SF$ = biosolids feed rate, t/d

86,400 = time conversion factor (number of seconds per day), s/d

EXAMPLE 3

If the following are given, determine the limit of lead concentration in the biosolids:

1. The dispersion factor is 3.4 µg/m$^3$/g/s
2. The control efficiency is 0.92
3. The biosolids feed rate is 12.86 t/d
4. The NAAQS for lead is 1.5 µg/m$^3$

SOLUTION

$$C = \frac{(0.1)(NAAQS)(86{,}400)}{(DF)(1-CE)(SF)} \qquad (4)$$

$$C = \frac{(0.1)(1.5)(86{,}400)}{(3.4)(1-0.92)(12.86)}$$

$$C = 3700 \text{ mg/kg}$$

Equation 4 requires determination of certain characteristics of the incineration operation such as control efficiency, the dispersion factor, and the feed rate (Figure 6.3). The permitting authority can provide guidance for developing information about an incinerator's control efficiency and dispersion factor, which are described briefly below.

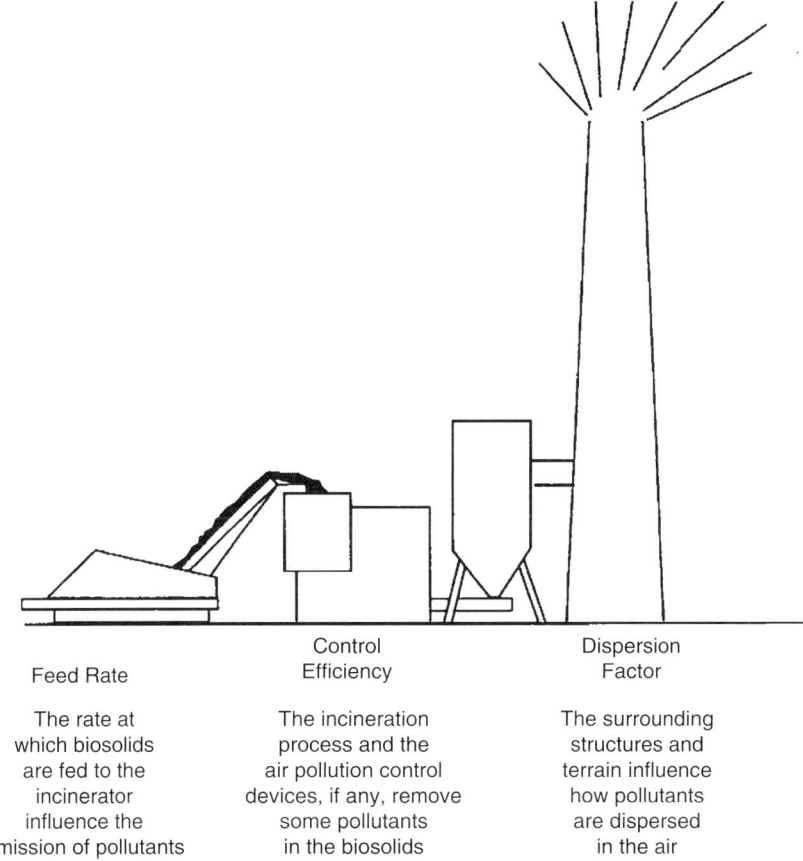

**Fig. 6.3.** Factors that affect the pollutant limits for biosolids fired in an incinerator. *Source*: US EPA (8).

Control efficiency measures the degree to which a biosolids incinerator furnace (in conjunction with an air pollution control system, if used), removes a particular pollutant. For example, if a quantity of biosolids fed to an incinerator contains 100 g of lead, and 1 g of lead is released from the stack, the incinerator has 99% control efficiency for lead.

A dispersion factor is the ratio of the increase in the concentration of a pollutant in the air at or beyond the property line of an incinerator site relative to the rate pollutants are emitted from the stack. The dispersion factor is determined through an air dispersion model, in which particular site conditions are considered, such as the type of terrain or adjacent buildings, whether an area is urban or rural, the temperature and velocity of the gas in the incinerator stack, the height of the stack, and the emission rate for the pollutant.

Control efficiency is determined through a performance test of the incinerator, as specified by the permitting authority. Generally, the permitting authority requires that the performance test be conducted under conditions that represent normal operating circumstances. The incinerator will be allowed to operate with more flexibility, however, if the performance test covers a broader range of operational parameters. The control efficiency determination is based on the concentrations of regulated metals in the biosolids, the concentrations of regulated metals in the incinerator air emissions, and documentation on the incinerator, as well as APCD operating conditions (e.g., biosolids feed rate, exhaust flow rate, combustion temperature, and biosolids characteristics), during the test.

The permitting authority should be consulted on which air dispersion model to use. Air dispersion models range from simple screening tools to complex computer models (34). Screening techniques, though inexpensive, tend to be more conservative in their predictions, resulting in higher estimates of pollutants emitted. Complex models require qualified air quality modelers to perform analyses, but result in more accurate estimations.

The incinerator operator usually determines the biosolids feed rate based on the design capacity of the incinerator and the rate at which biosolids are generated and must be disposed (35). The biosolids feed rate is needed to calculate biosolids pollutant limits in Eqs. 4 and 6. Feed rate itself can be determined based on either of the following (see Example 4):

1. The average daily design capacity for all biosolids incinerators within a site
2. The average daily amount of biosolids fired in all incinerators within the property line of a site for the number of days that the incinerator operates during a 365-day period.

EXAMPLE 4

Calculation of the biosolids feed rate for use in the pollutant limit calculations: A site has three incinerators. Their design capacities are as follows:

- Unit 1: 100 T/d
- Unit 2: 100 T/d
- Unit 3: 200 T/d

Part 503 allows the operator to choose one of two methods to calculate the biosolids feed rate, which is used in the pollutant limit calculations:

1. Method one: Design capacity for all incinerators:
   Calculate the total design capacity for all incinerators at the site:

   $$\text{Total capacity} = 100 \text{ T/d} + 100 \text{ T/d} + 200 \text{ T/d} = 400 \text{ T/d}$$

2. Method two: Average daily feed rate for all incinerators:

*Case 1:* For the first 20 days of the year, unit 1 operated at 50 T/d; for the first 100 days of the year, unit 2 operated at 50 T/d and unit 3 operated at 100 T/d. Calculate the total amount of biosolids fired in a 365-day period:

Unit 1: 50 T/d × 20 d = 1000 T/d (shut down 345 days)
Unit 2: 50 T/d × 100 d = 5000 T/d (shut down 265 days)
Unit 3: 100 T/d × 100 d = 10,000 T/d (shut down 265 days)
Total = 1000 T/d + 5000 T/d + 10,000 T/d = 16,000 T/d

Calculate the average daily amount of biosolids fired during the total number of days the incinerators operated during a 365-day period:

$$\text{Average} = \frac{16{,}000T}{100d} = 160\,\text{T/d}$$

*Case 2:* If the incinerators in the above example did not operate at the same time, but instead operated sequentially, the average would be based on the total number of days any incinerator at the site was operated, which is $20 + 100 + 100 = 220$ days. In that case, the average daily feed rate would be:

$$\text{Average} = \frac{16{,}000T}{220d} = 73\,\text{T/d}$$

For greater flexibility, the operator may want to consider using method one to calculate pollutant limits to have greater latitude in the amount of biosolids fed to the incinerator.

Many incinerator facility operators are finding that they are not limited by biosolids pollutant concentrations under the Part 503 rule. Figure 6.4 provides incinerator test data that are typical for most biosolids incineration facilities. The figure shows allowable metal concentration rates that are significantly higher than actual limits. Thus, incinerator operators should not expect to encounter difficulty meeting the more strict pollutant limits that were calculated based on incinerator design capacity.

Information about lead is also required to use Eq. 4. Emission standards have been set nationally for several substances. These limits, known as National Ambient Air Quality Standards (NAAQS, 40 CFR 50.12), protect human health and the environment from the possible harmful effects of pollutants in air. The manager of a biosolids incinerator must use the NAAQS for lead in the lead equation. The current NAAQS for lead is 1.5 µg/m$^3$; if the NAAQS for lead changes in the future, the number used in Eq. 4 also must change.

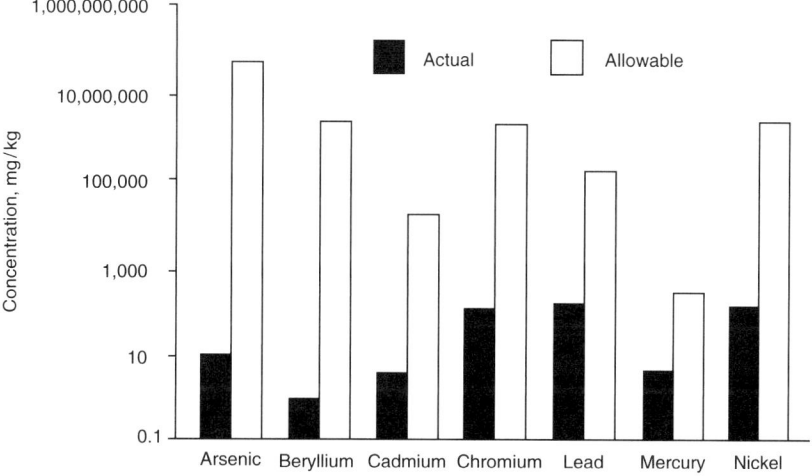

**Fig. 6.4.** Allowable vs. actual biosolids metal concentrations. *Source*: US EPA (8).

### 4.1.3. Arsenic, Cadmium, Chromium, and Nickel Pollutant Limits

As with lead, the pollutant limits for arsenic, cadmium, chromium, and nickel in biosolids fired in a biosolids incinerator are calculated using Eq. 6. Also, the control efficiencies, dispersion factor, and biosolids feed rates must be determined to calculate the limits for these four pollutants. Instead of using a NAAQS, however, as is done for lead, risk-specific concentrations (RSCs, the allowable increase in the average daily ground-level ambient air concentration for a pollutant at or beyond the property line of the site, µg/m$^3$) are used to calculate limits for arsenic, cadmium, chromium, and nickel:

$$C = RSC \frac{86,400}{(DF)(1 - CE)(SF)} \qquad (5)$$

where

$C$ = the pollutant limit (allowable daily concentration of arsenic, cadmium, chromium, or nickel in biosolids, dry-weight basis), mg/kg

$RSC$ = risk-specific concentration (the allowable increase in the average daily ground-level ambient air concentration for a pollutant at or beyond the property line of the site, µg/m$^3$

86,400 = conversion factor (s/d)

$DF$ = dispersion factor (based on an air dispersion model), µg/m$^3$/g/s

$CE$ = control efficiency for arsenic, cadmium, chromium, or nickel based on a performance test, fraction

$SF$ = biosolids feed rate, t/d

EXAMPLE 5: POLLUTANT LIMIT FOR ARSENIC

Given:

- RSC = 0.023 µg/m$^3$
- DF = 3.4 µg/m$^3$/g/s
- CE = 0.975
- Biosolids feed rate = 12.86 t/d

Then:

$$C = RSC \frac{86,400}{(DF)(1 - CE)(SF)} \qquad (5)$$

$$C = (0.023) \frac{86,400}{(3.4)(1 - 0.975)(12.86)}$$

$$C = 1820 \, \text{mg/kg}$$

If the dispersion factor were 0.6 instead of 3.4 µg/m$^3$/g/s, then the allowable concentration for arsenic would be 3.4/0.6, or 5.667 times greater at 10,300 mg/kg.

The RSCs, which are based on human health concerns, represent the allowable increase in the average daily ground-level ambient air concentrations of pollutants at or beyond the property line of the site where the biosolids incinerator is located. The RSCs for arsenic, cadmium, and nickel are listed in Table 6.17.

**Table 6.17**
**Risk-specific concentrations for arsenic, cadmium, and nickel limit exceedance**

| Pollutant | Risk-specific concentrations (RSC) ($\mu g/m^3$) |
|---|---|
| Arsenic | 0.023 |
| Cadmium | 0.057 |
| Nickel | 2.0 |

**Table 6.18**
**Risk-specific concentration (RSC) values to use when calculating the pollutant limit for chromium**

| Type of incinerator | RSC ($\mu g/m^3$) |
|---|---|
| Fluidized bed with wet scrubber | 0.65 |
| Fluidized bed with wet scrubber and wet electrostatic precipitator | 0.23 |
| Other types with wet scrubber | 0.064 |
| Other types with wet scrubber and wet electrostatic precipitator | 0.016 |

In contrast, the RSC for chromium is based on either (1) the type of incinerator used (Table 6.18), or (2) analytical sampling and Eq. 6, below. The analysis involves sampling stack gas to determine the ratio of hexavalent chromium to total chromium analytes.

$$RSC = \frac{0.0085}{r} \qquad (6)$$

where

$r$ = the decimal fraction of the hexavalent chromium concentration in the total chromium concentration measured in the exit gas from the biosolids incinerator stack. This is an analytical measurement based on an average of representative samples.

If measurements in biosolids for beryllium and mercury (where required) and lead, arsenic, cadmium, chromium, or nickel are higher than the pollutant limits derived as discussed above, the biosolids incinerator will be in violation until adjustments are made that allow the limits to be met. Such adjustments include, but are not limited to, improvements in biosolids quality through pretreatment efforts, reduction in the biosolids feed rate, or improved furnace operations or the addition of APCDs to improve the control efficiency. Better control efficiency will allow higher pollutant limits. If significant furnace or APCD improvements are made, however, the performance test must be repeated. Then the approved operating conditions under which emissions were in compliance during the performance test must be maintained whenever the incinerator is operating.

## 4.2. Total Hydrocarbons

Organic compounds that are emitted as a result of incomplete combustion or the generation of combustion by-products (e.g., benzene, phenol, vinyl chloride) can be present in biosolids incinerator emissions. Because these compounds can be harmful to the public health, the Part 503 rule regulates the emission of organic pollutants from biosolids incinerators through an operational standard that limits the amount of total hydrocarbons (THC)—or carbon monoxide (CO)—allowed in stack gas. The Part 503 rule as amended is structured to use THC or CO to represent all organic compounds.

### 4.2.1. Total Hydrocarbon and Carbon Monoxide Measurement

The THC concentration (or CO) is used to represent all organic compounds in the exit gas covered by the Part 503 rule. The US EPA does not require that biosolids themselves be monitored for THC (or CO), as is required for metals. Instead, the stack gas must be monitored for THC (or CO) because organic pollutants could be present due to incomplete combustion of organic compounds or the generation of by-products of combustion (36).

The Part 503 rule allows a monthly average concentration of up to 100 ppmv of THC (or CO) in the stack gas. Thus, an incineration facility operator firing biosolids must continuously monitor THC (or CO) levels in the stack gas to ensure that the monthly average concentration of THC (or CO) is at or below the limit. The monthly average THC (or CO) concentration is the arithmetic mean of the hourly averages; the hourly averages must be calculated based on at least two readings taken each hour that the incinerator operates in a day (i.e., in a 24-hour period).

The THC (or CO) must be measured using a flame ionization detector with a sampling line heated to 150°C or higher. The THC (or CO) concentration measurement taken from the stack must be corrected for moisture content and oxygen (as described below) before being compared to the 100 ppmv limit.

### 4.2.2. Correction for 0% Moisture

The THC (or CO) concentration in the stack gas must be corrected for 0% moisture, using Eq. 7, below. This correction is required so that all THC (or CO) emissions can be evaluated on a standardized basis. Once the correction factor for moisture is known, the original THC (or CO) concentration must be multiplied by this correction factor.

$$\text{Correction factor} = \frac{1}{1 - X} \qquad (7)$$

where
  $X$ = the decimal fraction of the % moisture in the biosolids incinerator exit gas

EXAMPLE 6
  If:

- $X = 0.12$
- Original THC measurement = 40 ppmv

What is the THC concentration corrected for 0% moisture?

SOLUTION

$$\text{Correction factor} = \frac{1}{1-X} \quad (7)$$

$$\text{Correction factor for moisture} = \frac{1}{1-0.12}$$

$$\text{Correction factor for moisture} = 1.14$$

Multiply the original THC (or CO) measurement 40 ppmv by the correction factor:

THC concentration corrected for 0% moisture = 40 ppmv × 1.14 = 46 ppmv

### 4.2.3. Correction to 7% Oxygen

The THC (or CO) concentration in the stack exit gas also must be corrected to 7% oxygen, using Eq. 9, below. This correction is required so that all THC (or CO) emissions can be evaluated on a standardized basis. Seven percent oxygen was chosen as the standard correction because this amount of oxygen, which is representative of 50% excess air, is frequently used for operational and measurement purposes. Once the correction factor for oxygen is known, the THC (or CO) concentration (which has already been corrected for moisture) must be multiplied by this value.

$$\text{Correction factor} = \frac{14}{21-Y} \quad (8)$$

where

$Y$ = the % oxygen concentration in the biosolids incinerator exit gas (dry volume basis)
$21$ = the % of oxygen in air
$14$ = the difference between the % oxygen in air (21%) and 7% oxygen

After being corrected to 7% oxygen and for 0% moisture, the THC (or CO) concentration may not be above 100 ppmv on a monthly average basis. If the monthly average THC (or CO) concentration in the stack gas measures above 100 ppmv, the biosolids incinerator is in violation until adjustments are made to meet the limits. These adjustments can include more careful control of furnace operations (and improvements of other procedures, if necessary) to reduce the amount of THC (or CO) released from the stack.

EXAMPLE 7
If:

- $Y = 10\%$
- $X = 0.12$
- Original THC measurement = 40 ppmv

What is the THC concentration corrected for 0% moisture and 7% oxygen?

SOLUTION

$$\text{Correction factor for oxygen} = \frac{14}{21 - Y} \qquad (8)$$

$$\text{Correction factor for oxygen} = \frac{14}{21 - 10}$$

Correction factor for oxygen = 1.27

The THC concentration corrected for 0% moisture = 46 ppmv (from Example 6)

The THC concentration corrected for moisture and oxygen = 46 × 1.27 = 58 ppmv

### 4.3. Management Practices for Biosolids Incineration

The management practices for biosolids incineration in the Part 503 rule cover the following (8):

1. Instrument operation and maintenance
2. Temperature requirements
3. Operation of air pollution control devices
4. Protection of threatened or endangered species

All but the last of these management practices are necessary to ensure that the limits set for arsenic, beryllium, cadmium, chromium, lead, mercury, nickel, and THC (or CO) are met. The required management practices are described below.

#### 4.3.1. Instruments Operation and Maintenance

Biosolids incinerator operators must use instruments to continuously measure and record certain information, including the following:

1. THC (or CO) in the stack exit gas
2. Oxygen in the stack exit gas
3. Information used to calculate moisture content in the stack exit gas
4. Combustion temperature in the furnace

Each of the instruments used for these measurements must be installed, calibrated, operated, and maintained according to guidance provided by the permitting authority. Examples of such instruments include extractive or in situ oxygen analyzers, thermocouples to measure the temperature of a saturated stream and combustion temperature, and dewpoint detectors (known as wet bulb/dry bulb detectors) to obtain information used to determine moisture content (37–40).

In addition, the instrument used for THC (or CO) measurements must utilize a flame ionization detector, utilize a sampling line heated to 150°C or higher at all times, and be calibrated using propane at least once every 24-hour operating period.

*4.3.2. Temperature Requirements*

The maximum combustion temperature allowed in the incinerator furnace is set by the permitting authority based on performance test data. A limit on combustion temperature is necessary to ensure that the incinerator is operating as it did during the performance test. If biosolids are incinerated at higher temperatures than the allowed maximum temperature, the control efficiency could change and the concentration of metals in the stack gas could increase. The incinerator would then be out of compliance until operated below the maximum allowed temperature or until shown to be in compliance with a new set of pollutant limits calculated using control efficiencies relevant to the new set of operating conditions.

*4.3.3. Air Pollution Control Devices*

Conditions for operating APCDs are determined by the permitting authority from the performance test. These conditions (e.g., gas flow rate and gas temperature) ensure that the APCDs are operating as efficiently as possible. If they are not operating properly, the control efficiency could change, which would affect the ability to meet pollutant limits. Therefore, these values must be within the range established during the performance test. Examples of operating parameters for APCDs that the permitting authority might set are shown in Table 6.19.

*4.3.4. Protection of Threatened or Endangered Species*

The final management practice for biosolids incinerators does not allow biosolids to be incinerated if a threatened or endangered animal or plant species or its critical habitat is likely to be adversely affected. Threatened and endangered species are listed in the Endangered Species Act (18). Critical habitat is defined as any place where a threatened or endangered species lives and grows during any stage of its life cycle. Any direct or indirect action in a critical habitat that diminishes the likelihood of survival and recovery of a listed species is considered destruction or adverse modification of a critical habitat.

## *4.4. Frequency of Monitoring Requirements for Biosolids Incineration*

The person firing biosolids in a biosolids incinerator must monitor at specified intervals for certain metals in the biosolids; for the THC (or CO) concentration, oxygen content, and information needed to determine moisture in the stack exit gas; for combustion temperature in the furnace; and for certain conditions of air pollution control device operation. These monitoring requirements are discussed below.

*4.4.1. Monitoring for Metals*

Biosolids must be monitored for the concentration of metals for which pollutant limits have been set, including arsenic, beryllium, cadmium, chromium, lead, mercury, and nickel. The permitting authority will determine how often the facility operator must monitor for beryllium and mercury. For the other metals (arsenic, cadmium, chromium, lead, and nickel), the minimum frequency for monitoring is based on the amount of biosolids incinerated (Table 6.20). The greater the amount of biosolids incinerated, the more frequently metals must be monitored.

**Table 6.19**
**Operating parameters for air pollution control devices**

| Operating parameter | Air pollution control device | Example of measuring instrument |
|---|---|---|
| Pressure drop | Venturi scrubber, impingement scrubber, mist eliminator,* fabric filter | Differential pressure gauge/transmitter |
| Liquid flow rate(s) | Venturi scrubber, impingement scrubber, mist eliminator, wet electrostatic precipitator (ESP) | Orifice plate with differential pressure gauge/transmitter |
| Gas temperature (inlet or outlet) | Venturi scrubber, impingement scrubber, dry scrubber, fabric filter, wet ESP | Thermocouple/transmitter |
| Liquid/reagent flow rate to atomizer | Dry scrubber (spray dryer absorber) | Magnetic flow meter |
| pH of liquid/reagent to atomizer | Dry scrubber (spray dryer absorber) | pH meter/transmitter |
| Atomized motor power (for rotary atomizer) | Dry scrubber (spray dryer absorber) | Watt meter |
| Compressed air pressure (for dual fluid flow) | Dry scrubber (spray dryer absorber) | Pressure gauge |
| Compressed airflow rate | Dry scrubber (spray dryer absorber) | Orifice plate with differential pressure gauge/transmitter |
| Opacity | Fabric filter | Transmissometer |
| Secondary voltage (for each transformer/rectifier) | Wet ESP | Kilovolt meters/transmitter |
| Secondary currents (for each transformer/rectifier) | Wet ESP | Milliammeters/transmitter |

*Types of mist eliminators include the wet cyclone, vane demister, chevron demister, and mesh pad.

### 4.4.2. Continuous Monitoring

As shown in Table 6.20, certain monitoring must be done continuously. Continuous monitoring is required for THC (or CO) concentrations, oxygen levels, and information used to calculate moisture content in the stack exit gas. Continuous monitoring also is required for combustion temperature in the furnace.

### 4.4.3. Monitoring Conditions in Air Pollution Control Devices

Certain operating conditions must be monitored in air pollution control devices, as discussed above. The specific conditions that must be monitored are based on the type of APCDs in place and the operating parameters that are important for maintaining

**Table 6.20**
**Monitoring frequency for biosolids incinerators**

| Pollutant/parameter | Amount of biosolids fired (dry-weight basis; T/yr) | Monitoring frequency |
|---|---|---|
| Arsenic, cadmium, chromium, lead, and nickel in biosolids | Greater than zero but less than 290 | Once per year |
| | Equal to or greater than 290 but less than 1500 | Once per quarter (four times per year) |
| | Equal to or greater than 1500 but less than 15,000 | Once per 60 days (six times per year) |
| | Equal to or greater than 15,000 | Once per month (12 times per year) |
| Beryllium and mercury in biosolids or stack exit gas | NA | As often as permitting authority requires |
| THC (or CO) concentration in stack exit gas | NA | Continuously; monthly averages reported, which is the arithmetic mean of hourly averages that include at least 2 readings/h |
| Oxygen concentration in stack exit gas | NA | Continuously |
| Information needed to determine moisture content in stack exit gas | NA | Continuously |
| Combustion temperature in furnace | NA | Continuously |
| Air pollution control device conditions | NA | As often as permitting authority requires |

the control efficiency demonstrated in the performance test. The ultimate monitoring frequency for APCDs will be specified in the permit.

### 4.5. Record-Keeping and Reporting Requirements for Biosolids Incineration

The person who incinerates biosolids must develop and keep the records for a minimum of 5 years. The record-keeping requirements include information on the pollutant limits, management practices, and monitoring requirements.

All Class 1 treatment works, treatment works serving a population of 10,000 or more, and treatment works with a 1 MGD or greater design flow have to report the information every February 19th to the permitting authority.

## 5. PATHOGEN AND VECTOR ATTRACTION REDUCTION REQUIREMENTS

Pathogens are disease-causing organisms, such as certain bacteria, viruses, and parasites. Vectors are organisms, such as rodents and insects, that can spread disease by carrying and transferring pathogens. Subpart D of the Part 503 rule covers alternatives for reducing pathogens in biosolids (including domestic septage), as well as options for reducing the potential for biosolids to attract vectors.

The alternatives concern the designation of biosolids as Class A or Class B in regard to pathogens. These classifications indicate the density (numbers/unit mass) of pathogens in biosolids where applicable. The requirements for land application or surface disposal of biosolids vary depending on the class of pathogen reduction achieved. Biosolids have to meet applicable requirements for both pathogen and vector attraction reduction to be in compliance with the rule. This chapter describes the pathogen alternatives and VAR options in the Part 503 rule. For more detail, the reader is referred to US EPA publications (41, 42).

The pathogen and VAR requirements apply to biosolids, including domestic septage, and their application to or placement on the land for beneficial use or disposal. Domestic septage applied to nonpublic contact sites (i.e., agricultural land, forests, and reclamation sites) is covered by a simplified portion of the rule that is explained in a separate US EPA guidance document (21).

Depending on how biosolids are used or disposed and which pathogen alternative and VAR option are relied on, compliance with the pathogen and vector attraction requirements is the responsibility of persons who do the following (8):

1. Generate biosolids that are either land applied or surface disposed
2. Derive a material from biosolids that are either land applied or surface disposed
3. Apply biosolids to the land
4. Place biosolids on a surface disposal site
5. Own or operate a surface disposal site

### 5.1. Pathogen Reduction Alternatives

The pathogen reduction alternatives ensure that pathogen levels in biosolids are reduced to levels considered safe for the biosolids to be land applied or surface disposed.

*Regulations and Costs of Biosolids Disposal and Reuse* 321

The rule includes criteria to classify biosolids as Class A or Class B with respect to pathogens. These classifications are based on the level of pathogens present in biosolids that are used or disposed.

If pathogens (*Salmonella* sp. bacteria, enteric viruses, and viable helminth ova) are below detectable levels, the biosolids meet the Class A designation. Biosolids are designated Class B if pathogens are detectable but have been reduced to levels that do not pose a threat to public health and the environment as long as actions are taken to prevent exposure to the biosolids after their use or disposal. When Class B biosolids are land applied, certain restrictions must be met at the application site; other requirements have to be met when Class B biosolids are surface disposed. The land application restrictions allow natural processes to further reduce pathogens in the biosolids before the public has access to the site. In general, Class A corresponds to the processes to further reduce pathogens (PFRP) designation, and Class B corresponds to the processes to significantly reduce pathogens (PSRP) designation.

### 5.1.1. Class A Pathogen Requirements

The rule lists six alternatives for treating biosolids so they can be classified as Class A with respect to pathogens. These alternatives are summarized in Table 6.21 and are discussed in detail below. Any one of these six alternatives may be met for the biosolids to be deemed Class A.

**Table 6.21**
**Summary of the six alternatives for meeting Class A pathogen requirements**

In addition to meeting the requirements in one of the six alternatives listed below, the requirements in Table 6.22 must be met for all six Class A alternatives

*Alternative 1: thermally treated biosolids*
  Biosolids must be subjected to one of four time-temperature regimes
*Alternative 2: biosolids treated in a high-pH–high-temperature process*
  Biosolids must meet specific pH, temperature, and air-drying requirements
*Alternative 3: biosolids treated in other processes*
  Demonstrate that the process can reduce enteric viruses and viable helminth ova; maintain operating conditions used in the demonstration after pathogen reduction demonstration is completed
*Alternative 4: biosolids treated in unknown processes*
  Biosolids must be tested for *pathogens—Salmonella* sp. or fecal coliform bacteria, enteric viruses, and viable helminth ova—at the time the biosolids are used or disposed of, or, in certain situations, prepared for use or disposal
*Alternative 5: biosolids treated in a PFRP*
  Biosolids must be treated in one of the processes to further reduce pathogens (PFRP) (see Table 6.24)
*Alternative 6: biosolids treated in a process equivalent to a PFRP*
  Biosolids must be treated in a process equivalent to one of the PFRPs, as determined by the permitting authority

**Table 6.22**
**Pathogen requirements for all Class A alternatives**

Either one of the following:
- The density of fecal coliform in the biosolids must be less than 1000 most probable numbers (MPN)/g total solids (dry-weight basis)
- The density of *Salmonella* sp. bacteria in the biosolids must be less than 3 MPN/4 g of total solids (dry-weight basis)

Either of these requirements must be met at one of the following times:
- When the biosolids are used or disposed
- When the biosolids are prepared for sale or giveaway in a bag or other container for land application
- When the biosolids or derived materials are prepared to meet the requirements for EQ biosolids

Pathogen reduction must take place before or at the same time as vector attraction reduction, except when the pH adjustment, percent solids vector attraction, injection, or incorporation options are met

Table 6.22 lists several requirements that must be met for all six of the Class A alternatives. Perhaps the most significant of the requirements is to avoid regrowth of bacteria as indicated by the results of a fecal coliform or Salmonella test.

### 5.1.1.1. ALTERNATIVE 1 FOR MEETING CLASS A: THERMALLY TREATED BIOSOLIDS

This alternative applies when specific thermal heating procedures (43, 44) are used to reduce pathogens. Equations are used to determine the length of heating time at a given temperature needed to obtain Class A pathogen reduction (i.e., reduce the pathogen content to below detectable levels). The equations take into consideration the solid-liquid nature of the biosolids being heated, along with the particle size and how particles are brought into contact with the heat. The equations also take into consideration that the internal structure of the mixture can inhibit mixing. For example, a safety factor is included in the equation for Regime C (Table 6.23) that adds more time for heating because less information is available about operational parameters that could influence the degree of pathogen destruction per unit of heat input. The rule identifies and provides equations for four different acceptable heating regimes.

The minimum indicated boundary conditions (i.e., solids content, mixing with the heat source, time of heating, and operating temperature) are specified in Table 6.23 for each of the four thermal heating regimes. The equation specified for a particular heating regime is used to calculate the actual time and temperature for operating the system within the boundaries of the applicable regime. In addition to the requirements for each regime, the requirements in Table 6.22 must be met.

## Table 6.23
### The four time-temperature regimes for Class A pathogen reduction under alternative 1

| Regime | Applies to: | Requirement | Time-temperature relationship* | |
|---|---|---|---|---|
| A | Biosolids with 7% solids or greater (except those covered by Regime B) | Temperature of biosolids must be 50°C or higher for 20 min or longer | $t = \dfrac{131{,}700{,}000}{10^{0.14T}}$ | (9) |
| B | Biosolids with 7% solids or greater in the form of small particles and heated by contact with either warmed gases or an immiscible liquid | Temperature of biosolids must be 50°C or higher for 15 s or longer | $t = \dfrac{131{,}700{,}000}{10^{0.14T}}$ | (9) |
| C | Biosolids with less than 7% solids | Heated for at least 15 s but less than 30 min | $t = \dfrac{131{,}700{,}000}{10^{0.14T}}$ | (9) |
| D | Biosolids with less than 7% solids | Temperature of sludge is 50°C or higher with at least 30 min or longer contact time | $t = \dfrac{50{,}070{,}000}{10^{0.14T}}$ | (10) |

*$t$ = time in d; $T$ = temperature in °C.

### EXAMPLE 8

Biosolids contain 10% solids and are heated with a biosolids dryer at 55°C. What is the required minimum time for achieving Class A pathogen status?

### SOLUTION

$$t = \frac{131{,}700{,}000}{10^{0.14T}}$$

$$t = \frac{131{,}700{,}000}{10^{0.14(55)}} = 2.6 \text{ d} = 63 \text{ h} \qquad (9)$$

where
  $t$ = time

The minimum time would be 63 hours if the operator followed Regime A in Table 6.23. Under Regime A the temperature cannot be lower than 50°C or the time shorter than 20 minutes.

### EXAMPLE 9

Biosolids contain 10% solids and are treated in a biosolids dryer for about 1.5 minutes (0.001 day). What is the required minimum temperature ($T$)?

SOLUTION

$$t = \frac{131{,}700{,}000}{10^{0.14T}} \quad (9)$$

$$0.001 = \frac{131{,}700{,}000}{10^{0.14(T)}}$$

$$0.001[10^{0.14(T)}] = 131{,}700{,}000$$

$$T = 79°C$$

The minimum temperature to achieve Class A pathogen status would be 79°C if the operator followed Regime B in Table 6.23. Under this regime, the temperature cannot be lower than 50°C or the time shorter than 15 seconds and the biosolids must be in the form of small particles (e.g., from a steam drier) in intimate contact with the drying unit. Otherwise, Regime A would apply.

#### 5.1.1.2. ALTERNATIVE 2 FOR MEETING CLASS A: BIOSOLIDS TREATED IN A HIGH pH-HIGH TEMPERATURE PROCESS

This alternative describes conditions of a specific temperature-pH process (45) that is effective in reducing pathogens to below detectable levels. The process conditions required by the regulation are as follows:

1. Elevating the pH to greater than 12 (measured at 25°C) for 72 hours or longer
2. Maintaining the temperature above 52°C for at least 12 hours during the period that the pH is greater than 12
3. Air drying to over 50% solids after the 72-hour period of elevated pH
4. Meeting all the requirements in Table 6.22

#### 5.1.1.3. ALTERNATIVE 3 FOR MEETING CLASS A: BIOSOLIDS TREATED IN OTHER PROCESSES

This alternative requires comprehensive monitoring of enteric viruses and viable helminth ova during each monitoring episode until demonstration has shown that the process achieves adequate reduction of pathogens. The presence of enteric viruses and viable helminth ova has to be shown in the biosolids prior to pathogen treatment to document the effectiveness of the treatment process. The tests and requirements are as follows:

1. Once shown to be present prior to treatment, the density of enteric viruses in the biosolids after pathogen treatment must be less than 1 plaque-forming unit (PFU)/4 g of total solids (dry-weight basis).
2. Likewise, the density of viable helminth ova in the biosolids after pathogen treatment must be less than $1/4$ g of total solids (dry-weight basis).
3. All the requirements in Table 6.22 must be met.

Alternative 3 is useful for demonstrating that a new process fully meets Class A pathogen requirements under the tested set of operating parameters. Subsequent testing for enteric viruses and viable helminth ova is unnecessary whenever the tested set of operating parameters has been met. It is important to realize that the tested set of operating parameters may have included ranges of values.

If no enteric viruses or viable helminth ova are present before treatment, then the tested batch of biosolids can be considered Class A. The tests, however, must be repeated during each subsequent monitoring episode until

1. pathogens are detected before the process and demonstrated to have been reduced to below detectable levels after the process, or
2. after 2 years of testing with no detection of pathogens before the process, the permitting authority modifies the monitoring requirements for enteric viruses and viable helminth ova.

As already mentioned, monitoring for fecal coliform or *Salmonella* sp. bacteria is always required in accordance with the requirements listed in Table 6.22.

#### 5.1.1.4. ALTERNATIVE 4 FOR MEETING CLASS A: BIOSOLIDS TREATED IN UNKNOWN PROCESSES

This alternative is used in situations where

1. a biosolids treatment process is unknown, or
2. the biosolids were treated in a process operating under less-stringent conditions than those under which the biosolids could qualify as Class A under any of the other alternatives.

This alternative requires that the biosolids be analyzed for *Salmonella* sp. bacteria, enteric viruses, and viable helminth ova at each of the following times:

1. When the biosolids are used or disposed
2. When biosolids are prepared for sale or for giveaway in a bag or other container for application to the land
3. When the biosolids are prepared to meet the EQ requirements

As in alternative 3, the required test results for this alternative are as follows:

1. The density of viruses in the biosolids must be less than 1 PFU/4 g of total solids (dry-weight basis).
2. The density of viable helminth ova in the biosolids must be less than $1/4$ g of total solids (dry-weight basis).
3. All the requirements in Table 6.22 must be met.

Although biosolids must meet the same pathogen test results as in alternative 3, alternative 4 requires testing of each batch of the biosolids that is used or disposed, rather than just monitoring the operating parameters, after the demonstration that the process reduces pathogens.

#### 5.1.1.5. ALTERNATIVE 5 FOR MEETING CLASS A: BIOSOLIDS TREATED IN A PROCESS TO FURTHER REDUCE PATHOGENS

This alternative states that biosolids are considered to be Class A if

1. they are treated in one of the PFRPs listed in Table 6.24, and
2. all requirements in Table 6.22 are met

**Table 6.24**
**Processes to further reduce pathogens (PFRPs)**

| Process | Conditions |
|---|---|
| 1. Composting | Using either the within-vessel composting method or the static aerated pile composting method, the temperature of the biosolids is maintained at 55°C or higher for 3 days. |
| | Using the windrow composting method, the temperature of the biosolids is maintained at 55°C or higher for 15 days or longer. During the period when the compost is maintained at 55°C or higher, the windrow is turned a minimum of five times. |
| 2. Heat drying | Biosolids are dried by direct or indirect contact with hot gases to reduce the moisture content of the biosolids to 10% or lower. Either the temperature of the biosolids particles exceeds 80°C or the wet bulb temperature of the gas in contact with the biosolids as the biosolids leave the dryer exceeds 80°C. |
| 3. Heat treatment | Liquid biosolids are heated to a temperature of 180°C or higher for 30 minutes. |
| 4. Thermophilic aerobic digestion | Liquid biosolids are agitated with air or oxygen to maintain aerobic conditions, and the mean cell residence time of the biosolids is 10 days at 55° to 60°C. |
| 5. Beta ray irradiation | Biosolids are irradiated with beta rays from an accelerator at dosages of at least 1.0 megarad at room temperature (20°C). |
| 6. Gamma ray irradiation | Biosolids are irradiated with gamma rays from certain isotopes, such as cobalt 60 and cesium 137, at room temperature (20°C). |
| 7. Pasteurization | The temperature of the biosolids is maintained at 70°C or higher for 30 minutes or longer. |

### 5.1.1.6. ALTERNATIVE 6 FOR MEETING CLASS A: BIOSOLIDS TREATED IN A PROCESS EQUIVALENT TO A PROCESS TO FURTHER REDUCE PATHOGENS

Under alternative 6, biosolids are considered to be Class A if

1. they are treated by any process determined to be equivalent to a PFRP by the permitting authority, and
2. all requirements in Table 6.22 are met.

The Part 503 rule gives the permitting authority responsibility for determining equivalency. To be equivalent, a treatment process must be able to consistently reduce pathogens to levels comparable to the reduction achieved by listed PFRPs. The process must be equivalent in its ability to achieve Class A status with respect to enteric viruses and viable helminth ova as long as it is operated under the same conditions that produced the required reductions.

**Table 6.25**
**Summary of the three alternatives, and their conditions, for meeting Class B pathogen requirements**

Alternative 1: the monitoring of indicator organisms
   Test for fecal coliform density as an indicator for all pathogens. The geometric mean of seven samples shall be less than 2 million MPNs/g of total solids or less than 2 million CFUs/g of total solids at the time of use or disposal.
Alternative 2: biosolids treated in a PSRP
   Biosolids must be treated in one of the processes to significantly reduce pathogens (PSRP) (see Table 6.26).
Alternative 3: biosolids treated in a process equivalent to a PSRP
   Biosolids must be treated in a process equivalent to one of the PSRPs, as determined by the permitting authority.

The US EPA's Pathogen Equivalency Committee (PEC) is available as a resource to provide recommendations on equivalency determinations to the permitting authority and guidance to the regulated community.

### 5.1.2. Class B Pathogen Requirements

Class B pathogen requirements can be met using one of three alternatives, as listed in Table 6.25 and described below. Unlike a Class A biosolids, in which pathogens are at levels below detectable limits, Class B biosolids may contain some pathogens. For this reason, the Class B requirements for land application of biosolids also include site restrictions that prevent crop harvesting, animal grazing, turf growing, and public access for a certain period of time (1 to 14 months) until environmental conditions have further reduced pathogens. Management practices rather than site restrictions prevent exposure to the pathogens in biosolids for surface disposed Class B biosolids.

#### 5.1.2.1. ALTERNATIVE 1 FOR MEETING CLASS B: THE MONITORING OF INDICATOR ORGANISMS

Alternative 1 requires that seven samples of treated biosolids be collected shortly before biosolids use or disposal, and that the geometric mean fecal coliform density of these samples be less than 2 million colony-forming units (CFUs) or most probable number (MPN) per gram of biosolids (dry-weight basis). The US EPA suggests that these seven samples be collected over a 2-week period. This approach uses fecal coliform density as an indicator of the average density of bacterial and viral pathogens.

The US EPA recommends that seven samples be taken over the 2-week period preceding use or disposal because the test methods used to determine fecal coliform density (membrane filter methods and the multiple tube dilution method) have poor precision, and biosolids quality can vary. Using at least seven samples should provide a sufficiently representative sampling of the biosolids.

#### 5.1.2.2. ALTERNATIVE 2 FOR MEETING CLASS B: BIOSOLIDS TREATED IN A PROCESS TO SIGNIFICANTLY REDUCE PATHOGENS

Under alternative 2, biosolids are considered to be Class B if they are treated in one of the PSRPs listed in Table 6.26.

**Table 6.26**
**Processes to significantly reduce pathogens (PSRPs)**

| Process | Conditions |
|---|---|
| 1. Aerobic digestion | Biosolids are agitated with air or oxygen to maintain aerobic conditions for a specific mean cell residence time at a specific temperature. Values for the mean cell residence time and temperature shall be between 40 days at 20°C and 60 days at 15°C. |
| 2. Air drying | Biosolids are dried on sand beds or on paved or unpaved basins. The biosolids dry for a minimum of 3 months. During 2 of the 3 months, the ambient average daily temperature is above 0°C. |
| 3. Anaerobic digestion | Biosolids are treated in the absence of air for a specific mean cell residence time at a specific temperature. Values for the mean cell residence time and temperature shall be between 15 days at 35°C to 55°C and 60 days at 20°C. |
| 4. Composting | Using the within-vessel, static aerated pile, or windrow composting methods, the temperature of the biosolids is raised to 40°C or higher and maintained for 5 days. For 4 hours during the 5-day period, the temperature in the compost pile exceeds 55°C. |
| 5. Lime stabilization | Sufficient lime is added to the biosolids to raise the pH of the biosolids to 12 after 2 hours of contact. |

Unlike the comparable Class A requirement, this alternative does not require microbiological monitoring for regrowth of fecal coliform or *Salmonella* sp. bacteria.

5.1.2.3. ALTERNATIVE 3 FOR MEETING CLASS B: BIOSOLIDS TREATED IN A PROCESS EQUIVALENT TO A PROCESS TO SIGNIFICANTLY REDUCE PATHOGENS

Under alternative 3, biosolids treated by any process determined to be equivalent to a PSRP by the permitting authority are considered to be Class B biosolids.

Part 503 gives the permitting authority responsibility for determining equivalency. The US EPA Pathogen Equivalency Committee is available as a resource to provide recommendations on equivalency determinations to the permitting authorities. As with Class A, the Class B equivalency determination can be made on either a site-specific or a national basis.

## 5.2. Requirements for Reducing Vector Attraction

The pathogens in biosolids pose a disease risk when they are brought into contact with humans or other susceptible hosts (plant or animal). Vectors, which include flies, mosquitoes, fleas, rodents, and birds, can transmit pathogens to humans and other hosts physically through contact or biologically by playing a specific role in the life cycle of the pathogen. Reducing the attractiveness of biosolids to vectors reduces the potential for transmitting diseases from pathogens in biosolids.

The rule contains 12 options, which are summarized in Table 6.27 and described below, for demonstrating reduced vector attraction for biosolids. (Note: Option 12

**Table 6.27**
**Summary of options for meeting vector attraction reduction**

| Option | Description |
|---|---|
| *Option 1* | Meet 38% reduction in volatile solids content |
| *Option 2* | Demonstrate vector attraction reduction with additional anaerobic digestion in a bench-scale unit |
| *Option 3* | Demonstrate vector attraction reduction with additional aerobic digestion in a bench-scale unit |
| *Option 4* | Meet a specific oxygen uptake rate for aerobically digested biosolids |
| *Option 5* | Use aerobic processes at greater than 40°C for 14 days or longer |
| *Option 6* | Alkali addition under specified conditions |
| *Option 7* | Dry biosolids with no unstabilized solids to at least 75% solids |
| *Option 8* | Dry biosolids with unstabilized solids to at least 90% solids |
| *Option 9* | Inject biosolids beneath the soil surface |
| *Option 10* | Incorporate biosolids into the soil within 6 hours of application to or placement on the land |
| *Option 11* | Cover biosolids placed on a surface disposal site with soil or other material at the end of each operating day (Note: only for surface disposal.) |
| *Option 12* | Alkaline treatment of domestic septage to pH 12 or above for 30 minutes without adding more alkaline material |

applies only to domestic septage.) These requirements are designed to either reduce the attractiveness of biosolids to vectors (options 1 through 8 and option 12) or prevent vectors from coming in contact with the biosolids (options 9 through 11).

*5.2.1. Option 1: Reduction in Volatile Solids Content*

Under this option, vector attraction is reduced if the mass of volatile solids in the biosolids is reduced by at least 38% during the treatment of the biosolids. This percentage is the amount of volatile solids reduction that is attained by anaerobic (46) or aerobic digestion (47) plus any additional volatile solids reduction that occurs before the biosolids leave the treatment works, such as through processing in drying beds (48) or lagoons (49), or by composting (50).

*5.2.2. Option 2: Additional Digestion of Anaerobically Digested Biosolids*

Frequently, biosolids have been recycled through the biological wastewater treatment section of a treatment works or have resided for long periods of time in the wastewater collection system. During this time, they undergo substantial biological degradation. If the biosolids are subsequently treated by anaerobic digestion for a period of time, they are adequately reduced in vector attraction. Because they will have entered the digester already partially stabilized, however, the volatile solids reduction after treatment is frequently less than 38%.

Under these circumstances, the 38% reduction required by option 1 might not be possible. Option 2 allows the operator to demonstrate vector attraction reduction by testing a portion of the previously digested biosolids in a bench-scale unit in the laboratory. Vector attraction reduction is demonstrated if after anaerobic digestion of the biosolids for an additional 40 days at a temperature between 30° and 37°C, the volatile solids in the biosolids are reduced by less than 17% from the beginning to the end of the bench test.

### 5.2.3. Option 3: Additional Digestion of Aerobically Digested Biosolids

This option is appropriate for aerobically digested biosolids that cannot meet the 38% volatile solids reduction required by option 1. This includes biosolids from extended aeration plants, where the minimum residence time of biosolids leaving the wastewater treatment processes section generally exceeds 20 days. In these cases, the biosolids will already have been substantially degraded biologically prior to aerobic digestion.

Under this option, aerobically digested biosolids with 2% or less solids are considered to have achieved vector attraction reduction if, in the laboratory after 30 days of aerobic digestion in a batch test at 20°C, volatile solids are reduced by less than 15%. This test is only applicable to liquid aerobically digested biosolids.

### 5.2.4. Option 4: Specific Oxygen Uptake Rate for Aerobically Digested Biosolids

Frequently, aerobically digested biosolids are circulated through the aerobic biological wastewater treatment process for as long as 30 days. In these cases, the biosolids entering the aerobic digester are already partially digested, which makes it difficult to demonstrate the 38% reduction required by option 1.

The specific oxygen uptake rate (SOUR) is the mass of oxygen consumed per unit time per unit mass of total solids (dry-weight basis) in the biosolids. Reduction in vector attraction can be demonstrated if the SOUR of the biosolids that are used or disposed, determined at 20°C, is equal to or less than 1.5 mg oxygen/h/g of total biosolids (dry-weight basis). This test is based on the fact that if the biosolids consume very little oxygen, their value as a food source for microorganisms is very low and therefore microorganisms are unlikely to be attracted to them. Other temperatures can be used for this test, provided the results are corrected to a 20°C basis. This test is only applicable to liquid aerobic biosolids withdrawn from an aerobic process.

### 5.2.5. Option 5: Aerobic Processes at Greater Than 40°C

This option applies primarily to composted biosolids that also contain partially decomposed organic bulking agents. The biosolids must be aerobically treated for 14 days or longer, during which time the temperature always must be over 40°C and the average temperature must be higher than 45°C. This option can be applied to other aerobic processes, such as aerobic digestion (47), but options 3 and 4 are likely to be easier to meet for the other aerobic processes.

### 5.2.6. Option 6: Addition of Alkaline Material

Biosolids are considered to be adequately reduced in vector attraction if sufficient alkaline material is added to achieve the following (45):

1. Raise the pH to at least 12, measured at 25°C, and without the addition of more alkaline material, maintain a pH of at least 12 for 2 hours.
2. Maintain a pH of at least 11.5 without addition of more alkaline material for an additional 22 hours.

The conditions required under this option are designed to ensure that the biosolids can be stored for at least several days at the treatment works, transported, and then used or disposed without the pH falling to the point where putrefaction occurs and vectors are attracted.

### 5.2.7. Option 7: Moisture Reduction of Biosolids Containing No Unstabilized Solids

Under this option, vector attraction is considered to be reduced if the biosolids do not contain unstabilized solids generated during primary treatment and if the solids content of the biosolids is at least 75% before the biosolids are mixed with other materials. Thus, the reduction must be achieved by removing water, not by adding inert materials.

It is important that the biosolids not contain unstabilized solids because the partially degraded food scraps likely to be present in such biosolids would attract birds, some mammals, and possibly insects, even if the solids content of the biosolids exceeded 75%.

### 5.2.8. Option 8: Moisture Reduction of Biosolids Containing Unstabilized Solids

The ability of any biosolids to attract vectors is considered to be adequately reduced if the solids content of the biosolids is increased to 90% or greater, regardless of whether this increase was for biosolids from primary treatment. The solids increase should be achieved by removal of water and not by dilution with inert solids. Drying to this extent severely limits biological activity and strips off or decomposes the volatile compounds that attract vectors.

The way dried biosolids are handled, including their storage before use or disposal, can create or prevent vector attraction. If dried biosolids are exposed to high humidity, the outer surface of the biosolids will gain in moisture content and possibly attract vectors. This should be properly guarded against.

### 5.2.9. Option 9: Biosolids Injection

Vector attraction reduction can be demonstrated by injecting the biosolids below the ground surface. Under this option, no significant amount of biosolids can be present on the land surface within 1 hour of injection, and if the biosolids are Class A with respect to pathogens, they must be injected within 8 hours after discharge from the pathogen-reducing process. The reason for this special consideration for Class A biosolids (assuming vector attraction has not been reduced by some other means) is that pathogens could regrow and Class A biosolids have no site restrictions to provide crop, grazing, and access protection.

Injection of biosolids beneath the soil places a barrier of earth between the biosolids and vectors. The soil removes water from the biosolids, which reduces the mobility and odor of the biosolids. Odor is usually present at the site during the injection process but quickly dissipates when injection is complete.

### 5.2.10. Option 10: Incorporation of Biosolids into the Soil

Under this option, biosolids must be incorporated into the soil within 6 hours of application to or placement on the land. Incorporation is accomplished by plowing or some other means of mixing the biosolids into the soil. If the biosolids are Class A with respect to pathogens, the time between processing and application or placement must not exceed 8 hours, the same as for injection under option 9.

*5.2.11. Option 11: Covering Biosolids*

Under this option, biosolids placed on a surface disposal site must be covered with soil or other material at the end of each operating day. Daily covering reduces vector attraction by creating a physical barrier between the biosolids and vectors. Covering also helps meet pathogen requirements by allowing environmental conditions to reduce pathogens.

*5.2.12. Option 12: Alkaline Treatment for Domestic Septage*

This option pertains only to vector attraction reduction for domestic septage. Under this option, the pH of domestic septage must be raised to at least 12 and remain at pH 12 or above for a minimum of 30 minutes during which no additional alkaline material may be added.

For details on biosolids management at local, state, and international levels, the reader is referred to the following references: in the U.S., Arizona, (51, 52), Florida (53), Georgia (54), Orange County, CA (55), and Pennsylvania (56); in Canada, British Columbia (57) and Greater Vancouver Regional District (58); and in Australia, Victoria (59) and Western Australia (60).

# 6. COSTS

This section presents a series of cost relationships for sludge disposal alternatives and describes briefly the process and type of information used in creating the cost relationships. In this context the term *sludge disposal alternative* is used to denote the combination of sludge treatment processes and sludge transport and ultimate disposal methodologies comprising a sludge management system (61).

The basic premises or conditions selected for development of the cost curves are as follows:

1. The variables to be considered in the relationships are as follows:
   a. Sewage treatment plant flow rates varying from 1 to 1000 MGD
   b. Two levels of treatment, primary and (activated sludge) secondary
   c. Sludge treatment processes incorporating incineration and anaerobic digestion
   d. Transport to ocean disposal by barging and to land disposal by truck, rail, and pipeline
   e. Land disposal by landfill and land spreading
2. The range of transport distances for the ocean barging and land disposal methodologies were selected to reflect the transport distances likely to be considered by cities, and were as follows:
   a. Barge transport to ocean disposal locations at distances of 15, 50, 80, 110, 150, and 180 miles from the barge loading station
   b. Land transport over distances of 20, 50, 100, and 150 miles
3. Barging transport costs were developed for two situations:
   a. Simple case, where all sludge generated in a metropolitan area can be loaded at a single barge loading station
   b. Complex case, where sludge is collected from a multiplicity of barge loading stations in a metropolitan area before the barge can be towed to sea
4. All cost relationships were developed using 1975 costs.

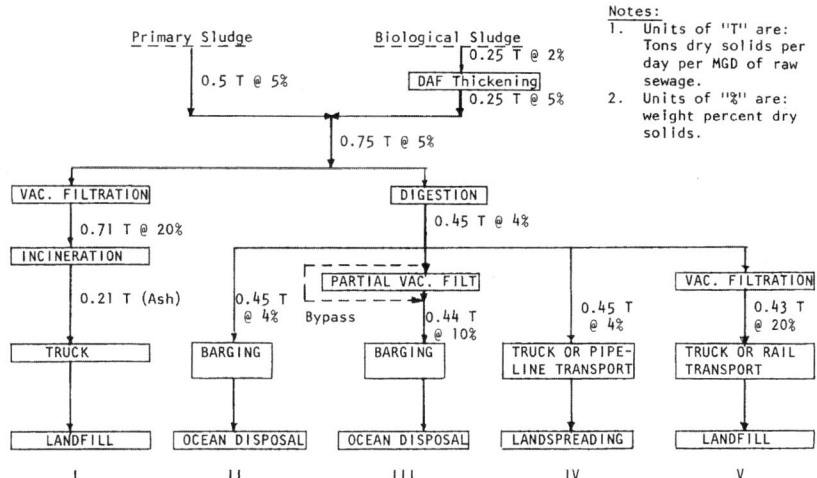

**Fig. 6.5.** Sludge disposal alternatives. *Source*: US EPA (61).

5. All costs can be converted to 2007 U.S. dollars (USD) by using U.S. Army Corps of Engineers Civil Works Construction Yearly Average Cost Index for Utilities (see appendix):

$$2007 USD = Cost \frac{539.74}{190.49}$$

Cost in $2007 USD = 2.83$ *Cost* from Figure 6.6 \hfill (11)

## 6.1. Description of Alternatives

The sludge disposal alternatives for which cost relationships were developed are shown in Figure 6.5. The basic alternatives are as follows:

1. *Alternative I*: Vacuum filtration of primary and thickened biological sludges to produce a sludge stream at 20% solids, followed by incineration and truck haul of the incinerator ash to a landfill site.
2. *Alternative II*: Digestion of primary and thickened biological sludges followed by barging to an ocean disposal site under simple and complex barging conditions.
3. *Alternative III*: Digestion of primary and thickened biological sludges, vacuum filtration of a portion of the digested sludge stream and blending of this portion with undewatered digested sludge to produce a sludge at 10% solids, followed by barging to an ocean disposal site under simple and complex barging conditions.
4. *Alternative IV*: Digestion of primary and thickened biological sludge, followed by tank truck or pipeline transport to a landspreading site.
5. *Alternative V*: Digestion of primary and thickened biological sludge, followed by vacuum filtration to produce a sludge stream at 20% solids, and truck or rail transport to a landfill site.

Subalternatives were defined in each alternative to account for sludge handling and disposal associated with primary and activated sludge secondary treatment as follows:

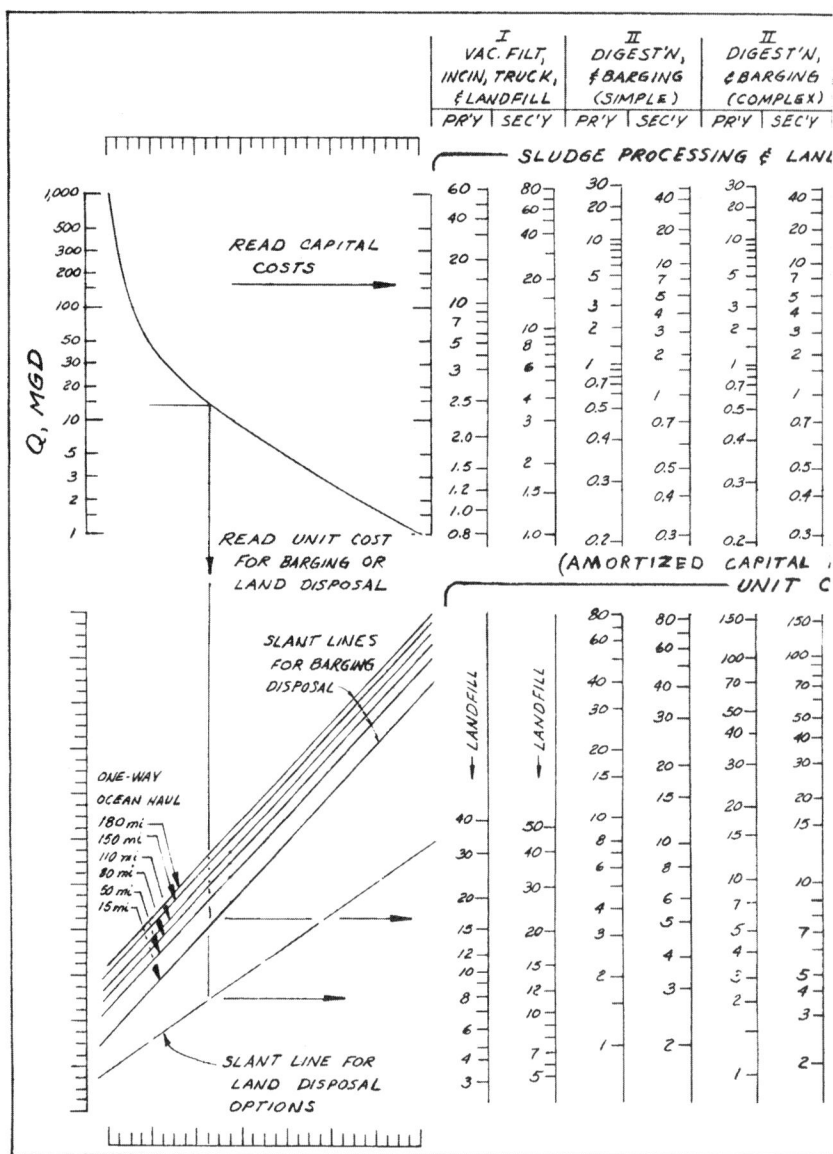

**Fig. 6.6** Sludge disposal cost curves. *Source*: US EPA (61).

1. The secondary treatment subalternatives within each alternative were cost-evaluated as shown in Figure 6.5 and described above, that is, inclusive of the biological sludge stream and the thickening unit process for this stream.
2. The primary treatment subalternatives were cost-evaluated exclusive of the biological sludge stream and the thickening unit process for this stream.

In developing material balances for the sludge flow through the unit process trains in each alternative, the normalized parameter selected was "t of dry solids/d/MGD of flow." The values of this parameter for the sludge stream at each point in the unit process

# Regulations and Costs of Biosolids Disposal and Reuse

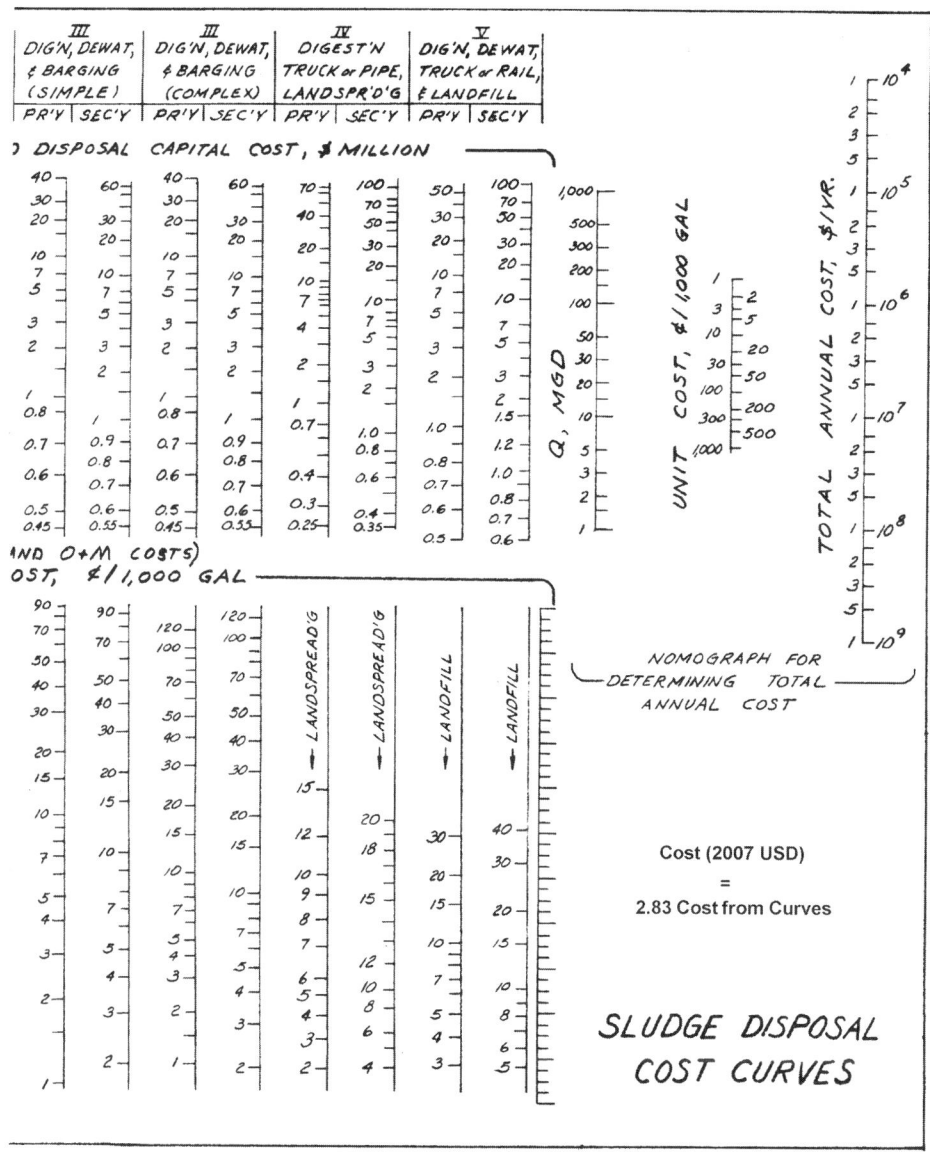

**Fig. 6.6** (*Continued*)

trains, and the corresponding solids concentrations of the sludge stream, are shown in Figure 6.5. These data were developed using the following assumptions:

1. Primary treatment generates 0.5 t dry solids/d/MGD at 5% solids in the sludge flow from the primary clarifiers.
2. Secondary treatment generates 0.5 t dry solids/d/MGD at 5% solids in the sludge flow from the primary clarifiers and 0.25 t dry solids/d/MGD at 2% solids in the biological sludge flow from the secondary clarifiers.
3. Thickening by dissolved-air flotation (DAF) increases the solids concentration in biological sludges from 2% to 5%.

4. Incineration results in destruction of 70% of the dry solids content of the sludge flow.
5. Digestion (anaerobic) results in destruction of 40% of the dry solids content of the sludge flow.
6. Dewatering to 20% solids can be achieved with vacuum filtration.
7. The maximum solids concentration of which sludge can be discharged (pumped) from a barge is 10%.

## 6.2. Cost Relationships

Some 288 sets of cost data were utilized in the development of a graphical presentation of capital cost and total annual (amortization plus operation/maintenance) costs as a function of flow rate, treatment level, transport distance, etc., as the end product.

The estimating conditions that were applied in the development of cost curves are as follows:

1. Economic factors: interest rate $= 5\frac{7}{8}\%$ and project life $= 20$ years
2. Sludge generation quantities: the assumption used is that secondary treatment results in the generation of 0.75 t dry solids/d/MGD, and primary treatment in 0.50 t dry solids/d/MGD; thus for any given flow rate, the sludge quantity generated for primary treatment is assumed to be equal to two thirds of the quantities generated for secondary treatment.
3. Landspreading: an application rate of 10 t dry solids/acre/yr

The cost relationships were developed as follows:

1. Capital costs are presented graphically for the sludge processing and land disposal related elements of each alternative; no capital costs are included in any alternative for the transport element, such that the capital costs for each alternative are distance independent.
2. Total annual costs for the land-based alternatives are presented graphically for the sludge processing land disposal related, and transport elements on a distance independent basis.

## 6.3. Sludge Disposal Cost Curves

A graphical presentation of the sludge disposal cost curves is presented in Figure . The elements of the cost curves are as follows:

1. A flow rate scale ranging from 1 to 1000 MGD.
2. Capital cost scales for the primary and secondary treatment subalternatives within each alternative, wherein capital costs are included for the sludge processing and land disposal elements within each subalternative.
3. Unit cost scales (in units of cents: USC/1000 gal) for each subalternative; unit cost is defined as the normalized total annual cost, inclusive of amortization and O/M costs, for the sludge processing, transport, and disposal elements of each alternative.
4. Slant lines for determining unit costs for barging disposal at one-way ocean haul distances of 15, 50, 80, 110, and 150 miles.
5. A slant line for determining unit costs for the land disposal alternatives as a function of haul distances between 20 and 150 miles.
6. A nomograph for determining total annual cost (USD/yr) as a function of flow rate and the unit cost for a given subalternative at that flow rate.

## 6.4. Procedure for Using the Diagram

1. Select flow rate, treatment level, alternative, and haul distance (haul distance if an ocean disposal alternative is selected).
2. Enter flow rate on the "Q, MGD" scale in the upper left hand corner of the graph.
3. Read capital cost by moving rightward horizontally across the capital cost scales; guide scales are provided at either side of the capital cost scales to assist lineup of the straight edge.
4. Read unit cost by reading horizontally from the "Q, MGD" scale to the deflector line; proceed vertically downward to the appropriate slant line; then read unit cost by moving rightward horizontally across the unit cost scales; guide scales are provided to assist lineup of the edge at each step.
5. Using the selected flow rate, the unit cost as determined above, and a straight edge, enter the "Nomograph for Determining Total Annual Cost" and read total annual cost directly.

## ACRONYMS

*AWSAR* = Annual Whole Sludge Application Rate
*APCD* = Air Pollution Control Devices
*APLR* = Annual Pollutant Loading Rate
*BLM* = Bureau of Land Management
*CAA* = Clean Air Act
*CFR* = Code of Federal Register
*CFU* = Colony-Forming Units
*CO* = Carbon monoxide
*CPLR* = Cumulative Pollutant Loading Rate
*CWA* = Clean Water Act
*EQ* = Exceptional Quality
*LEL* = Lower Explosive Limit
*MCL* = Maximum Contaminant Level
*MPN* = Most Probable Number
*MPRSA* = Marine Protection, Research, and Sanctuaries Act
*NEPA* = National Environmental Policy Act
*NAAQS* = National Ambient Air Quality Standards
*NESHAPs* = National Emission Standards for Hazardous Air Pollutants
*NPDES* = National Pollutant Discharge Elimination System
*PC* = Pollutant Concentration
*PEC* = Pathogen Equivalency Committee
*PFRP* = Processes to Further Reduce Pathogens
*PFU* = Plaque-Forming Unit
*POTW* = Publicly Owned Treatment Works
*PSRP* = Processes to Significantly Reduce Pathogens
*RCRA* = Resource Conservation and Recovery Act
*SOUR* = Specific Oxygen Uptake Rate
*THC* = Total hydrocarbons
*TSCA* = Toxic Substances Control Act

$t$ = English ton
$T$ = Metric Ton
US ACE = US Army Corps of Engineers
US EPA = US Environmental Protection Agency
US FWS = US Fish and Wildlife Service
USGS = US Geological Survey
US SCS = US Soil Conservation Service
VAR = Vector Attraction Reduction

## NOMENCLATURE

$APLR$ = annual pollutant loading rate, kg of pollutant/ha/yr
$AWSAR$ = annual whole sludge (biosolids) application rate, dry T of biosolids/ha/yr
$C$ = pollutant concentration (dry weight basis), mg/kg
$C$ = pollutant limit (allowable daily concentration of arsenic, cadmium, chromium, or nickel in biosolids, dry-weight basis), mg/kg
$C_{lead}$ = pollutant limit (allowable daily concentration of lead in biosolids, dry-weight basis), mg/kg
$CE$ = biosolids incinerator control efficiency, fraction
$DF$ = dispersion factor based on an air dispersion model, $\mu g/m^3/g/s$
NAAQS = National Ambient Air Quality Standard for lead, $\mu g/m^3$ (1.5 $\mu g/m^3$)
$r$ = the decimal fraction of the hexavalent chromium concentration in the total chromium concentration measured in the exit gas from the biosolids incinerator stack.
RSC = risk specific concentration (the allowable increase in the average daily ground-level ambient air concentration for a pollutant at or beyond the property line of the site, $\mu g/m^3$
$SF$ = biosolids feed rate, t/d
$t$ = time, d
$T$ = temperature, °C
$X$ = the decimal fraction of the % moisture in the biosolids incinerator exit gas
$Y$ = the % oxygen concentration in the biosolids incinerator exit gas (dry volume basis)

## REFERENCES

1. Federal Register, *Clean Water Act (CWA)*, 33 U.S.C. ss/1251 et seq. (1977), U.S. Government Public Law. Full text is at www.access.gpo.gov/uscode/title33/chapter26_.html, May (2002).
2. Federal Register, *Resource Conservation and Recovery Act (RCRA)*, 42 U.S. Code s/s 6901 et seq. (1976), United States Government, www.access.gpo.gov/uscode/title42/chapter82_.html, January (2004).
3. Federal Register, *Toxic Substances Control Act (TSCA)*, 15 U.S.C. s/s 2601 et seq. (1976), U.S. Government Public Law. Text at www.access.gpo.gov/uscode/title15/chapter53_.html, January (2004).

4. Federal Register, *Marine Protection, Research and Sanctuaries Act (MPRSA)*, U.S. Government Pub. L. 92–532, October 23, 1972, 86 Stat. 1052 and 1061, Titles I and II are codified at 33 U.S.C. 1401–1445, Title III is codified at 16 U.S.C. 1431–1445. Web: www.fws.gov/laws/lawsdigest/marprot.html, (1972–1996).
5. Federal Register, *National Environmental Policy Act (NEPA)*, US Government Pub. L. 91–190, 42 U.S.C. 4321–4347, January 1, 1970, as amended by Pub. L. 94-52, July 3, 1975, Pub. L. 94-83, August 9, 1975, and Pub. L. 97–258, § 4(b), September 13, 1982, http://www.nepa.gov/nepa/regs/nepa/nepaeqia.htm (1969–1982).
6. Federal Register, *Clean Air Act (CAA)*, 42 U.S.C. s/s 7401 et seq. (1970), U.S. Government, Public Laws. Full text is at www.epa.gov/airprogm/oar/caa/index.html, March (2006).
7. Federal Register, *The Standards for the Use or Disposal of Sewage Sludge (Federal Biosolids Rule)*, Title 40 of the Code of Federal Regulations (CFR), Part 503, *58 FR 9248 to 9404*, U.S. Government Public Law, February 19 (1993).
8. U.S. Environmental Protection Agency, *A Plain English Guide to the US EPA Part 503 Biosolids Rule*, US EPA/832/R-93/003, Washington, DC, September (1994).
9. U.S. Environmental Protection Agency, *Biosolids Recycling: Beneficial Technology for a Better Environment*, US EPA/832-R-93-009, Washington, DC, June (1994).
10. U.S. Environmental Protection Agency, *A Guide to the Biosolids Risk Assessment Methodology for the US EPA 503 Rule*, US EPA/832-B-93-005, Washington, DC, September (1995).
11. U.S. Environmental Protection Agency, *Guidance for Writing Permits for the Use or Disposal of Sewage Sludge*, Washington, DC (1994).
12. U.S. Environmental Protection Agency, *Biosolids Generation, Use, and Disposal in the United States*, US EPA530-R-99-009, Washington, DC, September (1999).
13. U.S. Environmental Protection Agency, *Technical Support Document for Land Application of Sewage Sludge, Volumes 1 and 2*, Washington, DC, NTIS: PB93-110583 and P1393-110575 (1993).
14. California State Water Resources Control Board, *General Waste Discharge Requirements for Biosolids Land Application, Chapter 14 Alternative Analysis*, Draft Statewide Program EIR, February (2004).
15. L. K. Wang, N. K Shammas, and Y. T. Hung (eds.), *Biosolids Treatment Processes*, 820 pp. Humana Press, Totowa, NJ (2007).
16. M. J. McFarland, *Biosolids Engineering*, McGraw-Hill, New York (2000).
17. Metcalf and Eddy, Inc., *Wastewater Engineering Treatment and Reuse*, 4$^{th}$ ed. McGraw-Hill, New York (2003).
18. U.S. Fish and Wildlife Service, *Threatened and Endangered Species System (TESS)*, U.S. Department of Interior, http://ecos.fws.gov/tess_public/StartTESS.do (2007).
19. U.S. Environmental Protection Agency, *Preparing Sewage Sludge for Land Application or Surface Disposal—A Guide for Preparers of Sewage Sludge on the Monitoring, Record Keeping, and Reporting Requirements of the Federal Standards for Use or Disposal of Sewage Sludge, 40 CFR Part 503*, US EPA/813-B-93-002a, Washington, DC, NTIS: PB94-102415, September (1993).
20. U.S. Environmental Protection Agency, *Land Application of Sewage Sludge: A Guide for Land Appliers on the Record Keeping and Reporting Requirements of the Federal Standards for the Use or Disposal of Sewage Sludge, 40 CFR Part 503*, US EPA/831-B-002b, Washington, DC, June (1994).
21. U.S. Environmental Protection Agency, *Domestic Septage Regulatory Guidance: A Guide to the US EPA 503 Rule*, US EPA/832-B-92-005, Washington, DC, September (1993).

22. U.S. Environmental Protection Agency, *Process Design Manual: Land Application of Sewage Sludge and Domestic Septage*, Cincinnati, OH (1995).
23. Federal Register, *Comprehensive Environmental Response, Compensation, and Liability Act (CERCLA or Superfund)* 42 U.S.C. s/s 9601 et seq. (1980), United States Government, Public Laws. Full text is at www.access.gpo.gov/uscode/title42/chapter103_.html, January (2004).
24. U.S. Environmental Protection Agency, *Process Design Manual: Surface Disposal of Sewage Sludge and Domestic Septage*, Cincinnati, OH (1995).
25. U.S. Environmental Protection Agency, *Surface Disposal of Sewage Sludge: A Guide for Owners/Operators of Surface Disposal Facilities on the Monitoring, Record Keeping, and Reporting Requirements for the Use or Disposal of Sewage Sludge, 40 CFR Part 503*, US EPA/831-B-93-002c, Washington, DC, June (1994).
26. U.S. Army Corps of Engineers, *Federal Manual for Identifying and Delineating Jurisdictional Wetlands* (1989).
27. U.S. Environmental Protection Agency, *Methods Manual for Compliance with the BIF (Boilers and Industrial Furnaces) Regulation*, US EPA/530/SW-91-010, December (1990).
28. L. K. Wang, J. R. Taricska, Y. T. Hung, J. E. Elgridge, and K. H. Li, Wet and dry scrubbing. In: *Air Pollution Control Engineering*, pp. 197–306. L. K. Wang, N. C. Pereira, and Y. T. Hung, (eds.), Humana Press, Totowa, NJ (2004).
29. C. J. Yuan and T. T. Chen, Electrostatic precipitators. In: *Air Pollution Control Engineering*, pp. 153–196. L. K. Wang, N. C. Pereira, and Y. T. Hung (eds.), Humana Press, Totowa, NJ (2004).
30. L. K. Wang, C. Williford, and W. Y. Chen, Fabric filtration. In: *Air Pollution Control Engineering*, L. K. Wang, N. C. Pereira, and Y. T. Hung (eds.), Humana Press, Totowa, NJ (2004).
31. L. K. Wang, W. Lin, and Y. T. Hung, Thermal oxidation. In: *Air Pollution Control Engineering*, pp. 59–96. L. K. Wang, N. C. Pereira, and Y. T. Hung (eds.), Humana Press, Totowa, NJ (2004).
32. U.S. Environmental Protection Agency, *Solid Waste Disposal Act*, Washington, DC. Text at www.epa.gov/cpg/pdf/rcra-6002.pdf, December (2002).
33. Federal Register, *Resource Conservation and Recovery Act (RCRA)*, 42 US Code s/s 6901 et seq. (1976), United States Government, Public Laws. Full text at www.access.gpo.gov/uscode/title42/chapter82_.html, January (2004).
34. U.S. Environmental Protection Agency, *Guideline on Air Quality Models (Revised) (Appendix W of 40 CFR Part 51*, US EPA/450/2-78-027R, Washington, DC, August (1993).
35. P. N. Kane, A biosolids technician training course with a "hands on" team approach using professionals from the field, *Journal of Extension*, 40, 2, April (2002).
36. U.S. Environmental Protection Agency, *Technical Support Document for Incineration of Sewage Sludge*, Washington, DC, NTIS: PB93-110617 (1993).
37. L. K. Wang, N. C. Pereira, and Y. T. Hung (eds.), *Air Pollution Control Engineering*, 504 pp. Humana Press, Totowa, NJ (2004).
38. L. K. Wang, N. C. Pereira, and Y. T. Hung (eds.), Advanced *Air and Noise Pollution Control*, 526 pp. Humana Press, Totowa, NJ (2005).
39. U.S. Environmental Protection Agency, *POTW Sludge Sampling and Analysis Guidance Document*, Washington, DC, NTIS: PB93-227957, August (1989).
40. U.S. Environmental Protection Agency, *Sewage Sludge Sampling Video*, 202-260-7786, Washington, DC (1993).

41. U.S. Environmental Protection Agency, *Environmental Regulations and Technology: Control of Pathogens and Vector Attraction in Sewage Sludge*, US EPA/625/R-92/013, Washington, DC, December (1992).
42. U.S. Environmental Protection Agency, *Technical Support Document for Reduction of Pathogens and Vector Attraction in Sewage Sludge*, Washington, DC, NTIS: P1393-110609 (1993).
43. L. K. Wang, C. Williford, W. Y. Chen, and N. K. Shammas, Low temperature thermal processes. In: *Biosolids Treatment Processes*, pp. 299–330. L. K. Wang, N. K. Shammas, and Y. T. Hung (eds.), Humana Press, Totowa, NJ (2007).
44. C. Williford, W. Y. Chen, L. K. Wang, and N. K. Shammas, High temperature thermal processes. In: *Biosolids Treatment Processes*, pp. 613–644. L. K. Wang, N. K. Shammas, and Y. T. Hung (eds.), Humana Press, Totowa, NJ (2007).
45. C. Williford, W. Y. Chen, N. K. Shammas, and L. K. Wang, Lime stabilization. In: *Biosolids Treatment Processes*, pp. 207–242. L. K. Wang, N. K. Shammas, and Y. T. Hung (eds.), Humana Press, Totowa, NJ (2007).
46. J. R. Taricska, D. A. Long, J. P. Chen, Y. T. Hung, and S. W. Zou, Anaerobic digestion. In: *Biosolids Treatment Processes*, pp. 135–176. L. K. Wang, N. K. Shammas, and Y. T. Hung (eds.), Humana Press, Totowa, NJ (2007).
47. N. K. Shammas and L. K. Wang, Aerobic digestion. In: *Biosolids Treatment Processes*, pp. 177–206. L. K. Wang, N. K. Shammas, and Y. T. Hung (eds.), Humana Press, Totowa, NJ (2007).
48. L. K. Wang, Y. Li, N. K. Shammas, and G. P. Sakellaropoulos, Drying beds. In: *Biosolids Treatment Processes*, pp. 403–430. L. K. Wang, N. K. Shammas, and Y. T. Hung (eds.), Humana Press, Totowa, NJ (2007).
49. L. K. Wang, Y. T. Hung, and J. P. Chen, Animal wastes treatment using anaerobic lagoons. In: *Biosolids Treatment Processes*, pp. 431–450. L. K. Wang, N. K. Shammas, and Y. T. Hung (eds.), Humana Press, Totowa, NJ (2007).
50. N. K. Shammas and L. K. Wang, Biosolids composting. In: *Biosolids Treatment Processes*, pp. 645–688. L. K. Wang, N. K. Shammas, and Y. T. Hung (eds.), Humana Press, Totowa, NJ (2007).
51. Arizona State Department of Environmental Quality, *Biosolids/Sewage Sludge Management Program*, Arizona Pollutant Discharge Elimination System Program (Azpdes), September (2003).
52. J. F. Artiola, *Biosolids Land Use in Arizona*, AZ1426, The University of Arizona, College of Agriculture and Life Sciences, Tucson, Arizona, November (2006).
53. Florida State Department of Environmental Protection, *Domestic Wastewater Residuals*, Tallahassee, FL (1998).
54. Georgia Environmental Protection Division, *Guidelines for Land Application of Sewage Sludge (Biosolids) at Agronomic Rates*, June (2006).
55. Orange County Code, *Domestic Wastewater Residual Management Ordinance of Orange County*, FL, Ordinance No. 99-15, 6-29-99, Orange County, FL (1999).
56. Pennsylvania State Department of Environmental Protection, *Operating a Biosolids Land Application Program*, http://www.dep.state.pa.us/dep/biosolids/landapplication.htm, February (2007).
57. GVRD, *The Biosolids Report*, Report No. 1, Greater Vancouver Regional District (GVRD), Burnaby, British Columbia, Canada, May (2000).
58. (GVRD Regulatory Framework for Biosolids Management in Canada, Greater Vancouver Regional District, Policy and Planning Department, Burnaby, BC, Canada, February (2006).

59. Australia Environmental Protection Agency, *Guidelines for Environmental Management: Biosolids Land Application*, Publication 943, Victoria, Australia, April (2004).
60. Western Australia Department of Environmental Protection, *Guidelines for Direct Land Application of Biosolids and Biosolids Products*, Perth, WA, February (2002).
61. U.S. Environmental Protection Agency, *Wastewater Sludge Utilization and Disposal Costs*, US EPA-430/9-75-015, Washington, DC, September (1975).
62. U.S. Army Corps of Engineers, *Civil Works Construction Cost Index System Manual*, 110-2-1304, 2000 Tables, Washington, DC, 44 pp., http://www.nww.usace.army.mil/cost (2007).

## APPENDIX

**U.S. Army Corps of Engineers Civil Works Construction Yearly Average Cost Index for Utilities (62)**

| Year | Index | Year | Index |
|---|---|---|---|
| 1967 | 100 | 1988 | 369.45 |
| 1968 | 104.83 | 1989 | 383.14 |
| 1969 | 112.17 | 1990 | 386.75 |
| 1970 | 119.75 | 1991 | 392.35 |
| 1971 | 131.73 | 1992 | 399.07 |
| 1972 | 141.94 | 1993 | 410.63 |
| 1973 | 149.36 | 1994 | 424.91 |
| 1974 | 170.45 | 1995 | 439.72 |
| 1975 | 190.49 | 1996 | 445.58 |
| 1976 | 202.61 | 1997 | 454.99 |
| 1977 | 215.84 | 1998 | 459.40 |
| 1978 | 235.78 | 1999 | 460.16 |
| 1979 | 257.20 | 2000 | 468.05 |
| 1980 | 277.60 | 2001 | 472.18 |
| 1981 | 302.25 | 2002 | 484.41 |
| 1982 | 320.13 | 2003 | 495.72 |
| 1983 | 330.82 | 2004 | 506.13 |
| 1984 | 341.06 | 2005 | 516.75 |
| 1985 | 346.12 | 2006 | 528.12 |
| 1986 | 347.33 | 2007 | 539.74 |
| 1987 | 353.35 | | |

# 7
# Engineering and Management of Agricultural Land Application

## Lawrence K. Wang, Nazih K. Shammas, and Gregory Evanylo

**CONTENTS**

INTRODUCTION
AGRICULTURAL LAND APPLICATION
PLANNING AND MANAGEMENT OF AGRICULTURAL LAND
  APPLICATION
DESIGN OF LAND APPLICATION PROCESS
OPERATION AND MAINTENANCE
NORMAL OPERATING PROCEDURES
EMERGENCY OPERATING PROCEDURES
ENVIRONMENTAL IMPACTS
LAND APPLICATION COSTS
PRACTICAL APPLICATIONS AND DESIGN EXAMPLES
GLOSSARY OF LAND APPLICATION TERMS
NOMENCLATURE
REFERENCES
APPENDICES

**Abstract** The controlled application of sewage sludge biosolids to cropland by subsurface injection or surface spreading is introduced in this chapter. Specifically, the land application process operation, design criteria, performance, application rates, staffing requirements, process monitoring, sensory observation, normal operating procedures, process control considerations, emergency operating procedures, safety considerations, 19 application and design examples, costs, and trouble shooting guide are presented and discussed in detail.

**Key Words** Agricultural land application • biosolids • septage • plant available nitrogen • agronomic rate • design examples • heavy metals • plant available nitrogen (PAN) • nitrogen • phosphorus • potassium • lime • fertilizer.

From: *Handbook of Environmental Engineering, Volume 7: Biosolids Engineering and Management*
Edited by: L. K. Wang, N. K. Shammas and Y. T. Hung © The Humana Press, Totowa, NJ

## 1. INTRODUCTION

Land application of biosolids is defined as the spreading, spraying, injection, or incorporation of liquid, semiliquid, or solid organic by-product of the wastewater treatment process, onto or below the surface of the land to take advantage of the nutrient supplying and soil property enhancing qualities of the residuals. These organic by-products are land applied to improve the structure of the soil and to supply nutrients to crops and other vegetation grown in the soil. These by-products are commonly applied to agricultural land (including pasture and range land), forests, reclamation sites, and, if properly treated, public contact sites (e.g., parks, turf farms, highway median strips, golf courses), lawns, and home gardens.

Biosolids contain significant concentrations of essential plant nutrients. The availability of these nutrients to vegetation at an application site depends on these materials' composition, processing, handling, and method of application, as well as a number of soil and climatic factors. Under most situations, the amount of these materials that can be applied to the soil is based on satisfying a nutrient requirement of the vegetation. This quantity is called the "agronomic" rate of application.

This chapter illustrates how nutrient management is achieved using the approach of the U.S. Environmental Protection Agency (US EPA), and how the agronomic rate of application on agricultural land is determined. Although the emphasis of this chapter is on application of biosolids on agricultural land, the same method of analysis can also be applied to livestock manure and other organic by-products.

### 1.1. Biosolids

Sewage sludge means any solid, semisolid, or liquid residue removed during the treatment of municipal wastewater or domestic sewage. Sewage sludge includes, but is not limited to, solids removed during primary, secondary, or advanced wastewater treatment, scum, septage, portable toilet pumpings, and sewage sludge products. Sewage sludge does not include grit or screenings, or ash generated during the incineration of sewage sludge. Septage means the liquid and solid material pumped from a septic tank, cesspool, or similar domestic sewage treatment system, or holding tank when the system is cleaned or maintained. Biosolids are solid, semisolid, or liquid materials, resulting from biological treatment of domestic sewage that have been sufficiently processed to permit these materials to be safely land-applied. The term *biosolids* was introduced by the wastewater treatment industry in the early 1990s and has been recently adopted by the US EPA to distinguish high-quality, treated sewage sludge from raw sewage sludge and from sewage sludge containing large amounts of pollutants. Although the term *biosolids* does not evoke the same negative connotation as does *sewage sludge*, the use of the term is appropriate when it makes the distinction described above (1–10). Manure is the waste discharged from livestock.

### 1.2. Biosolids Production and Pretreatment Before Land Application

Biosolids are produced primarily through biological treatment of domestic wastewater. Biosolids comprise the solids that are removed from the wastewater and further processed before the treated water is released into streams or rivers.

Thickening, digestion, stabilization, conditioning, dewatering, composting, and heat-drying processes are often employed additionally to improve the biosolids handling characteristics, increase the economic viability of land application, and reduce the potential for public health, environmental, and nuisance problems associated with land application practices. These processes control disease-causing organisms and reduce characteristics that might attract rodents, flies, mosquitoes, or other organisms capable of transporting infectious disease. Table 7.1 shows how various biosolids pretreatment processes affect the suitability of biosolids to be applied on land especially for agricultural use (2).

## 1.3. Biosolids Characteristics

The suitability of biosolids for land application can be determined by biological, chemical, and physical analyses. Biosolids' composition depends on wastewater constituents and treatment processes. The resulting properties determine application method and rate, and the degree of regulatory control required. Several of the more important properties of biosolids are discussed below.

Total solids (TS) include suspended and dissolved solids and are usually expressed as the concentration present in biosolids. The TS depends on the type of wastewater process and biosolids' treatment prior to land application. Typical solids contents of various biosolids are liquid (2% to 12%), dewatered (12% to 30%), and dried or composted (50%). Volatile solids (VS) provide an estimate of the readily decomposable organic matter in biosolids and are usually expressed as a percentage of total solids. The VS is an important determinant of potential odor problems at land application sites. A number of biosolids treatment processes, including anaerobic digestion, aerobic digestion, alkaline stabilization, and composting, can be used to reduce VS content and thus the potential for odor.

The pH is a measure of the degree of acidity or alkalinity of a substance. The pH of biosolids is often raised with alkaline materials to reduce pathogen content and attraction of disease-spreading organisms (vectors). High pH (greater than 11) kills virtually all pathogens and reduces the solubility, biological availability and mobility of most metals. Lime also increases the gaseous loss (volatilization) of the ammonia form of nitrogen (ammonia-N), thus reducing the N-fertilizer value of biosolids (2).

Pathogens are disease-causing microorganisms that include bacteria, viruses, protozoa, and parasitic worms. Pathogens can present a public health hazard if they are transferred to food crops grown on land to which biosolids are applied; contained in runoff to surface waters from land application sites; or transported away from the site by vectors such as insects, rodents, and birds. For this reason, federal and state regulations specify pathogen and vector attraction reduction requirements that must be met by biosolids applied to land. A partial list of pathogens that can be found in untreated sewage sludge, and the diseases or symptoms that they can cause is presented by an US EPA report (11).

Nutrients are elements required for plant growth that provide biosolids with most of their economic value. These include nitrogen (N), phosphorus (P), potassium (K), calcium (Ca), magnesium (Mg), sodium (Na), sulfur (S), boron (B), copper (Cu), iron (Fe), manganese (Mn), molybdenum (Mo), and zinc (Zn). Concentrations in biosolids can vary significantly (Table 7.2); thus, the actual material being considered for land application should be analyzed.

**Table 7.1**
**Effects of biosolids treatment processes on land application practices**

| Treatment process and definition | Effect on biosolids | Effect on land application practices |
|---|---|---|
| Thickening: Low force separation of water and solids by gravity, flotation, or centrifugation | Increases solids content by removing water | Lowers transportation costs |
| Digestion (anaerobic and aerobic): Biological stabilization through conversion of organic matter to carbon dioxide, water, and methane | Reduces the biodegradable content (stabilization by conversion to soluble material and gas; reduces pathogen levels and odor | Reduces the quantity of biosolids |
| Alkaline stabilization: Stabilization through the addition of alkaline materials (e.g., lime, kiln dust) | Raises pH; temporarily decreases biological activity; reduces pathogen levels and controls putrescibility and odor | High pH immobilizes metals as long as pH levels are maintained |
| Conditioning: Processes that cause biosolids to coagulate to aid in the separation of water | Improves sludge dewatering characteristics; may increase dry solids mass and improve stabilization | The ease of spreading may be reduced by treating biosolids with polymers |
| Dewatering: High force separation of water and solids; methods include vacuum filters, centrifuges, filter and belt presses, etc. | Increases solids concentration to 15% to 45%; lowers nitrogen and potassium concentrations; improves ease of handling | Reduces land requirements and lowers transportation costs |
| Composting: Aerobic, thermophilic, biological stabilization in a windrow, aerated static pile or vessel | Lowers pathogenic activity and converts sludge to humus-like material | Excellent soil conditioning properties; contains less plant available nitrogen than other biosolids |
| Heat drying: Use of heat to kill pathogens and eliminate most of the water content | Disinfects sludge, destroys most pathogens, and lowers odors and biological activity | Greatly reduces sludge volume and mass |

*Source*: US EPA (10).

Table 7.2
**Typical nutrient concentrations in biosolids from all biosolids treatment processes source**[a,b]

| Nutrient | No. of samples | Range | Mean |
|---|---|---|---|
| Total N (%) | 191 | <0.1–17.6 | 3.9 |
| Ammonia N (%) | 103 | 0.0005–6.7 | 0.65 |
| Nitrate N (%) | 45 | 0.0002–0.49 | 0.05 |
| Total P (%) | 189 | <0.1–14.3 | 2.5 |
| K (%) | 192 | 0.02–2.64 | 0.40 |
| Ca (%) | 193 | 0.1–25.0 | 4.9 |

[a]Nutrient concentrations are on a dried solids basis.
[b]Biosolids treatment processes include anaerobically and aerobically digested, lagooned, primary, tertiary, and unspecified biosolids.
*Source*: Sommers et al. (58).

Trace elements are found in low concentrations in biosolids. The trace elements of interest in biosolids are those commonly referred to as heavy metals. Some of these trace elements (e.g., copper, molybdenum, and zinc) are nutrients needed for plant growth in low concentrations, but all of these elements can be toxic to humans, animals, or plants at high concentrations. Possible hazards associated with a buildup of trace elements in the soil include their potential to cause phytotoxicity (i.e., injury to plants) or to increase the concentration of potentially hazardous substances in the food chain. Federal and state regulations have established standards for the following nine trace elements: arsenic (As), cadmium (Cd), copper (Cu), lead (Pb), mercury (Hg), molybdenum (Mo), nickel (Ni), selenium (Se), and zinc (Zn).

Organic chemicals are complex compounds that include man-made chemicals from industrial wastes, household products, and pesticides. Many of these compounds are toxic or carcinogenic to organisms exposed to critical concentrations over certain periods of time, but most are found at such low concentrations in biosolids that do not pose significant human health or environmental threats. Although no organic pollutants are included in the current U.S. federal biosolids regulations, further assessment of specific organic compounds is expected. Biosolids can be considered as waste to be disposed of or as a beneficial soil amendment. Ocean dumping, a disposal practice common in the 1980s, was banned due to concerns about excess nutrient loading in these waters (7). Currently, about half of the biosolids generated in Virginia are applied to land and the remainder is disposed of through landfilling or incineration (2–5).

### 1.4. Agricultural Land Application for Beneficial Use

As an alternative to disposal by landfilling or incineration, land application seeks to beneficially recycle the soil property-enhancing constituents in biosolids, which are ultimately derived from crops grown on agricultural land. Biosolids are about 50% mineral and 50% organic matter. The mineral matter includes plant nutrients, and organic matter is a source of slow release nutrients and soil conditioners. Land application returns those materials to the soil where they can contribute to further crop production (2–5).

Farmers can benefit from biosolids application by reducing fertilizer costs. The main fertilizer benefits are through the supply of nitrogen, phosphorus, and lime (where lime-stabilized biosolids are applied). Biosolids also ensure against unforeseen nutrient shortages by supplying essential plant nutrients that are rarely purchased by farmers because crop responses to their application are unpredictable. These include elements such as sulfur, manganese, zinc, copper, iron, molybdenum, and boron. Land application replenishes valuable organic matter, which occurs in less than optimum amounts in most soils. The addition of organic matter can improve soil tilth, the physical condition of soil as related to its ease of tillage, fitness as a seedbed, and its impedance to seedling emergence and root penetration. Other benefits imparted by the addition of organic matter to soil include the following:

1. Increases water infiltration into the soil and soil moisture-holding capacity
2. Reduces soil compaction
3. Increases the ability of the soil to retain and provide nutrients
4. Reduces soil acidification
5. Provides an energy source (carbon) for beneficial microorganisms

The addition of organic matter in biosolids to a fine-textured clay soil can help make the soil more friable and can increase the amount of pore space available for root growth and entry of water and air into the soil. In coarse-textured sandy soils, organic residues in biosolids can increase the water-holding capacity of the soil and provide chemical sites for nutrient exchange and adsorption.

Land application is usually less expensive than alternative methods of disposal. Consequently, wastewater treatment facilities and the public they serve benefit through cost savings. The recycling of nutrients and organic matter can be attractive to citizens concerned with environmental protection and resource conservation.

Land application of biosolids involves some risks, which are addressed through federal and state regulatory programs. Pollutants and pathogens are added to soil with organic matter and nutrients. Human and animal health, soil quality, plant growth, and water quality could be adversely affected if land application is not conducted in an agronomically and environmentally sound manner. In addition, nitrogen and phosphorus in biosolids, as in any fertilizer source, can contaminate ground and surface water if the material is overapplied or improperly applied. There are risks and benefits to each method of biosolids disposal and reuse.

### 1.5. U.S. Federal and State Regulations

The US EPA has developed the regulations and the standards for the use or disposal of sewage sludge (Title 40 of the Code of Federal Regulations [CFR], Part 503). The Part 503 rule establishes minimum requirements when biosolids are applied to land to condition the soil or fertilize crops or other vegetation grown in the soil. The Clean Water Act required that this regulation protect public health and the environment from any reasonably anticipated adverse effects of pollutants and pathogens in biosolids (6–25).

Federal regulations require that state regulations be at least as stringent as the Part 503 rule (7–9). Many state regulations prohibit land application of low-quality sewage sludge and encourage the application of biosolids that are of sufficient quality that they will not adversely affect human health or the environment (2–5, 26). Determination of biosolids quality is based on trace element (pollutant) concentrations and pathogen and vector attraction reduction. Appendix A summarizes the US EPA general requirements for land application of sewage sludge (including biosolids).

*1.5.1. Heavy Metal Limits*

The Part 503 rule prohibits land application of sewage sludge that exceeds the *ceiling concentration limits* (Table 7.3) for nine trace elements: arsenic, cadmium, copper, lead, mercury, molybdenum, nickel, selenium, and zinc. Sewage sludge exceeding the ceiling concentration limit for even one of the regulated pollutants is not classified as biosolids and hence cannot be land applied.

*Pollutant concentration limits* are the most stringent pollutant limits included in Part 503 for land application. Biosolids meeting pollutant concentration limits are subject to fewer requirements than biosolids meeting ceiling concentration limits. Results of the US EPA's 1990 National Sewage Sludge Survey (NSSS) (24) demonstrated that the mean concentrations of the nine regulated pollutants are considerably lower than the most stringent Part 503 pollutant limits (Table 7.3).

*The cumulative pollutant loading rate* (Table 7.3) is the total amount of a pollutant that can be applied to a site in its lifetime by all bulk biosolids applications meeting ceiling concentration limits. No additional biosolids meeting ceiling concentration limits can be applied to a site after the maximum cumulative pollutant loading rate is reached at

Table 7.3
Land applied biosolids pollutant limits and mean concentrations from the National Sewage Sludge Survey (NSSS)

| Pollutant | CCL[a,b] mg/kg | PCL[a,c] mg/kg | CPLR[a,d] lb/acre | NSSS[a] mg/kg |
|---|---|---|---|---|
| Arsenic (As) | 75 | 41 | 36 | 10 |
| Cadmium (Cd) | 85 | 39 | 35 | 7 |
| Copper (Cu) | 4300 | 1500 | 1340 | 741 |
| Lead (Pb) | 840 | 300 | 270 | 134 |
| Mercury (Hg) | 57 | 17 | 16 | 5 |
| Molybdenum (Mo) | 75 | e | e | 9 |
| Nickel (Ni) | 420 | 420 | 375 | 43 |
| Selenium (Se) | 100 | 100 | 89 | 5 |
| Zinc (Zn) | 7500 | 2800 | 2500 | 1202 |

[a] Dry weight basis.

[b] CCL (ceiling concentration limits) = maximum concentration permitted for land application.

[c] PCL (pollutant concentration limits) = maximum concentration for biosolids whose trace element pollutant additions do not require tracking (i.e., calculation of CPLR).

[d] CPLR (cumulative pollutant loading rate) = total amount of pollutant that can be applied to a site in its lifetime by all bulk biosolids applications meeting CCL.

[e] The February 25, 1994, Part 503 rule amendment deleted Mo PCL for sewage sludge applied to agricultural land but retained Mo CCL.

*Source*: US EPA (11, 24).

that site for any one of the nine regulated trace elements. Only biosolids that meet the more stringent pollutant concentration limits may be applied to a site once a cumulative pollutant loading rate is reached at that site.

In 1987 the US EPA established the most recent pretreatment guidelines on the development and implementation of local discharge limitations that require industries to limit the concentrations of certain pollutants, including trace elements and organic chemicals, in wastewater discharged to a treatment facility. An improvement in the quality of biosolids over the years has largely been due to pretreatment and pollution prevention programs (2, 3). Appendix B summarizes the US EPA's pollutant limits for the land application of sewage sludge (including biosolids).

### 1.5.2. Organic Chemicals

Part 503 does not regulate organic chemicals in biosolids because the chemicals of potential concern have been banned or restricted for use in the U.S., are no longer manufactured in the U.S., are present at low concentrations based on data from the US EPA's 1990 NSSS (24), or because the limit for an organic pollutant identified in the Part 503 risk assessment is not expected to be exceeded in biosolids that are land applied (25). Restrictions will be imposed for agricultural use if testing of certain toxic organic compounds verifies that biosolids contain levels that could cause harm to human health or the environment.

### 1.5.3. Pathogen Reduction

The U.S. federal and state regulations require the reduction of potential disease-causing microorganisms, called pathogens (e.g., viruses, bacteria, and parasitic worms), and vector (e.g., rodents, birds, insects that can transport pathogens away from the land application site) attraction properties. Biosolids intended for land application are normally treated by chemical or biological processes that greatly reduce the number of pathogens and odor potential in sewage sludge. Two levels of pathogen reduction, Class A and Class B, are specified in the regulations (3, 7).

The goal of Class A requirements is to reduce the pathogens (including *Salmonella* sp., bacteria, enteric viruses, and viable helminth ova) to below detectable levels. Class A biosolids can be land applied without any pathogen-related site restrictions. *Processes to further reduce pathogens* (PFRP) treatment, such as those involving high temperature, high pH with alkaline addition, drying, and composting, or their equivalent, are most commonly used to demonstrate that biosolids meet Class A requirements.

The goal of Class B requirements is to ensure that pathogens have been reduced to levels that are unlikely to cause a threat to public health and the environment under specified use conditions. *Processes to significantly reduce pathogens* (PSRP), such as digestion, drying, heating, and high pH, or their equivalent, are most commonly used to demonstrate that biosolids meet Class B requirements. Because Class B biosolids contain some pathogens, certain site restrictions are required. These are imposed to minimize the potential for human and animal contact with biosolids until environmental factors (temperature, moisture, light, microbial competition) reduce the pathogens to below detectable levels. As an example of waiting periods after land application, Table 7.4 summarizes the Class B biosolids application land use restrictions imposed by the Virginia Department of Health (26). The site restriction requirements in combination

**Table 7.4**
**Class B biosolids application land use restrictions**

| | |
|---|---|
| Root crops, where biosolids remain on land surface: | |
| ≥4 mos. prior to soil incorporation | Harvest 20 months after application |
| <4 mos. prior to soil incorporation | Harvest 38 months after application |
| Food crops that touch biosolids or soil | Harvest 14 months after biosolids application |
| Other food, feed, or fiber crops | Harvest 30 days after application |
| Turf | Harvest 1 year after application when the turf is placed on land with high potential for public exposure |
| Grazing animals | |
| Lactating (milking) animals | No grazing prior to 60 days after application |
| Nonlactating animals | No grazing prior to 30 days after application |
| Public access to land | |
| High access potential | Restricted to 1 year after application |
| Low access potential | Restricted to 30 days after application |

*Source*: Virginia Department of Health (26).

with Class B treatment are expected to provide a level of protection equivalent to Class A treatment. All biosolids that are land applied must, at a minimum, meet Class B pathogen reduction standards (3, 7).

*1.5.4. Vector Attraction Reduction*

The objective of vector attraction reduction is to prevent disease vectors such as rodents, birds, and insects from transporting pathogens away from the land application site. There are 10 options available to demonstrate that land-applied biosolids meet vector attraction reduction requirements. These options fall into either of the following two general approaches: (a) reducing the attractiveness of the biosolids to vectors with specified organic matter decomposition processes (e.g., digestion, alkaline addition); and (b) preventing vectors from coming into contact with the biosolids (e.g., biosolids injection or incorporation below the soil surface within specified time periods).

*1.5.5. Categories of Biosolids Quality*

The quality of biosolids (i.e., pollutant concentrations, pathogen levels, and vector attraction reduction control) determines which land application requirements must be met. There are three categories of biosolids quality that are discussed below and described in Table 7.5.

Biosolids that meet the Part 503 pollutant concentration limits (PCLs), Class A pathogen reduction, and a vector attraction reduction option that reduces organic matter are classified as exceptional quality (EQ) biosolids. In general, EQ biosolids can be applied as freely as any other fertilizer or soil amendment to any type of land.

*Pollutant concentration* (PC) biosolids meet the same low PCLs as EQ biosolids, but PC biosolids usually meet Class B rather than Class A pathogen reduction requirements. Biosolids meeting Class A pathogen reduction requirements plus one of the practices designed to prevent vectors from coming into contact with biosolids also are PC biosolids.

Table 7.5
Summary of requirements for different quality bulk biosolids

| Biosolids type | Ceiling concentration limit | Other pollutant limits | Pathogen class | Vector attraction reduction[a] | Siting restrictions | Track added pollutant | Required management practices |
|---|---|---|---|---|---|---|---|
| Exceptional quality (EQ) | Yes | Pollutant | A | Treatment options | No | No | No[b] |
| Pollutant concentration (PC) | Yes | Pollutant conc limits | A or B | Any option | No[a]/Yes[b] | No | Yes[c] |
| Cumulative pollutant loading rate (CPLR) | Yes | Cumulative pollutant loading rate | A or B | Any option | Yes | Yes | Yes |

[a] The eight vector attraction reduction treatment options that reduce the attractiveness of the biosolids to vectors by further decomposition of the volatile solids. Two additional management options (incorporation and injection) prevent vectors from coming into contact with the biosolids.
[b] EQ biosolids can be applied as freely as any other fertilizer or soil amendment to any type of land. Virginia requires additional record keeping for distribution of bulk quantities and specific labeling information for bagged products marketed under a registration filed with the Virginia Department of Agriculture and Consumer Services. EQ biosolids are exempt from Part 503 general requirements and management practices.
[c] Management practices are required when biosolids do not meet EQ criteria either because they meet vector attraction reduction through soil injection or incorporation.
Source: Evanylo (3).

*Cumulative pollutant loading rate* (CPLR) biosolids, unlike EQ or PC biosolids, require tracking of the cumulative metal loadings to ensure adequate protection of public health and the environment. Additional land application terminologies can be found in the glossary at the end of this chapter and in the literature.

### 1.5.6. Nutrients

The U.S. federal regulations specify that biosolids may only be applied to agricultural land at or less than the rate required to supply the nitrogen (N) needs of the crops to be grown. This agronomic rate is "designed (a) to provide the amount of N needed by the food crop, feed crop, fiber crop, or vegetation grown on the land; and (b) to minimize the amount of N in the biosolids that passes below the root zone of the crop or vegetation grown on the land to the ground water" (40 CFR 503.11 [B]). Agronomic rate may also be based on crop phosphorus (P) needs if it is determined that excessive soil P poses a threat to water quality.

### 1.5.7. Site Suitability and Location

Site physical characteristics that influence the land application management practices include topography; soil permeability, infiltration, and drainage patterns; depth to groundwater; and proximity to surface water. Federal, state, and local regulations, ordinances, or guidelines place limits on land application based on these physical characteristics. Potentially unsuitable areas for biosolids application include (a) areas bordered by ponds, lakes, rivers, and streams without appropriate buffer areas; (b) wetlands and marshes; (c) steep areas with sharp relief; (d) undesirable geology (fractured bedrock) if not covered by a sufficiently thick layer of soil; (e) undesirable soil conditions (rocky, shallow); (f) areas of historical or archeological significance; and (g) other environmentally sensitive areas, such as floodplains.

Many states have enacted regulations establishing site-specific management practice standards more demanding than the Part 503 rule. Such regulations define standards of practice to ensure that biosolids use does not compromise the public health or the environment. An example of regulations promulgated by a state that are more stringent than the federal regulations is the Virginia Biosolids Use Regulations (26), which specify minimum distances to land application areas from occupied dwellings, water supply wells or springs, property lines, perennial streams and other surface waters, intermittent streams/drainage ditches, improved roadways, rock outcrops and sinkholes, and agricultural drainage ditches (Table 7.6).

## 2. AGRICULTURAL LAND APPLICATION

### 2.1. Land Application Process

The land application operation and maintenance information in this chapter applies to controlled application of biosolids to cropland. The most appropriate application method for agricultural land depends on the physical characteristics of the biosolids and the soil, as well as the types of crops grown. Biosolids are generally land applied using one of the following methods: (a) sprayed or spread on the soil surface and left on the surface for pastures, range, and forest land; (b) incorporated into the soil after being surface applied; or (c) injected directly below the surface for producing row crops or other

Table 7.6
Minimum distances (ft) to land application area

| Adjacent feature | Surface application[a] | Incorporation | Winter[b] |
|---|---|---|---|
| Occupied dwellings | 200 | 200 | 200 |
| Water supply wells or springs | 100 | 100 | 100 |
| Property lines | 100 | 50 | 100 |
| Perennial streams and other surface water, except intermittent streams | 50 | 35 | 100 |
| Intermittent streams/drainage ditches | 25 | 25 | 50 |
| All improved roadways | 10 | 5 | 0 |
| Rock outcrops and sinkholes | 25 | 25 | 25 |
| Agricultural drainage ditches with slopes equal to or less than 2% | 10 | 5 | 10 |

[a] Not plowed or disked to incorporate within 48 hours.
[b] Application occurs on average site slope greater than 7% during period between November 16 of one year and March 15 of the following year.
*Source*: Virginia Department of Health (26).

vegetation. Biosolids application methods such as incorporation and injection can be used to meet Part 503 vector attraction reduction requirements. Alternatively, both liquid and dewatered biosolids may be applied to land without subsequent soil incorporation. Although there are many field variations to the land application operation, certain process requirements, limitations, design criteria, operating procedures, and monitoring methods must be established to optimize the process performance (1–15, 26–38).

The most common form of biosolids applied to agricultural land is that which has undergone conditioning and dewatering. Dewatering typically increases the solids content of liquid biosolids from less than 5% to between 25% and 30%, thus precluding the need to transport considerable quantities of water in liquid biosolids to the application site. Dewatered biosolids can be applied to cropland by equipment similar to that used for applying limestone, animal manures, or commercial fertilizer. Typically, dewatered biosolids are surface-applied and incorporated by chisel plowing, disking, or another form of tillage. Incorporation is not used when applying dewatered biosolids to forage or to the increasing amount of no-till land.

Liquid biosolids can be applied by surface spreading or subsurface injection. Surface methods include spreading by tractor drawn tank wagons, special applicator vehicles equipped with flotation tires, or irrigation systems. Surface application with incorporation in Virginia is limited to soils with less than a 7% slope. Biosolids are commonly incorporated by plowing or disking after the liquid has been applied to the soil surface and allowed to partially dry, unless minimum or no-till systems are being used.

Spray irrigation systems generally should not be used to apply biosolids to forages or row crops during the growing season, although a light application to the stubble of a forage crop following a harvest is acceptable. The adherence of biosolids to plant vegetation can have a detrimental effect on crop yields by reducing photosynthesis. In addition, spray irrigation increases the potential for odor problems and reduces the aesthetics at the application site.

# Engineering and Management of Agricultural Land Application

**Fig. 7.1.** Biosolids injection equipment. *Source*: US EPA (6, 18).

Liquid biosolids can also be injected below the soil surface using tractor-drawn tank wagons with injection shanks and tank trucks fitted with flotation tires and injection shanks. Both types of equipment minimize odor problems and reduce ammonia volatilization by immediate mixing of soil and biosolids. Injection can be used either before planting or after harvesting crops, but it is likely to be unacceptable for forages and sod production. Some injection shanks can damage the sod or forage stand and leave deep injection furrows in the field. Subsurface injection minimizes runoff from all soils and can be used on slopes of up to 15%. Injection should be made perpendicular to slopes to avoid having liquid biosolids run downhill along injection slits and pond at the bottom of the slopes. As with surface application, drier soil is able to absorb more liquid, thereby minimizing downslope movement. Despite the advantages with regard to odor and maintaining soil vegetative cover with injecting liquid biosolids, such a practice is much less common than surface application of dewatered biosolids.

Typical injector trucks are shown in Figure 7.1. Liquid application of biosolids form an application truck is shown in Figure 7.2. Figure 7.3 shows the application of liquid biosolids to a forest land.

## 2.2. Agricultural Land Application Concepts and Terminologies

To understand when and where to apply biosolids on agricultural land, certain common terms must be understood (12).

The *farm field* is the basic management unit used for all farm nutrient management, and is defined as "the fundamental unit used for cropping agricultural products." An *area*

**Fig. 7.2.** Liquid application of biosolids. *Source*: US EPA (6, 17).

**Fig. 7.3.** Application of liquid biosolids to forest land. *Source*: US EPA (6, 17).

*of cropland* that has been *subdivided* into several strips is not a single field. Rather, each strip represents an *individual field unit*. Individual fields that are managed in the same manner, with the similar yield goals, are called a *crop group*.

The cycle of crop *planting and harvesting periods*, not the calendar year, dictates the timing of biosolids land application activities. Winter wheat and perennial forage grasses are examples of crops that may be established and harvested in different calendar years. In many regions, biosolids are commonly applied in the fall or early winter, in anticipation of a crop that will be planted the following spring. Crop nutrient management practices are linked to crop nutrient uptake (crop growth) and nutrient removal at harvest time. Agricultural land application programs must be coordinated with the cropping cycle.

The basic time management unit is often called the *crop year* or *planting season*. The *crop year* is defined as the year in which a crop receiving the biosolids treatment is harvested. For example, fall applications of biosolids in 2006 intended to provide nutrients for a crop to be harvested in 2007 are earmarked for crop year 2007. Likewise, biosolids applied immediately prior to planting winter wheat in October 2006 should be identified as fertilizer intended for crop year 2007 because the wheat will be harvested in the summer of 2007. Similarly, if instead of wheat, the field is planted as corn in May 2007 that will be harvested in November 2007, applications of biosolids made in the fall of 2006 are credited to crop year 2007. Typically, biosolids applied January through June would be intended for a crop harvested in the same calendar year. Biosolids applied in the last 6 months of the calendar year usually fertilize crops harvested the next calendar year. This generalization does not always hold true. For example, biosolids may be applied in July on a grass forage crop or in preparation for a buckwheat crop that will be harvested before winter. Other common exceptions are likely in hot, humid sections of the U.S. (12). The first step in computing the agronomic rate is to establish the amount of nitrogen needed for a desired crop yield. The crop yield consists of crop removal rates and nutrient recommendations for proposed crops. This is also a good time to evaluate other primary nutrient crop removal amounts (phosphate and potash), though these elements are not normally regulated at the state or federal level.

Realistic yield goals can be obtained from agronomy guidelines (or equivalent) published by state land grant universities or cooperative extension. Yield goals are commonly based on soil productivity categorization of soils that take into account soil depth, drainage characteristics, or other important soil features. Tables 7.7 and 7.8 provide typical agronomic information commonly available from the state agronomy guides (2–5, 12).

Only a portion of the total nitrogen present in biosolids is available for plant uptake. This *plant available nitrogen* (PAN) is the actual amount of N in the biosolids that is available to crops during a specified period. The biosolids application rate is a field measurement determined for the particular application equipment.

*Total Kjeldahl nitrogen* (TKN) is the summation of ammonium nitrogen ($NH_4^+$-N), and organic nitrogen. *Total nitrogen* (TN) is the summation of ammonium nitrogen ($NH_4^+$-N), nitrate nitrogen ($NO_3^-$-N), nitrite nitrogen ($NO_2^-$-N), and organic nitrogen. Usually nitrite nitrogen is in negligible amount. Crops directly utilize nitrogen

Table 7.7
**Typical yield capabilities of select soils**

| Soil series | Crop productivity group | Corn grain (bu/acre) | Corn silage (ton/acre) | Wheat (bu/acre) | Soybeans (bu/acre) |
|---|---|---|---|---|---|
| Abbottstown | 4 | 100 | 17 | 40 | 30 |
| Albrights | 3 | 125 | 21 | 50 | 30 |
| Allenwood | 1 | 150 | 25 | 60 | 45 |
| Basher | 2 | 120–125 | 21 | 60 | 40 |

*Source*: Pennsylvania State University (36).

Table 7.8
**Nitrogen, phosphate, and potash removal from soil by various crops***

| Crop | Unit | N (lb removed per unit production) | $P_2O_5$ (lb removed per unit production) | $K_2O$ (lb removed per unit production) |
|---|---|---|---|---|
| Corn, grain | bu | 1.0 | 0.4 | 0.3 |
| Corn, silage (65% moisture) | ton | 7.0 | 3.0 | 9.0 |
| Soybeans, grain | bu | 3.8 (a) | 1.0 | 1.5 |
| Wheat, grain, and straw | bu | 1.5 | 0.7 | 1.4 |
| Wheat, grain | bu | 1.3 | 0.5 | 0.3 |

*Note*: This table is provided as an example only. Similar information is available for each state.
*Source*: Brandt and Martin (34).

in its inorganic forms, principally nitrate-N and ammonium-N. Biosolids nitrate-N concentrations, however, are typically less than 0.05%. This translates to less than one pound per dry ton of biosolids. Hence, this fraction is usually insignificant and is not included in most agronomic rate calculations. However, it is advisable to test the biosolids nitrate-N content before eliminating this factor.

Most nitrogen exists in biosolids as organic-N, principally contained in proteins, nucleic acids, amines, and other cellular material. These complex molecules must be broken apart through biological degradation for nitrogen to become available to crops. The conversion of organic-N to inorganic ammonium N is called *mineralization*.

The mineralization rate depends on soil factors such as temperature, moisture, pH, and availability of oxygen, as well as the inherent biodegradability of organic materials. Biosolids that are digested undergo some mineralization before ever reaching the farm field. Hence, the method and degree of biosolids treatment prior to application influences the amount of nitrogen easily released for plant uptake. Further microbial conversion of ammonium-N to nitrate-N occurs under aerobic conditions and is termed nitrification. Nitrification occurs rapidly in well-drained agricultural soils and results in a predominance of nitrate rather than ammonium N by early summer in most humid, temperate climatic regions of the U.S.

### Table 7.9
**Estimated biosolids mineralization rate factors ($F_{year}$): percent organic-N mineralized from field-applied biosolids ($F_{year}$)**

| Time after biosolids application (crop year) | Unstabilized and waste activated sludges | Lime stabilized or aerobically digested biosolids | Anaerobically digested biosolids | Composted biosolids |
|---|---|---|---|---|
| 0–1 | 40% | 30% | 20% | 10% |
| 1–2 | 20% | 15% | 10% | 5% |
| 2–3 | 10% | 8% | 5% | 3% |
| 3–4 | 5% | 4% | | |

Source: Virginia Department of Health (26).

### Table 7.10
**Biosolids mineralization factors ($K_m$): pounds of N mineralized per dry ton of biosolids per percent organic-N ($K_m$)**

| Time after biosolids application (crop year) | Unstabilized and waste activated sludges | Lime stabilized or aerobically digested biosolids | Anaerobically digested biosolids | Composted biosolids |
|---|---|---|---|---|
| 0–1 | 8.00 | 6.00 | 4.00 | 2.00 |
| 1–2 | 2.40 | 2.10 | 1.60 | 0.90 |
| 2–3 | 0.96 | 0.90 | 0.72 | |
| 3–4 | 0.44 | 0.42 | | |

Source: US EPA (19).

Organic-N in biosolids becomes available to crops (i.e., mineralized) over a period of several years. Because of the many influencing factors, we rely on estimates of mineralization. The mineralization factors shown in Tables 7.9 and 7.10 illustrate how the length of time since field application affects the amount of biosolids PAN (19, 26).

Ammonium-N in biosolids/manure can be significant, making up even half the initial PAN of biosolids. The ammonium-N in biosolids can vary widely depending on treatment and storage. Since ammonium-N is prone to volatilization (as ammonia gas, $NH_3$), the application method affects PAN. For instance, surface applied biosolids are expected to lose half of their ammonium-N. Conversely, direct subsurface injection or soil incorporation of biosolids within 24 hours minimizes volatilization losses (12). The conversion of ammonium-N to ammonia gas form ($NH_3$) is called *volatilization*. Tables 7.11 and 7.12 present estimates of biosolids volatilization losses of ammonium N to ammonia gas from Virginia regulations (26) and the US EPA (7, 11), respectively.

The sum of nitrogen from the readily available N (i.e., $NO_3^-$-N, $NH_4^+$-N not lost as $NH_3$) and organic N mineralized to $NH_4^+$ is plant-available and used to determine N application rate. Table 7.13 presents historical mean yields and potential biosolids nitrogen utilization of various crops growth on soils of different productivity groups in Virginia (4, 26).

### Table 7.11
### Estimated plant-available percentage of ammonia from biosolids

| Management practice | Available portion (%) | |
|---|---|---|
|  | Biosolids pH <10 | Biosolids pH >10 |
| Injection below surface | 100 | 100 |
| Surface application with: |  |  |
|   Incorporation within 24 hours | 85 | 75 |
|   Incorporation within 1–7 days | 70 | 50 |
|   Incorporation after 7 days | 50 | 25 |

*Source*: Virginia Department of Health (26).

### Table 7.12
### Biosolids volatilization factors ($K_v$) for losses of ammonium-N as ammonia gas

| Sewage sludge | Factor $K_v$ |
|---|---|
| Liquid and surface applied | 0.50–0.75 |
| Liquid and injected into the soil | 1.0 |
| Dewatered and applied in any manner | 0.50* |

*Use value obtained from state regulatory agencies, if available.
*Source*: US EPA (7, 11).

### Table 7.13
### Historical mean yields and potential biosolids nitrogen utilization of various crops growth on soils of different productivity groups

| Crop | Soil productivity mean yield | Group IIA N use (lb/ac) | Soil productivity mean yield | Group IIIB N use (lb/ac) |
|---|---|---|---|---|
| Corn |  |  |  |  |
|   Grain (yield = bu/ac) | 140 | 140–160 | 110 | 110–130 |
|   Silage (yield = tons/ac) | 19 | 140–160 | 16 | 110–130 |
| Soybean (yield = bu/ac) |  |  |  |  |
|   Early season | 40 | 140–160 | 35 | 110–130 |
|   Late season | 34 | 140–160 | 25 | 110–130 |
| Wheat (yield = bu/ac) |  |  |  |  |
|   Standard | 56 | 90 | 48 | 80 |
|   Intensive | 70 | 90 | 60 | 80 |
| Tall-grass hay (yield = tons/ac) | 3.5–4 | 250 | <3 | 200 |
| Pasture, fescue/orchard grass | a | 120 | a | 100 |
| Alfalfa (tons/ac) | 4–6 | 300 | <4 | 210 |

[a] Insufficient data to make a good estimate.
*Source*: (4, 26).

## 3. PLANNING AND MANAGEMENT OF AGRICULTURAL LAND APPLICATION

### 3.1. Planning

Biosolids nutrient management controls should be planned for activities before, during, and after land application, according to the National Biosolids Partnership (12).

#### 3.1.1. Planning Before Land Application

The following are the planning activities before land application of biosolids: (a) Confirm that the biosolids meet all pollutant, pathogen reduction, and vector attraction reduction requirements at the time proposed for application. Do not just rely on past history; a responsible representative must personally review the data to ensure that all is in order. (b) Confirm the N, P, and K content of the biosolids. If the material has been stored for greater than 6 weeks, nutrient content should be reevaluated. (c) Review the farm nutrient management plan for the crop(s) being planted in order to calculate the biosolids agronomic rate. (d) Access information on past biosolids applications in order to consider residual N when calculating the biosolids agronomic rate. (e) Calculate the "target" biosolids agronomic rate based on the nitrogen content of the biosolids, crop nitrogen need, and residual N from past biosolids applications. (f) Discuss the proposed biosolids application with the farm operator to confirm that the recycling program is consistent with the farm operator's intentions. Address any last minute changes on the farm operator's part. (g) Check that all regulatory approvals and notices have been completed.

#### 3.1.2. Planning During Land Application

The following are the planning activities during land application of biosolids: (a) Check the area applied versus the volume (or tonnage) of biosolids applied to confirm that the actual application rate is consistent with the target agronomic rate. This exercise should be performed daily. (b) Record the location (field, portion of field) where each load of biosolids is applied, the weather conditions, responsible parties involved, visits by regulators, and any unusual observations or complaints by neighbors.

#### 3.1.3. Planning After Land Application

The following are the planning activities after land application of biosolids: (a) Assemble and file all records documenting the application event. (b) Submit any required regulatory reports. (c) Provide pertinent information to the farm operator, particularly the biosolids nutrients applied. (d) Notify the farm operator (and specified regulatory officials as required) that land application activities have been completed.

### 3.2. Nutrient Management

#### 3.2.1. Nutrient Management Goal

The goal of nutrient management is to develop environmentally responsible strategies for field application of agricultural fertilizers. A sound nutrient management plan (NMP) provides a site-specific strategy for supplying necessary nutrients for crop growth while at the same time protecting local water quality. The Part 503 Rule limits land application of biosolids N to only the amount used by growing crops (12).

Nutrient sources include, but are not limited to, livestock and poultry manures, compost as fertilizer, commercially manufactured chemical fertilizers, biosolids, or combinations thereof. The Part 503 rule specifies that biosolids may not be applied at a rate greater than the agronomic rate. The practice of limiting biosolids applications to supply only as much N as will be consumed by the crop and removed during harvest, termed agronomic N rate, is not a new concept. Most state biosolids regulations have recognized this practice for decades.

Experience has shown that repeat applications of nutrients to the same farm field may eventually result in elevated levels of soil test phosphorus (P). When elevated soil test P is found, terms such as *high* or *excessive* are used in soil test reports to indicate that further addition of phosphate fertilizer will not increase crop yields. Such interpretations of soil fertility tests are based on agronomic/economic considerations, not on potential environmental risk posed by high soil test P levels.

Biosolids, like animal manure, are not a balanced fertilizer. The primary nutrients (nitrogen [N], phosphorus [P], and potassium [K]) required to achieve target crop yields do not match the amounts available from biosolids on a mass basis. For example, when biosolids applications are performed to meet crop-N need, P is typically overapplied. At the same time K is often underapplied. The degree to which P and K are mismatched to crop needs depends on the particular biosolids and the crop.

Current standard practice bases biosolids application on plant available nitrogen (PAN) content. The approach strives to ensure that at least two of the three primary nutrients, N and P, are present in the soil in sufficient quantities to achieve the desired crop yield.

### 3.2.2. Farm Identification Elements for Nutrient Management

Biosolids application can substantially offset, or even completely eliminate, the need for chemical fertilizers when careful and deliberate nutrient management is employed. There are four basic components to a voluntary biosolids nutrient management plan: (a) farm identification, (b) nutrient management plan summary, (c) nutrient allocation and use, and (d) restrictions (12).

The following is a list of required information in order to develop the first component, farm identification: (a) operator's name, address, telephone number, and signature (including a landowner consent form); (b) county where operation is located; (c) name(s) of adjacent streams; (d) indication of any special protection waters; (e) total acres of operation; (f) total cropland acres available for nutrient application; (g) total cropland acres planned for manure recycling, excluding biosolids and other organic-N nutrient sources; (h) total cropland acres planned for biosolids recycling, excluding manure and other organic-N nutrient sources; (i) total cropland acres to which biosolids and manure both will be applied; (j) number of animal equivalent units (AEUs) per acre receiving manure, if applicable; (k) name and certification number of nutrient management specialist, if applicable; (l) location maps showing outline of farm site and soil survey maps containing soil types and slopes with outline of farm site; and (m) farm maps of sufficient scale to show the field and operation boundaries and the areas where biosolids application is limited or restricted.

### 3.2.3. Nutrient Management Plan Summary Elements

A nutrient management plan summary should include the following elements: (a) manure management summary table, if applicable; (b) total manure generated on the farm site annually; (c) total manure used on the farm site annually; (d) total manure exported from the farm site annually; (e) biosolids management summary table; (f) total biosolids generated by contributing sources; (g) total amount of biosolids which could be recycled in accordance with the computed agronomic rate; (h) nutrient application rates by field or crop group; (i) general summary of excess manure utilization procedures; and (j) implementation schedule (12).

### 3.2.4. Nutrient Allocation and Use Elements

Although the methodology introduced in this chapter is applied to biosolids on agricultural lands, it can also be applied to manure or other nutrient-containing materials. Accordingly, the following nutrient allocation and use elements should be investigated and reported when possible: (a) amounts and various nutrient sources used on the operation; (b) the number of animals of each animal type, if applicable; (c) acreage and expected crop yields for each crop group; (d) the amount of nutrients necessary to meet expected crop yields; (e) residual N from legumes; (f) the nutrient content of the manure(s), if applicable; (g) the amount of PAN originating from manure(s), considering the application method and planned manure incorporation time (volatilization losses), if applicable; (h) the amount of PAN originating from past manure applications, if applicable; (i) the nutrient content of conventional fertilizers that will be used regardless of other N sources (e.g., starter fertilizer, herbicide carrier solutions, etc.); (j) the amount of PAN originating from conventional fertilizers; (k) the nutrient content of the biosolids; (l) the amount of PAN originating from biosolids, considering the biosolids treatment method, biosolids-N forms, and planned application method; (m) the amount of PAN originating from past biosolids applications; (n) planned manure application rate(s), if applicable; (o) target spreading periods for manure application, if applicable; (p) nitrogen balance calculation showing the biosolids agronomic rate for each management unit; and (q) winter manure spreading procedures, if applicable.

Table 7.14 provides a summary of conventional nitrogen fertilizers in terms that are helpful in discussions with farm operators. Table 7.15 illustrates the type of information needed when assessing nitrogen contributions from legumes. It is necessary to consult the state agronomy guide for guidance specific to the land application area (12).

### 3.2.5. Restrictions Elements

There are only four elements for restrictions for nutrient management plan: (a) frozen, snow covered, and saturated soil conditions; (b) slope constraints; (c) manure application isolation distances; and (d) biosolids application isolation distances and harvest waiting periods.

**Table 7.14**
**Conventional nitrogen fertilizer materials**

| Fertilizer | Total N, % | Available phosphoric acid, % | Soluble potash, % |
|---|---|---|---|
| Anhydrous ammonia ($NH_3$) | 82 | 0 | 0 |
| Urea ($NH_2$-CO-$NH_2$) | 46 | 0 | 0 |
| Ammonium nitrate ($NH_4NO_3$) | 33–34 | 0 | 0 |
| N solutions (UAN) (urea + $NH_4NO_3$ + water) | 28–32 | 0 | 0 |
| Ammonium sulfate $(NH_4)_2SO_4$ | 21 | 0 | 0 |
| Diammonium phosphate (DAP) $(NH_4)_2HPO_4$ | 18 | 46 | 0 |
| Monoammonium phosphate (MAP) $NH_4H_2PO_4$ | 11 | 52 | 0 |
| Ammonium polyphosphate | 10 | 34 | 0 |
| Potassium nitrate $KNO_3$ | 13 | 0 | 45 |

*Source*: Pennsylvania State University (36).

**Table 7.15**
**Residual nitrogen contributions from legumes for corn production***

| Previous crop | % Stand | Soil productivity group 1 nitrogen credit (lb/acre) |
|---|---|---|
| First year after alfalfa | | |
| | ≥50% | 120 |
| | 25–49% | 80 |
| | <25% | 40–45 |
| Second year after alfalfa | | |
| | <50% | 60 |
| First year after clover or trefoil | | |
| | ≥50% | 90 |
| | 25–49% | 60 |
| | <25% | 40 |
| First year after soybeans harvested for grain | | 1 lb N/bu soybeans |

*\*Note*: This table is proved as an example only. Similar information is available for each state.
*Source*: Virginia Department of Health (26).

## 4. DESIGN OF LAND APPLICATION PROCESS

### 4.1. Biosolids Application Rate Scenario

Design criteria for land application programs address issues related to application rates and suitable sites. Biosolids, site, and vegetative characteristics are the most important design factors to consider. Biosolids must meet regulatory requirements for stabilization and metals content. In addition, nutrient content and physical characteristics, such as percent solids, are used to determine the appropriate application rate for the crop that will be grown and the soil in which the crops will be grown. Site suitability is determined based on such factors as soil characteristics, slope, depth to groundwater, and proximity to surface water. In addition, many states have established site requirements to further protect water quality. Some examples from various states include (a) sufficient land to provide areas of nonapplication (buffers) around surface water bodies, wells, and wetlands; (b) depth from the soil surface to groundwater equal to at least 1 m; (c) soil pH in the range of 5.5 to 7.5 to minimize metal leaching and maximize crop growing conditions; and (d) site suitability is also influenced by the character of the surrounding area. While odors and truck traffic may not be objectionable in an agricultural area, both will adversely impact residential developments and community centers close to fields where biosolids are applied.

The type of vegetation to be grown is also a design consideration. Vegetation, like soil characteristics, will generally not exclude biosolids application since most vegetation benefits from the practice. However, the type of vegetation has impact on the choice of application equipment, the amount of biosolids to be applied, and the timing of applications. The amount of biosolids that may be applied to a site is a function of the amount of nutrients required by the vegetation and the amount of metals found in the biosolids.

Table 7.16 summarizes the application frequency, timing, and rates for various types of sites. Another factor to be considered in designing a land application program is the timing of applications. Long periods of saturated or frozen ground limit opportunities for application. This is an important consideration in programs using agricultural lands; applications must be performed at times convenient for the farmer and must not interfere with the planting of crops. Most application of biosolids to agricultural land occurs in the early spring or late fall. As a result, storage or an alternate biosolids management option must be available to handle biosolids when application is not possible. Forest lands and reclamation sites allow more leeway in the timing of applications. In some areas of the

Table 7.16
**Typical biosolids application scenarios**

| Type of site/ vegetation | Schedule | Application frequency | Application rate dry ton/acre |
|---|---|---|---|
| **Agricultural land** | | | |
| Corn | April, May, after harvest | Annually | 5 to 10 |
| Small grains | March–June, August, fall | Up to 3 times per year | 2 to 5 |
| Soybeans | April–June, fall | Annually | 5 to 20 |
| Hay | After each cutting | Up to 3 times per year | 2 to 5 |
| **Forest land** | Year round | Once every 2–5 years | 5 to 100 |
| **Range land** | Year round | Once every 1–2 years | 2 to 60 |
| **Reclamation sites** | Year round | Once | 60 to 100 |

*Source*: US EPA (6).

U.S., application can proceed year round. Application is most beneficial on agricultural land in late fall or early spring before the crop is planted.

Timing is less critical in forest applications when nutrients can be incorporated into the soil throughout the growing period. Winter application is less desirable in many locales. Range lands and pasture lands also are more adaptable to applications during various seasons. Applications can be made as long as ground is not saturated or snow covered and whenever livestock can be grazed on alternate lands for at least 30 days after the application. The timing of single applications in land reclamation programs is less critical and may be dictated by factors such as regulatory compliance schedules. Table presents the data of N, P, K, and other nutrient removal from soil by various crops (52).

The concentrations of trace elements (largely, heavy metals) in biosolids affect the suitability of land application and the application rate. Table 7.3 presents the US EPA ceiling concentration limits, pollutant concentration limits, and cumulative pollutant loading rates of heavy metals and the mean concentrations from the NSSS (11, 14). Table 7.18 presents possible trace element concentration in typical unamended and biosolids-amended soils and the time required to reach cumulative loading limits for the regulated trace elements (5). Actual U.S. federal regulations on the heavy metal (trace element) control is summarized in Appendix C, which is similar to Table 7.3, but includes both English and metric units.

### 4.2. Step-by-Step Procedures for Biosolids Application Rate Determination

The biosolids loading rate or application rate is a function of the biosolids characteristics, soil characteristics, crop, crop end use, and crop nutrient requirements. By-product and soil characteristics vary even for a given situation, and crop requirements vary widely. For this reason and for changing nutrient requirements based on crop rotation, there is no one general application rate.

The biosolids application rate is based on the estimated biosolids plant available nitrogen content and the nitrogen requirements of the crop. Nitrogen is present in aerobically digested biosolids in the organic, ammonium, and nitrate forms. The nitrate form (nitrate-nitrogen) is not present in anaerobically digested biosolids. Nitrogen is available for immediate plant use in the ammonium ($NH_4^+$) or nitrate ($NO_3^-$) forms. The availability of organic nitrogen to the crop depends on the mineralization rate and is normally available over a period of several years.

Usually, the biosolids application rate is first determined based on nitrogen requirements. This rate is then used to calculate phosphorus supply and is compared to soil and crop P needs. The priority trace elements are lead, zinc, copper, nickel, cadmium, arsenic, mercury, molybdenum, and selenium (1, 7). There are four basic steps involved in determining the biosolids agronomic rate (AR): (a) crop nitrogen fertilizer rate (CNFR) determination, (b) crop nitrogen deficit (CND) determination, (c) biosolids plant available nitrogen (PAN) determination, and (d) AR calculation.

The following subsections provide detailed step-by-step procedures for calculation of biosolids AR using all of the separate components listed above. In practice,

### Table 7.17 (a)
### Nutrient removal by selected field crops

| Crop | Yield | N | P | K | Ca | Mg | S | Cu | Mn | Zn |
|---|---|---|---|---|---|---|---|---|---|---|
| | | | | | | Nutrient removal, lb/acre | | | | |
| Barley (grain) | 60 bu | 65 | 14 | 24 | 2 | 6 | 8 | 0.04 | 0.03 | 0.08 |
| Barley (straw) | 2 tons | 30 | 10 | 80 | 8 | 2 | 4 | 0.01 | 0.32 | 0.05 |
| Corn (grain) | 200 bu | 150 | 40 | 40 | 6 | 18 | 15 | 0.08 | 0.10 | 0.18 |
| Corn (stover) | 6 tons | 110 | 12 | 160 | 16 | 36 | 16 | 0.05 | 1.50 | 0.30 |
| Cotton (seed + lint) | 1.3 tons | 63 | 25 | 31 | 4 | 7 | 5 | 0.18 | 0.33 | 0.96 |
| Cotton (stalk + leaf) | 1.5 tons | 57 | 16 | 72 | 56 | 16 | 15 | 0.05 | 0.06 | 0.75 |
| Oats (grain) | 80 bu | 60 | 10 | 15 | 2 | 4 | 6 | 0.03 | 0.12 | 0.05 |
| Oats (straw) | 2 tons | 35 | 8 | 90 | 8 | 12 | 9 | 0.03 | — | 0.29 |
| Peanuts (nuts) | 2 tons | 140 | 22 | 35 | 6 | 5 | 10 | 0.04 | 0.30 | 0.25 |
| Peanuts (vines) | 2.5 tons | 100 | 17 | 150 | 88 | 20 | 11 | 0.12 | 0.15 | — |
| Rye (grain) | 30 bu | 35 | 10 | 10 | 2 | 3 | 7 | 0.02 | 0.22 | 0.03 |
| Rye (straw) | 1.5 tons | 15 | 8 | 25 | 8 | 2 | 3 | 0.01 | 0.14 | 0.07 |
| Soybean (grain) | 50 bu | 188 | 41 | 74 | 19 | 10 | 23 | 0.05 | 0.06 | 0.05 |
| Soybean (stover) | 3 tons | 89 | 16 | 74 | 30 | 9 | 12 | — | — | — |
| Wheat (grain) | 60 bu | 70 | 20 | 25 | 2 | 10 | 4 | 0.04 | 0.10 | 0.16 |
| Wheat (straw) | 2.5 tons | 45 | 5 | 65 | 8 | 12 | 15 | 0.01 | 0.16 | 0.05 |
| Tobacco (burley) | 2 tons | 145 | 14 | 150 | — | 18 | 24 | — | — | — |
| Tobacco (flue cured) | 1.5 tons | 85 | 15 | 155 | 75 | 15 | 12 | 0.03 | 0.55 | 0.07 |

### Table 7.17 (b)
### Nutrient removal by selected hay field, fruit and vegetable crops

| Crop | Yield | N | P | K | Ca | Mg | S | Cu | Mn | Zn |
|---|---|---|---|---|---|---|---|---|---|---|
| | | | | | | Nutrient removal, lb/acre | | | | |
| | | | | | | pounds per acre | | | | |
| Alfalfa | 6 tons | 350 | 40 | 300 | 160 | 40 | 44 | 0.10 | 0.64 | 0.62 |
| Bluegrass | 2 tons | 60 | 12 | 55 | 16 | 7 | 5 | 0.02 | 0.30 | 0.08 |
| Coastal Bermuda-grass | 8 tons | 400 | 45 | 310 | 48 | 32 | 32 | 0.02 | 0.64 | 0.48 |
| Fescue | 3.5 tons | 135 | 18 | 160 | — | 13 | 20 | — | — | — |
| Orchard grass | 6 tons | 300 | 50 | 320 | — | 25 | 35 | — | — | — |
| Red clover | 2.5 tons | 100 | 13 | 90 | 69 | 17 | 7 | 0.04 | 0.54 | 0.36 |
| Soybean | 2 tons | 90 | 12 | 40 | 40 | 18 | 10 | 0.04 | 0.46 | 0.15 |
| Timothy | 4 tons | 150 | 24 | 190 | 18 | 6 | 5 | 0.03 | 0.31 | 0.20 |
| Apples | 500 bu | 30 | 10 | 45 | 8 | 5 | 10 | 0.03 | 0.03 | 0.03 |
| Cabbage | 20 tons | 130 | 35 | 130 | 20 | 8 | 44 | 0.04 | 0.10 | 0.08 |
| Peaches | 600 bu | 35 | 20 | 65 | 4 | 8 | 2 | — | — | 0.01 |
| Potato (sweet) | 300 bu | 40 | 18 | 96 | 4 | 4 | 6 | 0.02 | 0.06 | 0.03 |
| Potato (white) | 15 tons | 90 | 48 | 158 | 5 | 7 | 7 | 0.06 | 0.14 | 0.08 |
| Snap bean | 4 tons | 138 | 33 | 163 | — | 17 | — | — | — | — |
| Spinach | 5 tons | 50 | 15 | 30 | 12 | 5 | 4 | 0.02 | 0.10 | 0.10 |
| Tomatoes (fruit) | 20 tons | 120 | 40 | 160 | 7 | 11 | 14 | 0.07 | 0.13 | 0.16 |

*Source*: Mullins and Hansen (53).

**Table 7.18**
**Possible trace element concentration in typical unamended and biosolids-amended soils and the time required to reach cumulative loading limits for the regulated trace elements**

| Trace element | Typical background soil concentration range for non-contaminated[a] (mg/kg) | Theoretical soil concentration at US EPA cumulative loading limit[b] (mg/kg) | Time required to reach cumulative loading limit[c] (years) |
|---|---|---|---|
| Arsenic | 6–10 | 21 | 360 |
| Cadmium | 0.2–0.5 | 20 | 500 |
| Copper | 17–65 | 750 | 181 |
| Lead | 8–22 | 150 | 201 |
| Mercury | 0.06–0.15 | 9 | 320 |
| Nickel | 7–45 | 210 | 871 |
| Selenium | 0.3–0.4 | 50 | 1780 |
| Zinc | 19–82 | 1400 | 208 |

[a]*Source*: Evanylo (5).
[b]Theoretical maximum soil concentrations after application of the maximum allowable amount of that element.
[c]Assumes an annual application rate of 5 dry ton/acre of a biosolid with trace element concentrations equal to the means allowable concentrations.

this analysis must be repeated for each farm field contained in a land application program (1, 12).

*4.2.1. Determining Unit Nitrogen Fertilizer Rate and Crop Nitrogen Fertilizer Rate*

Table 7.7 presents yield capabilities of selected Pennsylvania soils (36), and Tables 7.8 and summarize N, $P_2O_5$, and $K_2O$ removal from soil by various crops. The data presented in Table 7.8 is termed the unit nitrogen fertilizer rate (UNFR) in the unit of lb-N per unit crop yield, where the units are either bushels or tons. (Note: 1 U.S. bushel = 1.2444 ft$^3$; 1 British bushel = 1.2843 ft$^3$; 1 ton [British ton] = 2000 lb; 1 T [metric ton] = 1000 kg). The sludge (biosolids/manure) AR is based on meeting crop needs without overapplication of nitrogen. The total CNFR can be calculated by Equation 1:

$$CNFR = (Yield)(UNFR) \qquad (1)$$

where
   CNFR = crop nitrogen fertilizer rate, lb N/acre
   Yield = crop yield, bu/acre or ton/acre harvested, as shown in Table 7.7
   UNFR = unit nitrogen fertilizer rate, lb N/unit crop yield, as shown in Table 7.8

*4.2.2. Determining Crop Nitrogen Deficit*

Crop nitrogen deficit (CND) equals the anticipated crop nitrogen fertilizer rate (CNFR) minus all past biosolids PAN (PANA) and nonbiosolids sources (PANS), in the unit of lb N/acre, as shown in Equation 2. Previous biosolids carry-over nitrogen is included in this calculation.

$$CND = CNFR - (PANA) - (PANS) \tag{2}$$
$$CND = CNFR - (PANT) \tag{2a}$$

where
  $CND$ = crop nitrogen deficit, lb N/acre
  $CNFR$ = crop nitrogen fertilizer rate, lb N/acre
  $PANA$ = crop year biosolids PAN applied in previous years, lb N/acre
  $PANS$ = crop year nonbiosolids PAN applied in previous years, lb N/acre
  $PANT$ = crop year total PAN applied in previous years, lb N/acre

Specifically, PAN contributions from all past and current planned nonbiosolids sources must be subtracted from the calculated CNFR to determine the CND that may be supplied by biosolids applications for a particular field crop. Nitrogen from all past sources (PANT) that must be considered include (a) manure-N, if applicable, including current and historical applications; (b) residual legume-N, if applicable, carryover from previous legume crops; (c) starter fertilizer-N, if applicable; (d) conventional N-containing chemical fertilizers; (e) biosolids organic-N carryover including nitrogen originating from the previous 3 years' applications; and (f) other nitrogen sources, such as land applied crop or food processing residuals, irrigation water, nitrogen-solution pesticide carriers, other nonconventional fertilizer materials, etc.

The PANA (i.e., mainly organic-N application per acre) carryover from past biosolids applications can be calculated using Equation 3, Table 7.10 or Table 7.19.

$$PANA = A(K_m)(Organic\text{-}N) \tag{3}$$

where
  $PANA$ = crop year biosolids PAN applied in previous years, lb N/acre
  $A$ = biosolids application per acre in previous years, dry ton/acre
  $Organic\text{-}N$ = organic nitrogen concentration in biosolids, %
  $K_m$ = biosolids crop year organic-N mineralization factor based on the method of biosolids treatment, lb/ton/% (see Tables 7.9, 7.10, and 7.19)

### Table 7.19
**Mineralization of sludge organic nitrogen**

| Years after sludge application | Mineralization rate, % | Annual nitrogen available during yr, lb N/ton sludge % Organic nitrogen in sludge, % by weight | | | | | | | | |
|---|---|---|---|---|---|---|---|---|---|---|
| | | 2.0 | 2.5 | 3.0 | 3.5 | 4.0 | 4.5 | 5.0 | 5.5 | 6.0 |
| 1 | 15 | 6.0 | 7.5 | 9.0 | 10.5 | 12.0 | 13.5 | 15.0 | 16.5 | 18.0 |
| 2 | 10 | 3.4 | 4.2 | 5.1 | 6.0 | 5.8 | 7.6 | 8.5 | 9.4 | 10.2 |
| 3 | 5 | 1.5 | 1.9 | 2.3 | 2.7 | 3.1 | 3.4 | 3.8 | 4.2 | 4.6 |
| 4 | 5 | 1.5 | 1.8 | 2.2 | 2.5 | 2.9 | 3.3 | 3.6 | 4.0 | 4.4 |
| 5 | 5 | 1.4 | 1.7 | 2.1 | 2.4 | 2.8 | 3.1 | 3.4 | 3.8 | 4.1 |

*Source*: US EPA (1).

### 4.2.3. Determining First-Year Plant Available Nitrogen ($PAN_{0-1}$)

Computing biosolids plant available nitrogen (PAN) should account for (a) the type of biosolids, (b) the method of biosolids application, (c) organic-N mineralization in subsequent growing seasons, and (d) both inorganic and organic contributions to PAN. Total nitrogen is the sum of nitrate nitrogen, nitrite nitrogen, organic nitrogen, and ammonia (all expressed as N). Note that for laboratory analysis purposes, total Kjeldahl nitrogen (TKN) is made up of both organic nitrogen and ammonia nitrogen.

The first year plant available nitrogen ($PAN_{0-1}$) in biosolids may be summarized by Equations 4 and 4a, in which all biosolids mass is based on dry solids:

$$N_{total} = N_{ammonium} + N_{nitrate} + N_{nitrite} + N_{organic} \tag{4}$$

$$PAN_{0-1} = (K_V)(N_{ammonium}) + N_{nitrate} + (F_{0-1})N_{organic} \tag{4a}$$

where

$PAN_{0-1}$ = first year plant available nitrogen in biosolids, lb N/ton of biosolids

$N_{total}$ = total nitrogen content in biosolids, lb nitrate-N/ton of biosolids

$N_{ammonium}$ = ammonium nitrogen content in biosolids, lb ammonium-N/ton of biosolids

$N_{nitrate}$ = nitrate nitrogen content in biosolids, lb nitrate-N/ton of biosolids

$N_{nitrite}$ = nitrite nitrogen content in biosolids, lb nitrate-N/ton of biosolids = 0 (assumed)

$N_{organic}$ = organic nitrogen content in biosolids, lb organic-N/ton of biosolids

$K_V$ = ammonium-N volatilization factor, based on the method of land application, as shown in Table 7.10

$F_{0-1}$ = biosolids first year organic-N mineralization factor based on the method of biosolids treatment (see Tables 7.9, 7.10, and 7.19)

### 4.2.4. Determining Biosolids Application Rate or Agronomic Rate

Determination of the AR involves five basics: (a) selecting a realistic crop yield goal; (b) determining N needs of this crop; (c) estimating residual N in the soil from past manures/legumes/biosolids; (d) determining the amount of supplemental N needed to meet the crop need; and (e) calculating the amount of biosolids necessary to supply this amount.

All of the above listed crop-N sources have been discussed in previous sections of this chapter. Note that the principal source for historical data is the farm operator. The agronomic rate is calculated using the first year PAN content of the biosolids intended to be recycled and the CND. Any change in either of these factors will impact the computed AR. Equation 5 describes the calculation of the AR:

$$AR = (CND)/(PAN_{0-1}) \tag{5}$$

where

$AR$ = agronomic rate, dry ton/acre

$CND$ = crop nitrogen deficit, lb N/acre

$PAN_{0-1}$ = first year plant available nitrogen in biosolids, lb N/ton of biosolids

*Engineering and Management of Agricultural Land Application*

### 4.2.5. Determining Maximum Allowable Biosolids Application

The determination of the lifetime allowable biosolids application rate is based on the total accumulated heavy metals for application of biosolids not meeting PCL. The total metals that can be applied are shown in Table 7.3. Using the information from Table 7.3, the maximum total biosolids application rate is the lowest of the following $R$ computations:

$$R_{Pb} = (\text{lb Pb/acre})/(\text{mg/kg Pb} \times 0.002) \tag{6}$$
$$R_{Zn} = (\text{lb Zn/acre})/(\text{mg/kg Zn} \times 0.002) \tag{7}$$
$$R_{Cu} = (\text{lb Cu/acre})/(\text{mg/kg Cu} \times 0.002) \tag{8}$$
$$R_{Ni} = (\text{lb Ni/acre})/(\text{mg/kg Ni} \times 0.002) \tag{9}$$
$$R_{Cd} = (\text{lb Cd/acre})/(\text{mg/kg Cd} \times 0.002) \tag{10}$$
$$R_{As} = (\text{lb As/acre})/(\text{mg/kg As} \times 0.002) \tag{11}$$
$$R_{Hg} = (\text{lb Hg/acre})/(\text{mg/kg Hg} \times 0.002) \tag{12}$$
$$R_{Mo} = (\text{lb Mo/acre})/(\text{mg/kg Mo} \times 0.002) \tag{13}$$
$$R_{Se} = (\text{lb Se/acre})/(\text{mg/kg Se} \times 0.002) \tag{14}$$

where

$R_{Pb}$ = biosolids application rate based on lead content, ton biosolids/acre
$R_{Zn}$ = biosolids application rate based on zinc content, ton biosolids/acre
$R_{Cu}$ = sludge application rate based on copper content, ton sludge/acre
$R_{Ni}$ = sludge application rate based on nickel content, ton sludge/acre
$R_{Cd}$ = sludge application rate based on cadmium content, ton sludge/acre
$R_{As}$ = sludge application rate based on arsenic content, ton sludge/acre
$R_{Hg}$ = sludge application rate based on mercury content, ton sludge/acre
$R_{Mo}$ = sludge application rate based on molybdenum content, ton sludge/acre
$R_{Se}$ = sludge application rate based on selenium content, ton sludge/acre

### 4.2.6. Determine Phosphorus Balance

$$P_{balance} = (BAR_{design}) \times (P_{content})F - P_{required} \tag{15}$$

where

$BAR_{design}$ = biosolids application rate selected for design, ton biosolids/acre
$P_{content}$ = phosphorus content in biosolids, lb P/ton biosolids (or lb $P_2O_5$/ton biosolids)
$P_{required}$ = phosphorus requirement on land, lb P/acre (or lb $P_2O_5$/acre)
$P_{balance}$ = positive value shows the excess lb P/acre; negative value shows the needed lb P/acre. (or lb $P_2O_5$/acre)
$F$ = availability factor = 0.5 (assumed for design)

### 4.2.7. Determination of Potassium Balance

$$K_{balance} = (BAR_{design}) \times (K_{content})F - K_{required} \tag{16}$$

where

$BAR_{design}$ = biosolids application rate selected for design, ton biosolids/acre
$K_{content}$ = potassium content in biosolids, lb K/ton biosolids (or lb $K_2O$/ton biosolids)
$K_{required}$ = potassium requirement on land, lb K/acre (or lb $K_2O$/acre)
$K_{balance}$ = positive value shows the excess lb K/acre; negative value shows the needed lb K/acre (or lb $K_2O$/acre)
$F$ = availability factor ($F = 1.0$ for potassium in biosolids)

There is no specific limit on K, and generally there will be a K deficiency (i.e., $K_{balance}$ = negative value) unless more K is added over that contained in biosolids.

### 4.3. Simplified Sludge Application Rate Determination

The land application system should provide safe and beneficial biosolids use. In general, biosolids application rates computed by nitrogen balance result in adequate crop phosphorus but inadequate potassium. Table 7.20 presents a simplified approach for calculating application rates for various crops based on the nitrogen requirements, assuming 70 lb organic nitrogen/ton of sludge with a mineralization rate of 15-10-5 (N-P-K), and an ammonium nitrogen content of 30 lb/ton of sludge.

Table 7.20
Simplified application rate determination

| Crop* | Application yr | | | | |
|---|---|---|---|---|---|
| | 1 | 2 | 3 | 4 | 5 |
| | Ton/acre of sludge | | | | |
| Alfalfa | 9.1 | 7.8 | 7.4 | 7.0 | 6.6 |
| Orchard grass | 7.4 | 6.3 | 6.0 | 5.6 | 5.4 |
| Clover grass | | | | | |
| Corn (grain) | 4.2 | 3.6 | 3.4 | 3.2 | 3.0 |
| (stover) | 1.7 | 1.5 | 1.4 | 1.3 | 1.2 |
| Sorghum (grain) | 3.0 | 2.5 | 2.4 | 2.2 | 2.1 |
| (stover) | 3.2 | 2.7 | 2.6 | 2.4 | 2.3 |
| Corn silage | 5.9 | 5.0 | 4.8 | 4.5 | 4.3 |
| Oats (grain) | 2.0 | 1.7 | 1.6 | 1.5 | 1.4 |
| (straw) | 0.86 | 0.74 | 0.70 | 0.66 | 0.62 |
| Soybeans (grain) | 6.0 | 5.1 | 4.8 | 4.6 | 4.3 |
| (straw) | 2.1 | 1.8 | 1.7 | 1.6 | 1.5 |
| Wheat (grain) | 3.6 | 3.0 | 2.9 | 2.7 | 2.6 |
| (straw) | 1.0 | 0.89 | 0.84 | 0.79 | 0.75 |
| Barley (grain) | 2.7 | 2.3 | 2.2 | 2.1 | 2.0 |
| (straw) | 1.0 | 0.84 | 0.80 | 0.75 | 0.71 |

*Source*: US EPA (1).

## 5. OPERATION AND MAINTENANCE

### 5.1. Operation and Maintenance Process Considerations

Land application systems generally use uncomplicated, reliable equipment. Operations include pathogen reduction processing, dewatering, loading of transport vehicles, transfer to application equipment, and the actual application. Operations and maintenance considerations associated with pathogen reduction processing are discussed elsewhere (41). The other operations require labor skills of heavy equipment operators, equipment maintenance personnel, and field technicians for sampling, all normally associated with wastewater treatment facilities. In addition, the biosolids generator is responsible for complying with state and local requirements as well as federal regulations. The biosolids manager must be able to calculate agronomic rates and comply with record keeping and recording requirements. In fact, the generator and land applier must sign certification statements verifying accuracy and compliance. The generator should also allocate time to communicate with farmers, landowners, and neighbors about the benefits of biosolids recycling. Control of odors, along with a viable monitoring program, is most important for public acceptance (6).

### 5.2. Process Control Considerations

Control of biosolids land application involves determination of application rate by close monitoring of biosolids and soil conditions and determination of crop nutrient requirements. The operation may change substantially after each year of operation. For example, biosolids application rates may be lower each year due to residual nitrogen. The rate may be reduced after several years of application due to phosphorus buildup. Added to this variability is crop rotation, which the farmer may practice periodically.

Process control steps required are proper rate setting, as described previously, and daily control of actual biosolids quantities applied. The actual biosolids application rate is varied by changing the number of passes made by the truck over the site. The field should be marked with numbered stakes to aid the equipment operators in proper application.

### 5.3. Maintenance Requirements and Safety Issues

Maintenance requirements are mainly cleaning and equipment service. The cleaning operation includes daily flushing of the injectors and periodic flushing of the tanks. Truck and equipment preventative maintenance schedules are specified in the manufacturer's data.

Safety is related to vehicle and equipment operation. Generally, the highest potential for accidents is when equipment is being backed or trailers are being connected or disconnected from tractors. All drivers should be given a thorough driver training course including classroom and practice operation. All should be required to pass a driving test specially designed for this operation.

The only safety measures necessary beyond the usual common sense is to require a spotter to assist drivers when backing trailers at the plant and to ensure that truck tires are at adequate pressure and not excessively worn.

## 6. NORMAL OPERATING PROCEDURES

### 6.1. Startup Procedures

The startup procedures include a daily check of trucks for oil level, fuel level, battery condition, radiator water level, lights, and turn signals. The injector(s) should be checked for flushing and lubrication after the previous use. Solids content of the biosolids should be determined in order to set the biosolids application rate. The total biosolids application rate should be determined and provided to operating personnel along with an application plan. If the biosolids have very high moisture content, the site may have to be covered more than once with rest periods between applications to prevent ponding.

### 6.2. Routine Land Application Procedures

Biosolids are transferred to the site(s) and applied according to the predetermined plan. The operator should be alert for ponding or other signs of problems. A record of the biosolids application should be prepared. The operator should show the covered areas each day as well as the number of passes and resulting biosolids application rate. These records enable the farmer to determine additional fertilizer requirements and the future sludge application rates, and provide plant personnel with a record of the biosolids application.

### 6.3. Shutdown Procedures

At the end of the day the truck and applicator should be washed to remove any remaining solids and serviced. Tillage may be required if the biosolids were surface applied rather than injected; however, sites permitted for pasture and hayland, no-till crop production, or forestland do not require incorporation.

## 7. EMERGENCY OPERATING PROCEDURES

### 7.1. Loss of Power or Fuel

Loss of electrical power does not affect the field or transport operations for biosolids application, but there may be an impact on the characteristics of the biosolids. The nature of this impact depends on the type of processes involved. Most likely the solids content decreases. Under these circumstances the solids concentration should be determined for each load of biosolids.

Adequate provisions must be made to pump liquid biosolids from the holding tank to the transport truck at the sewage treatment plant. If the trucks are equipped with diesel engines and the fuel runs out, the entire fuel system must be bled to remove air prior to starting the engine.

### 7.2. Loss of Other Biosolids Treatment Units

Other treatment units that directly impact the land application operation are those required for stabilization and concentration or dewatering. If the stabilization process is not operating properly, biosolids characteristics change. If the concentration or dewatering process is not operating properly, the biosolids moisture content will be high and a greater volume must be handled. In either case, the biosolids application rate must be changed to account for the change in the biosolids characteristics.

## 8. ENVIRONMENTAL IMPACTS

Despite many positive impacts on the environment, land application can have negative impacts on water, soil, and air if not practiced correctly (6–9).

Negative impacts to water result from the application of biosolids at rates that exceed the nutrient requirements of the vegetation. Excess nutrients in the biosolids (primarily nitrogen compounds) can leach from the soil and reach groundwater. Runoff from rainfall may also carry excess nutrients to surface water. However, because biosolids are a slow release fertilizer, the potential for nitrogen compounds to leach from biosolids amended soil is less than that posed by the use of chemical fertilizers. In areas fertilized by either biosolids or chemicals, these potential impacts are mitigated by proper management practices, including the application of biosolids at agronomic rates. Maintenance of buffer zones between application areas and surface water bodies and soil conservation practices will minimize impacts to surface water.

Federal regulations contain standards related to all metals of concern, and application of biosolids, which meets these standards, should not result in the accumulation of metals to harmful levels. Stringent record keeping and reporting requirements on both the federal and state level are imposed to prevent mismanagement.

Odors from biosolids applications are the primary negative impact to the air. Most odors associated with land application are a greater nuisance than threat to human health or the environment. Odor controls focus on reducing the odor potential of the biosolids or incorporating them into the soil. Stabilization processes such as digestion can decrease the potential for odor generation. Biosolids that have been disinfected through the addition of lime may emit ammonia odors but they are generally localized and dissipate rapidly. Biosolids stabilization reduces odors and usually results in an operation that is less offensive than manure application.

Overall, a properly managed biosolids land application program is preferable to the use of conventional fertilizers for the following reasons:

1. Biosolids are a recycled product, use of which does not deplete nonrenewable resources such as phosphorus.
2. The nutrients in biosolids are not as soluble as those in chemical fertilizers and are therefore released more slowly.
3. Biosolids appliers are required to maintain setbacks from water resources and are often subject to more stringent soil conservation and erosion control practices, nutrient management, and record keeping and reporting requirements than farmers who use only chemical fertilizers or manures.
4. Biosolids are closely monitored.
5. The organic matter in biosolids improves soil properties for optimum plant growth, including tilth, friability, fertility, and water holding capacity. They also decrease the need for pesticide use.

A joint policy statement of the U.S. Department of Agriculture, the U.S. Food & Drug Administration, and the US EPA states, "The use of high quality biosolids coupled with proper management procedures, should safeguard the consumer from contaminated crops and minimize any potential adverse effect on the environment" (6).

## 9. LAND APPLICATION COSTS

It is difficult to estimate the cost of land application of biosolids without specific program details. For example, there is some economy of scale due to large equipment purchases. The same-size machine might be needed for a program that manages 10 dry tons of biosolids/d as one managing 50 dry tons/d; the cost of that machine can be spread over the 10 or 50 dry tons, greatly affecting average costs per dry ton. One source identified costs for land application varying from $60 to $290/dry ton (27). This range reflects the wide variety in land application methods as well as varying methods to prepare biosolids for land application. For example, costs for programs using dewatered biosolids include an additional step, whereas costs for programs using liquid biosolids do not reflect the cost of dewatering. They do, however, include generally higher transportation costs.

Despite the wide range of costs for land application programs, several elements must be considered in estimating the cost of any biosolids land application program: (a) purchase of application equipment or contracting for application services; (b) transportation; (c) equipment maintenance and fuel; (d) loading facilities; (e) labor; (f) capital and operation and maintenance of stabilization facilities; (g) ability to manage and control odors; (h) dewatering (optional); (i) storage or alternate management option for periods when application is not possible due to weather or climate; (j) regulatory compliance, such as permit applications, site monitoring, and biosolids analyses; and (k) public education and outreach efforts.

Land must also be secured. Some municipalities have purchased farms for land application; others apply biosolids to privately held land. Some operating costs can be offset through the sale of the biosolids material. Since the biosolids reduce the need for fertilizers and pH adjustment, farmers sometimes pay to have biosolids applied to their lands.

## 10. PRACTICAL APPLICATIONS AND DESIGN EXAMPLES

### 10.1. Biosolids Pretreatment Before Agricultural Land Application

Discuss the necessity of having sludge stabilization before application of biosolids on agricultural land.

*Solution*

The US EPA's 40 CFR Part 503, *Standards for the Use and Disposal of Sewage Sludge* (the Part 503 rule), requires that wastewater solids be processed before they are land applied. This processing is referred to as "stabilization" and helps minimize odor generation, destroys pathogens (disease causing organisms), and reduces vector attraction potential. There are several methods to stabilize wastewater solids: (a) adjustment of pH, or alkaline stabilization; (b) digestion; (c) composting; and (d) heat drying.

The Part 503 rule defines two types of biosolids with respect to pathogen reduction, Class A and Class B, depending on the degree of treatment the solids have received. Both types are safe for land application, but additional requirements are imposed on Class B materials. These are detailed in the Part 503 rule and include such things as restricting public access to the application site, limiting livestock grazing, and controlling crop-harvesting schedules. Class A biosolids (biosolids being treated so that there are no detectable pathogens) are not subject to these restrictions.

Vectors such as flies, mosquitoes, rodents, and birds can transmit diseases directly to humans or play a specific role in the life cycle of a pathogen as a host. Vector attraction reduction refers to processing that makes the biosolids less attractive to vectors, thereby reducing the potential for transmitting diseases.

## 10.2. Advantages and Disadvantages of Biosolids Land Application

Discuss the applicability, advantages, and disadvantages of applying biosolids on agricultural land based on real case histories.

*Solution*

Land application is well suited for managing solids from any size wastewater treatment facility. As the method of choice for small facilities, it offers cost advantages, benefits to the environment, and value to the agricultural community. However, biosolids produced by many major metropolitan areas across the country are also land applied. For example, biosolids from the Blue Plains Wastewater Treatment Facility serving the District of Columbia and surrounding communities in Virginia and Maryland have been land applied since the plant began operation in 1930. The cities of Philadelphia, Chicago, Denver, New York, Seattle, and Los Angeles all land apply at least part of their biosolids production. Land application is most easily implemented where agricultural land is available near the site of biosolids production, but advances in transportation have made land application viable even where hauling distances are greater than 1000 miles. For example, Philadelphia hauls dewatered biosolids 250 miles to reclaim strip mines in western Pennsylvania, and New York City ships some of its biosolids over 2000 miles to Texas and Colorado.

Land application offers several advantages as well as some disadvantages that must be considered before selecting this option for managing biosolids. Land application is an excellent way to recycle wastewater solids as long as the material is quality-controlled. It returns valuable nutrients to the soil and enhances conditions for vegetative growth. Land application is a relatively inexpensive option, and capital investments are generally lower than other biosolids management technologies. Contractors can provide the necessary hauling and land application equipment. In addition, on-site spatial needs can be relatively minor depending on the method of stabilization selected.

Although land application requires relatively less capital, the process can be labor intensive. Even if contractors are used for application, management oversight is essential for program success. Land application is also limited to certain times of the year, especially in colder climates. Biosolids should not be applied to frozen or snow covered grounds, while farm fields are sometimes not accessible during the growing season. Therefore, it is often necessary to provide a storage capacity in conjunction with land application programs. Even when the timing is right (for example, prior to crop planting in agricultural applications), weather can interfere with the application. Spring rains can make it impossible to get application equipment into farm fields, making it necessary to store biosolids until weather conditions improve. Another disadvantage of land application is potential public opposition, which is encountered most often when the beneficial use site is close to residential areas. One of the primary reasons for public concern is odor. In worst-case situations, municipalities or counties may pass ordinances that ban or restrict the use of biosolids in areas close to dwellings. However, many successful programs have gained public support through effective communications, an absolutely essential component in the beneficial use of biosolids.

### 10.3. Design Worksheet for Determining the Agronomic Rate

A design worksheet for determining the agronomic rate has been prepared by the US EPA (7).

*Solution*

The US EPA worksheet for determining the agronomic rate is presented in Table 7.21.

### 10.4. Calculation for Available Mineralized Organic Nitrogen

Assume that anaerobically digested biosolids with a 3% organic nitrogen content (dry weight basis) was applied to the site at a rate of 5 T/hectare (ha) in 2006. The following year, 2007, 3 T/ha of biosolids (same organic nitrogen content as in 2006) was applied to the same site. It is now 2008, and you want to calculate the PAN from previous biosolids applications.

*Solution*

The worksheet and the calculations are both presented in Table 7.22 (7). Here 1 ha = 2.471 acres. 1 T = 1 metric ton = 1 mt = 1000 kg.

### 10.5. Risk Assessment Approach Versus Alternative Regulatory Approach to Land Application of Biosolids

Study the publications of G.K. Evanylo of the Virginia Cooperative Extension (2–5), and discuss the risk assessment approach versus alternative regulatory approaches to the land application of biosolids.

*Solution*

RISK ASSESSMENT APPROACH

The risk assessment process was the most comprehensive analysis of its kind ever undertaken by the US EPA. The resultant U.S. federal regulations Part 503 was designed to provide "reasonable worst case," not absolute, protection to human health and the environment. The initial task of the 10-year risk assessment process was to establish a range of concentrations for trace elements and organic compounds that had the greatest potential for harm based on known human, animal, and plant toxicities. Maximum safe accumulations for the chemical constituents in soil were established from the most limiting of 14 pathways of exposure (Table 7.23), which included risks posed to human health, plant toxicity, and uptake, effects on livestock or wildlife, and water quality impacts. A total of 200 chemical constituents were screened by the US EPA (24),

**Table 7.21**
**Design worksheet for determining the agronomic rate**

Key to Symbols and Abbreviations

$NH_4^+$-N = Ammonium nitrogen content of the sewage sludge obtained from analytical testing of the sewage sludge, kg/mt (dry weight basis)

Kv = Volatilization factor estimating ammonium nitrogen remaining after atmospheric losses

Org-N = Organic nitrogen content of the sewage sludge obtained from analytical testing or determined by subtracting $NH_4^+$-N from TKN, kg/mt (dry weight basis)

$NO_3^-$-N = Nitrate nitrogen content of the sewage sludge obtained from analytical testing, kg/mt (dry weight basis)

$F_{0-1}$ = Mineralization rate for the sewage sludge during the first year of application, in percent of organic nitrogen expressed as a fraction (e.g., 20% = 0.2).

Helpful Conversions

mg/kg = lb/ton × 500
kg/ha = lb/acre × 1.12
kg/ha = tons/acre × 2242
mt/ha = tons/acre × 2.24

1. Total available nitrogen from sewage sludge
   a. Ammonium nitrogen  _____ kg/mt
      Calculated with the following formula: analytical result for $NH_4^+$-N (kg/mt) × $k_v$
   b. Mineralized organic nitrogen for first year of application  _____ kg/mt
      Calculated with the following formula: Org-N × $F_{0-1}$
   c. Nitrate nitrogen  _____ kg/mt
      Use analytical result for $NO_3^-$-N  _____ kg/mt
   d. Total  _____ kg/mt
2. Available nitrogen in the soil  _____ kg/ha
   (Use whichever is greater a or b)
   a. Soil test results of background nitrogen in soil
   b. Estimate of available nitrogen from previous sewage sludge applications
3. Nitrogen supplied from other sources (optional, but recommended):
   a. Nitrogen from supplemental fertilizers (if appropriate)  _____ kg/ha
   b. Nitrogen from irrigation water (if appropriate)  _____ kg/ha
   c. Nitrogen from previous crop (unless #2 is based on soil testing)  _____ kg/ha
   d. Other (if appropriate) (specify): \_\_\_\_\_  _____ kg/ha
   e. Total (add a, b, c, d, if available)  _____ kg/ha
4. Total nitrogen available from existing sources  _____ kg/ha
   Add 2 and 3e
5. Available nitrogen loss to denitrification (optional) (check with regulatory authority before using this site-specific factor)  _____ kg/ha
6. Adjusted nitrogen available  _____ kg/ha
   Subtract 5 from 4
7. Total nitrogen requirement of crop (obtain information from agricultural extension agents or other agronomy professionals)  _____ kg/ha
8. Supplemental nitrogen needed from sewage sludge  _____ kg/ha
   Subtract 4 or 6 from 7
9. Agronomic loading rate  _____ mt/ha
   Divide 8 by 1

TKN, total Kjeldahl nitrogen.
*Source*: US EPA (7).

and 50 of these were selected for further evaluation, using the criteria above and the availability of data for a preliminary risk assessment. Twenty-three of the 50 constituents were identified as warranting consideration for regulation based on the risk assessment. No regulatory limits were set for the 13 trace organic compounds in this group because the US EPA risk assessment showed that the safe levels were considerably higher than the observed concentrations in biosolids. The 503 rule was

**Table 7.22**
**Calculation for available mineralized organic nitrogen**

The organic nitrogen in sewage sludge continues to decompose and release mineral nitrogen through the mineralization process for several years following its initial application. This residual nitrogen from the previously applied sewage sludge must be accounted for as part of the overall nutrient budget when determining the agronomic rate for sewage sludge. Residual nitrogen can be determined through soil analysis or calculated using the following procedure. These calculations must be done for each yearly sewage sludge application unless soil analysis is performed prior to land application (see example calculations).

Instructions: Complete a separate table for each year sewage sludge was land applied. Note that most do not calculate beyond the third year because the values become negligible. Sum the values of mineralized Org-N (column d) from each table for the particular calendar year you're trying to determine Org-N available. (See example below.)

| a. Year[1] | b. Starting Org-N[2] (kg/ha) | c. Mineralization | d. Mineralized Org-N[3] (kg/ha) | e. Org-N Remaining[5] (kg/ha) |
|---|---|---|---|---|
| 0–1 (year sewage sludge was applied) | | | | |
| 1–2 (1st year after) | | | | |
| 2–3 (2nd year after) | | | | |

[1] Begin with year sewage sludge is applied, and continue for 2 more years.

[2] In the first year, this equals the amount of Org-N initially applied. In subsequent years, it represents the amount of org-N remaining from the previous year (i.e., column e).

[3] The org-N content of the initially applied sewage sludge continues to be mineralized, at decreasing rates, for years after initial application.

[4] Multiply column b and column c.

[5] Subtract column d from column b.

### Example

Assume that anaerobically digested sewage sludge with a 3% org-N content (dry weight basis) was applied to the site at a rate of 5 mt/ha in 2006. The following year, 2007, 3 mt/ha of sewage sludge (same org-N content as in 2006) was applied to the same site. It is now 2008, and you want to calculate the available nitrogen from previous sewage sludge applications.

In 2006, the org-N in the sewage sludge applied = (0.03) (5 mt/ha) (1000 kg/mt) = 150 kg/ha.
In 2007, the org-N in the sewage sludge applied = (0.03) (3 mt/ha) (1000 kg/mt) = 90 kg/ha.

Calculate the available nitrogen from 2006 and 2007 in the following manner (assume anaerobically digested sewage sludge).

**Table 7.22** (*Continued*)

| a. Year[1] | b. Starting Org-N (kg/ha) | c. Mineralization | d. Mineralized Org-N (kg/ha) | e. Org-N Remaining (kg/ha) |
|---|---|---|---|---|
| 2006 Sewage Sludge | | | | |
| 0–1 (first application-2006) | 150 | 0.2 | 30 | 120 |
| 1–2 (2007) | 120 | 0.1 | 12 | 108 |
| 2–3 (2008) | 108 | 0.05 | 5.40 | 102.60 |
| 2007 Sewage Sludge | | | | |
| 0–1 (first application-2007) | 90 | 0.2 | 18 | 72 |
| 1–2 (2008) | 72 | 0.1 | 7.2 | 64.80 |
| 2–3 (2009) | 64.8 | 0.05 | 3.24 | 61.56 |

To determine the total mineralized organic nitrogen available in 2008 from the sewage sludge applied in 2006 and 2007, add the mineralized Org-N value in the 2008 row of column d of the table for the 2006 sewage sludge to the mineralized Org-N value in the 2008 row of column d of the table for the 2007 sewage sludge (i.e., $5.40 + 7.2 = 12.6$ kg/ha).

Total mineralized Org-N available in 2008 from previous sewage sludge 12.6 kg/ha.

*Source*: US EPA (7).

then limited to 10 trace elements (arsenic, cadmium, chromium, copper, lead mercury, molybdenum, nickel, selenium, and zinc). Chromium was subsequently dropped on a court challenge because the risk assessment had shown a very low risk level for this metal.

The most limiting pathway for each of the nine regulated trace elements was used to develop pollutant concentration limits and lifetime loading rate standards. For example, the greatest risk to a target organism from lead (Pb) is a child directly ingesting biosolids that have been applied to soil. The pollutant limits are therefore based on estimates of childhood soil consumption that the US EPA considered conservative (i.e., they predict a greater impact on human health than is likely to occur). Ingestion of biosolids is the most limiting pathway for five of the trace elements (As, Cd, Pb, Hg, and Se), phytotoxicity was most limiting for three trace elements (Cu, Ni, and Zn), and feed consumption by animal was the most limiting for Mo (2–5).

Under Part 503, the cumulative pollutant loading rate (CPLR) limits established by the US EPA for eight trace elements would allow the concentrations of these elements to increase to levels that are 10 to 100 times the normal background concentrations in soil (Table 7.18). The time that it would take for each of the eight elements to reach its cumulative loading limit when biosolids with typical trace element concentrations (2–5, 24) are applied annually at a rate of 5 dry tons/acre is presented in Table 7.18. These are conservative estimates where agronomic loading rates are normally applied once every 3 years, not annually. The CPLR limits were developed to ensure that soil metals never reach harmful levels. Future applications of biosolids to the site would be prohibited if the cumulative loading limit for any of the eight trace elements was reached (5).

**Table 7.23**
Exposure pathways used in the risk assessment of land application (5)

| Pathway | Description of highly exposed individual |
| --- | --- |
| 1. Sludge > soil > plant > human | Human (except home gardener) lifetime ingestion of plants grown in sludge-amended soil |
| 2. Sludge > soil > plant > human | Human (home gardener) lifetime ingestion of plants grown in sludge-amended soil |
| 3. Sludge > human | Human (child) ingesting sludge |
| 4. Sludge > soil > plant > animal > human | Human lifetime ingestion of animal products (animals raised on forage grown on sludge-amended soil) |
| 5. Sludge > soil > animal > human | Human lifetime ingestion of animal products (animals ingest sludge directly) |
| 6. Sludge > soil > plant > animal | Animal lifetime ingestion of plants grown on sludge-amended soil |
| 7. Sludge > soil > animal | Animal lifetime ingestion of sludge |
| 8. Sludge > soil > plant | Plant toxicity due to taking up sludge pollutants when grown in sludge-amended soils |
| 9. Sludge > soil > organism | Soil organism ingesting sludge-soil mixture |
| 10. Sludge > soil > predator | Predator of soil organisms that have been exposed to sludge-amended soils |
| 11. Sludge > soil > airborne dust > human | Adult human lifetime inhalation of particles (dust) [e.g., tractor driver tilling a field] |
| 12. Sludge > soil > surface water > human | Human lifetime drinking surface water and ingesting fish containing pollutants in sludge |
| 13. Sludge > soil > air > human | Human lifetime inhalation of pollutants in sludge that volatilize to air |
| 14. Sludge > soil > groundwater > human | Human lifetime drinking well water containing pollutants from sludge that leach from soil to groundwater |

ALTERNATIVE REGULATORY APPROACH: BEST AVAILABLE TECHNOLOGY

An alternative to the risk assessment approach, termed *best available technology* (BAT) approach, limits contaminants in biosolids to concentrations attained by the best current technology (e.g., industrial pretreatment and separation of sanitary, storm, and industrial sewerage). The BAT approach is more restrictive of land application than risk assessment (i.e., lower pollutant concentrations can be attained using the best available technology than are permitted under the risk assessment approach). Biosolids are more likely to be landfilled or incinerated under this approach than under risk assessment (5).

ALTERNATIVE REGULATORY APPROACH: NONCONTAMINATION APPROACH

The US EPA Part 503 takes the position that all biosolids management options incur some risk, and that these risks can be evaluated so that regulations governing use and management options can be developed to reduce risk to acceptable (safe) levels. There are some who believe that the application of any biosolids that would cause an increase in the soil concentration of any pollutant is unacceptable. This is called the noncontamination approach. According to this approach, any addition of a pollutant to the soil must be matched by removal of that pollutant so that no long-term buildup occurs in the soil. This is the most restrictive of approaches to the land application of biosolids and is favored by those who believe that any increase in pollutant concentration in the soil is undesirable, regardless of what risk assessment demonstrates. Although this approach reduces to zero any environmental risks from land application of biosolids, it diverts more biosolids to landfills or incinerators, thereby increasing the environmental risks associated with disposal and reducing recycling of nutrients and organic matter (2–5).

Each approach for regulating contaminants in biosolids has its technical and scientific foundation, but the approach selected is based primarily on legislative mandates and policy decisions (5).

## 10.6. Tracking Cumulative Pollutant Loading Rates on Land Application Sites

Introduce the worksheet that has been prepared by the US EPA.

*Solution*

The definition and environmental engineering significance of the CPLR can be found from Table 7.3 and Appendix B.

The US EPA worksheet for tracking CPLR is presented in Table 7.24 (7).

## 10.7. Management of Nitrogen in the Soils and Biosolids

Study the US EPA report (7) and discuss the managerial strategy for controlling nitrogen in the soils and sewage sludge.

*Solution*

Nitrogen exists in the soils and biosolids in three basic forms:

1. Organic nitrogen: This refers mainly to carbon-based compounds such as proteins and amino acids. Little of this form is available to plants and must largely be converted to inorganic nitrogen by soil microorganisms. Mineralization is the conversion of organic N to inorganic N in the form of ammonium. Mineralization rates of organic N in biosolids varies for different climatic regimes and soils. The rate of N mineralization decreases with time and is typically not calculated as for longer than 3 years after biosolids application.
2. Inorganic nitrogen (ammonium-N, nitrite-N, and nitrate-N): Plants readily assimilate nitrate and ammonium ions. The soil microbes and plants compete for this inorganic N. Rapidly growing soil microbes can immobilize or "tie up" the ammonium and nitrate in the soil by converting it to the organic form, and may temporarily deplete the available N in the soil for plant uptake when the C/N ratio of a soil amendment is wide (i.e., >25 : 1). The positively charged ammonium ions are adsorbed by clay and organic matter so that little of this form is leached. Nitrification is the process whereby soil microbes convert ammonium to nitrate. Nitrate is very mobile and readily leached. Nitrite is usually not present in significant concentrations.

## Table 7.24
### Management worksheet for tracking CPLR on land application sites

| Pollutant | Cumulative pollutant loading rates (CPLRs) (kg/ha) 100% | Cumulative pollutant loading rates (CPLRs) (kg/ha) 90% | Calculation for determining cumulative loading[1] | | | | | | | | |
|---|---|---|---|---|---|---|---|---|---|---|---|
| | | | Concentration in sewage sludge (mg/kg) (dry weight) | × | Sewage sludge application rates (T/ha) (from Item 2) | × | 0.001 (conversion factor) | + | Amount of pollutant applied since July 20, 1993 (kg/ha)[2] | = | Total amount of pollutant applied to date (kg/ha) |
| Arsenic | 41 | 37 | ___ | × | ___ | × | 0.001 | + | ___ | = | ___ |
| Cadmium | 39 | 35 | ___ | × | ___ | × | 0.001 | + | ___ | = | ___ |
| Chromium | 3000 | 2700 | ___ | × | ___ | × | 0.001 | + | ___ | = | ___ |
| Copper | 1500 | 1350 | ___ | × | ___ | × | 0.001 | + | ___ | = | ___ |
| Lead | 300 | 270 | ___ | × | ___ | × | 0.001 | + | ___ | = | ___ |
| Mercury | 17 | 15 | ___ | × | ___ | × | 0.001 | + | ___ | = | ___ |
| Nickel | 420 | 378 | ___ | × | ___ | × | 0.001 | + | ___ | = | ___ |
| Selenium | 100 | 90 | ___ | × | ___ | × | 0.001 | + | ___ | = | ___ |
| Zinc | 2800 | 2520 | ___ | × | ___ | × | 0.001 | + | ___ | = | ___ |

[1] Use the following equations to convert from English system units (i.e., tons per acre) to metric system units (i.e., metric tons per hectare):

- To convert from tons per acre to metric tons per hectare, multiply tons per acre by 2.2421.
- To convert from acres to hectares, multiply the number of acres by 0.4047.
- To convert from tons to metric tons, multiply the number of tons by 0.9072.

[2] Land appliers are prohibited from applying CPLR sewage sludge to a site if CPLR sewage sludge was previously applied to the site after July 20, 1993, and the amounts of the pollutants regulated under Part 503 in the previously applied sewage sludge are unknown.

*Source*: US EPA (7).

3. Gaseous nitrogen (nitrogen gas, ammonia gas): Nitrogen gas is present in the soil atmosphere (air) and is a source of N for legumes, which can convert this to plant usable ammonium ion ($NH_4^+$). Under anaerobic conditions and readily available carbon source, microorganisms can convert nitrate to nitrogen gas and nitrous oxide ($N_2O$) in a process termed denitrification. Under alkaline conditions, ammonium ions lose a hydrogen ion and become ammonia, which readily volatilizes as ammonia gas ($NH_3$).

Plants use only a portion of the total nitrogen in biosolids. Some of the nitrate and ammonium is lost to the atmosphere by denitrification and volatilization, and some of the organic nitrogen becomes available over time as the mineralization process converts the organic forms to ammonium and nitrate. Some of the nitrate is lost through leaching. The goal when designing the agronomic rate for an application site is to supply the necessary amount of nitrogen needed for the crops or vegetation to produce the desired harvest yield while minimizing leaching of the nitrogen below the root zone. The rates of mineralization, plant uptake, volatilization, and denitrification are dependent on many factors and vary from site to site and at the same site.

Predicting how much biosolids are needed to provide the nitrogen sufficient to meet crop yield goals and minimize leaching of nitrogen below the root zone requires consideration of numerous factors. The following are some factors that influence the amount of sewage sludge that can be applied:

1. Total nitrogen content in biosolids and the concentrations (or percentage of the total nitrogen) of the various nitrogen forms in the biosolids are influenced by the types of processing operations. Anaerobic digestion (30 days or longer) produces biosolids that have high concentrations of organic N and ammonium but little nitrate (oxygen is required to proceed from ammonia to nitrate). Aerobically digested biosolids have higher levels of nitrate than anaerobically digested sewage sludge, but are still composed largely of organic and ammonium forms of N. Dewatering reduces the concentrations of nitrate and ammonium N, which are soluble in biosolids liquid fraction.
2. The mineralization rate at the application site is affected by how well the sewage sludge was stabilized in the digester. Poor stabilization results in more organic nitrogen for mineralization. Good stabilization converts most of the organic nitrogen into readily available inorganic N, leaving only that which is relatively inert and resistant to further mineralization (this sewage sludge may have a low mineralization rate).
3. The mineralization rate is also influenced by soil temperature and texture. Higher temperature increases the metabolic rate of microorganisms; thus, mineralization rates are typically higher in warmer than in colder periods and regions. Mineralization and nitrification are increased by aeration; thus, coarser textured soils, which facilitate gas exchange and are more likely than fine-textured soils to be well-drained have higher rates of N mineralization and nitrification,
4. The amount of ammonium lost through volatilization to the atmosphere is affected by soil/biosolids pH: The fraction of $NH_4/NH_3$ in the gaseous ammonia phase increases with pH. Volatilization losses of ammonia are reduced as biosolids are more thoroughly mixed with soil after application. Volatilization occurs rapidly, with the greatest loss occurring within the first week, if biosolids are left on the soil surface. Incorporation of surface-applied biosolids immediately after application or injection of liquid biosolids can greatly reduce the loss.

5. The amount of ammonia lost through volatilization is decreased when biosolids are incorporated into moist soils or when rainfall occurring immediately after surface application transports the soluble ammonia into the soil.
6. The amount of nitrate that is lost to the atmosphere by denitrification is affected by factors that contribute to anaerobic conditions and by the metabolic rate of the denitrifying microorganisms. The factors are the following:
    a. Soil moisture: Saturated soils have fewer pore spaces occupied by oxygen, thus creating anaerobic conditions that favor the growth of denitrifying microorganisms.
    b. Soil type: Fine-textured soils are more likely to become anaerobic than coarse-textured soils, thus increasing the potential for denitrification even when soil is not saturated.
    c. Carbon source: An abundant source of readily oxidizable carbon will increase the denitrification rate.
    d. Nitrate levels: Denitrification will occur rapidly where nitrate levels provide a sufficient source of nitrogen for the microorganisms.

## 10.8. Converting Dry Tons of Biosolids per Acre to Pound of Nutrient per Acre

The following equations are used to convert dry ton of biosolids per acre to pound of nutrient per acre when percent content of a nutrient is known:

$$(LB_{organic})/acre = (D_{ton}/acre)(PC_{organic})(20) \tag{17}$$
$$(LB_{nitrate})/acre = (D_{ton}/acre)(PC_{nitrate})(20) \tag{18}$$
$$(LB_{nitrite})/acre = (D_{ton}/acre)(PC_{nitrite})(20) \tag{19}$$
$$(LB_{ammonium})/acre = (D_{ton}/acre)(PC_{ammonium})(20) \tag{20}$$
$$(LB_{TKN})/acre = (D_{ton}/acre)(PC_{TKN})(20) \tag{21}$$
$$(LB_P)/acre = (D_{ton}/acre)(PC_P)(20) \tag{22}$$
$$(LB_K)/acre = (D_{ton}/acre)(PC_K)(20) \tag{23}$$

where

$LB_{organic}$ = weight of organic nitrogen, lb
$LB_{nitrate}$ = weight of nitrate nitrogen, lb
$LB_{nitrite}$ = weight of nitrite nitrogen, lb
$LB_{ammonium}$ = weight of ammonium nitrogen, lb
$LB_{TKN}$ = weight of total Kjeldahl nitrogen, lb
$LB_P$ = weight of phosphorus, lb
$LB_K$ = weight of potassium, lb
$D_{ton}$ = dry tons of biosolids applied
$PC_{organic}$ = Percent of organic nitrogen, %
$PC_{nitrate}$ = Percent of nitrate nitrogen, %
$PC_{nitrite}$ = Percent of nitrite nitrogen, %
$PC_{ammonium}$ = Percent of ammonium nitrogen, %
$PC_{TKN}$ = Percent of total Kjeldah nitrogen, %
$PC_P$ = Percent of phosphorus, %
$PC_K$ = Percent of potassium, %
20 = a conversion factor (1% = 20 lb/ton = 20 lb/2000 lb)

Total Kjeldahl nitrogen (TKN) is the summation of ammonium nitrogen ($NH_4^+$-N) and organic nitrogen (organic-N).

For the purpose of illustration, determine the biosolids organic nitrogen applied, $(LB_{organic})$/acre, in the year of 2007, if the total biosolids applied = 2.5 dry ton/acre, and biosolids organic nitrogen content = 4.5%.

*Solution*

$$(LB_{organic})/acre = (D_{ton}/acre)(PC_{organic})(20)$$
$$= (2.5 \text{ dry tons/acre})(4.5)(20)$$
$$= 225 \text{ lb organic-N/acre}$$

## 10.9. Converting Percent Content to Pound per Dry Ton

The following equations can be used for converting percent content to pound per dry ton using the conversion factor of 1% = 20 lb/dry ton:

$$N_{organic} = (PC_{organic})(20) \quad (24)$$
$$N_{nitrate} = (PC_{nitrate})(20) \quad (25)$$
$$N_{ammonium} = (PC_{ammonium})(20) \quad (26)$$
$$N_{nitrite} = (PC_{nitrite})(20) \quad (27)$$
$$N_{total} = (PC_{total})(20) \quad (28)$$
$$N_{TKN} = (PC_{TKN})(20) \quad (28a)$$

where

20 = a conversion factor (% = 20 lb/ton = 20 lb/2000 lb)

Explain how the above equations are derived. Assuming aerobically digested biosolids contain 4% organic nitrogen; determine the amount of organic nitrogen (N) per dry ton of biosolids.

*Solution*

$$20 \text{ lb/dry ton} = (20 \text{ lb})/(\text{dry } 2000 \text{ lb}) = 1/100 = 1\%$$
$$PC_{organic} = 4\%$$
$$N_{organic} = (PC_{organic})(20) = (4)(20) = 80 \text{ lb organic N/dry ton}$$

## 10.10. Calculating Net Primary Nutrient Crop Need

The anticipated wheat yield for a field consisting of primarily Basher soils is 60 bushels per acre (Table 7.7), with a unit nitrogen fertilizer rate (UNFR) of 1.3 lb of nitrogen per bushel (Table 7.8). Determine the resultant crop nitrogen fertilizer rate (CNFR), in lb N/acre.

*Solution*

Equation 1 is used for calculating CNFR:

$$CNFR = (Yield)(UNFR) \quad (1)$$

where
  Yield = crop yield = 60 bu/acre harvested from Table 7.7
  UNFR = unit nitrogen fertilizer rate = 1.3 lb N/bu crop yield, as shown in Table 7.8
  CNFR = crop nitrogen fertilizer rate

$$\text{CNFR} = (60\,\text{bu/acre})\,(1.3\,\text{lb N/bu}) = 78\,\text{lb N/acre}$$

### 10.11. Calculating the Components of Plant Available Nitrogen in Biosolids

For all practical purposes, total nitrogen (TN) in soil or nutrient-containing materials is the summation of *ammonium nitrogen* ($NH_4^+$-N), *nitrate nitrogen* ($NO_3^-$-N), and organic nitrogen because nitrite nitrogen occurs in negligible amount. Crops directly utilize nitrogen in its inorganic forms, principally nitrate-N and ammonium-N.

Analytical methods (39) determine only total Kjeldahl nitrogen (TKN), ammonium nitrogen ($NH_4^+$-N), nitrate nitrogen ($NO_3^-$-N), and nitrite nitrogen ($NO_2^-$-N). Organic nitrogen (organic-N) can be calculated by subtracting ammonium N from TKN as in Equation 29c, below. The following equations are used to calculate plant available N:

$$TN = (NH_4^+\text{-}N) + (NO_3^-\text{-}N) + (NO_2^-\text{-}N) + (\text{Organic-}N) \tag{29}$$
$$TN = TKN + (NO_3^-\text{-}N) + (NO_2^-\text{-}N) \tag{29a}$$
$$TKN = (NH_4^+\text{-}N) + (\text{Organic-}N) \tag{29b}$$
$$(\text{Organic-}N) = TKN - (NH_4^+\text{-}N) \tag{29c}$$

$$N_{\text{total}} = N_{\text{ammonium}} + N_{\text{nitrate}} + N_{\text{nitrite}} + N_{\text{organic}} \tag{4}$$
$$N_{\text{TKN}} = N_{\text{ammonium}} + N_{\text{organic}} \tag{30}$$
$$N_{\text{organic}} = N_{\text{TKN}} - N_{\text{ammonium}} \tag{30a}$$

where
  TN = total nitrogen, mg/kg, %, or lb/ton
  TKN = total Kjeldahl nitrogen, mg/kg, %, or lb/ton
  $NH_4^+$-N = ammonium nitrogen, mg/kg, %, or lb/ton
  $NO_3^-$-N = nitrate nitrogen, mg/kg, %, or lb/ton
  $NO_2^-$-N = nitrite nitrogen, mg/kg, %, or lb/ton
  Organic-N = organic nitrogen, mg/kg, %, or lb/ton

The units of $N_{\text{total}}$, $N_{\text{TKN}}$, $N_{\text{ammonium}}$, $N_{\text{nitrate}}$, $N_{\text{nitrite}}$, and $N_{\text{organic}}$ are all lb N per ton of sludge.

The following is an example to illustrate how the components of PAN in biosolids can be calculated using the above equations. Lime stabilized biosolids have a nitrate nitrogen concentration of 1000 mg/kg, an ammonium nitrogen concentration of 2000 mg/kg, a nitrite nitrogen concentration of 0 mg/kg, and a total Kjeldahl nitrogen concentration of 27,000 mg/kg, all on a dry weight basis. The biosolids contain 17.6% dry solids. Determine the components of PAN in lb N/ton of dry biosolids.

*Engineering and Management of Agricultural Land Application*

*Solution*

Since nitrate nitrogen concentration, ammonium nitrogen concentration, nitrite nitrogen concentration, and TKN concentration are 1000 mg/kg, 2000 mg/kg, 0 mg/kg, and 27,000 mg/kg, respectively, then $PC_{nitrate}$, $PC_{ammonium}$, $PC_{nitrite}$, and $PC_{TKN}$ are 0.1%, 0.2%, 0%, and 2.7%, respectively. Equations 25 to 28a and Equation 4 are used for the following calculations:

$$N_{nitrate} = (PC_{nitrate})(20) = (0.1)(20) = 2 \text{ lb N/ton} \quad (25)$$
$$N_{ammonium} = (PC_{ammonium})(20) = (0.2)(20) = 4 \text{ lb N/ton} \quad (26)$$
$$N_{nitrite} = (PC_{nitrite})(20) = (0)(20) = 0 \text{ lb N/ton} \quad (27)$$
$$N_{TKN} = (PC_{TKN})(20) = (2.7)(20) = 54 \text{ lb N/ton} \quad (28a)$$
$$N_{TKN} = N_{ammonium} + N_{organic} \quad (4)$$
$$N_{organic} = N_{TKN} - N_{ammonium} = 54 - 4 = 50 \text{ lb N/ton}$$

## 10.12. Calculating the First Year $PAN_{0-1}$ from Biosolids

### 10.12.1. Determining the First-Year $PAN_{0-1}$ from Lime-Stabilized Biosolids

Lime stabilized sludge (i.e., biosolids) has a nitrate nitrogen concentration of 1000 mg/kg, an ammonium nitrogen concentration of 2000 mg/kg, a nitrite nitrogen concentration of 0 mg/kg, and a total Kjeldahl nitrogen (TKN) concentration of 27,000 mg/kg, all on a dry weight basis. The calcium carbonate equivalent (CCE) and pH of the lime stabilized biosolids are 40% and >10, respectively. The biosolids contain 17.6% dry solids. The biosolids will be surface-applied and disked into the soil within 24 hours in the mid-Atlantic region of the U.S. Determine the first year $PAN_{0-1}$ from the lime stabilized biosolids.

*Solution*

From the design example in Section 10.11, the following parameters have been calculated:

$N_{ammonium}$ = ammonium nitrogen content in biosolids
= 4 lb ammonium-N/ton of biosolids
$N_{nitrate}$ = nitrate nitrogen content in biosolids = 2 lb nitrate-N/ton of biosolids
$N_{organic}$ = organic nitrogen content in biosolids = 50 lb organic-N/ton of biosolids

The $PAN_{0-1}$ (first year plant available nitrogen in biosolids, lb N/ton of biosolids) can be calculated using Equation 4 when the ammonium nitrogen volatilization factor ($K_v$) and the biosolids first year organic nitrogen mineralization factor ($F_{0-1}$) are also known. Here,

$K_V$ = ammonium-N volatilization factor, based on the method of land application
= 75% (from Table 7.12).
$F_{0-1}$ = biosolids first year organic-N mineralization factor based on
the method of sludge treatment = 30% (from Table 7.9).

$$PAN_{0-1} = (K_V)(N_{ammonium}) + N_{nitrate} + (F_{0-1})N_{organic} \quad (4)$$
$$= (0.75)(4 \text{ lb N/ton}) + (2 \text{ lb N/ton}) + (0.3)(50 \text{ lb N/ton}) = 20 \text{ lb N/ton}$$

## 10.12.2. Determining the First-Year $PAN_{0-1}$ from Aerobically Digested Biosolids

An aerobically digested biosolids has a nitrate nitrogen concentration of 0%, an ammonium nitrogen concentration of 1%, a nitrite nitrogen concentration of 0%, and a TKN concentration of 5%, all on a dry weight basis. The biosolids will be surface-applied as a liquid. Determine the first year $PAN_{0-1}$ from the aerobically digested biosolids.

*Solution*

Step 1. Determine the amount of nitrogen per dry ton of biosolids.

$$N_{nitrate} = (PC_{nitrateN})(20) = (0)(20) = 0 \text{ lb N/ton}$$
$$N_{ammonium} = (PC_{ammoniumN})(20) = (1)(20) = 20 \text{ lb N/ton}$$
$$N_{nitrite} = (PC_{nitriteN})(20) = (0)(20) = 0 \text{ lb N/ton}$$
$$N_{TKN} = (PC_{TKN})(20) = (5)(20) = 100 \text{ lb N/ton}$$
$$N_{organic} = N_{TKN} - N_{ammonium} = 100 - 20 = 80 \text{ lb N/ton}$$

Step 2. Determine the first year PAN from biosolids using Equation 4.

$K_V$ = ammonium-N volatilization factor, based on the method of land application = 50% = 0.5 (from Table 7.12)

Assume that the biosolids pH is <10, and incorporation into soil is after 7 days.

$F_{0-1}$ = biosolids first year organic-N mineralization factor based on the method of sludge treatment = 30% = 0.3 (Table 7.9)

$$PAN_{0-1} = (K_V)(N_{ammonium}) + N_{nitrate} + (F_{0-1})N_{organic}$$
$$= (0.5)(20 \text{ lb N/ton}) + (0 \text{ lb N/ton}) + (0.3)(80 \text{ lb N/ton}) = 34 \text{ lb N/ton}$$

## 10.13. Calculating Biosolids Carryover Plant Available Nitrogen

### 10.13.1. Single Previous Biosolids Application

The PANA (i.e., mainly organic-N applied per acre) carryover from past biosolids applications can be calculated using Equation 3 and Table 7.10. The following information is given for a biosolids land application operation: (a) aerobically digested biosolids; (b) applied 2 years ago; (c) biosolids application rate = 5.1 dry tons per acre; and (d) organic nitrogen content = 4.0%. Determine the biosolids PANA from previous biosolids application using the $K_m$ factor in Table 7.10.

*Solution*

$A$ = biosolids application per acre in previous years = 5.1 dry ton/acre
Organic-N = organic nitrogen concentration in sludge = 4.0%
$K_m$ = biosolids crop year organic-N mineralization factor based on the method of sludge treatment
= 0.9 lb/ton/% (see Table 7.10 for crop year = CY = 2 to 3 years)

$PANA$ = crop year biosolids PAN applied in previous years
= the amount of biosolids N carry-over available
= $A(K_m)$ (Organic-N) = (5.1)(0.9)(4.0) = 18.4 lb N/acre

It should be noted that Table 7.19 can also be used for determination of PANA. For instance, when CY = 2 to 3 years, Organic-N = 4%, the annual nitrogen available during the crop year = (5.8 + 3.1)/2 = 4.45 lb N/ton. Therefore, PANA = (5.1 ton/acre) (4.45 lb N/ton) = 22.7 lb N/acre, which is close to 18.4 lb N/acre.

### 10.13.2. Multiple Previous Biosolids Applications

Past biosolids applications are as follows:

| Data from records | Previous year, 2006 | Two years ago, 2005 | Three years ago, 2004 |
|---|---|---|---|
| Total biosolids applied (dry ton/acre) | 2.5 | 3.0 | 4.0 |
| Biosolids Org-N content (%) | 4.3 | 4.6 | 4.9 |
| Biosolids Org-N applied* (lb N/acre) | 215 | 276 | 392 |

*Biosolids Org-N lb/acre = (dry ton/acre) × (Org-N%) × 20.

Determine the PANA from past biosolids applications using the biosolids organic-N mineralization $K_m$ shortcut factor method (12).

*Solution*

The calculations and the answer are presented in Table 7.25.

### 10.14. Calculating Nitrogen-Based Agronomic Rate

The following technical information is given: (a) Planned crop = corn, silage; (b) predominant soil series = Allenwood; (c) target crop yield (based on soil productivity group) = 25 ton/acre; (d) unit nitrogen fertilizer rate (UNFR) = 7.0 lb N/ton yield; (e) previous year legume crop and yield = soybeans with 45 bu/acre; (f) starter fertilizer usage = 100 lb/acre of 11-52-0; (g) historical manure usage information: manure type = dairy; frequency of application = 5 out of last 10 years; typical manure application rate = 10 wet ton/acre; typical manure-N content = 10 lb N/wet ton; (h) past biosolids applications:

| Data from records | Previous year, 2006 | Two years ago, 2005 | Three years ago, 2004 |
|---|---|---|---|
| Total biosolids applied (dry ton/acre) | 2.5 | 3.0 | 4.0 |
| Biosolids Org-N content (%) | 4.3 | 4.6 | 4.9 |
| Biosolids Org-N applied (lb N/acre) | 215 | 276 | 392 |

Biosolids characteristics are given: (a) biosolids stabilization method = aerobic digestion; (b) biosolids application method = dewatered and surface applied; (c) biosolids organic-N content = 4.9%; (d) biosolids ammonium-N content = 0.1%; (e) biosolids nitrate-N content = 0.0%; (f) biosolids solids content = 20%.

**Table 7.25**
**Biosolids organic-N mineralization $K_m$ "shortcut" factor method for determination of PAN applied (PANA) from past biosolids applications**

| | T | U | V | W | X | Y | Z |
|---|---|---|---|---|---|---|---|
| | Crop year (CY) | No. of years prior to plan yr | Equivalent years since application | $K_m$ factor (from Table 7.10) | Total applied biosolids (dry ton/acre) | Biosolids organic-N content (%) | Mineralized PANA from past biosolids (lb/acre) |
| 1 | (CY) 2004 | 3 | 3–4 | $K_m$ (3–4) value 0.42 | 4.0 | 4.9 | W1 × X1 × Y1 = 8.2 = Z1 |
| 2 | (CY) 2005 | 2 | 2–3 | $K_m$ (2–3) value 0.90 | 3.0 | 4.6 | W2 × X2 × Y2 = 12.4 = Z2 |
| 3 | (CY) 2006 | 1 | 1–2 | $K_m$ (1–2) value 2.1 | 2.5 | 4.3 | W3 × X3 × Y3 = 22.6 = Z3 |
| 4 | Plan CY 2007 | | | | | Total | Z1 + Z2 + Z3 = 43.2 = Z4 |

Total carryover N from biosolids applied in three previous years equals (cell Z4).
*Source*: National Biosolids Partnership (12).

### Table 7.26
### Crop nitrogen fertilizer rate (CNFR) and crop nitrogen deficit (CND) determinations

| | Units | Enter value |
|---|---|---|
| Step 1: CNFR determination | | |
| 1a. Planned crop | Crop name | Corn, silage |
| 1b. Predominant soil series (optional) | Soil series name | Allenwood |
| 1c. Soil productivity group (optional) | Soil group no. | Group I (Table 7.7) |
| 1d. Target crop yield | bu/acre or ton/acre | 25 ton/acre (Table 7.7) |
| 1e. Unit nitrogen fertilizer rate (UNFR) | lb N/acre | 7.0 (Table 7.8) |
| 1f. Crop nitrogen fertilizer rate (CNFR) | lb N/acre | 175 (1d × 1e) |
| Step 2: CND determination | Units | Enter value |
| 2a. PAN from legumes | lb N/acre | 45 (Table 7.25) |
| 2b. PAN from conventional fertilizers | lb N/acre | 11 = 100 × 11% |
| 2c. PAN from recent or panned livestock manure applications | lb N/acre | 0 |
| 2d. PAN from historical livestock manure applications | lb N/acre | 15 (given) |
| 2e. PAN from past biosolids applications (PANA) | lb N/acre | 43.2 (Table 7.25) |
| 2f. PAN from other sources | lb N/acre | 0 |
| 2g. Total PAN (PANT from above) | lb N/acre | 114.2 (sum of above) |
| 2h. Crop nitrogen deficit (CND) | lb N/acre | 175 − 114.2 = 61 |

*Source*: National Biosolids Partnership (12).

Determine the nitrogen-based agronomic rate (AR) given the above conditions.

*Solution*

The four steps outlined below and in Tables 7.26 and 7.27 for calculating the agronomic rate (AR) of agricultural land application of biosolids are established by the National Biosolids Partnership (NBP) (12). This important design example illustrates how the nutrient management of biosolids and septage can be scientifically achieved using the US EPA method (1, 7, 11). It is suggested by the NBP that the US EPA method be adopted and tailored for local cropping practices, and the regulatory requirements in each state (2–5, 12, 29, 41). The following are the four steps of tabulated calculations:

Step 1. Calculating CNFR (Table 7.26)
Step 2. Calculating CND (Table 7.26)
Step 3. Calculating biosolids PAN (Table 7.27)
Step 4. Calculating AR (Table 7.27)

In step 4, the calculated AR is 2 dry ton/acre (equivalent to 10 wet ton/acre), as demonstrated in the AR calculation in Table 7.27. The AR in dry tons per acre is converted to wet tons per acre by dividing dry ton/acre by the solids content (in decimal form). In this case, 2 dry ton/acre is equivalent to 10 wet ton/acre when the solids given content is 20%.

**Table 7.27**
**Biosolids PAN and agronomic determinations**

| Step 3: biosolids PAN determination | Units | Enter value |
|---|---|---|
| 3a. Biosolids stabilization method | | Aerobic |
| 3b. Biosolids application method | | Surface |
| 3c. Biosolids organic-N PAN when $F_{0-1} = 0.3$ (Table 7.9) | lb N/dry ton | 29.4 (= 4.9 × 20 × 0.3) |
| 3d. Biosolids ammonium-N, PAN, when $Kv = 0.5$ (Table 7.12) | lb N/dry ton | 1.0 (= 0.1 × 20 × 0.5) |
| 3e. Biosolids nitrate-N, PAN | lb N/dry ton | 0.0 (= 0.0 × 20) |
| 3f. Total biosolids $PAN_{0-1}$ | lb N/dry ton | 30.4 (= 3c + 3d + 3e) |
| Step 4: calculate AR | | |
| 4a. Agronomic rate (AR) | Dry ton/acre | 2.0 (= 2h/3f) |
| | Wet ton/acre | 10.0 (= 2/20%) |

*Source*: National Biosolids Partnership (12).

## 10.15. Calculating the Required Land for Biosolids Application

The recommended minimum amount of nitrogen needed by a typical corn crop to be grown in the state of New Jersey is 120 lb/acre/yr. Biosolids containing 3% nitrogen could be applied at up to 5.4 dry tons/acre if used to supply all the nitrogen needed by the crop (i.e., no other nitrogen fertilizers used). A city in New Jersey produces 10 dry tons of biosolids a day. Determine approximately the area of cornfield that will be needed for the agricultural land application.

*Solution*

    1% = 20 lb/ton
    3% nitrogen content in biosolids = 3(20 lb/ton) = 60 lb N/ton biosolids
    Minimum PAN required for corn crop = 120 lb N/acre/yr
    Minimum agricultural land application rate = (120 lb N/acre/yr)/(60lb N/ton)
        = 2 ton/acre/yr
    Maximum allowable land application rate = 5.4 dry ton/acre/yr
    Biosolids production rate = 10 dry ton/d
    Minimum area of land required for biosolids application = (10 ton/d)/
        (5.4 ton/acre/365 d) = 676 acres of corn field

## 10.16. Calculating the Nitrogen-Based and the Phosphorus-Based Agronomic Rates for Agricultural Land Application

Applying biosolids to meet the phosphorus, rather than the nitrogen, needs of the crop is a conservative approach for determining annual biosolids application rates (1, 2–5). Supplemental nitrogen fertilization will be needed to optimize crop yields (except for nitrogen-fixing legumes) if biosolids application rates are based on a crop's phosphorus needs. The phosphorus in biosolids is estimated to be about half as available for plant uptake as the phosphorus normally applied to soils in commercial inorganic fertilizers.

# Engineering and Management of Agricultural Land Application

The phosphorus balance and the phosphorus-based agronomic rate of biosolids for land application can be determined by Equations 16 and 31, respectively.

$$(AR_P) = (P_{required})/[(P_{content})F] \tag{31}$$

where

$AR_P$ = phosphorus-based agronomic rate, dry ton/acre
$P_{content}$ = phosphorus content in biosolids, lb P/ton biosolids (or lb $P_2O_5$/ton biosolids)
$P_{required}$ = phosphorus requirement for the land, lb P/acre (or lb $P_2O_5$/acre)
$F$ = a factor of availability = 0.5 (assuming 50% will be available)

It should be noted that Equation 31 is another version of Equation 16 assuming $P_{balance} = 0$ and $F = 0.5$ for practical applications. The units of $P_{content}$ and $P_{required}$ in the above equation must be compatible (i.e., both based on P, or both based on $P_2O_5$).

$P_{required}$ is the phosphorus fertilizer recommendation for the harvested crop, or the quantity of phosphorus removed by the crop. It is assumed that only 50% of $P_{content}$ will be available as the plant available phosphorus (PAP). Two conversion factors are used in determining the phosphorus-based agronomic rate:

1% on dry basis = 20 lb/ton = 10 kg/T (where 1 T = 1 mt = 1 metric ton = 1000 kg)
1 lb P = 2.3 lb $P_2O_5$
1 kg P = 2.3 kg $P_2O_5$

An example prepared by G.K. Evanylo of Virginia Tech (4) is presented here for showing how the phosphorus-based AR can be calculated for agricultural land application.

Lime-stabilized biosolids have a pH >10, a calcium carbonate equivalent (CCE) of 40%, a nitrate nitrogen concentration of 1000 mg/kg (0.1%), an ammonium nitrogen concentration of 2000 mg/kg (0.2%), a total nitrogen concentration of 27,000 mg/kg (2.7%), and a total phosphorus concentration of 21,000 mg/kg (2.1%), all on a dry weight basis (percent dry solids is 17.6%). Corn for grain is to be grown on a Kempsville sandy loam soil that has a pH of 6.2, high Ca, Mg and K soil test ratings, and a low P soil test rating. The biosolids will be surface-applied and disked into the soil within 24 hours.

What should be the nitrogen-based agronomic rate and the phosphorus-based agronomic rate of the lime-stabilized biosolids? (40)

*Solution*

STEP 1. NITROGEN-BASED AGRONOMIC RATE

The estimated yield potential of corn grown on a Basher soil is 120 bu/acre (41) and the N rate permitted is 110 to 130 lb/acre according to Tables 7.7 and 7.13, which is adopted from Virginia Biosolids Use Regulations (26).

The nitrogen components of PAN in the biosolids have been calculated in previous section as follows:

$$PAN_{0-1} = (K_V)(N_{ammonium}) + N_{nitrate} + (F_{0-1})N_{organic}$$
$$= (0.75)(4 \text{ lb N/ton}) + (2 \text{ lb N/ton}) + (0.3)(50 \text{ lb N/ton}) = 20 \text{ lb N/ton}$$

The nitrogen-based agronomic rate (7 dry ton/acre) is obtained by dividing the adjusted fertilizer nitrogen rate (140 lb N/acre) by the calculated $PAN_{0-1}$ (20 lb N/dry ton).

$$AR = \text{nitrogen-based agronomic rate}$$
$$= (140 \text{ lbs N/acre})/(20 \text{ lb N/dry ton}) = 7 \text{ dry ton/acre}$$

STEP 2. PHOSPHORUS-BASED AGRONOMIC RATE

$$P_{content} = \text{phosphorus content in biosolids} = 2.1\%$$
$$= 2.1 \times 20 \times 2.3 \text{ lb } P_2O_5/\text{ton biosolids}$$
$$= 96.6 \text{ lb } P_2O_5/\text{ton biosolids}$$
$$F = \text{factor of availability} = 0.5 \text{ (assuming 50\% will be available)}$$
$$P_{required} = \text{phosphorus requirement for the land} = 120 \text{ lb } P_2O_5/\text{acre}$$

Note: A given local site-specific data (41):

$$(AR_P) = \text{phosphorus-based agronomic rate, dry ton/acre}$$
$$= (P_{required}/[(P_{content})F] = (120)/[(96.6)0.5]$$
$$= 2.5 \text{ dry ton/acre (Note: It is about one third of the nitrogen-based AR)}$$

## 10.17. Calculating the Lime-Based Agronomic Rate for Agricultural Land Application

Application rates for lime-stabilized or -conditioned biosolids may be computed by determining the biosolids calcium carbonate equivalent (CCE) (4, 42–44). The CCE provides a direct comparison of the liming value of the biosolids with calcium carbonate limestone, which is the basis for soil testing liming requirements. Biosolids conditioned or stabilized with lime may have a CCE of between 10% and 50% on a dry weight basis. The agronomic lime rate for biosolids is determined from Equation 32:

$$(AR_L) = (L_{required})/(L_{content}) \tag{32}$$

where

$AR_L$ = lime-based agronomic rate, dry ton/acre

$L_{content}$ = lime content in sludge in terms of CCE, % (in decimal form; for instance, 40% = 0.4)

$L_{required}$ = lime requirement for the land in terms of CCE, ton CCE/acre

Determine the lime-based agronomic rate for the same lime-stabilized biosolids introduced previously.

*Solution*

It is known that the CCE of the lime stabilized biosolids is 40%. The coarse-textured Kempsville soil is permitted 0.75 tons limestone/acre according to the Virginia Biosolids

Use Regulations (26). Thus, the rate of lime-stabilized biosolids to provide 0.75 tons CCE/acre can be calculated by Equation 32 as follows:

$L_{content}$ = lime content in biosolids in terms of CCE = 40% = 0.4
$L_{required}$ = lime requirement for the land in terms of CCE = 0.75 ton CCE/acre
$(AR_L)$ = lime-based agronomic rate, dry ton/acre
   = $(L_{required})/(L_{content})$ = (0.75)/(0.4) = 1.9 tons biosolids/acre

In summary, the N-based, P-based, and lime-based agronomic rates for the examples presented in Section 10.17 and this section are 7, 2.5, and 1.9 dry tons/acre, respectively. The most limiting is the lime-based agronomic rate; thus, 1.9 dry tons or 10.8 wet tons (1.9 dry ton/acre divided by 0.176 dry ton/wet ton) should be the appropriate agronomic rate.

## 10.18. Calculating Potassium Fertilizer Needs

The amounts of plant-available nitrogen, phosphorus, potassium, and other nutrients added by biosolids should be calculated once the design application rate ($BAR_{design}$) has been determined. Supplemental fertilizers should be applied if the amount of any nutrients in the biosolids is less than that recommended (1, 4, 7).

The amount of potassium applied in biosolids can be calculated from biosolids composition data (as done earlier for P) according to Equation 17:

$$K_{balance} = (BAR_{design}) \times (K_{content})F - K_{required} \qquad (17)$$

All of the potassium in biosolids can be assumed to be readily plant-available because potassium is a soluble element. The availability factor $F$ is assumed to be 100% or 1.0 for potassium in biosolids.

In case the nitrogen-based agronomic rate (AR) has been chosen for agricultural land application of biosolids, then $BAR_{design} = AR$. $K_{balance}$ will be a negative value showing the needed lb K/acre (or lb $K_2O$/acre).

It has been known that the nitrogen-based agronomic rate ($AR$ = 7 dry ton/acre) is to be chosen to be for actual operation of land application. A biosolids containing 0.52% K is to be applied to a wheat field, which has a potassium fertilizer recommendation of 135 lb $K_2O$/acre. Determine the additional $K_2O$ needed (i.e., a negative $K_{balance}$ value) for the wheat field.

*Solution*

$BAR_{design}$ = biosolids application rate selected for design = nitrogen-based $AR$
   = 7 dry ton biosolids/acre
$K_{content}$ = potassium content in sludge = 0.52% K = 0.52 × 20 lb K/ton biosolids
   = 0.52 × 20 × 1.2 lb $K_2O$/ton dry biosolids
   = 12.48 lb $K_2O$/ton dry biosolids

$K_{\text{required}}$ = potassium requirement for the wheat field = 135 lb K$_2$O/acre
$F$ = availability factor ($F = 1.0$ for potassium in biosolids)
$K_{\text{balance}}$ = positive value shows the excess lb K$_2$O/acre;
　　　negative value shows the needed lb K$_2$O/acre
　　= $(BAR_{\text{design}}) \times (K_{\text{content}})F - K_{\text{required}}$
　　= (7 dry ton /acre) × (12.48 lb K$_2$O/ton)(1.0) − 135 lb K$_2$O/acre
　　= −48 lb K$_2$O/acre (a negative value shows the needed lb K$_2$O/acre)

## 10.19. Biosolids Land Application Costs and Cost Adjustment

Figure 7.4 shows the construction cost and operation and maintenance cost of biosolids land application for municipal biological wastewater treatment plants (45). The energy required to apply biosolids to land is approximately 1.2 million BTU/dry ton biosolids or 58,000 BTU/wet ton biosolids at 5% solids, excluding transport of biosolids to the site. The Engineering News Record (ENR) Construction Cost Index (CCI) of the cost data in Figure 7.4 is documented to be 2476 (45). Determine the approximate current construction cost of a new land application facility for a typical 10-MGD (million gallons per day) municipal wastewater treatment plant in the U.S. The following are

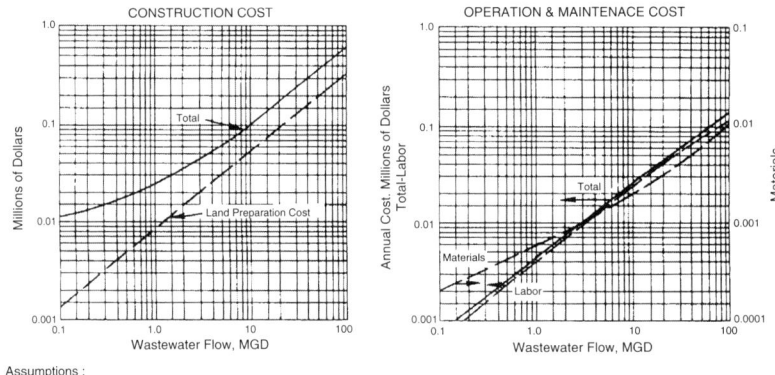

**Fig. 7.4.** Construction and operation and maintenance costs of biosolids land application. *Source*: US EPA (45).

the assumed conditions: (a) the municipal wastewater treatment plant (WWTP) uses anaerobic digestion producing 900 lb of 4% biosolids per MG (million gallons) of wastewater; (b) the biosolids application rate is 11 dry ton/acre/yr; (c) the design flow of the WWTP ($Q_{DESIGN}$) is 10 MGD; (d) the current month is April 2007; and (e) the service lives of the land application infrastructure and the land application equipment are 30 years and 10 years, respectively. Also determine the number of monitoring wells that will be required for the new land application infrastructure.

*Solution*

To use the cost data presented in Figure 7.4, an agricultural or environmental engineer must first calculate the effective wastewater flow ($Q_E$) using Equation 33.

$$Q_E = Q_{design}(10 \text{ ton/acre/yr})/(AR_{design}) \tag{33}$$

where

$Q_E$ = effective wastewater flow, MGD
$Q_{design}$ = municipal WWTP design wastewater flow = 10 MGD
$BAR_{design}$ = biosolids application rate selected for design = 11 ton biosolids/acre/yr

Therefore,

$$Q_E = Q_{design}(10 \text{ ton/acre/yr})/(BAR_{design}) = (10)(10)/(11) = 9.1 \text{ MGD}$$

The construction cost curves in Figure 7.4 will yield the following cost data: (a) $50,000 for the land preparation cost (ENR CCI = 2476); and (b) $90,000 for the total construction cost (ENR CCI = 2476)

The ENR publishes both a CCI and building cost index (BCI) that are widely used in the construction industry. Its Web site contains an explanation of the indexes methodology and a complete history of the 20-city national average for the CCI and BCI in the U.S. Both indexes have materials and labor component. According to the April 2007 issue of ENR (46), the CCI in April 2007 is 7864.70. Equations 34 and 36 and Equations 35 and 37 can be used for estimating the current and future costs, respectively:

$$COST_{current} = COST_{past}(CCI_{current})/(CCI_{past}) \tag{34}$$
$$COST_{future} = COST_{past}(CCI_{future})/(CCI_{past}) \tag{35}$$
$$COST_{current} = COST_{past}(BCI_{current})/(BCI_{past}) \tag{36}$$
$$COST_{future} = COST_{past}(BCI_{future})/(BCI_{past}) \tag{37}$$

Using Equation 34, one can then determine the current construction costs for a new land application site:

$$COST_{current} = \text{land preparation cost}$$
$$= (\$50,000)(7864.70)/(2476) = \$158,819$$
$$COST_{current} = \text{total land application infrastructure construction cost}$$
$$= (\$90,000)(7864.70)/(2476) = \$285,874$$

Similarly all past, current, and future costs of both construction and operation and maintenance can be calculated using Equations 34 to 37 when the cost indexes of past, current, or future are known (46). It is important to note that many other cost indexes and cost models can also be adopted for the intended cost conversation or updating (40, 47–59).

Since the design capacity of the municipal WWTP is 10 MGD, eight monitoring wells will be needed according to Figure 7.4 from the US EPA.

## 11. GLOSSARY OF LAND APPLICATION TERMS

Agricultural land: Land on which a food, feed, or fiber crop is grown. This includes range land or land used as pasture.

Agronomic rate: The whole sludge application rate designed to (1) provide the amount of nitrogen needed by a crop or vegetation grown on the land and (2) minimize the amount of nitrogen in the biosolids that passes below the root zone of the crop or vegetation grown on the land to the ground water.

Annual pollutant loading rate (APLR): The maximum amount of a pollutant that can be applied to a unit area of land during a 365-day period. This term describes pollutant limits for biosolids that is given away or sold in a bag or other container for application to the land.

Annual whole sludge application rate: The maximum amount of biosolids on a dry weight basis that can be applied to a land application site during a 365-day (1-year) period.

Area of cropland: An area of cropland that has been subdivided into several strips is not a single field. Rather, each strip represents an individual field unit.

Bagged biosolids: Biosolids that are sold or given away in a bag or other container.

Biosolids: Solids, semisolids, or liquid materials resulting from biological treatment of domestic sewage that has been sufficiently processed to permit these materials to be safely land-applied. The term *biosolids* was introduced by the wastewater treatment industry in the early 1990s and has been recently adopted by the US EPA to distinguish high-quality, treated sewage sludge from raw sewage sludge and from sewage sludge containing large amounts of pollutants.

Bulk biosolids: Biosolids that are not sold or given away in a bag or other container for application to the land.

Ceiling concentration limits (CCL): The maximum concentrations of the nine trace elements allowed in biosolids to be land applied. Biosolids exceeding the ceiling concentration limit for even one of the regulated pollutants is not classified as biosolids and hence cannot be land applied.

Class I sludge management facility: Publicly owned treatment works (POTWs) required to have an approved pretreatment program under 40 CFR 403.8(a), including any POTW located in a state that has elected to assume local pretreatment program responsibilities under 40 CFR 403.10(e). In addition, the regional administrator or, in the case of approved state programs, the regional administrator in conjunction with the state director, has the discretion to designate any treatment works treating domestic sewage (TWTDS) as a Class I sludge management facility.

Crop group: Individual farm fields that are managed in the same manner, with the similar yield goals, are called a c*rop group*

Crop nitrogen deficit (CND): Anticipated crop nitrogen fertilizer rate (CNFR) minus all past plant available nitrogen (PAN) sources (PAN-past) and current planned non-biosolids PAN sources (PAN-plan), in the unit of lb N/acre. Previous biosolids carryover nitrogen is included in this calculation.

Crop nitrogen fertilizer rate (CNFR): A rate (lb N/acre) that is equal to the yield multiplied by the unit nitrogen fertilizer rate (UNFR; lb N/unit crop yield). The yield is the crop harvested or crop yield (bu/acre or ton/acre).

Crop year: The year in which a crop receiving the biosolids/manure treatment is harvested. For example, fall applications of biosolids/manure in 2000 intended to provide nutrients for a crop to be harvested in 2001 are earmarked for crop year 2001. Likewise, biosolids/manure applied immediately prior to planting winter wheat in October 2000 should be identified as fertilizer intended for crop year 2001 because the wheat will be harvested in the summer of 2001. The basic time management unit is also referred to as the *planting season*.

Crop yield: The crop harvested in the unit of bu/acre or ton/acre.

Cumulative pollutant loading rate (CPLR): The total amount of pollutant that can be applied to a site in its lifetime by all bulk biosolids applications meeting the CCL. It is the maximum amount of an inorganic pollutant that can be applied to an area of land. This term applies to bulk biosolids that is land applied.

Domestic septage: Either a liquid or solid material removed from a septic tank, cesspool, portable toilet, type III marine sanitation device, or similar treatment works that receives only domestic sewage. This does not include septage resulting from treatment of wastewater with a commercial or industrial component.

Exceptional quality (EQ) biosolids: Biosolids that meet the most stringent limits for the three biosolids quality parameters. In gauging biosolids quality, the US EPA determined that three main parameters of concern should be considered: (a) pollutant levels; (b) the relative presence or absence of pathogenic organisms, such as *Salmonella* and *Escherichia coli* bacteria, enteric viruses, or viable helminth ova; and (c) the degree of attractiveness of the sewage sludge to vectors, such as flies, rats, and mosquitoes, that could potentially come in contact with pathogenic organisms and spread disease. Given these three variables, there can be a number of possible biosolids qualities. *Exceptional quality*, a term that does not appear in the Part 503 regulation, is used to describe biosolids that meet the highest quality for all three of these biosolids quality parameters (i.e., ceiling concentrations and pollutant concentrations in 503.13 for metals, one of the Class A pathogen reduction alternatives, and one of the sewage sludge processing vector attraction reduction options 1 through 8).

Farm field: The basic management unit used for all farm nutrient management, defined as "the fundamental unit used for cropping agricultural products."

Feed crops: Crops produced primarily for consumption by animals. These include, but are not limited to, corn and grass. For a crop to be considered a feed crop, it has to be produced for consumption by animals (e.g., grass grown to prevent erosion or to stabilize an area is not considered a feed crop).

Fiber crops: Crops, such as flax and cotton, that were included in Part 503 because products from these crops (e.g., cotton seed oil) may be consumed by humans.

Food crops: Crops consumed by humans. These include, but are not limited to, fruits, grains, and vegetables.

Forest land: Tract of land thick with trees and underbrush.

Heavy metals: Trace elements are found in low concentrations in biosolids. The trace elements of interest in biosolids are those commonly referred to as heavy metals. Some of these trace elements (e.g., copper, molybdenum, and zinc) are nutrients needed for plant growth in low concentrations, but all of these elements can be toxic to humans, animals, or plants at high concentrations. Possible hazards associated with a buildup of trace elements in the soil include their potential to cause phytotoxicity (i.e., injury to plants) or to increase the concentration of potentially hazardous substances in the food chain. Federal and state regulations have established standards for the following nine trace elements: arsenic (As), cadmium (Cd), copper (Cu), lead (Pb), mercury (Hg), molybdenum (Mo), nickel (Ni), selenium (Se), and zinc (Zn).

Indicator organism: A nonpathogenic organism (e.g., fecal coliform) whose presence implies the presence of pathogenic organisms. Indicator organisms are selected to be conservative estimates of the potential for pathogenicity.

Individual field unit: An area of cropland that has been subdivided into several strips is not a single field. Rather, each strip represents an individual field unit.

Land application: The spreading, spraying, injection, or incorporation of liquid or semiliquid organic substances, such as biosolids, livestock manure, compost, and other types of waste, onto or below the surface of the land to take advantage of the soil-enhancing qualities of the substances. These substances are land applied to improve the structure of the soil. It is also applied as a fertilizer to supply nutrients to crops and other vegetation grown in the soil. The liquid or semiliquid organic substances are commonly applied to agricultural land (including pasture and range land), forests, reclamation sites, public contact sites (e.g., parks, turf farms, highway median strips, golf courses), lawns, and home gardens.

Land application site: An area of land on which biosolids are applied to condition the soil or to fertilize crops or vegetation grown in the soil.

Manure: Any wastes discharged from livestock

Mineralization: The conversion of organic-N to inorganic-N forms. Most nitrogen exists in biosolids/manure as organic-N, principally contained in proteins, nucleic acids, amines, and other cellular material. These complex molecules must be broken apart through biological degradation for nitrogen to become available to crops.

Nutrients: Elements required for plant growth that provide biosolids with most of their economic value. These include nitrogen (N), phosphorus (P), potassium (K), calcium (Ca), magnesium (Mg), sodium (Na), sulfur (S), boron (B), copper (Cu), iron (Fe), manganese (Mn), molybdenum (Mo), and zinc (Zn).

Pasture: Land on which animals feed directly on feed crops such as legumes, grasses, or grain stubble.

Pathogens: Disease-causing microorganisms that include bacteria, viruses, protozoa, and parasitic worms. Pathogens can present a public health hazard if they are transferred to food crops grown on land to which biosolids are applied; contained in runoff to surface waters from land application sites; or transported away from the site by vectors such as insects, rodents, and birds.

pH: A measure of the degree of acidity or alkalinity of a substance. The pH of biosolids is often raised with alkaline materials to reduce pathogen content and attraction of disease-spreading organisms (vectors). High pH (greater than 11) kills virtually all pathogens and reduces the solubility, biological availability, and mobility of most metals. Lime also increases the gaseous loss (volatilization) of the ammonia form of nitrogen (ammonia-N), thus reducing the N-fertilizer value of biosolids

Plant available nitrogen (PAN): The amount of N in the biosolids/manure that is available to crops during a specified period. Only a portion of the total nitrogen present in biosolids is available for plant uptake.

Planting and harvesting periods: A cycle, which is not the calendar year, that dictates the timing of biosolids land application activities. Winter wheat and perennial forage grasses are examples of crops that may be established and harvested in different calendar years.

Planting season: The basic time management unit; also see *crop year*.

Pollutant concentration limit (PCL): The maximum concentration of heavy metals for biosolids whose trace element pollutant additions do not require tracking (i.e., calculation of CPLR (cumulative pollutant loading rate). The PCL is the most stringent pollutant limit included in U.S. Federal Regulation Part 503 for land application. Biosolids meeting pollutant concentration limits are subject to fewer requirements than biosolids meeting ceiling concentration limits.

Preparer: The person who generates biosolids during the treatment of domestic sewage in a treatment works. Also the person who derives a material from sewage sludge.

Public contact site: Land with a high potential for contact by the public, including public parks, ball fields, cemeteries, nurseries, turf farms, and golf courses.

Range land: Open land with indigenous vegetation.

Reclamation site: Drastically disturbed land, such as strip mines and construction sites, that is reclaimed, possibly with the use of biosolids.

Septage: The liquid and solid material pumped from a septic tank, cesspool, or similar domestic sewage treatment system, or holding tank when the system is cleaned or maintained.

Sewage sludge: The solid, semisolid, or liquid residue generated during the treatment of domestic sewage in a treatment works. Sewage sludge includes, but is not limited to, domestic septage, scum, and solids removed during primary, secondary, or advanced wastewater treatment processes. The definition of sewage sludge also includes a material derived from sewage sludge (i.e., sewage sludge whose quality is changed either through further treatment or through mixing with other materials).

Total Kjeldahl nitrogen (TKN): The summation of ammonium nitrogen ($NH_4^+$-N), and organic nitrogen (organic-N).

Total nitrogen: The summation of ammonium nitrogen ($NH_4^+$-N), nitrate nitrogen ($NO_3^-$-N), nitrite nitrogen ($NO_2^-$-N), and organic nitrogen (organic-N). Usually nitrite nitrogen is in negligible amount. Crops directly utilize nitrogen in its inorganic forms, principally nitrate-N and ammonium-N.

Total solids (TS): Suspended and dissolved solids, usually expressed as the concentration present in biosolids. TS depends on the type of wastewater process and biosolids' treatment prior to land application. Typical solids contents of various biosolids are liquid (2% to 12%), dewatered (12% to 30%), and dried or composted (50%).

Trace elements: Elements found in low concentrations in biosolids. The trace elements of interest in biosolids are those commonly referred to as heavy metals.

Treatment works: Federally owned, publicly owned, or privately owned device or system used to treat, recycle, or reclaim either domestic sewage or a combination of domestic sewage and industrial waste of a liquid nature.

Treatment works treating domestic sewage: Publicly owned treatment works (POTWs) or other sewage sludge or wastewater treatment system or device, regardless of ownership, used in the storage, treatment, recycling, and reclamation of municipal or domestic sewage, including land dedicated for the disposal of sewage sludge.

Unit nitrogen fertilizer rate (UNFR): A rate in lb-N per unit crop yield, where the unit can either bushel or ton. (Note: 1 bu (U.S. bushel) = 1.2444 ft$^3$; 1 British bushel = 1.2843 ft$^3$; 1 t (British ton) = 2000 lb; 1 T (metric ton) = 1000 kg)

Vector attraction: Characteristics (e.g., odor) that attract birds, insects, and other animals that are capable of transmitting infectious agents.

Vectors: Rodents, birds, and insects that can transport pathogens away from the land application site.

Volatile solids (VS): An estimate of the readily decomposable organic matter in biosolids, usually expressed as a percentage of total solids. VS is an important determinant of potential odor problems at land application sites.

Volatilization: Ammonium-N in biosolids can be significant, making up even half the initial PAN of biosolids. The ammonium-N of biosolids can vary widely depending on treatment and storage. Since ammonium-N is prone to volatilization (as ammonia gas, $NH_3$), the application method affects PAN. For instance, surface applied biosolids are expected to lose half of their ammonium-N. Conversely, direct subsurface injection or soil incorporation of biosolids within 24 hours minimizes volatilization losses. The conversion of ammonium-N to ammonia gas form ($NH_3$) is called *volatilization*.

Yield: The crop harvested, in units of bu/acre or ton/acre.

## NOMENCLATURE

$A$ = sludge application per acre in previous years, dry ton/acre
$AR$ = agronomic rate, dry ton/acre
$AR_L$ = lime-based agronomic rate, dry ton/acre
$AR_P$ = phosphorus-based agronomic rate, dry ton/acre
$BAR_{design}$ = sludge application rate selected for design, ton sludge/acre
$CND$ = crop nitrogen deficit, lb N/acre
$CNFR$ = crop nitrogen fertilizer rate, lb N/acre

$D$ = the concentration of the pollutant in the sewage sludge on a dry weight basis, in mg/kg

$D_{ton}$ = dry tons biosolids applied

$E_m$ = the effective mineralization factor for the growing season portion of the year

$F_{0-1}$ = sludge first year organic-N mineralization factor based on the method of sludge treatment

$K_{balance}$ = positive value shows the excess lb K/acre; negative value shows the needed lb K/acre

$K_{content}$ = potassium content in sludge, lb K/ton sludge

$K_m$ = sludge crop year organic-N mineralization factor based on the method of sludge treatment, lb/ton/%

$K_{required}$ = potassium requirement on land, lb K/acre

$K_V$ = ammonium-N volatilization factor, based on the method of land application

$L_{content}$ = lime content in sludge in terms of CCE, % (in decimal form; for instance, 40% = 0.4)

$L_{required}$ = lime requirement for the land in terms of CCE, ton CCE/acre

$LB_{ammonium}$ = weight of ammonium nitrogen, lb

$LB_K$ = weight of potassium, lb

$LB_{nitrate}$ = weight of nitrate nitrogen, lb

$LB_{nitrite}$ = weight of nitrite nitrogen, lb

$LB_{organic}$ = weight of organic nitrogen, lb

$LB_P$ = weight of phosphorus, lb

$LB_{total}$ = weight of total nitrogen, lb

$LB_{TKN}$ = weight of total Kjeldahl nitrogen, lb

$NH_4^+$-N = ammonium nitrogen, mg/kg, %, or lb/ton

$NO_3^-$-N = nitrate nitrogen, mg/kg, %, or lb/ton

$NO_2^-$-N = nitrite nitrogen, mg/kg, %, or lb/ton

$N_{ammonium}$ = ammonium nitrogen content in sludge, lb ammonium-N/ton of sludge

$N_{nitrate}$ = nitrate nitrogen content in sludge, lb nitrate-N/ton of sludge

$N_{organic}$ = organic nitrogen content in sludge, lb organic-N/ton of sludge

$N_{TKN}$ = total Kjeldahl nitrogen content in sludge, lb TKN/ton of sludge

$N_{total}$ = total nitrogen content in sludge, lb N/ton of sludge

Organic-N = organic nitrogen concentration in sludge, %

PAN = plant available nitrogen

PANA = crop year biosolids PAN applied in previous years, dry ton/acre

PANS = crop year non-biosolids PAN applied in previous years, lb N/acre

PANT = crop year total PAN applied in previous years, lb N/acre

$PAN_{0-1}$ = first year plant available nitrogen in sludge, lb N/ton of sludge

$PC_{ammonium}$ = percent of ammonium nitrogen, %

$PC_K$ = Percent of potassium, %

$PC_{nitrate}$ = percent of nitrate nitrogen, %

$PC_{nitrite}$ = percent of nitrite nitrogen, %

$PC_{organic}$ = percent of organic nitrogen, %

$PC_P$ = percent of phosphorus, %

$PC_s$ = percent of solids, %

$PC_{TKN}$ = percent of total Kjeldah nitrogen, %

$P_{balance}$ = positive value shows the excess lb P/acre; negative value shows the needed lb P/acre.

$P_{content}$ = phosphorus content in sludge, lb P/ton sludge

$P_{required}$ = phosphorus requirement on land, lb P/acre

$R$ = rate of application, lb/acre

$R_{As}$ = sludge application rate based on arsenic content, ton sludge/acre

$R_{Cd}$ = sludge application rate based on cadmium content, ton sludge/acre

$R_{Cr}$ = sludge application rate based on chromium content, ton sludge/acre

$R_{Cu}$ = sludge application rate based on copper content, ton sludge/acre

$R_{Hg}$ = sludge application rate based on mercury content, ton sludge/acre

$R_{Mo}$ = sludge application rate based on molybdenum content, ton sludge/acre

$R_{Ni}$ = sludge application rate based on nickel content, ton sludge/acre

$R_{Pb}$ = sludge application rate based on lead content, ton sludge/acre

$R_{Se}$ = sludge application rate based on selenium content, ton sludge/acre

$R_{Zn}$ = sludge application rate based on zinc content, ton sludge/acre

TKN = total Kjeldahl nitrogen

TN = total nitrogen

UNFR = unit nitrogen fertilizer rate, lb N/unit crop yield

$W$ = the concentration of the pollutant in the sewage sludge on a wet basis in mg/L

Yield = crop yield, bu/acre or ton/acre harvested

## REFERENCES

1. U.S. Environmental Protection Agency, *Sludge Handling and Conditioning*. Technical Report US EPA-430/9-78-002, Washington, DC (1978).
2. G. K. Evanylo, *Agricultural Land Application of Biosolids in Virginia: Production and Characteristics of Biosolids*. Publication 452-301. Virginia Cooperative Extension, Virginia Polytechnic Institute and State University, Blacksburg, VA (1999).
3. G. K. Evanylo, *Agricultural Land Application of Biosolids in Virginia: Regulations*. Publication 452-302. Virginia Cooperative Extension, Virginia Polytechnic Institute and State University, Blacksburg, VA (1999).
4. G. K. Evanylo, *Agricultural Land Application of Biosolids in Virginia: Managing Biosolids for Agricultural Use*. Publication 452-303. Virginia Cooperative Extension, Virginia Polytechnic Institute and State University, Blacksburg, VA (1999).
5. G. K. Evanylo, *Agricultural Land Application of Biosolids in Virginia: Risks and Concerns*. Publication 452-304. Virginia Cooperative Extension, Virginia Polytechnic Institute and State University, Blacksburg, VA (1999).
6. U.S. Environmental Protection Agency, *Biosolids Technology Fact Sheet—Land Application of Biosolids*, US EPA-832-F-00-064, Washington, DC, September (2000).
7. U.S. Environmental Protection Agency, *Land Application of Sewage Sludge*. US EPA-831-B-93-002b, Washington, DC, December (1994).
8. U.S. Environmental Protection Agency, *Emerging Technologies for Biosolids Management*, US EPA-832-R-06-005, Washington, DC, September (2006).
9. U.S. Environmental Protection Agency, *NPDES Part II Standard Conditions*. Washington, DC, www.epa.gov/ne/npdes/permits/generic/partIIfinal2007.pdf (2007).

10. U.S. Environmental Protection Agency, *Use and Disposal of Municipal Wastewater Sludge*. US EPA/625/10-84/003, Cincinnati, OH (1984).
11. U.S. Environmental Protection Agency, *Process Design Manual: Land Application of Sewage Sludge and Domestic Septage*, US EPA/625/R-95/001, Washington, DC (1995).
12. U.S. Environmental Protection Agency, NACWA and WEF. Biosolids nutrient management—calculating agronomic rate of application. Chapter 13. *National Manual of Good Practice for Biosolids*. National Biosolids Partnership, Alexandria, VA, January (2005).
13. U.S. Environmental Protection Agency, *Amendments to the Standards for the Use or Disposal of Sewage Sludge (40 Code of Federal Regulations Part 503)*, Washington, DC (1995).
14. U.S. Environmental Protection Agency, *Biosolids Recycling: Beneficial Technologies for a Better Environment*. Technical Report US EPA-832-R-94-009, Washington, DC (1994).
15. U.S. Environmental Protection Agency, *Standards for the Use or Disposal of Sewage Sludge (40 Code of Federal Regulations Part 503)*, Washington, DC (1993).
16. U.S. Environmental Protection Agency, *National Pretreatment Program: Report to Congress*. Technical Report US EPA-21-W-4004, Washington, DC (1991).
17. U.S. Environmental Protection Agency, *Sewage Sludge Management Primer, Technology Transfer Series*, Cincinnati, OH (1986).
18. U.S. Environmental Protection Agency, *Environmental Regulations and Technology Use and Disposal of Municipal Wastewater Sludge*. Technical Report US EPA-625/10-84-003, Cincinnati, OH (1984).
19. U.S. Environmental Protection Agency, *Process Design Manual Land Application of Municipal Sludge*. Technical Report US EPA-625/1-83-016, Cincinnati, OH (1983).
20. U.S. Environmental Protection Agency, *Interagency Policy on Beneficial Use of Municipal Sewage Sludge*, Washington, DC (1981).
21. U.S. Environmental Protection Agency, *Technical Support Document for Reduction of Pathogens and Vector Attraction in Sewage Sludge*. US EPA 822-R-93-004, Washington, DC (1992).
22. U.S. Environmental Protection Agency, *Survival and Transport of Pathogens in Sludge-amended Soil: A Critical Review*. US EPA/600/2-87/028, Cincinnati, OH (1987).
23. U.S. Environmental Protection Agency, Guidelines for carcinogen assessment; Guidelines for estimating exposure; Guidelines for mutagenicity risk assessment; Guidelines for health assessment of suspect developmental toxicants; and Guidelines for health risk assessment of chemical mixtures. *Federal Regulations*. Vol. 51, No. 185 (1986).
24. U.S. Environmental Protection Agency, National Sewage Sludge Survey (NSSS)—availability of information and data, and anticipated impacts on proposed regulations. *Federal Regulations, 55, p. 218*, Washington, DC (1990).
25. U.S. Environmental Protection Agency, *Technical Support Document for Land Application of Sewage Sludge, Volume I*. US EPA –822-R-93-900-9, Washington, DC (1992).
26. Virginia Department of Health, *Biosolids Use Regulations*. 12VAC-5-585-10et seq.32.1-164.5 of the Code of Virginia (1997).
27. M. S. Byerly, *Land Application of DAF Sludge*. Georgia Tech Research Institute, Atlanta, GA (2007).
28. R. G. O'Dette, Determining the most cost effective option for biosolids and residuals management. *Proceedings of the 10th Annual Residuals and Biosolids Management Conference: 10 Years of Progress and a Look toward the Future*. Water Environment Federation, Alexandria, VA (1996).

29. W. E. Sopper, E. M. Seaker, and R. K. Bastian (eds.), *Land Reclamation and Biomass Production and Municipal Wastewater and Sludge*. The Pennsylvania State University Press, University Park, PA (1982).
30. Water Environment Federation, *National Outlook—State Beneficial Use of Biosolids Activities*. Washington, DC (1997).
31. B. D. Knezek, and P. H. Miller (eds.), *Applications of Sludges and Wastewaters on Agricultural Land: A Planning and Educational Guide*, North Central Regional Research Publication 235, Cooperative Extension Service, Ohio State University, Columbus, OH (1980).
32. American Society of Agronomy, *Olsen Bicarbonate Test Methods of Soil Analysis, Part 2*, Madison, Wisconsin 53771, p. 1044 (1965).
33. T. L. Jones-Lepp and R. Stevens. Pharmaceuticals and personal care products in biosolids & sewage sludge. *Analytical and Bioanalytical Chemistry*, Vol. 387, No. 4, February (2007).
34. R. C. Brandt, and K. M. Martin. *The Food Processing Residual Management Manual*. Pennsylvania Department of Environmental Protection, Harrisburg, PA. Pub. No. 2500-BK-DER-1649 (1994).
35. M. M. Deitzman, S. Mostaghimi, T. A. Dillaha, and C. D. Heatwole. Tillage effects on P losses from sludge amended soils, *Journal Soil and Water Conservation*, 247–251 (1989).
36. Pennsylvania State University, College of Agricultural Sciences, *Soil Fertility Management Agronomy Guide*, Report No. 5.5M1096NV0, University Park, PA (1996).
37. Pennsylvania State Conservation Commission, *Pennsylvania's Nutrient Management Act Program Technical Manual*, Harrisburg, PA (1998).
38. L. Sommers, Chemical composition of sewage sludges and analysis of their potential use as fertilizers. *Journal of Environmental Quality*, 6, 225–239 (1977).
39. Pennsylvania State University, *Land Application of Sewage in Pennsylvania*, Pennsylvania Cooperative Extension, University Park, PA (1998).
40. L. S. Clesceri, A. E. Greenberg, and A. D. Eaton (eds.), *Standard Methods for the Examination of Water and Wastewater*, 20th ed., American Public Health Association, Water Environment Federation and American Water Works Association (2007).
41. N. K. Shammas and L. K. Wang, Land application of biosolids. In: *Biosolids Treatment Processe*, L. K. Wang, N. K. Shammas, and Y. T Hung (eds.), Humana Press, Totowa, NJ, pp. 705–746. (2007).
42. T. W. Simpson, S. J. Donohue, G. W. Hawkins, M. M. Monnett, and J. C. Baker, *The Development and Implementation of the Virginia Agronomic Land Use Evaluation systems (VALUES)*, Department of Crop and Soil Environmental Sciences, Virginia Tech., Blacksburg, VA (1993).
43. A. C. Chao, L. K. Wang, and M. H. S. Wang, *Use of the Ames Mutagenecity Bioassay as a Water Quality Monitoring Method*, PB88-168422/AS, National Technical Information Service, Springfield, VA (1988).
44. L. K. Wang, M. H. S. Wang, and V. Renak, *Determination of Solids and Water Content of Highly Concentrated Sludge Slurries and Cakes*, PB85-182624/AS, National Technical Information Service, Springfield, VA (1985).
45. U.S. Environmental Protection Agency, *Inspectors Guide for Evaluation of Municipal Wastewater Treatment Plants*, US EPA-430/9-79-010, Washington, DC (1979).
46. U.S. Environmental Protection Agency, *Innovative and Alternative Technology Assessment Manual*, US EPA-430/9-78-009, Washington, DC (1980).
47. *Engineering News Record (ENR)*, McGraw-Hill, New York, http://construction.ecnext.com/coms2/summary_0249-235734_ITM, April 7 (2007).

48. M. H. S. Wang and L. K. Wang, *Electrical Energy Consumption and Heating Requirements of Municipal Wastewater Treatment Plants*, PB82-183393, National Technical Information Service, Springfield, VA (1982).
49. J. C. Wang, D. B. Aulenbach, L. K. Wang, and M. H. S. Wang, Energy models and cost models for water pollution control, *Clean Production—Environmental and Economic Perspectives*, K. B. Misra (ed.), Springer, New York, 685–720 (1996).
50. U.S. Environmental Protection Agency, *Survival and Transport of Pathogens in Sludge-amended Soil: A Critical Review*. US EPA/600/2-87/028, Cincinnati, OH (1987).
51. J. E. Smith, Jr., P.D. Millner, W. Jakubowski, N. Goldstein, and R. Rynk. Contemporary perspectives on Infectious disease agents in sewage sludge and manure. *Proceedings of the Workshop on Emerging Infectious Disease Agents and Issues Associated with Sewage Sludge, Animal Manures, and Other Organic By-Products*, Cincinnati, OH, June (2001).
52. *Compost Science and Utilization*. JG Press, Inc., Emmaus, PA. (Four issues per year).
53. G. L. Mullins and D. J. Hansen, Basic soil fertility. In: K. C. Haering and G. K. Evanylo. *The Mid Atlantic Nutrient Management Handbook*. MAWP 06-02, pp. 53–93 (2006). http://www.mawaterquality.org/themes/nutrient_management/manmh2006.htm.
54. J. T. Gilmour, Carbon and nitrogen mineralization during co-utilization of biosolids and composts. In: *Beneficial Co-utilization of Agricultural, Municipal and Industrial By-products*, S. Brown, J. S. Angle, and L. W. Jacobs (eds.), pp. 89–112. Kluwer Academic Publishers, Dordrecht, Netherlands (1998).
55. J. T. Gilmour and M.D. Clark. Nitrogen release from wastewater sludge: A site specific approach. *Journal Water Pollution Control Federation*, 60, 494–498 (1988).
56. J. T. Gilmour, C.G. Cogger, L.W. Jacobs, S.A. Wilson, G.K. Evanylo, and D.M. Sullivan. *Estimating Plant-Available Nitrogen in Biosolids*. Project 97-REM-3, Water Environment Research Foundation, Alexandria, VA (2000).
57. L. W. Jacobs, Process design for agricultural land application sites. In: *Process Design Manual—Land Application of Sewage Sludge and Domestic Septage*, US EPA/625/R-95/001, pp. 63–93, U.S. Environmental Protection Agency, Washington (1995).
58. D. C. Sommers, C. F. Parker, and G. J. Meyers, *Volatilization, Plant Uptake and Mineralization of Nitrogen in Soils Treated with Sewage Sludge*. Technical Report 133, Purdue University Water Resources Research Center, West Lafayette, IN (1981).
59. U.S. Environmental Protection Agency, Process design for agricultural utilization. In: *Process Design Manual—Land Application of Municipal Sludge*, US EPA-625/1-83-016, pp. 6-1 to 6–48, Cincinnati, OH (1983).
60. U.S. Environmental Protection Agency, Process design for agricultural land application sites. In: *Process Design Manual—Land Application of Sewage Sludge and Domestic Septage*, US EPA/625/R-95/001, pp. 63–93, Washington, DC (1995).

# APPENDIX A

## US EPA General Requirements for Land Application of Sewage Sludge (7)

| | |
|---|---|
| (a) | No person shall apply sewage sludge to the land except in accordance with the Part 503 land application requirements. |
| (b) | No person shall apply bulk sewage sludge that is non-EQ for pollutants (i.e., subject to cumulative pollutant loading rates in 503.13(b)(2)) to agricultural land, forest, a public contact site, or a reclamation site if any of the cumulative pollutant loading rates in 503.13(b)(2) has been reached. |
| (c) | No person shall apply domestic septage to agricultural land, forest, or a reclamation site during a 365-day period if the annual application rate in 503.13(c) has been reached during that period. |
| (d) | The person who prepares bulk sewage sludge that is applied to agricultural land, forests, areas where the potential for contact with the public is high (i.e., public contact site) or a reclamation site shall provide the person who applies the bulk sewage sludge written notification of the concentration of total nitrogen (as N on a dry weight basis) in the bulk sewage sludge. |
| (e)(1) | The person who applies sewage sludge to the land shall obtain information needed to comply with applicable Part 503 requirements. |
| (e)(2)(i) | Before bulk sewage sludge that is subject to cumulative pollutant loading rates (CPLRs) in 503.13(b)(2) is applied to the land, the person who proposes to apply the bulk sewage sludge shall contact the permitting authority for the state in which the bulk sewage sludge is being applied, to determine whether, bulk sewage sludge subject to the cumulative pollutant loading rates in 503.13(b)(2) has been applied to the site since July 20, 1993. |
| (ii) | If bulk sewage sludge subject to CPLRs has not been applied to the site since July 20, 1993, the cumulative amount of each pollutant listed in Table 2 of 503.13 may be applied to the site in accordance with 503.13(a)(2)(i). |
| (iii) | If bulk sewage sludge subject to CPLRs in 503.13(b)(2) has been applied to the site since July 20, 1993, and the cumulative amount of each pollutant applied to the site since that date is known, the cumulative amount of each pollutant applied to the site shall be used to determine the additional amount of each pollutant that can be applied to the site in accordance with 503.13(a)(2)(i). |
| (iv) | If bulk sewage sludge subject to CPLRs in 503.13(b)(2) has been applied to the site since July 20, 1993, and the cumulative amount of each pollutant applied to the site since that date is not known, sewage sludge subject to CPLRs may no longer be applied to the site. |
| (f) | A person who prepares bulk sewage sludge shall provide the person who applies the bulk sewage sludge notice and necessary information to comply with applicable Part 503 requirements. |
| (g) | When the person who prepares sewage sludge gives the material to another person who prepares sewage sludge, the person who provides the sewage sludge shall provide to the person who receives sewage sludge notice and necessary information to comply with the applicable Part 503 requirements. |
| (h) | The person who applies bulk sewage sludge to the land shall provide the owner/leaseholder of the land on which the bulk sewage sludge is applied notice and necessary information to comply with applicable Part 503 requirements. |

**Appendix A** (*Continued*)

(i) Any person who prepares bulk sewage sludge that is applied to land in a state other than the state in which the bulk sewage sludge is prepared, shall provide written notice, prior to the initial application of bulk sewage sludge to the land application site by the applier, to the permitting authority for the state in which the bulk sewage sludge is proposed to be applied. The notice must include:

  (1) The location by either street address or latitude/longitude, of each land application site.
  (2) The approximate time period in which the bulk sewage sludge will be applied to the site.
  (3) The name, address, telephone number, and National Pollutant Discharge Elimination System permit number (if appropriate) for the person who prepares the bulk sewage sludge.
  (4) The name, address, telephone number, and National Pollutant Discharge Elimination System permit number (if appropriate) for the person who will apply the bulk sewage sludge.

(j) Any person who applies bulk sewage sludge subject to the cumulative pollutant loading rates in 503.13(b)(2) to the land shall provide written notice, prior to the initial application of the bulk sewage sludge to the application site by the applier, to the permitting authority for the state in which the bulk sewage sludge will be applied, and the permitting authority shall retain and provide access to the notice. The notice must include:

  (1) The location, by either street address or latitude/longitude, of each land application site.
  (2) The name, address, and National Pollutant Discharge Elimination System permit number (if appropriate) of the person who will apply the bulk sewage sludge.

# APPENDIX B

## US EPA Pollutant Limits for the Land Application of Sewage Sludge (7)

| Pollutant | Concentration limits | |
|---|---|---|
| | Ceiling concentrations (Table 1 of 40 CFR 503.13) (milligrams per kilogram, dry weight) | Pollutant concentrations (Table 3 of 40 CFR 503.13) monthly average (milligrams per kilogram, dry weight) |
| Arsenic | 75 | 41 |
| Cadmium | 85 | 39 |
| Chromium | 3000 | 1200 |
| Copper | 4300 | 1500 |
| Lead | 840 | 300 |
| Mercury | 57 | 17 |
| Molybdenum* | 75 | — |
| Nickel | 420 | 420 |
| Selenium | 100 | 36 |
| Zinc | 7500 | 2800 |

| Pollutant | Loading rates | | | |
|---|---|---|---|---|
| | Cumulative pollutant loading rates (Table 2 of 40 CFR 503.13) | | Annual pollutant loading rates (Table 4 of 40 CFR 503.13) | |
| | (kilograms per hectare, dry weight) | (pounds per acre, dry weight) | (kilograms per hectare per 365-day period, dry weight) | (pounds per acre per 365-day period, dry weight) |
| Arsenic | 41 | 37 | 2.0 | 1.8 |
| Cadmium | 39 | 35 | 1.9 | 1.7 |
| Chromium | 3000 | 2677 | 150 | 134 |
| Copper | 1500 | 1339 | 75 | 67 |
| Lead | 300 | 268 | 15 | 13 |
| Mercury | 17 | 15 | 0.85 | 0.76 |
| Molybdenum* | — | — | — | — |
| Nickel | 420 | 375 | 21 | 19 |
| Selenium | 100 | 89 | 5.0 | 4.5 |
| Zinc | 2800 | 2500 | 140 | 125 |

*Note: The pollutant concentration limit, cumulative pollutant loading rate, and annual pollutant loading rate for molybdenum were deleted from Part 503 effective February 19, 1994. The US EPA may reconsider establishing these limits at a later date.

# APPENDIX C

## US EPA Sample Format for Providing Notice and Necessary Information (7)

*This form is to assist compliance with the bulk sewage sludge notification requirements (503.12(f). if the sewage sludge meets the exceptional quality requirements, however, then the notification requirements do not apply.*

### Part I - To be completed by PREPARERS of sewage sludge

A. Please provide pollutant concentrations

| Pollutant | Concentration (mg/kg) measured dry weight (indicate monthly average or instantaneous value) | Pollutant concentrations (mg/kg) (Table 3 40 *CFR* 503.13) (monthly average) | Ceiling concentrations* (Table 1 in 40 *CFR* 503.13) (Instantaneous maximum) |
|---|---|---|---|
| Arsenic | | 41 | 75 |
| Cadmium | | 39 | 85 |
| Chromium | | 1200 | 3000 |
| Copper | | 1500 | 4300 |
| Lead | | 300 | 840 |
| Mercury | | 17 | 57 |
| Molybdenum | | N/A | 75 |
| Nickel | | 420 | 420 |
| Selenium | | 36 | 100 |
| Zinc | | 2800 | 7500 |
| TKN | | N/A | N/A |
| $NH_4^+$-N | | N/A | N/A |
| $NH_3$-N | | N/A | N/A |

*Sewage sludge may not be land applied if any pollutant concentrations in any sample exceed these values.

B. Pathogen reduction (40 *CFR* 503.32)—please indicate the level achieved

 ☐ Class A    ☐ Class B

C. Vector attraction reduction (40 *CFR* 503.33)—please indicate the option performed

 ☐ Option 1   ☐ Option 2   ☐ Option 3   ☐ Option 4

 ☐ Option 5   ☐ Option 6   ☐ Option 7   ☐ Option 8

 ☐ No vector attraction reduction options were performed

D. Certification

I certify, under penalty of law, that this document and all attachments were prepared under my direction or supervision in accordance with a system designed to assure that qualified personnel properly gather and evaluate the information submitted. Based on my inquiry of the person or persons who manage the system or the persons directly responsible for gathering the information, the information submitted is, to the best of my knowledge and belief, true, accurate, and complete. I am aware that there are significant penalties for submitting false information, including the possibility of fine and imprisonment for knowing violations.

| A. Name and Official Title *(type or print)* | B. Area Code and Telephone Number |
|---|---|
| C. Signature | D. Date Signed |

*(Continued)*

## Appendix C (*Continued*)

| Part II - To be completed by LAND APPLIERS of sewage sludge |
|---|

*Land appliers using non-EQ sludge should provide the following information, when applicable to the landowner/leaseholder.*

A. If the pollutant levels in the sewage sludge do not meet the **pollutant concentration** limits in Table 3 of 40 *CFR* Part 503, then the land applier should provide the landowner with the following information:

1. Location of land application site _____
2. Number of hectares where the bulk sewage sludge was applied _____
3. Date and time bulk sewage sludge was applied _____
4. Amount of bulk sewage sludge applied in metric tons, dry weight _____
5. Record the amount of each metal and nitrogen applied and appropriate units (i.e., **kilograms per hectare, pounds per acre**):

| Units | Arsenic | Cadmium | Chromium | Copper | Lead | Mercury | Nickel | Selenium | Zinc | Available Nitrogen |
|---|---|---|---|---|---|---|---|---|---|---|
|  |  |  |  |  |  |  |  |  |  |  |

B. If a Class B pathogen reduction alternative was used (see Part I), then the following site restrictions must be met. Please check the boxes for the site restrictions met if any.

   A. Food crops that may touch the sewage sludge/soil mixture cannot be harvested before the end of the following waiting period:
   
   ☐ 1. If harvested parts are totally **above** the land wait to harvest for **14 months** after the application of sewage sludge.
   
   ☐ 2. If harvested parts are **below** the land surface and the sewage sludge remains on top of the soil for 4 months or longer before the field was plowed, wait to harvest for **20 months** after the initial application of sewage sludge.
   
   ☐ 3. If harvested parts are below the land surface, and the sewage sludge was incorporated into the soil within 4 months of being applied, wait to harvest for **38 months** after the initial application.
   
   B. ☐ Food crops that do not touch sludge/soil mixture, feed crops, and fiber crops cannot be harvested for 30 days after sewage sludge
   
   C. ☐ Animals cannot be grazed on the land for 30 days after application of the sewage sludge.
   
   D. ☐ If harvested turf is used for a lawn or other purpose where there is a high potential for public exposure, then the turf cannot be harvested for 1 year after the application of the sewage sludge to the land.
   
   E. ☐ Public access to land with a high potential for public exposure (e.g., parks, playgrounds, golf courses) will be restricted for 1 year after the application of the sewage sludge.
   
   F. ☐ Public access to land with a **low potential** for public exposure (e.g., private property, remote or restricted public lands) will be restricted for 30 days after the application of the sewage sludge.

C. If the preparer did not perform any of the vector attraction reduction Options 1-8 (see Part I), then either Option 9 or 10 must be performed by the land applier. Please indicate if Option 9 or 10 was performed. Check appropriate box.

   ☐ Option 9-Subsurface Injection    ☐ Option 10-Incorporated (plowed) into the Soila    ☐ N/A

D. CERTIFICATION

   I certify, under penalty of law, that this document and all attachments were prepared under my direction or supervision in accordance with a system designed to assure that qualified personnel properly gather and evaluate the information submitted. Based on my inquiry of the person or persons who manage the system or the persons directly responsible for gathering the information, the information submitted is, to the best of my knowledge and belief, true, accurate and complete. I am aware that there are significant penalties for submitting false information, including the possibility of fine and imprisonment for knowing violations.

| A. Name and Official Title *(type or print)* | B. Area Code and Telephone Number |
|---|---|
| C. Signature | D. Date Signed |

# 8
# Landfilling Engineering and Management

### Puangrat Kajitvichyanukul, Jirapat Ananpattarachai, Omotayo S. Amuda, Abbas O. Alade, Yung-Tse Hung, and Lawrence K. Wang

**CONTENTS**

INTRODUCTION
REGULATIONS AND POLLUTANT STANDARDS FOR BIOSOLIDS
  LANDFILLING
TYPES OF BIOSOLIDS FOR LANDFILLING
REQUIREMENTS OF BIOSOLIDS CHARACTERISTICS FOR LANDFILLING
BIOSOLIDS TREATMENT FOR LANDFILLING
DESIGN OF BIOSOLIDS LANDFILLING
CASE STUDY AND EXAMPLE
REFERENCES

**Abstract** Landfilling of biosolids is classified as one type of surface disposals presented in the U.S. Environmental Protection Agency under subpart C of Title 40 of the Code of Federal Regulations (CFR), Part 503. This method offers the simplest solution to biosolids handling by concentrating the material in a single location and provides several benefits. To meet the requirement for biosolids landfilling, the treatment process to improve characteristics of sewage sludge corresponding to the Part 503 regulation is necessary. In this chapter, characteristics of biosolids for landfilling and the treatment methods are provided. The design criteria for biosolids landfilling are also given.

**Key Words** Sludge landfill • area fill • handling • design • limitations.

## 1. INTRODUCTION

Landfilling of biosolids is classified as one type of surface disposals presented in the U.S. Environmental Protection Agency (US EPA) under subpart C of Title 40 of the Code of Federal Regulations (CFR), Part 503 (1). However, the term *biosolids* is not defined in this document. Definition of biosolids first appeared in US EPA (1995) (2) as the primarily organic solids product yielded by municipal wastewater treatment processes that can be beneficially recycled (whether or not they are currently being recycled). The surface disposal subpart of the regulation deals with biosolids that are placed on an area of land where only biosolids are placed for final disposal. It does not include biosolids that are placed on land for either storage (generally less than 2 years) or treatment (e.g., lagoon treatment for pathogen reduction). The regulated surface disposal method involves landfilling of biosolids in monofills (biosolids-only landfills), dedicated surface disposal practices, disposal in permanent piles or mounds, and disposal in impoundments or lagoons (1). Among these four techniques, monofilling of biosolids is possibly the most common surface disposal technique covered by the Part 503 biosolids rule. It typically involves excavating a trench without a liner, placing the biosolids in the trench, and then backfilling the trench to return the soil to its original contours. Other techniques are less commonly used (3). Several previous works have also shown that the addition of biosolids can benefit the operation of a landfill by increasing organic matter decomposition rate, resulting in more efficient methane gas recovery and improving the quality of leachate collected from the landfill (1).

To meet the requirement for biosolids landfilling, treatment to improve characteristics of sewage sludge is necessary. The biosolids standard for landfilling can follow the sludge criteria for use as a fertilizer or soil conditioner. These standards include meeting metals limitations, pathogen reduction, and vector requirements as appeared in the Part 503 biosolids rule. If the biosolids landfill is properly constructed, operated, and maintained, the risk of its releasing pollutants and pathogens can be minimized.

This chapter describes the disposal of biosolids by the landfilling method. Regulations related to biosolids landfill are reviewed. Types of sewage sludge from wastewater treatment plants, treatment processes for sewage sludge, and requirements of biosolids in landfilling are discussed. The design of landfill and some existing biosolids landfill sites are provided. Some relevant design calculations in biosolids landfilling are also included.

## 2. REGULATIONS AND POLLUTANT STANDARDS FOR BIOSOLIDS LANDFILLING

Sewage sludge management and its usages have been presented in several US EPA documents. The surveys of sewage sludge characteristics by the US EPA (4) have led to identifying and characterizing the risks from chemical pollutants in sewage sludge. Accordingly, the chemical pollutants concentration and pathogen standards were established in the Part 503 rule and have been used as the major requirement of biosolids for land application including landfilling. The Part 503 regulations establish allowable concentrations of metals, as well as appropriate management practices to ensure that metals found within land-applied biosolids will not endanger human health or the environment. The chemical standards, cumulative pollutant loading rates, pollutant concentration limits, and annual pollutant loading rates for 10 inorganic chemicals were originally established (5). These chemicals are arsenic, cadmium, chromium, copper, lead, mercury, molybdenum, nickel, selenium, and zinc. The risk-based concentration limit for dioxins, a category of organic compounds that includes 29 specific congeners of polychlorinated dibenzo-$p$-dioxins polychlorinated dibenzofurans, and coplanar polychlorinated biphenyls (PCBs) were issued in 1999 to amend the Part 503 rule for land-applied biosolids (6). In the 40 CFR Part 503 biosolids rule, land application limits

## Table 8.1
**Heavy metals limits for land application according to the Part 503 biosolids rule**

| Pollutant | Ceiling concentration limits (mg/kg)[a] | Cumulative pollutant loading rates (kg/ha) | High-quality pollutant concentration limits (mg/kg)[b] | Annual pollutant loading rates (kg/ha/yr) |
|---|---|---|---|---|
| Arsenic | 75 | 41 | 41 | 2.0 |
| Cadmium | 85 | 39 | 39 | 1.9 |
| Copper | 4300 | 1500 | 1500 | 75 |
| Lead | 840 | 300 | 300 | 15 |
| Mercury | 57 | 17 | 17 | 0.85 |
| Molybdenum | 75 | c | c | c |
| Nickel | 420 | 420 | 420 | 21 |
| Selenium | 100 | 100 | 100 | 5.0 |
| Zinc | 7500 | 2800 | 2800 | 140 |

*Note*: All values are on dry weight basis.
[a] Absolute values.
[b] Monthly averages.
c: The February 25, 1994 Part 503 Rule Amendment kept only the ceiling concentration limit and deleted all other limits for Mo.
*Source*: US EPA (2).

for nine heavy metals (arsenic, cadmium, copper, lead, mercury, molybdenum, nickel, selenium, and zinc) were set. Four sets of limits are provided: the ceiling limits for land application, more stringent high-quality pollutant concentration limits, the cumulative loading limits, and the annual limits for bagged products not meeting the high-quality pollutant concentration limits. These limits are shown in Table 8.1. Of the various metals found in biosolids, cadmium and lead are of the greatest concern to human health. Table 8.2 summarizes the characteristics of solids and biosolids generated from wastewater treatment steps.

To apply on land, biosolids must meet the pollutant ceiling concentrations and cumulative pollutant loading rates or pollutant concentration limits. Biosolids that meet pollutant concentration limits can be applied to sites without tracking cumulative loading limits. However, biosolids that meet the ceiling limits but not the pollutant concentration limits can only be applied until the amount of metals on the site have accumulated up to the cumulative limits.

Besides heavy metals, pathogens are also considered major pollutants that may be present in biosolids and cause human health problem. Pathogens are microorganisms that cause disease. Pathogens associated with sewage and biosolids include bacteria, viruses, protozoa, and helminth. Considering pathogen limits, the preliminary set of risk assessments (7, 8) for viruses, bacteria, and parasites was conducted, and the operational standards for pathogens in biosolids were established. The pathogen reduction requirements are of major concern for biosolids application to the land, as the presence of pathogens (including enteric viruses, bacteria, parasites, and viable helminth ova) in biosolids should be at levels that are unlikely to pose a threat to public health and the environment under specific use conditions.

Table 8.2
**Summary of characteristics of solids and biosolids generated from each wastewater treatment step**

| Step | Processes | Characteristics of solids or biosolids |
|---|---|---|
| Screening and grit removal | Screening, grit chamber | Suspended solids or solid wastes |
| Primary wastewater treatment | Primary sedimentation | Biosolids containing 3% to 7% solids, easily to thicken or dewater |
| Secondary wastewater treatment | Biological treatment process (suspended growth or fixed growth system) | Biosolids containing 0.5% to 2.0% solids, difficult to thicken and dewater |
| Tertiary wastewater treatment | Chemical precipitation, nitrogen-phosphorus removal, chlorination-dechlorination | Biosolids with varying water-absorbing characteristics, containing chemicals in composition of biosolids, chemicals in biosolids may include lime, polymers, iron, or aluminum salts |

*Source*: US EPA (3).

The Part 503 biosolids rule divides biosolids into Class A and Class B biosolids in terms of pathogen levels. Class A applies to biosolids that are sold or given away in any container for application to land or applied to lawn or home garden. Class B applies to biosolids that are placed on surface disposal. This class was developed from the 1979 40 CFR 257 regulations for processes to significantly reduce pathogens (PSRP) (9). In the initial development of those requirements, a process to significantly reduce pathogens (PSRP) was defined as a process that reduces pathogenic viruses, *Salmonella* bacteria, and indicator bacteria (fecal coliform). Thus, Class B is the minimum level of pathogen reduction for land application. In comparison, Class A biosolids are treated to a greater degree than Class B biosolids and are safe for direct human contact (e.g., bagged compost). Class B biosolids are treated to a safe level but do require compliance with land and crop use restrictions. The goal of these standards is to reduce the presence of pathogens (including enteric viruses, bacteria, parasites, and viable helminth ova) in biosolids to levels that are unlikely to pose a threat to public health and the environment under specific use conditions.

According to the requirements specified in the 40 CFR Part 503 biosolids rule, "The Standards for the Use or Disposal of Sewage Sludge," any biosolids that are land applied shall contain pathogens and metals that are below specified levels to protect the health of humans, animals, and plants. The vector (e.g., flies and rodents) attraction should be reduced. In addition, Part 503 also specifies that the biosolids are applied at an "agronomic rate," which is the biosolids application rate designed to provide the amount of nitrogen needed by the crop or vegetation and to minimize the amount of nitrogen in the biosolids that passes below the root zone of the crop to the ground water. The specific management practices, monitoring frequencies, and record keeping and reporting are also required.

As regulated by the Part 503 biosolids rule, biosolids can be placed on an area of land where only biosolids are placed for final disposal. It does not include biosolids that are placed on land for either storage (generally less than 2 years) or treatment (e.g., lagoon treatment for pathogen reduction). Biosolids landfilling is one type of surface disposal specified in Part 503. In general, the landfills for municipal solid waste (MSW) have to comply with the requirement of the Part 258 landfill rule and follow the specified design, operation, and closure of municipal solid waste landfills (10). For biosolids landfilling, Part 503 requires that the site meet certain locational restrictions similar to the site restrictions in the Part 258 landfill rule. For example, the surface disposal unit must meet management requirements similar to those for municipal solid waste landfills. These management practices include requirements for runoff collection, leachate collection and disposal (if the unit is lined), vector control, methane monitoring, and groundwater monitoring or certification, and restrictions on public access, growing of crops, and grazing of animals. However, the Part 503 biosolids rule regulates specifically requirements for the surface disposal of biosolids. Provisions for closure and postclosure care must be made, and a plan for leachate collection (if the unit is lined), must be developed. In addition, if the surface disposal unit is unlined, the biosolids must meet concentration limits on arsenic, chromium, and nickel. For landfilling, the biosolids treatment processes that are approved for Class A and B biosolids are discussed in Sections 4.1 and 4.2, below. The further details of design, and management of biosolids landfilling are provided in Section 6.

To dispose biosolids in MSW landfill, biosolids must meet the Part 258 landfill rule requirement. In general, biosolids must not meet the definition of hazardous wastes under the Resource Conservation and Recovery Act (RCRA) or the Toxicity Characteristic Leachate Procedure test. In addition, biosolids cannot contain more than 50 parts per billion of PCBs (11). It is also stated that dewatering biosolids to about 20% solids is required.

To use biosolids as the daily MSW landfill cover, the biosolids should meet the limits on metals, vector attraction reduction, and other management requirements of Part 503. In addition, dewatering biosolids to achieve soil-like characteristics is needed. With this requirement, biosolids can be used as the daily cover for the MSW solid waste landfill at the end of each day or more frequently as needed to control disease vectors, fires, odors, blowing litter, and scavenging. Biosolids can also be used as part of a final MSW landfill cover, which must meet the Part 258 landfill rule cover criteria for permeability, infiltration, and erosion control.

## 3. TYPES OF BIOSOLIDS FOR LANDFILLING

*Biosolids* is the new term for what had previously been referred to as *stabilized sewage sludge*. In fact, biosolids are primarily organic treated wastewater residues from municipal wastewater treatment plants that are suitable for recycling as a soil amendment. *Sewage sludge* currently is the term that is used to refer to untreated primary and secondary organic solids. The differences between these two terms are the stabilization properties of sewage sludge from municipal wastewater treatment plant. These properties lead to the improvement of biosolids so that they have soil-like characteristics and can be recycled as a soil amendment. Other types of sludges (such as industrial oil and gas field wastes) that cannot be beneficially recycled as soil amendment are not referred to as biosolids (1, 2).

Generally, the factors that influence the quantity and characteristics of the biosolids generated from municipal wastewater treatment plants include the following:

- Composition of the wastewater: The characteristics of the biosolids produced can change annually, seasonally, or even daily because of variations in the incoming wastewater composition and variations in treatment processes (3).
- Type of wastewater treatment used: Both types of wastewater treatment and pretreatment used affect the characteristics of biosolids, which in turn can affect the types of biosolids treatment chosen. Higher degrees of wastewater treatment can increase the total volume of biosolids generated. Higher levels of treatment also can increase the concentrations of contaminants in biosolids (3).
- Type of subsequent treatment applied to the biosolids: Different biosolids treatments such as conditioning, thickening, stabilization, dewatering, thermal conversion, drying, and other processes cause different characteristics of biosolids. Further details in each biosolids treatment method can be found in Section 4.

Biosolids are produced primarily through biological treatment of domestic wastewater. Biosolids are the solids that are removed from the wastewater and further processed before the treated water is discharged into natural water. Generally, the solids that can be removed by several steps in wastewater treatment processes include screenings and grit, naturally floating materials called scum, and the solids removed from primary and secondary clarifiers called sewage sludge. These solids can be stabilized and become biosolids. However, biosolids can come from wastewater processes that involve the addition of chemicals to precipitate the solids (such as ferric chloride, alum, lime, or polymers). These chemicals can be concentrated in the resulting biosolids. For example, alum (as aluminum hydroxide) adsorbs phosphorus or causes trace metals (e.g., cadmium) to precipitate out of the wastewater, and these chemicals become the major composition of the resulting biosolids.

Typically, a wastewater treatment plant involves primary, secondary, and tertiary treatment systems. First, the solids are separated from the raw wastewater by primary treatment. Then dissolved biological matter is progressively converted into a solid mass by microorganisms in the secondary treatment system. Finally, the treated water may be disinfected chemically or physically. The final effluent can be discharged into natural water or recycled for irrigation or groudwater discharge. Biosolids can be obtained from all treatment steps with different characteristics. The types of wastewater treatment and resulting types of biosolids are as follows:

1. Screening and grit removal: Grit removal is an important part of any wastewater treatment process. This unit removes any solids that can cause serious damage to pipes and downstream equipment. Any material in the wastewater that cannot pass through these openings accumulates on the screen and is automatically removed by a mechanical brush and a continuous flow of water. Biosolids from this unit are heavy, inorganic, sandlike solids that can be handled as a solid waste and nearly always landfilled. This material is excluded

from the definition of biosolids and from the 40 CFR Part 503 regulation governing the use of disposal of biosolids (3).
2. Primary wastewater treatment: Primary wastewater treatment is used to remove any material that can be easily separated from the raw wastewater. It usually involves gravity sedimentation to remove suspended solids prior to secondary treatment. The typical materials that are removed from this unit are biosolids that include fats, oils, and greases (FOG); sand; gravel and rocks; larger settleable solids including human waste; and floatable materials. Biosolids from this unit usually contain 3% to 7% solids, and their water content can be easily reduced by thickening or dewatering (3).
3. Secondary wastewater treatment: In wastewater treatment plant, biological content in raw wastewater will be degraded in secondary wastewater treatment. This unit relies on microorganism activity. Secondary wastewater treatment systems are classified as fixed film or suspended growth. In the fixed-film treatment process, the biomass grows on media and the wastewater passes over its surface. In contrast, the biomass in a suspended growth system is well mixed with the wastewater. As a result, biosolids from this process are mainly the sewage sludge with a low solids content in the range of 0.5% to 2.0%. In comparison to biosolids from primary wastewater treatment, the obtained biosolids are more difficult to thicken and dewater (3).
4. Tertiary wastewater treatment: Tertiary (advanced) wastewater treatment is usually applied when high effluent quality is needed. The effluent quality from this process is higher than that produced with secondary treatment. The tertiary wastewater treatment includes nitrogen-phosphorus removal, disinfection, and biological and chemical precipitation. More than one tertiary treatment process may be used at any treatment plant. Several types of chemicals such as lime, polymers, iron, and aluminum salts are also applied in these systems. Consequently, characteristics of biosolids from this unit varied depending on the chemical used. For example, high-level lime precipitation produces alkaline biosolids (3).

## 4. REQUIREMENTS OF BIOSOLIDS CHARACTERISTICS FOR LANDFILLING

As landfill is one type of surface disposal, biosolids subject to disposal by landfilling have to meet the US EPA requirements in subpart C of 40 CFR Part 503. Biosolids landfilling is similar to land application in that it entails the placement of sewage sludge on the land. The major difference between the two is that in the case of landfilling, sewage sludge is placed on the land for the purpose of final disposal, without regard to the soil-enhancing qualities of the sewage sludge (12). The Part 503 regulations require that biosolids placed in a landfill meet either Class A or Class B pathogen reduction requirements. Levels of pathogen reduction of each classification are described below.

### 4.1. Class A Pathogen Requirements

Class A pathogen requirements are mainly concerned with reducing the pathogen levels to below the regulated level at the same time as or before the vector attraction reduction requirements are met. Two types of pathogens are fecal coliform and *Salmonella*. The Class A pathogen criteria require that both treatment-process control requirements and prescribed densities of either fecal coliform or *Salmonella* are satisfied as follow (2):

**Table 8.3**
Guidelines for temperature and time for pathogens reduction

| Total solids | Temperature | Time | Equation, $D$ = time in days, $t$ = temp in °C | Notes |
|---|---|---|---|---|
| ≥7% | ≥50 °C | ≥20 min | $D = \dfrac{131{,}700{,}000}{10^{0.14t}}$ | No heating of small particles by warmed gases or immiscible liquid |
| ≥7% | ≥50 °C | ≥15 s | $D = \dfrac{131{,}700{,}000}{10^{0.14t}}$ | Small particles heated by warmed gases or immiscible liquid |
| <7% | >50 °C | ≥15 s to <30 min | $D = \dfrac{131{,}700{,}000}{10^{0.14t}}$ | |
| <7% | ≥50 °C | ≥30 min | $D = \dfrac{50{,}070{,}000}{10^{0.14t}}$ | |

*Note*: Temperature calculated using the appropriate equation must never be less than 50°C. The time values are not used in the calculations, but are provided to indicate the prescribed duration that temperature must be maintained.
*Source*: US EPA (2).

- The fecal coliform density must be less than 1000 most probable number (MPN) per gram (g) of total solids (TS), and that must be satisfied immediately after the treatment process is completed.
- The *Salmonella* density must be less than 3 MPN per 4 g of TS, and that must be satisfied immediately after the treatment process is completed.
- In addition, one of the following treatment processes described below must be met. The goal of these processes is to reduce pathogen densities below specified detection limits for three types of organisms: *Salmonella* sp. (<3 MPN per 4 g TS), enteric viruses (<1 plaque-forming unit [PFU] per 4 g TS), and helminths (<1 viable organism per 4 g TS) (5).

The listed treatment processes to reduce pathogen densities include the following:

- Temperature and time process: This process is based on the time and temperature relationship that is related to pasteurization studies. An increased sewage-sludge temperature must be maintained for a prescribed period according to the guideline in Table 8.3 (2).
- Alkaline treatment process: The requirement of this process is to maintain the pH of the sewage sludge greater than 12 for at least 72 hours. During this process, the temperature of the sewage sludge must be greater than 52°C for at least 12 hours. In addition, after the 72-hour period, the sewage sludge must be air dried to at least 50% total solids (2).
- Prior test for enteric virus and viable helminth ova: The analysis of enteric viruses and viable helminth ova in sewage sludge before pathogen-reduction processing is required.

If the densities of enteric virus are less than 1 PFU/4 g TS and the viable helminth ova are less than one per 4 g TS, the sewage sludge is considered Class A biosolids with respect to enteric virus and viable helminth ova until the next monitoring event (2).

If the sewage sludge is analyzed before pathogen-reduction processing and is found to have densities of enteric virus greater than or equal to 1 PFU/4 g TS or viable helminth ova of more than 1 per 4 g TS and is tested again after processing and found to have densities of enteric virus of less than 1 PFU/4 g of TS and viable helminth ova less than one per 4 g of TS, the sewage sludge is considered Class A biosolids when the treatment process is operated under the same conditions that successfully reduced enteric virus and helminth ova (2).

- Posttest for enteric virus and viable helminth ova process: If the sewage sludge is not analyzed before pathogen-reduction processing for enteric viruses and viable helminth ova, the sewage sludge must meet the enteric virus and viable helminth ova levels noted below for Class A at the time the sewage sludge is used, disposed of, or prepared for sale or giveaway in a bag or container or when the biosolids meets exceptional quality requirements and pollutant concentration limits (2):
  - The density of enteric viruses must be less than 1 PFU/4 g TS
  - The density of viable helminth ova must be less than one per 4 g TS

- Process to further reduce pathogens (PFRP) and PFRP equivalent: To obtain a Class A biosolid rating, the process must reduce *Salmonella* species or fecal coliforms to below the Class A criteria, and must operate under the specified conditions used in the application demonstration to the US EPA Pathogen Equivalency Committee. These processes include composting, heat drying, heat treatment, thermophilic aerobic digestion process, irradiation process, and pasteurization. The requirements of each process are described in Table 8.4. Details of each process are described in the next section.

## 4.2. Class B Pathogen Requirements

Class B is the minimum level of pathogen reduction for land application and surface disposal. However, if the biosolids are placed in a landfill as daily cover, they are not required to achieve the Class B qualification. The goal of Class B is to reduce pathogens to levels protective of human health and the environment, but not to undetectable levels.

Class B sewage sludge must meet one of the following pathogen requirements:

- The geometric mean of at least seven separate samples at the time of use or disposal must be calculated and analyzed for fecal coliforms during each monitoring period. The geometric mean of the densities of these samples should meet the following criteria:
  - Less than 2,000,000 MPN per gram of total dry solids (2,000,000 MPN/g TS), or
  - Less than 2,000,000 colony-forming units (CFU) per gram of total dry solids (2,000,000 CFU/g TS).

- The sewage sludge must be treated by a process to significantly reduce pathogens (PSRP) or PSRP equivalent process. The requirements of each process are described in Table 8.5. Details of each process are described in the next section.

## 4.3. Other Biosolids Characteristics for Landfilling

### 4.3.1. Reduction of Vector Attraction

Vector attraction reduction is concerned mainly with the method used to reduce the potential for the spreading of infectious disease agents by vectors, such as flies,

Table 8.4
Description of processes to further reduce pathogens (PFRP)

| Pathogen treatment processes | Description |
| --- | --- |
| Composting | |
| (a) Vessel or static-aerated-pile composting methods | Maintain the temperature of the sewage sludge at 55°C or higher for 3 days |
| (b) Windrow methods | Maintain the temperature of the sewage sludge at 55°C or higher for 15 days or longer; during this period, a minimum of five windrow turnings are required |
| Heat drying process | Dry the sewage sludge by direct or indirect contact with hot gases to reduce the moisture content of the sewage sludge to 10% or lower |
| | Either the temperature of the sewage-sludge particles must exceed 80°C or the wet bulb temperature of the gas in contact with the sewage sludge leaving the dryer must exceed 80°C |
| Heat treatment process | Heat liquid sewage sludge to a temperature of 180°C or higher for 30 min |
| Thermophilic aerobic digestion process | Agitate liquid sewage sludge with air or oxygen to maintain aerobic conditions |
| | The mean cell residence time for the sewage sludge must be 10 days at 55°C to 60°C |
| Beta-ray irradiation process | Irradiate the sewage sludge with beta rays from an accelerator at a dose of at least 1.0 megarad at room temperature |
| Gamma-ray irradiation process | Irradiate the sewage sludge with gamma rays from certain isotopes, such as cobalt 60 and cesium 137, at a dose of at least 1.0 megarad at room temperature |
| Pasteurization process | Maintain the temperature of the sewage sludge at 70°C or higher for 30 min or longer |

*Source*: US EPA (2).

rodents, and birds. There are two classifications, long-term and short-term stabilization, to decrease vector attraction:

- Long-term stabilization is defined as the biological degradation of the organics, and it results in a reduction of vector attraction.
- Short-term stabilization is defined as the inhibition of biological activity before application. These processes must be demonstrated at the time of use to ensure that the criteria are satisfied. The treatment should be completed before land application so that further reaction does not occur in the field.

Vector attraction reduction processes include aerobic and anaerobic digestion, aerobic process (e.g., composting), alkaline adjustment, and drying process. Details of each technique for vector attraction reduction are listed in Table 8.6.

**Table 8.5**
**Description of processes to significantly reduce pathogens (PSRP)**

| Pathogen treatment processes | Description |
| --- | --- |
| Aerobic digestion | Agitate the sewage sludge with air or oxygen to maintain an aerobic condition for a mean cell residence time and temperature between 40 days at 20°C and 60 days at 15°C |
| Anaerobic digestion | Treat the sewage sludge in the absence of air for a specific mean cell residence time at a specific temperature between 15 days at 35°C to 55°C and 60 days at 20°C |
| Air drying | Dry the sewage sludge on sand beds or in paved or unpaved basins for a minimum of 3 months. During 2 of the 3 months, the ambient average daily temperature must be above 0°C |
| Lime stabilization | Add sufficient lime to the sewage sludge to raise the pH to 12 after 2 h of contact |
| Composting | Compost the sewage sludge using either within-vessel, static-aerated-pile, or windrow composting methods. Raise the temperature of the sewage sludge to 40°C or higher for 5 days; for 4 h at some point during each of the 5 days, the temperature in the compost pile must exceed 55°C |

*Source*: US EPA (2).

*4.3.2. Physical and Chemical Characteristics of Biosolids*

The suitability of biosolids for landfilling can also be determined by biological, chemical, and physical analyses. Several of the more important properties of biosolids are as follows (13):

- Total solids (TS) include suspended and dissolved solids, which are usually expressed as the concentration present in biosolids. The values of total solids are varied owing to the type of wastewater process and biosolids treatment prior to landfill. Typical solids contents of various biosolids processes are as follows: liquid (2% to 12%), dewatered (12% to 30%), and dried or composted (50%).
- Volatile solids (VS) are an estimate of the readily decomposable organic matter in biosolids, which is usually expressed as a percentage of the total solids. A number of treatment processes, including anaerobic digestion, aerobic digestion, alkaline stabilization, and composting, can be used to reduce VS content and thus the potential for odor.
- The pH is a measure of the degree of acidity or alkalinity of a substance. The pH of biosolids is often raised with alkaline materials to reduce pathogen content and the attraction of disease-spreading organisms (vectors). High pH (greater than 11) kills virtually all pathogens and reduces the solubility, biological availability, and mobility of most metals.
- Nutrients are elements required for plant growth that provide biosolids with most of their economic value. Biosolids generally have lower nutrient content than commercial fertilizers. Biosolids typically contain 3.2% nitrogen, 2.3% phosphorus, and 0.3% potassium.

**Table 8.6**
**Details of processes for vector attraction reduction**

| Vector attraction reduction processes | Description |
|---|---|
| Aerobic digestion | Mass of volatile solids (VS) should be reduced by 38% or more |
| | If 38% VS cannot be achieved, vector attraction reduction can be demonstrated by further digesting a portion of the digested sewage sludge with a solids content of 2% or less in a bench scale unit for an additional 30 days at 20°C and achieving a further VS reduction of less than 15% |
| Anaerobic digestion | Mass of volatile solids (VS) should be reduced by 38% or more |
| | If 38% VS cannot be achieved, vector attraction reduction can be demonstrated by further digesting a portion of the digested sewage sludge in a bench scale unit for an additional 40 days at 30° to 37°C or higher and achieving a further VS reduction of less than 17% |
| Composting | Temperature is kept at greater than 40°C for at least 14 days and the average temperature during this period is greater than 45°C |
| Alkaline stabilization | pH is raised to at least 12 by alkali addition and, without the addition of more alkali, remains at 12 or higher for 2 hours and then at 11.5 or higher for an additional 22 hours (when pH is measured at 25°C) |
| Drying | Total solids (TS) is at least 75% when the sewage sludge does not contain unstabilized primary solids and at least 90% when unstabilized primary solids are included |
| | Blending with other materials is not allowed to achieve the total solids % |
| Injection | Liquid sewage sludge (or domestic septage) is injected beneath the surface with no significant amount of sewage sludge present on the surface after 1 hour |
| | Sewage sludges that are Class A for pathogen reduction must be injected within 8 hours of discharge from the pathogen reduction process |

*Source*: US EPA (2).

However, concentrations of nutrients in biosolids can vary significantly; thus, the actual material being considered for land application should be analyzed (14).

- Trace elements (e.g., copper, zinc, molybdenum, boron, calcium, iron, magnesium, and manganese) are found in low concentrations in biosolids. The trace elements of interest in biosolids are those commonly referred to as heavy metals. Possible hazards associated with a buildup of trace elements in the soil include their potential to cause phytotoxicity (i.e., injury to plants) or to increase the concentration of potentially hazardous substances in the food chain.
- Organic chemicals are complex compounds that include man-made chemicals from industrial wastes, household products, and pesticides. Many of these compounds are toxic or carcinogenic to organisms exposed to critical concentrations over certain periods of time, but most are found at such low concentrations in biosolids that the US EPA concluded they do not pose significant human health or environmental threats. Although no organic pollutants are included in the current federal biosolids regulations, further assessment of several specific organic compounds (e.g., dioxins, PCBs) is being conducted.

- Odors of biosolids are unpleasant. Biosolids may have their own distinctive odor depending on the type of treatment they have been through. Some biosolids may have only a slightly musty ammonia odor. Others have a stronger odor that may be offensive to some people. Much of the odor is caused by compounds containing sulfur and ammonia, both of which are plant nutrients.

### 4.4. Analytical Methods in Determining Biosolids Characteristics

To ensure that biosolids have the proper properties for landfilling, the analytical methods for each characteristic are specified in Part 503. Methods for analyzing inorganic pollutants are also described in the following documents: from the US EPA, *A Guide to the Biosolids Risk Assessments for the US EPA Part 503 Rule* (2), *Environmental Regulations and Technology: Control of Pathogens and Vector Attraction in Sewage Sludge* (6), and *Test Methods for Evaluating Solid Waste* (15); from the American Society for Testing and Materials, *Standard Practice for Recovery of Viruses from Wastewater Sludge* (16); from the American Public Health Association, *Standard Methods for the Examination of Water and Wastewater* (17); and other documents (18, 19).

## 5. BIOSOLIDS TREATMENT FOR LANDFILLING

Biosolids characteristics are the major factor in determining a municipality's choice of use or disposal methods. Moreover, biosolids management, including the handling method, transportation, and cost, is also affected by these characteristics. As regulated on Part 503, biosolids must meet certain requirements for pathogens, vector attraction reduction, and metal content. Consequently, the treatment of biosolids is needed to improve biosolids characteristics to meet the regulated requirements. Some biosolids treatment processes increase the volume or mass of the biosolids, while others reduce the biosolids mass. Biosolids treatment technology provides the primary mechanism for pathogen reduction and the necessary stabilization to reduce the attraction of biosolids as a food source for vectors. In addition, the treatment process also reduces odors and the related public nuisance and public health concerns.

The treatment of biosolids can be initiated before the wastewater enters the wastewater treatment plant. The incoming wastewater streams can be monitored to ensure their recyclability and compatibility with the treatment plant process. When the wastewater reaches the plant, the sewage passes through physical, chemical, and biological processes. The solids can be separated and removed from the clean wastewater. The removable solids can be treated with lime to raise the pH level to eliminate odors. In addition, the obtained biosolids can be sanitized by adding chemicals to control pathogens and other organisms capable of transporting disease. After wastewater treatment processes, the obtained solids might be recycled and applied as fertilizer to improve and maintain productive soils and stimulate plant growth. The quality and characteristics of biosolids for meeting the requirement are dependent not only on the treatment process but also on the composition of wastewater as well. However, most biosolids need further treatment processes after passing the wastewater treatment plant. Physical and chemical processes are often employed additionally to improve the biosolids handling characteristics, increase the economic viability of land application, and reduce the potential for public health, environmental, and nuisance problems associated with land application practices (3). These processes stabilize wastewater treatment solids to control pathogens, including enteric viruses, bacteria, parasites, and viable helminth ova, and to reduce characteristics that might attract rodents, flies, mosquitoes, or other organisms capable of transporting infectious disease (3). The type and extent of the processes used to treat wastewater will affect directly the characteristics of he biosolids. Several biosolids treatments include conditioning, thickening, stabilization, dewatering, thermal conversion, drying, and other processes.

## 5.1. Conditioning

Conditioning is the process that causes coagulation of biosolids and improves sludge dewatering characteristics. This method may increase dry solids mass and enhance stabilization for biosolids. Two types of conditioning methods used to improve the biosolids characteristics are chemical and physical conditionings.

Chemical conditioning can be done by adding chemicals to the biosolids. The chemicals used are ferric chloride ($FeCl_3$), lime ($CaO$), and organic polymers. The chemical additives work as coagulants, which can improve sludge dewatering. Polymers are the most frequently used chemicals in treating biosolids.

Physical conditioning is based on the heating method. The conditioning temperature of biosolids is at 350° to 390°F. The heating process is done for 30 minutes, during which the pathogens are destroyed and the moisture is removed. Using this method, the dewaterability can be improved greatly.

Recently, several innovation technologies in conditioning have been proposed. These emerging techniques are electrocoagulation, ultrasonic disintegration, and enzyme addition, as well as a combination of conditioning processes. In ultrasonic disintegration, acoustic waves are applied to solids prior to digestion to attain extremely high pressures and temperatures within the biosolids (20). This results in the implosion of gas bubbles, which produces shear stresses that break up surfaces of bacteria, fungi, and other cellular matter. The full-scale technology has been used at the Bad Bramstedt Sewage Works, Germany; in Kävlinge, Sweden (2002); and at the Mangere Wastewater Treatment Plant, New Zealand (2005) (20).

Electrocoagulation was initiated for mining and metals industry applications in the early 1900s. Electrical current is used to dissolve a sacrificial anode and thereby introduce chemically reactive aluminum into the wastewater stream. The negatively charged ions and particles in water will be attracted to the positively charged aluminum ions (20). The agglomeration particles obtained no longer remain in suspension. These particles can also be removed by a dissolved air flotation technique.

For enzyme addition, the objective of this method is to degrade organic material to increase the dewaterability of biosolids and to reduce odors and aid in digestion processes (20). Enzymes used in this method are specialized nutrients (i.e., humic acids, amino acids) and aerobic and anaerobic bacteria cultures. The mixtures of enzymes are added to thickening and digestion systems that produce enzymes specially engineered to degrade organic materials, converting them into carbon dioxide and water (20).

## 5.2. Thickening

The objective of thickening is to increase the concentration and reduce the volume of biosolids. Thickening enhances the treatment processes that follow such as stabilization, dewatering, and drying. In general, there are two basic types: gravity thickeners and flotation thickeners. Gravity thickeners are usually settling tanks with or without mechanical thickening devices. This method can produce solids contents in sludges of up to 8.0% for primary sludges and up to 2.2% for activated sludge. Lime and polyelectrolytes are also used as chemical additives to enhance efficiency. For flotation thickening, the sludge solids rise to the surface where they are collected by a dissolved-air flotation process. This method is suitable for activated sludge treatment where solids contents of 4% or higher are obtained.

Several technology innovations have been integrated in the thickening process to enhance its ability in increasing solids concentrations. These techniques are anoxic gas flotation, membrane thickeners, and recuperative thickeners. These technologies can help reduce chemicals and increase the efficiency of subsequent processes such as digestion. Anoxic gas flotation is used to thicken the contents of an anaerobic digester to reduce the volume of solids to dewater and transport. This technology was used for over a year at the Salmon Creek Plant in Burien, King County, Washington (20). Volatile solids were increased from approximately 50% to 71%, resulting in a 34% reduction in the volume of solids to be hauled off-site. In addition, odors from the belt presses and the amount of polymer required for dewatering were both reduced. Membrane thickeners are normally used in an aerobic environment to achieve separation of liquid from biomass. The types of membranes applied for this purpose include tubular, hollow-fiber, spiral wound, plate and frame, and pleated cartridge filters. The membrane thickeners are operating in several locations throughout the U.S. Full-scale facilities are in use in Dundee, Michigan, and Fulton County, Georgia, among other locations (20). The last type of innovative thickener process is the recuperative thickener. It can reduce biosolids volume, enhance biosolids destruction and gas production, and reduce dewatering and beneficial use/disposal costs (20). In this process the digested biosolids are removed from the anaerobic digestion process, thickened, and returned to the anaerobic digestion process. This technology allows for the use of existing biosolids process equipment and does not have additional capital costs (20).

## 5.3. Stabilization

The stabilization process reduces the pathogen levels, eliminates offensive odors, and reduces, inhibits, or eliminates the potential for putrefaction that leads to odor production. In addition, this process is also focused on reducing attraction to vectors. Several methods of stabilization are available, including alkali stabilization, anaerobic digestion, aerobic digestion, composting, and heat drying. The selection of the method depends on the level of sewage sludge treatment and the required characteristics for Class A and Class B biosolids. Table 8.7 presents the application of each stabilization method for use or disposal. Details of each method in stabilization process are described below.

### 5.3.1. Alkali Stabilization

The objective of alkali stabilization is to improve the structure characteristics of the stabilized biosolids. Stabilization of sludge is possible through the addition of alkali. This method can be achieved by raising pH to 12 or higher by alkali addition. However, without the addition of more alkali, the pH should be maintained at 12 or higher for

**Table 8.7**
**Stabilization processes for use and disposal methods**

| Treatment process | Use or disposal method |
| --- | --- |
| Aerobic or anaerobic treatment (digestion) | Produces biosolids used as a soil amendment and organic fertilizer on pasture and row crops, forests, and reclamation sites; additional treatment, such as dewatering, also can be performed |
| Alkaline treatment | Produces biosolids useful for land application and for use as daily landfill cover |
| Composting | Produces highly organic, soil-like biosolid with conditioning properties for horticultural, nursery, and landscape uses |
| Heat-drying/palletizing | Produces biosolids for fertilizers generally used at a lower rate because of higher cost and higher nitrogen content |

*Source*: US EPA (3).

30 minutes. Lime is most commonly used as the chemical for this process. Other alkaline materials can be used, including cement kiln dust, lime kiln dust, Portland cement, and fly ash. The high pH of alkali-stabilized sludge tends to immobilize heavy metals in the sludge. In addition, this method temporarily decreases biological activity. Pathogen levels can be reduced and the putrescibility can be controlled. Owing to the temporally effects of pH, this method cannot permanently halt the decomposition of biosolids. Leachate generation and the release of gas, odors, and heavy metals can occur.

### 5.3.2. Digestion (Anaerobic and Aerobic)

This method entails biological stabilization of biosolids through conversion of some of the organic matter to water, carbon dioxide, and methane. Digestion can be divided into anaerobic and aerobic processes. Anaerobic digestion involves the stabilization of sludge in a closed tank to reduce the organic content, mass, odor (and the potential to generate odor), and pathogen content of biosolids. Anaerobic bacteria in the digester convert organic solids to carbon dioxide, methane, and ammonia. Anaerobic digestion is typically operated at about 35°C (95°F), but also can be operated at higher temperatures (greater than 55°C [131°F]) to further reduce the solids and the pathogen content of the stabilized biosolids (3).

In the aerobic process, the stabilization of sludge is handled in an open or closed vessel or lagoon using aerobic bacteria to convert the organic solids content to carbon dioxide, water, and nitrogen. This process is mostly operated under the high temperature (i.e., higher than 55°C [131°F]) of aerobic digestion (3). This process can produce biosolids with lower pathogen levels and higher solids content.

Both anaerobic and aerobic digestion can reduce the volatile and biodegradable organic content. Pathogen levels and putrescibility and odor are also diminished. The volume of biosolids may be reduced by concentrating solids into a denser sludge. Consequently, the sludge quantity for landfilling is reduced. This process is mostly applied to biosolids before landfilling.

## 5.3.3. Composting

Composting is an aerobic process involving biological stabilization in which organic matter is decomposed by microorganisms. This method starts by mixing dewatered biosolids with a bulking agent (e.g., wood chips, municipal yard trimmings, rice hulls, etc.) and allowing the biosolids mixture to decompose in the presence of air for a period of time. In this process, the bulking agent is added to lower the moisture content of the biosolids mixture, increase porosity, and add a source of carbon. Product derived from composting is a humus-like material with excellent soil conditioning properties. In addition, during the process the temperature can increase to 55° to 60°C, and the pathogens are destroyed (3). The biosolids compost can be ready in about 3 to 4 weeks of active composting followed by about 1 month of the curing process. This process can be operated in a variety of systems, such as in a windrow, aerated static pile, or vessel.

Composting is most likely not appropriate for landfilling due to its cost. Generally, this method is used to create a sludge suitable for land application rather than surface disposal.

## 5.3.4. Heat Drying

Heat drying is a biosolids treatment process using heat from direct or indirect dryers to evaporate water from sewage sludge. As noted on a biosolids technology fact sheet (21), compared to other biosolids improvement methods this technique is ideal for producing Class A biosolids. This process is one of several methods that can reduce the volume and improve the quality of wastewater biosolids. The superb characteristic of biosolids resulting from this process is low water content (the typical heat dried product is at least 90% solids, compared to 15% to 30% solids commonly produced by mechanical dewatering). Consequently, the transportation costs are lower and sludge storage is not complicated. Moreover, this method increases the sludge calorific value, so that biosolids can be easily incinerated without any additional fuel. However, the major disadvantages include a requirement of substantial capital investment and a large amount of energy to operate this process. Heat drying systems can require 1400 to 1700 British Thermal Units (BTU) per pound of water evaporated (21). The most important feature of a heat-drying system is the dryer, which can be classified as direct, indirect, or other. Direct and indirect dryers typically have been widely used for drying biosolids. In direct dryers, the biosolids come into contact with hot gases, which cause evaporation of moisture. The rotary dryer is the most typical dryer used in biosolids treatment. In indirect dryers, the heating medium is used to enhance the drying process. The heating medium can be oil or steam. In this process, the biosolids remain separated from the heating medium by metal walls. The evaporation of moisture occurs when the biosolids contact the metal surface heated by the hot medium. Indirect dryers include hollow-flight dryers, steam dryers, and tray dryers.

## 5.4. Dewatering

Dewatering is aimed at decreasing the biosolids volume by reducing the water content of biosolids and increasing the solids concentration. Consequently, the end products from this process are more easily disposed of by landfill, incineration, or other means. Normally, sewage sludge must be treated by adding chemicals prior to dewatering. For this purpose, chemicals such as ferric chloride, lime, or polymers are added to facilitate the separation of solids by aggregating small particles into larger masses. Then, in the thickening process, the moisture bound to biosolids is removed to concentrate the biosolids. By the dewatering process, the liquid biosolids are converted to a damp cake, which is easier to handle and transport. The dewater processes used in improving sludge characteristics are gravity thickening, air drying, vacuum filters, centrifuges, and belt filter press. Details of each dewatering process are as follows:

- Belt filter press: This process employs single or double moving belts to dewater sludges. The three stages in the belt filter press are chemical conditioning of the feed sludge, gravity drainage to a non-fluid consistency, and shear and compression dewatering of the drained sludge. A belt filter press typically produces a cake solids concentration in the 20% to 32% range (3).
- Air drying: This process employs large land areas in which to place biosolids and long periods of time in which to allow biosolids to dry through evaporation and drainage. Normally, this process uses sludge drying beds. These beds are very susceptible to climatic conditions such as precipitation, air temperature, relative humidity, and wind velocity. Products from this process typically contain a solids concentration as high as 45% to 90% (3).
- Vacuum filter: This process relies on mechanical dewatering systems, which can be of the drum type, belt type, string discharge type, or coil type. Dewatering of biosolids involves rotating a drum submerged in a vat of biosolids and applying a vacuum from within the drum, drawing water into the drum, and leaving the solids or filter cake on the outer drum filter medium (3). The dewatered biosolids are scraped off the filter. Chemicals such as $FeCl$ are the coagulant most commonly used in this process. Biosolids from this process normally have a 12% to 22% solids content (3).
- Centrifugation: Dewatering of biosolids by this process can occur by applying a cylindrical vessel (drum or bowl) at high speed to generate a fast rotation in separating the biosolids from liquid. The major factor playing an important role in separating biosolids from liquid is the difference in density. Centrifugation can result in a 25% to 35% solids content (3).

## 6. DESIGN OF BIOSOLIDS LANDFILLING

### 6.1. Landfilling Application for Biosolids

Landfilling is one type of surface disposal, according to the Part 503 biosolids rule. This method is possibly the most common surface disposal regulated in this document. In fact, the application of biosolids in landfilling can be done in several ways, such as disposal in a monofill (a landfill that accepts only wastewater treatment plant biosolids) or in a co-disposal landfill (a landfill that combines biosolids with municipal solid waste), and land application as materials for the final cover in landfill. However, the Part 503 biosolids rule regulates the surface disposal of biosolids as monofill other than landfilling of biosolids with municipal solid waste in an municipal solid waste landfill, which is required to meet the Part 258 landfill rule. In addition, the applications of biosolids that are placed on land for either storage (generally less than 2 years) or treatment (e.g., lagoon treatment for pathogen reduction) are also not included in Part 503.

Generally, most biosolids are utilized for land applications such as fertilizer, lawns and gardens, potted growing media, soil amendments, composting, or palletizing. An application of biosolids for landfill cover or as aggregate is accounted for by land application and not within the disposal. As reported by the US EPA (3), only a small

*Landfilling Engineering and Management* 433

amount of biosolids are believed to be disposed of in monofills, piles, or surface disposal sites. Among these surface disposals, monofill is probably the most widely used method. This section mainly addresses biosolids monofill; however, the application of biosolids in co-disposal landfill is also included.

## 6.2. Biosolids Monofill

Generally, landfill can be defined as burying of waste to minimize nuisance conditions and public health problems. Monofill is one type of landfill that accepts only a single type of waste for disposal. For biosolids, monofill is considered when land application or other beneficial reused is not possible. For example, biosolids have odorous material that may create a public nuisance or contain high concentration of metals or other toxins (22). As noted in the US EPA biosolids technology fact sheet, US EPA 832-F-03-012, there are advantages and disadvantages in using landfill for biosolids. The advantages are as follows (22):

- Landfilling is suitable for biosolids with high concentrations of metals or other toxics.
- Landfills may require a smaller land area than does land application.
- Landfilling improves the packing of solid waste and increases biogas production.
- Landfills may be the most economical biosolids management solution, especially for malodorous biosolids.

The disadvantages of biosolids landfilling compared to other methods are to be considered before making a decision. These disadvantages are as follows (22):

- Landfilling biosolids eliminates their reuse potential and is contrary to the US EPA national beneficial reuse policy.
- Landfilling requires extensive planning, including the selection of a proposed landfill site, and operation, closure, and postclosure care of the site.
- Operation, maintenance, and postclosure care of landfills are labor intensive.
- Landfill sites have a potential for groundwater contamination from leachate.
- Decomposition of biosolids in a landfill produces methane gas, which must be collected and reused or disposed of by flaring or venting. Energy can be recovered through methane capture systems to offset the cost of the necessary collection system.
- Landfills have a potential for odor generation.

To apply in landfill, it is required that biosolids should be stabilized prior to allowing them to be deposited in monofill and should contain 15% or greater solids content. In some case, soil may be mixed with biosolids to increase the solids concentration to this level (22). Also, biosolids must pass a paint filter test due to the regulatory prohibition of materials containing free liquids. Details of the paint filter test are described in US EPA publication SW-846, Test Methods for Evaluating Solid Waste, Physical/Chemical Methods, Method 9095. In addition, a toxicity characterization leaching procedure (TCLP) must be performed to verify that the biosolids are nonhazardous. Details of the TCLP method are also defined in US EPA SW-846, method 1311.

### 6.2.1. Site Selection

Site selection is one of several crucial aspects in landfilling that must be considered so as to reduce the impact on the environment. There are several potential environmental impacts associated with landfilling of biosolids: groundwater contamination from leachate produced in the landfill; the emission of gases that can perturb air quality in the neighboring area; greater traffic volume; greater land use; and risks to public health, aesthetics, wildlife, and habitats of endangered species. Selecting an appropriate landfill site can reduce these impacts. However, specific impacts vary among landfill locations. The difference in environmental impacts for each site cannot be simply quantified. In assessing a possible location for landfilling, the fundamental criterion that should be considered is that the site must be free of potentially crucial problems such as landslides, subsidence, and flooding.

Selection of appropriate site for landfill is discussed in the siting criteria established by either Part 503 (for monofills) or Part 258 regulations (for co-disposal landfills). Requirements are as follows (22):

- A landfill shall not be developed if it is likely to adversely affect a threatened or endangered species.
- The landfill cannot be located in a wetland unless a special permit is obtained.
- A landfill cannot restrict the flow of a 100-year flood event.
- The landfill must not be located in a geologically unstable area.
- The landfill must be located 60 m (200 ft) or more from a fault area that has experienced displacement in Holocene time.
- If the landfill is located in a seismic impact zone, it must be designed to resist seismic forces.
- The landfill must be located at least 300 m (1000 ft) from an airport runway.

The first step in the site selection process is identification of a number of potential sites that comply with the regulated criteria. Several sites have to be compared to obtain the most appropriate one. Analyses must be carried out to determine the deficiencies of each site. The natural environmental impact of each site has to be addressed. The most appropriate site is the one that has no or minimal impact on the environment and public health.

### 6.2.2. Methods of Biosolids Landfilling

Methods of biosolids landfilling can be divided into three types:

1. Trench method: This method is suitable for areas where the water table is not near the surface. In this method, the biosolids are filled in the trench. Monofill trenches can be either narrow (typically less than 3 m wide) or wide (typically greater than 3 m wide), depending on the solids content of the biosolids. It is suggested that narrow trenches are suitable for disposal of biosolids with a solids content less than 20%, and wide trenches are appropriate for disposal of biosolids with a solids content of 20% or more (22). Trench depths and widths are variable. A schematic showed a cross section and site plan of a landfill using the trench method is illustrated in Figures 8.1 and 8.2. For the wide trench, equipment selection depends on site-specific constraints. The monofill trench is generally not used at sites that require a liner because of the potential of damage to the liner during trench excavation (22).
2. Area method: Biosolids are placed in a natural or excavated depression, or they are mixed with soil and placed on top of the existing layer. This method is appropriate for areas where the terrain is unsuitable for the excavation of trenches. It is particularly suited to areas where bedrock or ground water are shallow. There are several categories of area method (23, 24):
    a. Area fill mound: Biosolids are mixed with a bulking agent, which is usually soil. The mixture is transported to the filling site or area where it is stacked in approximately 6-foot-high mounds. A 3-foot thickness of cover material is then applied; this may be increased to 5 feet if additional mounds are applied on top of the first mounds. The appropriate soil/sludge bulking ratio is 0.5–2:1. This soil cover thickness may also

# Landfilling Engineering and Management

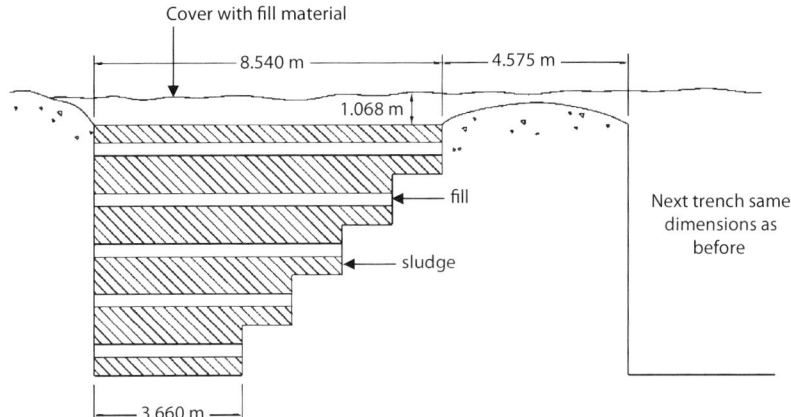

**Fig. 8.1.** Landfill trench section. *Source*: US EPA (23).

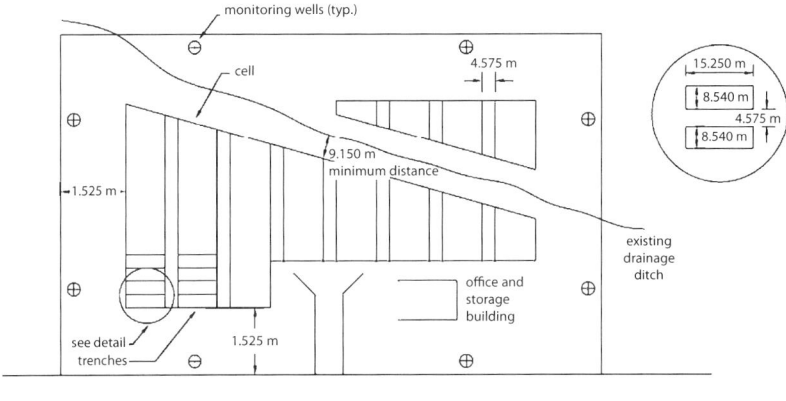

**Fig. 8.2.** Landfill site plan. *Source*: US EPA (23).

depend on the solid content of the sludge, the requirement for mound stability, and the bearing capacity based on the number of lifts and the equipment weight. Lightweight equipment with swamp pad tracks is appropriate for area fill mound operations; heavier wheel equipment is appropriate in transporting bulking material to and from stockpiles (1).

b. Area fill layer: Biosolids are mixed with soil on- or off-site. The mixture is then spread evenly in consecutive layers of 0.5 to 3 feet thick. Interim cover may be applied to an indefinite height before the final cover is applied. Lightweight equipment with swamp pad trucks and heavier wheel equipment are appropriate for area fill layer and hauling soil, respectively. A very mild slope should be constructed to prevent sludge from flowing downhill. Otherwise, layering could be performed on a mildly slopping terrain, if the sludge solids content is high or sufficient bulking soil is employed.

c. Diked containment: This is a four-sided containment area with level ground on which dikes are constructed around the four sides. Dikes could also be constructed in such a way that the containment area is placed at the top of the hill and the steep slope is used as containment on one or two sides. The remaining sides would then house the dikes.

3. Ramp method: This method is applied to the areas with the shape of canyons, ravines, dry borrow pits, and quarries. Biosolids are spread and compacted along a slope of the terrain. The technique to place and compact biosolids varies with the geometry of the site.

## 6.3. Design Criteria

### 6.3.1. Area Requirement

After obtaining the landfill site, the landfill design starts by estimating the average biosolids quantity. The biosolids characteristics of sludge unit weight and solid content percentage have to be determined for use in this calculation. Once the biosolids quantity is known, the area required for the landfill site and its probable life span should be calculated. Generally, a site should provide 10 to 20 years of operational capacity (22). In addition, other facilities should be designed with the proper manner. For example, a runoff collection system should be designed to contain a 25-year, 24-hour storm, which occurs once every 25 years and lasts for 24 hours.

### 6.3.2. Landfill Size Estimation

Design criteria for landfill operations and soil layer depths for a typical site using the trench method are presented in Table 8.8.

Generally, the depths of landfill are required for safety to prevent contamination of adjacent areas and for ease of operation with conventional equipment. Elevation of trenches to 15 to 20 feet above the impervious layer effectively controls the leachate (25). It is also suggested that application rates for trenches less than 3 m in width be approximately 2270 to 10,580 $m^3$/hectares (ha). For wide trench ranges, the typical application rates are 6000 to 27,000 $m^3$/ha (26, 27).

### 6.3.3. Landfill Liner

Landfill liner is used as the barrier to protect the groundwater from the contamination of leachate. However, in some cases the liner is not required for biosolids landfilling. In the Part 503 regulations, the liner is required for a monofill if the concentration of the

Table 8.8
Design criteria for landfill operations and soil layer depths for a typical site

| Design criteria | Size |
| --- | --- |
| Trench width | |
|    Narrow trench | <3 m |
|    Wide trench | >3 m |
| Trench length | 15 m |
| Sludge layer | 60 cm |
| Intermediate fill layer | 30 cm |
| Top fill cover | 1–1.5 m |
| Distance between trenches | 4.5 m |
| Distance from drainage ditch | 9 m |

Source: US EPA (24).

*Landfilling Engineering and Management* 437

**Table 8.9**
**Maximum pollutant concentration for surface disposal of biosolids**

| Distance from boundary of active biosolids unit to property line (m) | Pollutant concentration (dry weight basis) | | |
|---|---|---|---|
| | Arsenic (mg/kg) | Chromium (mg/kg) | Nickel (mg/kg) |
| 0 to less than 25 | 30 | 200 | 210 |
| 25 to less than 50 | 34 | 220 | 240 |
| 50 to less than 75 | 39 | 260 | 270 |
| 75 to less than 100 | 46 | 300 | 320 |
| 100 to less than 125 | 53 | 360 | 390 |
| 125 to less than 150 | 62 | 450 | 420 |
| Equal to or greater than 150 | 73 | 600 | 420 |

*Source*: US EPA (1).

pollutants (arsenic, chromium, and nickel) in biosolids exceeds the regulated criteria. The maximum concentrations of these heavy metals in biosolids are shown in Table 8.9.

If liners are required, three types can be used in monofill: low permeability soil such as clay, geosynthetic clay liners (GCL), and geomembrane liners. Liners are regulated as shown in Part 503; a maximum hydraulic conductivity of $1 \times 10^{-7}$ cm/s is required for a monofill liner (22).

For the application of biosolids in co-disposal landfill, as shown in the Part 258 regulations, a composite liner is required. The top component of the liner must consist of a minimum 30-mil flexible membrane liner. At least a 60-cm-thick layer of compacted soil with a hydraulic conductivity of $1 \times 10^{-7}$ cm/s is required for the bottom component. The flexible membrane component must be installed in direct and uniform contact with the compacted soil (22).

### 6.3.4. Gas Collection Requirement

During the degradation of biosolids in landfill, the natural gas can be generated. The natural gas primarily consists of methane and carbon dioxide. Methane is explosive at concentrations of 5% to 15%. For carbon dioxide, the mixing of this gas with water can cause acidic groundwater. To protect the environment from the emission of these gases, either passive or active gas collection can be applied.

A passive gas collection consists of perforated collection pipes and header pipes. This system collects and releases gases to the atmosphere. Active systems consist of a network of gas wells drilled into the biosolids with a blower to collect the gas. This system is normally used to recover energy from methane. In addition, this technique is also used to control odors at the landfill site.

### 6.3.5. Landfill Operation and Maintenance

Performance of a landfill is measured as a function of disposal without causing nuisance to the surrounding environment (25). The landfill system should be operated in a way that prevents odors and vector habitants. To do this, there should be no runoff or change in natural drainage.

In landfill operating and maintenance, biosolids can be directly dumped into the containment by haul vehicles through an access made at the top of the dikes. During filling, an interim cover of 1 to 3 feet may be applied, and at the end of the filling a final cover of 3 to 5 feet should be applied. In addition, to achieve stability and soil-bearing capacity, biosolids can be applied with a bulking agent, which is usually soil. Excess moisture is absorbed by the solids from the sludge, and this increases its workability.

In applying the area method to a biosolids landfill, the area fill mound is applied to stabilized the sludge, and it facilitates land utilization; however, this method has the major disadvantage of entailing higher manpower and equipment requirements due to the constant operation of pushing and stacking the slumping mounds. An area fill layer is also applied to stabilized sludge, which involves lower manpower and equipment requirements. Efficient land use, suitability for stabilized and unstabilized sludge, and lower soil requirements are the advantages of dike containment over other categories of sludge landfilling.

*6.3.6. Limitation of Landfilling*

Landfill is considered a very reliable biosolids disposal method. For the handling of biosolids, landfill disposal also offers the simplest solution by concentrating the material in a single location (13). However, the limitations of landfill are that rainfall could cause mounds to slump, and operation becomes difficult in wet and freezing weather.

The risk of landfill disposal of biosolids also occurs when buried organic wastes undergo anaerobic decomposition, which produces methane gas. Methane, a greenhouse gas, has been implicated in possible global warming when released to the atmosphere (13). In addition, in older landfills that do not have synthetic liners or in newer landfills whose liners leak, there can be the threat of pollution to local groundwater from chemicals and nutrients from the landfills. The potential importance of the organic matter and plant nutrients in the biosolids is lost with landfilling (13). Many communities in the United States have made a positive environmental decision to embark on biosolids recycling, even in cases where landfilling is less expensive than land application (13).

## 7. CASE STUDY AND EXAMPLE

### 7.1. Future Trends in Biosolids Landfilling

Quantities of biosolids generation have been projected as illustrated in the US EPA document, "Biosolids Generation, Use, and Disposal in the United States" (3). An estimation of the percentages of biosolids that were used or disposed of in 2000 and 2005, and a projection of the percentages for 2010 are given in Table 8.10. The beneficial use of biosolids is expected to continue in the future, while the trend toward disposal is expected to decrease slightly. By 2010, the biosolids disposal might account for only about 30% of all biosolids generated.

Several factors such as regulatory influences, voluntary improvements in biosolids quality, and the resulting increase in biosolids use are based on the decreasing of biosolids disposal. Regulatory influences include the increased restrictions on landfilling in the Part 503 biosolids rule, the Part 258 landfill rule, and various state requirements, which also have driven up the costs of these disposal methods (10). As biosolids

**Table 8.10**
**Projections of use and disposal of biosolids in 2000, 2005, and 2010**

| | Beneficial use | | | | Disposal | | | |
|---|---|---|---|---|---|---|---|---|
| Year | Land application | Advanced treatment | Other beneficial use | Total | Surface disposal/ landfill | Incineration | Other | Total |
| 1998 | 41% | 12% | 7% | 60% | 17% | 22% | 1% | 40% |
| 2000 | 43% | 12.5% | 7.5% | 63% | 14% | 22% | 1% | 37% |
| 2005 | 45% | 13% | 8% | 66% | 13% | 20% | 1% | 34% |
| 2010 | 48% | 13.5% | 8.5% | 70% | 10% | 19% | 1% | 30% |

*Source*: US EPA (3).

landfilling declines, an increase in the use of biosolids for landfill cover and composting can be expected.

The application of biosolids as a part of landfill material, especially landfill cover, is done in several areas in the U.S. The sanitary landfill in Tuscarawas County, Ohio, has planned to use biosolids from a wastewater treatment plant as an alternative daily cover material (3). Following the Ohio Administrative Code (OAC) rule 3745-27-19(F)(3)(a), it is required that the biosolids material proposed for use must provide protection comparable to 6 inches of soil and is protective of human health and the environment. The state requires a 180-day trial period in which biosolids are used as an alternative to 6 inches of soil for daily cover at the facility. The landfill monitoring report discusses the following issues:

- The number of days the biosolids were used
- Weather conditions at the time of placement of the biosolids (evaluated for effectiveness as daily cover)
- Number of days odors were detected
- Presence of vectors, rodents, and insects
- Any increase in blowing litter
- Method(s) for controlling fire during use of alternative daily cover material

After applying biosolids as daily cover in landfill, the records must be made available for state inspection.

## 7.2. Calculation Examples

*Example 1*

Let the criteria in monofill design be as follow:

Trench length is 15 m.
Depth of bottom sludge layer is 0.6 m.
Depth of intermediate sludge layer is 0.3 m.
Depth of top sludge layer is 0.6 m.
Depth of top fill cover is 1 m.

If the monofill contains five sludge layers and the widths of each layer (from bottom to top) are 3.5, 4.5, 5.5, 6.5, and 7.5 m, find the trench capacity of sludge.

If the sludge unit weight is $1000 \text{ kg/m}^3$, find the total weight of biosolids to be placed in landfill.

SOLUTION

1. Calculate the volume of sludge filling in monofill for each layer:

   Volume of biosolids filled in bottom layer = $15 \text{ m} \times 0.6 \text{ m} \times 3.5 \text{ m} = 31.5 \text{ m}^3$
   Volume of biosolids filled in 2nd layer = $15 \text{m} \times 0.3 \text{ m} \times 4.5 \text{ m} = 20.25 \text{ m}^3$
   Volume of biosolids filled in 3rd layer = $15 \text{ m} \times 0.3 \text{ m} \times 5.5 \text{ m} = 24.75 \text{ m}^3$
   Volume of biosolids filled in 4th layer = $15 \text{ m} \times 0.3 \text{ m} \times 6.5 \text{ m} = 29.25 \text{ m}^3$
   Volume of biosolids filled in 5th layer = $15 \text{ m} \times 0.6 \text{ m} \times 7.5 \text{ m} = 67.5 \text{ m}^3$

2. Find trench capacity:

   Trench capacity = $31.5 + 20.25 + 24.75 + 29.25 + 67.5 \text{ m}^3$
   Trench capacity = $173.25 \text{ m}^3$

3. Find total weight of biosolids to be filled:

   Volume of biosolids = $173.25 \text{ m}^3$
   Sludge unit weight = $1000 \text{ kg/m}^3$
   Total weight of biosolids to be filled = $173.25 \text{ m}^3 \times 1000 \text{ kg/m}^3$
   Total weight = $173.25 \times 10^3$ kg or $173.25$ metricTons

*Example 2*

For scheduling purposes, the capacity of each trench should be expressed as days of operation. Determine the days of operation using information from Example 1, if the daily haul rate is 30 tons/day.

SOLUTION

From the previous example, total weight of biosolids to be placed in one trench is 173.25 tons and the daily haul rate is 30 tons/day. The days of operation or fill time period can be calculated as follow:

   Fill time = total sludge weight (tons)/daily haul rate (tons/day)
   Fill time = 173.25 metricTons/30 metricTons/day
   Fill time = 5.78 days

This means that a new trench must be prepared every 5 days.

*Example 3*

Let the total area of monofill be $3000 \text{ m}^2$, which can be divided into two parts: trench and infrastructure, and a utilities areas. The infrastructure and facilities areas normally include office buildings, sludge storage, roads, drainage system, etc. If 30% of the monofill is to be the infrastructure and utilities areas, provide the maximum capacity of biosolids to be filled in this site. Use the design criteria from Example 1.

SOLUTION

1. Find the filled area for one trench. The trench area is obtained from Example 1:

   Trench area = width $\times$ length = $15 \text{ m} \times 7.5 \text{ m} = 37.5 \text{ m}^2$

2. Find the trench area in the monofill:

$$\text{Total area} = 3000\,\text{m}^2$$
$$\text{Trench area} = 0.70 \times 3000\,\text{m}^2 = 2100\,\text{m}^2$$
$$\text{Infrastructure and utilities area} = 0.30 \times 3000\,\text{m}^2 = 900\,\text{m}^2$$

3. Number of trenches in this monofill:

$$\text{Number of trenches} = 2100\,\text{m}^2/3.75\,\text{m}^2$$
$$\text{Number of trenches} = 56\,\text{trenches}$$

4. Maximum capacity of biosolids to be filled in this monofill:

$$\text{From Example 1, trench capacity} = 173.25\,\text{m}^3/\text{trench}$$
$$\text{Maximum capacity} = 173.25\,\text{m}^3/\text{trench} \times 56\,\text{trenches}$$
$$\text{Maximum capacity} = 9702\,\text{m}^3$$

Total weight of biosolids to filled in this monofill is 9702 metric Tons.

## REFERENCES

1. U.S. Environmental Protection Agency, *The Standards for the Use or Disposal of Sewage Sludge*, US EPA 822/Z-93/001, Washington, DC (1993).
2. U.S. Environmental Protection Agency, *A Guide to the Biosolids Risk Assessments for the US EPA Part 503 Rule*, US EPA 832/B-93-005, Washington, DC (1995).
3. U.S. Environmental Protection Agency, *Biosolids Generation, Use, and Disposal in the United States*, US EPA 530-R-99-009, Washington, DC (1999).
4. U.S. Environmental Protection Agency, *National Sewage Sludge Survey: Availability of Information and Data, and Anticipated Impacts on Proposed Regulations*, Washington, DC (1990).
5. National Research Council, *Biosolids Applied to Land: Advancing Standards and Practices*, National Academy Press, Washington, DC (2002).
6. U.S. Environmental Protection Agency, *Environmental Regulations and Technology: Control of Pathogens and Vector Attraction in Sewage Sludge*, US EPA/625/R-92/013, Washington, DC (1999).
7. U.S. Environmental Protection Agency, *Pathogen Risk Assessment Feasibility Study*, US EPA 600/6-88/003, Cincinnati, OH (1985).
8. U.S. Environmental Protection Agency, *Environmental Regulation and Technology: Control of the Pathogens in Municipal Wastewater Sludge for Land Application*, Cincinnati, OH (1989).
9. U.S. Environmental Protection Agency, *Criteria for Classification of Solid Waste Disposal Facilities and Practices*, Washington, DC (1979).
10. U.S. Environmental Protection Agency, *Landfill Rule and Follow the Design, Operation and Closure of Municipal Solid Waste Landfills*, Washington, DC (1994).
11. U.S. Environmental Protection Agency, *Disposal of Polychlorinated Biphenyls (PCBs)*, Washington, DC (1998).
12. U.S. Environmental Protection Agency, *US EPA Land Application of Sewage Sludge: A Guide for Land Appliers on the Requirements of the Federal Standards for the Use or Disposal of Sewage Sludge*, US EPA/831-B-93-002b, Washington, DC (1994).
13. G. K. Evanylo, *Production and Characteristics of Biosolids, Agricultural Land Application of Biosolids in Virginia*, Department of Crop and Soil Environmental Sciences, Virginia Tech, Publication Number 452-301, Virginia (1999).

14. L. Sommers, Chemical composition of sewage sludges and analysis of their potential use as fertilizers. *Journal of Environmental Quality*, 6, 225–239 (1977).
15. U.S. Environmental Protection Agency, *Test Methods for Evaluating Solid Wastes, Physical/Chemical Methods (SW-846)*, Washington, DC (1986).
16. American Society for Testing and Materials, *Standard Practice for Recovery of Viruses from Wastewater Sludge*, Annual Book of ASTM Standards, Section 11, Water and Environmental Technology, Philadelphia (1992).
17. American Public Health Association, *Standard Methods for the Examination of Water and Wastewater*, 18th ed., American Public Health Association, Washington, DC (1992).
18. B. A. Kenner, and H. A. Clark, Detection and Enumeration of *Salmonella* and *Pseudomonas aeruginosa*, *Journal of Water Pollution Control Federation*, 46, 2163–2171 (1974).
19. W. A. Yanko, *Occurrence of Pathogens in Distribution and Marketing Municipal Sludge*, US EPA 600/1-87-014, NTIS in Springfield, VA-NTIS PB 88-154273/AS, Washington, DC (1987).
20. U.S. Environmental Protection Agency, *Emerging Technologies for Biosolids Management*, US EPA 832-R-06-005, Washington, DC (2006).
21. U.S. Environmental Protection Agency, *Biosolids Technology Fact Sheet: Heat Drying*, US EPA 832-F-06-029, Washington, DC (2006).
22. U.S. Environmental Protection Agency, *Biosolids Technology Fact Sheet: Use of Landfilling for Biosolids Management*, US EPA 832-F-03-012, Washington, DC (2003).
23. U.S. Environmental Protection Agency, *Sludge Handling and Conditioning*, US EPA 430/9-78-002, Washington, DC (1978).
24. U.S. Environmental Protection Agency, *Innovative and Alternative Technology Assessment Manual*, US EPA 430/9-78-009, Washington, DC (1980).
25. U.S. Environmental Protection Agency, *Sludge Landfilling—Area Fill*, US EPA Fact Sheet 6.1.4, Washington, DC (2003).
26. U.S. Environmental Protection Agency, *Process Design Manual for Municipal Sludge Landfills*, US EPA 625/1-78-010, Cincinnati, OH (1978).
27. L. K. Wang, N. K. Shammas, and Y. T. Hung (eds.), *Biosolids Treatment Processes*. Human Press, Totowa, NJ, 820 pp. (2007).

# 9
# Ocean Disposal Technology and Assessment

## Kok-Leng Tay, James Osborne, and Lawrence K. Wang

**CONTENTS**

  INTRODUCTION
  CONVENTION ON THE PREVENTION OF MARINE POLLUTION
    BY DUMPING OF WASTES AND OTHER MATTER—LONDON
    CONVENTION 1972
  WASTE ASSESSMENT GUIDANCE
  WASTE ASSESSMENT AUDIT
  WASTE CHARACTERIZATION PROCESS AND DISPOSAL PERMIT
    SYSTEM
  DISPOSAL SITE SELECTION
  DISPOSAL SITE MONITORING
  LAND-BASED DISCHARGES OF WASTES TO THE SEA:
    ENGINEERING DESIGN CONSIDERATIONS
  MARINE POLLUTION PREVENTION (THE CITY OF LOS ANGELES
    BIOSOLIDS ENVIRONMENTAL MANAGEMENT SYSTEM)
  OCEAN DISPOSAL TECHNOLOGY ASSESSMENT AND CONCLUSIONS
  NOMENCLATURE
  REFERENCES

**Abstract** The current, managed process of disposal at sea, practiced by most nations worldwide, is in sharp contrast to practices, extending even as late as 1970, involving indiscriminant dumping, based largely on ignorance and an attitude that the ocean had unlimited resources and an unending capacity to absorb impact. If done in an environmentally acceptable manner, disposal at sea can be an attractive option, since it is generally less expensive than other waste management options such as waste treatment or land disposal. The London Convention 1972 and the London 1996 Protocol have essentially banned the disposal at sea of all waste and other matter other than the wastes listed in the Annexes to the convention, namely, dredged material, sewage sludge, ships and platforms, fish wastes, organic wastes of natural origin, inert inorganic geologic material and bulky wastes. Ocean disposal should be the last option if the wastes can be recovered for beneficial use. For instance, ocean disposal of nutrient-rich biosolids has been terminated in the U.S. in favor of land application and compost, in which biosolids are used as a soil amendment, fertilizing material, or compost conditioner. This chapter discusses these important topics: international convention on marine pollution prevention and control, waste assessment guidance, waste assessment audit, waste characterization process and resource recovery, an ocean disposal permit system, disposal site selection and monitoring, and land-based discharges of wastes to the sea. A design example is provided, and the biosolids Environmental Management System (EMS) of the City of Los Angeles for ocean pollution prevention is discussed.

**Key Words** Ocean disposal • wastes • biosolids • London Convention • marine pollution control • waste assessment guidance • waste assessment audit • resource recovery • permit system • disposal site selection • monitoring

## 1. INTRODUCTION

The world's oceans and estuaries provide opportunities for the disposal of a wide range of materials generated by human activities. The current, managed process of disposal at sea, practiced by most nations worldwide, is in sharp contrast to practices, extending even late into the last century, involving indiscriminant dumping, based largely on ignorance and on the attitude that the ocean had unlimited resources and an unending capacity to absorb impact. For example, it was widely assumed that the ocean's assimilative capacity would adequately absorb wastes without creating significant environmental and human health risks (1–4). If done in an environmentally acceptable manner, disposal at sea can be an attractive option, since it is generally less expensive than other waste management options such as waste treatment or land disposal. Over the years, many types of waste have been dumped at sea as a result of human activities. These wastes include dredged material from harbor and shipping channel dredging; industrial wastes from land-based discharges of manufacturing or processing operations; raw sewage or sewage sludge (biosolids) from sewage treatment plants; drilling fluids and cuttings from offshore hydrocarbon exploration and development; seabed mining wastes from offshore activities; surplus munitions; chemical warfare agents; and radioactive wastes from nuclear power production, medical and research activities.

International agreements for the prevention of marine pollution identify at least five sources of marine pollution related to discharges, dumping, or release of wastes at sea (5):

- Pollution caused by dumping at sea: disposal of wastes at sea by ships that have been loaded on land for the purpose of dumping the material at sea
- Land-based sources of pollution: discharges of wastes from land
- Vessel-generated pollution: operational discharges of wastes from ships
- Pollution caused by seabed activities: discharges or release of material directly from exploration, exploitation, and associated offshore processing of seabed minerals
- Pollution from or through the atmosphere: the release of material into the atmosphere by man-made activities on land, vessels, or aircraft, which subsequently deposits in the ocean

Concern for the effect of waste dumping on marine life and its potential linkage to human health risk began to attract international attention in the early 1970s. In response, several international conferences were convened to adopt multilateral conventions to protect the marine environment by controlling, restricting, or prohibiting the release, discharge, and dumping of wastes at sea. One of these conventions was the Convention on the Prevention of Marine Pollution by Dumping of Wastes and Other Matter (London Dumping Convention, 1972) adopted on November 13, 1972, at a conference organized by the British government in London. This convention applies to wastes or other materials loaded for the purpose of dumping at sea. It entered into force in 1975 after it was ratified by 25 nations, including some of the leading industrial countries such as Canada, the United States, the United Kingdom, France, and Germany (6–9).

Other conventions have also been adopted over the past 30 years to address the different sources of marine pollution. The International Convention for the Prevention of Pollution from Ships (MARPOL [marine pollution] 73/78) addresses the routine operational discharges from ships. It has set criteria for the discharge of oil in bilge water, the discharge of residues of liquid cargo carried in bulk, treatment and management of sewage from ships, at-sea disposal of garbage, and prevention of air pollution from ships. Marine pollution resulting from mining of the sea bed for sand or minerals is controlled by the International Sea Bed Authority operating under the umbrella of the United Nations Law of the Sea Convention (UNCLOS), while marine pollution from or through the atmosphere is regulated internationally by both MARPOL (as noted above) and the Convention on Long-range Transboundary Air Pollution.

In 1995, the Global Program of Action for the Protection of the Marine Environment from Land-Based Activities (GPA), a nonbinding agreement to address land-based sources of marine pollution, was signed by 108 countries. The GPA addresses concerns over land-based discharges of sewage, heavy metals, radioactive materials, persistent organic pollutants, radioactive substances, oils and hydrocarbons, nutrients and seabed mobility, and the impacts of habitat destruction. A number of countries have put national programs of action in place to implement the GPA, and the program has been carried out through the United Nations Regional Seas Program.

There are currently no international conventions dealing with management and treatment of discharges from the offshore oil and gas industry, which releases a significant quantity of wastes to the marine environment (10, 11). An international conference held in the Netherlands in 1998 examined the need for international controls, and determined that the industry is adequately managed through regional and national regulatory means and that there was no pressing need at this time to develop international agreements. As a result of the conference, the United Nations Environment Program (UNEP) established an Offshore Oil and Gas Forum on the Internet to provide up–to-date information on industry practices.

This chapter focuses on activities that are generally referred as "ocean disposal" by the international conventions (i.e., the loading of wastes and other matter for deliberate disposal at sea from ships and other structures at sea). It also briefly discusses management of waste discharge from land-based sources of marine pollution, and engineering design considerations for submarine outfall systems.

## 2. CONVENTION ON THE PREVENTION OF MARINE POLLUTION BY DUMPING OF WASTES AND OTHER MATTER—LONDON CONVENTION 1972

Since 1972, the number of countries that have become party to the London Dumping Convention, 1972 (LC72) (changed to the London Convention [LC] in the early 1990s) has increased to 79. Numerous resolutions have been adopted through regular consultative meetings of contracting parties, ad hoc scientific groups, and expert groups to improve the implementation of the convention. The most significant resolutions include the following:

- The 1993 decision on banning ocean dumping of low-level radioactive wastes, phasing out ocean dumping of industrial wastes by December 31, 1995, and banning incineration of wastes at sea
- The development and implementation of the LC 1996 protocol, which replaces the 1972 convention.

The 1996 protocol represents a major change of approach in regulating use of the oceans as a depository for wastes. It places emphasis on progressively reducing the need to use the sea for waste dumping, and states that contracting parties shall prohibit the dumping of all wastes or other matter with the exception of those materials specifically enumerated in Annex 1 of the protocol. The precautionary approach was adopted in the protocol to address uncertainties in relation to assessment of impacts on the marine environment by ocean disposal of wastes. Under this approach, the contracting parties are required to take appropriate preventative measures when there is reason to believe that the dumping of wastes or other matter is likely to cause harm, even when there is no conclusive evidence to prove a causal relation between inputs and their effects (12, 13).

In addition, a generic waste assessment guidance (WAG), which applies to all classes of material that may be considered for ocean disposal, was developed to guide contracting parties in evaluating applications for disposal of wastes at sea in a manner consistent with the provisions of the LC72 and its 1996 protocol. Complementary waste-specific guidelines have also been developed for each group of wastes, including dredged material, sewage sludge, fish offal, vessels proposed for disposal at sea, platforms and other man-made objects, inert geologic materials, organic wastes, and bulky materials.

To meet their international commitments, the LC contracting parties are required to establish appropriate administrative authorities to enforce the provisions of the convention. Not all countries have complied by developing appropriate legislation to meet the provisions; some countries have not yet ratified the 1996 protocol, while others, including Canada, Australia, the United Kingdom, France, and Spain, have developed and implemented the necessary regulatory processes to comply with the protocol.

## 3. WASTE ASSESSMENT GUIDANCE

In Canada, which is an example of a nation that has a regulatory process to implement the LC protocol, an ocean disposal permitting system is administered under the Canadian Environmental Protection Act (CEPA) of 1999 (14). This legislation incorporates the pollution prevention and waste management principles of the LC Waste Assessment

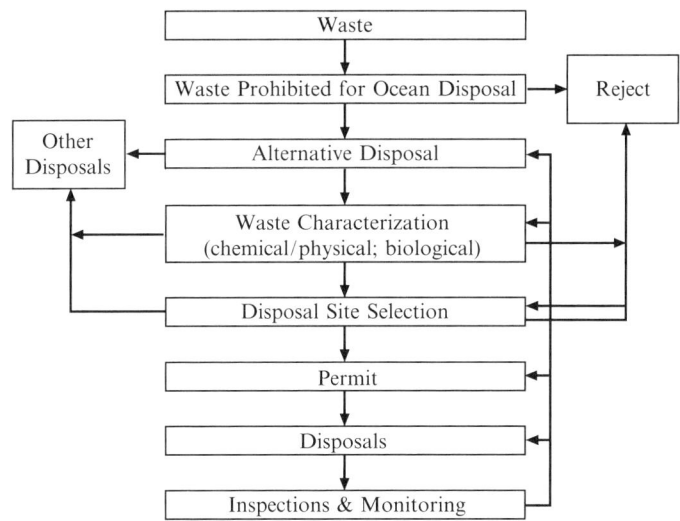

Fig. 9.1. The waste assessment framework.

Guidance (WAG). A simplified Waste Assessment Framework adapted by the Canadian Disposal at Sea Program from the LC Waste Assessment Guidelines is shown in Figure 9.1, and demonstrates a systematic waste assessment process for disposal of wastes at sea (9, 15). Although this process was developed for the assessment of wastes proposed for disposal at sea, it can also be used for the assessment of other waste streams, such as land-based discharges of wastes to the sea.

Under the London Convention 1996 protocol, certain categories of waste can be considered for disposal, and all others are prohibited. Categories of waste that can be considered for disposal at sea are the following:

- Dredged material
- Sewage sludge (i.e., biosolids)
- Fish waste, or material resulting from industrial fish processing operations
- Vessels and platforms or other man-made structures at sea
- Inert, inorganic geological material
- Organic material of natural origin
- Bulky items of iron, steel, concrete, or comparable materials for which the concern is physical impact, and which are generated in locations, such as small islands with isolated communities, having no practical access to other disposal options.

## 4. WASTE ASSESSMENT AUDIT

The LC Waste Assessment Guidance (WAG) recommends that prior to choosing disposal at sea as the appropriate management approach, alternate methods of disposal should be examined. The following aspects of the project should be evaluated to allow assessment of alternatives to dumping of wastes at sea (LC 1996 protocol):

- The types, amounts, and hazardous nature of the wastes
- The sources of the wastes and their production process

- The feasibility of waste reduction/prevention techniques:
  - Product reformulation
  - Clean production technologies
  - Process modification
  - Input substitution
  - On-site, closed-loop recycling

In addition, proponents are required to demonstrate that appropriate consideration has been given to the following hierarchy of waste management options, ranked in order of increasing environmental impact:

- Re-use
- Off-site recycling
- Destruction of hazardous constituents
- Treatment to reduce or remove the hazardous constituents
- Disposal on land, into air, and into water

Ocean disposal shall not be permitted if it is determined that appropriate opportunities exist to reuse, recycle, or treat the wastes without undue risks to human health or the environment or disproportionate costs.

Both the Canadian ocean disposal permitting process under the Canadian Environmental Protection Act (CEPA) and the U.S. approach under Section 103 of the U.S. Marine Protection, Research, and Sanctuaries Act (MPRSA) (described below) have provisions for assessment of wastes to be disposed at sea.

In the Canadian experience, a good example of the successful application of a waste audit is the reuse and recycling of fish wastes from fish processing plants and aquacultural activities. Since the adoption of the waste assessment framework in the Canadian Disposal at Sea permit system, applicants for fish waste disposal permits are required to conduct a waste audit to identify environmentally friendly options before applying for an ocean disposal permit. This process has reduced the number of permits issued by Environment Canada for ocean disposal of fish wastes in the three Canadian maritime provinces (Nova Scotia, New Brunswick, and Prince Edward Island) from more than 50 permits per year to only one per year, and all fish wastes from these provinces are now composted or recycled in fish meal plants. Disposal-at-sea permits for fish wastes are still issued, however, in some isolated communities in Newfoundland and Labrador, where practical options of reprocessing fish wastes are not available.

The beneficial use of clean dredged material from channel dredging is another example of waste reduction options (16). This includes the use of clean sand removed from channel deepening operations to replenish beaches and the use of dredged material from harbors to reclaim wetland or create crab and lobster habitat (17, 18).

In the United States, MPRSA (19) requires that applicants for an MPRSA permit must evaluate dredged material disposal alternatives, including an examination of potential beneficial uses of the proposed dredge material and a consideration of alternative disposal options before selecting the ocean disposal option. Contaminated dredged material is not permitted for open sea disposal. However, if the contaminants can be removed or reduced from the dredged material, the dredge spoils can also be used for beneficial purposes. The U.S. Environmental Protection Agency (US EPA) and the

U.S. Army Corps of Engineers (ACE) have done extensive work in beneficial uses of dredged material, including habitat restoration, beach nourishment, aquaculture, parks and recreation, agriculture, forestry, horticulture, strip mine reclamation, landfill cover, shoreline stabilization and erosion control, construction and industrial use, and material transfer (20).

## 5. WASTE CHARACTERIZATION PROCESS AND DISPOSAL PERMIT SYSTEM

### 5.1. Assessment of Material for Disposal

If ocean disposal is the environmentally preferable and practical alternative, a waste characterization process is required to characterize the chemical, physical, and biological properties of the wastes. Each LC contracting party is required to develop a national action list of contaminants and associated lowest and highest acceptable concentrations for the screening of wastes to assess their potential effects on human health and the marine environment. When selecting substances for consideration in an action list, the contracting parties are advised to give priority to toxic, persistent, and bioaccumulative substances. When dealing with contaminants, several action levels are available, including a screening action level to identify wastes that are acceptable for ocean disposal, and an upper action level to reduce or eliminate acute or chronic effects on human health or on sensitive marine organisms.

In Canada, applying for and obtaining a disposal-at-sea permit involves a level of self-assessment, including a waste characterization component. A permit application consists of a completed application form, a nonrefundable application fee of $2500 (Canadian Dollars), and supporting environmental assessment documents. To ensure that the public is well informed about the proposed operations, applicants are required to publish a notice of projects in local newspapers before submitting the application (9).

A tiered assessment process based on the precautionary approach is used for waste characterization under the Canadian Disposal at Sea Program. The process uses chemical screening levels to assess suitability of the material in terms of lower action levels. When chemical screening levels are exceeded, biological (toxicity) tests are used to determine the upper action levels, above which ocean disposal is prohibited (9, 21). The set of chemical criteria used by Environment Canada to make regulatory decisions based on lower action levels for wastes proposed for ocean disposal includes cadmium (Cd), mercury (Hg), polychlorinated biphenyls (PCBs), and polynuclear aromatic hydrocarbons (PAHs), which are among the contaminants on the blacklist of the LC Annex I.

For contaminants on the "gray list" of the Convention Annex I, including lead (Pb), zinc (Zn), and copper (Cu), no lower action levels are assigned; however, Environment Canada has developed and published a set of marine sediment quality criteria for these and many other contaminants on the gray list (22). Section 127 of CEPA (1999) requires that Environment Canada take into account any other factors (including levels of these and other contaminants) that are necessary for assessment of the waste before issuing an ocean disposal permit. This legislative requirement allows Environment Canada to request analysis of any contaminants that are likely to be present in the wastes proposed for disposal at sea. For example, ocean disposal proponents may be required to analyze

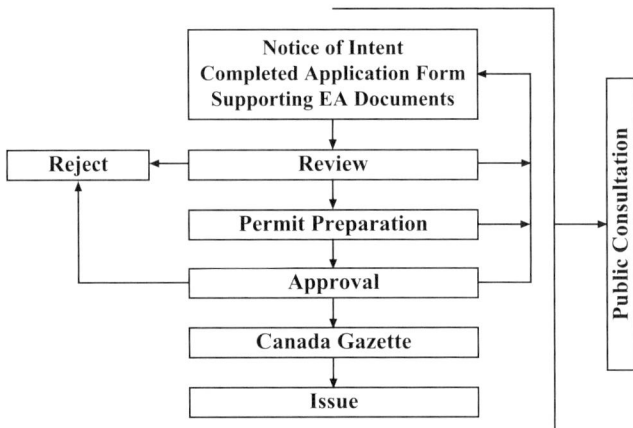

Fig. 9.2. Ocean disposal permit review process. EA, Environmental Assessment.

and report the concentrations of trubutyltin (TBT) in dredged material from shipyards where TBT paints had been used (use of the chemical is now banned).

Permit applications are reviewed by a regional committee, the Regional Ocean Disposal Advisory Committee (RODAC), and by other interested parties (the public, environmental groups, aboriginal interests, and other federal and provincial government agencies). RODAC consists of scientists and managers from Environment Canada and Fisheries and Oceans Canada. Figure 9.2 shows a typical permit review process. The public is actively encouraged to submit comments or concerns about the proposed ocean disposal activities throughout the application process. To assist proponents requiring ocean disposal of dredged material, Environment Canada has published a series of guidance documents, including "Users Guide to the Application Form for Ocean Disposal" (15).

In the United States, ACE is the permitting authority for dredged material disposal under MPRSA (23). Dredged material assessment is conducted by ACE, subject to US EPA review and concurrence. A national guidance, "Evaluation of Dredged Material Proposed for Ocean Disposal—Testing Manual," more commonly known as the "Green Book," was published by the US EPA to assist implementation of the regulations under MPRSA. As in the Canadian approach, the "Green Book" uses a tiered approach, which includes both chemical and biological assessments, with an emphasis on the direct assessment of biological impact. The Tier I evaluation helps to identify the information needed and to determine appropriate tiers and tests necessary to collect the information, as well as the identification of contaminants of concern. Tiers I, II, and III are intended to suffice for almost all evaluations, while Tier IV is intended only for extremely rare situations (24).

## 5.2. Chemical Screening

In Canada, up to 95% of disposal-at-sea permit applications can be screened by reviewing chemical screening values. The approach permits rapid decisions by regulators to approve or reject applications for ocean disposal, and is effective because it is enforceable by law, providing a generic and consistent national benchmark. Furthermore, chemical screening criteria are easy to understand, and are generally well accepted by regulators, environmental managers, and, industrial groups, the latter because the approach saves time and money. There are several inherent weaknesses to the approach:

- Scientific validity of the regulated values: Chemical screening levels are not truly cause-and-effect values (25), and are sometimes generated with very conservative safety factors. For some contaminants that have very low screening values, use of this chemical screening approach can be overprotective, ecologically invalid, and costly.
- Long list of chemicals: It is impossible to analyze all chemicals in the dredged material.
- Lack of flexibility: Regulated values do not provide flexibility for regulators to approve permit applications for wastes containing chemicals with concentrations slightly exceeding the lower regulated levels (e.g., a permit application with 0.61 mg/kg of cadmium is considered unacceptable, while a permit application with 0.60 mg/kg is acceptable under the Canadian Disposal at Sea Program).
- Not site specific: In some cases (e.g., cadmium), concentrations of a chemical in the dredged material exceed the regulated level because of high natural background levels at the dredged site. Furthermore, use of chemical screening levels does not consider target receptors, exposure pathways, characteristics of the receiving environment, or socioeconomic impact.
- Susceptible to sampling and method errors: Since the chemical screening method relies solely on comparing concentrations of contaminants to regulated criteria, poor sampling design, small sample numbers, improper analytical methods, and inadequate quality assurance/quality control (QA/QC) in laboratory practice can result in poor decision making.

Guidance documents for sampling, and field and laboratory performance have been developed in the United States and Canada to address these issues (24, 26). Standard reference samples for marine sediments developed by several countries including the United States and Canada are available for sediment chemical analyses when testing dredged material.

## 5.3. Biological Testing

Under the Canadian ocean disposal legislation (CEPA), proponents are encouraged to follow a tiered assessment approach by conducting biological effects assessment when their wastes fail the chemical screening process. Similarly, biological tests are required under Tiers II and III of the US EPA Green Book. The biological approach includes performing biological test methods recommended in the guidance documents published by Environment Canada and in the US EPA Green Book. These methods include the following:

- A 10-day amphipod acute lethality test on whole sediment (24, 27)
- A bacterial bioluminescence test (Microtox solid-phase) (28)
- An echinoid fertilization test on sediment pore water (29)
- A bedded sediment bioaccumulation test (30)

United States Ocean Dumping Regulations under MPRSA state that "trace contaminants are not defined in terms of numerical chemical limits, but rather in terms of persistence, toxicity, and bioaccumulation that will not cause an unacceptable adverse impact after dumping." By this definition of trace contaminants, biological end points are regarded, in a sense, as analytical instruments for determining the environmentally adverse consequences of any contaminants present. Biological tests are a direct measurement of the bioavailability of the contaminants. They measure the toxicity of contaminants even when their presence is unknown or unmeasured (24).

As for the chemical screening approach, guidance documents have also been developed to guide the performance of biological tests and proper use of the tests for making regulatory decisions for ocean disposal of wastes. The most common issues in using biological tests for regulatory decision making are as follows:

- The selection of appropriate tests, such as pore water or solid-phase tests
- The selection of suitable test species
- Proper use of control and reference samples
- Consideration of confounding factors, such as particle size, ammonia, and sulfide
- Interpretation of the data generated by the tests

When using biological tests to assess wastes such as dredged material, it is important to ensure that a battery of properly selected test methods, including lethal and sublethal, pore water, and solid-phase tests, are used. The different end points provide useful information on the type of toxicity effects (lethal or sublethal) and the potential sources of toxicity (contaminants in pore water or bound to particles) in the sediments (31, 32). The effect of these differences in interpretation of test results are illustrated by the results of two sediment assessment studies conducted by Environment Canada (33). More than half of the sediment samples collected from Sydney and Halifax harbors, two major harbors in Nova Scotia, Canada, were toxic to amphipods. However, only the Halifax harbor samples were toxic to the sea urchin, as indicated by the results of a fertilization bioassay (Table 9.1). Since the amphipod method is a solid-phase test while the sea urchin method is a pore water test, it was suggested that the difference between the sediment samples and biological response was that the contaminants in the Sydney harbor sediments were mostly particle-bound, while those in the Halifax harbor were both particle- and pore-water-bound. If the sea urchin test was the only test used for permit assessment, the result would have led to a decision allowing the ocean disposal of dredged material from Sydney harbor but not for the dredged material from Halifax harbor, yet PAH concentrations in the former are many times higher.

When selecting for biological tests, it is important to know the sensitivities to different types of sediment of each test species. For example, at least six different species of marine amphipod can be used in the amphipod lethality test. Some species are more sensitive to particle size changes (Figure 9.3); some are more sensitive to ammonia or sulfide (Figure 9.4) (34); and some are more tolerant of lower salinity. Thus, if an inappropriate species is selected for a test, the test results can be misleading, and an unsound decision based on false-positive results can be made.

Selection of control and reference sites is also critical in the biological testing process. Confounding factors such as particle size (Figure 9.5), sulfide and ammonia tolerance (Figures 9.6 and 9.7), and inappropriate selection of reference sites, can interfere with interpretation of the test results (24, 34).

Dredged material criteria in both the U.S. and Canada deem sediments unacceptable for ocean dumping if they fail prescribed toxicity tests or if chemicals of concern are accumulated in the tissues of test organisms in laboratory exposures (15, 24, 35). The use of the bioaccumulation end point for sediment assessment, however, can also be

Table 9.1
Polynuclear aromatic hydrocarbons (PAHs) concentrations and results of biological tests conducted on sediment samples from Sydney and Halifax harbors, Nova Scotia, Canada

| Station | PAHs (mg/kg) | Amphipod (% survival) | Sea urchin ($IC_{50}$, % pore water) |
|---|---|---|---|
| Sydney Harbor[1] | | | |
| 1 | 246.4 | 56 | >100 |
| 2 | 132.18 | 75 | >72* |
| 3 | 199.97 | 48 | >100 |
| 4 | 87.28 | 67 | >100 |
| 5 | 35.35 | 67 | >100 |
| 6 | 16.94 | 81 | >100 |
| 7 | 33.67 | 74 | >100 |
| 13 | 195.41 | 27 | >100 |
| 23 | 4.77 | 97 | >100 |
| 26 | 20.98 | 83 | >100 |
| Halifax Harbor[2] | | | |
| 1 | 0.025 | 92 | 83 |
| 2 | 0.06 | 91 | 35 |
| 3 | 61.79 | 69 | 39 |
| 4 | 21.79 | 89 | 66 |
| 5 | 20.69 | 76 | 90 |
| 6 | 19.42 | 34 | 91 |
| 7 | 17.93 | 21 | 0 |

$IC_{50}$, concentration that inhibits 50%.
[1]Zajdlik et al. (33).
[2]Environment Canada unpublished data.

questionable, and in particular has been criticized as being overly protective. O'Connor (35), in comparing criteria for land application of sewage sludge and ocean disposal of dredged material, pointed out that because of the different uses of the bioaccumulation end point, environmental regulations in the U.S. have allowed much higher levels of chemical contamination in sewage sludge to be used on land than in dredged material to be dumped at sea. Disposal at sea is not permitted for a dredged material if bioaccumulation of chemicals in test organisms exceeds the U.S. Food and Drug Administration (FDA) action levels for the protection of human health, while criteria for land application of sewage assume that bioaccumulation will occur. The later criteria were derived from a lengthy risk assessment that considered 14 pathways for chemicals to migrate from sludge-amended soil to plants, animals, and humans. O'Connor suggested that perhaps a similar recognition that marine organisms and humans get only part of their food requirements from sources exposed to dredged material would enhance the existing assessment method, which uses the bioaccumulation end point for dredged material. (35)

Use of the bioaccumulation end point for sediment assessment has been a discussion topic in recent workshops and conferences organized by the US EPA. The interpretation criteria for bioaccumulation tests are still under debate (36, 37). Other criteria, such as

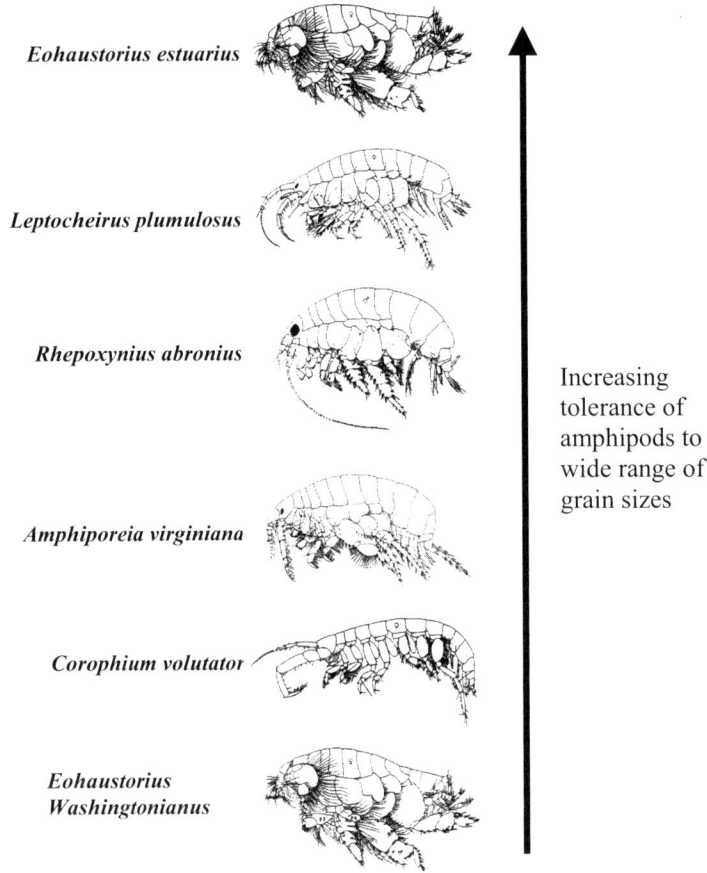

**Fig. 9.3.** Comparative tolerance of six species of amphipod to changing of grain size in artificial sediments (Environment Canada, unpublished data).

the use of tissue indicators, may be developed in future. In an effort to demonstrate this possibility, Tay et al. (38) showed that clams, *Macoma balthica*, exposed to contaminated sediments using a method identical to the US EPA bioaccumulation test method, exhibited cellular lesions that are correlated with PAHs and PCBs. They proposed using the suite of histopathologic and histochemical biomarkers employed in their study to complement the bioaccumulation end point for predicting the health of the test organisms when assessing dredged material.

### 5.4. Ecological and Human Health Risk Assessment

In recent years, the use of risk assessment as a decision-making tool for environmental assessment projects has become more popular. In the 2000 LC consultation meeting, representatives of the United States submitted a proposal for the use of ecological and human health risk assessment for the evaluation of wastes for disposal at sea. Although no decision was made by the convention, it is clear that site-specific ecological and human health risk assessment is more flexible than the two-tiered chemical and

# Ocean Disposal Technology and Assessment 455

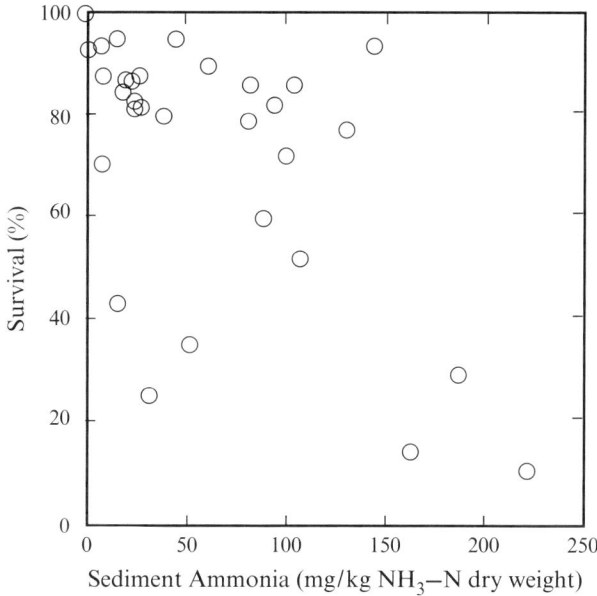

**Fig. 9.4.** Relationship between percent survival of *E. washingtonianus* and sediment ammonia (34).

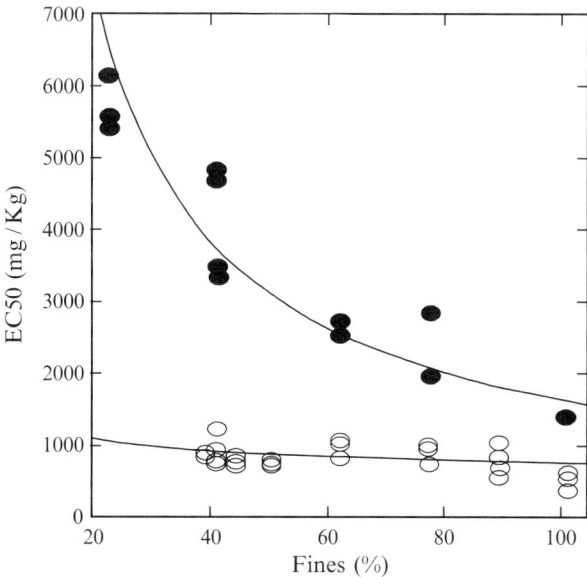

**Fig. 9.5.** Effects of particle size on Microtox solid-phase $EC_{50}s$ of clean sand/clay mixtures (●–●) and sand/clay mixtures with a fixed amount (40%) of National Research Council of Canadian Marine Sediment Reference Sample HS-6 (O–O) (34).

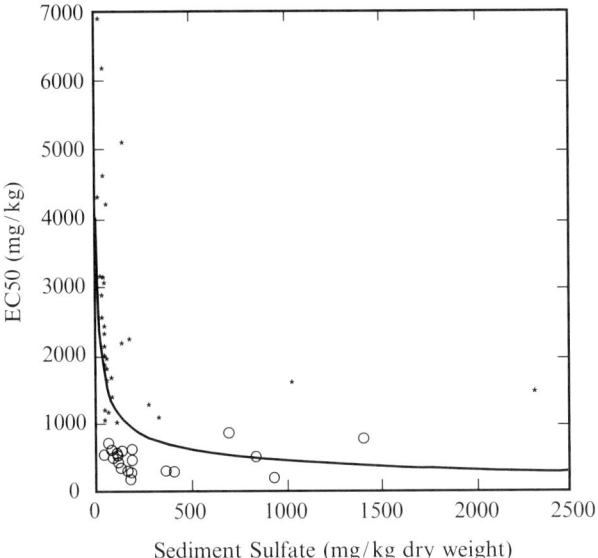

**Fig. 9.6.** Relationship between Microtox solid-phase $EC_{50}$ and sediment sulfide. Spearman correlation coefficient $r = -0.6$; $p = .001$. * = nontoxic, o = toxic (according to Canadian Ocean Disposal Guidelines) (34).

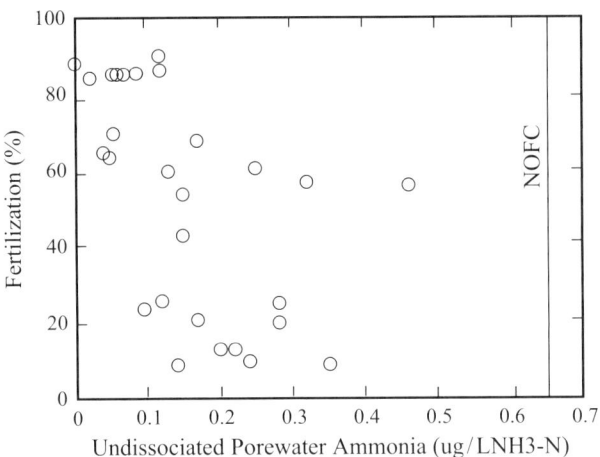

**Fig. 9.7.** Relationship between fertilization success of *L. pictus* and undissociated pore water ammonia (34). NOEC, no observable effect concentration.

biological approach. It is also more ecologically valid and sometimes cost-effective. Risk assessment requires users to make assumptions and generate uncertainties, a process that sometimes is not "user friendly" and is more time-consuming. To achieve their own objectives, some users can adopt uncommon processes to make assumptions and calculate risks. If used improperly, the approach can be underprotective, ecologically invalid, and can lead to long-term environmental problems.

## 5.5. Water Quality Issues

Water quality data can be important if the proposed wastes for ocean disposal contain mostly silt and clay or high levels of fecal coliform bacteria. High levels of suspended particulate material (SPM) caused by disposal of fine dredged material at ocean disposal sites can be a problem for nearby aquaculture sites, shellfish growing areas, fishing grounds, coastal recreation, and residential areas (39–43). If the dredged material also contains a high level of fecal coliforms, and if dumping is conducted at or near aquaculture and shellfish growing sites, it can cause closure of these sites and create significant risks to human consumers. Disposal of nutrient-rich material such as dredged material with high concentrations of fertilizers or fish wastes may trigger eutrophication in sheltered areas and damage important coastal resources.

Tier II of the U.S. dredged material assessment process requires determination of compliance with applicable water quality criteria (WQC). The Tier II water column evaluations are conducted using a numerical model recommended by the "Green Book." The numerical model is a screening tool that assumes that all contaminants are released into the water column during the disposal process. Water column bioassays, such as those using fish, bivalves, and crustaceans, may be used if there are no applicable marine WQCs for the contaminants of concern, or synergistic effects between certain contaminants are suspected (24).

Despite the above concerns, research has shown that dredged material disposal at disposal sites may not increase the incidence of fecal coliforms (44); and in most cases, turbidity is only an aesthetic concern, having limited direct impact to biota (45). It remains important, however, to take into consideration water quality issues when assessing wastes proposed for disposal at sea, to protect the important near-shore resources and recreational areas.

## 6. DISPOSAL SITE SELECTION

In the Canadian Disposal at Sea Program, the requirements for supporting data on the physical, chemical, and biological properties of the sites are minimal for disposal sites that have been used in the past for the same type of wastes, especially when the sites have been used very recently. For any new ocean disposal site, proponents are required to conduct a detailed site evaluation, including an assessment of the size and capacity of the site and its location with respect to sensitive areas, amenities, and other uses of the sea. Guidance for dumpsite selection can be found in a report of the Joint Group of Experts on the Scientific Aspects of Marine Environmental Protection (46). The proposed use of a site is a determining factor in site selection. For example, a disposal site located in an open area subjected to strong currents and wave action may be more suitable for fish waste disposal, while a sheltered, nondispersive site may be more suitable for dredged material disposal.

In addition to requirements under CEPA, prior to issuing a disposal-at-sea permit, Environment Canada must ensure that the proposed activities have been reviewed in accordance with the requirements of the Canadian Environmental Assessment Act (CEAA), the federal environmental assessment legislation in Canada (47). The CEAA requirements are similar to those of CEPA (1999) but there are additional obligations, including the need to address the following factors:

- Cumulative environmental effects
- Health and socioeconomic conditions
- Physical and cultural heritage concerns
- Current use of lands and resources for traditional purposes by aboriginal persons
- Environmental effects of malfunctions or accidents that may occur
- Technical and economic feasibility of mitigation measures

Selection of appropriate dump sites will need to take into consideration the above factors to ensure regulatory compliance.

Since 1990, the Canadian Disposal at Sea Program has included a requirement for the sampling and assessment of biological species, community diversity measures, and abundance and biomass of the benthic biota at proposed ocean disposal sites (48). Benthic biota are useful biological indicators for disposal site assessment because they are typically sessile organisms and likely to have prolonged exposure to the disposal site environment (49). By studying the changes in benthic community structure, we can predict the impacts of the disposal activity on the aquatic environment. The effects of exposure of benthos to disposed material at ocean disposal sites have been well studied (50–53).

## 7. DISPOSAL SITE MONITORING

Disposal site monitoring is an essential component of ocean disposal programs in both Canada and the U.S., and to other signatories of the London Convention. It is conducted to confirm predictions of the assessments and to refine the assessment practices of the permit system. The LC contracting parties are required to submit annual reports to the convention on their monitoring activities as part of their commitment. The ACE has done extensive monitoring of ocean disposal sites in the U.S., one example of which is the Disposal Area Monitoring System (DAMOS) administrated by the New England District of the ACE.

The Canadian Disposal at Sea Program has instituted a cost recovery system to pay for monitoring activities, in keeping with the user-pay principle, an initiative of the Canadian federal government to recover costs for some of the services it provides. Disposal at sea permit holders are required to pay $470 (Canadian Dollars) for every 1000 cubic meters of dredged or excavated material that is authorized for disposal at sea. The fees support Environment Canada's annual disposal site monitoring activities at various ocean disposal sites. Results of this program are summarized annually and submitted to the London Convention as part of the Canada's international commitment. Disposal site monitoring is currently focused on sites used for disposal of dredged material and inert geological material, together which account for 95% of all material disposed at sea in Canada.

Guidance documents have been developed by Environment Canada for conducting ocean disposal site monitoring to ensure a high quality of information collection and national consistency (15, 54). Most of the monitoring conducted in Atlantic Canada has been jointly carried out by Environment Canada and Natural Resources Canada

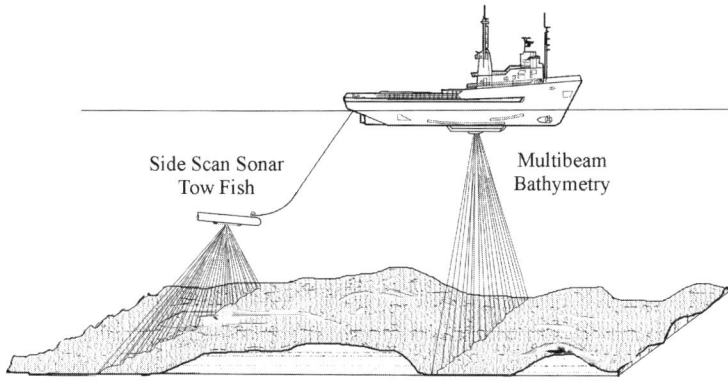

**Fig. 9.8.** Side-scan sonar and multibeam.

through study programs that involved integrated geophysical, geochemical, and biological monitoring. Techniques have included acoustic surveys (multibeam, side-scan sonar, and sub-bottom profiler surveys); current meters; seabed erosion measurement; sediment and benthic biota sampling; and bottom photographs and video imaging. The ACE uses a comparable suite of techniques to monitor its ocean disposal sites. Some of the technical approaches used in Canadian and U.S. monitoring are presented in additional detail below.

### 7.1. Acoustic Geophysical Surveys

#### 7.1.1. Side-Scan Sonar

High-resolution, acoustic images of the disposed material at ocean disposal sites can be produced using a side-scan sonar system (120 and 330 kHz). The system collects back-scattered signals emitted from a towed transducer deployed about 50 m behind the survey vessel (Figure 9.8) and is capable of resolving objects as small as about 0.15 m. A survey consists of running a set of survey lines with predesigned spacing to provide a full mosaic of the sea bottom. Data collected from survey lines are then combined to produce a side-scan sonar mosaic. Rocks and hard bottom reflect strong signals and appear as dark areas, while soft and muddy bottom, which absorb sonar energy, appear as lighter areas in the acoustic images.

#### 7.1.2. Multibeam Bathymetry

Instead of using single-beam bathymetric equipment that covers only a small percentage of the total sea-floor area, Environment Canada and Natural Resources Canada use multibeam bathymetric equipment to survey all dredged material disposal sites (Figure 9.8). Bathymetric data are collected using a multibeam bathymetry system that uses a 300-kHz transducer with 127 beams and a beamwidth of $1.5° \times 1.5°$. It provides a depth resolution of 1 cm with an accuracy of 5 cm root mean square (RMS) and a positional accuracy of 1 to 2 m. For deeper water, a Simrad EM1002 system that uses a 95-kHz transducer with 121 beams and a beamwidth of $2°$ is employed. The deep water system provides coverage to a depth of 1200 m in deep water. Survey lines can

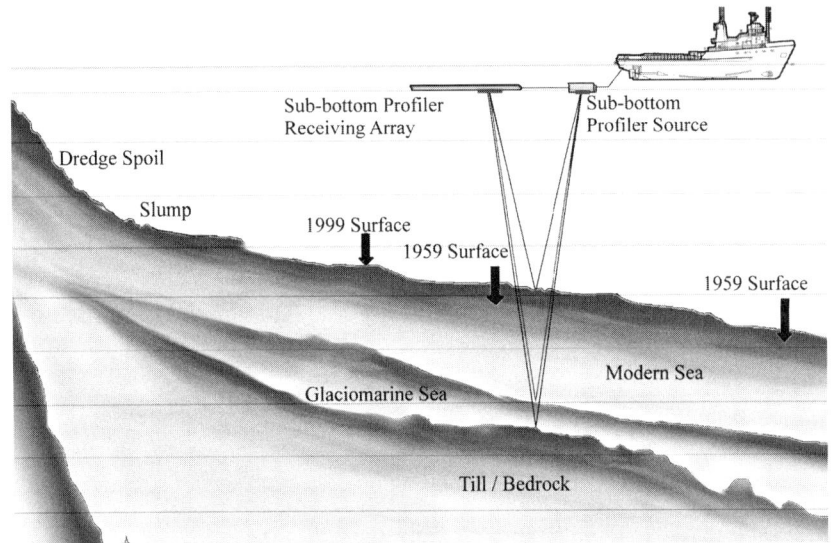

**Fig. 9.9.** Sub-bottom profiler.

be run at various spacing throughout the survey area to provide 200% coverage in water depth greater than about 20 m (18). The technology allows the production and rendering of detailed 3D images of the seabed showing geological and disposal site features in considerable detail.

*7.1.3. Sub-Bottom Profiler*

A high-resolution sub-bottom profiler system has been used to map the thickness and vertical cross section of the disposed material at the ocean disposal sites. The system uses an electrodynamic (boomer) sound source to produce a repeatable impulse-like output. The sound penetrates the seabed and reflects off sediment layers, providing a vertical resolution, normally of 0.25 m or better (Figure 9.9).

## 7.2. Currents and Sediment Transport Survey

As part of monitoring programs at disposal sites, currents have been measured by current meters deployed at selected locations in the vicinity of disposal sites. Sediment transport can be measured by different methods including the benthic annular flume known as a "sea carousel," designed to measure in situ sediment erosion rates (55, 56), and the use of artificial fluorescent sediment tracers to monitor sediment movement (57).

## 7.3. Chemical and Biological Sampling

Ekman, Van Veen, and Shipek grab samplers and Benthos gravity corers are used to collect samples for grain size and chemical analyses. Samples for analysis of benthic communities are typically sieved through a fine-mesh sieve, fixed in formalin, and stored in alcohol before the animals are removed ("sorting") and identified, typically to species. For sediment bioassays, grab samples from each sampling station are taken, pooled, and mixed thoroughly, and subsamples are stored at low temperature until the test.

Species indices including species diversity, richness and evenness, abundance and biomass, as well as community composition and characteristics are commonly used for the assessment of disposal impacts to benthic biota (58). These can be used together with statistically based monitoring designs such as control-impact comparisons as outlined in monitoring protocols for Environment Canada's successful national Pulp and Paper Environmental Effects Monitoring Program, to assess the near field and far field impacts of the disposal (59). Habitat mapping using bottom images generated by the acoustic survey, combining bottom photographs and video images, has also been used successfully to assess bottom habitat changes at ocean disposal sites (60).

The REMOTS$^{TM}$ system has been used by the ACE Disposal Area Monitoring System (DAMOS) for routine disposal site monitoring for over 20 years. It is a sampling technique using equipment designed to obtain undisturbed, vertical, cross-sectional photographs (in-situ profiles) of the upper 15 to 20 cm of the sea floor. Images obtained by REMOTS yield a suite of standard parameters, including sediment particle size, camera prism penetration depth, surface boundary roughness, depth distribution of redox potential, infaunal successional stage, and organism-sediment index. Using this tool, the DAMOS program has mapped the distribution of thin (<20 cm) dredged material layers, delineated benthic disturbance gradients, and monitored the process of benthic recolonization following the disposal of dredge material at the disposal sites, to assess disposal impacts on benthic biota and monitor their recovery (61–63).

## 7.4. Case Studies

### 7.4.1. Black Point Dredged Material Ocean Disposal Site, Saint John, New Brunswick, Canada

Monitoring at the Black Point ocean disposal site is difficult due to regional hydrographic conditions. The disposal site is located in a high-energy area with exposure to the Bay of Fundy and Gulf of Maine, which is affected by the outflow of the Saint John River, and the enormous Bay of Fundy tides. Based on the transport pattern in this area, it was predicted that the dredged material disposed of at the site would be dispersed by the strong outwardly directed current (55, 64). However, after 60 years of dredged material disposal, data obtained by the acoustic geophysical surveys (multibeam, side-scan sonar, and sub-bottom profiler) show a significant buildup of disposed materials within a 1-km radius of the disposal site buoy. The area occupied by the dumped material is about 2.8 km$^2$ (Figure 9.10), and the volume represents approximately 15% to 20% of the total volume disposed at the site since 1959. Currents and erosion rate measurements and repeated multibeam surveys indicate that the site has been subjected to significant bottom current activity, erosion, and smoothing. The sub-bottom profiling data and historical bathymetric data suggest that up to 10 m of material has accumulated in the vicinity of the disposal site buoy. Within the area of heavy disposal activity, the most recent spoil deposits, containing blocky cohesive clay material, can be readily distinguished from the smoother seabed areas (Figure 9.11). Sediment grain size data, bottom photos, and video were used to ground-truth the acoustic information.

Currents at the Black Point disposal site are predominantly east-west and the resultant transport is east on the flood tide and west on the ebb tide (65), moving materials in

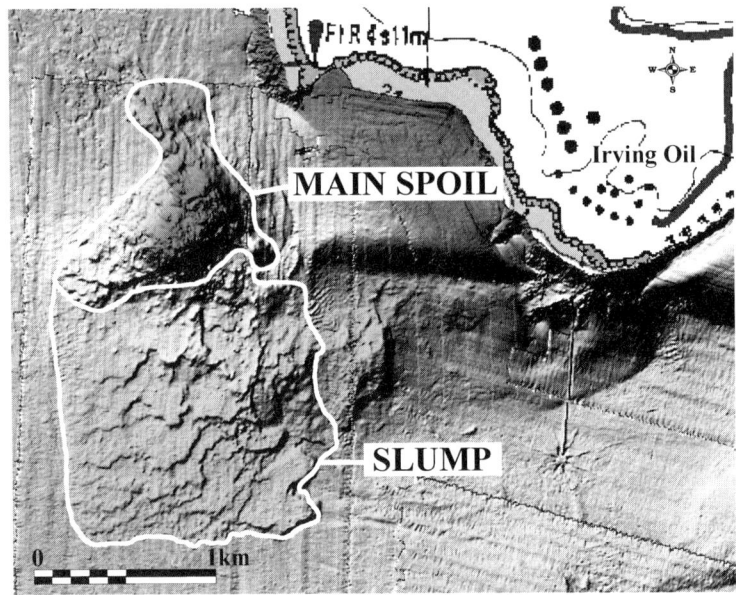

**Fig. 9.10.** Multibeam image of the Black Point Disposal Site showing the dumped materials (2.8 km$^2$).

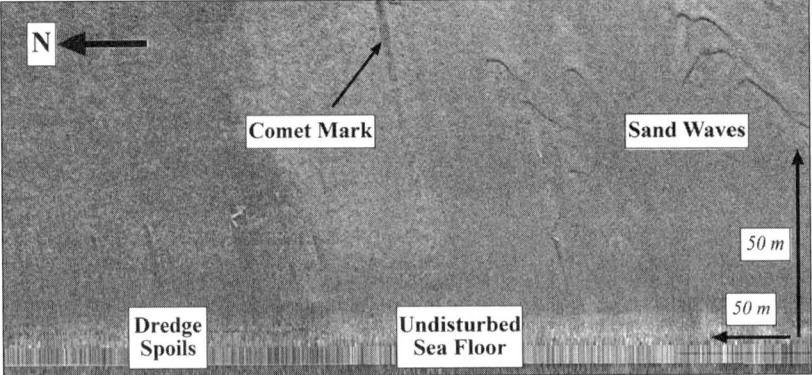

**Fig. 9.11.** Side-scan sonar image showing the down slope limit of the debris flow south of the Black Point disposal site.

suspension from the site strongly to the east or the west (66, 67). As a particle moves either west or east due to the alternating current regime, it has a chance of coming out of suspension and entering the seabed on either side of the disposal site. The enrichment of vanadium—a contaminant of crude oil that is likely to have originated in the dredged areas of inner Saint John harbor—in sediments west of the disposal site supports the current prediction results indicating the instability of the disposal site (60).

When acoustic surveys are used in a integrated manner with other physical measurements such as currents and erosion measurements, underwater photos and video, and sediment grain size analysis, they can provide integrated real-time information of the zone of impact at ocean disposal sites and serve as a basis for the design of chemical and biological monitoring programs at the sites.

Contaminants of concern for dredged material disposal at the Black Point disposal site have been shown to be below regulated levels under most circumstances. Toxicity test results of the dumped materials using a battery of standard toxicity tests demonstrate limited toxicity. Apart from changing the biological community through physical disturbance where dredged spoil has been regularly dumped, there have been no indications of significant changes in biological communities with time around the disposal site (58).

### 7.4.2. Amherst Cove Dredged Material Ocean Disposal Site, Prince Edward Island, Canada

The Amherst Cove disposal site is located 2 to 3 km from the Confederation Bridge, a 13-km structure connecting the two Atlantic Canada provinces, Prince Edward Island and New Brunswick, across the Northumberland Strait. The site was used for the disposal of 473,000 cubic meters of dredged materials from 59 pier base locations and a jetty required for the construction of the bridge. Dredged materials were disposed in a pattern to enhance the existing habitat for crustaceans (lobster and rock crab) at the site. Acoustic surveys, grain size study, and bottom photos and video show the materials are mostly large boulders and large pieces of rock (Figure 9.12). Preliminary examination of the physical and biological data indicates a slight improvement in lobster habitat and a significant improvement in rock crab habitat at the disposal site (68).

## 8. LAND-BASED DISCHARGES OF WASTES TO THE SEA: ENGINEERING DESIGN CONSIDERATIONS

Land-based sources of marine pollution by direct discharges of wastes from land to the sea represent one of the five main sources of marine pollution identified by the International Maritime Organization (IMO) that need to be controlled and regulated. Although technically land-based disposal is another form of ocean disposal, it is not internationally regulated by the London Convention or its 1996 protocol and the activity is normally described by the LC contractual parties as "land-based discharges to the sea," not "ocean disposal."

Most countries have national and local legislation in place to control the discharge of waste from land to the sea. In particular, many countries have industry-specific regulations for industries such as metal mining, petroleum refining, pulp and paper production, etc. For those countries that have no such legislation, management of these waste streams should follow the Global Plan of Action for Protection of the Marine Environment from Land-Based Activities (GPA). For countries that are part of a UNEP Regional Seas network, the basis for control of land-based activities may already have been laid. In the United States, discharges by outfalls to the sea are regulated by the Clean Water Act (CWA) (69), and in Canada they are regulated by the Fisheries Act (70).

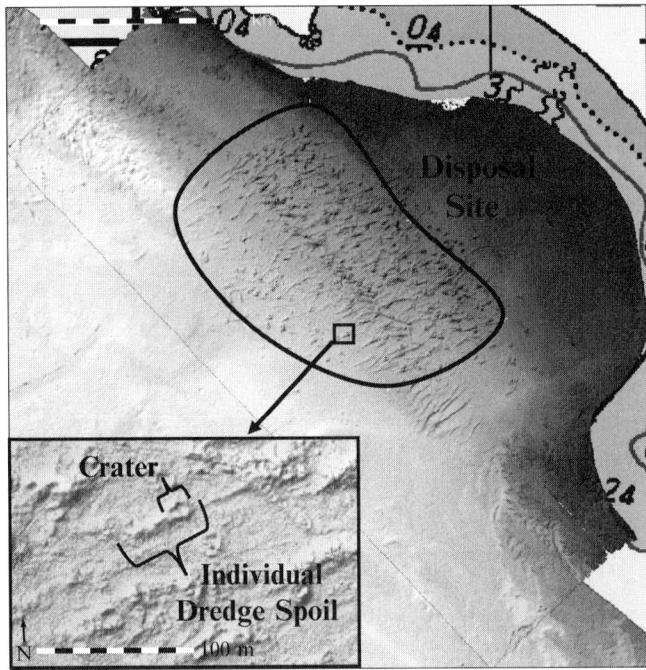

**Fig. 9.12.** Multibeam image of the Amherst Cove disposal site. The discrete isolated mounds of dredged materials are visible within the boundary of the disposal site. Enlargement of a section of the image shows individual dumped material (10 to 15 m across and many are over 100 m long) with crater located at the centre of most of the large mounds.

Commonly, discharges to the sea are through pipelines and involve offshore outfalls. This chapter provides engineers with the information necessary to design marine outfalls for the wastes in question. In particular, the outfalls are for wastes that have been adequately treated to reduce contaminants. Although discharges of wastes from land-based activities via diffuser pipes is not regulated under the LC, the waste assessment procedures outlined in earlier sections for wastes proposed for ocean disposal can also be used to assess the potential environmental impact of the discharges from land-based sources to the marine environment.

### 8.1. Ocean Outfall System

A submarine outfall system for land-based discharges of waste consists of a long section of pipe to transport the wastewater or waste sludge some distance from shore, and a diffuser section to dilute the waste with seawater. Diffusers provide initial dilution of a waste in a waterway (Figure 9.13). An outfall design must meet applicable receiving water standards. Near the end of the outfall, the liquid waste is released in a simple stream or jetted through a manifold or multiple-point diffuser. At this end point, the liquid waste mixes with surrounding seawater and rises to the surface where it drifts as a liquid waste field in accordance with the prevailing ocean currents. This drift or movement with the currents is termed advection (71). At the same time, the field is

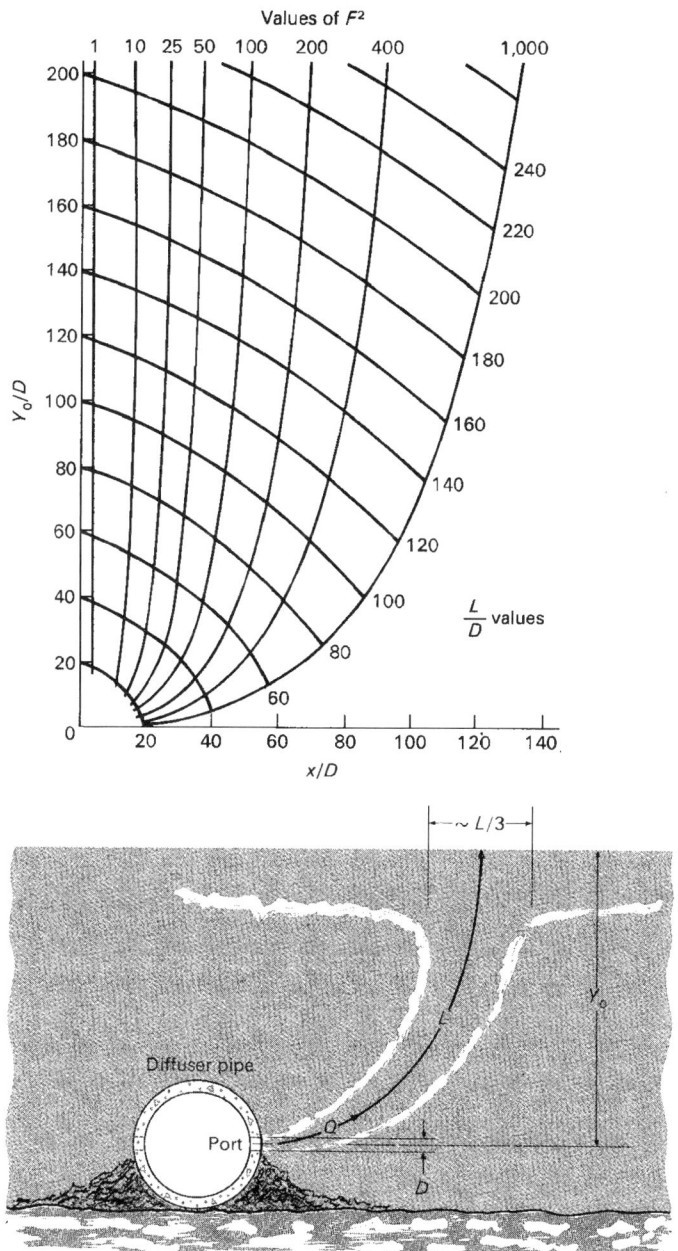

**Fig. 9.13.** Outfall diffuser system and jet mixing analysis (71, 74–76).

also diffusing outward into the surrounding water. Since the initial dilution from an efficient diffuser is so large that the reduction in dissolved oxygen (DO) is negligible, only coliform bacteria, floatable solids, nutrient, and toxicity requirements will govern the design and location of most outfalls. The following subsections introduce the types of dilution and design parameters, and provide a design example.

## 8.2. Initial Dilution

The velocity of the jet causes turbulent mixing with the surrounding water when waste is discharged from a single- or multiple-port diffuser. If the waste is of lower specific gravity than the dilution water, the mixing jet will bend upward and may eventually reach the water surface. The ocean is vertically stratified with an upper layer of warm water riding over colder water, generally observed during warm summer months. It is possible to dilute the waste sufficiently with cold water so that the specific gravity of the waste-cold water mixture is greater than that of the warm upper layer. In this situation the waste plume will remain submerged under the upper layer. The dilution is a function of many parameters, and can be estimated from Equation 1:

$$D1 = (Vbd)/Q \qquad (1)$$

where
$D1$ = initial dilution due to mixing the liquid waste and sea water
$V$ = current velocity, m/s or m/h (ft/s or ft/h)
$b$ = effective width of diffuser system, m (ft)
$d$ = average depth of the liquid waste field, m (ft)
$Q$ = liquid waste flow rate, $m^3/s$ or $m^3/h$ ($ft^3/s$ or $ft^3/h$)

## 8.3. Dispersion Dilution

A uniform waste-seawater mixture is formed above the diffuser section after initial dilution. The waste field then begins to move in response to the prevailing current. As it moves, the outer edge of the field entrains seawater as a result of turbulent mixing. The waste field begins to diffuse outward and takes on the shape of a plume. Dilution due to eddy diffusion after initial dilution can be calculated by Equation 2:

$$D2 = C_0/C_t \qquad (2)$$

where
$D2$ = dilution due to eddy diffusion after initial dilution
$C_t$ = maximum pollutant concentration at time $t$, mg/L
$C_0$ = pollutant concentration after initial dilution, mg/L

## 8.4. Decay Dilution

Bacterial decay increases with time as portions of the plume reaching the shore or critical area become more concentrated. The decay rate of the liquid waste, therefore, is another significant factor in waste dilution. Accurate estimation of the number of coliform bacteria requires taking into account their reduction due to die-off, flocculation, and sedimentation. Bacterial decay includes mortality as well as flocculation and sedimentation, and is most commonly assumed to follow first-order kinetics. The dilution due to waste decay can be determined by Equation 3:

$$D3 = C_0/C_t = \exp[(2.3\ X)/(T_{90} \times 60 \times V)] \qquad (3)$$

where
$D3$ = dilution due to waste decay
$C_0$ = bacterial concentration after initial dilution, mg/L

$C_t$ = bacterial concentration at time $t$, mg/L
$X$ = distance along plume centerline, m (ft)
$T_{90}$ = the time to achieve a 90% reduction in bacterial numbers, h
$V$ = the velocity in m/min (ft/min)

## 8.5. Outfall Design Criteria

### 8.5.1. Velocity of Liquid Waste in Pipeline

An outfall conveys the liquid waste to the diffuser section. Its size is determined by the velocity, headloss, structural considerations, and economics of the situation. To avoid excessive headloss, waste velocities of 0.6 to 0.9 m/s (2 to 3 ft/s) at the average flow rate ($V_{\text{pipe}}$) are normally recommended. Lower velocities will not be a problem provided the waste has received preliminary treatment to reduce the quality of settleable solids. Velocities higher than 2.4 to 3.0 m/s (8 to 10 ft/s) should be avoided because of excessive headloss.

### 8.5.2. Diffuser Orientation, Size, and Length

The diffuser section should be oriented perpendicular to the prevailing ocean current. If the currents are not predominant in any one direction, a Y- or V-shaped diffuser is commonly used. The size of diffuser ports varies from 7.6 to 23 cm (3 to 9 in). Port diameters of less than 7.6 cm (3 in) are being used on the fourth White's Point outfall of the Los Angeles County Sanitation districts, in Los Angeles, California. One meter of the diffuser length is required for each 1000 m³/d of average liquid waste flow rate ($Q$) transported by the pipeline.

### 8.5.3. Diffuser Alignment, Spacing, and Numbers

Diffuser ports are normally aligned horizontally, with ports alternating from side to side in order to avoid interference between adjacent jet plumes as they rise toward the surface. A rule of thumb is that ports should be spaced ($S_d$) between 2.5 and 4.5 m (8 and 15 ft) on centers. A number of different sizes of ports along the diffuser section may be required. The port diameters ($D_p$) chosen during the preliminary design represent an average value. The final hydraulic analysis will determine the range of port sizes required to discharge the liquid waste flow uniformly along the length of the diffuser section ($L_d$). The following design equations are used:

$$L_d = Q/1000 \text{ m}^3/\text{d/m} \tag{4}$$
$$N = L_d/S_d \tag{5}$$
$$A_{\text{tp}} = Q_p/V_p \tag{6}$$
$$A_p = A_{\text{tp}}/N \tag{7}$$
$$D_p = (A_p \times 4/3.14)^{0.5} \tag{8}$$
$$A_{\text{pipe}} = Q/V_{\text{pipe}} \tag{9}$$
$$D_{\text{pipe}} = (A_{\text{pipe}} \times 4/3.14)^{0.5} \tag{10}$$

where
$L_d$ = total diffuser length, m
$Q$ = average liquid waste flow rate in the pipeline, m³/d
$N$ = number of diffusers

$S_d$ = diffuser spacing, m
$A_{tp}$ = total area of diffuser ports, m$^2$
$A_p$ = area per diffuser port, m$^2$
$D_p$ = diffuser port diameter, m
$A_{pipe}$ = pipeline cross-sectional area, m$^2$
$V_{pipe}$ = liquid waste velocity in pipeline, m/s
$D_{pipe}$ = diameter of pipeline, m

## 8.6. Design Example

The design of an ocean outfall system is illustrated in this subsection. Frequently, dredged materials have been disposed of in the ocean through a pipeline (72, 73). The readers are referred to other sources (74–76) for design criteria and procedures. The following are the conditions for designing an ocean outfall system for discharging a liquid waste to the ocean:

1. Peak flow rate of liquid waste, $Q_p = 92{,}000\ \text{m}^3/\text{d}$
2. Average flow rate of liquid waste, $Q = 46{,}000\ \text{m}^3/\text{d}$
3. Bottom slope = 12.5 m/1000 m
4. Coliform concentration in untreated wastewater = 1,000,000/mL
5. Critical onshore current = 6 m/min
6. Diffuser length requirement = 1 m diffuser length per 1000 m$^3$/d of average flow rate of liquid waste.
7. Diffuser spacing ($S_d$) = 3 m
8. Port discharge velocity at peak flow ($V_p$) = 5 m/s
9. Diffuser configuration = diffusers are V-shaped with each side being 20 m in length, and are offset at an angle of 45 degrees with respect to the shoreline
10. Time to achieve a 90% reduction in bacterial numbers ($T_{90}$) = 4 h
11. Velocity in pipeline at average flow rate = 0.75 m/s

Solution:

1. Determine the total diffuser length:

$$L_d = Q/1000\ \text{m}^3/\text{d/m}$$
$$L_d = (46{,}000\ \text{m}^3/\text{d})/(1000\ \text{m}^3/\text{d/m}) = 46\ \text{m} \tag{4}$$

2. Determine the number of diffusers:

$$N = L_d/S_d$$
$$N = 46\ \text{m}/3\ \text{m} = 15.3\ (\text{use 16 ports}) \tag{5}$$

3. Determine the total area of ports:

$$A_{tp} = Q_p/V_p$$
$$A_{tp} = (92{,}000\ \text{m}^3/\text{d})/[(5\ \text{m/s})(86{,}400\ \text{s/d})]$$
$$= 0.213\ \text{m}^2 \tag{6}$$

4. Determine the area per diffuser port:

$$A_p = A_{tp}/N$$
$$A_p = 0.213\ \text{m}^2/16 = 0.0133\ \text{m}^2/\text{port} \tag{7}$$

5. Determine the diffuser port diameter:

$$D_p = (A_p \times 4/3.14)^{0.5}$$
$$D_p = (0.0133 \, m^2 \times 4/3.14)^{0.5}$$
$$= 0.1298 \, m \text{ (use } 0.130 \, m\text{)} \tag{8}$$

Decision: use 16 ports with a 130-mm port diameter.

6. Determine the cross-sectional area of pipeline:

$$A_{pipe} = Q/V_{pipe}$$
$$A_{pipe} = (46,000 \, m^3/d)/[(0.75 \, m/s)(86,400 \, s/d)]$$
$$= 0.713 \, m^2 \tag{9}$$

7. Determine the pipeline diameter:

$$D_{pipe} = (A_{pipe} \times 4/3.14)^{0.5}$$
$$D_{pipe} = (0.713 \, m^2 \times 4/3.14)^{0.5}$$
$$= 0.953 \, m \tag{10}$$

Decision: use a 1 m (1000 mm) diameter pipe.

## 9. MARINE POLLUTION PREVENTION (THE CITY OF LOS ANGELES BIOSOLIDS ENVIRONMENTAL MANAGEMENT SYSTEM)

The City of Los Angeles, California, recovers residues, known as biosolids, from its wastewater treatment plants. The Hyperion Treatment Plant (HTP) and the Terminal Island Treatment Plant (TITP) are responsible for managing the biosolids that are produced from the wastewater processing. Strict quality control procedures and regulatory compliance with the U.S. federal and California local laws are followed for the production and use of biosolids. Farmers use some of the biosolids as a soil amendment and fertilizing material and some are mixed with green materials to produce compost (79). The biosolids produced at the HTP were disposed in the ocean from 1957 to 1987, as shown in Table 9.2. Between 1987 and 1989, biosolids were primarily disposed in landfills. The city started an extensive beneficial reuse program in 1989, which continues today. The city received national awards from the US EPA for rapid conversion from disposal to beneficial use of biosolids in 1989 and an outstanding 100% beneficial reuse in 1994. In 2003, special recognition and awards were received from the Association of Metropolitan Sewerage Agencies (AMSA) and the US EPA for the city's Exceptional Quality Biosolids Program. In addition, a video produced by the city entitled "Where Does it Go?" received a National Environmental Achievement Award for Public Information and Education, Video Category from AMSA.

The biosolids Environmental Management System (EMS) is a program developed by the National Biosolids Partnership (NBP) to improve the quality of biosolids management programs nationwide and to promote public acceptance of biosolids use and disposal practices. The EMS addresses the management aspects of the city's biosolids

Table 9.2
The City of Los Angeles disposal method of biosolids in wet tons per year

| Year | Ocean disposal | Landfill | Chemical fixation | Land application | On-site energy recovery | Compost | Total for year | Cost |
|---|---|---|---|---|---|---|---|---|
| 1987 | 259,500 | 163,558 | 0 | 0 | 0 | 0 | 423,058 | $7,986,675.00 |
| 1988 | 0 | 359,111 | 74,990 | 1248 | 508 | 0 | 435,857 | $19,248,716.00 |
| 1989 | 0 | 137,773 | 104,072 | 185,994 | 5562 | 0 | 433,401 | $17,486,784.00 |
| 1990 | 0 | 0 | 142,893 | 287,819 | 3228 | 352 | 434,292 | $14,232,575.00 |
| 1991 | 0 | 0 | 78,623 | 190,583 | 5786 | 107,228 | 382,220 | $13,158,226.00 |
| 1992 | 0 | 0 | 76,340 | 123,736 | 4482 | 148,587 | 353,145 | $12,262,285.00 |
| 1993 | 0 | 0 | 24,800 | 153,680 | 3593 | 143,215 | 325,288 | $11,022,563.00 |
| 1994 | 0 | 0 | 0 | 190,482 | 5021 | 104,353 | 299,856 | $9,286,662.00 |
| 1995 | 0 | 0 | 0 | 191,712 | 3764 | 71,176 | 266,652 | $7,566,419.00 |
| 1996 | 0 | 0 | 0 | 228,753 | 2257 | 51,577 | 282,587 | $6,778,887.00 |
| 1997 | 0 | 0 | 0 | 283,397 | 243 | 17,138 | 302,775 | $6,012,903.00 |
| 1998 | 0 | 0 | 0 | 229,843 | 0 | 44,267 | 274,110 | $5,563,702.00 |
| 1999 | 0 | 0 | 0 | 259,571 | 0 | 38,190 | 297,761 | $5,850,783.00 |
| 2000 | 0 | 0 | 0 | 287,603 | 0 | 1806 | 289,409 | $7,300,000.00 |
| 2001 | 0 | 0 | 0 | 276,961 | 0 | 2049 | 279,010 | $7,490,000.00 |
| 2002 | 0 | 0 | 0 | 288,764 | 0 | 1480 | 290,244 | $6,639,000.00 |
| 2003 | 0 | 0 | 0 | 261,954 | 0 | 1672 | 263,626 | $6,972,706.00 |
| 2004 | 0 | 0 | 0 | 248,235 | 0 | 1714 | 249,949 | $6,755,000.00 |
| 2005 | 0 | 0 | 0 | 255,450 | 0 | 1964 | 257,414 | $6,767,457.00 |
| 2006 | 0 | 0 | 0 | 238,608 | 0 | 2608 | 241,216 | $6,674,348.00 |
| Total | 259,500 | 660,442 | 501,718 | 4,184,393 | 34,444 | 739,376 | 6,381,870 | $185,055,691.00 |

program and encourages public participation and communication. In September 2003, the city's EMS program was verified by an independent third-party auditor and became the second agency in the nation to be admitted to the NBP EMS program. The certification was retained in 2004 and 2006 after the program was verified by an independent auditor. The city received the NBP EMS platinum certification status in October 2006. This status designates that the city has maintained the highest standards possible for biosolids management and environmental stewardship. Table 9.2 documents the accomplishment of the city of Los Angeles (79).

## 10. OCEAN DISPOSAL TECHNOLOGY ASSESSMENT AND CONCLUSIONS

In the United States, federal legislation aimed at controlling water pollution first appeared in 1899, and was strengthened during each decade since the 1950s. Thousands of municipal sewage treatment systems, or Publicly Owned Treatment Works (POTWs), were built, although ocean disposal of biosolids and dredged materials is still permitted (77–79).

Since the London Convention came into force in 1975, major steps have been taken by the parties to the convention to eliminate or reduce the risk to the marine environment and human health from waste disposal at sea. The banning of ocean disposal of industrial and radioactive wastes is one of the many significant achievements of the convention. The adoption of the precautionary approach and the development of the waste assessment guidance by the convention provide the basis for signatories to follow a consistent approach to maximize the protection of the marine environment and human health when making decisions to permit disposal of wastes and other materials at sea.

The ocean is without boundary. To protect and preserve our marine environment, all contracting parties to the global conventions for the protection of marine environment should take the necessary steps to implement the conventions by developing legislation and environmental regulations, and, where necessary, follow up with strict enforcement to ensure regulatory compliance.

The approach taken by international conventions in protecting the marine environment has focused either on ocean-based or land-based sources—a medium-by-medium approach. This approach remains an issue for the contracting countries who structured their legislation and regulations under individual conventions. This issue has been extensively discussed by Champ et al. (2) and Wolfe (4) in the late 1980s. The practical outcome of this situation in Canada results, for example, in ocean dumping of wastes by loaded barges being regulated by the Canadian Environmental Protection Act (CEPA), while discharges of wastes into coastal waters from outfalls or pipes are regulated by the Fisheries Act. Similarly, waste disposal in the United States is regulated under the MPRSA, while the waste discharges into coastal waters are regulated by the U.S. Clean Water Act. This situation can lead to the following management problems (4):

- The different criteria, requirements, and restrictions imposed by the various laws may lead to inconsistencies in waste-management procedures.

- Application of various regulations individually may prohibit (or encourage) disposal of a particular waste in one medium without regard to the environmental effects associated with the discharge or disposal into other media.
- Regulating waste disposal using a "medium-by-medium" approach may result in shifting the risk posed by individual classes of wastes to the medium of least regulation, rather than to the medium of least risk.
- Acts or regulations created by this approach are difficult for the general public to understand and follow.

Consideration of multiple media in waste management was proposed and discussed by environmental managers and scientists in the late 1980s (2). To this day, it remains a challenging issue to the policy makers and environmental managers.

The London Convention 1972 and the London 1996 Protocol have essentially banned the disposal at sea of all waste and other matter other than the wastes listed in the Annexes to the convention, namely, dredged material, sewage sludge, ships and platforms, fish wastes, organic wastes of natural origin, inert inorganic geologic material and bulky wastes. Ocean disposal should be the last option if the wastes can be recovered for beneficial use. For instance, ocean disposal of nutrient-rich biosolids has been terminated in the U.S. in favor of land application and compost in which biosolids are used as a soil amendment, fertilizing material, or compost conditioner (79).

The best management approach, however, is the promotion of pollution prevention, which will reduce and eventually eliminate the sources of wastes that are entering the marine environment. Key sources where improvements can be made by pollution prevention include disposal at sea, land-based discharges, ship source pollution, seabed activities, and emissions from land-based industries. These improvements will surely lead to improved marine environmental quality and reduced risk to human health in our oceans in the future.

## NOMENCLATURE

$A_p$ = area per diffuser port, m$^2$
$A_{pipe}$ = pipeline cross-sectional area, m$^2$
$A_{tp}$ = total area of diffuser ports, m$^2$
$b$ = effective width of diffuser system, m (ft)
$C_0$ = pollutant concentration (or bacterial concentration) after initial dilution, mg/L
$C_t$ = maximum pollutant concentration (or bacterial concentration) at time t, mg/L
$d$ = average depth of the liquid waste field, m (ft)
$D1$ = initial dilution due to mixing the liquid waste and sea water
$D2$ = dilution due to eddy diffusion after initial dilution
$D3$ = dilution due to waste decay
$D_p$ = diffuser port diameter, m
$D_{pipe}$ = diameter of pipeline, m
$L_d$ = total diffuser length, m
$N$ = number of diffusers
$Q$ = average liquid waste flow rate in the pipeline, m$^3$/d; or in m$^3$/s or m$^3$/h (ft$^3$/s or ft$^3$/h)

$S_d$ = diffuser spacing, m
$T_{90}$ = the time to achieve a 90% reduction in bacterial numbers, h
$V$ = current velocity, m/s or m/h (ft/s or ft/h); or m/min (ft/min) m/min (ft/min)
$V_{pipe}$ = liquid waste velocity in pipeline, m/s
$X$ = distance along plume centerline, m (ft)

## REFERENCES

1. E. D. Goldbert, The oceans as waste space: the argument. *Oceanus*, 24 (1), 2–9 (1981).
2. M. A. Champ, M. A. Conti, and P. K. Park, Multimedia. Risk assessment and ocean waste management. In: *Oceanic Processes in Marine Pollution, Vol. 3: Marine Waste Management: Science and Policy*, M. A. Champ and P. K. Park (eds.), Robert E. Krieger, Malabar, FL, pp. 3–24 (1989).
3. K. S. Kamlet, Ocean dumping regulations: issues and approaches. In: *Oceanic Processes in Marine Pollution, Vol. 3: Marine Waste Management: Science and Policy*, M. A. Champ and P. K. Park (eds.), Robert E. Krieger, Malabar, FL, pp. 111–121 (1989).
4. D. A. Wolfe, Urban wastes in coastal waters: assimilative capacity and management. In: *Oceanic Processes in Marine Pollution, Vol. 5: Urban Wastes in Coastal Marine Environments*, D. A. Wolfe and T. P. O'Connor (eds.), Robert E. Krieger, Malabar, FL, pp. 3–22 (1988).
5. International Maritime Organization (IMO), *London Dumping Convention: The First Decade and Beyond—Provisions of the Convention on the Prevention of Marine Pollution by Dumping Wastes and Other Matter, 1972 and Decisions Made by the Consultative Meeting of Contracting Parties (1975–1988)*. London, UK (1991).
6. I. W. Duedall, B. H. Ketchum, P. K. Park, and D. R. Kester, Global inputs, characteristics, and fates of ocean-dumped industrial and sewage wastes: an overview. In: *Wastes in the Ocean, Vol. 1: Industrial and Sewage Wastes in the Ocean*, I. W. Duedall, B. H. Ketchum, P. K. Park, and D. R. Kester (eds.), Wiley-Interscience, New York, pp. 3–45 (1983).
7. M. K. Nauke, Obligations of contracting parties to the London Dumping Convention. In: *Oceanic Processes* in *Marine Pollution, Vol. 3: Marine Waste Management: Science and Policy*, M. A. Champ and P. K. Park (eds.), Robert E. Krieger, Malabar, FL, pp. 123–136 (1989).
8. W. C. Holton, No safe harbor. *Environmental Health Perspectives*, 106 (5), 228–233 (1998).
9. K. L. Tay, Controlling the disposal of waste at sea in Canada. In: *Proceedings of The Sixth Mainland-Taiwan Environmental Protection Conference*. National Sun Yat-Sen University, Kaohsiung, Taiwan, pp. 989–993 (1999).
10. J. M. Osborne, and K. R. Karr, *Environmental Code of Practice for Treatment and Disposal of Wastes Discharges from Offshore Oil and Gas Operations*, Report EPS 1/PN/2, Environment Canada. Ottawa, Ontario, Canada (1990).
11. S. Patin, *Environmental Aspects of Offshore Oil and Gas Industry*. EcoMonitor, New York (1999).
12. United Nations Environment Program, *1992 Rio Declaration on Environment and Development, Principle 15* (1993).
13. B. Thorne-Miller, The LDC, the precautionary approach, and the assessment of wastes for sea-disposal. *Maritime Pollution Bulletin*, 24 (7), 335–339 (1992).
14. Canadian Environmental Protection Act, *Statutes of Canada 1999, Chapter 33, Part VI, Canada Gazette*, Ottawa, Ontario, Canada (1999).

15. Environment Canada, *Users Guide to the Application Form for Ocean Disposal, Environmental Protection Series*, report EPS 1/MA/1, Ottawa, Ontario, Canada (1995).
16. U.S. Army Corps of Engineers, *Beneficial Uses of Dredged Material, Engineer Manual*, EM 110–2–5026, Washington, DC (1986).
17. L. A. Murray, Progress in England and Wales on the development of beneficial uses of dredged material. In: *Proceedings of the Second International Conference on Dredging and Dredged Material Placement, Lake Buena Vista, Florida, 13–16 November, 1994*, E. C. McNair, Jr. (ed.), American Society of Civil Engineers, New York, pp. 644–653 (1994).
18. D. R. Parrott, *Monitoring of the Offshore Disposal Site in the Middle Shoal Area, Cape Breton Island, Nova Scotia, 19–30 October 2002*, Environment Canada, Dartmouth, Nova Scotia, Canada, 60 pp. (2003).
19. U.S. Congress, *The Marine Protection, Research and Sanctuaries Act (MPRSA) of 1972 ("The Ocean Dumping Act")*, Public Law 92–532, U.S. Code. 86 Stat. 1052. 33 U.S.C. 1401 et seq. as amended, Government Printing Office, Washington, DC (1972).
20. U.S. Environmental Protection Agency, *Evaluating Environmental Effects of Dredged Material Management Alternatives: A Technical Frameworks*, US EPA 542–B-92–008, Washington, DC (1992).
21. Environment Canada. *Technical Guidance on Biological Monitoring for Ocean Disposal*, EVS Environment Consultants, EVS Project No. 3/047–52, Ottawa, Ontario, Canada (1995).
22. Canadian Council of Ministers of Environment, *Canadian Environmental Quality Guidelines*, Winnipeg, Manitoba, Canada (2003).
23. E. Erdheim, US marine waste disposal policy. In: *Wastes in the Ocean, Vol. 6: Nearshore Waste Disposal*, B. H. Ketchum, J. M. Capuzzo, W. V. Burt, I. W. Duedall, P. K. Park, and D. R. Kester (eds.), Wiley-Interscience, New York, pp. 421–460 (1985).
24. U.S. Environmental Protection Agency, *Evaluation of Dredged Material Proposed for Ocean Disposal*, US EPA 503/8–91/001, Washington, DC 214 pp. (1991).
25. P. M. Chapman, Why are we still emphasizing chemical screening-level numbers? *Maritime Pollution Bulletin*, 40 (6), 465–466 (2000).
26. Environment Canada, *Guidance Document on Collection and Preparation of Sediments for Physicochemical Characterization and Biological Testing, Environmental Protection Series*, report EPS 1/RM/29, Ottawa, Ontario, Canada (1994).
27. Environment Canada, *Reference Method for Determining Acute Lethality of Sediment to Marine or Estuarine Amphipods*, report EPS 1/RM/35, Environment Canada, Ottawa, Ontario, Canada (1998).
28. Environment Canada, *Biological Test Method: Solid-Phase Reference Method for Determining the Toxicity of Sediment Using Luminescent Bacteria (Vibrio fischeri)*, EPS 1/RM/42, Environment Canada, Ottawa, Ontario, Canada (2002).
29. Environment Canada, *Biological Test Method: Fertilization Assay Using Echinoids (Sea Urchins and Sand Dollars)*, report EPS 1/RM/27, Environment Canada, Ottawa, Ontario, Canada (1992).
30. U.S. Environmental Protection Agency, *Guidance Manual: Bedded Sediment Bioaccumulation Tests*, US EPA/600/R-93/183, Washington, DC (1993).
31. C. J. True and A. A. Heyward, Relationships between Microtox test results, extraction methods, and physical and chemical compositions of marine sediment samples. *Toxicologic Assessment*, 5, 29–45 (1990).
32. K. L. Tay, K. G. Doe, S. I. Wade, D. A. Vaughan, R. E. Berrigan, and M. J. Moore, Sediment bioassessment in Halifax Harbour. *Environmental Toxicology and Chemistry*, 11, 1567–1581 (1992).

33. B. A. Zajdlik, K. G. Doe, and L. M. Porebski, *Report on Biological Toxicity Tests Using Pollution Gradient Studies—Sydney Harbour*, Marine Environment Division, Environment Canada, EPS 3/AT/2, Ottawa, Ontario, Canada (2000).
34. K. L. Tay, K. G. Doe, A. J. MacDonald, and K. Lee, The influence of particle size, ammonia and sulfide on toxicity of dredged materials for ocean disposal. In: *Microscale Testing in Aquatic Toxicology—Advance, Techniques and Practice*, P. G. Wells, K. Lee, and C. Blaise (eds.), CRC Lewis, Florida, pp. 559–574 (1998).
35. T. P. O'Connor, Comparative criteria: land application of sewage sludge and ocean disposal of dredged material. *Maritime Pollution Bulletin*, 36 (3), 181–184 (1998).
36. T. S. Bridges and D. M. Moore, *Summary of a Workshop on Interpreting Bioaccumulation Data Collected During Regulatory Evaluations of Dredged Material*, U.S. Army Corps of Engineers, Waterways Experiment Station, Miscellaneous Paper D-96-1 (1996).
37. U.S. Environmental Protection Agency, *Proceedings of the National Sediment Bioaccumulation Conference*, US EPA 823-R-98-002, Washington, DC (1998).
38. K. L. Tay, S. J. Teh, K. Doe, K. Lee, and P. Jackman, Histopathologic and histochemical biomarker responses of baltic clam, *Macoma balthica*, to contaminated Sydney Harbour sediment, Nova Scotia, Canada. *Environmental Health Perspectives*, 111 (3), 273–280 (2003).
39. K. Essink, Ecological effects of dumping of dredged sediments; options for management. *Journal of Coastal Conservation*, 5, 69–80 (1999).
40. D. J. Wildish and J. Power, Avoidance of suspended sediments by smelt as determined by a new "single fish" behavioral bioassay. *Bulletin of Environmental Contamination and Toxicology*, 34, 770–774 (1985).
41. Seakem Oceanography Limited, *Assessment of the Effects of Suspended Dredge Material on Aquaculture Organisms*. Report submitted to Environment Canada, Dartmouth, Nova Scotia, Canada, January (1990).
42. J. Grant and B. Thorpe, Effects of suspended sediment on growth, respiration, and excretion of the soft-shell clam (*Mya arenaria*). *Canadian Journal of Fishery and Aquatic Science*, 48, 1285–1292 (1991).
43. G. Radenac, P. Miramand, and J. Tardy, Search for impact of a dredged material disposal site on growth and metal contamination of *Mytilus edulis* (L.) in Charente-Maritime (France). *Maritime Pollution Bulletin*, 34 (9), pp. 721–729 (1997).
44. J. A. Babinchak, J. T. Graikoski, S. Dudley, and M. F. Nitkowski, Effect of dredge spoil deposition on fecal coliform counts in sediments at a disposal site. *Applied Environmental Microscopy*, 34 (1), 38–41 (1977).
45. R. Engler, L. Saunders, and T. Wright, Environmental effects of aquatic disposal of dredged material. *The Environmental Profession*, 13, 317–325 (1991).
46. GESAMEP, *Scientific Criteria for the Selection of Waste Disposal Sites at Sea*, Reports and Studies 16, Group of Experts on the Scientific Aspects of Marine Environmental Protection (GESAMEP) (1982).
47. Canadian Environmental Assessment Act, *Statutes of Canada 1992, Chapter 37, Canada Gazette*, Queen's Printer for Canada, Ottawa, Ontario, Canada (1992).
48. Environment Canada, *An Overview of Biological Community Data in the Ocean Disposal Permit Review Process, Atlantic Region—Current Practice, Statistical Validity, and Biological Communities*. Ocean Disposal Report 10, ISBN: 0-662-27742-2, Environment Canada, Dartmouth, Nova Scotia, Canada (1999).
49. C. D. Levings, E. P. Anderson, and G. W. O'Connell, Biological effects of dredged-material disposal in Alberni Inlet. In: *Wastes in the Ocean, Vol. 6: Nearshore Waste Disposal*, B. H.

Ketchum, J. M. Capuzzo, W. V. Burt, I. W. Duedall, P. K. Park, and D. R. Kester (eds.), Wiley-Interscience, New York, pp. 131–158 (1985).
50. R. D. Roberts, M. R. Gregory, and B. A. Foster, Developing an efficient macrofauna monitoring index from an impact study—a dredge spoil example. *Maritime Pollution Bulletin*, 36 (3), 231–235 (1998).
51. M. Harvey, D. Gauthier, and J. Munro, Temporal changes in the composition and abundance of the macro-benthic invertebrate communities at dredged material disposal sites in the Anse à Beaufils, Baie des Chaleurs, Eastern Canada. *Maritime Pollution Bulletin*, 36 (1), 41–55 (1998).
52. S. D. Smith and M. J. Rule, The effects of dredge-spoil dumping on a shallow water soft-sediment community in the Solitary Islands Marine Park, NSW, Australia. *Maritime Pollution Bulletin*, 42(11), pp. 1040–1048 (2001).
53. S. G. Bolam and H. L. Rees, Minimizing impacts of maintenance dredged material disposal in the coastal environment: a habitat approach. *Environmental Management*, 32 (2), 171–188 (2003).
54. A. Chevier and P. A. Topping, *National Guidelines for Monitoring at Ocean Disposal Site*, Environment Canada, Ottawa, Ontario, Canada (1998).
55. K. L. Tay, K. G. Doe, A. J. MacDonald, and K. Lee, *Monitoring of the Black Point Ocean Disposal Site, Saint John Harbour, New Brunswick, 1992–1994*, Ocean Dumping Report #9, Environment Canada, Dartmouth, Nova Scotia, Canada (1997).
56. C. L. Amos, M. Hughes, A. Robertson, B. Wile, and K. L. Tay, *Seabed Stability Monitoring at Dump Site B of Saint John Harbour, New Brunswick, Using Sea Carousel, Geological Survey of Canada*, internal technical report, Bedford Institute of Oceanography, Dartmouth, Nova Scotia, Canada (1993).
57. U.S. Army Corps of Engineers, *Disposal Area Monitoring System (DAMOS)—An Investigation of Sediment Dynamics in the Vicinity of Mystic River CAD Cells Utilizing Artificial Sediment Tracers*, DAMOS Contribution 150 (2003).
58. Envirosphere Consultants Limited, *Monitoring Seabed Animal Communities in Outer Saint John Harbour and at the Black Point Ocean Disposal Site—2001 Survey and Comparison to Previous Studies*, technical report submitted to Environment Canada, Dartmouth, Nova Scotia, Canada, December (2002).
59. Environment Canada, *Pulp and Paper Technical Guidance for Aquatic Environmental Effects Monitoring*, EEM/1998/1, Ottawa, Ontario, Canada (1998).
60. Envirosphere Consultants Limited, *Environmental Monitoring at the Black Point Ocean Disposal Site: Assessing Long-term Impacts of Dredge Spoil Disposal in Saint John Harbour, New Brunswick*, report submitted to Environment Canada under contract agreement (2003).
61. J. D. Germano, D. C. Rhoads, L. F. Boyer, C. A. Menzie, and J. Ryther, Jr., REMOTS® Imaging and side-scan sonar: efficient tools for mapping seafloor topography, sediment type, bedforms, and benthic biology. In: *Oceanic Processes in Marine Pollution, Vol. 4, Scientific Monitoring Strategies for Ocean Waste Disposal Hood*, D. W. Hood, D. W. A. Schoener, and P. K. Park (eds.), Robert E. Krieger, Malabar, FL, pp. 39–48 (1989).
62. D. C. Rhoads and J. D. Germano, *The Use of REMOTS® Imaging Technology for Disposal Site Selection and Monitoring*, ASTM STP 1087, 1916 Race Street, Philadelphia, PA 19103, pp. 50–64 (1990).
63. U.S. Army Corps of Engineers, *Post-Storm Monitoring Survey at the New London Disposal Site Seawolf Mound*, DAMOS Contribution 149, New England District, 81 pp. (2002).

64. L. P. Hildebrand, *An Overview of Environmental Quality in Saint John Harbour, New Brunswick, Environmental Protection Service Report Series*, Environment Canada, Atlantic Region, Halifax, Nova Scotia, EPS-5–AR-81–1 (1980).
65. HydroSimTec Consultant, *Analysis of S4 Data, Black Point Dumping Site*, report No. 010001 submitted to Geological Survey of Canada, Atlantic Region, Bedford Oceanography Institute, Dartmouth, Nova Scotia, Canada, March (2001).
66. HydroSimTec Consultant, *Numerical Simulation of Tidal Circulation and Sediment Transport for Black Point Dumping Site*, report No. 010002 submitted to Geological Survey of Canada, Atlantic Region, Bedford Oceanography Institute, Dartmouth, Nova Scotia, Canada, September (2001).
67. HydroSimTec Consultant, *Numerical Simulation of Dredge Material Dumping and Sediment Dispersion for Black Point Dumping Site*, report No. 030001 submitted to Geological Survey of Canada, Atlantic Region, Bedford Oceanography Institute, Dartmouth, Nova Scotia, Canada, May (2003).
68. M. Comeau, *Dredged Material Deposited at Amherst Cove, PEI, During the Construction of the Confederation Bridge: An Assessment of Habitat Enhancement for Lobster*, report submitted under the MOU between Environment Canada and Fisheries and Oceans Canada (2003).
69. U.S. Congress, *The Federal Water Pollution Control Act Amendments (FWPCA) of 1972*, Public Law 92–500, in U. S. Code. Government Printing Office, Washington, DC, 86 Stat. 816. 33 U.S.C. 1251 et seq. as amended (1972).
70. Fisheries Act, *Statutes of Canada 1992, C. 51, Canada Gazette*. Queen's Printer for Canada, Ottawa, Ontario, Canada (1992).
71. Metcalf and Eddy, Inc., *Wastewater Engineering: Treatment, Disposal, and Reuse*, 4th ed. McGraw-Hill, New York (2003).
72. L. K. Wang, Dredging operations and waste disposal. In: *Water Resources and Natural Control Processes*, 1st ed., L. K. Wang and N. C. Pereira (eds.). Humana Press, Totowa, NJ, pp. 447–492 (1986).
73. L. K. Wang and C. Huang, Dredging and environmental risk assessments. In: *Handbook of Environmental Engineering* Series, 2nd ed., L. K. Wang, et al. (eds.). Humana Press, Totowa, NJ (2009).
74. G. Abraham, *Jet Diffusion in Stagnant, Ambient Fluid*. Publication No. 29. Delft Hydraulics Laboratory, New York (1963).
75. E. A. Pearson, *An Investigation of the Efficiency of Submarine Outfall Disposal of Sewage and Sludge*, California Water Pollution Control Board, CA, Publication No. 14 (1956).
76. M. E. Burchett, G. Tchobanoglous, and A. J. Burdoinn, A practical approach to submarine outfall calculations. *Public Works*, 98, 5 (1967).
77. Sacramento Regional County Sanitation District, *Biosolids Fact Sheet: Historical Use of Biosolids* (2005).
78. New England Biosolids Residuals Association, *History of Bioslids Use*. Tamworth, NH, http://www.nebiosolids.org/history.html, May (2007).
79. City of Los Angeles, *The City of Los Angeles Biosolids Environmental Management System (EMS)*, www.lacity.org/san/biosolidsems/index.htm (2007).

# 10
# Combustion and Incineration Engineering

## Walter R. Niessen

**CONTENTS**

INTRODUCTION TO INCINERATION
PROCESS ANALYSIS OF INCINERATION SYSTEMS
INCINERATION SYSTEMS FOR MUNICIPAL SOLID WASTE
THERMAL PROCESSING SYSTEMS FOR BIOSOLIDS
ECONOMICS OF INCINERATION
AN APPROACH TO DESIGN
APPENDIX: WASTE THERMOCHEMICAL DATA
NOMENCLATURE
REFERENCES

**Abstract** Thermal processes (drying, pyrolysis, and combustion) provide powerful means to reduce the volume and sanitize municipal solid wastes (MSW) and the residual solids of wastewater treatment. Solids removed by wastewater treatment processes include screenings and grit, floating materials (scum), and the concentrated solids from primary and secondary clarifiers (sewage sludge). This chapter reviews the basic technology and analysis tools relating to thermal processes, and presents the application of these tools in the management of both municipal solid wastes and sewage sludge (biosolids).

**Key Words** Sewage sludge • municipal solid waste • biosolids • combustion • incineration • pyrolysis • drying • air pollution • emissions • municipal waste combustors • multiple hearth furnace • fluidized bed • mass burning • refuse derived fuel • rotary dryer • tray dryer • indirect dryer • direct dryer.

## 1. INTRODUCTION TO INCINERATION

Solving the problem of volume is often the key to successful solid waste and sludge management. The low bulk density of solid waste or sludge requires costly storage containers at the point of generation, greatly affects the cost and difficulty of collection and handling, and is the primary factor in setting the cost and scale of landfill and other ultimate disposal operations. Incineration is often an attractive processing step to reduce the volume of solid waste and sludge. Properly, incineration should not be considered as an ultimate disposal method since it cannot function without a landfill or other means to receive its solid residues.

In addition to volume reduction, incineration also offers the potential advantages of detoxification of combustible toxins and sanitation of pathologically contaminated material (especially important for biosolids), reduction of environmental impact related to the leaching of organic material from raw refuse or biosolids landfills, and provision of an energy resource through the use of a boiler enclosure. Extension of the scope of "incineration" to include gasification reactors introduces a new product: a "synthesis gas" composed largely of carbon monoxide (CO) and hydrogen ($H_2$) that has use as a chemical feedstock or as a clean fuel usable in diesel, turbine or fuel cell energy conversion systems.

The disadvantages of incineration include high capital and operating cost relative to many other waste management options, operating problems owing to the sensitivity of the process to changes in the character of the waste, staffing problems made more critical in light of the complexity of the process, potential secondary environmental impacts such as air and noise pollution and discharge of highly polluted wastewater, adverse public reactions in many cases, and technical risk reflecting the complexity of incineration systems.

## 2. PROCESS ANALYSIS OF INCINERATION SYSTEMS

The designer or analyst of an incineration system faces a formidable challenge. Incineration processes are complex, involving the interplay of chemical reactions, fluid flow, and heat transfer in a nonisothermal, nonhomogeneous, reacting system. This already complex physicochemical process is made less tractable to rigorous analysis by the ever-changing nature of the waste and the irregular pattern of major and minor adjustments of key parameters owing to the action of the plant operators or automatic control instrumentation.

This pessimistic picture, however, should not lead the engineer to abandon theoretical analysis as a powerful tool in design and problem-solving tasks related to incineration. In this chapter and in other texts (1, 2), the interested student or practicing engineer is introduced to analytical tools that facilitate the understanding and insight to cope with this important sector of environmental engineering.

The analytical methods presented here draw heavily on the disciplines of chemical and mechanical engineering, beginning with a review of the fundamentals of process analysis (heat and material balances, chemical kinetics, and equilibrium). This is followed by several topics relating to the fundamental processes occurring in burning, gasifying, and pyrolyzing systems.

### 2.1. Stoichiometry

Stoichiometry is the discipline of tracking matter (particularly the elements), partitioned in accord with the laws of chemical combining weights and proportions, and energy (in all its forms) in either batch or continuous processes. Application of the laws of conservation of mass and energy, supplemented by consideration of chemical kinetic and equilibrium relationships, provides great insight into the behavior of incineration systems.

# Combustion and Incineration Engineering

**Table 10.1**
**Values of the universal gas constant R for ideal gases**

| Energy | Pressure (P) | Volume (V) | Mols (n) | Temperature (T) | Gas constant (R) |
|---|---|---|---|---|---|
| – | atm | $m^3$ | kg mol | °K | $0.08205 \dfrac{m^3 \text{ atm}}{\text{kg mol °K}}$ |
| – | kPa | $m^3$ | kg mol | °K | $8.3137 \dfrac{\text{kPa } m^3}{\text{kg mol °K}}$ |
| kcal | – | – | kg mol | °K | $1.9872 \dfrac{\text{kcal}}{\text{kg mol °K}}$ |
| joules (abs) | – | – | g mol | °K | $8.3144 \dfrac{\text{joules}}{\text{g mol °K}}$ |
| ft-lb | psia | $ft^3$ | lb mol | °R | $1545.0 \dfrac{\text{ft lb}}{\text{lb mol °R}}$ |
| BTU | – | – | lb mol | °R | $1.9872 \dfrac{\text{BTU}}{\text{lb mol °R}}$ |
| – | atm | $ft^3$ | lb mol | °R | $0.7302 \dfrac{ft^3 \text{ atm}}{\text{lb mol °R}}$ |

In the paragraphs below, it will be advantageous to use the kilogram mol (or kilogram atom)—the molecular (atomic) weight expressed in kilograms—as the unit quantity. The advantage arises because one molecular (atomic) weight (mol) of any compound (element) contains the same number of molecules (atoms) and, for gases, occupies approximately the same volume (at similar pressures and temperatures). The approximation holds exactly if the gases are "ideal," an assumption acceptably accurate for gases at atmospheric pressure and elevated temperatures.

### 2.1.1. Gas Laws

The relationship between absolute pressure $P$, absolute temperature $T$, and volume $V$ for ideal gases is given by

$$PV = nRT \quad (1)$$

where $n$ is the number of mols of the gas and $R$ is the universal gas constant (Table 10.1).

EXAMPLE 1

To generate carbon dioxide for process use, 3000 kg/d of a waste containing 80% carbon, 7% ash, and 13% moisture is to be burned. The combustion gases leave the furnace at 1000°C and pass through a gas cooler, exiting at 80°C. How many kilogram-mols and how many kilograms of $CO_2$ will be formed per day? How many cubic meters of $CO_2$ are produced per day at the furnace outlet and at the gas cooler outlet at 1.04 atm?

The number of mols of carbon (atomic weight $= 12$) in the waste is $(3000)(0.80)(1/12) = 200$. Noting that with complete combustion each mol of carbon yields 1 mol of $CO_2$, 200 mol/d of $CO_2$ are produced. The weight flow of $CO_2$ (molecular weight $= 44$) is $200(44) = 8800$ kg/d. From $PV = nRT$,

$$V = \frac{nRT}{P} = \frac{200(0.08206)T}{1.04} = 15.78T$$

at 1000°C (1273 K), $V = 20{,}090 \, m^3$; at 80°C (353 K), $V = 5570 \, m^3$.

## 2.1.2. Material Balances

A material balance is a quantitative expression of the law of conservation of matter:

$$\text{Input} = \text{Output} + \text{Accumulation} \qquad (2)$$

This expression is always true for elements flowing through combustion systems but is often not true for compounds participating in combustion reactions.

The basic data used in calculating material balances can include analyses of fuels, waste, gases in the system, etc. (e.g., see Appendix) and some rate data (usually feed rate). Coupled with these data are fundamental relationships that prescribe combining proportions in molecules (e.g., two atoms of oxygen to one of carbon in carbon dioxide) and those that indicate the course and heat effects of chemical reactions.

Balances on elements in the fuel or waste allow one to calculate the amount of air theoretically required to completely oxidize the carbon, hydrogen, sulfur, etc. (recognizing that a portion of the oxygen required may be supplied by the oxygen contained in the material being burned). This quantity of air (known as the *theoretical* or *stoichiometric* air requirement) is often insufficient in a practical combustor, and excess air (expressed as a percentage of the stoichiometric air quantity) is usually supplied. For example, an incinerator operating at 50% excess air denotes a combustion process to which 1.5 times the stoichiometric air requirement has been supplied.

EXAMPLE 2

We are burning a waste with the following composition: 75.0% carbon, 6.2% hydrogen, 2.4% sulfur, 2.1% oxygen, 0.5% nitrogen, and 1.6% ash. The waste is burned with 50% excess air. The combustion air is at 15.5°C and 70% relative humidity. Calculate the flue gas composition and the quantity of combustion air that is required. The sequence of computations is shown in Table 10.2.

Several elements of the analysis on Table 10.2 should be noted:

- *Line 1*: Carbon is assumed to burn completely to carbon dioxide. In practice, some carbon may be incompletely burned (forming carbon monoxide), and some may end up as unburned carbon char in solid residues or as part of the particulate matter, leaving in the effluent gas as soot or char fragments.
- *Line 2*: Available hydrogen in the waste (other than the hydrogen in bound moisture) increases the amount of combustion air, but does not appear in the Orsat analysis (lines 16 and 17). Available hydrogen is presumed to react with oxygen (in the waste or supplied with the combustion air) to form water except for the hydrogen that preferentially reacts with organic chlorine to form hydrogen chloride (HCl).
- *Line 3*: Sulfur in the waste as sulfide or organic sulfur increases the amount of combustion air required in burning to $SO_2$. Inorganic sulfates may leave as ash or be reduced to $SO_2$. If selective analysis is not used for $SO_2$ (line 17), it is usually reported out as carbon dioxide. A small fraction (1% to 3%) of the $SO_2$ may be further oxidized to $SO_3$.
- *Line 4*: Oxygen in the waste reduces the amount of required combustion air.
- *Line 12*. Moisture entering as the humidity in the combustion air can be seen to be small and is often neglected. However, checking this assumption is prudent; especially in hot, humid locales.

Although this problem considered only waste components of C, H, O, N, and S, the analyst should review waste composition thoroughly and consider the range of possible secondary reactions:

**Table 10.2**
**Calculations for Example 2**

| Line | Component | kg | Atoms or mols[a] | Combustion product | Theoretical mols of O$_2$ required | Moles formed in stoichiometric combustion | | | | | |
|---|---|---|---|---|---|---|---|---|---|---|---|
| | | | | | | CO$_2$ | H$_2$O | SO$_2$ | N$_2$ | O$_2$ | Total |
| 1 | Carbon, C | 75.0 | 6.245 | CO$_2$ | 6.245 | 6.245 | 0.0 | 0.0 | 0.0 | 0.0 | 6.245 |
| 2 | Hydrogen, H$_2$ | 6.2 | 3.075 | H$_2$O | 1.538 | 0.0 | 3.075 | 0.0 | 0.0 | 0.0 | 3.075 |
| 3 | Sulfur, S | 2.4 | 0.075 | SO$_2$ | 0.075 | 0.0 | 0.0 | 0.075 | 0.0 | 0.0 | 0.075 |
| 4 | Oxygen, O$_2$ | 2.1 | 0.066 | — | (0.066) | 0.0 | 0.0 | 0.0 | 0.0 | 0.0 | 0.000 |
| 5 | Nitrogen, N$_2$ | 0.5 | 0.018 | N$_2$ | 0.0 | 0.0 | 0.0 | 0.0 | 0.018 | 0.0 | 0.018 |
| 6 | Moisture, H$_2$O | 12.2 | 0.678 | H$_2$O | 0.0 | 0.0 | 0.678 | 0.0 | 0.0 | 0.0 | 0.678 |
| 7 | Ash | 1.6 | N/A | — | 0.0 | 0.0 | 0.0 | 0.0 | 0.0 | 0.0 | 0.000 |
| 8 | Total | 100.0 | | | 7.792 | 6.245 | 3.753 | 0.075 | 0.018 | 0.0 | 10.091 |
| 9 | Mols of nitrogen in stoichiometric air[b] (79/21) (7.792) | | | | | | | | 29.312 | | 29.312 |
| 10 | Mols of nitrogen in excess air (0.5) (79/21) (7.792) | | | | | | | | 14.656 | | 14.656 |
| 11 | Mols of oxygen in excess air (0.5) (7.792) | | | | | | | | | 3.896 | 3.896 |
| 12 | Mols moisture in combustion air[c] | | | | | | 0.713 | | | | 0.713 |
| 13 | Total mols in flue gas | | | | | 6.245 | 4.466 | 0.075 | 43.985 | 3.896 | 58.695 |
| 14 | Volume (mol) percent in wet flue gas | | | | | 10.64 | 7.61 | 0.13 | 74.97 | 6.64 | 100.0 |
| 15 | Orsat (dry) flue gas analysis, mols | | | | | 6.245 | | 0.075 | 43.985 | 3.896 | 54.201 |
| 16 a. | With selective SO$_2$ testing, volume percent | | | | | 11.52 | N/A | 0.14 | 81.15 | 7.190 | 100.0 |
| 17 b. | With alkaline CO$_2$ testing only, volume percent | | | | | 11.66 | N/A | N/A | 81.15 | 7.190 | 100.0 |

[a] The symbol in the component column shows whether these are kg-mol or kg-atom.
[b] Throughout this chapter, dry combustion air is assumed to contain 21.0% O$_2$ by volume and 79.0% N$_2$.
[c] Calculated as follows: (0.008/18)[(29.377 + 14.688)(28) + (3.905)(32)] based on the basis of 0.008 kg water vapor per kg bone-dry air; found in standard psychrometric charts and tables.

- *Carbon monoxide*: Carbon monoxide (CO) is formed in appreciable quantities in grate-fired systems burning solid wastes that do not incorporate air jets over the fire to add oxygen and thoroughly mix the off-gases from the grate's gasification zone.
- *Chlorine*: Chlorine appearing in the waste as inorganic salts will, most likely, remain in the ash as the salt. Some chlorides may volatilize to some degree. Organic chlorine, however, bonds with hydrogen from the waste and forms HCl. Similar behavior is seen for organic fluorine compounds such as Teflon©.
- *Metals*: Metals usually burn to the oxide, although, in burning solid wastes, a large fraction of massive metal feed (e.g., tin cans, sheet steel, etc.) remains unoxidized.
- *Thermal decomposition*: Some compounds may decompose at combustor temperatures. Carbonates, for example, may dissociate to form an oxide and $CO_2$, and sulfides may "roast" to form the oxide and release $SO_2$.

In many instances, the analyst is called upon to evaluate an operating waste disposal system. In such studies, accurate data on the flue gas composition are readily obtainable and offer a low-cost means to characterize the operation and the feed waste.

One important combustor and combustion characteristic that can be immediately computed from the Orsat (dry gas) flue-gas analysis is the percentage of excess air where $O_2$, $N_2$, etc., are the volume percentages of the gases on a dry basis.

$$Percentage\ excess\ air = \frac{[O_2 - 0.5(CO + H_2)]\ 100}{0.266 N_2 - O_2 + 0.5(CO + H_2)} \qquad (3)$$

## EXAMPLE 3

The flue gas from a waste incinerator burning a low ash hydrocarbon waste believed to have little or no nitrogen or oxygen has an Orsat analysis (using alkaline $CO_2$ absorbent) of 11.6% $CO_2$, 7.2% $O_2$, and the rest nitrogen and inerts. From these data, calculate the weight ratio of hydrogen to carbon in the waste, the percent of carbon and hydrogen in the dry waste, the kilograms of dry air used per pound of dry waste, the percent of excess air used, and the mols of exhaust gas discharged from the unit per kilogram of dry waste burned. (Note that this example is derived from Example 2.)

Basis: 100 mol dry exhaust gas

| Component | Mols | Mol $O_2$ |
|---|---|---|
| $CO_2 (+SO_2)$ | 11.6 | 11.6 |
| $O_2$ | 7.2 | 7.2 |
| $N_2$ | 81.2 | – |
| Total | 100.0 | 18.8 |

Considering all $N_2$ to have come from the combustion air, a total of $81.2 \times (21/79) = 21.6$ mol $O_2$ entered with the $N_2$. The difference, $21.6 - 18.8 = 2.8$ mol $O_2$ may be assumed to have been consumed in burning hydrogen.

$H_2$ Burned: $2(2.8) = 5.6$ mol      11.2 kg
C Burned: $12(11.6)$ mol      +139.2 kg
    Total      150.4 kg

## Combustion and Incineration Engineering

a. Weight ratio of hydrogen to carbon: $(11.2/139.2) = 0.08$.
b. Percent (by weight) C in dry fuel: $(139.2/150.4)(100) = 92.55$.
c. Kilogram of dry air per kilogram of dry waste.

First, calculate the weight of air resulting in 1 mol dry exhaust gas from a nitrogen balance:

$$\frac{1}{100}(81.2 \, mol \, N_2)(^1/_{0.79} \, mol \, N_2/mol \, air)(29 \, kg \, air/mol) = 29.81 \, kg \, air/mol \, dry \, exhaust \, gas$$

then, $29.81(100/150)$ mol dry exhaust gas/kg waste $= 19.87$ kg dry air/kg dry waste

d. Percent excess air:

The oxygen *necessary* for combustion is: $11.6 + 2.8 \quad = 14.4 \, mol$
The oxygen *unnecessary* for combustion $\quad = 7.2 \, mol$
The total oxygen $\quad = 21.6 \, mol$

Note that the *necessary* oxygen increases and the *unnecessary* oxygen decreases if incompletely burned components (such as CO) are present.

The percent excess air (or oxygen) may be calculated as:

$$\frac{(100)(unnecessary)}{total - unnecessary} = \frac{100(7.2)}{21.6 - 7.2} = 50\% \quad (4a)$$

$$\frac{(100)(unnecessary)}{necessary} = \frac{100(7.2)}{21.6 - 7.2} = 50\% \quad (4b)$$

$$\frac{(100)(total - necessary)}{necessary} = \frac{100(21.6 - 14.4)}{14.4} = 50\% \quad (4c)$$

e. Mols of exhaust gas per kilogram of dry waste:

Noting that 5.6 mol water vapor must be added to the dry gas flow,

$$(100 + 5.6)/150.4 = 0.702 \, mol/kg \, waste.$$

Lessons learned by comparing of the results of Example 3 with the "true" situation from Example 2 are as follows:

- Waste analysis data are important in calculating combustion air requirements for design.
- Waste moisture data are necessary to determine total flue gas rates.
- Insight into the nature of the waste can be gained from stack gas analysis.

If data are available, all data—for both fuel and flue gas—should be used to cross-check for consistency.

### 2.1.3. Heat Balances

A heat balance is a quantitative expression of the law of conservation of energy. In waste incineration, five energy quantities are of prime interest:

- *Chemical energy*: the heat of chemical reaction, especially the heat of combustion
- *Latent heat*: the heat effect of changes in state, especially the heat of vaporization of moisture
- *Sensible heat*: the heat content (enthalpy) related to the temperature of materials
- *Useful heat*: the heat available for use, especially the sensible heat available to generate steam
- *Heat loss*: the heat lost through furnace walls by conduction, convection, and radiation

In analyzing incineration systems, the heat of combustion of the waste is perhaps the seminal variable defining the size, burning capacity, air supply, air pollution control system design, fan capacity and horsepower, energy recovery potential, and on and on. Ideally, laboratory determinations of the heat of combustion are available. Often, however, the analyst is left with a component analysis ($X\%$ paper, $Y\%$ wood, etc.) or, at best, an ultimate analysis. Component analysis can be used to develop the overall ultimate analysis of the mixed waste by combining appropriate proportions of the ultimate analysis of each component (1). One can then synthesize the mean heat of combustion using estimation formulas.

The three heating value estimation relationships that follow were developed to estimate the heat of combustion ($\Delta H_c$) of the combustible fraction of industrial and municipal wastes (kcal/kg) on a moisture and ash-free (MAF) basis. One uses the weight percent of hydrogen (H), oxygen (O), sulfur (S), etc. on a dry, ash-free basis. Substitute the percent (not the decimal percent) value for each in the following equations:

Chang equation (3):

$$\Delta H_c = 8561.11 + 179.72\,H - 63.89\,S - 111.17\,O - 90.00\,Cl - 66.94\,N \quad (5a)$$

Modified Dulong equation (4):

$$\Delta H_c = 78.31\,C + 359.32\,(H - O/8) + 22.12\,S + 11.87\,O + 5.78\,N \quad (5b)$$

Boie equation:

$$\begin{aligned}\Delta H_c =\ & 83.22\,C + 275.48\,H - 25.8\,O + 25.0\,S + 15.0\,N \\ & + 9.4\,Cl + 18.5\,F + 65.0\,P + 12.2\,Fe\end{aligned} \quad (5c)$$

The Chang equation, the modified Dulong equation, and the Boie equation have been tested against one another for the prediction of the heat of combustion of 150 pure organic compounds where laboratory data were available to test the accuracy of prediction. In this comparison (1), the average error relative to the laboratory value was as follows: for Chang, 1.48%; for Dulong, 5.54%; and for Boie, 11.38%. Chang's equation was clearly superior for this task. The Boie equation, however, was originally developed and is well regarded for estimation of the heat of combustion of mixed wastes (especially high cellulosic material) such as refuse or wood. The modified Dulong equation is generally best for fuel-like, high carbon/hydrogen materials such as coal, peat, or lignite.

In heat of combustion and sensible heat calculations, 15.6°C (60°F) is often used as a reference point for "zero energy." Most values of heat of combustion reported in the American and British incineration literature are the higher heating value (HHV), which includes the latent heat of vaporization of the water formed in combustion (10,520 kcal/kg-mol at 20°C). See the Appendix for HHV values for refuse and refuse components. The lower heating value (LHV) is often reported in the literature of mainland Europe and the Far East and does not include the latent heat. By international agreement, the joule has been selected as the preferred energy unit. One kcal is equivalent to 4190.02 joules.

Combustion and Incineration Engineering

The sensible heat content ($\Delta h$) at a temperature $T$ may be calculated relative to the reference temperature $T_o$ by:

$$\Delta h = \int_{T_o}^{T} Mc_p^o \, dT \; kcal/kg \, mol \tag{6}$$

where $Mc_p^o$ is the molar heat capacity (kcal kg-mol$^{-1}$°C$^{-1}$, which is numerically equal in units of Btu lb-mol$^{-1}$°F$^{-1}$). The calculation may be carried out using an empirical equation describing the functional relationship of $Mc_p^o$ on temperature (1). Also, one may use a graphical presentation (Figure 10.1) of the average molal heat capacity ($Mc_{p,avg}^o$)

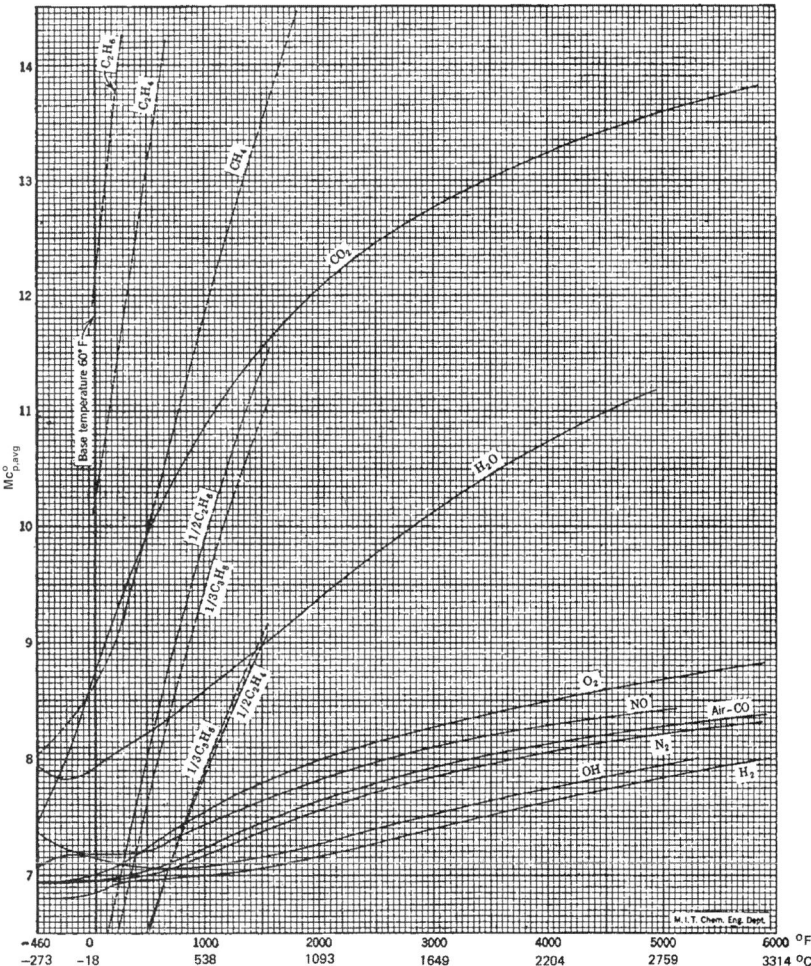

**Fig. 10.1.** Average molal heat capacity of fuel and combustion gases, ($Mc_{p,avg}^o$), at zero pressure between 60°F (15.5°C) and abscissa temperatures. (Courtesy of Massachusetts Institute of Technology Chemical Engineering Department.)

between a reference temperature of 60°F (15.6°C) and the abscissa temperature. Thus:

$$\Delta h = (T - T_o) M c^o_{p,ave} \quad (7)$$

EXAMPLE 4

If the 100 kg of waste described in Example 2 has a heat of combustion of 7500 kcal/kg (HHV) and the combustion air is preheated to 300°C, what is the temperature of the flue gases? How much steam can be generated if the gases are cooled to 180°C (about 350°F) in a boiler?

Assume 5% heat loss in the furnace and 5% in the boiler, and 570 kcal/kg enthalpy change from boiler feed water to product steam at 204°C. Basis: One hour of operation.

The total combustion air supplied to the system is 29.377 + 14.688 + 3(3.905) + 0.604 = 56.384 mol (see Table 10.2). From Figure 10.1, the heat content of the preheated air at 300°C is:

$$(56.384)(7.08)(300 - 15.5) = 113,572 \text{ kcal}$$

Therefore, the total energy impact is

$$7500(100) + 113,572 = 863,572 \text{ kcal energy addition}$$

To find the exit temperature of the combustion chamber and the steaming rate, it is useful to construct a plot of the heat content of the gas stream as a function of temperature, computed as shown in Table 10.3 and presented in Figure 10.2.

The flows of thermal energy are then:

| Energy flows | kcal | Temperature, °C |
|---|---|---|
| Energy into system = Heat of combustion | 750,000 | 15.5 |
| Air preheat | 113,572 | 300 |
| Total | 863,572 | 1630* |
| Heat loss (5%) from combustion chamber | (43,180) | |
| Energy into boiler | 820,392 | 1575 |
| Heat loss (5%) from boiler | (41,020) | |
| Heat loss out stack | (117,615) | 180 |
| Net energy into steam | 661,757 | 204 |

*The theoretical (adiabatic) flame temperature for this system (the temperature of the products of combustion assuming no heat loss).

For feedwater (at 100°C and 15.8 atm) changing to saturated steam at 15.8 atm, the enthalpy change is 567.9 kcal/kg, so that the resulting steaming rate for a burning rate of 1100 kg/h is:

$$\frac{661,757}{567.9} = 1165 \text{ kg/hour}$$

### Table 10.3
### Computation of heat content of flue gases from combustion of waste at 50% excess air

| | | | | | | |
|---|---|---|---|---|---|---|
| A. | Assumed temp., °C | 180 | 500 | 1000 | 1500 | 2000 |
| B. | A − 15.5°C | 164.5 | 484.5 | 984.5 | 1484.5 | 1984.5 |
| C. | $Mc^o_{p,avg} N_2$[a] | 6.93 | 7.16 | 7.48 | 7.74 | 8.10 |
| D. | $Mc^o_{p,avg} O_2$[a] | 7.17 | 7.48 | 7.87 | 8.13 | 8.26 |
| E. | $Mc^o_{p,avg} H_2O$[a] | 8.06 | 8.53 | 9.22 | 9.85 | 10.42 |
| F. | $Mc^o_{p,avg} CO_2$[a] | 9.67 | 10.64 | 11.81 | 12.57 | 12.91 |
| G. | Ash[b] | 0.2 | 0.2 | 0.2 | 0.2 | 0.2 |
| H. | Energy in $N_2$ 43.985(B)(C) | 50,109 | 152,250 | 324,480 | 507,925 | 699,850 |
| I. | Energy in $O_2$ 3.896(B)(D) | 4594 | 14,230 | 30,565 | 47,430 | 65,350 |
| J. | Energy in $H_2O$ 4.466(B)(E) | 5921 | 18,125 | 39,870 | 64,340 | 91,265 |
| K. | Energy in $CO_2$ 6.245(B)(F) | 9937 | 32,540 | 73,210 | 116,910 | 162,210 |
| L. | Energy in Ash 1.6(B)(G) +85[c] | 53 | 290 | 450 | 610 | 770 |
| M. | Latent Heat in $H_2O$ 4.466 (10,595)[d] | 47,320 | 47,320 | 47,320 | 47,320 | 47,320 |
| N. | Total (H + I + J + K + L + M)[e] | 117,880 | 264,689 | 514,600 | 781,800 | 1,057,501 |
| O. | kcal/mol gas | 2009 | 4512 | 8772 | 13,326 | 18,025 |

[a] *Source*: Fig. 10.1 (kcal/kg mol °C).
[b] Specific heat (typical) of the ash (kcal/kg °C) for solid or liquid.
[c] The latent heat of fusion of the ash (85 kcal/kg) is added at temperatures greater than 800°C, the assumed ash fusion temperature.
[d] Latent heat of vaporization at 15.5°C of free water in waste and from combustion of hydrogen in waste (kcal/kg mol). Total heat content of gas stream (kcal).
[e] Total heat content of gas stream (kcal).

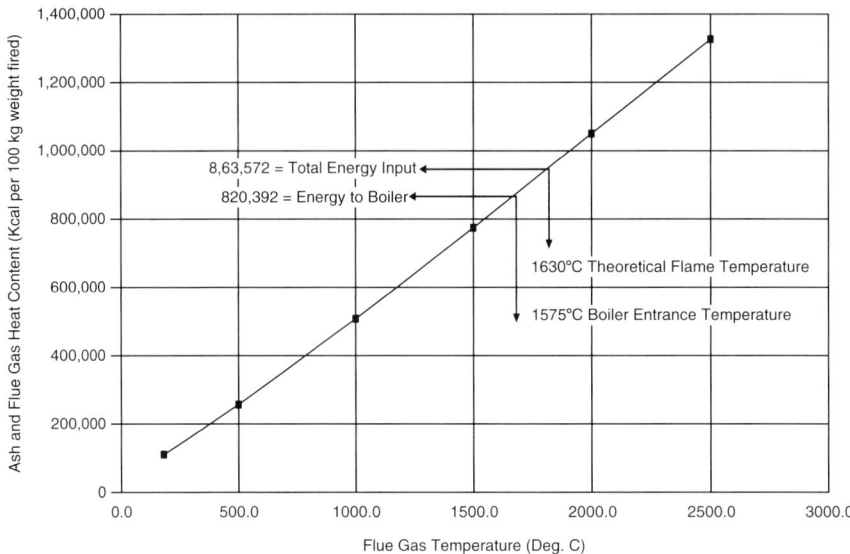

**Fig. 10.2.** Energy content of ash and flue gases as a function of flue gas temperature.

### 2.1.4. Equilibrium

No chemical reactions go to completion. Always, some fractions of the reactants remain in the reaction mass. For the gas phase reaction:

$$aA + bB \leftrightarrow cC + dD \tag{8}$$

where the reactant and product concentrations are expressed as partial pressures $P_A$, $P_B$, ..., the equilibrium constant $K_p$, which is a function (only) of temperature, is given by

$$K_p = \frac{P_C^c P_D^d}{P_A^a P_B^b} \tag{9}$$

where the units of $K_p$ depend on the stoichiometric coefficients a, b, c, and d. If $(c + d - a - b)$ is zero, $K_p$ is dimensionless. If the total is nonzero, $K_p$ will have the units of pressure raised to the appropriate integer or fractional power. Figures 10.3 and 10.3a show the temperature dependence of reactions of interest. Note that when solid carbon is a product or reactant, no partial pressure term for carbon is entered into the mathematical formulation.

EXAMPLE 5

At the furnace outlet temperature in Example 4 and at a total pressure of 1 atm, what is the emission rate of nitric oxide (NO) formed by the following reaction:

$$\frac{1}{2}O_2 + \frac{1}{2}N_2 \leftrightarrow NO \tag{10}$$

| Component | Mols | Partial pressure (atm) |
|---|---|---|
| NO | $x$ | $x/58.695$ |
| $N_2$ | $44,083 - 0.5x$ | $(44.083 - 0.5x)/58.695$ |
| $O_2$ | $3905 - 0.5x$ | $(3.905 - 0.5x)/58.695$ |
| Total[a] | 58.695 | 1.00 |

[a] For this reaction, the total number of mols does not change.

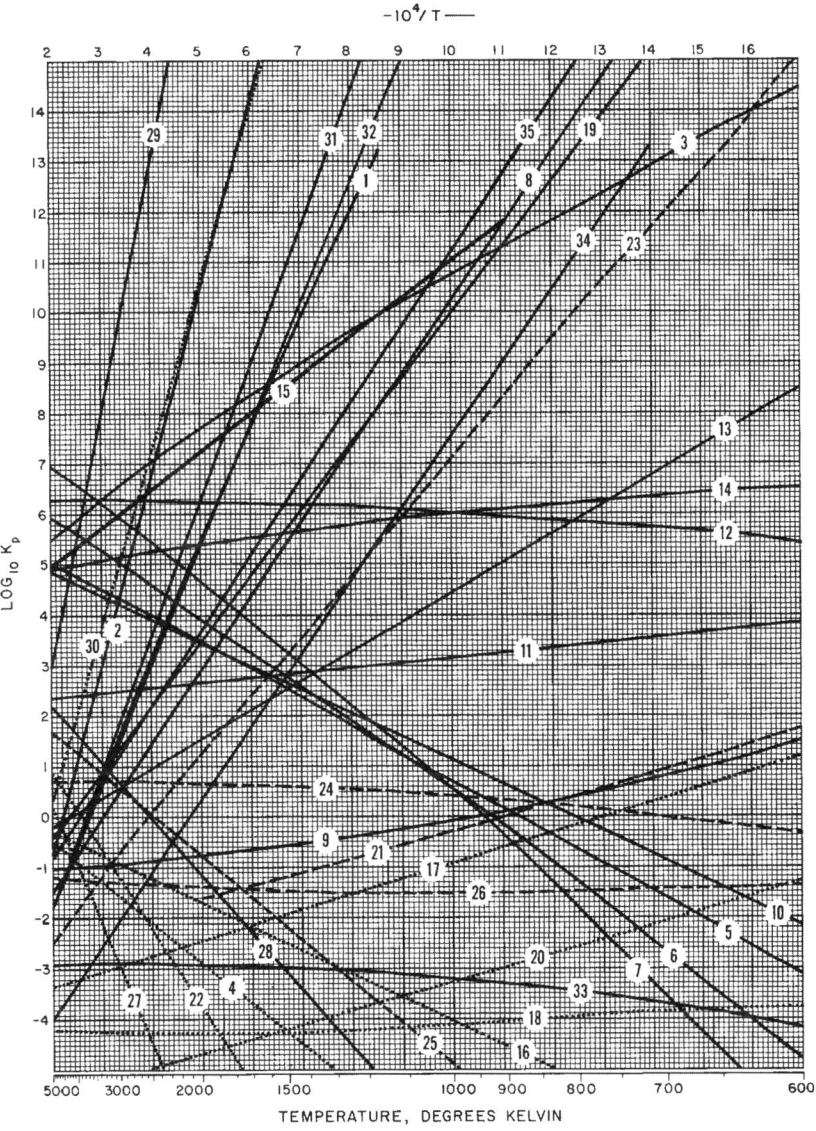

**Fig. 10.3.** Equilibrium constants of combustion reactions (partial pressure in atm). (Courtesy of Massachusetts Institute of Technology Chemical Engineering Department.)

### A. Carbon Reactions

| | |
|---|---|
| $\frac{1}{2}C_2(g) \longrightarrow C(s)$ | 8. $CO + \frac{1}{2}O_2 \longrightarrow CO_2$ |
| $C(g) \longrightarrow C(s)$ | 9. $CO + H_2O \longrightarrow CO_2 + H_2$ |
| $C + \frac{1}{2}O_2 \longrightarrow CO$ | 10. $CH_4 \longrightarrow C + 2H_2$ |
| $C + \frac{1}{2}N_2 \longrightarrow \frac{1}{2}C_2N_2$ | 11. $\frac{1}{2}C_2H_4 \longrightarrow C + H_2$ |
| $C + 2H_2O \longrightarrow CO_2 + 2H_2$ | 12. $HCHO \longrightarrow CO + H_2$ |
| $C + H_2O \longrightarrow CO + H_2$ | 13. $\frac{1}{2}C_2H_2 \longrightarrow C + \frac{1}{2}H_2$ |
| $C + CO_2 \longrightarrow 2CO$ | 14. $\frac{1}{3}C_3O_2 + \frac{1}{3}H_2O \longrightarrow CO + \frac{1}{3}H_2$ |

### C. Nitrogen and Oxygen Reactions

| | |
|---|---|
| $O_3 \longrightarrow \frac{3}{2}O_2$ | 18. $NO + \frac{1}{2}N_2 \longrightarrow N_2O$ |
| $\frac{1}{2}N_2 + \frac{1}{2}O_2 \longrightarrow NO$ | 19. $H_2 + \frac{1}{2}O_2 \longrightarrow H_2O$ |
| $NO + \frac{1}{2}O_2 \longrightarrow NO_2$ | 20. $\frac{1}{2}N_2 + \frac{3}{2}H_2 \longrightarrow NH_3$ |

### C. Sulfur Reactions

| | |
|---|---|
| $\frac{1}{2}S_2(g) \longrightarrow S(\ell)$ | 24. $\frac{1}{3}SO_2 + \frac{2}{3}H_2S \longrightarrow \frac{1}{2}S_2(g) + \frac{2}{3}H_2O$ |
| $SO_2 \longrightarrow SO + \frac{1}{2}O_2$ | 25. $H_2S \longrightarrow HS + \frac{1}{2}H_2$ |
| $SO_2 + 3H_2 \longrightarrow H_2S + 2H_2O$ | 26. $CO + H_2S \longrightarrow COS + H_2$ |

### D. Radical Reactions

| | |
|---|---|
| $C + \frac{1}{2}N_2 \longrightarrow CN$ | 31. $2O \longrightarrow O_2$ |
| $CH_4 \longrightarrow CH_3 + \frac{1}{2}H_2$ | 32. $2H \longrightarrow H_2$ |
| $2N \longrightarrow N_2$ | 33. $\frac{1}{2}H_2 + O_2 \longrightarrow HO_2$ |
| $N + O \longrightarrow NO$ | 34. $OH + O \longrightarrow HO_2$ |
| | 35. $OH + \frac{1}{2}H_2 \longrightarrow H_2O$ |

**Fig. 10.3a.** Reaction equilibria shown in Figure 10.3.

From Figure 10.3 at 1575°C, $\log K_p = 1.9 (K_p = 79.43)$ where

$$K_p = \frac{P_{O_2}^{1/2} P_{N_2}^{1/2}}{P_{NO}}$$

At equilibrium, then:

$$(3.905 - 0.5x)^{1/2}(44.083 - 0.5x)^{1/2} = 79.43x$$

Solving this equation gives $x = 0.164$ mol NO at equilibrium, or 0.279 mol% or 2794 ppm. In practice, however, kinetic limitations usually result in NO concentrations substantially below those predicted by equilibrium alone. Note that the total "$NO_x$ story" involves more than the "thermal $NO_x$" described by this equilibrium expression. See, therefore, other texts (1) that discuss the $NO_x$ generation in more detail to include, for example, $NO_x$ that derives from the nitrogen in the waste chemistry ("fuel nitrogen" $NO_x$).

### 2.1.5. Kinetics

All chemical reactions proceed at a finite rate depending on the concentration of the reactants, the static pressure (for reactions in the gas phase), and, importantly, the temperature. At combustion temperatures, reactions are usually very fast. Exceptions of importance are the oxidation reactions for carbon monoxide (CO), soot (carbon), and chlorinated hydrocarbons. Because of their importance as air pollutants, the reaction rate behavior (chemical kinetics) of CO and soot burning are discussed here.

#### 2.1.5.1. KINETICS OF CARBON MONOXIDE OXIDATION

Carbon monoxide (CO) is an important air pollutant, a poisonous gas in high concentrations, and represents unreleased (wasted) fuel energy if found in stack gases. The rate expression by Hottel et al. (5) for the rate of change of the CO mol fraction (fco) with time is given by:

$$\frac{-df_{CO}}{dt} = 12 \times 10^{10} \exp-\left[\frac{16,000}{RT}\right] f_{O_2}^{0.3} f_{CO} f_{H_2O}^{0.5} \left[\frac{P}{R'T}\right]^{1.8} \quad (11)$$

where $f_{CO}$, $f_{O_2}$, and $f_{H_2O}$ are the mol fractions of CO, $O_2$, and water vapor, respectively, $T$ is the absolute temperature (K), $P$ is the absolute pressure (atm), $t$ is the time in seconds, and R and R' are the gas constants expressed as 1.986 cal g mol$^{-1}$ K$^{-1}$ and 82.06 atm cm$^3$ g mol$^{-1}$ K$^{-1}$, respectively.

The term $(-16,000/RT)$ is the heart of the kinetic expression, functionally providing a strong sensitivity to temperature through the exponentiation of the ratio of 16,000 (the Arrhenius activation energy) to the kinetic energy of the molecules as scaled by the absolute temperature.

It is instructive to note that the reaction rate is dependent on the water vapor concentration, a reflection of the key role of hydrogen (H) and hydroxyl (OH) free radicals in the complex series of fundamental combustion reactions that, summed together, result in the overall reaction written as $2CO + O_2 \rightarrow 2CO_2$. Indeed, bone-dry CO is very difficult to burn whereas even a trace of moisture is sufficient to assist in ignition and to facilitate rapid combustion.

#### 2.1.5.2. KINETICS OF SOOT OXIDATION

When carbon-bearing wastes are burned, the existence of regions where the oxygen concentration falls to zero often results in the formation of soot (finely divided carbon). The high optical density of such black smoke can lead to violation of opacity regulations applying to stack discharges and creates system problems by fouling boiler tube surfaces, reducing the collection efficiency of electrostatic precipitators, etc.

Soot burnout is relatively slow in comparison to many other combustion reactions, owing in part to the slower pace of heterogeneous reactions and the possibility of diffusion limitations (viz., diffusion of oxygen to the surface of the soot particle).

For spherical particles, Field et al. (6) suggest that the rate of carbon consumption $q$ (g cm$^{-2}$ s$^{-1}$) is related to the oxygen partial pressure in atmospheres ($P_{O2}$,) by:

$$q = \frac{P_{O_2}}{1/k_s + 1/k_d} \tag{12}$$

$$k_d = \frac{4.335 \times 10^{-6} T^{0.65}}{d} \tag{13}$$

$$k_s = 0.13 \exp\left[(-35,700/R)\left(\frac{1}{T} - \frac{1}{1600}\right)\right] \tag{14}$$

where $k_s$ is the kinetic rate constant for the consumption reaction and $k_d$ is the diffusional rate constant for particles of diameter $d$ (cm) at a temperature $T$ (K) and where R is the gas constant (1.986 cal g-mol$^{-1}$ K$^{-1}$).

For a particle of initial diameter $d_o$ and an assumed specific gravity of 2, the time $t_b$ in seconds to completely burn out the soot particle is given by:

$$t_b = \frac{1}{p_{O_2}}\left[\frac{d_o}{0.13 \exp\left[\left(\frac{-35,700}{R}\right)\left(\frac{1}{T} - \frac{1}{1600}\right)\right]} + \frac{d_o^2}{8.67 \times 10^{-6} T^{0.75}}\right] \tag{15}$$

## 2.2. Thermal Decomposition (Pyrolysis)

The thermal decomposition or pyrolysis of carbonaceous solids in the absence of air or under limited air supply conditions occurs in most burning systems. Several solid-waste processing systems currently under advanced development exploit this process to effect gasification of refuse. Each produces a low heat-content gas stream containing volatilized water; a mixture of CO, hydrogen, and hydrocarbons; and a solid char, which often is burned completely in a specialized region of the "pyrolyzer."

Both physical and chemical changes occur in solids undergoing pyrolysis. The most important physical change is a softening effect, resulting in a plastic mass, followed by resolidification. Cellulosic materials increase in porosity and swell as volatiles are evolved.

As the cellulose pyrolysis begins (at about 200°C), complex, partially oxidized tars are evolved. As the temperature increases, these products further degrade, forming simpler, more hydrogen-rich gaseous compounds and solid carbon. The solid residue approaches graphitic carbon in chemical composition and physical structure.

The rate-controlling step in pyrolysis can be either the heat transfer rate into the solid or the chemical reaction rate. Below 500°C, the pyrolysis reactions appear rate-controlling for waste pieces less than 1 cm in size. Above 500°C, pyrolysis reactions are fast and both heat transfer and product diffusion are rate-limiting. For pieces larger than 5 cm, heat transfer probably dominates for all temperatures of practical interest.

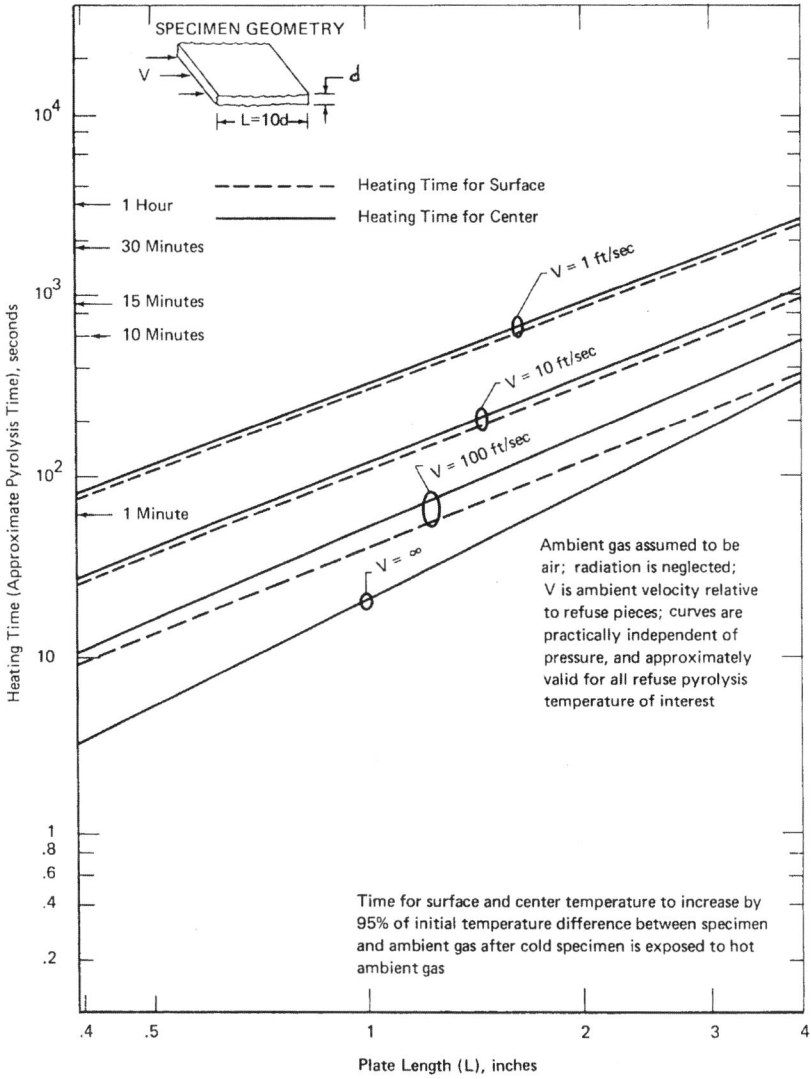

**Fig. 10.4.** Radiative and convective heating time for a thin plate.

### 2.2.1. Pyrolysis Time

The time required for pyrolysis of most wastes may be estimated by assuming that the rate is controlled by the rate of heating. Neglecting energy absorption or generation by reaction, Figures 10.4 and 10.5 facilitate estimation of the time for the center temperature of plates and spheres to rise by 95% of the initial temperature difference between specimen and surroundings. A thermal diffusivity of $3.6 \times 10^{-4}$ m$^2$/h has been assumed, roughly equal to that of paper or wood (7). The heating time at infinite cross-flow velocity ($V_\infty$) corresponds to radiant heating.

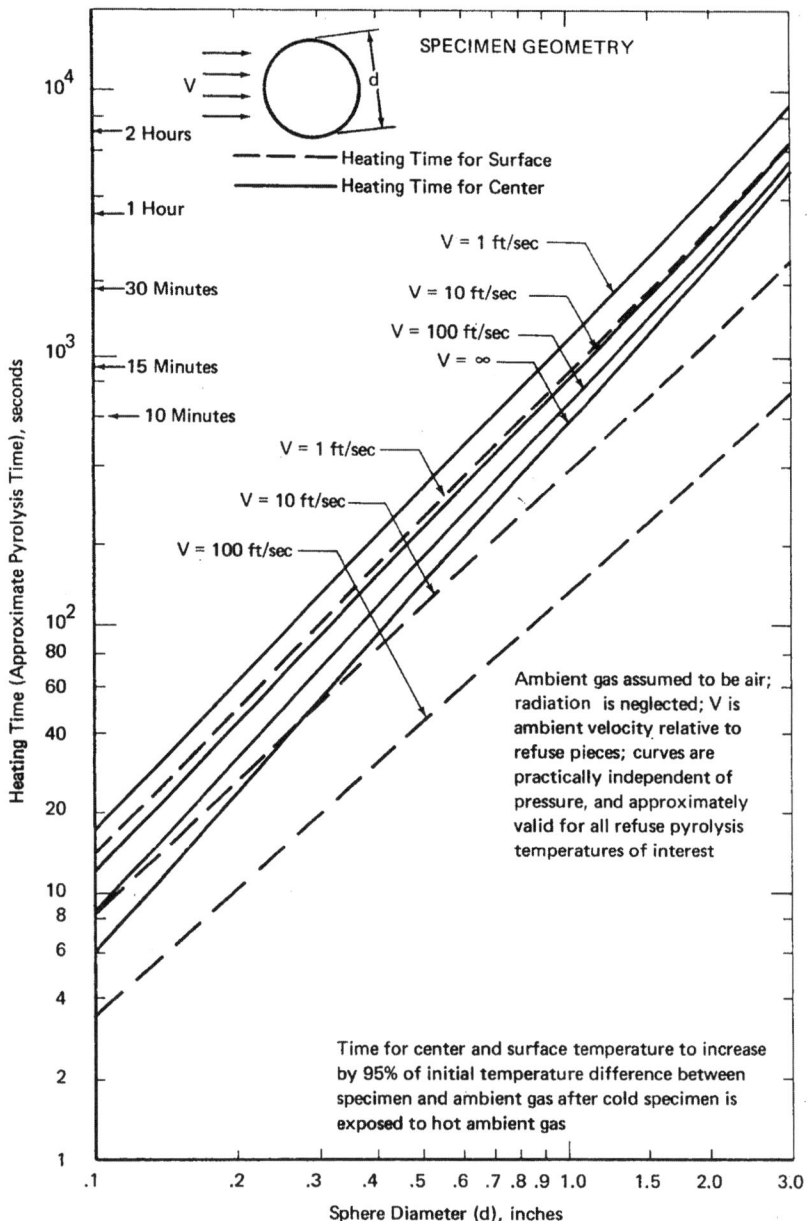

**Fig. 10.5.** Radiative and convective heating time for a sphere.

### 2.2.2. Pyrolysis Products

The pyrolysis products include ash and carbonaceous char in the solid phase; liquids (at room temperature), including $H_2O$, alcohols, aldehydes, and ketones (e.g., methanol, 2-methyl-1-propanol, 1-pentanol, 3-pentanol, 1,3-propanediol, and 1-hexanol), acetic and other acids; and gases, including $CO$, $CO_2$, $H_2$, and a variety of low molecular weight hydrocarbons.

The distribution between these products is known to be related to the heating rate and the ultimate temperature based on laboratory experiments with muffle furnaces. One

### Table 10.4
**Yields of pyrolysis products from different refuse components (weight % of refuse)**[a]

| Component | Gas | Water | Other liquid | Char (ash-free) | Ash |
|---|---|---|---|---|---|
| Cord hardwood | 17.30 | 31.93 | 20.80 | 29.54 | 0.43 |
| Rubber | 17.29 | 3.91 | 42.45 | 27.50 | 8.85 |
| White pine sawdust | 20.41 | 32.78 | 24.50 | 22.17 | 0.14 |
| Balsam spruce | 29.98 | 21.03 | 28.61 | 17.31 | 3.07 |
| Hardwood leaf mixture | 22.29 | 31.87 | 12.27 | 29.75 | 3.82 |
| Newspaper I | 25.82 | 33.92 | 10.15 | 28.68 | 1.43 |
| II | 29.30 | 31.36 | 10.80 | 27.11 | 1.43 |
| Corrugated box paper | 26.32 | 35.93 | 5.79 | 26.90 | 5.06 |
| Brown paper | 20.89 | 43.10 | 2.88 | 32.12 | 1.01 |
| Magazine paper I | 19.53 | 25.94 | 10.84 | 21.22 | 22.47 |
| II | 21.96 | 25.91 | 10.17 | 19.49 | 22.47 |
| Lawn grass | 26.15 | 24.73 | 11.46 | 31.47 | 6.19 |
| Citrus fruit waste | 31.21 | 29.99 | 17.50 | 18.12 | 3.18 |
| Vegetable food waste | 27.55 | 27.15 | 20.24 | 20.17 | 4.89 |
| Mean values | 24.25 | 23.50 | 22.67 | 24.72 | 11.30 |

[a] Refuse was shredded, air-dried, and pyrolyzed in a retort at 815°C (8).

### Table 10.5
**Percent yields of pyrolysis products from refuse at different temperatures by weight of refuse combustibles**

| Temperature °C | Gases | Liquid (including water) | Char |
|---|---|---|---|
| 480 | 12.33 | 61.08 | 24.71 |
| 650 | 18.64 | 59.18 | 21.80 |
| 815 | 23.69 | 59.67 | 17.24 |
| 925 | 24.36 | 58.70 | 17.67 |

From ref. (9).

suspects that another parameter affecting the degree of char gasification relates to the degree to which moisture, often evaporated and exhausted early in or prior to the test, is permitted to contact the char and thus engage in the water gas reaction:

$$H_2O + C \leftrightarrow CO + H_2 \qquad (16)$$

Table 10.4 shows the yield of pyrolysis products from different substrates. Tables 10.5 through 10.8 show the effect of ultimate temperature and heating rate on product mix. Table 10.9 shows the significant differences in gas composition and heat content for differing feed materials, and Table 10.10 shows the distribution in the products of the elements comprising a mixed municipal refuse. The yield of liquid (including water) is approximately 50% to 60% of air-dried, ash-free refuse, decreasing with increasing ultimate pyrolysis temperature. The heat content of the liquid per pound of refuse decreases as the pyrolysis temperature increases.

For typical refuse, gas yield ranges from 15% to 35% of the air-dried feed, with the yield decreasing and then increasing as the ultimate pyrolysis temperature is raised from

**Table 10.6**
Effect of heating rate on yields of pyrolysis products and heating value of the newspaper

| Time taken to heat to 815°C (min) | Yield of air-dried newspaper, wt% | | | | Heating value of gas, kcal/kg of newspaper |
|---|---|---|---|---|---|
| | Gas | Water | Other liquid | Char (ash-free) | |
| 1  | 36.35 | 24.08 | 19.14 | 19.10 | 1136 |
| 6  | 27.11 | 27.35 | 25.55 | 18.56 | 792 |
| 10 | 24.80 | 27.41 | 25.70 | 20.66 | 671 |
| 21 | 23.48 | 28.23 | 26.23 | 20.63 | 607 |
| 30 | 24.30 | 27.93 | 24.48 | 21.86 | 662 |
| 40 | 24.15 | 27.13 | 24.75 | 22.54 | 627 |
| 50 | 25.26 | 33.23 | 12.00 | 28.08 | 739 |
| 60 | 29.85 | 30.73 | 9.93  | 28.06 | 961 |
| 71 | 31.10 | 28.28 | 10.67 | 28.52 | 871 |

From ref. (8).

**Table 10.7**
Calorific value of pyrolysis gases obtained by pyrolyzing refuse at different temperatures

| Temperature, °C | Gas yield per kg of refuse combustibles* $m^3$ | Calorific value | |
|---|---|---|---|
| | | Gas, kcal/$m^3$ | Refuse combustibles, kcal/kg |
| 480 | 0.118 | 2670 | 316 |
| 650 | 0.173 | 3346 | 581 |
| 815 | 0.226 | 3061 | 692 |
| 925 | 0.211 | 3124 | 661 |

From ref. (9).
* At 15°C, 1 atm.

480° to 925°C. In general, 1 kg of refuse combustibles yields 0.125 to 0.185 $m^3$ of gas with a calorific value of about 3000 kcal/$m^3$.

The solid char formed from refuse pyrolysis is an impure carbon, similar to coal in proximate analysis. Chars formed at 480° and 925°C are comparable to bituminous and anthracite, respectively. Char yields range from 17% to 32% of the air-dried, ash-free feed, decreasing with an increasing heating rate and ultimate temperature. The char heating value is around 6600 kcal/kg (air-dried) and decreases slowly as the ultimate pyrolysis temperature increases.

### 2.2.3. Decomposition Kinetics

Pyrolysis of cellulose appears to follow a two-step process. The first step involves breaking of the C-O-C bond to yield a mixture of sugar-like molecules, which subsequently degrade by further breaking of C-O-C bonding.

Studies by Kanury (10) using an X-ray technique to monitor density changes during the pyrolysis of wooden cylinders provides useful insight into pyrolysis kinetics.

**Table 10.8**
**Composition of pyrolysis gases obtained by pyrolyzing refuse to different temperatures**

| Temperature, °C | Gas composition, volume % | | | | | |
|---|---|---|---|---|---|---|
| | $H_2$ | $CH_4$ | CO | $CO_2$ | $C_2H_4$ | $C_2H_8$ |
| 480 | 5.56 | 12.43 | 33.50 | 44.77 | 0.45 | 3.03 |
| 650 | 16.58 | 15.91 | 30.49 | 31.78 | 2.18 | 3.06 |
| 815 | 28.55 | 13.73 | 34.12 | 20.59 | 2.24 | 0.77 |
| 925 | 32.48 | 10.45 | 35.25 | 18.31 | 2.43 | 1.07 |

From ref. (9).

Kanury's data showed the pyrolysis reaction to follow:

$$\frac{d\rho}{dt} = -10^6 (\rho - \rho_c) \exp\left(\frac{-19{,}000}{RT}\right) \quad (17)$$

where $\rho$ is the instantaneous density (g/cm$^3$) and the subscript c denotes char, $t$ is time (min), $R$ is the gas constant (1.987 cal/mol K), and $T$ is the absolute temperature (K).

Kanury's data showed little reaction up to 350°C, but rapid reaction above this point. Shivadev and Emmons (11), studying the pyrolysis of filter paper, show an ignition-like, rapid increase in reaction rate above 407 ± 15°C, in rough agreement with Kanury.

Scrap automobile tires were subjected to pyrolytic conditions using derivative thermogravimetric (DTG) and thermogravimetric analysis. In their investigations, Kim et al. (11) studied the pyrolysis kinetics and mechanisms for the compositional components of two different sections of scrap tire rubbers: sidewall and tread. They found that the breakdown of each of the compounds comprising the tires followed an irreversible, one-step decomposition mechanism.

The tires used in their experiments involved tread (comprised of two types of styrene-butadiene rubber) and sidewalls (a mixture of natural rubber and polybutadiene rubber). The resulting kinetic constants are shown in Table 10.11. The decomposition left a residue approximating 34% of the initial weight: approximately 28% the carbon black originally in the tire compounds and not lost in the pyrolysis event, plus about 6% ascribed to char residues of the thermal decomposition.

### 2.3. Mass Burning

In mass-burning incinerators, solid wastes are burned in a relatively thick bed. In an idealized conceptualization of the bed processes (after ignition down to the grate line):

- Complete combustion is occurring at and near the grate, consuming the oxygen in the air supplied under the grate to form $CO_2$ and $H_2O$.
- As the gases pass upward, $CO_2$ and $H_2O$ react with char to form CO and $H_2$ in an endothermic reaction that, to a degree, is described by the water-gas shift equilibrium.
- Above this point, the only reaction that occurs is thermal pyrolysis of refuse in the essentially inert hot gases from below.

In the idealized mass-burning model described above, it was postulated that in thick beds the upper regions could behave as a true pyrolyzer. Evidence from coal and refuse

**Table 10.9**
Produced pyrolysis gas analysis

| Waste material | Gas analysis (dry basis), volume percent | | | | | | | | Heating value[a] | |
|---|---|---|---|---|---|---|---|---|---|---|
| | $H_2$ | $CO_2$ | $CH_4$ | $CO$ | $C_2H_2$ | $C_2H_4$ | $C_2H_6$ | $C_3H_8$ | Btu/scf | kcal/scm |
| MSW[b] | 44.47 | 15.78 | 6.96 | 24.76 | 4.97 | 1.49 | 0.66 | 0.91 | 421 | 6750 |
| Sawdust[c] | 29.32 | 12.13 | 11.04 | 43.79 | 3.12 | 0.36 | 0.36 | NM[e] | 398 | 6380 |
| Chicken manure | 35.91 | 29.50 | 8.31 | 21.37 | 2.22 | NM | 0.61 | NM | 308 | 4940 |
| Cow manure[d] | 31.07 | 20.60 | 7.70 | 38.06 | 1.86 | NM | 0.31 | NM | 328 | 5260 |
| Animal fat | 11.57 | 27.63 | 18.12 | 14.72 | 25.05 | NM | 2.91 | NM | 683 | 10,950 |
| Tire rubber | 33.81 | 15.33 | 29.09 | 5.67 | 12.94 | NM | 3.17 | NM | 661 | 10,600 |
| PVC plastic | 41.02 | 19.06 | 14.51 | 20.76 | 4.02 | 0.21 | 0.43 | NM | 412 | 6600 |
| Nylon | 45.38 | 6.03 | 15.47 | 34.64 | 0.0 | NM | 0.0 | NM | 403 | 6460 |
| Bituminous coal | 46.88 | 11.68 | 16.63 | 21.71 | 2.08 | NM | 1.01 | NM | 435 | 6980 |
| Sewage sludge (digested) | 47.01 | 22.88 | 11.22 | 15.57 | 3.12 | NM | 0.21 | NM | 360 | 5770 |

[a] scf, standard cubic feed (60°F, 1 atm); scm, standard cubic meter (15°C, 1 atm).
[b] Average of five tests.
[c] Average of three tests.
[d] Average of two tests.
[e] NM = Not measured.
From ref. (13).

### Table 10.10
**Dry-basis yields from pyrolysis of refuse in weight percent**

|  | C, wt% | H, wt% | O, wt% | N, wt% | S, wt% | Ash, wt% | Total, wt% |
|---|---|---|---|---|---|---|---|
| Feed composition | 30.85 | 3.84 | 22.32 | (0.4) | (0.1) | 42.49 | 100.00 |
| CO | 8.01 | – | 10.68 | | | | 18.69 |
| $CO_2$ | 4.32 | – | 11.52 | | | | 15.84 |
| H | – | 2.05 | | | | | 2.05 |
| $CH_4$ | 2.25 | 0.76 | | | | | 3.01 |
| $C_2H_2$ | 3.22 | 0.27 | | | | | 3.49 |
| $C_2H_4$ | 0.95 | 0.16 | – | | – | | 1.11 |
| $C_2H_6$ | 0.43 | 0.11 | – | | | | 0.54 |
| $C_3H_6$ | (0.52) | (0.09) | – | | | | 0.61 |
| $C_3H_8$ | (0.35) | (0.08) | – | | | | 0.43 |
| Liquids | 3.45 | (0.32) | (0.12) | (0.1) | – | – | 3.99 |
| Ash | – | | | – | – | 42.49 | 42.49 |
| Char | 7.35 | | | (0.3) | (0.1) | – | 7.75 |
| Pyrolysis product totals | 30.85 | 3.84 | 22.32 | (0.4) | (0.1) | 42.49 | 100.00 |

From ref. (14). Parentheses indicate estimated values.

### Table 10.11
**Kinematic parameters of scrap tire materials (27)**

| Material | ln A (min$^{-1}$) | ΔE (kJ/g mol) |
|---|---|---|
| Processing oils in natural and polybutadiene rubber | 7.84 | 48.0 |
| Processing oils in styrene-butadiene rubber | 7.56 | 43.3 |
| Natural rubber | 38.20 | 207.0 |
| Butadiene rubber | 34.08 | 215.0 |
| Styrene-butadiene rubber | 24.02 | 148.0 |

beds would tend either to discount the existence of the pyrolysis zone or, more probably, to suggest that this zone does not appear in beds of practical thickness (15, 16). Thus, the off-gas from a bed would be a mixture of gases from both burning and gasification zones.

Studies on the off-gas from refuse burning in a municipal incinerator (16) have confirmed the earlier data of Kaiser (17) and the hypothesis (7) that the off-gas composition is controlled by the water-gas shift equilibrium. This equilibrium describes the relative concentration of reactants in the following:

$$H_2 + CO_2 \leftrightarrow CO + H_2O$$

$$K_p = \frac{p_{H_2O} p_{CO}}{p_{CO_2} p_{H_2}} \qquad (18)$$

The importance of this equilibrium in mass burning is the incremental gasification potential given to the underfire air. In tests at Newton, Massachusetts (16), refuse and off-gas stoichiometry were studied. Average refuse was given the mol ratio formula: $CH_{1.585}O_{0.625}(H_2O)_{0.655}$. For this formula, and assuming that the water-gas shift equilibrium holds, over 1.5 times as much refuse can be gasified as would be predicted for stoichiometric combustion to $CO_2$ and $H_2O$. Burning rate data showed rates 1.7 to 2.1 times that corresponding to stoichiometric combustion with the combustion air supplied beneath the refuse bed. Entrainment of air from above the refuse bed was assumed to occur to account for relative burning rates in excess of the 1.5 factor due to the water gas shift reaction. A second result coming from the water-gas shift reaction is that a definite and relatively large combustion-air requirement will necessarily be placed on the overfire volume. The air requirement for the CO, $H_2$, distilled tars, and light hydrocarbons can indeed be as much as 30% to 40% of that expended in gasification, thus creating a need for effective overfire air injection and mixing.

## 2.4. Suspension Burning

In suspension burning, a particle of refuse is suddenly thrust into an environment of hot gases and intense radiative flux. The particle undergoes rapid drying and ignition while airborne, and proceeds to burn in an oxygen-rich atmosphere. Depending on the particle shape and weight, the velocity of the gas medium, and the geometry and dimensions of the combustion chamber, the particle may be partially or entirely burned while still suspended in the gas stream.

In general, the chemistry and heat transfer environment of the furnace and the details of the particle characteristics (moisture content, thermal and mass diffusivities, shape factors, etc.) are poorly defined, so that a rigorous analysis is difficult. Even for the somewhat simpler case of pulverized coal combustion, many simplifying assumptions are required to predict flame length, minimum air requirements, etc. (6).

For refuse, the second and third stages of the combustion process (heat-up of the dry solid and subsequent pyrolysis) may be analyzed using Figures 10.4 and 10.5.

## 2.5. Air Pollution from Incineration

Air pollutant emissions are central points of public concern and regulatory scrutiny for incinerators. Since the discovery of significant dioxin emissions from incinerator furnaces in the late 1970s, the fraction of incinerator plant capital cost related to compliance with air emission control has risen to more than 35% of the total investment. Often, resolution of public concern and reaching closure on the key permitting steps are the pacing and controlling events in the implementation of incineration facilities. Understanding the relationships between design and operating factors and air emissions and the relative effectiveness of alternative control technologies is critical.

Specifically, one must understand the relationship between emissions and the following:

- The specific chemical and physical characteristics of the wastes being burned
- The design features of the combustor
- The operating conditions in the combustor
- The control effected by the air pollution control device(s)

This knowledge allows a designer to set limits on the types or relative firing rate of the wastes burned and to configure the hardware design and operating strategy of the unit to ensure that air quality impacts remain within permit limits.

Air contaminants generated when burning fossil fuels or wastes have significance in three areas:

- Obtaining air emission permits from regulatory agencies
- Establishing the specifications for air pollution control systems
- Establishing the basic design, suggesting modifications to existing designs, or interpreting problems in emission minimization or control

Many air pollutants are emitted from combustion processes. Historically, the primary pollutant of concern was inorganic particulate matter. The total emission rate of particulate matter (total suspended particulate, TSP) is predominantly relatively inert "ash"—a mixture of benign compounds primarily composed of silicon, aluminum, calcium, iron aluminum, and oxygen. However, this portion of combustor emissions can also include the "heavy metals" or "air toxics" (especially lead, mercury, chromium, cadmium, arsenic, nickel) and other elements that may have significant toxic, carcinogenic, and other health effects.

Also, the inorganic TSP includes an important portion of the small particle size material denoted as "$PM_{2.5}$." $PM_{2.5}$ is the respirable fraction under 2.5 μm in mass mean diameter that has been shown to have an important role in both the onset and aggravation of asthma and other respiratory diseases. Data on fine particulate matter show sulfates and nitrates to be the most abundant species in atmospheric aerosols with sulfates being the predominant contributor to $PM_{2.5}$ (18). Inert particulate emissions are related to the fraction of ash in the feed and the fluid flow and other physical processes that can elutriate and convey the material from the combustion zone. A large fraction of the total $PM_{2.5}$ found in the atmosphere is formed following discharge of the flue gases from gaseous $SO_x$ and $NO_x$ emissions that are $PM_{2.5}$ precursors (secondary $PM_{2.5}$).

A second category of emission includes the combustible solids, liquids, and gases. A portion of these combustibles can be a fraction of the raw waste originally fed to the unit. Beyond this, there is a complex mixture of products of incomplete combustion (PICs) including carbonaceous soot and char; carbon monoxide; hydrocarbons and other volatile organic compounds (VOCs); and representatives of many classes of carbon-hydrogen-oxygen-nitrogen-halogen compounds such as benzene-soluble organic matter (BSO), polycyclic organic matter (POM) (e.g., benzo-($\alpha$)-pyrene), and a variety of polyhalogenated hydrocarbons (PHH), including the isomers and congeners comprising the families of polychlorinated and polybrominated dibenzo furans (PCDF, PBDF), p-dibenzo dioxins (PCDD, PBDD), and biphenyls (PCB, PBB). Net emissions of combustible pollutants such as the PICs arise partly from the waste chemistry (contributing "building blocks"), partly from failures in the combustion process (generating the PICs), and partly from successes in the combustion process (destroying a portion of the PICs).

Some of the organic or inorganic compounds emitted from incinerators exhibit or are suspected to exhibit significant adverse health effects. Some of the compounds react in the atmosphere (especially, under the influence of ultraviolet radiation) to generate ozone and a spectrum of eye-irritating oxygenated reaction products. With the great strides in sampling and analysis technology in recent years, the emission of these compounds can be quantified, regulated, and used as a basis for fines and penalties and, of vital importance, for continuation of permission to operate. In general, the net concentration of these pollutants in the atmosphere following normal dilution and dispersion following emission from the stack is significantly below the threshold where significant health effects are probable.

There are several pollutants, emitted as both gases and solids, where the emission level is directly related to fuel chemistry. These include the sulfur oxides, the halogens and hydrogen halides (especially HCl), trace elements, and radioactive elements. Others, such as the nitrogen oxides, are emitted at rates related to both fuel chemistry and the combustion process.

A detailed analysis of the emission process and emission factor estimation and detailed discussion and quantification of the mechanisms by which pollutants arise and may be controlled are available elsewhere (1). This knowledge provides guidance from which the waste characteristics and the incinerator design features and operating conditions can be used to quantitatively estimate uncontrolled emission rates for many of the important pollutants. Such calculations and the physical principles that may underlie them also provide the designer and operator (both have a role in most cases) with tools to minimize the emission rates.

### 2.5.1. Mineral Particulate

The great majority of the emissions of mineral particulate are associated with the carryover of mineral matter introduced with the waste. The most important mechanism leading to these emissions is the entrainment of ash fragments by the flow of air and combustion products passing as underfire air through the burning refuse mass.

Important particulate characteristics include size, size distribution, shape, density, stickiness, corrosivity, reactivity, and toxicity. Of these, particle size distribution (PSD) has the most important impact on air pollution control and can be expressed as a particle count and as a mass distribution. Most commonly, the greatest mass is associated with the larger particles, but the greater number of particles, visual impact (opacity due to light scattering), surface area, and difficulty in capture is found with the smaller particles.

The Gaussian or "normal" distribution function is often used to describe many data populations. However, the log normal distribution function (1) is usually superior to characterize particles from industrial sources. In this case, a plot of the log of the particle size against frequency produces a straight line on log-probability paper.

Undergrate air velocities typically range from $0.05\,sm^3/s/m^2$ of grate area to $0.5\,sm^3/s/m^2$. An analysis of the entrainment process (1, 7) suggests that this range of velocities would entrain particles ranging in maximum diameter from 70 to 400 μm at a mean temperature of the entraining fluid of about 1100°C. The importance of the underfire air entrainment mechanism on emissions is supported by the observed range of particle sizes of fly ash, the increases in particulate emission with undergrate air velocity, and the similarity in the chemical analyses of fly ash and that of the refuse fine ash (1).

Eberhardt and Mayer (19) and Nowak (20) report that 10% to 15% of the refuse ash can be expected to appear in the gases leaving the furnace. Data from the U.S. Public Health Service (PHS) on an experimental incinerator (21) showed emission rates from 8% to 22% of that in the refuse for underfire air rates comparable to those used in municipal incinerators.

The relationship between underfire air rates and emissions were correlated by the PHS as:

$$W = 4.35 \, v^{0.543} \qquad (19)$$

where $W$ is the emission factor (kg/ton of refuse burned) and $v$ is the undergrate air flow (sm$^3$s$^{-1}$m$^{-2}$). Data on three municipal incinerators obtained by Walker and Schmitz (22) showed general agreement with Equation 19 but scattered ±20%.

Small effects on emission rate are also shown by incinerator size (emissions increasing as size increases, perhaps owing to the natural convection effects on the velocity field in the furnace); burning rate relative to design capacity [Rehm (23) noted a 30% reduction in furnace emissions with a 25% reduction in throughput]; grate type (lower emission being associated with grates where sizable air openings pass a large fraction of the fine ash out of the flow as siftings); and several still lesser effects (1).

### 2.5.2. Combustible Solids, Liquids, and Gases

By definition, the appearance of combustible pollutants (soot, char, CO, and a spectrum of hydrocarbons) in the effluent of a combustion system reflects an inadequacy in combustion efficiency. This indicates any or all of the following:

- Inadequate residence time to complete combustion
- Inadequate temperature levels to speed combustion reactions to completion
- Inadequate oxygen concentrations in intimate conjunction with the combustibles to allow oxidation to proceed to completion, often indicating incomplete mixing rather than an oxygen deficiency in the overall flow.

Complete combustion is a basic objective of almost all combustion systems. In most real systems, however, the full attainment of this goal is either impossible or impractical. Thus, some unburned combustible will always be present in the effluent gases. In the 1970s and afterward, one saw several forces in the environmental area that focused attention on incompletely burned species:

- Air pollution aspects of the basic pollutants (particulate, $NO_x$, $SO_2$ and CO) were well understood. Air quality criteria (denoting the relationship between the long-term average concentration of the pollutant and human health effects) were set. The academic and the regulatory community were seeking new areas of endeavor.
- Analytical equipment and sampling/analysis techniques were rapidly evolving. It was now possible to identify and quantify the specific chemistry of complex organic species in flue gases. The combustor's "chemistry set" produces a spectrum of products: a broad range of polynuclear, oxygenated, halogenated, and otherwise medically significant species. The concentrations were small but the classes of compounds (POMs, halogenated organic compounds etc.) were provocative.
- Methodologies were emerging to evaluate the health risk of exposure to subacute levels of pollutants. Further, the public had seized on health risk as a new scale against which to evaluate projects.

These factors fostered increasing awareness and concern by regulatory authorities throughout the world over the health implications of combustible emissions. This concern emphasized the importance of careful design and operations to minimize both generation and survival of these pollutants.

The incomplete combustion of carbon-containing fuels or waste materials forms a spectrum of chemical species. Due to the refractive index and color of carbon particles and to the typically small particle size, the simplest product of incomplete combustion (carbon itself) contributes importantly to the opacity of the effluent. Small particles have a greater light scattering power for a given mass loading than coarse particles. Carbonaceous soot can be amorphous in character but is more often graphitic. This is unfortunate since graphitic carbon is more difficult to oxidize than the amorphous material.

Carbon-hydrogen compounds comprise the second class of combustible pollutants. These compounds include methane, ethane, acetylene, and other simple straight and branch-chained aliphatic compounds as well as complex saturated and unsaturated ring compounds. The health significance of these pollutants varies greatly. The subclass of complex aromatic compounds (POMs) includes compounds such as the carcinogenic benz-α-pyrene.

Carbon-hydrogen-oxygen compounds comprise the third class of pollutants. These pollutants range from simple compounds such as CO and formaldehyde to complex organic acids, esters, alcohols, ethers, aldehydes, ketones, and so forth. These compounds are often associated with odorous emissions and show very low minimum detection concentration limits. In particular, aldehydes that are formed as PICs are often major contributors to the burnt smell.

The fourth class of pollutants includes the carbon-hydrogen-nitrogen compounds. These include the PICs formed from combustion of amines, N-ring compounds, many proteins, and other chemical species. Some also incorporate oxygen in the molecule. These compounds are especially significant due to their participation in the formation of fuel nitrogen $NO_x$ and their contribution to odor.

Carbon-hydrogen-oxygen-halogen compounds are the fifth class of pollutants. These pollutants include chlorinated solvents, fluorinated and chlorinated polymers, and many other environmentally significant PIC-derived chemical compounds. Importantly, this class includes the several congeners and isomers of polychlorinated dibenzo-p-dioxin, dibenzo-furan, and biphenyl compounds. The chlorinated solvents and polymers contribute to the formation of the halogen acids (HF, HCl, HBr, and HI) that, under U.S. law, often triggers a requirement for acid gas control. Members of the polychlorinated dioxins, furans, and biphenyls are under intense scrutiny by the regulatory community and the general public worldwide. Carcinogenicity and other significant health effects are often associated with these materials.

The incompletely burned or PIC pollutants associated with almost every criterion of air quality include the following:

- Solid particulate or aerosols that contribute to atmospheric haze and solids fallout
- Photochemically reactive compounds that participate in the reactions leading to smog
- Compounds recognized as injurious to plants (e.g., ethylene) or animal life respiration (e.g., carbon monoxide), known to cause cancer in human beings (e.g., benz-α-pyrene), or known to promote adverse health effects in animals similar to that of pesticides (e.g., halogenated biphenyls).

Because of the health-related impact of these pollutants, their control assumes an importance out of proportion to the weight emitted.

The solution of the combustible pollutants problem is generally associated with the attainment of intense mixing in the hot regions of the furnace and provision of sufficient residence time for combustion reactions to be completed. The soot and CO kinetics discussed above allow estimation of the order of magnitude of time and temperature required after an adequate degree of mixedness is attained.

### 2.5.3. Acid Gases

Sulfur dioxide and hydrogen halides are emitted in proportion to the concentration of elemental sulfur, metal sulfides, organic sulfur-bearing compounds (mercaptans, sulfides, disulfides, etc.), and halogen-bearing organic compounds. In the incineration environment, oxidation of the sulfur or sulfur compounds and reactions of the halogens with hydrogen sources (e.g., $-OH$ or $-OH_2$ radicals) forms $SO_2$ and the appropriate halogen acid (HF, HCl, HBr, or HI). Estimation of their importance as pollutants, therefore, is to some extent dependent on the results of a comprehensive waste analysis for the area in question.

Sulfur oxides arise primarily from the oxidation of sulfur or sulfides, or from organic or inorganic sulfur-based acids. Owing to some absorption by alkaline fly ash, only about 70% of this sulfur will appear as $SO_2$ or $SO_3$ in the flue gases, by analogy with coal burning plants (24). About 97% of the sulfur appears as $SO_2$ and 3% as $SO_3$.

Hydrogen chloride arises most importantly from the incineration of municipal refuse components such as polyvinyl chloride (PVC, 59% chlorine) and polyvinylidene chloride (Saran, 73.2% chlorine). Conversion of the organically derived chlorine to hydrogen chloride in the furnace gases is almost quantitative. PVC has been posed as one of the primary sources of the halogen content of the polychlorinated dioxin compounds found in incinerator effluent. However, dioxin emission data from incinerators fired with known quantities of PVC showed no correlation between the HCl concentration in the flue gases and the dioxin emissions (25, 26).

### 2.5.4. Nitrogen Oxides

Nitrogen oxides (NO, $NO_2$) are formed in all air-oxidized combustion reactions, although nitrogenous fuels produce significantly higher concentrations than fuels barren of nitrogen. The fixation of nitrogen with oxygen occurs by the following overall chain reaction mechanism after Zeldovitch (27).

Equilibrium considerations lead to the following relationship for the overall reaction:

$$N_2 + O_2 \leftrightarrow 2\,NO \tag{20}$$

$$K_p = \frac{(NO)^2}{(N_2)(O_2)} = -21.9 \exp\left(\frac{-43{,}400}{RT}\right) \tag{21}$$

The rate of formation of NO is significant only at temperatures in excess of 1000°C due to kinetic limitations, and doubles for every 40°C increase in flame temperature. Thus, $NO_x$ emissions are encouraged by high flame temperatures (e.g., by air preheat) and high excess air. $NO_x$ can be reduced by the following:

- Water or steam injection or flue gas recirculation (to lower flame temperature)
- Operation at low excess air (to reduce oxygen concentrations)

- Staged combustion where the fuel is partially burned, heat is withdrawn through boiler surfaces, and then the rest of the required air for combustion is added (the overall effect is to lower the peak temperatures attained after the combustion gases contain a net excess of oxygen)
- Selection of burner designs ("low-NO$_x$ burners") that reduce the combustion intensity (volumetric burning rate), produce relatively long diffusion flames, or produce either two-stage combustion or low-temperature gas recirculation

A kinetic evaluation by Soete, reported by Bowman (28), can be solved iteratively to estimate the decimal fraction $Y$ of fuel nitrogen $N_f$ converted to NO. For $[N_f]_o$ equal to the NO concentration for 100% conversion of $N_f$ to NO and $[O_2]$, the oxygen concentration (gmol/cm$^3$) and temperature $T$ in degrees K, the following can be developed as a basis to estimate fuel-nitrogen related NO$_x$ generation:

$$Y = \left[ \frac{2}{1/Y - \{[2.5 \times 10^3 [N_f]_o]/[T \exp(-3{,}150/T)[O_2]]\}} \right] - 1 \qquad (22)$$

Liquid injection vortex type incinerators were studied by Kiang (29), who developed a *dimensional* empirical relationship between the percent of fuel nitrogen converted and several key design and operational parameters:

$$\log_e [Y] = 2.9 - 1.027 S - 0.4665 \log_e [percent\ O_2] - \frac{158.2452}{T} \qquad (23)$$

where

$Y$ is the percent of fuel nitrogen converted to NO (as a percent)
$S$ is the mean residence time (seconds)
Percent $O_2$ is the dry percent oxygen (as a percent)
$T$ is the temperature (degrees F)

Kiang found that the formation of NO was not related to the specific chemical form in which the nitrogen is found in the fuel (amine, cyanide, amide, nitroso group, etc.) nor to the partial pressure of water vapor in the incinerator flue gases.

Based on work by the PHS on fossil fuels (30), an estimate of the nitrogen oxides emission (expressed as NO$_2$) for refuse combustion as related to the heat release rate (kcal/hour) is given by:

$$\text{kg NO}_x/\text{hr} = [(\text{kcal/hour})/2.26 \times 10^6]^{1.18} \qquad (23)$$

### 2.5.5. Air Toxics

The term *air toxics* has been applied to a spectrum of organic and inorganic compounds found in stack emissions. The U.S. Environmental Protection Agency (US EPA) defines air toxics as pollutants known or suspected of causing serious health concerns such as cancer or birth defects. The US EPA's list of air toxics as of the year 2000 is shown in Table 10.12.

The health effects triggering inclusion in the air toxics list are summarized in Table 10.13. In many (if not most) cases, the concentration of these compounds in the stack gases is small; they are micropollutants. Thus, after dispersion and dilution of the stack gases by the atmosphere, the resulting average ambient air concentrations and

**Table 10.12**
**Air toxics**

| | |
|---|---|
| Acetaldehyde | Ethylene oxide |
| Acrolein | Formaldehyde |
| Acrylonitrile | Hexachlorobenzene |
| Arsenic compounds | Hydrazine |
| Benzene | Lead compounds |
| Beryllium compounds | Manganese compounds |
| 1,3-butadiene | Mercury compounds |
| Cadmium compounds | Methylene chloride |
| Carbon tetrachloride | Nickel compounds |
| Chloroform | Polychlorinated biphenyls (PCBs) |
| Chromium compounds | Polycyclic organic matter |
| Coke oven emissions | Quinoline |
| Dioxin/furans* | 1,1,2,2-tetrachloroethane |
| Ethylene dibromide | Perchloroethylene |
| Propylene dichloride | Trichloroethylene |
| 1,3-dichloropropene | Vinyl chloride |
| Ethylene dichloride | Diesel particulate matter* |

* Health effect results not yet available.
*Source*: US EPA (2000).

**Table 10.13**
**Health effect classifications**

| Health effect class | Characteristic impact |
|---|---|
| Toxic | Kills cells or the organism by poisonous nature; often, a threshold dose (mg/kg of body weight) characterizes the onset of impacts |
| Carcinogenic | Causes cancer; often, the dose-response function is characterized as a log-log correlation between dosage and the probability of observing cancerous cells |
| Mutagenic | Causes mutation of cells |
| Teratogenic | Causes changes in cells during prenatal development; may lead to birth defects or deformity |

associated inferred health risk consequences are often acceptable. However, since these pollutants are directly related to human health effects, they receive focused attention by both regulatory agencies and the general public.

Incineration of wastes is viewed, not unexpectedly, as an important potential source of air toxics emissions since inevitably the mix of diverse materials called "waste" often involves contamination by trace metals. Further, wastes may include a variety of toxic organic compounds targeted for disposal or toxics formed as by-products of industrial preparation reactions. Incomplete combustion results in emission of the compounds and "daughters" with similar chemical characteristics. Finally, due to moment-to-moment variation of the feed and other parameters, the incineration environment can degrade from time to time thus allowing a "puff" of toxic material to escape. The combination of these possibilities leads to intense scrutiny of incineration regarding air toxics.

**Table 10.14**
**Volatility temperatures for several air toxics metals (31)**

| Metal | Volatility temperature (°C) | Metal | Volatility temperature (°C) |
|---|---|---|---|
| Mercury | 16 | Antimony | 660 |
| Arsenic | 32 | Barium | 849 |
| Thallium | 138 | Silver | 904 |
| Cadmium | 216 | Beryllium | 1216 |
| Lead | 627 | Chromium | 1610 |

Trace elements of concern (especially mercury, arsenic, beryllium, selenium, lead, cadmium, zinc, and hexavalent chromium) are emitted in proportion to their concentrations in the waste fired. The simplest approach to estimating metal emission rates is to assume that the relative proportions of the metal in the emitted particulate matter is the same as that of the metal in the total inorganic fraction of the feed. There are several potential weaknesses in this approach:

- Some metals are trapped in relatively massive bottom ash and are never emitted to any significant degree. The chromium in large pieces of stainless steel is an example of this.
- Some metals are emitted strongly relative to their concentration in the ash. These "enrichment factors" (1) between the metal content of the bulk ash and that in the content in the fly ash can exceed 200-fold. Mercury and its compounds are examples. Because of the high volatility of mercury, almost 100% of this metal is emitted from the furnace. Other metals show lesser but often significant degrees of emission enhancement. Barton et al. (31) ranked several key metals by their "volatility temperature," the temperature where the sum of the vapor pressures of all species containing the metal present at equilibrium and weighted by their relative concentrations is $10^{-6}$ atmospheres (Table 10.14).

In a few instances, the balance between these and other factors influencing emissions has been studied and estimates can be made of the partitioning of feed metals between bottom ash and fly ash emissions (1).

Of the several air toxics, perhaps none has led to greater scrutiny nor impacted on regulatory acceptability than the several compounds from the family of polychlorinated dibenzo-p-dioxins (CDD) and furans (CDF). These complex chlorinated compounds included molecules with different degrees of chlorination (congeners) and, for a given congener, molecules where the chlorine molecules are located in different positions about the oxygenated benzene ring structure (isomers). Each of the congeners and isomers express their health effects to a different degree as shown in Table 10.15.

## 2.6. Fluid Mechanics in Furnace Systems

Furnace fluid mechanics are complex, with jet and buoyancy driven flows interacting in swirling, recirculating eddies, and all the while traversing complex geometrical sections. In a few instances, the flow field is laminar, but most situations are turbulent. Further, combustion of the fuel gases rising from beds of pyrolyzing waste results in heat release (and temperature change) of the gases. All of these phenomena are time and spatially variant. A basic understanding of several of these features of furnace flow, particularly the behavior of jets and of buoyancy effects, is of great assistance in the design and analysis of incineration systems.

**Table 10.15**
**Toxicity equivalence factors (TFEs) for specific PCDD and PCDF compounds**

| | Equivalency factor (TEF$_i$) | | | |
|---|---|---|---|---|
| Positions chlorinated | International Eadon (32) | US EPA (33) | BGA (34) | Nordic (35) |
| Dioxins | | | | |
| 2,3,7,8 Tetra CDD | 1.0 | 1 | 1 | 1 | 1 |
| 1,2,3,7,8 Penta CDD | 0.5 | 1 | 0.2 | 0.1 | 0.5 |
| 1,2,3,4,7,8 Hexa CDD | 0.1 | 0.03 | 0.04 | 0.1 | 0.1 |
| 1,2,3,6,7,8 Hexa CDD | 0.1 | 0.03 | 0.04 | 0.1 | 0.1 |
| 1,2,3,7,8,9 Hexa CDD | 0.1 | 0.03 | 0.04 | 0.1 | 0.1 |
| 1,2,3,4,6,7,8 Hepta CDD | 0.01 | 0 | 0 | 0.01 | 0.01 |
| 1,2,3,4,6,7,8,9 Octa CDD | 0.001 | 0 | 0 | 0.001 | 0.001 |
| Furans | | | | |
| 2,3,7,8 Tetra CDF | 0.1 | 0.33 | 0.1 | 0.1 | 0.1 |
| 1,2,3,7,8 Penta CDF | 0.01 | 0.33 | 0.1 | 0.1 | 0.01 |
| 2,3,4,7,8 Penta CDF | 0.5 | 0.33 | 1 | 0.1 | 0.5 |
| 1,2,3,4,7,8 Hexa CDF | 0.1 | 0.01 | 0.01 | 0.1 | 0.1 |
| 1,2,3,4,8,9 Hexa CDF | 0.1 | 0.01 | 0.01 | 0.1 | 0.1 |
| 1,2,3,6,7,8 Hexa CDF | 0.1 | 0.01 | 0.01 | 0.1 | 0.1 |
| 2,3,4,6,7,8 Hexa CDF | 0.1 | 0.01 | 0.01 | 0.1 | 0.1 |
| 1,2,3,4,6,7,8 Hepta CDF | 0.01 | 0 | 0.001 | 0.01 | 0.01 |
| 1,2,3,4,7,8,9 Hepta CDF | 0.01 | 0 | 0.001 | 0.01 | 0.01 |
| 1,2,3,4,6,7,8,9 Octa CDF | 0.001 | 0 | 0 | 0.001 | 0.001 |
| Other PCDD/PCDF | 0 | 0 | 0–0.01 | 0.001–0.01 | 0 |

BGA: Bundesgesunheitsamt (German Federal Health Office).

*2.6.1. Jet Behavior*

The behavior of jets in furnaces is of particular importance in incinerator design. This importance reflects the function of jets in the following:

- The controlled addition of mass to contribute to the oxidation process, to act as a thermal sink to temper gas temperatures (air jets) or to convey refuse into the incinerator (suspension burning).

The controlled addition of momentum to promote mixing (turbulence) of the furnace gases to assist in attaining complete combustion (air or steam jets).

The behavior and design of jets in combustion situations is covered in more detail elsewhere (1, 16). The simplest jet system (the round, isothermal, turbulent subsonic jet) is reviewed here since it embodies many of the basic characteristics important in understanding the behavior of the sidewall air jets commonly used in mass-burning incinerators. The complexities of cross flow, combustion, nonisothermal flow, and buoyancy effects should be considered, however, in any final design calculations (1, 36–38). The round jet (Figure 10.6) shows three characteristic flow regions:

- The *mixing region* adjacent to the nozzle and extending about 4 to 5 nozzle diameters from the discharge plane, which contains, as a distinguishing feature, an undisturbed flow near the axis of the jet (the "potential core") with a relatively flat velocity profile. The potential

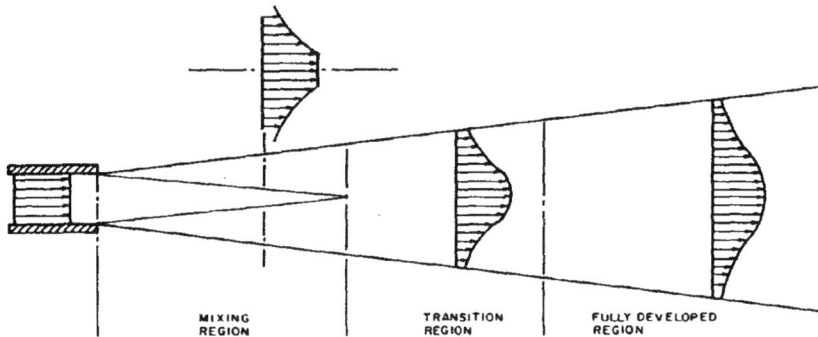

**Fig. 10.6.** Regions in jet flow.

core is surrounded by a flow field with a high-velocity gradient where the rapidly moving jet gases mix with the surrounding fluid.
- The *transition region* extending 4 to 8 diameters downstream where the radial velocity profile acquires a stable shape.
- The *fully developed flow region* where the normalized velocity profile about the jet axis remains of constant or self-preserving shape.

Important jet characteristics include (1) the change in centerline velocity and concentration (of nozzle fluid) with distance, (2) the shape of the radial velocity and concentration profiles in the fully developed region, and (3) the rate of entrainment of ambient fluid into the jet.

The following functional relationships (6) describe these characteristics using the nomenclature $u$ and $c$ for mean velocity and concentration, $\rho$ for density, $x$ for distance from the nozzle, $d_o$ for the effective nozzle diameter, $r$ for the radial distance from the jet centerline, and $m$ for the mass flow rate. As for the subscripts, $o$ denotes nozzle conditions, $x$ denotes conditions at a distance $x$ from the nozzle, $a$ denotes conditions in the ambient fluid, and $m$ denotes conditions on the jet centerline. Note that the "effective nozzle diameter" may be different from the physical orifice dimension. For example, if the jet issues from a circular hole in a sheet steel plenum, the effective nozzle diameter is about 60% of the diameter of the hole: the diameter of the *vena contracta*.

- Velocity
$$\frac{\overline{u_m}}{\overline{u_o}} = 6.3 \left(\frac{\rho_o}{\rho_a}\right)^{1/2} \frac{d_o}{(x + 0.6 d_o)} \tag{25}$$

$$\frac{\overline{u}}{\overline{u_m}} = \exp\left[-96 \left(\frac{r}{x}\right)^2\right] \tag{26}$$

- Concentration
$$\frac{\overline{c_m}}{\overline{c_o}} = 5.0 \left(\frac{\rho_o}{\rho_a}\right)^{1/2} \frac{d_o}{(x + 0.8 d_o)} \tag{27}$$

$$\frac{\overline{c}}{\overline{c_m}} = \exp\left[-57.5 \left(\frac{r}{x}\right)^2\right] \tag{28}$$

- Entrainment
$$\frac{\dot{m}_x}{\dot{m}_o} = 0.23 \left(\frac{\rho_a}{\rho_o}\right)^{1/2} \left(\frac{x}{d_o}\right) \tag{29}$$

These functional relationships indicate that the jet flow expands inside a cone-shaped envelope. Defining the boundary as that corresponding to a velocity one-half of that on the jet centerline, a cone of half-angle 4.85° is defined. The corresponding half-angle for concentration is 6.2° (6).

In the presence of strong cross-flow velocities such as the gas flow arising from grate-fired incinerators, the direction of the sidewall jet is "bent" in the direction of the cross-flow.

$$N_{\text{Re}} = \frac{\rho_o \bar{u}_o d_o}{\mu_o} \tag{30}$$

Dimensional analysis suggests that the coordinates of the dimensionless jet axis ($x/d_o$ and $y/d_o$) should depend on the ratio of momentum fluxes in the external and the jet flow as characterized by the parameter $M$ and the Reynolds number ($N_{\text{Re}}$), calculated using the following equations:

$$M = \frac{\rho_a \, u_1^2}{\rho_0 \, u_0^2} \tag{31}$$

For turbulent jets in the Reynolds number range above $10^4$, correlations of experimental data suggest that the Reynolds number effects are negligible and that $M$ is the controlling parameter. Patrick (39) developed a relationship describing the trajectory of jets injected normal to the cross-flow (with the jet centerline defined by the concentration of jet fluid) as:

$$\frac{y}{d_0} = 1.0 M^{1.25} \left(\frac{x}{d_0}\right)^{2.94} \tag{32}$$

Data show that the concentration axis shows a greater deflection than the velocity axis. This is due, perhaps, to the asymmetry of the external flow around the partly deflected jet as shown in Figure 10.7.

The relationships have utility in predicting the flow behavior of sidewall air jets. Indeed, the Bituminous Coal Research (BCR) method of overfire air jet design (38) is based on an assumed jet penetration depth (clearly, somewhat less than the width of the chamber) corresponding to a centerline velocity of 2.5 m/s as calculated using Equation 25. More elaborate jet design methodologies were developed by Ivanov (36) as described by Niessen (1).

### 2.6.2. Buoyancy

Since furnace gases are hot, buoyancy effects can result in substantial flow acceleration. Although often overlooked in furnace analysis, these effects can be of sufficient magnitude as to cause severe erosion damage or to greatly change the velocity field, thus influencing the penetration distance of sidewall jets.

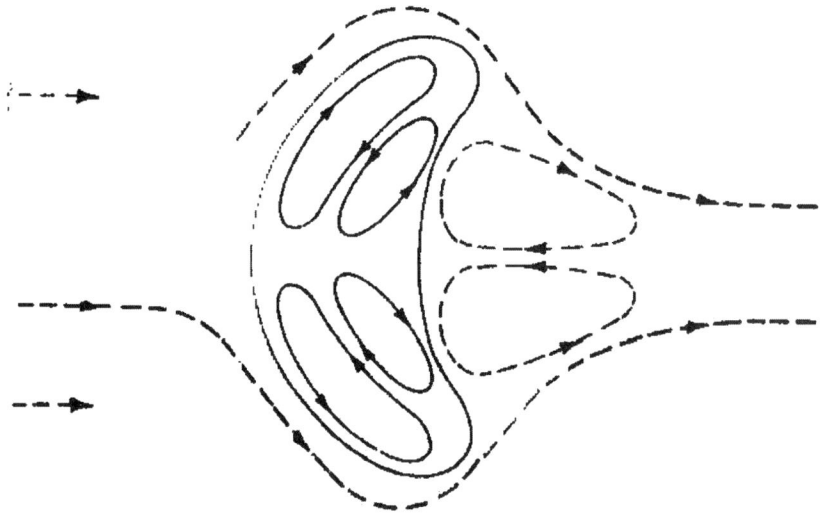

**Fig. 10.7.** Jet cross-section and circulation patterns for round jets in cross-flow.

The acceleration of gases from an initial velocity $u_o$, elevation $y_o$, and static pressure $P_o$ is described by Bernoulli's equation:

$$u^2 = u_o^2 + \frac{2(P_o - P)}{\rho_o} g_c - 2g(y - y_o) \tag{33}$$

where $\rho_o$ is the density of the gas and $g_c$ is the acceleration of gravity (9.807 m/s$^2$). This equation has utility in estimating the buoyant acceleration of gases rising from the grate of an incinerator.

CASE 1

A well-sealed incinerator has a zone of hot gases arising from the grate and a cold zone of stagnant or slowly moving gases above the residue quench tank. The change in static pressure $(P_o - P)$ for the cold gas, as the flow exits through a vertical outlet flue, is also experienced by the hot gas. Noting that the ratio of absolute temperatures is the inverse of the ratio of densities, writing Equation 33 for both flows and combining yields:

$$(u^2)_{hot} = (u_o^2)_{hot} + 2g(y - y_o)\left(\frac{T_{hot}}{T_{cold}} - 1\right) \tag{34}$$

CASE 2

For many older furnaces, the furnace is leaky, so that there is an interaction between the hot and cold gases and the ambient atmosphere, with both gas streams accelerating, but with the acceleration of the hot zone being more pronounced. In this case:

$$(u^2)_{hot} = (u_o^2)_{hot} + 2g(y - y_o)_{hot}\left(\frac{T_{hot}}{T_a} - 1\right) \tag{35a}$$

$$(u^2)_{cold} = (u_o^2)_{cold} + 2g(y - y_o)_{cold}\left(\frac{T_{cold}}{T_a} - 1\right) \tag{35b}$$

EXAMPLE 6

In a large, well-sealed furnace, 6000 m³/min of gases leave the burning refuse bed at a temperature of 1100°C (1373 K) at an elevation of 12.5 m and at a velocity of 1.2 m/s. At the end of the furnace, 25 m³/min of quench tank vapors leave the furnace at 300°C (573 K) at 0.1 m/s.

The two gas flows leave through a vertical outlet flue at the top of the chamber (elevation 17.5 m, with an area of 65 m²). Estimate the average velocity through the flue as well as the possible peak velocity due to buoyancy effects. Neglect the flow area for the cold gases:

1. Mean velocity

$$\bar{v} = \frac{\text{Volumetric flow rate}}{\text{Flue area}} = \frac{6{,}000}{65} = 92.21 \text{ m/min } or \text{ } 1.54 \text{ m/sec}$$

2. Buoyancy effects (Eq. 34)

$$(u^2)_{hot} = 1.2^2 + 2(9.807)(17.5 - 12.5)\left(\frac{1373}{573} - 1\right) = 11/76 \text{ m/sec}$$

At this velocity, the gas needs only 6000/(11.76 × 60) or 8.5 m² of duct area. Thus, the flow cross section shrinks as the gas accelerates, exiting as a high-velocity stream on the side of the flue nearest the grate with a slowly moving mass of gas filling the remainder of the flue.

## 3. INCINERATION SYSTEMS FOR MUNICIPAL SOLID WASTE

The period from 1960 through 1990 marked the halcyon years of municipal incineration in the United States. The most significant driving forces behind the growth in installed incinerator capacity were as follows:

- Increasing urbanization leading to increased waste generation in the cities and increasing difficulty in locating landfill sites (the principal competitive waste management approach) near to the city centers
- Substantial increases in the per-capita waste generation pattern of urban Americans
- Expectations (not realized) that energy cost increases, as driven by oil prices and concomitant growth in the revenue realized in selling the byproduct electricity that provides a major mitigation of incineration cost, would continue

Incinerators handling large quantities of solid waste exhibit wide variation in design. These variations reflect local conditions, scale of operation, the state of technology, and, not unimportantly, the personal experiences and prejudices of the design engineer. In a study of incineration practice in the U.S. (7), over 20 major engineering firms and equipment vendors were asked to identify those design parameters that they believed to be broadly reflected in U.S. practice. Only one such parameter (the burning rate per unit area of grate) was identified.

Until the 1970s, incinerators in the U.S. were generally designed through a technical collaboration between the public works department of the owner city or county, their consulting engineer, and the major component vendors (especially, the grate, fan and refractory manufacturers). In most cases, the incinerators were operated by staff of the public works departments of the owner city. The 1970s saw the emergence of a new paradigm: a system vendor who competed to design, construct, and operate the incinerator as a service to the owner community.

The incinerators operating in the early 1960s and earlier were relatively simple refractory chambers containing a grate to translate and mix (stoke) the waste. The combustion chamber was followed by very rudimentary air pollution control devices. The primary objective of these units was to achieve 80% to 90% volume reduction. Air pollution requirements focused on coarse particulate matter ("blackbirds") and were not stringent. Usually, simple water sprays and settling chambers were sufficient to meet code. Although incineration was always a capital-intensive alternative to landfill, most of the cost was associated with civil works: foundations, buildings, chimney structures, etc. The technological risk for this kind of system was not high, although poor designs could lead to problems with refractory degradation, smoking, and high maintenance expense.

In 1970 the Clean Air Act was passed. This act required a significant upgrade in the sophistication and cost for air pollution control. The high air dilution of the leaky refractory furnaces (some operated at as high as 800% excess air at the stack) made addition of the new control systems (electrostatic precipitator technology was "borrowed" from coal-burning power plants) prohibitively expensive. The desire to minimize flue gas volume and the increasing value of electrical energy suggested a new incinerator design concept: the waterwall boiler generating superheated steam. This design eliminated infiltration air and, being inherently self-cooled, operation at low overall excess air levels was possible without undue slagging of the walls. Passing the steam through a turbine generated electricity and produced a new revenue stream to mitigate the burgeoning cost of the incinerator.

However, concurrent with these technological changes was a great increase in risk. In years past, minor design mistakes could be made and the "fixes" absorbed into the operating budget. Now a prospective incinerator owner community would have to consider plant investments often in excess of $100 million and operating budgets that depended on a stable revenue stream from the electrical credit over a 20-year bond repayment life. Retaining a system vendor to design, build, and operate the facilities under a 20-year contract as a service to the host community appeared as a reasonable approach to contain and reallocate these risks. Importantly, this new project development and implementation concept recast the plant operations staff to one with standards approaching those of power utility operators. The working relationship between the community and the system vendor was codified and detailed in a comprehensive contract document—the service agreement. As one impact of these changes, the role of the consulting engineer shifted from one of being the engineering designer to project planning, preparing performance specifications for competitive procurement, providing technical assistance in financing, and securing permits. In other instances, entrepreneurial system vendors took the lead in developing projects.

The service agreement often goes beyond a simple documentation of a contract to provide waste incineration services. Since the lifetime of the agreement is often 15 to 20 years, many of the circumstances defining the nature of the service, economic factors, environmental requirements, and other important parameters will change. Thus, the agreement defines the set of reference system characterizations that were the basis of the original procurement and indicates methods and guidelines with which to update the cost or performance basis from the baseline. The characterizations include a "reference waste" composition and heat content; unit costs for labor, utilities, taxes, and chemicals; environmental requirements; and daily and annual processing rates and energy recovery targets.

The new, system vendor-dominated incineration business employs a wide variety of designs to do the same job. This individuality reflects both the growth of incineration technology in recent years and the large number of basic design parameters that are somewhat flexible and can be bent to the prejudices of the design firm. The systems offered by the system vendors can be divided into two broad categories: mass burn technology and refuse-derived fuel (RDF) technology. Mass burn technology is the approach representing both the most successful and frequently implemented method for combustion of MSW. Mass burn implies the combustion of unprocessed solid waste. In contrast, RDF technology is based on combustion of a prepared, refuse-based fuel. The goal in RDF systems is to process the refuse to a relatively homogeneous material in order to achieve the degree of combustion control and low excess air operation found in coal-burning systems. However, by avoiding the considerable technical risks and cost of preprocessing the waste (except, perhaps, to remove oversize "bulky waste" and certain hazardous or undesirable wastes such as auto batteries) mass burn strategy gains a clear economic and reliability advantage. Nonetheless, although mass burn technology dominates the market in the United States and Europe, both approaches have their strong points and their advocates.

The performance objectives of a municipal waste incineration system are as follows:

- To process each normal operating day not less than the quantity of waste with an analysis and heat content specified in the service agreement
- To process the minimum weekly, monthly, and yearly quantity of waste specified in the service agreement
- To consistently operate within the emission limits and other legal constraints of all applicable environmental regulations to include restrictions on the concentrations or mass rates of air or water pollutants, sound pressure levels, and the maintenance of specified system operating parameters within designated limits
- To protect the health and well-being of incinerator employees and of the neighboring commercial and residential community
- To protect the capital investment reflected in the equipment, buildings, roads, etc. comprising the incineration facility such that the useful operating life and maintenance and operating expenses of the incinerator are not adversely impacted
- To meet any production guarantees regarding residue quality and quantity; export rates of power, steam or other energy-related products; or other commercial promises

The achievement of these objectives is strongly supportive of a healthy plant operation, good customer relations, and good financial performance.

Many of the most critical performance objectives are highly dependent on the characteristics of the waste and, most importantly, on the heat content. This importance derives from the fact that, in essence, an incinerator is a system to process heat. Although the capacity of incinerator plants are most often described in mass-based terms, the actual metric of capacity is intrinsically associated with their maximum heat release rate (the maximum continuous rating, MCR) and not a mass throughput rate (except as the mass rate, multiplied by the waste heat content, is equivalent to a heat release

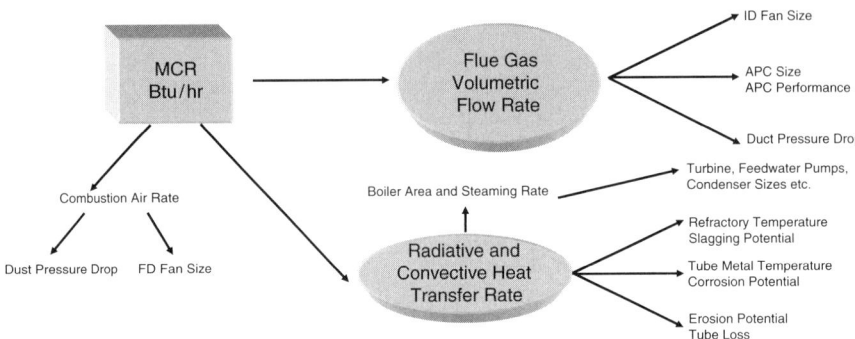

**Fig. 10.8.** Maximum continuous rating (MCR).

rate). Unfortunately, many municipal clients believe that their contract relationship with the incinerator operator is a commitment to process a given mass of material (e.g., 800 tons per day) rather than to process a specified number of millions of BTUs per day. Experience has shown that if this potential misperception is not clearly addressed in the service agreement, changes in waste heat content over the contract life will lead to customer dissatisfaction and even lawsuits.

Why is heat release the real variable? Figure 10.8 illustrates the process and hardware connections that spring from the MCR heat release parameter. Heat release rate, because of the approximate equivalence between heat release and combustion air quantity, is strongly related to the volumetric flow rate of combustion air and of the products of combustion. Thus, the heat release rate sets the size and horsepower of the forced draft and induced draft fans and the air pollution control system, and sizes the ductwork and dampers, pressure drops, etc. throughout the flow system. Also, the heat release rate, for a given combustion chamber, strongly impacts on the heat transfer rates (both convective and radiative), which affects the temperature of surfaces in boilers and on refractory walls. Thus, exceeding the design heat release rate can result in overheating of critical system components. All of these factors illustrate why incinerator capacity is quite properly equated to the MCR rather than the tons fed.

The firing diagram shown in Figure 10.9 provides a concise, graphical statement of the operating process envelope of a 800 ton per day incineration system. Specifically, the area bounded by the dotted lines represents the combinations of mass feed rate and refuse heat content that are supported by the referenced incineration furnace. For all points within the dotted area, the furnace can meet its design mass disposal rate and still remain at a technically sound fraction of MCR and the physical throughput limitations.

Let us consider the various elements of the boundary of the operating zone:

- Maximum heat release: The horizontal top line of the zone is the MCR. Heat release rates over this limit unduly stress the equipment or exceed design limits for fans, air pollution control equipment, etc. Also, in waterwall boiler systems, operation above this heat release rate may lead to boiler tube failures, tube erosion, etc. contributing to unscheduled outage.
- 50% of MCR: The horizontal bottom line of the zone is set at 50% of the MCR. While somewhat arbitrary, burning at less than half of the design heat release is often accompanied by poor mixing within the furnace (increasing CO and hydrocarbon pollutant emissions), degradation in residue quality, furnace control problems, draft control problems, etc.

Combustion and Incineration Engineering 519

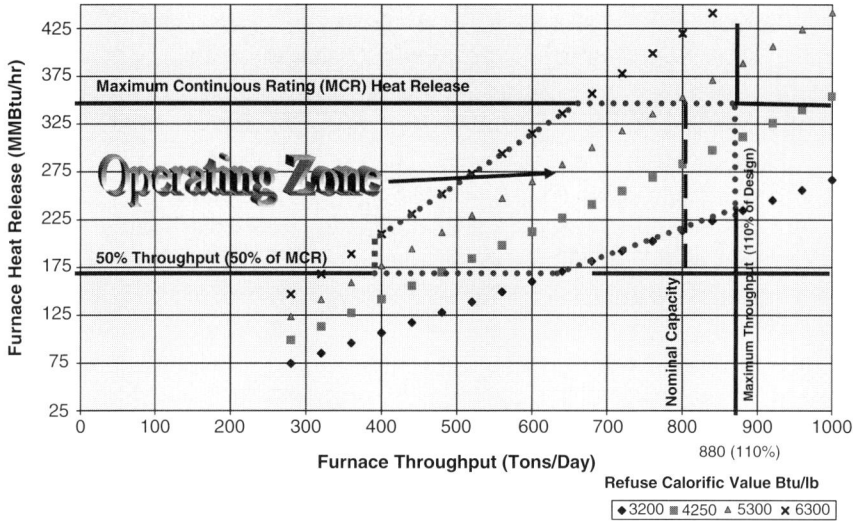

**Fig. 10.9.** 800 TPD incinerator firing diagram.

- 110% of maximum throughput: The vertical right-most boundary of the zone is set at 110% of the design mass throughput. This is a reasonable estimate of the maximum feed rate that can be accommodated by the structural strength and materials handling capabilities of the grate and the physical dimensions and capacity of the residue and fly ash handling systems.
- 50% of maximum throughput: The vertical, left-most boundary of the zone is set at 50% of the design mass throughput, reflecting the constraint that as the throughput drops from the design level, it becomes more likely that the grate will be exposed to furnace radiation. Also, the breakdown in the performance of the solids materials handling equipment becomes more likely.
- Maximum rate of highest heat content refuse: The sloped top boundary of the zone is set by the heat release throughput line for the highest plausible refuse heat content. This line intersects the MCR line at the design throughput line. This is the maximum heat content refuse used as the basis of design in setting the MCR. Note that for this heat content refuse, the system can just meet the design throughput rate (often equal to the minimum rate set in the service agreement) and stay within the MCR.
- Maximum rate of lowest heat content refuse: The sloped bottom boundary of the zone is set by the heat release throughput line for the lowest heat content refuse that intersects the 50% MCR line and extends to the design capacity limit.

Since the operating zone described in the firing diagram is a simple, graphical statement of the maximum operational capabilities of the incineration system, there are merits to including the diagram as part of the service agreement.

Thus an incinerator can be and is many things. The subsections that follow outline very briefly the principal options in incinerator design. Any one topic would justify a chapter or a book in its own right if the topic were to be explored at a level of detail fully supporting design decision-making.

### 3.1. Receipt and Storage

The system used for the receipt and storage of raw refuse is of major concern to the design engineer. Since this portion of the plant interfaces with the entire refuse collection system and in some cases with the individual citizen, the traffic patterns and dumping areas should be carefully thought out. Consideration must be given to the rapid processing of incoming vehicles (especially since refuse deliveries are seldom equally spread through the day) and to the safety of all parties concerned.

In larger mass-burning plants, refuse is received and stored in a pit below ground level. A traveling bridge crane with a clamshell, "orange-peel," or grapple-type bucket is used to pile and mix the refuse as well as to feed the incinerator furnace. The orange-peel and grapple-type bucket are the means most common in new plants. The capacity of the pit is generally equivalent to the quantity of refuse that can be burned in 2 to 3 days and, most certainly, from the time of the last receipt Friday evening until waste receipts begin again on Monday. Good pit design should provide for drainage, for fire control, and with means to recover and remove problem wastes that are discovered after dumping.

The crane operator is more than a cog in the materials handling system. The operator should be concerned with mixing of refuse to even out variations in refuse character, setting aside problem wastes (mattresses, engine blocks, refrigerators, and other "white goods"), moving refuse away from the dumping wall to allow uninterrupted dumping during peak receipt periods, keeping the feed hoppers filled, and planning the refuse withdrawals to effect a systematic cleaning of the pit. The latter responsibility is important for sanitation, good housekeeping, minimization of housefly nuisance (by processing refuse in a shorter time period than the larval–pupal cycle of the housefly), and the elimination of odors from decomposing refuse.

In some cases, a paved "tipping floor" serviced by a front-end loader or dozer blade-equipped vehicle is used to receive and charge refuse. In such cases particular attention must be given to frequent cleaning to avoid the hazard of a slippery and unsanitary floor area. Special concern should also be devoted to the selection of rugged tires for refuse handling vehicles (to avoid rapid wastage or blowouts) and in using heavy-duty radiators (to avoid dust clogging and overheating).

## 3.2. Charging

Most solid wastes usually have a low bulk density and include a substantial fraction of cellulose (wood, paper, cardboard). The low bulk density of these materials requires relatively large storage space and often involves rehandling of the waste (stacking) to increase storage capacity. Here is a summary of the materials handling options and problem areas in solid waste management facilities:

| Step | Small plant | Large plant | Problem areas |
| --- | --- | --- | --- |
| Receipt | Manual scale | Automatic scale | Delays, queues |
| Storage | Floor dump | Pit | Cleanout, fires |
| Reclaim | Front-end loader | Crane/orange peel | Mixing |
| Feeding | Ram | Chute, ram | Jamming |
| Support | Refractory hearth | Metal grates | Overheating |
| Residue handling | Quench tank, belt | Quench tank, drag conveyor or ram | Jamming |

*Combustion and Incineration Engineering* 521

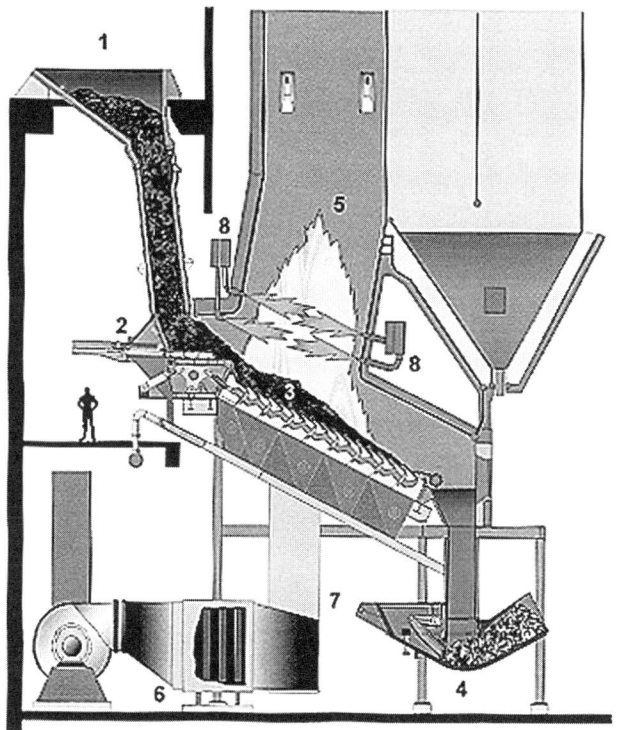

**Fig. 10.10.** Martin© municipal waste combustor. 1, feed hopper; 2, feeder; 3, Martin© reverse-acting grate; 4, Martin© discharger; 5, radiant furnace; 6, steam-heated air-preheater; 7, injection of underfire combustion air; 8, injection of overfire combustion air. (Courtesy of Ogden Energy Group, Inc.)

Feeding solid waste to a furnace presents special problems. It is critical to introduce the new waste using means that do not also allow large quantities of ambient air into the furnace. Such uncontrolled flows upset the draft, overly cool or quench the combustion so that unburned or incompletely burned material (products of incomplete combustion, PICs) are formed, and may result in rapid cooling of the refractory with consequent spalling. Also, combustion systems operate best with steady heat release rates. Uniform heat release results in uniform gas and refractory temperatures, and stable furnace draft (no "puffs" that push furnace gases into the operator's work space), and are easier for the control system to follow.

In almost all large continuous mass-burning incinerator furnaces (>250 tons/day), charging is effected using a rectangular gravity-fed chute discharging to a feeding grate (Figure 10.10, item 1). The chutes are often water-cooled and are usually designed with a slight taper, opening toward the discharge, and with well-ground butt welds to avoid bridging. The base of the chute is generally fitted with a hydraulic ram feeder (Figure 10.10, item 2) to ensure positive and controllable feed of waste to the combustor. In small facilities, wastes are often dumped on a tipping floor and pushed about or stacked using a front-end loader.

## 3.3. Enclosures

A combustor consists of one or more of the following chambers (or zones):

| Chamber | Function |
|---|---|
| Primary | Drying and gasification of solids, evaporation of liquids |
| Secondary | Burnout of soot, CO, residual organic matter |
| Afterburner | Temperature maintenance at levels where burnout will be achieved |
| Quench/cool | Gas conditioning to reduce temperature and/or humidify ahead of air pollution control |

These chambers may be constructed in several ways depending on the size and basic concept (energy recovery or not) of the combustor:

*Refractory*: Formed from brick, special shapes or castable refractory and primarily appearing only in smaller and older incinerators. Key properties include abrasion, impact and corrosion (fluxing) resistance, refractoriness, and insulating qualities.

*Water Wall*: Comprised of water-filled steel tubes. The metal tubes comprising the wall panels are welded together with a narrow steel strip between the individual tubes to form a continuous, gas-tight membrane or water-wall. Key properties include abrasion resistance (along the grate-line of mass burning units) and corrosion resistance (attack from deposits and gases).

The chamber geometry is important as it affects the bulk gas flow patterns and heat re-radiation within the enclosure. Baffles, "bull-noses," and other configurations are used to effect gross mixing, but induced turbulence using air jets is usually critical to the attainment of satisfactory mixing.

The floor of the furnace includes a refractory hearth (in small units) or, in all new and large plants, a metal grate or stoker that moves the refuse from under the charging chute and along the length of the incinerator's primary chamber. Refractory material (usually of silicon carbide composition for its abrasion resistance and high thermal conductivity) is commonly installed at the grate line, often extending 3 to 10 feet above the grate. Frequently, the refractory cladding extends to about 30 feet above the grate to protect the waterwall tubes from accelerated metal wastage due to the changing oxidizing and reducing character of the gases arising from the burning refuse bed that sweep across the sidewalls.

The containment of combustion processes in the older refractory-lined chambers limited the peak temperatures that could be tolerated both to avoid degradation of the refractory and to control accumulation of slag on the walls. Also, air in-leakage cooled and quenched combustion along the sidewalls. This usually meant that refractory incinerators were operated at high levels of excess air (high flue gas volumetric flow rates) with consequent penalties in the cost of air-pollution control devices, fans, and stacks.

The waterwall boiler design resulted in a substantial reduction in overall excess air. As an often more important benefit, the withdrawal of energy as steam (usually as an intermediate to electricity) or hot water produced an important revenue stream to mitigate the burgeoning capital and operating cost of incineration facilities. When steam generation is the desired objective, the incinerator enclosure design changes to maximize the fraction of the refuse heat content transferred to the water. Also, banks of tubes are immersed in the hot gas flow to recover heat convectively.

To better understand the system, let us follow the course of the water entering the boiler plant:

1. Water treatment: Raw water, containing dissolved minerals and suspended matter, would be an unsatisfactory feed to a boiler. To avoid scale buildup or corrosion from these contaminants, the water must be treated. The level of treatment increases with the severity of the water-side environment, as characterized by the temperature and pressure of the product steam. Treatment methods include filtration, softening, distillation, and ion exchange plus addition of chemicals.
2. De-aeration: Water also contains dissolved gases (air components, $CO_2$, etc.), which would accumulate in the boiler. To remove most of the dissolved gases, the treated water is heated with steam or electricity to the atmospheric boiling point in a de-aerator. The water leaving the de-aerator is ready for introduction into the boiler using the feed-water pumps to raise the pressure to the boiler's working level.
3. Boiler: At the point of introduction into the boiler, the feed water is treated, de-aerated, and perhaps somewhat preheated such that its temperature is in the range of 100° to 200°C. In passing the water through the boiler, it is desirable to optimize the temperature difference between the water and the hot combustion gases (maximum heat transfer rate) to minimize the required amount of heat transfer area (capital cost) while still extracting the maximum amount of heat from the combustion gases. In larger boilers, this will include such components as the following:
   a. Radiant boiler: These components are water walls or, in some cases, banks of tubes exposed to the combustion zone between which the hot flue gases flow. Heat transfer rates are very high, and radiant energy transport from the incandescent refuse bed, flame, or hot gases is the predominant means of heat transfer. The gas temperature leaving the radiant boiler zone is still quite hot, ranging between 1000° and 1100°C.
   b. Convection boiler: The convection boiler consists of one or more banks of tubes (or "passes") between which the hot flue gases flow, and where the water (from the economizer) is evaporated. Flow through the tubes may result from buoyancy effects (a natural convection boiler) or pumps (a forced convection boiler) and is two-phase, containing both liquid water and steam. Heat transfer efficiency and boiler cost factors favor following the radiant boiler directly with the superheater; the heat recovery element where the gaseous steam generated in the boiler is further heated as a means of increasing its power generation potential. However, incinerators constructed in this configuration in the mid-1970s experienced severe corrosion of the superheater tubes. It was found that the accelerated corrosion was due to the deposition of chloride-containing ash on the superheater under conditions where the fireside (outside) tube metal temperature exceeded 315° to 370°C. While one can mitigate the corrosion by using more exotic metals or with special protective treatment of the tubes, it has been shown to be more cost-effective to precede the superheater with a portion of the convection boiler section such as to drop the gas temperature below 870°C. With the lower gas temperature driving force, the peak metal temperature in the superheater stays below the high corrosion threshold. The remainder of the convection section is, then, installed following the superheater.
   c. Steam drum: The steam drum device consists of one or more large accumulators with disengagement space and mechanical devices to separate the gaseous steam from the

liquid water. The latter is recirculated to the convection or radiant boiler sections. The product steam, in thermodynamic equilibrium with liquid water, is saturated at the temperature and pressure of the steam drum contents.

d. Superheater: The superheater is composed of a radiantly or convectively heated tube bank where, at (roughly) constant pressure, the steam is further heated to produce dry steam with heat content greater than that of the saturated steam. Such superheated steam conditions are often advantageous, since they will produce more mechanical energy in a turbine than will saturated steam or will tolerate moderate heat losses without condensation in, say, a steam-distribution system pipeline.

e. Economizer: This component consists of one or more banks of tubes between which the hot flue gases flow and convectively transfer heat to the treated, de-aerated feed water. The feed water is usually not heated to the point where evaporation occurs. The economizer is located in the part of the boiler where the flue gas temperature is the lowest. Consequently, one must be careful that the dew point of the acidic flue gas components (both $H_2SO_4$ from oxidation of the $SO_3$ formed from a portion of the $SO_2$ generated in burning sulfur-bearing waste materials, and HCl and other halogen acids) is always safely below the temperature of the feed water so that condensation on the carbon steel tubes does not occur. Algorithms to calculate the dew point from the concentration of acid gases and moisture in the flue gases can be found in the literature (1).

f. Air preheater: In some designs, the last step in heat withdrawal from the incinerator flue gases takes place to preheat the combustion air. Since incinerator flue gases often contain considerable moisture, one must be careful that condensation (and associated corrosion from the acidic components in the flue gas) does not occur during times when the ambient air is cold.

### *3.4. Grates and Hearths*

All large municipal-scale incineration furnaces employ one of a variety of grates that stoke or mix the refuse during the combustion process in various ways depending on the type of grate or stoker. In smaller units, a refractory hearth or fixed array of static grate bars support the burning refuse. There are many different types of hearth or grate designs, each of which has its own special features. The more common concepts are described below.

#### *3.4.1. Stationary Hearth*

These incinerator furnace systems that operate without grates include the stationary hearth and rotary kiln types. The stationary hearth is usually a refractory floor to the furnace, and may have openings for the admission of air under a slight pressure below the burning material on the hearth (underfire air). In the absence of underfire air ports, air is admitted along the sides or from the top of the furnace (overfire air). It is usually necessary to provide manual stoking in order to achieve a reasonable degree of burnout. Stationary hearth furnaces are used for many smaller commercial and industrial incinerators. They are also used in crematories and for hospital wastes, when assisted with auxiliary gas or oil burners to maintain the furnace temperature above 650° to 900°C.

#### *3.4.2. Rotary Kiln*

Rotary kiln incinerators have been used at both industrial and municipal installations for disposal of combustible solids, liquids, and gaseous wastes. While municipal rotary kilns normally handle only solid wastes, industrial kilns are generally designed for both solids (often including drums) with provision for firing liquid wastes. There has been little use of the rotary kiln for municipal incinerator furnaces except to provide additional residence time for improved residue burnout *after* the burning of refuse on a multiple-grate system.

The rotary kiln is a cylindrical, horizontal, refractory-lined shell that is mounted at a slight (usually less than 3%) incline. Rotation of the shell (usually at less than 2 RPM) causes mixing and "opening" of the waste, thus improving contact with the combustion air and improving combustion efficiency. Variable speed control is preferred to tailor residence time to waste characteristics. The length to diameter ratio of the combustion chamber normally varies between 2:1 and 10:1, and the peripheral speed of rotation is normally in the range of 0.3 to 1.5 m per minute. Kiln loading is usually in the range of 3% to 12% of the cross-sectional area. Combustion temperatures vary according to the characteristics of the material being incinerated but normally range from 815°C to 1650°C. Residence times of the solids vary from 20 minutes to hours depending on the nature of the residue, the kiln incline, the length of the slope, and the rotational speed (1). Kiln gases experience residence times of up to 3 seconds since, often, a large secondary combustion chamber is provided.

Although the rotary kiln approaches the waste management ideal of an "omnivore" (capable of processing all types of waste and in varying proportions), its inherent characteristics also limit its utility:

- Size: Kiln systems are most compatible with large waste disposal requirements. Although units are available rated as low as 750,000 kcal/h, the small units are more costly than the modular combustion unit (MCU) of equivalent capacity. A design limit of 800,000 kcal per hour per square meter of cross-sectional area is recommended so that even small capacity units are relatively large.
- Stratification: Kiln systems are prone to stratification of the gas flow. Induced jet mixing and a large secondary chamber are required. A design basis of 1 second mean residence time at 900°C (minimum) is recommended for the secondary chamber.
- Particulate entrainment: As the load rotates, a portion of the ash is lifted and then drops through the gas stream. Since all of the air for combustion must be introduced at the feed end of the kiln, gas velocities over the bed are also higher than for other systems. To minimize entrainment, gas velocities should be limited to 3 to 4 m/s. Measured uncontrolled particulate loading is much higher for kilns than for modular combustion units or other nonfluidized systems.
- Maintenance: Because of the severe abrasion, refractory life is usually limited to about 2 years. Shorter life is common when slag attack occurs or when the hardness of the superduty brick linings is not matched to that of the insulating brick used on the shell. Under the latter conditions, abrasion loss at the interface is rapid. In many plants, castable refractory is preferred, especially near the discharge seal where thermal spalling is common and frequent patching is required.

Selection of a rotary kiln as an industrial waste incineration system is usually based on its ability to incinerate a diverse range of waste types. Liquids, sludge, and solids can all be accommodated by the unit within the operating parameters dictated by design. Liquids can be directly pumped to the kiln where they are injected through burner nozzles located in the front wall. Automatic control modulates waste fuel flow to the burners based on desired kiln exit temperature. A continuous pilot flame is usually provided to eliminate the possibility of flame failure during waste feeding. Liquids can be fed to the kiln from a variety of containers. Direct feed from on-site storage tanks is the most common method with the ability also existing to pump from tank trucks, small portable tanks, or steel drums.

Solids and semisolids can be fired in drums (usually but not necessarily limited to fiber packs), which are fed mechanically to the kiln through an airlock system. The feed rate of drums to the kiln may be under either automatic or manual control. Shutdown of the feed system usually occurs when there is a flame detected in the feed chute or temperatures in the feed lock exceed a preset minimum.

As the kiln rotates, the burned out material and ash travel through the kiln and fall into a water-filled quench trough. The water quenches the ash, which settles to the bottom of the trough. A drag chain conveyor removes the ash from the trough and conveys it to a closed bottom container for eventual landfill disposal.

The hot gases pass from the rotary kiln through a mixing chamber and into a secondary combustion chamber. Waste liquids may be burned directly in the secondary chamber as a supplement to, or replacement for, purchased fuel to maintain temperatures high enough (900° to 1100°C) to ensure complete burnout of distilled waste components. The secondary combustion chamber provides additional retention time (usually at least 1 second) and should include secondary air injection to induce mixing. The burners in this chamber are temperature controlled and have a continuous pilot flame to ensure uninterrupted combustion.

The hot gases pass from the secondary combustion chamber to a quench chamber, where they are cooled and large particles of ash are removed. The incoming gases are quenched through the use of water sprays, usually under manual control with automatic bypass in the event of water failure. The quench chamber is used to cool gases for introduction to a downstream scrubber.

A Venturi scrubber, typically with a variable throat opening, is usually installed downstream of the quench chamber. The cooled gases pass from the quench chamber to the high-efficiency Venturi scrubber. Recycle water backed up by makeup water is injected into the gas approach to the Venturi throat and is also sprayed directly into the throat. A remotely controlled damper can vary the cross-sectional area across the throat to achieve a range of pressure drops. Submicron particles in the gases are removed in this high-efficiency scrubber. Gaseous and liquid contaminants are likewise absorbed in the scrubber liquid. Water flow through the scrubber must be maintained at an optimum rate. Below this rate, scrubbing efficiency drops off rapidly. Above this rate, the efficiency increases only slightly, if at all. The scrubbed gases then pass through a liquid-gas separator; the entrained water containing the pollutants removed from the gas is separated from the gases and flows back to a recycle water tank. The gases proceed to an induced draft fan and out through the stack to the atmosphere. The induced draft (ID) fan is usually equipped with an adjustable inlet vane to regulate flow through the system. Water discharged from the quench chamber, scrubber, and demister usually flows to a recycle water tank where it is held for reuse.

As with any refractory-lined furnace system, the rotary kiln incinerator offers the best results when it is run 24 hours a day, 7 days a week on a continuing, steady-state basis.

### 3.4.2.1. Transportable Configurations

The cleanup of abandoned sites involving fixed quantities of hazardous wastes (such as the Superfund sites in the U.S.) is often found to be appropriate for the broad waste-type tolerance of the rotary kiln. However, the limited quantity of material available in such sites makes a permanent incinerator installation unfeasible. In such instances, the transportable rotary kiln has shown itself appropriate and cost-effective. Following on the successful performance of a "proof-of-principle" demonstration transportable kiln by the US EPA, a number of such systems have been deployed for the cleanup of small hazardous waste sites.

### 3.4.2.2. O'Conner Combustor Rotary Kiln Configuration

In the 1980s, a new rotary kiln design emerged—the O'Conner combustor. The kiln was fabricated of boiler tubes with a space (<0.5 cm) between each tube. At each end of the kiln, the tubes are fitted to a torroidal steam drum with the same diameter as the kiln. At the discharge end, a number of tubes are assembled in a radial pattern to a central rotating seal. Air is forced through the space between the tubes so, unlike the conventional rotary kiln, heat release is distributed along the length of the kiln. Several O'Conner combustor municipal refuse incineration plants were constructed in the U.S., but the lead system vendor offering the concept withdrew from the municipal marketplace in the early 1990s.

### 3.4.3. Stationary Grates

Stationary grates have been used in small incinerator furnaces for a longer time than any grate system except the stationary hearth. The stationary grate is composed of cast metal or fabricated metal grates with, perhaps, provision for rotating the grate sections to permit dumping of the ash residue. Although some stoking action can be obtained by shaking the grate, stationary grates normally require manual stoking.

### 3.4.4. Mechanical Grates: Batch Operations

Mechanically operated grates installed in batch-type furnaces were a natural evolution from the stationary grate furnaces. Although batch-type incineration furnaces have given way to continuous furnaces for new, large installations, many of the new small-capacity incinerators still utilize batch-fed furnaces, either with stationary or intermittently operated grates or without grates, the latter in small commercial and industrial installations.

### 3.4.4.1. Cylindrical Furnace Grates

In the circular batch furnace, the grates form annuli inside the vertical cylindrical walls of the furnace. A solid grate or "dead plate" covers the central area of the annulus. A hollow rotating hub with extended rabble arms rotates slowly above the circular dead plate to provide mechanical stoking or mixing. The rotating hub is covered with a hemispherical cone, and one or more consecutively smaller cones are stacked on top of the first one. Forced air (called "cone air") for combustion is supplied through the hub to the hollow rabble arms, and thence through openings in the arms to the space just above the dead plate. Additional cone air is supplied to each of the cones in order to cool the metal. The annular grate area is divided into pairs of keystone-shaped segments;

each pair is arranged to open downward for dumping the ash residue into the ash hopper below. These segmental grates are either hand-operated or hydraulically operated.

#### 3.4.4.2. RECTANGULAR BATCH FURNACE GRATES

Mechanically operated grates in rectangular batch-operated incinerator furnaces include reciprocating (pusher) grates and rocking grates. The grates are installed in a slightly inclined position from the horizontal, with the lower end of the incline at the ash discharge point. With these grates, the furnace is fed intermittently through an opening in the top and at the higher end of the grate, and the fresh refuse is deposited over the bonfire of previously ignited refuse.

As the burning continues, the grates are operated under manual control to move the burning bed of refuse toward the discharge, with manual control, ideally to prevent the discharge of residue that has not been completely burned. In some instances, a dump grate is installed at the ash discharge point to hold back ash residue that is still burning, with manual operation of the dump grate after the accumulated ash has been completely burned.

### 3.4.5. Mechanical Grates: Continuous Operations

Mechanical constant-flow grates are used in most of the newer continuous-burning incinerators. The constant-flow grate feeds the refuse continuously from the refuse feed chute to the incinerator furnace, provides movement of the refuse bed and ash residue toward the discharge end of the grate, and does some stoking and mixing of the burning material on the grates. Underfire air passes upward through the grate to provide oxygen for the combustion processes, while at the same time cooling the metal portions of the grate to protect them from oxidation and heat damage. Typical grate designs correspond to an average heat release rate of $13,500 \, \text{kcal/m}^2/\text{min}$. Clearly, the actual rate in different portions of the grate differs widely from this average.

#### 3.4.5.1. RECIPROCATING GRATE

The reciprocating or pusher grate is, by far, the most frequently used grate system in modern incineration plants. The illustration in Figure 10.11 shows a common design strategy where alloy grate bars are installed stepwise in rows with no or a slight downward incline toward the discharge. In the flat or downward inclining approach, the rows of grate bars move alternately to push the refuse from the feed chute, through the combustion area to the ash hopper. Additional stoking and mixing (breaking open packed refuse masses) may be obtained by providing a drop-off (Figure 10.11, item 5), so the refuse tumbles from one grate section (step) to the next. Up to four grate sections are commonly included in this type of grate for a continuous flow incinerator.

In another design, (the Martin© reverse-acting reciprocating grate system shown in Figure 10.10), the grate surface is inclined at an especially steep angle. In the Martin system, the grate bars (Figure 10.10, item 3) push "uphill," and due to the steep incline, the lower level of refuse is forced back against the gravity flow direction, a concept leading to vigorous stoking and internal "turbulence" within the bed, which breaks up the refuse mass and facilitates combustion.

Combustion and Incineration Engineering

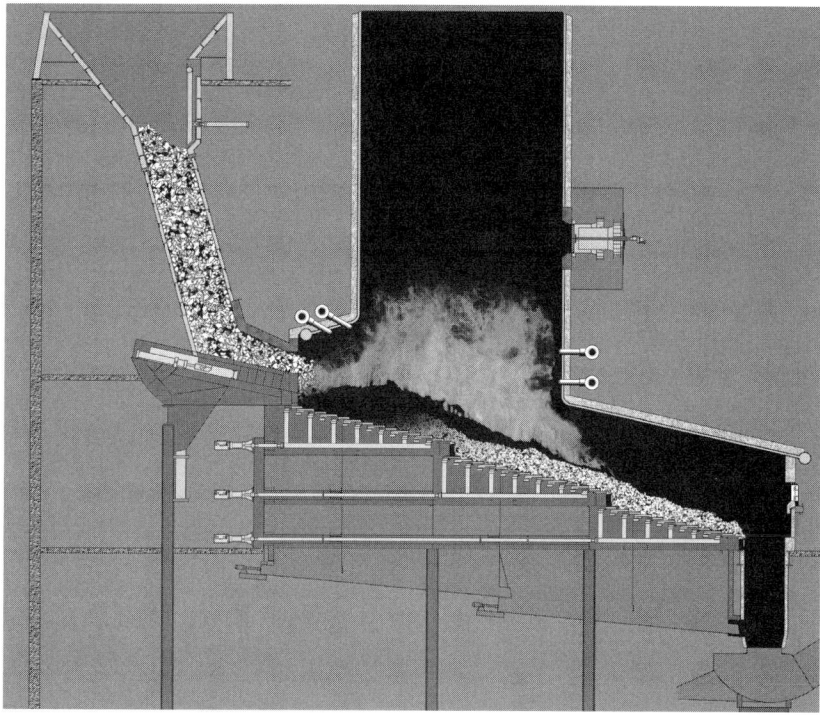

**Fig. 10.11.** Detroit stoker reciprocating grate furnace system. 1, refuse charging hopper; 2, refuse charging throat (refractory lined or water cooled); 3, hydraulic charging ram; 4, grates (high Cr-Ni alloy); 5, vertical drop-off (lined with alloy tuyeres to admit air); 6, overfire air jets. (Courtesy of Detroit Stoker, Inc.)

#### 3.4.5.2. ROCKING GRATE

The rocking grate was a common design in the 1950s and 1960s in the U.S. This grate system also slopes downward from the feed toward the discharge end, with up to four or more grate sections installed in series, and with or without a drop-off or spill-off between grate sections. The rocking grate includes a multiplicity of grate sections or segments that are approximately quarter-cylindrical and include openings for undergrate air. These grates are arranged as successive "stair treads" with risers of less than 3 cm. Alternate rows of grate sections are rotated approximately 90 degrees about the edge toward the discharge of the grates, with the grate face rising up into the burning mass and thus breaking it up and thrusting it forward toward the discharge.

#### 3.4.5.3. TRAVELING GRATE

The traveling grate was widely used in the 1950s and 1960s in the U.S. continuous flow mass burning incinerator furnaces. This grate type is still in use as the burnout grates in spreader stoker refuse-derived fuel (RDF)-burning boiler furnaces. There are two types of traveling grate stokers: the chain grate and the bar grate.

For continuous flow systems, both convey the refuse from the gravity feed chute through the incinerator furnace to the ash residue discharge, much as a conveyor belt.

Because the traveling grate stoker does not stoke or mix the fuel bed as it conveys, incinerator traveling-grate stokers are often cascaded in two, three, and even four or more units with spill-offs of a meter or more between grate sections. For use with a spreader stoker for RDF, the mechanical or pneumatic feeder "flings" the RDF material out over the burning refuse mass on the grate. The unburned fraction then falls to the grate and is drawn back toward the boiler face incorporating the feeder mechanism, dumping the ash through an air seal just below the feeder.

### 3.5. Combustion Air

The combustion air supply is critical not only to meet the stoichiometric requirements to oxidize the waste, but also to dry and gasify solids, atomize liquids, and induce mixing in the flow field. The air supply system is usually characterized as one or more of the following:

- Underfire air: supplied under the grates of mass-fired combustion units. Preheat may be used to assist in drying wet wastes, although only a fraction of the latent heat load for drying can be related to the heat content of the underfire air. However, if only a portion of the waste can be dried and subsequently ignited, heat release within the bed can provide the needed additional energy. Underfire air is also critical to cool the metal alloy grate bars, which otherwise would oxidize and degrade at the high temperatures (>1100°C) achieved in the burning mass.

    As discussed above, the gasification process (both full oxidation and partial oxidation to gaseous "fuel gases") is driven by the underfire air. In mass burn systems, the underfire air rate is often close to the stoichiometric requirement. Air demands along the grate are uneven, leading to the use of several underfire air plenums to control the underfire air distribution along the grate. There is a concentration of gasification and air demand (far in excess of air supply) near the feed end of the grate (say, at 15% to 30% of the way along the grate) and limited air utilization in the discharge (ash cooling) zone.
- Atomization air: a term applicable to liquid waste burners relating to the (high pressure) air stream used to atomize the liquid waste. Steam may also be used if available.
- Primary air: a term usually applied to gas or liquid waste burners and relating to the air stream entering with the waste. Usually the design of the air supply system is such as to induce intense mixing of fuel and air. Swirling flows are often generated to increase mixing. The inducement of swirl results in the generation of a counterrotating axial flow moving back toward the burner tip, which carries both heat and reactive free radicals to mix with, dry, and ignite the incoming material to facilitate rapid initiation of combustion.
- Secondary air or overfire air: the air supplied to the gas stream following gasification. This air is usually supplied at high pressure (velocity) and in carefully researched locations and directions to induce mixing. Design features of secondary air jets include the stoichiometric implications (How much excess air will result from the introduction of the flow?), penetration (Will the flow penetrate the flow of gases from the primary zone without adverse impingement effects on the opposite wall or the fuel bed?), coverage (Will all of the primary zone flow be mixed by the secondary flow?), and turndown effects (Will the desired features of secondary flows be maintained adequately as the overall system moves off design conditions?).

    In smaller incinerators, secondary air is often added at low velocity through slots or small openings in a bridge wall separating the primary chamber from the secondary chamber. In the latter case, mixing is dependent more on the shape of the chamber and changes in direction of the main gas stream than on the energy carried by the air jets. Low velocity

Combustion and Incineration Engineering 531

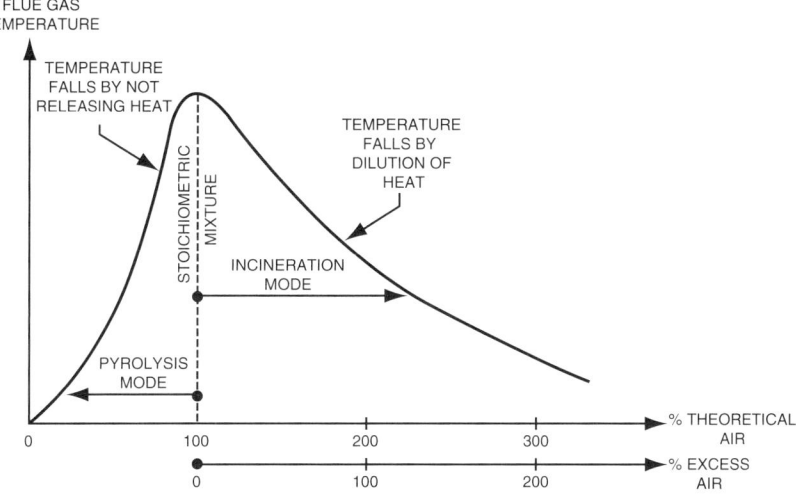

**Fig. 10.12.** Effect of excess air on flame temperature.

combustion air can also be admitted through furnace openings as a result of the negative pressure or draft within the furnace chamber. The quantity of combustion air admitted through such openings can be controlled by dampers or by the door opening.
- Tertiary air: additional air supply beyond secondary, which is used to cool the flue gases by dilution. Due to the high cost of air pollution control, this option for gas cooling is seldom employed.

Addition of air to a combustion system acts broadly in three ways:

- The air oxidant chemistry (oxygen content) interacts with the fuel value of the waste to release heat and increase the temperature, evaporate free water, and pyrolyze waste.
- The mass and specific heat of air components moderates the temperature.
- The momentum of the air flow acts to mix the furnace gases, thus facilitating complete combustion.

Figure 10.12 illustrates both the chemical effect (heat release is almost directly related to the proportion of the stoichiometric oxygen that is added) and the tempering effect. To the left of the stoichiometric mixture line, air addition increases the mean off-gas temperature by facilitating the release of heat. To the right of the stoichiometric line, the temperature progressively decreases by "dilution" of the heat (now fully released) by increasing the total flue gas mass rate.

### 3.6. Flue Gas Conditioning

Flue gas conditioning is defined as the cooling of the flue gas after it has left the combustion zone to permit discharge to mechanical equipment such as dry air-pollution control devices and fans or a stack. In general, cooling to 230° to 370°C is necessary if the gas is discharged to mechanical equipment, although cooling to 315° to 600°C is adequate for discharge to a refractory-lined stack.

Both wet and dry methods are used for cooling (or tempering) incinerator flue-gas streams. The following subsections discuss the technical and economic features of several embodiments of these methods.

*3.6.1. Cooling by Water Evaporation*

In wet methods, water is introduced into the hot gas stream and evaporation occurs. The degree of cooling is controlled by (1) the amount and droplet size of the water that is added to the gas and (2) the residence time of the gas in the water atmosphere. Currently, two types of wet cooling are used: the wet-bottom method and the dry-bottom method. Generally, the dry-bottom method is preferred.

In the dry-bottom method, only enough water is added to cool the gas to the desired temperature, and the system is designed and operated to ensure complete evaporation. A conditioning tower 9 to 36 m high is required and fine, high-pressure spray nozzles are used. Booster pumps are necessary to raise the water pressure to ensure fine atomization; water pressures from 6 to 36 bar are common in such systems. Alternatively, atomization can be effected using compressed air or steam. Control is usually accomplished with a temperature controller measuring the outlet flue gas temperatures and modulating the flow of water to the conditioning tower sprays.

The advantages of the dry-bottom system are that it minimizes water consumption, eliminates water pollution and discharge problems (the wastewater stream is generally quite acidic and contains numerous toxic metals), produces a dry effluent gas (free from entrained water), and reduces the volume of the flue gas. The disadvantages of the system are that it is expensive to design and install, power consumption is high, control is somewhat complex, and the small orifices of the atomizing nozzles make them susceptible to plugging. Also, the dry-bottom system is more costly than the wet-bottom system because of the need for water filtration and the maintenance required for the high-pressure pumps, nozzles, and control systems.

The wet-bottom method involves the flow of large quantities of water (much more than is required for cooling the flue gas). The water is supplied by coarse sprays operated at relatively low pressures. The excess water falls to the bottom of the cooling zone and is rejected or recycled. With recycling, the water approaches the wet bulb temperature of the flue gas.

The equipment used in this system consists of several banks of sprays, each with several nozzles with relatively large openings (over 0.5 cm), located in the flue leading to the stack or air pollution control equipment. Line water pressure is adequate for satisfactory operation. The system is generally controlled by measuring the gas temperature downstream of the sprays and modulating water flow, either manually or automatically.

The advantages of the wet-bottom system are that it is relatively simple, reliable, and inexpensive to design and install. Also, there is a reduction in total gas volume during cooling. A disadvantage of the wet-bottom system is that much more water (greater than 100% excess) is used than is necessary for cooling the gas. The excess water is acidified in use and is contaminated with particulate and dissolved solids. Also, the flue gas leaving the spray chamber may carry entrained water droplets or wet particulate

# Combustion and Incineration Engineering

matter. These moist particles can cause operating problems with the air pollution control (APC) and fan equipment owing to fly ash adherence and accumulation.

### 3.6.2. Cooling by Heat Withdrawal

The second method of gas conditioning (assuming that the incinerator itself is not a boiler system) uses a convection water tube boiler in which heat is removed from the flue gas to generate steam or hot water. The equipment consists of a convection-tube waste-heat boiler and perhaps an economizer. In smaller units, a fire-tube boiler is common and low in cost. Most of the conventional boiler auxiliary equipment is required such as boiler-feed water pumps, steam drums, and boiler water treatment facilities.

The advantages of this system are that heat is recovered and that the shrinkage in flue gas is greater than with any other method discussed. No water is added to the system during cooling, which may or may not be desirable. The disadvantages of the method are that the system is expensive to design and install, the boiler installation is complex to operate, corrosion and erosion problems with the boiler tubes may occur, sticking of fly ash on the boiler tubes can occur, and reliable markets for the steam must be found or still greater investment is required for air- or water-cooled condensers.

## 3.7. Air Pollution Control

Concern of the public about air pollution impacts on the contiguous environment is probably the major factor affecting the decision of a community government to construct and operate an incinerator. The most noticeable forms of air pollution are fly ash, smoke, odors (from the stack as well as other areas), noxious gases, and dust. Another class of pollutants (the air toxics) is emitted in small quantities but is important because of the health impact potential. All emanate from an incinerator at times. However, stringent regulatory emission standards throughout the world (Table 10.16) and a strong response from the waste to energy vendor community in terms of system design/operating features and add-on control equipment make the level of modern incinerator emissions well below that which would present a significant health impact. Indeed, one often finds that an incinerator, viewed as a power plant, has an emission pattern (mass of pollutants per kilowatt hour) with significantly less impact than the fossil fuel fired power plants it replaces (40).

Although the flue gases from incinerators contain a number of pollutants, air pollution control (APC) equipment installed on these units is primarily directed at particulate removal. Note, however, that particulate control addresses a number of pollutants:

- Total suspended particulate (the inorganic ash materials elutriated from the combustion process, tarry aerosols of high molecular weight polynuclear hydrocarbons and soot arising due to incomplete combustion)
- Heavy metals such as cadmium, arsenic, lead, chromium, and nickel that are either part of the elutriated particulate or vaporized in the combustion zone and, later, condense (often on the fine particulate ash particles)
- Sulfur oxides, HCl, and other acid gases that have reacted with injected hydrated lime or limestone to form solid products ($CaSO_4$, $CaCl_2$)
- Mercury and some of the heavy hydrocarbons (including dioxins and furans) that have been adsorbed onto activated carbon that has been injected into the flue gas stream

Table 10.16
International municipal waste combustor air emission codes (ca. 2003): all values corrected to 0 deg C, 101.3 kPa; 11% vol O$_2$ dry unless otherwise indicated

| Pollutant mg/Nm$^3$ | Hong Kong best practicable means requirements | Germany TALuit 1986 | Germany 17 BimSCHV 1990 & France 10/10/96 | Netherland BLA 93 | Italy | EC Draft 20.8.94 | Greece Turkey Rumania Bulgaria | Sweden | Russia | U.S. | Taiwan (r) | Japan |
|---|---|---|---|---|---|---|---|---|---|---|---|---|
| Particulate | 50 | 10 | 10/30 (a) | 5 (e) | 5 | 10 | 30 | 10 | 10 | 24 (h,i) | 70 | 80–250 |
| Opacity | (v) | – | – | – | 10% | – | – | – | 10% | 10% (i,j) | 10% | – |
| C-Org | 20 | 20 | 10/20 (a) | 10 | 10 | 10 | – | – | – | – | – | – |
| HCl | 50 | 50 | 10/60 (a) | 10 (e) | 10 | 10 | 50 | 60 | 10 | 41 (i,j) | 40 ppmdv | (f) |
| HF | 2 (y) | 2 | 1/4 (a) | 1 (e) | 1 | 1 | 2 | 1 | 1 | – | 7.3 (a) | – |
| SO$_2$ | 250 (x) | 100 | 50/200 (a,b) | 40 (e) | 50 | 50 | 150 | 200 | 50 | 85 (k,m) | 80 ppmdv | – |
| CO | 100 | 100 | 50/100 (a) | 50 (f) | 50 | 50 | 100 | – | 100 | 125 (n) | 120 ppmdv | – |
| NO$_2$ | 400 | 500 | 200/400 (a) | 70 (e) | 150 | 200/NH3<10 | – | 100 | 200 | 200 (l) | 180 ppmdv | (f) |
| Dioxin ng/Nm$^3$ | 0.1 | – | 0.1 | 0.1 | 0.1 | 0.1 | – | 0.1 | 0.1 | 0.18 (i,o) | 0.1 | – |
| Hg | 0.2 (u) | 0.2 | 0.05 | 0.05 | 0.05 | 0.05 | 0.2 | 0.03 | 0.01 | 0.08 (q) | 0.3 | – |
| Cd | 0.2 (u) | – | 0.05 (c) | 0.05 | 0.05 | 0.05 | 0.2 | 0.001 | 0.01 | 0.01 mg/dscm | 0.3 | – |
| Total heavy metal | 5 (w) | 1.0/5.0 | 0.5 (d) | 0.5 (g) | 0.5 | 0.5 | – | – | 0.5 | 0.1 mg/dscm | 2 (p) | – |
| Effluent from APC | – | yes until 1989 | No | No | No | Yes but stringent quality limits | – | Yes | No | – | – | Yes |

Notes:

a. (daily average)/(1/2 hr average); b. SO$_2$ + SO$_3$; c. (Cd + Tl); d. (Sb, As, Pb, Co, Cr, Cu, Mn, V, Sn, Ni, Se, Te); e. Short-term average; f. 1-hour average; g. (Sb, As, Pb, Co, Cr, Cu, Mn, Sn, Nl); h. Metals as particulate; i. Stack test once per year; j. 95% reduction or 25 ppmdv @ 20 deg C, 101.3 kPa, 7% vol O$_2$ dry; k. 80% reduct/on or 50 ppmdv @ 20 deg C, 101.3 kPa, 7% vol O$_2$ dry; l. 6-minute average; m. Daily geometric average; n. 4-hour average for mass burn combustor; o. TEC presented in table above standard is 30 ng/dscm total dioxins; p. Total lead and lead compounds; q. 0.08 dscm or 85% control; r. All values corrected to 10% O$_2$ dry, 0 deg C 101.3 kPa; s. Total fluorides as F; t. Limit varies with district based in part on stack height; u. Total CD + Hg; v. Less than Rigelman shade 1; w. Nickel and arsenic and their compounds 1 mg/Nm$^3$; x. H$_a$S, 5 mg/Nm$^3$; y. Fluorine and it compounds 10 mg/Nm$^3$, HBr. 5 mg/Nm$^3$, phosphorus 5 mg/Nm$^3$.

# Combustion and Incineration Engineering

The process of selecting an optimal APC technology is complex. Defining the meaning of "optimal" requires consideration of several technical, economic, and other criteria:

- Pollutant removal efficiency (relative to all regulated pollutants and giving attention to the potential generation of new pollutants in the course of abating others)
- Capital investment:
  - Basic APC device and installation
  - Fans, pumps, and piping
  - Foundations and structural support
  - Ductwork to incinerator and to fan and stack
  - Insulation
  - Instrumentation and control systems
  - Electrical supply, motor control centers
  - Water supply, sewer, and portion of treatment system
  - Working capital in spare parts
  - Allowance for device replacement (useful life)
- Operating costs:
  - Operating labor
  - Special Costs for training in operations and maintenance
  - Operating parts and supplies
  - Absorbents, reagents, and similar consumables
  - Fuel
  - Electricity
    - Fan power
    - Pump power
    - Device power (especially for electrostatic precipitator but also for shakers, air compressors, etc.)
  - Water
  - Maintenance labor
  - Maintenance parts and supplies
  - Wastewater treatment and disposal
  - Ash treatment and disposal
  - Contract maintenance expenses (e.g., for instrumentation maintenance)
- Impact on incinerator availability (reliability) and capacity
- Operability and maintainability (including consideration of compatibility with general plant operating and maintenance labor skills, maintenance facilities and tools)
- Compatibility with layout limitations
- Compatibility with other regulations (noise, odor, illumination, visible plume, icing of adjacent roadways or electrical wires)

The detailed design and selection of APC systems is a major technical area in itself. Presenting such information is beyond the scope and objectives of this book. Details of design and performance with special application to incineration systems are presented elsewhere (1). A number of devices have been used over the years for particulate control, ranging in particulate removal efficiency from 5% to 15% (in the 1960s) to modern systems achieving upward of 99%. In light of current U.S. federal particulate emission standards, control efficiencies in excess of 98% are generally required.

At one time, simple settling chambers or expansion chambers were used in the breeching and flue gas ducts, and many of the older installations simply used refractory baffles across the breechings extending downward from the roof or upward from the bottom of the breeching to require the flue gases to pass under and over such baffles to settle the larger particles. In some instances, a coarse spray of water was directed into the flue gases and toward the baffles, with most of the water falling to the floor of the chamber without vaporization. The wet floor and baffles improved particulate removal by preventing re-entrainment of settled ash into the flue gas stream. At best, however, such systems only attained a control efficiency of 20% to 35%, far below modern requirements.

In the 1970s, a few plants installed high-energy scrubbers or cyclones, but the dominant move was to the dry electrostatic precipitator (ESP) that controlled particulate to the more stringent New Source Performance Standards arising from the Clean Air Act of 1970. The economics of the ESP were helped greatly by the emergence of the gas-tight water-wall boiler design that limited excess air (flue gas volume). In fact, the initial movement of incinerator construction to water-wall designs in 1970 to 1973 was primarily driven by the need to reduce flue gas volume, not to recover energy. By 1970, the ESP had already become dominant for European facilities.

In the mid/late 1970s, the regulatory spotlight in the U.S. shifted to the polychlorinated dioxin/furan compounds. The proper control technology for these materials focuses on upgrading the combustion environment. Burn out the precursors and there will be little or no dioxin. The data on emission factors suggests that combustion control, by itself, is an effective solution and that the dioxin problem is now well in hand.

In parallel with the dioxin/furan controls, the requirements for back-end treatment underwent a major change, extending the control requirement to include acid gases, most importantly HCl and $SO_2$. The solution to the requirement for acid gas control was the combination of the semidry absorber with lime slurry addition followed by a fabric filter.

Alkaline reagents show their highest utilization (low stoichiometric ratio of reagent to acid gas neutralization requirement) for a given level of control if the temperatures are low and if reaction time is long. These two process features led to the development, testing, and implementation of the spray dryer absorber–fabric filter control concept for application to resource recovery systems. In this technology, a large chamber is added following the boiler when the temperature has been reduced to, say, 230°C. At the top of the chamber, a water slurry of hydrated lime is injected through a spray dryer atomizer. The gas temperature and water quantities are kept in balance such that evaporation is rapid and complete in, say, 2 seconds or less. A conservative margin is added to the mean residence time to ensure a dry particulate as input to the bag house or ESP. Conservative specifications call for 18 seconds in a downflow configuration and 8 seconds in an upflow reactor (41).

In most facilities, pebble lime (CaO) is received in truck-load quantities and stored in a carbon steel silo. The total storage capacity depends on local supply availability and reliability, but is often a minimum of 14 days. The lime is slaked to $Ca(OH)_2$ using a minimum of two slakers, each designed for 100% capacity. Dilution water is added to the slurry in accord with the needs for exhaust temperature control for the spray dryer system.

Mechanical rotary atomizers have been used to produce a mist of finely divided droplets. Two-fluid nozzles are also used to form the fine droplets. Since the spray pattern from the two-fluid nozzles is narrower than that of the mechanical atomizers, the subsequent evaporation chamber can be designed with a smaller diameter.

The injected alkaline slurry and associated water evaporation has two effects: cooling of the flue gases from about 300°C to about 150°C, and the generation of a distributed field of acid absorbent particles (a mixture of CaO and Ca(OH)$_2$). Following the spray dryer chamber, the gas can be passed to an ESP or, more conventionally, a fabric filter. The ESP effects removal of the solid reactants (and other flue gas particulate). The fabric filter option accomplishes particulate removal, but also provides another opportunity for acid gas contact as the acid gases pass through the dust cake. About 15% to 20% of the removal occurs on the filter cake (42). The second contact opportunity not only increases reagent utilization and increases the ultimate removal efficiency but provides reserve alkalinity to respond to spikes in acid gas concentration. The spray dryer absorber–fabric filter combination has achieved above 95% HCl removal and above 80% SO$_2$ removal while accomplishing general particulate control to levels well below regulatory requirements.

In the early 1990s, attention moved to mercury and to the nitrogen oxides. Mercury can be controlled to an acceptable degree through injection of activated carbon ahead of the fabric filter. Injection of alkali sulfides (e.g., solutions of sodium sulfide) was also effective to bind mercury vapor but operational hazards with the sulfide makes the carbon option preferable. Activated carbon injection was also shown to be effective in reducing dioxin/furan and other high molecular weight hydrocarbons (PNH, PCBs) to levels below those achieved by combustion control. Modern plants also address nitrogen oxides control to the 50% to 65% level using ammonia or urea solution injection into the flue gases in the superheater region (selective, noncatalytic reduction [SNCR] technology).

Therefore, at present, optimum control for municipal waste combustors involves high-quality combustion controls and overfire air mixing to minimize CO, unburned hydrocarbons, soot (an important precursor to dioxin/furan species), and other PICs; urea/ammonia injection for NO$_x$ control; activated carbon injection to adsorb mercury and its compounds as well as dioxins, PNH, and other high molecular weight compounds; a semidry spray dryer absorber for acid gas capture; and a fabric filter. This combination of operating practice and control equipment was followed with a tall stack to effect dispersion and avoid building downwash.

The direction of U.S. regulatory agencies in future years is difficult to predict. However, one might speculate that greater control will be required for metals and perhaps for nitrogen oxides. This could lead to the use of a dry followed by a wet electrostatic precipitator (replacing the fabric filter) with reheat ahead of catalytic, ammonia-based NO$_x$ reduction. The fabric filters used in incineration systems generally use fiberglass bags, often with Teflon B coatings and thin film polytetrafluoroethylene membranes to enhance collection of very fine particles. Some plants with this APC flowsheet are already operating in Europe. One notes, however, that this APC train is considerably more costly than that in present use and that the ground-level pollutant concentration gains that are obtained are quite modest.

## 3.8. Special Topics

### 3.8.1. Heat Recovery

As fossil fuel and other energy costs increase, the economics of energy recovery from waste incineration improve in kind. For solid waste incinerator applications, however, particular attention must be given to the areas of corrosion, slagging, and other tube-fouling problems; irregularity of heat release in comparison to fossil fuel boilers; and reliability issues. One should carefully consider whether the recovered energy value exceeds the incurred capital and operating expense. Also, the decision maker should thoughtfully reflect on the impacts that energy recovery functions will cause in the most important and critical objective of the incineration system: reliable waste management.

Recovery of energy from the hot flue gases from waste incineration can be effected using several different types of equipment. Selecting among the alternatives depends on the scale of energy recovery, the sophistication of the work force, and the severity of the steam conditions (temperature and pressure) required. Alternatives include the following:

Water tube boiler
    Water wall primary furnace (radiant boiler)
    Convection tube bank (mixed radiant and convection boiler)
    Superheater tube bank (convection)
    Convection tube bank (convection)
    Economizer (feed water heating by convection)
Waste heat boiler
    Generally convection-based water tubes
Fire tube boiler
    Gases pass inside tubes immersed in a large, pressurized tank; these boilers are simple and low in cost but are limited to producing relatively low pressure saturated steam
Hot-water boiler
    Simple in design and produce high temperature hot water useful for heating or, with a secondary "step-down" heat exchanger, lower temperature general purpose hot water

#### 3.8.1.1. Energy Markets

Energy markets are important to justify incorporating energy recovery into an incinerator design. Adding a boiler increases capital cost and decreases system reliability. Both adverse effects impact on the ability of the system to carry out its primary function: cost effective and reliable waste management.

Energy markets may be characterized in four ways: (1) The size of the market, (2) the energy type, (3) the reliability with which energy will be used (the market stability over the year), and (4) the reliability of revenues.

1. Market size: The best markets for waste-derived energy are large markets such that fluctuations in energy recovery or outright outages from plant shutdowns are readily

absorbed by other energy generators. Large markets of this type include electric utilities (serving an effectively infinite market and well backed-up) and major steam users (large steam-intensive manufacturing plants such as paper mills). Smaller users are often concerned about the reliability of energy supply and may demand backup, which negates the "avoided investment" value to the waste-derived energy supply and may even require a shadow workforce or warm-running to allow quick response to outages, thus negating a large portion of the savings by reduced workforces and maintenance expense.

2. Market type: The ideal market is a steam user. This avoids the need to install energy conversion equipment (turbo-generators), and the inefficiency in energy conversion for incineration plants that often operate at inferior steam conditions compared to a utility generating plant. Unfortunately, steam customers using the quantity of steam and the steam conditions (pressure and temperature) that are most compatible with an incinerator and that are located near to the incinerator plant (to minimize the acquisition of easements and the legal, cost, and energy loss impacts of long transmission lines) are often difficult to find.

Electrical generation has the advantage of being infinite in extent and assuredly continuing. The primary disadvantage is that there exists a highly efficient competitor (the utility itself) also generating electricity such that energy credits are minimal. Also, the capital investment in the energy recovery system may not be recoverable with a capacity credit associated with the sale of electricity unless the utility capacity is limited and the addition of the incinerator generating capacity has worth. Wheeling of your electrical energy product over utility lines is conceptually possible (for a fee) to tie in to a more attractive retail electrical customer. Often, issues of reliability of supply and legal challenges from the utility make this alternative unattractive. Wheeling is not a mandated service under present U.S. federal law and may be denied.

3. Market reliability: The solid waste problem (usually) knows little seasonality. Also, as with all combustion systems, continuous operation is strongly recommended to avoid temperature cycling of the equipment and refractory. Thus, the ideal energy market is 24 hours a day, 7 days a week, and 52 weeks a year. Although true for electricity, most steam markets fall short of this goal. Large steam users such as paper mills are the closest. Seven-day industrial operations of lesser size may be acceptable for much of the year. However, white-collar office buildings, school or office buildings, and other energy markets primarily using the steam for space conditioning show strong and unfavorable diurnal and seasonal patterns.

4. Revenue reliability: The financial lifetime of incineration systems is long. Therefore, energy marketing agreements will extend over many cycles of base energy cost (the cost of the reference fuel that is often specified in energy contracts and used to scale the unit value of the plant's energy product, such as the cost of No. 2 distillate oil in New York). In some cases, contract terms at fixed prices may be obtained over relatively long times but usually at levels that are significantly lower than the more risky floating rates. On the other hand, if the unit value of energy floats, the project revenue stream is uncertain. This usually requires some kind of backing for the energy-related revenue stream if bonds (where the payment of principle and interest is based in part on the energy revenues) are used to finance the project. Also, most steam customers present a measure of risk that, in a time frame much shorter than the incinerator life, they may go out of business, cut back operations, or significantly improve their energy utilization patterns such that the steam market weakens or disappears altogether. Since such an evolution is always possible and is in the best interests of the steam user, few steam clients would agree to steam purchase agreements that limit their flexibility to make such changes. Thus, the risk of cessation of energy revenues hangs heavy over the heads of the incinerator operations.

### 3.8.1.2. CORROSION ISSUES AND ENERGY RECOVERY

Energy recovery components add to the unreliability of incineration systems. Clearly, part of the unreliability comes simply by the addition of another element in series with the incineration system that must work to make the incinerator operable. Further, however, the incineration process (as contrasted with the burning of fossil fuels) creates corrosive, fouling and erosive conditions in the flow path, which significantly degrade the availability and working life of boiler components, thus increasing outage frequency and severity and maintenance expense.

Corrosion of water-wall and tube metal surfaces is perhaps the most serious technical problem in the design of refuse-fired boilers. In the operation of a boiler using wastes as fuels (and for conventional liquid or solid fuels as well), metal wastage owing to corrosion and erosion and tube fouling owing to the buildup of deposits have presented serious problems to the system designer and operator. Detailing the nature and cures for such problems is beyond the scope of this chapter and is still a matter of intense study and speculation. Several basic concepts, however, merit qualitative description:

- Low-temperature corrosion: Condensation of moisture from flue gases occurs in regions of the energy recovery system where the surface temperature exposed to the flue gases falls below the dew point. This can occur with a too-low feed water temperature in the economizer or when cold ambient air overcools the surfaces used to transfer heat to incoming combustion air in the air heater. Condensation can also occur in "dead zones" in the air pollution control system and ducting where the circulation of (hot) flue gas is slow in corners or other low-flow zones such that heat loss to the ambient significantly reduces the metal surface temperature. Also, condensation can occur in an uninsulated stack when high winds and cold ambient conditions result in wall cooling below the dew point. The presence of mineral acids (especially sulfur trioxide or sulfuric acid) in the flue gas increases the dew point considerably above 100°C. If condensation occurs, the resulting metal wastage rate, accelerated by the presence of soluble chlorides or acids, can become unacceptably high. Clearly, this corrosive mechanism is always operative during boiler startup and shutdown unless fossil fuels are fired to warm up the unit. Algorithms to estimate the acid dew points are available in the literature (1).

    The cure for this type of corrosion is straightforward, namely, to design the system so as to avoid dead zones and to maintain metal temperatures safely above the dew point of the flue gases. Furthermore, frequency and duration of cool-downs should be minimized.
- High-temperature corrosion: In regions of the furnace where the flue gases are above, say, 870°C, a variety of mechanisms for chemical attack of the metal tube surfaces can become operative. Chlorine, appearing in the flue gases as hydrochloric acid (e.g., from the combustion of chlorinated hydrocarbon wastes or PVC), or in salts such as sodium or potassium chloride, has been shown to participate in corrosive attack of metal tubes. Sulfur, appearing as the dioxide or trioxide or as sulfates, appears to slow the rate of attack of the metal by chlorides (43). Although the exact mechanism of attack is still in question, it is clear that fireside metal temperature is the single most useful parameter with which to judge the potential for rapid metal wastage. The flue gas temperature, however, is a second, though less direct variable at gas temperatures of interest.

    Data reported by staff at the Battelle Laboratories (43–49) indicate that a maximum fireside metal temperature of about 200°C should give long (say, 15 years) carbon-steel boiler-tube service for systems burning 100% municipal refuse. Allowing for a 30°C temperature drop across the tube wall (a reasonable average for boiler tubes containing liquid water and thus experiencing a high heat transfer rate on the inner wall, but about half of that experienced by superheater tubes containing gaseous steam), this corresponds to a maximum (saturated) steam pressure of 30 bar. For higher pressures and temperatures, increased wastage must be accepted or more costly tube metal alloys must be used. The problems of metal wastage are of special concern for superheater surfaces. The lower heat-transfer rates of the steam (compared to liquid water) results in higher fireside tube temperatures and greatly accelerated corrosion. Some relief from this problem has been found by inserting a portion of the convective boiler surface ahead of the superheater to lower the gas temperature and the use of rammed silicon carbide type refractory coatings on the tubes (50). The latter solution comes at the cost of lowered heat transfer rates and increased investment and maintenance expense.

- Oxidizing and reducing corrosion: In the combustion of highly non-ideal fuels such as raw municipal refuse and especially for mass-burning systems, the gases rising from the grate fluctuate in composition between oxidizing (having an excess of oxygen) and reducing (having an absence of oxygen and the presence of carbon monoxide, hydrogen, hydrocarbon gases, etc.). Tube metal surfaces exposed to such changing gas compositions are subject to rapid wastage. During the oxidizing period, the surface metal oxidizes to, say, $Fe_2O_3$. When the gas switches to the reducing side, the oxide is reduced to the metal but the metal thus formed is not adherent and flakes off. The process repeats and net metal loss occurs. Flaking of the weak, reduced metal structure is accelerated by the "shot blasting" effect of erosion from particulate matter entrained in the gas stream.

   In mass-burning systems, the bed processes always produces reducing gases and, consequently, sidewalls and radiant tube banks are particularly prone to this type of attack. Protection of the sidewalls with refractory up to a distance of about 10 m above the grate line has been used successfully to cure the sidewall corrosion problem. Introduction of sufficient secondary air above the fire and stimulation of high levels of turbulence can greatly assist the burnout of reducing gases prior to their entry into the initial convective tube banks.

- Abrasion (erosion) wastage: The mechanical erosion of tube surfaces by fast-moving fly ash particles can rapidly lead to tube failure. The fly ash from municipal refuse combustion has been shown to be particularly abrasive, more than that from, for example, most coal ash. This problem can be mitigated by designs that keep the velocity of the flue gases between the tubes below, say, 3.5 to 4.5 m/s; by coating the tubes with refractory (or letting slag build up to a degree); and by careful design of tube bank geometry and flow patterns. In general, however, these remedies lead to larger, more costly boiler facilities.

- Slagging: Slag buildup is not directly responsible for tube wastage. Indeed, slag accumulations act to protect metal tubes from erosive metal losses. Slag accumulation, however, reduces heat transfer rates and increases the pressure drop and gas velocity (erosion rate) through the boiler flues. Also, important corrosion reactions occurring under reducing conditions (in the absence of oxygen) have been shown to occur within the slag layer so that slag buildup cannot be used to infer a lack of chemical attack.

   Slagging can be avoided by designs that maintain fireside metal temperatures below the range where the slag becomes tacky. For municipal refuse, this range is approximately 600° to 700°C. Commonly, the boiler passes are equipped with soot blowers that periodically use jets of steam or compressed air or blasts of metal shot to dislodge adherent slag. Note, however, that cleaning the tube surface can also result in the removal of coatings that were performing a protective role with respect to erosive or corrosive tube attack. Thus, following soot blower activation, wastage will be initiated, typically at very high rates, until a protective coating of slag or corrosion products is reestablished. Table 10.17 summarizes the corrosion and slagging problems in boiler systems that are commonly experienced by incinerators.

**Table 10.17**
**Corrosion regimes in incinerator boiler systems**

| Component | System slagging | Chloride corrosion | $H_2SO_4$ corrosion | HCl corrosion | Oxidizing-reducing corrosion |
|---|---|---|---|---|---|
| Sidewall | • | • | | | • |
| Superheater | • | • | | | • |
| Convection boiler | | • | | | |
| Economizer | | | • | • | |
| Air heater | | | • | • | |
| Stack | | | • | • | |

*3.8.2. Burning in Suspension*

Although the mass burn concept inherently avoids the cost and problems of handling and processing raw refuse, the combustion process on the grate is paced by the drying of the incoming refuse and mechanical opening of the mass to expose the combustible material to air. These steps tend to reduce the combustion intensity (kcal/hr/m$^3$), thus requiring a larger and more costly combustion space. Further, the combustion, grate cooling, and air flow requirements of mass burning systems lead to operation at a relatively high overall excess air level within the furnace (more than 85% excess even in modern facilities). High excess air operation increases the size of the boiler, the air pollution control systems, the induced draft fan, and the stack. If the waste is particularly wet (e.g., a watermelon or wet paper), one may find essentially raw waste in the ash discharge conveyor.

These problems can be mitigated by first processing the waste to reduce the particle size; separating "heavies" such as rocks, china, metal objects, and the like; and then mechanically or pneumatically injecting the processed refuse-derived fuel (RDF) into a furnace. Depending on the trajectory and the thermal environment into which the waste is injected, the waste may burn partially or entirely while it is airborne. Ultimate burnout (if required) may require a "burnout grate." In practice, this is accomplished with either a spreader stoker-fired system (sometimes called semi-suspension burning) or a true suspension-fired system such as is used for powdered coal.

In either case, the refuse is received and stored and then subdivided using a hammer-mill, rasp, or other device. If it is desired to reduce the quantity of residue, to minimize problems owing to the buildup of slag deposits or to recover unincinerated ferrous metal, the shredded refuse may be processed to (1) reduce massive residue content, (2) reduce (somewhat) fine residue content, (3) increase mean heat content, (4) recover ferrous metal using a belt magnet, and (5) simplify and reduce problems with transport of the processed refuse. The processed refuse is then stored for later retrieval or fed directly to the furnace using either belt or other mechanical conveyors, or a pneumatic feeder.

3.8.2.1. SPREADER STOKER

In a spreader stoker-fired system, the raw waste is first processed to a 10- to 15-cm top-size RDF, which is then bunkered. The RDF is then reclaimed from storage and projected into the incinerator enclosure using a rotating, vaned "flinger," or is pneumatically blown in. The trajectory of an entering refuse particle carries it over a flat traveling grate stoker moving the bed of burning residue back toward the firing face. The burned-out residue is discharged just below the firing chutes. Hot gases rising from the burning refuse bed and radiation from airborne burning particles rapidly heat the incoming refuse, drying and igniting it. The refuse particle may be entirely burned out while suspended in the gas or be swept out of the furnace. If the particle is large and heavy, it falls to the grate to aid in the drying and ignition of incoming refuse. Both burning patterns are desirable to provide ignition and to protect the grate. Thus, extra-fine shredding is avoided, a benefit since shredding costs rise rapidly as product top-size decreases (1). Due to the improved contact between waste particles and the combustion air, the combustion air levels can be reduced to approximately 50% excess air (overall).

Experience in spreader stoker firing of RDF has been generally acceptable, although most plants have experienced extensive startup problems associated with waste processing and RDF storage, reclaim, and feeding. These problems have resulted in increased capital costs (often more than 20% increase over the initial budget allocation) and losses in the startup period due to unrealized power revenue and waste processing fees.

A second problem that has shown itself with the spreader stoker systems is an increase in the dioxin generation in these plants in comparison to the mass burn facilities. The underlying mechanism underlying the increase appears to be the higher concentration of partially charred organic matter (carryover from the injection process) in the flue gases leaving the high-temperature zones. This graphitic char reacts with inorganic chlorine salts in the presence of certain heavy metals that act as catalysts to form dioxin species in the downstream zones of the incinerator.

### 3.8.2.2. Suspension Burning

In a suspension-fired furnace, the RDF is processed to a smaller (say, 5-cm top-size) refuse fragment than is the case for the spreader stoker. This significantly increases the processing cost (1). The RDF is pneumatically blown into the system and almost all burning occurs while the refuse particle is suspended in the hot furnace gases. To ensure that this occurs, tall furnaces (similar in dimensions to anthracite coal fired boilers) are used or a high degree of refuse subdivision is required. Often, a small burnout grate is provided in the floor of the furnace to increase the retention time for larger wood fragments or exceedingly wet refuse materials that survived the subdivision processing steps.

Suspension burning is well adapted to burning several fuels, and indeed most suspension burning of refuse (up to 20% of the total heat release rate) has been in co-firing applications with coal in large (>100 MW) steam electric plants. Long-term experience involving co-firing with coal has demonstrated the following:

- No significant reduction in the performance of the electrostatic precipitators used for air pollution control
- No marked increase in boiler corrosion or tube fouling when burning upgraded RDF with precombustion removal of a large portion of glass and "heavies" using an air classification step following shredding
- Rapid burnout of most refuse particles, but a clear requirement for a burnout grate
- Significant problems with refuse feeding systems owing to the tendency of the RDF to bridge, jam, and hang up and to the high abrasiveness of the refuse

Combustion air for suspension burning, without main grates, requires a different consideration from combustion air for a grate system. In suspension burning, the air that conveys the shredded refuse into the chamber may be half or all of the theoretical air required for combustion; however, it must be sufficient to convey and inject the shredded refuse into the furnace. The technology of Combustion Engineering Inc (Winsor, CT). involves addition of air at points in the corners of the furnace chamber targeted tangential to an imaginary cylinder in the center of the boiler pass. This creates a cyclonic action, with the burning mass in the center of the rotating cyclone and the air injection surrounding the cyclonic flame. If the suspension burning system includes an auxiliary grate at the bottom of the furnace chamber for completing the burnout of the ash residue, a small amount of underfire air is desirable through this grate.

### 3.8.3. *Residue Processing and Disposal*

Municipal solid waste includes inert materials that cannot be destroyed in the combustion process. Also, the incineration process is inherently imperfect so that some potentially combustible material is dried, heated, and carbonized but the desired next step (gasification of the char) is not achieved. Further, some material simply "falls between the cracks" (siftings) and leaves the hot combustion environment substantially unburned. These three components comprise *bottom ash*, the inevitable residue of municipal solid waste incineration operations.

In addition, there is *fly ash*, the fine ash that becomes airborne in the primary chamber and either settles in the ducts and devices of the incinerator or ultimately becomes the inlet loading of particulate matter to the air pollution control system. Also, the fly ash includes refuse constituents that volatilize in the high-temperature zones of the furnace and subsequently condense on particulate (often the small diameter particles that present a large surface area). These may include heavy metals and high molecular weight hydrocarbons with a significant health effect.

The presence of ash imposes several technical and economic stresses on the incineration operation and the incineration business:

- Since ash is a solid and cannot simply be drained from the incineration system, costly high-maintenance devices are needed to remove the solids from the combustor and to handle the ash stream.
- Ash (especially the smaller particles in the fly ash) is a concentrate of toxic elements such as lead, nickel, and mercury as well as elements that are both carcinogenic and toxic, such as cadmium, hexavalent chromium, and arsenic.
- Ash constitutes a waste stream of the incinerator, and a place must be found to get rid of it. This generates a continuing operating cost for both transport to its disposal site and for the disposal itself. Potentially, landfill disposal leaves the incinerator firm with a liability for groundwater contamination and other adverse short- and long-term consequences of residue disposal.
- Ash hazards (real or imagined) have emerged in many countries as a significant concern among the public and the regulatory agencies. These concerns can be addressed, but they can be an impediment to project implementation.
- Ash (especially bottom ash) is not an especially desirable material. It is quite variable in its properties including both large clinkers and fine dusts, it may include both massive and wire-form metals and ceramic and stony materials, and it exhibits a variety of colors, mechanical

strengths, and other physical and chemical properties. Other than by the extraction of ferrous metal (easy to accomplish with a simple magnetic separator), processing the residue to adjust its properties to meet the demands of the marketplace can be quite costly in comparison to the modest revenue stream that can be expected.

All of these factors can be important in making the environmental assessments, developing the operating strategy, and carrying out the economic analysis concerned with municipal solid waste incineration. Understanding the practical technical alternatives, the legitimate environmental concerns and the realistic economic factors associated with ash management is an important element of design, operations, and business planning for refuse incineration.

The residue from the mass burn incineration of municipal refuse contains partially oxidized metal, glass and ceramics, inert mineral matter, and some unburned and partially burned organic material. Ideally, the combustible fraction is small. Practically, however, wet refuse components, incompletely burned heavy wood pieces, unbroken compacted masses, and the like will be found in the residue conveyor. The residue is most commonly trucked to a landfill for disposal. The weight of residue usually comprises about 20% of the refuse fired (dry basis), but its volume (assuming reasonably efficient burning) is less than 10% of the original refuse volume. The requirements for landfills receiving incinerator residue do not differ greatly from those for raw refuse, although gas formation is quite limited. The leachate characteristics are predominantly related to the metal and salt constant rather than to the organic matter. This latter characteristic is important since without the pH drop due to the organic acids formed during biodegradation of organic matter in the fill, the solution rate of the metals in the residue is very slow.

Incinerator residue may be readily processed by use of, say, a 2-cm-opening trommel screen and magnetic separator to produce three products: (1) ferrous metal, (2) mixed glass and (primarily) inorganic fines, and (3) oversize.

- The ferrous metal product, though inferior to the ferrous metal recovered from raw refuse, can be sold on the scrap metal market. The primary reasons for the lower quality of this material include its partially oxidized state, the mechanical trapping of impurities by the collapse of cans and other light gauge metal in processing, and the alloying of copper, tin, and other metals that occurs in the incineration environment.
- The mixed glass and fines product is useful as a fill and has been used in Baltimore (with further size separation) as an aggregate for road topping ("Glasphalt").
- The oversize fraction, importantly containing the unburned combustible, can be reburned or landfilled.

### 3.8.3.1. Ash Conveyance and Discharge

Except for the case where shredded solid waste is burned in suspension, most solid incineration systems require means to convey ash through the furnace and to discharge it from the system when it is burned out. In mass burn systems, conveyance through the system is accomplished using moving grates. The normal gravity flow effects movement of both waste and ash through kilns. In the smaller modular systems (described below), a ram pushes waste into the system and, thereby, displaces ash. In a few smaller batch-type systems, the residue accumulates in the furnace until it exceeds some set volume limit or until the feed is exhausted. Then, after cooling, the residue is shoveled out.

Continuous removal of ash creates a problem if, even intermittently, a relatively large gate, "bomb bay," or other opening is made through which the residue dumps out. In such circumstances, great volumes of cold ambient air may be pulled into the system, thus greatly disturbing the balance of pressure in the furnace, often temporarily cooling and quenching combustion and overloading the air pollution control device. In modern systems, these instances, however short, are unacceptable.

For granular residues, double flapper valves activated by gravity or (preferably) by sequenced pneumatic cylinders will allow solids to escape without disturbing the furnace. For larger residue objects (continuously or on occasion), the residue chute can be fitted with a normally open gate and mated to a removable container. The gate is closed to change containers, and careful attention must be given to the design and to the operating practice to avoid overheating, overloading, or jamming the gate.

More commonly, residue is dumped into a water-filled quench tank fitted with a drag chain conveyor or pusher ram (see Figure 10.10, item 4). The water both stops combustion and forms an air seal. In systems with the pusher ram, the ram, periodically actuated, pushes the residue up a ramp, which allows for some drain-back of water. The residue then falls onto a vibrating or belt conveyor. In most plant designs, a single residue conveyor is provided, extending along a bank of several incinerator furnaces. Often, provision is made to insert a chute between the ram discharge and the conveyor to allow the ash to bypass the conveyor and dump on the floor. Ash is then picked up and hauled off using a front-end loader and a small dumpster container or truck. This bypass chute is used to deal with emergencies or for conveyor maintenance.

Since residue removal is critical to continuous operation, drag chain systems, which are susceptible to jamming and other outages, are usually fitted with twin tanks, each with its own conveyor. A deflection plate is used to direct the residue into one or the other quench tank.

### 3.8.3.2. Ducts, Boilers, and Breeching

Generally, solids incinerator flue gases contain a high concentration of fly ash. To avoid buildup in ducts, boilers, and breeching, it is appropriate to install hopper bottoms for ash withdrawal. Long horizontal runs should be avoided unless frequent downtime for clean-out is anticipated. Even then, consideration should be given to providing clean-out doors and other means to remove the accumulated solids.

### 3.8.3.3. Residue Properties

The residue must be regarded as a source of secondary pollution, or, if nothing else, as a new waste requiring disposal. For many wastes, the residue is enriched in heavy metals (especially fly-ash solids), which may be of significance in establishing acceptable means of ultimate disposal. In recent years, concern has been expressed by environmental and regulatory groups regarding the hazards of ash handling and disposal.

Handling, storage, and transportation practices, as they might affect worker health and safety, and ultimate (landfill) disposal of the ash, as it might impact on groundwater, have been under scrutiny. In some states, regulations have been promulgated to require ash testing, ash planning, and other steps to address concerns that have focused on heavy metals and dioxins.

Residue data are available in the literature (1, 51) and in the appendix.

### 3.8.4. Pyrolysis and other Gasification Systems
#### 3.8.4.1. EARLY PROCESS DEVELOPMENTS

Earlier in this chapter, the basic concept of pyrolysis was described. This subsection describes the hardware concepts that have been employed to carry out refuse pyrolysis on a commercial scale. In each method to process refuse by pyrolysis, the objective was the same, that is, to convert a heterogeneous, hard-to-handle material into gaseous and liquid fuel products suitable for firing in conventional combustors. Affecting such a fuel conversion step resolves in part a critical step in exploiting refuse energy content to reduce solid waste disposal costs, and thus lead to the marketing of the energy resource.

It may be significant to observe that all of the systems were studied by relatively large, sophisticated firms, each with extensive experience in design and operation of combustion systems and associated air pollution control equipment. Nonetheless, all of the processes, after the investment of years of effort and many millions of dollars in both equipment and manpower, have been shut down. Commercial as well as process failures drove the decisions to cease development, and the failures reflect the especially difficult challenges of handling and processing solid waste while achieving the superior environmental performance demanded of facilities that operate in, near, and for communities.

The problem facing the designer of a system operating in the pyrolysis mode is to find a way to apply heat to refuse sufficient to raise its temperature 4300°C or more without adding oxygen, which will burn the pyrolysis products. This has been accomplished in three ways: (1) indirect heating, (2) zoned partial combustion, and (3) fluid-bed flash pyrolysis. Each of these approaches was studied in the 1960 to 1970 time period, the halcyon years of waste processing process development.

*3.8.4.1.1. Indirect Heating*   Using this approach, refuse is placed inside a metal cylinder that is heated on the outside surface by burning the pyrolysis gases. Such a processing method has been used by Pan American Resources of Albuquerque, New Mexico. Their 40-ton/d unit (the Lantz Converter) uses a rotating kiln design with an airlock seal for the feed. Pyrolysis temperatures range between 480° and 820°C depending on the refuse characteristics. Test work, initiated in 1968 at a Ford Motor Company assembly plant in San Jose, California, was not promising and the operation was discontinued. The Destrugas system, developed in Copenhagen, Denmark, is also reported to use a retort-type, vertically oriented cylindrical system.

*3.8.4.1.2. Zoned Partial Combustion*   In general, pyrolysis systems seek to produce a gaseous product suitable for transmission over some (limited) distance and useful as a fuel in existing boilers or gas turbine combustors. Thus, the char fraction represents a waste product. The zoned partial combustion approach makes use of the heat of combustion of the char to produce a hot gas that, when passed over or through the incoming refuse, produces the thermal environment necessary for pyrolysis. Careful control of the addition of oxidizing gas (air or pure oxygen) limits the combustion to the char fraction, thus maximizing the product gas yield and heating value (kcal/m$^3$).

The Monsanto Landgard© system, constructed during the 1970s in Baltimore, Maryland, with a design capacity of 910 ton/d, was an embodiment of zoned partial combustion using a rotary kiln as the pyrolysis vessel. Fuel oil was burned in the discharge zone to add additional pyrolysis heat. Excess air (relative to the fuel oil) acted to burn the char. The hot, oxygen-deficient combustion products then passed up the kiln, effecting pyrolysis and feed drying. The Monsanto system encountered severe problems in control and with air pollution. The latter may have been associated with an inadequate design of the scrubber system, although it was suggested that volatilization of salts in the extremely hot char-burning zone may have produced such an abundance of submicron particulate that almost any air pollution control device would have been overwhelmed. In any event, limited combustion capacity, air problems, and other operating problems ultimately led to the shutdown of the operation.

A second type of zoned partial combustion system uses a vertical shaft furnace configuration. Combustion occurs in the lower region of the furnace and the hot gases pass upward through the incoming feed plug. The Union Carbide Purox© system effects combustion with pure oxygen, thus avoiding dilution of the product gas with the nitrogen associated with air-derived oxygen. The Andco Torrax© system uses air for combustion, but preheats the air using a portion of the product gas. These systems were demonstrated at the 100- to 180-ton/d scale, but continuing problems with sidestream treatment (the tarry water condensed from the product gas), maintenance, on-line availability, and capacity led to shutdown of the operations.

*3.8.4.1.3. Flash Pyrolysis*   The third approach to commercialized pyrolysis processes refuse that has been thoroughly prepared (coarse shredding, drying, air classification, and final shredding of the combustibles to a very fine size), in a fluid bed. Fluidizing gases, preheated by burning the pyrolysis char, rapidly heat the incoming refuse. The rapid heating rate maximizes the yield of liquid products. A 180-ton/d San Diego, California, demonstration plant was tested in the 1970s by its developer, the Garrett Division of Occidental Petroleum. Operating problems and fuel preparation problems ultimately required shutdown of the facility.

3.8.4.2. GASIFICATION

The strong concern of many environmental action groups and of the regulatory community has put the spotlight on air emissions from municipal waste combustors (MWCs) fostering vigorous opposition to proposed plants and to the adoption of increasingly stringent emission requirements. One response in this adversarial environment has been to upgrade the processes and enhance the control technology for "conventional" mass burn and RDF-type MWC systems. An alternative approach is to develop altogether new thermal processing technologies that are inherently low in emissions yet still have the target of accepting the wide range of feedstocks comprising MSW. One such class of new, environmentally benign technologies are those based on the gasification of refuse coupled with intensive cleanup of the product gas prior to its use as a fuel.

In its simplest embodiment, gasification occurs by simple heating of organic material to temperatures of the order of 400° to 600°C, whereupon complex molecules break into shorter chain species that are gases under the extant conditions. This is pyrolysis-type gasification as described above. A second strategy used to effect gasification exposes the organic material to hot steam that heats the material, inducing pyrolysis, and also acts as an oxidant that, via the water gas and water gas shift reactions, yields a gaseous fuel product with minimum char residual and without addition of nitrogen. The key reactions include the following:

$$\text{Water gas reaction:} \quad C + H_2O \leftrightarrow CO + H_2 \quad (35)$$

$$\text{Water gas shift:} \quad CO + H_2O \leftrightarrow CO_2 + H_2 \quad (36)$$

Gasification processes applied to MSW are often (but not exclusively) fed waste previously subjected to recycling and refined to an RDF. The organic fraction is heated with limited or no air to yield a gaseous product stream with substantial heat content. This intermediate product gas can then be cleaned of metals and other particulate matter and of HCl, HF, ammonia, hydrogen sulfide, and other gaseous contaminants. The gas can then be used as a chemical feedstock or can be burned in a gas engine or gas turbine to generate electricity. The gas cleanup effort need only deal with the relatively small quantity of product gas stream from the gasification reactor. This contrasts with the scale of cleanup for conventional incinerator flue gases that have been greatly expanded in volume by the addition of almost twice the theoretical quantity of combustion air and associated high dilution with nitrogen and excess oxygen. As a consequence, the equipment and operating cost for environmental emissions control applied to gasification systems can be substantially lower.

Further, the ultimate combustion process and energy conversion takes place with relatively high-quality, clean fuels comprised, importantly, of CO and $H_2$ rather than a flue gas derived from MSW and thus bearing corrosive particulate, acid gases, and moisture. Combustion of a "clean gas" allows use of high-efficiency technology for energy conversion (high kWh/BTU) and produces very low emission rates of dioxins, acid gases, and other problematic pollutants.

Gasification is an old technology, with roots in charcoal making, in the reduction of iron and other metal ores, in the manufacture of city gas, and the like. MSW gasification is a more recent application of the technology and, early in the new millennium, is still in the advanced developmental stage with only a few plants, worldwide, based on this technology that have continuous operating experience under commercial conditions. Thus, it is reasonable to expect that process or equipment deficiencies or difficulties will appear in the technologies still under development.

A study of emerging gasification technologies conducted in 1996 identified three that appeared near to commercialization: TPS Termiska AB, Nykoping, Sweden, a process developed by Battelle Laboratories, Columbo, OH, and a process by Thermoselect Inc. Malvern, PA (52). All involve gasification of MSW (Termiska and Battelle with an RDF feed stream and Thermoselect with raw, unprocessed MSW) followed by fuel gas cleanup and gas combustion for the generation of electricity. The descriptions of the processes given below are intended to highlight the technical features and characteristics of these three examples of MSW gasification.

*3.8.4.2.1. General* The handling of MSW is one of the major challenges to successful implementation of any MSW system. Since heat transfer to the feed waste is a basic step in gasification, almost all gasification technologies require processing of raw MSW to an RDF. The more uniform and highly subdivided character of RDF increases the rate of gasification processes and assists in achieving high conversion efficiency.

Two of the three processes discussed use fluid bed technology. A classical, bubbling fluid bed combustor involves a cylindrical or rectangular chamber containing coarse sand or similar bed material through which gas is passed at a rate that causes the sand bed to expand and move in a turbulent, roiling motion. If the gas velocity is further increased, a fraction of the particles will be blown out of the bed. One can then interpose a medium- to low-efficiency particulate collector such that the larger particles will be captured and returned to the bed for continued processing. Often, an array of vertically oriented steel channels is mounted in the duct leaving the freeboard. These channels (known as "U-beams") intercept the coarser solids. The U-beams are followed by a somewhat more efficient particulate capture system such as a cyclone that collects 100% of particles with greater than a 30-μm aerodynamic diameter. This embodiment of the suspended combustion concept is called a circulating fluid bed.

The advantages of the fluid bed environment include the uniformity of chemical and temperature environment brought about by the mixing effects and thermal inertia of the dynamic motion of the sand and the effectiveness of the circulating sand both to carry heat to incoming material to dry and to abrade feed particles. The abrasion removes the ash layer that protects the unreacted core material, and reduces the particle size to facilitate combustion.

Product gas can be characterized by its relative heat content class:

| | | |
|---|---|---|
| Low heat content gas: | 950 to 2800 kcal/Nm$^3$ | 100 to 300 BTU/sft$^3$ |
| Medium heat content gas: | 2800 to 5700 kcal/Nm$^3$ | 300 to 600 BTU/sft$^3$ |
| High heat content gas: | >5700 kcal/Nm$^3$ | >600 BTU/sft$^3$ |

The basic objective of gasification-based processes is to convert a heterogeneous solid fuel with handling and pollutant-emissions problems into a combustible gas containing the maximum remaining heating value. In a way, this is an extension of the RDF preparation to produce a high form value—gaseous fuel. In many cases, the combustible gas is burned in a gas engine or turbine combustor to generate electricity. Where warranted, heat recovery from the exhaust of the engine or turbine can be passed to a boiler to produce steam, and the steam, in turn, is converted to a second quantity of electricity using a conventional steam turbine/generator.

*3.8.4.2.2. TPS Termiska Processor–Gasification by Partial Combustion* The TPS Termiska Processor (Thermal Processes) is a Swedish firm. The TPS gasifier is composed of a bubbling fluid bed into which RDF or RDF pellets are fed. Secondary air addition partway up the furnace transforms the bed aerodynamic balance such that smaller, lighter particles are blown from the circulating bed. Heavy, still-burning "chunks" remain in the dense bubbling fluid bed until they are consumed. Ground dolomitic lime is added in a second bed to catalyze the breakdown of high molecular weight hydrocarbons into lighter products. The product gas may be cleaned to generate a fuel gas suitable for use in a gas engine or turbine or can be burned directly in a boiler or process furnace. The flowsheet is shown in Figure 10.13.

Extensive development and testing work using MSW and other biomass feed stocks were carried out in the late 1990s. Semi-works scale facilities using the TPS technology were tested using wood biomass in the early years of the new millennium.

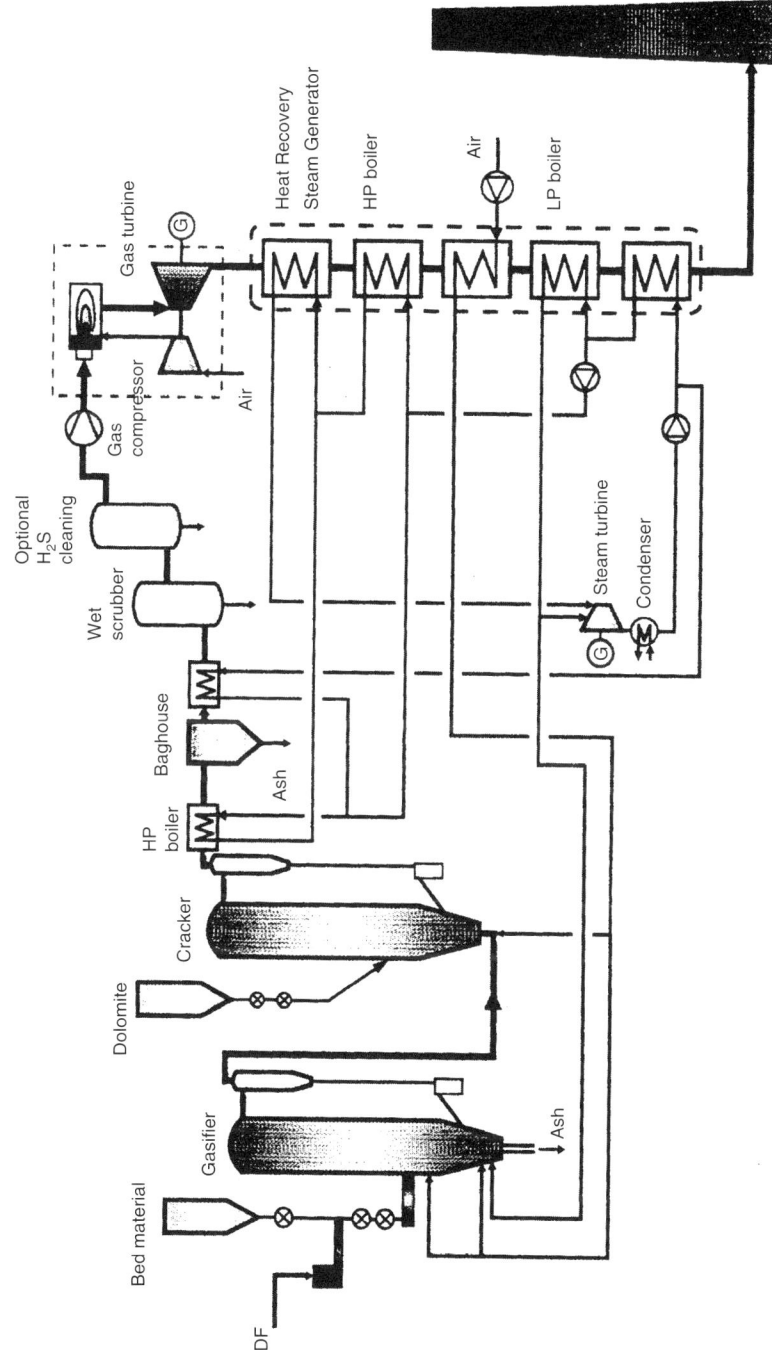

**Fig. 10.13.** Termiska processor refuse gasification system flowsheet.

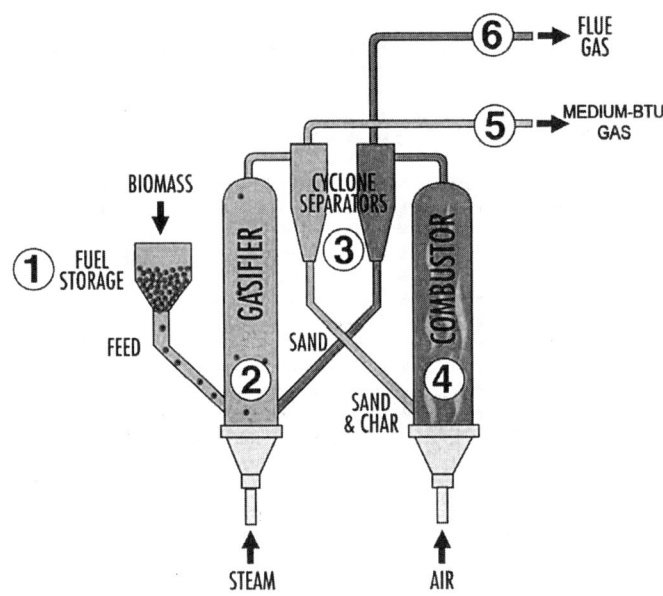

**Fig. 10.14.** SilvaGas© refuse gasification system flowsheet.

*3.8.4.2.3. SilvaGas Process—Gasification by Pyrolysis and Steam Reforming* The SilvaGas Process was originally developed by Battelle Memorial Institute of Columbus, Ohio, as the Battelle High Throughput Gasification System (BHTGS). The process uses a pair of interconnected, circulating fluidized bed (CFB) reactors. In a CFB, relatively small particles of solid material are introduced into a fast-moving gas stream. After a few seconds of gas-solid contact time (during which heating, reacting, or oxidation processes occur), the solids are carried out of the chamber and are captured using a simple cyclone device. In the first high-throughput CFB of the BHTGS, refuse-derived fuel (RDF) or other biomass feedstock is gasified to a medium-heating-value gas using steam (without oxygen) as the fluidizing medium and hot sand as the heat source. Residual char and sand collected in the cyclone separator that follows the CFB is burned with air in the second CFB. The sand (now heated by the combustion process) is collected following the second bed and is then circulated back to the first bed as its source of heat. The gasified RDF fuel gas is cleaned and passed to a combustor (a boiler, or a gas engine or gas turbine) from which electrical energy or shaft power can be extracted. The char combustor off-gas is scrubbed and discharged after energy recovery. A schematic of the process is shown in Figure 10.14.

Battelle's development efforts began with laboratory studies and theoretical analysis in 1977. More comprehensive process development activities were begun in 1980 with the construction of 6- and 10-in diameter process research units (PRUs). The PRU investigations continued through the mid-1980s. The PRU tests logged over 22,000 operating hours on a variety of feedstocks. The tests demonstrated the technical feasibility of the gasification process and provided the basis for generating a detailed process conceptual design. Based on this design, capital and operating costs estimates for commercial plants were also prepared. Testing of a highly prepared RDF was conducted in 1989 in a

10-inch inside diameter, 22.7-foot-high gasifier associated with a 40-inch, 11.5-foot-high combustor.

Battelle subsequently licensed its BHTGS for the North American market to Future Energy Resources Corporation (FERCO) in Atlanta, Georgia. The license is presently owned by Silva Gas Corporation, also located in Atlanta, GA. A 25-MW commercial-scale demonstration was carried out at Burlington Electric's McNeil Generating Station in Burlington, Vermont, using whole tree wood chips as the energy source. The Vermont Gasification Project (VGP), capable of processing 500 tons per day of wood chips, demonstrated the commercial scale viability of the technology. The product gas was burned in the existing McNeil boilers. Based on the success of the VGP plant, SilvaGas is currently focused on project development efforts of SilvaGas plant sizes equivalent to 10 to 50 MW in size (200 to 1000 tons per day of feed material). The U.S. Department of Energy and private investors have provided over $64 million in funding to develop the SilvaGas process.

The various stages of the process are as follows:

- RDF preparation: The raw waste is received and stored using conventional floor dump and front-end loader means. In larger plants, pit and crane technology could be used. Waste is then processed using horizontal shaft hammer mills, air separators, and similar conventional RDF processing equipment to produce a 2-inch top-size RDF feed material. The RDF product is then stored in a live-bottom bin that continually recirculates the RDF to avoid felting entanglements that often result in an irregular feed rate.

    Typically raw MSW contains many materials such as the residual glass and metals remaining in the waste that do not provide any fuel value for energy recovery. To prepare the MSW for use as a fuel for the SilvaGas process, the glass and metal materials must be removed. After the glass and metals are removed from MSW, the remaining material should meet at least the American Society for Testing Materials (ASTM) classification of RDF-3: shredded fuel derived from MSW processed for the removal of metal, glass, and other entrained inorganic material; particle size of this material is such that 95% by weight passes through a 2-inch-square mesh screen. The resulting fuel to the gasifier is assumed to be 480 tons per day (TPD) on a dry basis (654 TPD wet). Interim storage of the RDF is provided by a live bottom bin or similar device (Figure 10.14, label 1) to act as an accumulator.
- Gasifier CFB: The SilvaGas system employs a CFB gasifier to provide high throughputs of biomass material. Superheated steam is used as the fluidizing medium to react with the incoming RDF organic matter. On entering the CFB, the RDF first dries and then is rapidly heated to facilitate reactions between the steam and the RDF organic material to form the product fuel gas (Figure 10.14, label 5). Heat for the endothermic drying, heating and pyrolysis steps is supplied to the reactor by introducing sand that has been heated to 1000°C (1800°F). Fluidizing gases enter the gasifier at a level below the RDF feed entry port and an L-valve sand recycle entry.

    The product gas and the sand and char from incompletely consumed RDF organic matter exit at the top of the gasifier at 1500°F and enter gasifier cyclone separator systems (Figure 10.14, label 3). The primary cyclone separates the sand and char from the product gas. The product gas, with a small amount of sand, char and ash is then directed to the secondary cyclone separator to complete the gas cleaning by mechanical dust collection.
- Char combustor CFB: Heat necessary for the gasification reactions is provided by using a stream of circulating sand which passes between the gasifier CFB (Figure 10.14, label 2) and a high throughput CFB combustion reactor (Figure 10.14, label 4). A small amount of char is produced as a result of the gasification reactions (typically 20% of the feed material).

The heat of combustion of this char, released in the combustion reactor, provides the energy to reheat the circulating sand.

The combustor, a bubbling fluidized bed with a refractory lining, is designed to minimize heat losses. Sand enters the combustor through a closed chute line from the gasifier cyclone. This line enters through the top of the combustor and extends downward into the fluidized bed. The sand bed is returned to the gasifier from the combustor by an L-valve. The L-valve provides the necessary seal between the combustor and gasifier environments.

- Gas cleanup and cooling

    Product gas: The SilvaGas process provides a cooled, clean, 4000 to 5000 kcal/m$^3$ (500- to 600-BTU/sft$^3$ (standard cubic feet)) product gas. The product gas is a mixture of CO and $CO_2$, $H_2$, low molecular weight hydrocarbons, and water vapor plus some higher molecular weight "tars" and some uncollected sand and other inorganic particles and vapors. Because the gas is reducing in nature, sulfur compounds in the waste appear as $H_2S$ rather than $SO_2$. Organic halogens in the waste appear as hydrochloric acid (HCl) and hydrofluoric acid (HF). All of the contaminate materials must be reduced in concentration to meet the environmental requirements of the fuel gas user. These may vary widely, and therefore the technology for gas cleanup is necessarily tailored to the energy market technology.

    In general, however, cooling is the first step in gas treatment. Cooling from the CFB discharge temperature of 815°C (1500°F) to approximately 315°C (600°F, 60°F, 1 atm.) provides energy to preheat the combustion air or to warm air with which to pre-dry the RDF stream (dry RDF yields a more efficient conversion of the chemical energy in the waste into product gas fuel value).

    The second step in product gas cleanup involves a wet scrubber. The condensed, organic phase scrubbed from the product gas is separated from the water, in which it is insoluble, and injected into the combustor CFB. Also, the wet scrubber would remove much of the particulate matter and, if maintained alkaline, would collect the acid gases $H_2S$, HCl, and HF. The final gas treatment train, however, has not been settled pending discussions with the ultimate gas user.

    Combustor gas: Exhaust gases from the combustor at 1000°C (1800°F) pass through a cyclone separator, which discharges the hot, coarse, separated sand particles directly back into the gasification fluidized bed. The flue gases then are further cleaned and cooled by a waste-heat recovery system before being exhausted to the atmosphere. Waste heat recovery can include the production of steam for the gasification CFB, electrical generation, or turbine drives on the fluidized beds or for RDF drying.

    In pilot tests with RDF, the SilvaGas process tests produced gaseous emissions from the char combustor that were in compliance with US EPA's New Source Performance Standards (NSPS) for MWCs. Wastewater from the process contained only trace quantities of organic materials. At Battelle's test site, the outlet of a simple, industrial treatment system—a sand filter followed by a simple charcoal filter—showed wastewater to be within US EPA's drinking water standards.

    Chlorine was not measured during the RDF testing. However, subsequent proprietary Battelle data indicate that chlorine in the waste stream is converted completely to HCl in the gasifier and not to chlorinated organic materials such as dioxins and furans. There is a small concentration of HCl present in the gas, most of which is removed by the wet scrubber.

*3.8.4.2.4. Thermoselect®: Gasification of Raw MSW by Pyrolysis* The Thermoselect process embodies a fully developed method of gasifying MSW without apparent adverse impact on the environment. The residues of the gasification process and the gas cleanup system can be converted into potentially commercially useful by-products such that there is no net waste to landfill. A standard design has been developed around a nominal 400 t/d furnace. Larger facility capacities are offered by adding multiples of the standard

**Table 10.18**
**Typical syngas composition**

| Average syngas composition | Volume percent |
|---|---|
| Carbon monoxide (CO) | 34–39 |
| Hydrogen ($H_2$) | 32–35 |
| Carbon dioxide ($CO_2$) | 22–27 |
| Nitrogen ($N_2$) | 3–4 |
| Methane ($CH_4$) | <0.1 |
| Other | <0.6 |

**Table 10.19**
**Installations of thermoselect process units**

| Location | Start year | Capacity mg/day* | Capacity tons/day* |
|---|---|---|---|
| Fondotoce, Italy | 1992 | 1 × 95 | 1 × 105 |
| Chiba, Japan | 1999 | 2 × 150 | 2 × 165 |
| Karlsruhe, Germany | 2000 | 3 × 240 | 3 × 264 |
| Tokushiki Yoshino, Japan | 2003 | 2 × 60 | 2 × 66 |
| Matsu (Shimokita Area) Japan | 2003 | 2 × 70 | 2 × 77 |
| Sainokuni City, Japan | 2004 | 2 × 200 | 2 × 220 |
| Kyokuto, Japan | 2005 | 1 × 95 | 1 × 105 |
| Kurashiki City, Japan | 2005 | 3 × 185 | 3 × 203 |
| Isahaya, Japan | 2005 | 3 × 100 | 3 × 110 |

* Number of processing modules × capacity/furnace, e.g., 2 × 200 means two furnaces, each of 200 mg/day.

modules. Thermoselect technology is made available in the U.S. through Interstate Waste Technologies of Malvern, Pennsylvania.

The Thermoselect system processes commingled MSW and clean commercial and industrial waste and converts them into what are stated to be environmentally safe and marketable products including a cleaned synthetic gas (syngas), vitrified solid granulates, metal pellets, a zinc salt concentrate, and elemental sulfur. If a full "zero discharge" concept is desired, no net liquid effluents need be discharged into the environment since all process water can be treated and recycled. In addition, the process is intended to minimize the formation or emissions of particulates, nitrogen oxides, and other pollutants.

When a waste feed containing 50% organic matter, 25% organic, and 25% water at 4472 BTU/lb is processed, a syngas of about 224 BTU/$ft^3$ and having the average composition shown in Table 10.18 is produced.

Gasification is achieved at a high temperature. The products of gasification are then held at high temperatures for more than 4 seconds—a relatively long residence time. Data indicate that this combination of time and temperature destroys the complex organic compounds produced in the gasification process and yields a product gas that, substantially, has reached chemical equilibrium. The raw gas is cleaned in an air pollution control/gas purification system, which removes acid gases, hydrogen sulfide, particulates, and volatile heavy metals. Air emissions result only from the combustion of

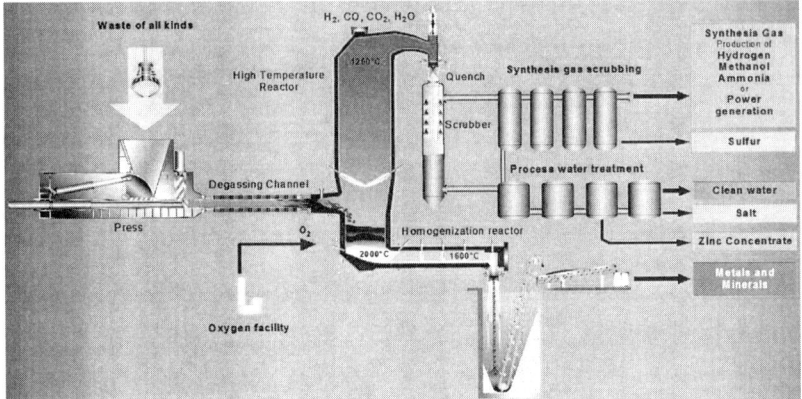

**Fig. 10.15.** Thermoselect© refuse gasification system flowsheet.

the cleaned syngas in the production of steam in boilers or other means for the generation of electric power.

The original Thermoselect development and demonstration facility was located at Fondotoce, Italy. The equipment consisted of one process line with a nominal capacity of 4.4 t/h, housed in a low-level building with two relatively short stacks. Normal operation was in progress during an inspection visit by Camp Dresser and M-Kee Inc. (CDM) staff in 1966 and included the delivery of municipal and bagged industrial wastes by truck. No odors or noise was observed either inside or outside the plant. Similar operation was noted during a subsequent inspection by CDM of the Karlsruhe, Germany, facility in 2001.

Commercial plants constructed and operating elsewhere are listed in Table 10.19. The third and fourth columns show the number of processing modules and their capacities. Of the nine plants shown, the Karlsruhe and the Fondotoce plants have been shut down. Recently, a project has been proposed for Puerto Rico. Figure 10.15 is a schematic of the Thermoselect gasification system. The various stages of this process are as follows:

- Compaction: The raw waste is dropped by grapple from the waste pit into the housing of a compactor that compresses commingled waste to less than 10% of its original volume, thereby removing the trapped air in the original loose material. As the process calls for additional feed, a gate opens and the compactor pushes the plug of waste though an unheated transition section into the degasification channel.
- Degassing/pyrolysis of the organic fraction of the waste: The degasification channel is externally heated to a temperature of 600°C (1100°F). Volatile components in the waste vaporize as the waste moves to the next stage. The heated vapors include steam from water that is evaporated from the solid waste. As the waste plugs are pushed down the degassing channel, they receive radiant heat from the next stage. The temperature in this area approaches 800°C (1470°F). At this transition point between the degassing channel and the next stage, identified as the high temperature chamber (HTC), the waste plugs have been greatly reduced in physical size due to the loss of volatile components; the nonvolatile organic portion has been to a high degree carbonized; and the inorganic portion of the waste has remained virtually unaffected and is part of the carbon matrix.
- High-temperature gasification: In the degasification channel, carbon and water vapor react to form hydrogen and carbon monoxide. These fuel gases move with the other gases into the upper section of the HTC, which is maintained at a temperature of 1200°C (2200°F). This

high temperature, combined with a residence period approaching 4 seconds and turbulence (provided in this section through a proprietary oxygen introduction technique), destroys any remaining organic compounds. The resulting hot gases exit the HTC and are immediately water quenched in a spray chamber to below 70°C (158°F).

The char and inorganic material fill the lower section of the HTC and, at this point, gaseous oxygen is introduced at a controlled rate. This produces a temperature of 2000°C (3600°F), which melts the inorganic fractions (glass products and various metals) that flow from the lower HTC into the homogenization stage where the melt is prepared for removal from the process.

- Homogenization chamber: The metal and mineral from the HTC enter this stage by gravity. Oxygen is introduced to burn out any remaining carbon particles. To maintain the melt, gas burners may be used. The molten metal and mineral streams are quenched in a water bath. The mineral stream cools and forms a vitrified granulate, and the metal mix freezes as the flow enters the water bath to form metal alloy pellets. The mix of granulate and metal pellets are recovered using a drag chain conveyor. The vitrified mineral granulate meets the US EPA's TCLP (Toxic Characteristic Leaching Procedure) leaching standards.

The vitrified mineral product is not dusty and its metal content remains bound in a glass matrix and does not dissolve in water. This black, glass-like mineral product can be used as the following:

> Raw material components for making clinker brick
> A substitute for cement analogous to the use of anthracite fly-ash
> An additive for concrete
> A filler for bituminous mixtures
> A filler and anti-frost layer in underground engineering
> Mineral fiber and heat insulation fibers
> Decorative pavers and blocks for the building industry

The gaseous environment in the homogenization chamber reduces the metal oxides and causes typical alloy-forming metals such as nickel, chromium, and copper to pass into an iron-rich metal melt. Since this melt has a very low concentration of high vapor pressure components, such as mercury, zinc, cadmium, lead and arsenic, it can be directly used for metallurgical purposes.

The severe duty imposed on the homogenizer section results in a requirement for periodic replacement. The developer includes a spare homogenizer in the basic plant capitalization such that an exchange with a spare section can be performed with minimum line outage. The replacement period is 6 months, and Thermoselect has found that cooling and removal of the spent unit, positioning of the refreshed unit, and restart can be accomplished over a weekend.

- Gas cooling and gas separation from process water: The hot gases in the upper section of the HTC exit are rapidly water quenched to below 70°C (158°F). The syngas and sulfur compounds (present as hydrogen sulfide or $H_2S$ rather than $SO_2$ because the gases are reducing in nature) are separated from the quench water and passed through successive scrubber stages: an acid wash at 60°C (140°F), desulfurization and a base wash at 40°C (104°F); and finally a cooling state bringing the gas temperature to 4°C (40°F) to remove water vapor. The gases are passed through a coke filter and warmed to ambient temperature prior to use. If appropriate, the syngas can be compressed for distribution and regulation.

The cleaned syngas can serve as an energy source for the production of electricity using diesel engines or as a fuel to a steam boiler. The syngas can also be used as a chemical feed stock to form methanol or benzene. During emergency conditions (as when the energy market is unavailable), the gasifier unit can be shut down. In the coast-down period, the

Thermoselect facility is equipped with an emergency flare chamber to safely burn off the syngas.

The sulfur removal system converts hydrogen sulfide ($H_2S$) to sulfur using an iron complex via a well-proven, proprietary process. The resulting iron complex is regenerated (52) using oxygen from the air in an adjoining stage.

- By-products of the process: The process water solutions generated from the gas cleaning sections are subject to conventional chemical treatment. Heavy metal hydroxides are removed as a solid concentrate. Elemental sulfur and a concentrate containing heavy metal hydroxides are collected in addition to the syngas, vitrified mineral product, and metal alloy pellets.

The metal and vitrified mineral granulates collected from the homogenization chamber can be density separated when in molten form, but are more easily handled in granulate and metal pellet form, which can then be magnetically separated into vitrified mineral product and metal alloy pellets. The metal pellets consist of iron (>90%), with considerable amounts of copper (3% to 5%), nickel (0.6%), chromium (0.3%), tin (0.4%), and phosphorus (0.1%). Concentrations of heavy metals that find their way into these by-products are at acceptable levels.

The air pollution emissions of the Thermoselect process are uniquely low and essentially are those of a gas turbine burning the CO and hydrogen fuel gas. Thermoselect's goal is for essentially no solid effluents. Thus, the vitreous material is to be sold for use as a fill or concrete aggregate, the metal pellets are to be sold to the scrap metal markets, and the zinc hydroxide concentrate produced in treating the wastewater from the scrubber is to be sold to the zinc recycle market.

### *3.8.5. Modular Incineration Systems for Municipal and Commercial Wastes* (<100 ton/d)

In the 1960s, a number of manufacturers, recognizing the increasing emphasis on smoke abatement, began producing incinerators that limited (or starved) the combustion air supply. This design strategy kept flue gas temperatures high and minimized the fuel consumption of afterburner devices. In such units, refuse is charged in batches, and after 10 to 12 hours of charging, the unit is allowed to burn down. Residue is raked out after the burn-down (8 to 12 hours) and the process is repeated. These incinerators are known as modular combustion units (MCUs).

The starved air units consist of a cylindrical or elliptical cross-section primary chamber incorporating underfire air slots in the hearth-type floor. Overfire air is also supplied. The quantity of underfire air is limited to avoid quenching combustion and to minimize the elutriation of particulate matter. The reduced air supply to the primary has a strongly beneficial effect on the entrainment of particulate matter. The slight adverse effect on residue burnout (more carbon char is developed than under fully oxidizing condition) can be resolved in systems where a prolonged burnout period (say, the second and third shifts) is available and, in any event, is not a serious drawback when compared to the significant reduction in particulate emission.

All combustion air is provided by forced draft fans. In many such incinerators, the proportion of underfire to overfire air is regulated from a temperature sensor (thermocouple) in the exhaust flue (higher exit temperatures increase overfire air and decrease underfire air). Most units incorporate a gas- or oil-fired afterburner that is energized whenever the exit gas temperature falls below the set point temperature (usually 650° to

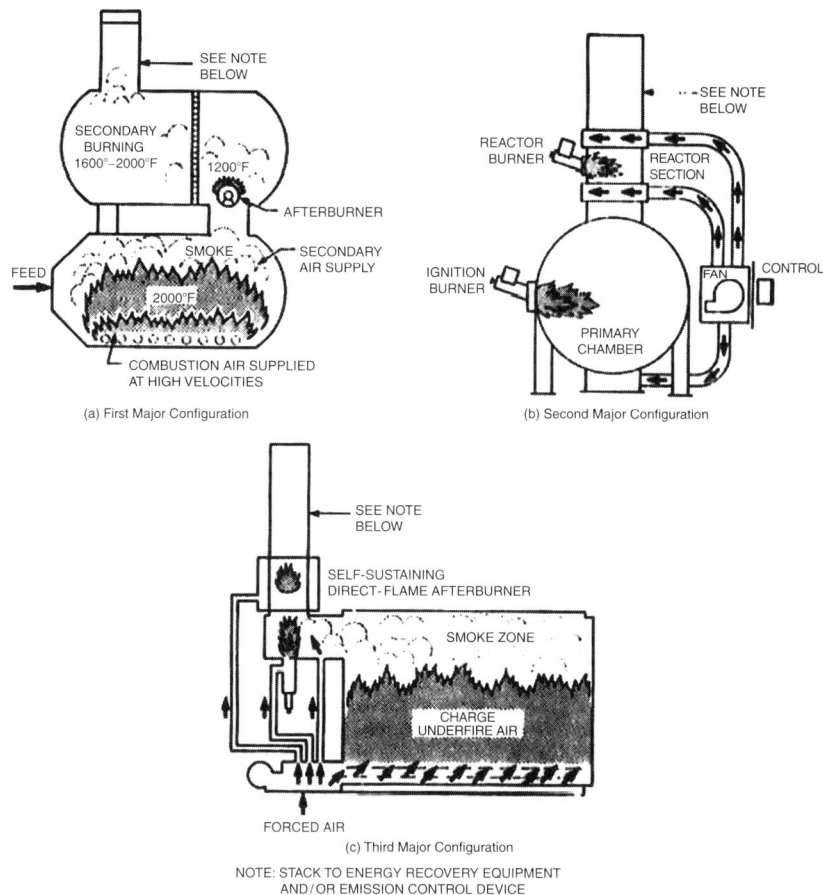

**Fig. 10.16.** Modular combustion unit (MCU) designs.

750°C). The afterburner is mounted either in a separate, secondary combustion chamber or in a refractory-lined stack. Typical configurations are shown in Figure 10.16.

The MCU incineration systems for solid waste and, in some cases, sludge, liquid, and gaseous wastes are a still evolving product. Many of the major manufacturers of these units entered the market in the 1940s and 1950s when the emergence of the shopping center and the burgeoning growth of packaging stimulated sales of relatively small (45 to 225 kg/h) and simply designed volume reduction aids.

Starved air incinerators are available in capacities from 90 to 900 kg/h (200 to 2000 lb/h) and, for the smaller units and if properly operated and maintained, can meet many federal, state, and local particulate air pollution codes when fed with typical office and plant trash (principally paper, cardboard, and wood). For the larger, municipal scale units, current and more restrictive codes for acid gases and mercury require addition of the conventional dry scrubber and fabric filter system with carbon injection. The modular combustion units can be equipped with a boiler (usually of the fire-tube type) for the recovery of heat. The larger units can also be equipped with automatic feed and residue removal systems thus increasing the daily throughput and somewhat stabilizing the steam generation rate.

Because of the limited air supply to the primary furnace (about 80% of theoretical), the combustion gases leaving the primary furnace are incompletely burned. The gases leaving the primary contain soot and carbon monoxide, hydrogen, and a mixture of hydrocarbons ranging from methane to tarry aerosols (smoke). Addition of air at approximately 50% excess overall and the use of a small (say, 65,000 kcal/h) burner in the secondary chamber allows combustion to be completed. Because of the operation at low excess air, a higher waste moisture content can be tolerated and energy recovery efficiency is enhanced. The secondary combustion chamber is vital to successful operation and, depending on the throughput of the unit, takes on many forms.

For industrial wastes, vendors of MCUs have developed numerous useful auxiliary equipment including ram loaders, drag chain conveyors for continuous ash removal, and burners suitable for readily pumpable liquid wastes. Generally, the vendor catalogues indicate no change in the hardware configuration between systems to burn the same feed rate of municipal waste (at 2500 kcal/kg) and industrial waste (at, say, 3900 kcal/kg). This inconsistency with normal combustion system experience (where heat release per unit area or volume are key design parameters) suggests that some caution should be exercised when selecting such units. Of particular importance is ensuring that the secondary chamber has sufficient volume to provide adequate retention time at elevated temperatures. This is important for industrial wastes that contain a high concentration of soot-forming materials such as styrene, rubber, or polyethylene. Where no air pollution control is provided, complete burnout of soot and smoke is critical.

The MCU systems have or are being constructed in capacities up to 100 tons per day (24-hour continuous operation). Batteries of the larger units can be installed to serve smaller communities or regions. The facility design is usually simple with a floor dump and refuse storage area serving two or more incinerators. A typical facility cross-section is shown in Figure 10.17. The energy market for these smaller facilities is often steam-oriented rather than electricity.

Field tests by Camp Dresser and McKee, Inc. in a 25 ton per day continuous ash removal MCU showed that the mean residence time of the ash residue in such units is approximately 10 hours. Because of the long residence time, slow drying or slow burning wastes can be incinerated in such units, limited primarily by the overall system heat balance. Such wastes include dewatered sewage sludge and thick-section plastic.

## 4. THERMAL PROCESSING SYSTEMS FOR BIOSOLIDS

### 4.1. Introduction

It is an interesting observation that the names of many of the foulest and most detested things, the adjectives that evoke the greatest disgust, the verbs that describe the vilest actions, and the adverbs that shape actions into their darkest forms begin with the letter "s." It is not surprising, therefore, that the residuum of sewage treatment is one of these "s" words: sludge. The assignment of a negative connotation to sludge is not only related to its origins or its often difficult properties as an engineering material, but to the increasing political, environmental, technical, and economic problems that it presents to

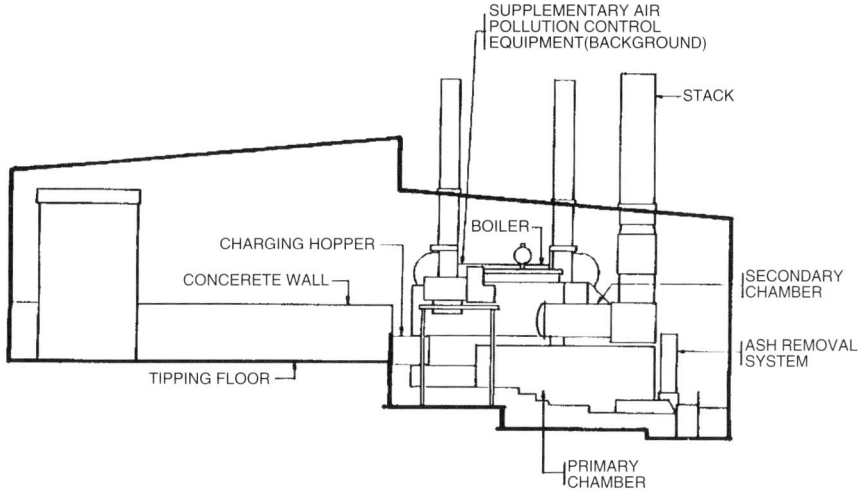

**Fig. 10.17.** Modular combustion unit (MCU) facility cross section.

the professional biosolids manager. More recently, the noun *sludge* has been changed to *biosolids*, which is not only less offensive but also has broadened in its meaning to incorporate the several other solid and semisolid residuals from wastewater treatment: grit, scum, skimmings, and screenings.

Since the signing of the Water Pollution Control Act in 1961 and its substantial amendments in 1972, the biosolids management community has told the (true but no longer interesting) story of the problems occasioned by the burgeoning quantity of biosolids produced as the level of wastewater treatment has increased. In recent years, say, since 1980, this story has been expanded to include dialogue regarding the growing significance of biosolids quality as it affects its impact on the environment. However, professional biosolids managers now play the lead in a story that is exploding into the headlines and is shouted across council chambers as public and regulatory concern, focused and reinforced by interest groups, impact on the shrinking number of disposal options that are open to the managers.

Most importantly, the public and political clamor is saying "No!" to more and more of the methods traditionally used in biosolids management. In desperation, the old approaches are being refined, updated, and equipped with more elaborate and effective pollution control systems to meet upgraded regulations and public expectations. Further, where landfills are used for ultimate disposal, these elements of the biosolids management system are rapidly approaching class I quality in their sophistication (and cost).

The broad area of thermal processing offers numerous alternatives for the processing of biosolids. Some of the techniques truly constitute disposal in that the total material constituting the original biosolids solids pass into another's hands. Other techniques reduce volume, increase biological stability, improve landfill structural characteristics, or otherwise aim to mitigate the management problem, but still result in a residue material requiring ultimate disposal.

## 4.2. Objectives and General Approach

This section reviews biosolids thermal processing technologies from several perspectives in order to assist the biosolids management professional in understanding the breadth of assistance that can be obtained from thermal processing systems—ultimately, an aid in decision making. It covers the full range of concepts from simple (and not-so-simple) drying, through the several types of thermal destruction by pyrolysis and conventional combustion to high-temperature slagging systems.

Each of the techniques is discussed from several standpoints: the process, the controlling thermodynamic principles, operating characteristics, environmental characteristics, a summary features presentation, and a comment on the current state of development. Where the technologies involve basic combustion processes, the principles and analysis methods discussed previously are applicable and are not repeated here.

The process description is brief in cases where the technology is well known and well documented elsewhere. Where useful, references for additional reading are given. The principal content includes the major items of equipment (their function and operating ranges) and the energy and material flows through the system.

The discussion of thermodynamics and combustion focuses on the heat balance for the system and how it is affected by biosolids and system parameters. Parameters of particular interest in this regard are the following:

- Percent solids: Perhaps the most important single parameter characterizing biosolids is the percent solids. In most wastewater treatment literature, percent solids (rather than percent moisture) is the preferred descriptor characterizing the water content of the waste stream. Note that, for biosolids, moisture content is high (say, when compared to conventional solid wastes) and the treatment process involves progressive separation of more and more water. The percent solids focuses on the goal of maximal moisture removal.

    For most publicly owned treatment plants (POTWs), the combination of gravity and mechanical dewatering seldom produces better than a 27% solids cake, and therefore the burning of biosolids is more the burning of water than of the organic biomass.

    Further, in many plants, the percent solids is allowed to be the dependent variable in the plant with detention times, effluent quality, and almost all other process variables being held to narrow tolerances. From the standpoint of thermal processing systems, this means that the most important process variable is out of control. Clearly, in view of the increases in fuel expense, the increase in environmental impact, and other adverse consequences of uncontrolled swings in operating conditions that are often associated with such a plant operating strategy, the decision to implement a thermal processing system should be the occasion to seriously reevaluate plant wet-end operating priorities. The performance of several classes of dewatering technology are summarized in Table 10.20.

    The solids in biosolids fall into two categories: the combustible and the ash. Combustible includes the organic biomass and other organic matter (scum, leaves, etc.) that is oxidized and driven off in the gas phase in the course of combustion processes. In this category are also the (usually) trace amounts of pesticides and other toxic organic compounds. Often, the combustible content is equated to volatile solids (VS), a biosolids characterization variable often reported in the literature. One must be cautious, however, since a portion of the mass associated with lime and ferric hydroxide and some other inorganic biosolids conditioning chemicals is often included in the determination of VS.

    The ash component of the biosolids includes the relatively inert inorganic matter associated with the wastewater flow (grit, silt, and sand, etc.) but also includes the insoluble toxic

**Table 10.20**
**Typical biosolids dewatering effectiveness levels**

| Level | Percent solids |
|---|---|
| Gravity settling | |
|     Clarifier | 0.20–4 |
|     Thickener | 3–8 |
|     Hydrocyclones | 3–8 |
|     Biosolids drying bed | 85+ |
| Mechanical dewatering | |
|     Vacuum filter | 14–23 |
|     Belt filter | 16–34 |
|     Filter press | 30–45 |
|     Centrifuge (conventional) | 14–23 |
|     Twin-roll nip press | 15–25 |
|     Centrifuge (high "g") | 23–35 |

metal compounds that can be important regarding environmental impact. The ash fusion temperature is an important property of the ash. If too low (e.g., due to the presence of ferric phosphate or other fluxes), the ash can melt together into a "clinker" at combustor operating temperatures. Clinker formation can damage the refractory, block passages, and cause other operating and maintenance problems in combustion systems.

The moisture content of biosolids significantly affects thermal processes due to the high latent heat absorption associated with the evaporation of water and the high sensible heat absorbed in raising the temperature of water vapor to combustion temperatures. The heat of evaporation of water at standard atmospheric pressure is about 539 Kcal/kg (970.3 BTU per pound). The heat capacity of water vapor (Kcal/kg-°C, BTU/pound-°F) at ambient temperatures is about 20% higher than that of air and increases rapidly as the temperature increases above ambient (Figure 10.1). Clearly, these thermal loads are important, stealing from the ability of the inherent heat of combustion of the biosolids to meet acceptable combustion temperatures without use of purchased fuel.

- Heat of combustion: The higher heat of combustion of biosolids combustible (4400 to 7200 kcal per dry kg [8000 to 13,000 BTU per dry pound] is roughly comparable to that of peat. The heat of combustion is somewhat elevated by the presence of excess oils and greases. Often, biosolids heat content is reported in Kcal/kg (Btu/lb) VS but this is not desirable if the VS content varies much from the combustible content due to high calcium or ferric hydroxide content.

The heat of combustion can be estimated from the ultimate chemical analysis of the biosolids using the Boie relationship (drawn from the DuLong equation (see Equation 5). It should be noted that heat of combustion can be expected to vary over time. Data indicate that digested biosolids fall into the low end of the heat of combustion range and heat treated (e.g., Zimpro thermally conditioned biosolids fall into the high range). Typical values if the higher heating value (HHV) are shown in Table 10.21.

- Sooting (smoking) tendency: Although not an unfailing rule, the tendency of a material to form soot under less-than-optimal combustion conditions appears to be generally related to the ratio between the unbound hydrogen and the carbon content of a material. This ratio is calculated as:

$$\text{Unbound Hydrogen–to–Carbon Ratio} = \frac{\%Hydrogen - \%Oxygen/8}{\%Carbon} \quad (36)$$

**Table 10.21**
**Typical higher heating values for wastewater treatment solids**

| Biosolids type | Typical HHV (kcal/kg – dry basis) |
| --- | --- |
| Raw primary sludge | 5500–7000 |
| Activated (secondary) sludge | 4700–5500 |
| Anaerobically digested primary sludge | 3050 |
| Raw, primary sludge, $FeCl_3$–lime conditioned | 3900 |
| Biological trickling filter solids | 4700–5500 |
| Grease and scum | 9250 |
| Fine screenings | 4300 |
| Ground garbage | 4550 |
| High organic grit | 2200 |

**Table 10.22**
**Comparative sooting tendency of waste materials (1)**

| Waste material | Unbound H-to-C ratio |
| --- | --- |
| Mixed paper | 0.0065 |
| Softwood | 0.0191 |
| Linoleum | 0.0625 |
| Polyurethane | 0.0641 |
| Polystyrene | 0.0913 |
| Waxed paper cartons | 0.0927 |
| Polyvinyl chloride | 0.1200 |
| Cooking fats, scum | 0.1325 |
| Rubber | 0.1333 |
| Oils, paints | 0.1343 |
| Polyethylene | 0.1677 |

For biosolids, this ratio is about 0.0032, and Table 10.22 shows comparisons with other materials. It can be seen that biosolids are not inherently a smoky burning material. Note, however, that including greases (scum) with the biosolids adds material of considerably greater smoking tendency. Thus "no-problem" systems can acquire emission problems.

- Percent excess air: The exact air quantity to meet the oxidation needs of the biosolids chemistry (two oxygen atoms for each carbon, etc.) is called the theoretical air or stoichiometric air requirement. Numerous combustion texts (e.g., 1) show the method of calculation of this quantity. Air supplied to the system above the stoichiometric amount is called excess air and usually is added (1) as part of air flow used to mix the flue gases (inducing turbulence to ensure that, everywhere, the gases are above the stoichiometric ratio), and (2) to temper combustion temperatures to avoid overheating of the furnace or ash fusion (clinkering).

As the air supplied to a combustion system increases from zero to the stoichiometric point, the temperature of the resulting flue gases increases steadily. This increase reflects the close coupling between the quantity of oxygen supplied and the release of heat. Above the stoichiometric point, the temperature steadily declines reflecting a dilution of the heat (now, theoretically, 100% released) with excess air.

Auxiliary fuel is required if, with air quantities that meet process requirements for mixing and adequate oxidation of combustible gases, the heat of combustion of the biosolids itself is unable to bring the flue gases above 800°C (1500°F). The fuel adds sufficient heat to evaporate all liquid water and to heat the flue gases to the level ensuring burnout of smoke and odor. Clearly, the more excess air that has been used with the biosolids, the larger the quantity of gas to be heated and the greater the amount of fuel (per ton of biosolids) that is required.

- Energy parameter: The energy parameter (EP) has been shown as a useful variable to combine in a single term the heat and material balance for the combustion of biosolids or other fuels or wastes. The EP is calculated as follows:

$$EP = (1 - S) \times 10^6 / (S)(B)(V) \tag{37}$$

where
   $S$ = decimal percent solids;
   $V$ = decimal percent volatiles
   $B$ = heat of combustion (BTU/lb volatile solids)

The advantage in using the EP instead of, for example, percent solids to correlate thermochemical calculations, is that EP correlations are usually linear, whereas over broad ranges, the percent solids correlations are strongly curved. It can be noted that for a given percent excess air, the EP collapses the heat effects of water evaporation, flue gas heating, and waste-derived energy supply into a single term.

The operating characteristics of interest here include the relative sophistication needed of operators, the sensitivity of the process to upsets upstream in dewatering or other plant operations, sensitivity of facility availability to the quality and frequency of maintenance; energy efficiency, needs for utilities support (especially water supply), and the like.

The primary environmental characteristics involved with thermal systems involve air emissions: opacity, odor, excessive particulate, and air toxics.

The problem of opacity is usually more associated with organic aerosols (e.g., from greases) and finely divided soot particles than with excessive mass emissions. That is due to the high effectiveness of small aerosol or soot particles to scatter light—the fundamental process underlying an appearance of plume opacity. Therefore, the causes of and cures for this type of problem rest more with improving combustion than with upgrading (particulate control oriented) air pollution control systems.

Odor problems reflect a combination of (1) process characteristics that release or generate odorous compounds; (2) deficient control processes (scrubbers, biofilters, combustion environments, etc.); and (3) inadequate plume dispersion (short stacks, low discharge velocity, unfavorable topography, and unfavorable meteorology). Clearly, the distance to residential dwellings, commercial enterprises, or recreational areas is a nontechnical but key aspect affecting the consequences of paying inadequate attention to odor issues.

All of the processes discussed here have the potential to release odors. The odor character is most often a sewage odor (usually sulfide-based odorant chemistry) but may also include odorants from solvents or other chemical families (volatile organic compounds [VOCs]) or amine-based fishy odors generated in low-temperature thermal processing of high-nitrogen compounds that are stripped from the wastewater solids. Burnt odors (often quite offensive odors formed by partial oxidation of proteins at temperatures high enough to support primitive oxidation reactions but too low for full combustion at, say, 400° to 550°C [750° to 1000°F] or greater) can arise in some dryer operations. Burnt odors often involve ketone chemistry.

For well-mixed combustion systems, control of all of these odor chemistries is readily affected at temperatures over 750°C (1400°F) in the presence of excess oxygen. If scrubbers are selected, conventional hypochlorite or permanganate systems are adequate for sulfides but acidic scrubbers are needed for amines. Scrubber systems show poor to no removal of the partially oxidized organic odorants. Biofilters are suitable for low concentration sulfide odors and many other odor chemistries where organic compounds are present to which biofilter organisms can be "trained."

Particulate control for most new incinerators is affected with a combination of a Venturi scrubber and a tray scrubber. Older units used the tray scrubber alone with mixed results. Tray scrubbers are quite adequate in attaining the US EPA control technology-based New Source Performance Standard (NSPS) for biosolids incinerators (1.3 pounds of particulate per dry ton burned) but can be ineffective at low pressure drop levels (10 to 15 inches of water) if they must collect submicron particulate enriched with the heavy metals and organics (collectively spoken of as "air toxics"). The US EPA's research and policy evaluations reassessed the original standards and promulgated special standards based on processing rate and biosolids heavy metal content (an environmental burden standard) in addition to the requirements for stack gas concentrations of (presumably benign) "particulate." The emission standards for dryer processes are usually drawn from "process weight" regulations that specify limits on the mass emitted per unit mass processed rather than from the US EPA incinerator NSPS codes. If heavy metal emission (air toxics) is high, newer incinerators have turned to wet electrostatic precipitators to reliably provide high levels of control. When organic emissions become problematical, afterburners that ensure combustion temperatures in excess of 875°C are necessary.

The features and problem areas presented by each technology are also discussed. The final section of each technology discussion notes the state of development: Is the process a commercial reality or still an experiment? In many cases, the technology is fully operational in several plants, so this issue is unimportant. Occasionally, however, a word of caution is appropriate. Since most biosolids management systems are implemented by public agencies with public funds where there is considerable political cost for failure and a lack of real compensation for risk taking, this factor can and should assume major significance in technology selection.

### 4.3. Low-Range (Ambient, 100°C) Drying Processes

Drying processes aim at stabilization of the biosolids by a drastic reduction in moisture content (say, to less than 15%) and at preparation for possible disposal as a product: a low assay (nitrogen, phosphorus, and potassium [NPK] content) fertilizer, a fertilizer base (to be enhanced by addition of high NPK chemicals), or a tilth-enhancing soil amendment. After making the considerable investment in capital and operating cost to upgrade the biosolids, the decision maker must consider seriously the question: What do I have? Is it still a waste? Or a marginal product with characteristics just barely meeting market requirements? Or a salable product that is welcomed in the marketplace and can be sold at a profit to generate a reliable revenue stream to offset production expense?

The product characteristics that affect marketability include the following:

- NPK value (a minimum of 4% available, slow-release nitrogen is preferred)
- Freedom from dustiness and odor in handling and storage
- Limited trash content (plastic bits, cigarette filters)
- Limited contaminants (cadmium or other heavy metals etc.) that limit applicability
- Reliability of supply
- High bulk density (reducing transportation cost)

### 4.3.1. Drying Beds

Since drying is energy intensive and generates a product of limited marketability (i.e., high energy costs could make economically untenable), concepts that use no fuel at all (solar) are of interest. Drying beds are the most widely used method of municipal biosolids dewatering, but only a fraction of these produce a biosolids product greater than 75% solids. However, for small communities in climates with favorable rainfall-to-evaporation characteristics, the process is low cost, simple, and the product is acceptable (on a giveaway basis) to the community or to local farmers.

Sand drying beds are the oldest and most common types of drying bed. The beds are rectangular with vertical sidewalls. Bed dimensions range from 4.5 to 45 m (15 to 150 feet) and include 10 to 23 cm (4 to 9 in) of sand over 20 to 46 cm (8 to 18 in) of graded gravel or crushed stone. Plastic or vitrified clay underdrains slope to a sump. Variations include paved systems with spaced sand-covered drains (smaller than the sand beds but capable of supporting mechanical equipment for biosolids removal and maintenance) and wedge-wire bed systems that show high rates and acceptable maintenance.

In all cases, thickened and sometimes chemically conditioned biosolids are run into the bed and allowed to drain and dry. In dry climates, the biosolids approach 95% solids. Biosolids removal is often performed manually (especially for sand beds), but some plants use mechanical means. The main sidestream from the process is the bed underdrain liquor that is recycled to the adjacent wastewater treatment plant. Since no purchased fuel is used, the process has excellent energy statistics. In most instances, however, the ultimate dewatering level is only to the 60% to 70% solids level. Sand beds are very simple to operate and maintain. Their often-erratic product quality reflects more climatic variation and loading effects than do process variables under operator control.

The primary environmental characteristic of the drying bed is the potential for odor. While the biosolids applied to drying beds are generally anaerobically digested, from time to time offensive odors are emitted. The very nature of the process makes gaining control of an odor episode almost impossible. Dispersal of masking agents or counter-odorant materials is the only mitigating technique that is available.

The clear benefits of the sand bed derive from its simplicity, limited operator attention and energy consumption, and relative insensitivity to changes in biosolids character. The problem areas include the large land requirements, sensitivity to climate (both rainfall and freezing) and, frequently, odor complaints. This latter problem generally mandates the use of digested or chemically stabilized biosolids as the feed. The product is of limited utility (being often dusty and of low bulk density) and is often land filled.

### 4.3.2. Direct-Fired Systems

Direct fired dryer systems, where the biosolids moisture is driven off by contact with hot combustion gases, represent a major departure from sand beds or similar approaches to biosolids management. Capital investment, operator sophistication, sensitivity of the process to biosolids variation, and other factors now become important. Further, since a direct-fired system directly contacts the biosolids with relatively large volumes of hot gas, a large and significant odor control problem (and cost) must be addressed either by application of odor control equipment or by other means.

Gases fed to direct dryers at 200° to 300°C (400° to 600°F) emerges with a typical sludge or $H_2S$-based smell. Experience has shown that this odor is controllable with a conventional hypochlorite-type, chemically oxidizing scrubber. At progressively higher high dryer gas temperatures, the odor character shifts through an intermediate "fishy" smell in the 300° to 400°C (600° to 750°F) range, to a burnt protein smell, which includes numerous pyrolysis-derived or partial combustion products that are not well-controlled with scrubbers; afterburner control devices are required. For energy efficiency, a high-efficiency regenerative afterburner design is preferred. In such units, the incoming cool gas from the dryer is preheated by passing through a bed of hot refractory. Firing fuel raises the temperature of the preheated gas to incineration temperatures of, say, 900°C (1650°F). The burned-out exhaust gases then pass through a second refractory bed, preheating it. From time to time, the gas flow path is switched. Clearly, it is critical that the off-gas is clean enough to avoid plugging of the refractory media.

The two major direct dryer concepts used for biosolids include the flash dryer where a portion of the biosolids (or, if desired or needed, purchased fuel) is burned to generate the hot gases for drying and the rotary pelletizing dryer where burning purchased fuel generates hot gases for drying.

#### 4.3.2.1. FLASH DRYER

Flash-drying is the rapid dewatering of biosolids by direct contact with hot gases. Disintegration of the incoming biosolids to expose surface area and facilitate evaporation is a key feature of this approach. The process was first applied to wastewater biosolids in 1932 at the Chicago Sanitary District. The flash-dryer systems used in the U.S. are primarily those of the Raymond Division of Combustion Engineering Inc. Wellsville, NY. A schematic of the Raymond system is shown in Figure 10.18 (53).

The process is most easily envisioned in three segments. In the first, mechanically dewatered biosolids at, say, 15% to 25% solids is blended to a 60% solids intermediate product by mixing with recycled dry biosolids in a double paddle mixer, pug mill, or similar device. The mixing is carried out to improve the materials handling characteristics for pneumatic transport and, importantly, to "jump" over the 40% to 50% solids range where the biosolids are very sticky and tend to adhere and build up on hot surfaces. The 60% solids blended product is then fed to a cage mill: a fanlike device where the biosolids are mixed with hot furnace gases at 700°C (1300°F). In the cage mill, the biosolids moisture flashes off to produce a 95% solids product that is carried out of the mill by the gas flow. The solids are separated from the gases using a simple large-diameter cyclone device. A portion of the dried biosolids is recycled to the mixer, and the remainder may be either burned to provide the drying heat or drawn off as a product.

# Combustion and Incineration Engineering

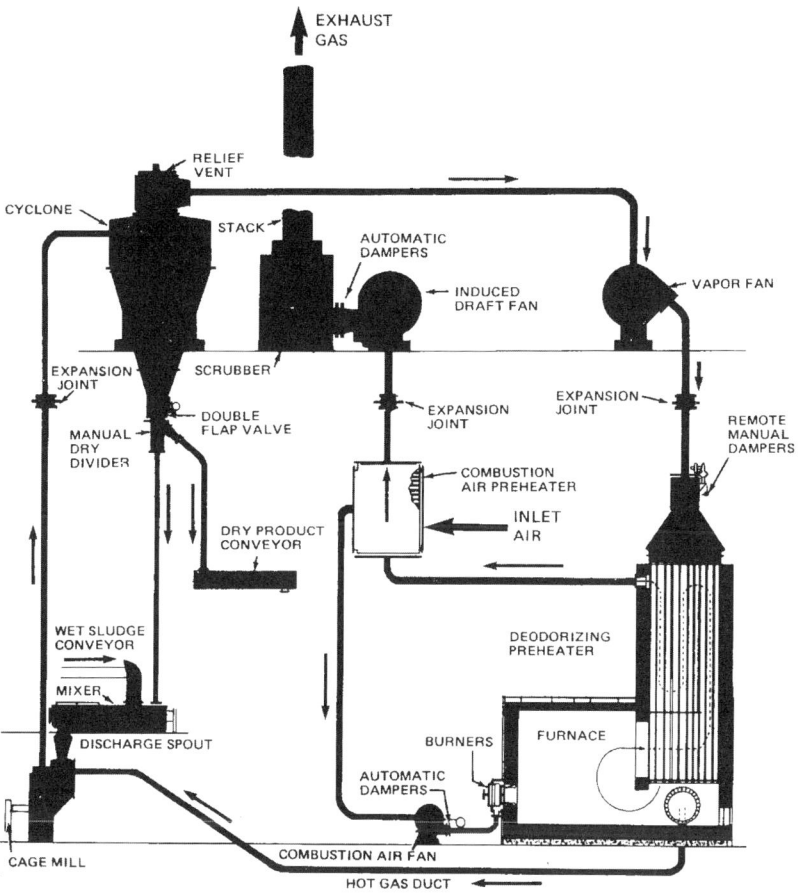

**Fig. 10.18.** Raymond flash dryer. (Courtesy of Raymond Division, Combustion Engineering Inc.) (53).

The second element of the flash dryer system involves the incineration or furnace process. Gas, oil, coal, or the dried biosolids are burned with preheated air. The combustion takes place at high temperatures, but the net furnace off-gas temperature is tempered by the recycle of cool gases from the cyclone to yield a net flue gas temperature of 700°C (1300°F). These gases are split with a portion returning to the cage mill and the remainder exhausted to atmosphere.

The third element of the flash drying system is the effluent gas treatment facility, which includes the deodorizing preheater that accepts cool gas from the cyclone that, through heat exchange with the combustion gases in the furnace, is preheated to about 600°C (1100°F). The deodorized gases leaving the furnace, having attained a minimum temperature of 650°C (1200°F), pass through a combustion air preheater and to a wet scrubber for particulate removal.

Ideally, the leaving gas temperature for the system is 105°C (220°F) to 150°C (300°F) reflecting the energy conservation principles built into the flow system. This results in a net heat rate of 1225 to 1335 kcal/kg (2200 to 2400 BTU/lb) of water evaporated in bringing the product biosolids to about 5% moisture (95% solids). As noted, either fossil fuels or the biosolids itself can be used as the heat source; the decision as to which fuel to use depends on the marketability of the dried biosolids product and the cost of the fossil fuel.

The flash dryer clearly is a complex flow system with several heat exchangers and numerous materials handling processes. Not surprisingly, these systems are subject to the severe abrasion of the dried biosolids (especially the pug mill and the cyclone metalwork) and fouling of heat exchange surfaces. Materials flow balance can be a problem leading to overflows at, say, the pug mill (a housekeeping problem). Also, plating of sticky biosolids on the cage mill leads to imbalance and bearing problems. The biosolids product made in this device is extremely dusty. This product characteristic creates a potential dust explosion problem in handling, storage, and transport, and detracts from the marketability of the material.

If dried biosolids is not burned as a fuel in the odor incineration system, the environmental problems of the flash dryer are primarily the air emissions from the fuel ($SO_2$, $NO_x$, and, possibly, minor amounts of particulate matter), particulate from the carryover through the cyclone of biosolids solids, and, if the deodorization process is inadequate, odor emissions. The latter should be susceptible to abatement through maintenance of sufficient temperatures in the deodorizing flow stream but at an energy penalty.

If biosolids is burned (in part) as the deodorizing fuel, one can be relatively assured of substantially complete burnout of biosolids organic materials (including trace amounts of pesticides and, most likely, PCBs). However, the combustor operates with a high flame temperature (the combined effects of dry biosolids and preheated air), which suggests potential problems with heavy metal volatilization. Niessen (54) reported that with a flame temperature in excess of 1050°C (1900°F) a large fraction of several heavy metals would vaporize and, without extraordinary efforts at air pollution control (e.g., gas cooling followed by fabric filtration or wet electrostatic precipitation), excessive metal emissions could be experienced.

The primary feature of the flash dryer is the ability to use biosolids energy to provide the drying interchangeably with fossil fuel. Thus, the product stream can be matched to the market; if the market price is high, burn fuel and sell biosolids; if low, burn biosolids and minimize manufacturing cost. As noted in the introduction, however, the dusty nature and inherently low NPK value of flash-dried biosolids reduces the marketability of the product so that careful attention must be given to finding secure markets for the material. Product pelletization after the flash dryer can be considered (with an additional capital and operating cost) to upgrade marketability. The maintenance problems and costs of this dryer concept create operating headaches for the owning organization, and this characteristic should be carefully considered before going forward. The flash dryer system is well-proven technology and is in use in the U.S., Europe, and elsewhere.

#### 4.3.2.2. ROTARY DRYER/PELLETIZER

The biosolids product generated in the flash dryer discussed above and in the indirect dryers discussed below have a significant marketing difficulty: the dustiness of the product. The rotary direct dryer produces a hard particle with limited dust and excellent materials handling and flow properties. In a rotary dryer, wet sludge is fed and hot

## Combustion and Incineration Engineering

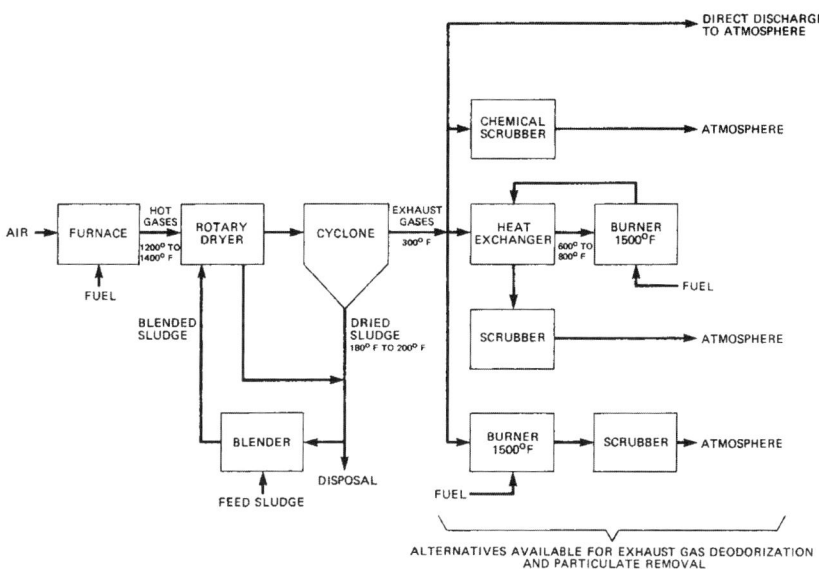

**Fig. 10.19.** Flowsheet of direct-fired rotary dryer (53).

gases passed either countercurrent or cocurrent. A characteristic of biological sludge is the formation of a sticky consistency at about 40% solids. If this condition forms within the dryer, a "ring" will develop. Thus, the wet sludge from dewatering is always blended with dry material to "jump" over the 40% sticky zone. The sludge is fed, then, at about 50% solids and dried the rest of the way in the dryer. The rotary dryer produces a spherical pelletized product at from 93% to 98% solids, which, using a triple–pass screen, is sized to the 4- to 6-mm-diameter range. Undersize and (crushed) oversize are recycled and blended with the incoming wet sludge cake. A typical flowsheet for a conventional rotary dryer is shown in Figure 10.19 (53).

Single-pass dryers, composed of a single, rotating tube, are more common in Europe. A single-pass design is usually equipped with lifters to elevate and distribute the solids through the gas stream. In the U.S., the triple-pass system composed of three, coaxial tubes is dominant. Flow of the solids in the single-pass configuration is facilitated by sloping the dryer shell (much as is done in a rotary kiln). For the triple-pass concept, the solids are fed to the innermost tube and are moved along by the air stream (thus, only cocurrent flow is possible). At the end of the first traverse, the solids fall to the second tube. The air flow follows and keeps moving the particles along. The process repeats at the end of the second traverse. For the triple-pass system, since the motive action requires operation within a tight range of gas-to-solids relative velocities, the gas flow rate and system dimensions are more narrowly defined and are closely coupled. The multipass system, therefore, is more limited in turn-down than the single-pass unit since a reduction in gas flow may lead to cessation in solid movement through the system.

The selection of inlet gas temperatures is driven by the heat balance on the dryer. First, it must be recognized that a direct dryer system is sized by the gas flow. Second, the evaporative capacity (for a given gas flow) scales directly to the temperature drop of the drying gas between the feed point and the discharge; typically, the discharge is at about 90°C (190°F). Thus, the capital cost per unit of productivity for the dryer drops as the inlet gas temperature increases. It is the play-off between the fall in dryer capital cost (and an increase in the thermal efficiency) as gas temperatures increases and the increased cost in pollution control (scrubber and afterburner) that sets the design point. Typically, the inlet gas temperature is set in excess of 425°C (800°F).

The rotary pelletizing concept in its several embodiments has become the most attractive alternative in biosolids drying. Particularly in Florida but elsewhere, too, this process is gaining attention as a means to produce a product that can be shipped offsite, not simply sent elsewhere for disposal. Although limited by the inherently low fertilizer value assay of biosolids, the material appears marketable as a soil amendment or for use as a carrier with fortification by chemical fertilizers to a more marketable assay.

A second class of variations in rotary pelletizers concerns the hot gas source and temperature. Conventionally, fossil fuels are direct fired in a furnace with excess air to produce the hot gases for biosolids drying. However, a gas turbine can be used to generate electricity (serving the treatment plant and biosolids dryer complex) with the off-gas (at about 700°C or 1300°F) used for all or a portion of the dryer needs. The gas turbine approach is used in Milwaukee, Wisconsin, in the making of Milorganite, one of the oldest and best known rotary dryer biosolids products. The integration of pelletizing with a gas turbine in Milwaukee's system uses a 650°C (1200°F) gas inlet temperature and would be expected to have odor problems. Their very high stack, which ensures good dispersion under almost all meteorological conditions, provides an important mitigation that enables the system to avoid the use of an afterburner odor control system.

A third concept takes the gases from the dryer, cleans and dehumidifies them in a water scrubber and condenser operation, and recycles them to a heat exchanger where they are reheated indirectly by very hot direct fired furnace gases to produce the medium-temperature dryer feed gas. A bleed stream (only 5% to 10% of the total flow) is drawn off and incinerated for odor control. Air pollution control includes treatment for particulate (usually a Venturi/tray tower scrubber combination) and, importantly, systems for odor control. The latter can include chemically oxidizing systems (chlorine or permanganate-based scrubbers) or thermal oxidation (afterburners).

As with all drying concepts, energy efficiency is vital. For the turbine off-gas systems, heat rates (based on the sensible heat of the turbine gas) range from 830 to 1000 kcal/kg (1500 to 1800 BTU/lb) of water evaporated. Adding an afterburner after the dryer for odor control pushes the upper limit to about 1600 kcal/kg (2900 BTU/lb) of water. If the turbine can be used as an afterburner and the recirculation mode is selected, the overall heat rate drops to 760 kcal/kg (1370 BTU/lb) water. Direct fired systems such as those in Figure 10.19 range from 975 to 1160 kcal/kg (1750 to 2100 BTU/lb) of water without an afterburner and reach 1330 kcal/kg (2400 BTU/lb) with an afterburner.

Rotary dryer systems are relatively easy to operate and maintain. As with the flash dryer, abrasion in the system (dryer body and flights, screw conveyors, and in the pug mill) is the principal area of concern.

The principal environmental concern with rotary pelletizing systems involves odor emissions. As the dryer inlet gas temperature is increased (to increase the dryer evaporative capacity and improve investment capital utilization), the off-gas odor character changes. At inlet temperatures in the 300° to 325°C (400° to 600°F) range, odors are characteristic of the sewage, that is, sulfide based. As the temperature increases above 325°C (600°F), the biosolids begin to show evidence of thermal cracking and new and more refractory odor chemistry appears, that is, partially oxidized and organic. Unlike the sulfide-based odor, the oxidized materials are not particularly susceptible to hypochlorite scrubbing. Thus, the potential exists to create an odor nuisance. Further, the high political price of creating an odor problem suggests the importance of ensuring positive odor abatement under any conceivable operating condition. This leads to the frequent recommendation for afterburner-like devices for this service. These systems, if designed with heat recuperation features, offer good energy efficiency within themselves and high odor destruction potential, but add significantly to the investment cost.

The primary benefit of the pelletizing dryer concept is the recognizable market for the product. This allows for the potential to recover a portion of the investment made in upgrading the biosolids. The primary disadvantage is the mirror image of the benefit; a system justified on the basis of product revenues may become economically untenable if fuel costs escalate rapidly or if markets become either saturated or constrained (as, for example, through restrictions on pellet use due to health risk concerns relating to the heavy metal content of the biosolids).

The gas recirculation concept is relatively new and creates some potential problems in gas flow control and draft balance. However, with recirculation, the net vent gas stream is very small (5% to 10% of the total), and vigorous odor control concepts (such as an afterburner) become economically acceptable. Full integration of the recirculating system with a turbine (including using the turbine as the afterburner) is as yet untested.

*4.3.3. Indirect-Fired Systems*

Indirect drying results in greatly reduced odor problems in comparison to those of direct-fired systems. Indirect systems dry the biosolids by contact with hot surfaces within closed chambers. The water vapor is condensed using a barometric condenser. The surface is usually heated by condensing steam but recirculating hot oil is also used. The primary technologies used are the torus disk dryer, the hollow flight dryer, and the tray dryer. Thin-film scraped dryers have been used (e.g., in Dieppe, France) to partially dry thickened sludge from 6% to 8% solids to a product with about 20% solids (i.e., a dewatering function) using "free" steam from an adjacent municipal refuse incinerator, but in the more general case high purchased fuel cost and a limited product dryness range limit the applicability of this approach.

4.3.3.1. PORCUPINE, BEPEX, AND OTHER DRYERS

Several dryer manufacturers have attempted to dry biosolids using steam-heated dryers. In the hollow flight dryer, biosolid feed (usually blended to 50% to 60% solids using a pug mill) is passed through a horizontal steam-heated shell. Steam-heated, hollow blades (such as the cut-flight Bethlehem Porcupine© Bethlehem, PA, dryer or the Komline-Sanderson paddle dryer) move and masticate the drying biosolids and contribute useful heat exchange surface. The Stord© Herlev, Denmark or Bepex© Minneapolis, MN torus disk systems use a shaft fitted with a series of hollow, steam-filled disks that rotate inside a steam-heated shell. A drawing of a Bepex© unit is shown in Figure 10.20.

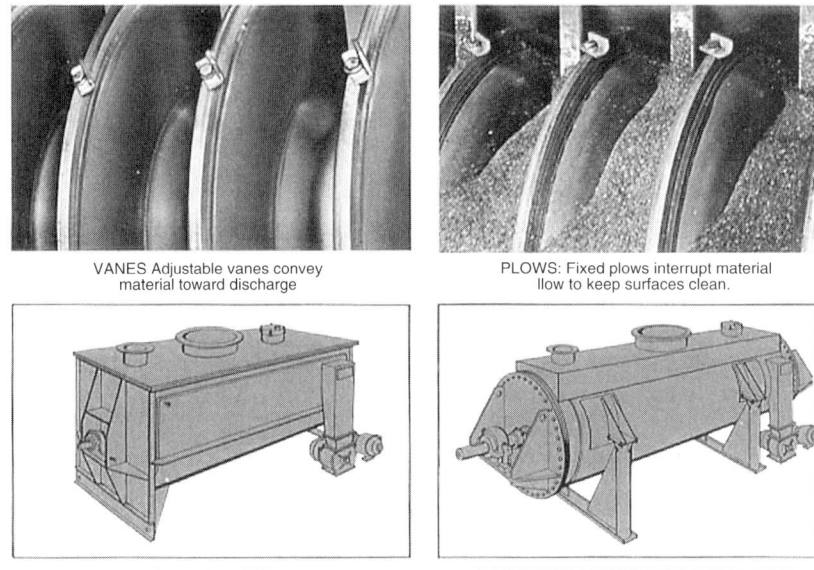

**Fig. 10.20.** Bepex© biosolids dryer. (Courtesy of Bepex Inc.)

Small metal vanes on the periphery of the disks give a slight but significant directionality to the motion of the material being dried, moving it through the unit. Water vapor (highly odorous and containing both organic oils and ammonia) rising from the drying mass is condensed in a barometric condenser and recycled to the head works of the POTW. All of the technologies mentioned above produce a relatively dusty product at 96% to 98% solids content. Postdryer pelletization is often utilized to reduce dustiness and thereby improve marketability.

#### 4.3.3.2. PELLETECH DRYER

The Pelletech© Baltimore, MD system is a tray dryer configured as a stack of trays mounted inside an insulated cylindrical shell. Each tray is heated with steam or recirculating oil. A central shaft rotates slowly, and arms that project out over the trays plow the biosolids first to the outside wall, where they fall through drop-holes on the periphery to the tray below, and then toward the center, where the biosolids fall through the annular space around the shaft. This device is shown in Figure 10.21. The kneading action of the plows and the drying action on the hot trays act to progressively build a shell on the small dry "seeds" that are part of the feed. This forms the biosolids into 5- to 10-mm spherical pellets. The product is screened on a double-deck screen. The "overs" are crushed and blended with the "unders" for recycle, and the "middlings" comprise the product stream.

Since the principal heat loss from the system is the water vapor at about 120° to 150°C (250° to 300°F), the thermal efficiency of the indirect dryer is high: about 900 kcal/kg (1600 BTU/lb) water evaporated (assuming 80% efficiency in the steam-raising boiler).

Although thermodynamically attractive, the indirect system is often plagued with many of the problems of biosolids drying: abrasion in the dry areas, corrosion from

# Combustion and Incineration Engineering

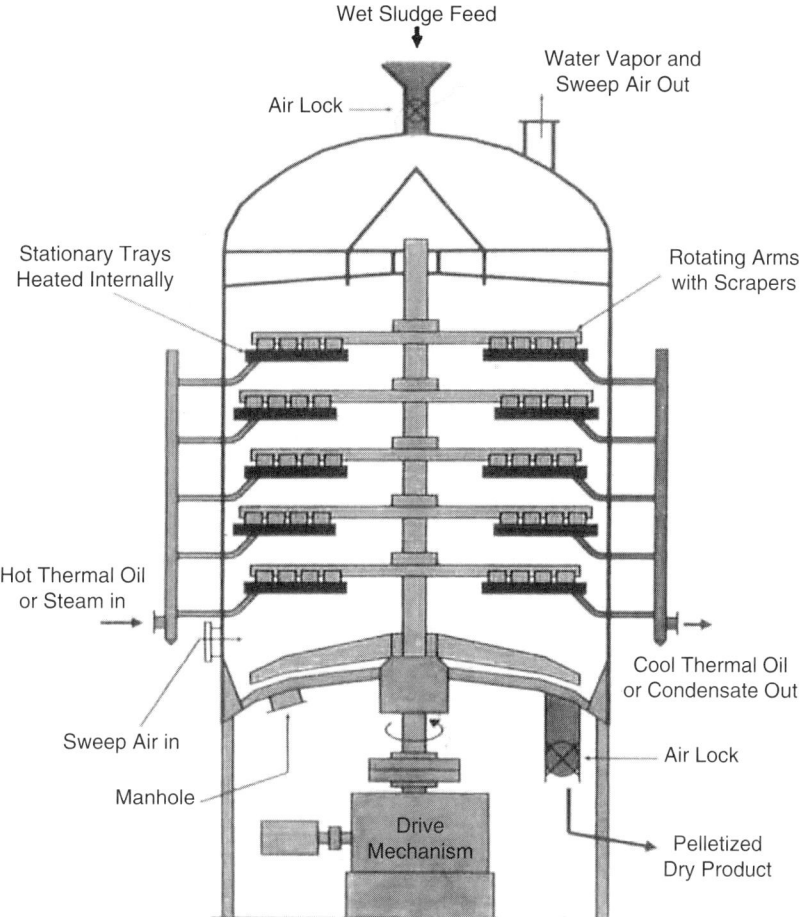

**Fig. 10.21.** Cross section of Pelletech© biosolids drying system. (Courtesy of Envirogro Inc.)

ferric chloride (if used as a biosolids conditioning aid) in the wet areas, and problems with sticking to the heat exchange surfaces if the biosolids in the unit fall into the tacky 40% to 50% solids range.

From an environmental standpoint, the indirect dryer is potentially superior; there is no emission of pollutant gases, particulate, or heavy metals, and there is inherent containment of odors within the unit. Note, however, that for indirect dryers, a small air flow is maintained through the unit to sweep out the steam (about 10% of the steam weight flow). This steam/air stream is passed to an indirect or direct condenser, and usually the final gas stream (small in volume but highly odorous) is incinerated for odor control. Often, the boiler used to generate the dryer steam can use all of the purged air as combustion air. This economically affects the afterburner function without new capital investment or control-related fuel expense. The off-gas from these units is very offensive, and gas containment and odor control are critical.

Other than the operating problems indicated above, the remaining problems revolve about the marketability of the dried product. Dusty products from the Stord or Bepex technologies do present a marketing problem.

Indirect rotary dryers have had only limited use in the U.S. The Porcupine system in Harrisburg, Pennsylvania, experienced problems with biosolids bonding to the paddle blades (insulating them). The Stord units in Chittenden County, Vermont, and Pittsburgh, Pennsylvania, have had some problems with odor, with buildup of biosolids on the heat transfer surfaces, and with capacity shortfalls.

Sometimes, drying is affected ahead of incineration. The dryer augments mechanical dewatering to reduce or eliminate the fuel requirements for sludge incineration. This application may be seen at the Pittsburgh, Pennsylvania, wastewater plant where steam dryers are used to reduce sludge moisture content ahead of fluidized bed incinerators. Steam is generated "free" by cooling the off-gas from the fluidized bed incinerators. Also, at the Hyperion plant in Los Angeles, the Carver-Greenfield multiple effect evaporation (MEE) process produces a bone-dry sludge powder subsequently incinerated in a two-bed fluid bed incinerator. The first bed is run air starved in a pyrolysis mode and the second bed is fully oxidizing. The City of Youngstown, Ohio, installed a steam dryer to partially dry sludge to eliminate fuel use in incineration.

### 4.4. Mid-Range (250° to 1000°C or 300° to 1800°F) Combustion Processes

This class of thermal processing includes conventional and several specialized incineration systems. Although capital intensive and incurring high operating cost, incineration offers the advantages of complete destruction of biosolids organic matter, which, in the past, promised easy disposal of the ultimate residue. Recent concerns regarding the leaching of heavy metals from ash make this assumption less confident from the standpoints of permitting and public acceptance.

As the temperature of organic materials is raised, one reaches a temperature above, say 300°C (550°F) where breakdown of the organic chemicals comprising the material is initiated. From the standpoint of heavy metal emissions, limiting the maximum temperature attained by the solids to, say, below 650°C (1200°F) avoids completely the heavy metal emissions, excepting for the special case of mercury and its compounds. Further, if this temperature control is achieved by limiting combustion air (the so-called pyrolysis or starved air mode), fuel use is minimized and essentially no $NO_x$ is formed, and organic or pyritic sulfur in the waste is released as $H_2S$. Although the off-gas from such a unit needs afterburning, the fuel value of the CO, hydrogen, and hydrocarbons in the furnace off-gas provide a portion of the afterburner energy. Clearly, the afterburning process will produce $NO_x$ and the $H_2S$ oxidizes to $SO_2$.

Alternatively, one can supply more air and operate throughout in the conventional, fully oxidizing incineration mode. Here, the systems are simpler, and considerably more operating experience exists but secondary problems such as fused clinker formation can arise (since operating temperatures can be high and, for certain biosolids, ash chemistries can exceed the ash fusion temperature). Further, heavy metal volatilization can take place.

#### 4.4.1. General

By far the most common biosolids incinerators (whether pyrolysis or full combustion) are either of the multiple hearth furnace (MHF) or fluidized bed (FB) types. Other than the obvious difference in overall combustion chemistry, the hardware used for pyrolysis

Combustion and Incineration Engineering 577

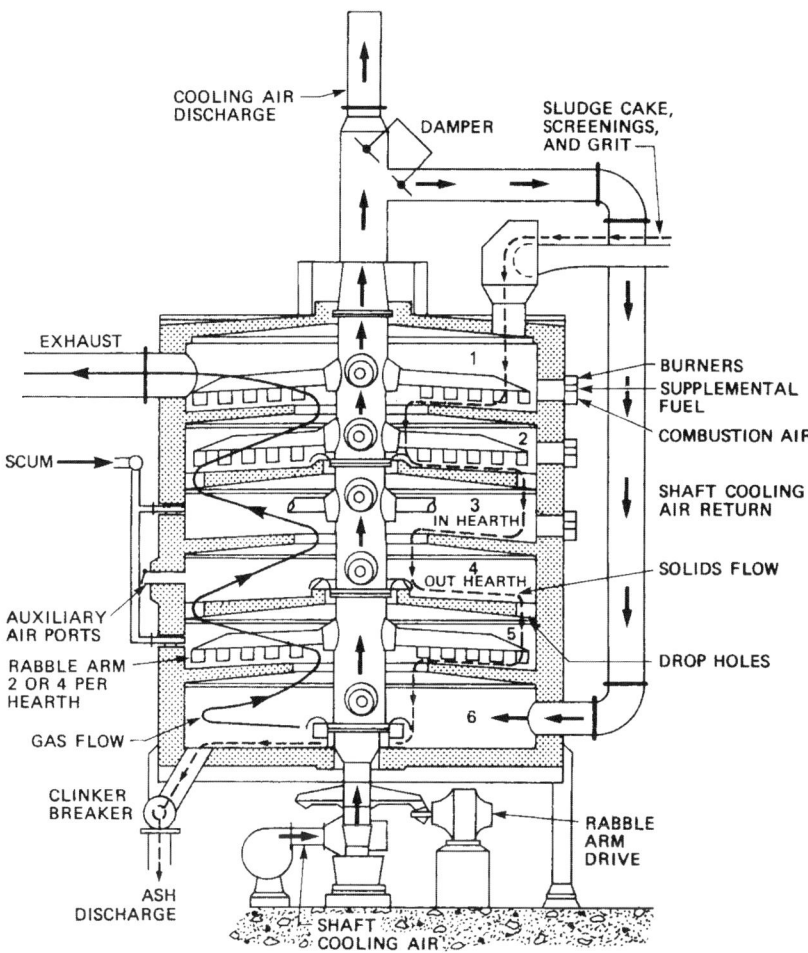

**Fig. 10.22.** Cross section of multiple hearth biosolids incinerator (53).

or full combustion mode of operation is essentially the same. Thus, for brevity it is appropriate to discuss the hardware for each alternative and then to discuss the specific differences brought about by the operating mode.

#### 4.4.1.1. MULTIPLE HEARTH SYSTEMS

Multiple hearth furnaces (MHFs) are the most common furnaces used for conventional biosolids incineration. By limiting the air supply, they may be operated in the pyrolysis (starved air mode) to give the desirable operating characteristics of pyrolysis within a system well proven in conventional service. Importantly, they also can be configured to switch over to the full combustion mode as a process backup.

The MHF system is designed for continuous operation. Startup and holding energy requirements are too costly and problematic for the equipment to operate intermittently. The system is composed of a stack of from 5 to 11, nearly flat, cylindrical, refractory hearths (Figure 10.22). The hearths are enclosed in a refractory lined shell.

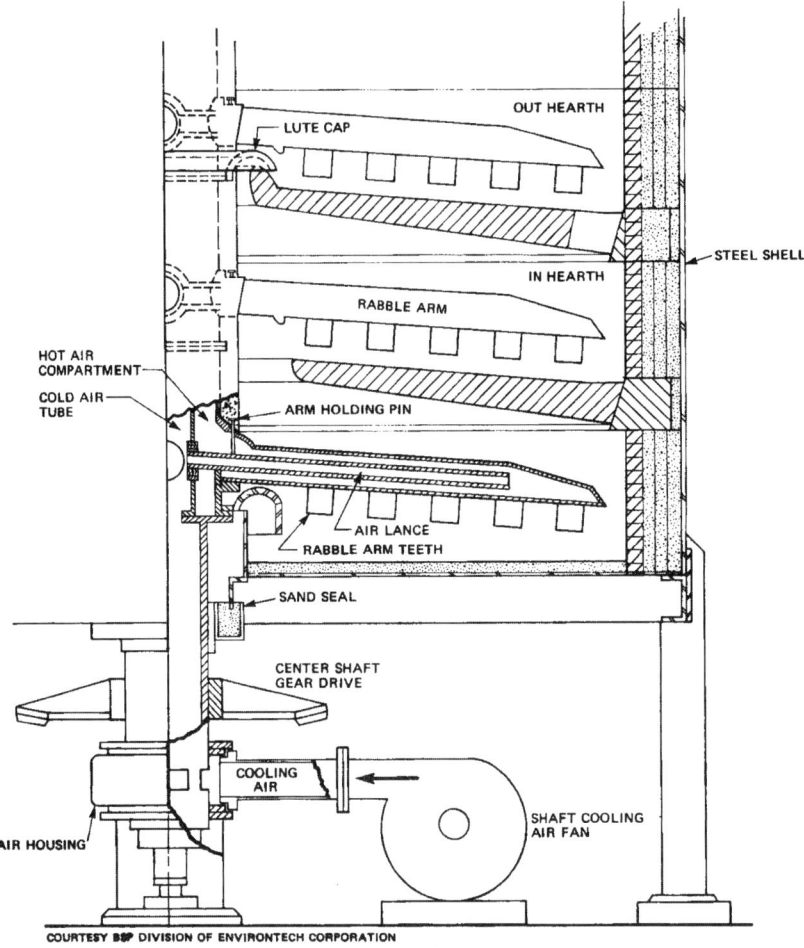

**Fig. 10.23.** Detail of rabble arm of multiple hearth biosolids furnace. (Courtesy of BSP Div. Envirotech Corp.) (53).

A rotating shaft is mounted in the center of the furnace. The center shaft is equipped with arms that are cantilevered out over each hearth (rabble arms). The refractory clad and air cooled arms (Figure 10.23) (53) are fitted with a series of rectangular plows along their length. The pitch of the plows is adjustable to modify the rate at which the biosolids mass is advanced per rotation of the center shaft.

Biosolids are fed to the outer edge of the top (or, with a bypass, the second) hearth, where it is plowed (rabbled) toward the center, where it drops to the hearth below through the annular space between the hearth and the center shaft. The plows on the second working hearth are pitched in reverse to that on the first hearth, and now rabble the biosolids toward the periphery. Spotted around the periphery are drop holes leading to the next hearth. And the process continues down through the furnace. The rabbling process acts not only to move the material but also to cut, furrow, and open the surface. Indeed, it is estimated that the effective biosolids surface area for drying and combustion is about 130% of the plan area of the hearth.

## Combustion and Incineration Engineering

The upper hearths are drying hearths whereon the incoming biosolids are dried until, at some point, all or a portion of the matter is dry enough to burn/pyrolyze. This heat release from combustion of the biosolids, in combination with energy from burners mounted on most hearths, provides the hot gases used for drying on the upper hearths. The burning process is concluded in two phases: gasification/carbonization and char burnout. Subsequently, the hot ash is cooled with a countercurrent flow of incoming air for heat recovery.

The MHF is a large furnace with considerable thermal inertia. Thus, it can absorb substantial swings in feed biosolids quantity and quality without producing unrecoverable upsets. The behavior of an MHF is complex, with each hearth's processes reflecting the contribution from below of lower hearth off-gases and burners; the contribution from above of biosolids that are in a changing position along the reaction sequence from cold, wet raw feed to ash cooling; and the hearth biosolids' own contribution of heat absorption (drying) or heat release (combustion/pyrolysis), as perturbed by the effects of rabbling speed and burner heat flux and mixing. However, with experience, many operators have learned to bring the units operating in the full combustion mode to a stable condition with a steady and under-control sequence of processes progressing down through the furnace.

The importance of feed stability to satisfactory performance cannot be overstated. Because of the staged character of the MHF process, cycling in feed rate or biosolids percent solids results in the generation of several combustion zones down through the furnace. This can result in the discharge of still-burning biosolids into the residue conveyors with resulting equipment damage. However, with stable feed, the MHF system can be a stable operating system with few operating problems. If the combustion zone temperatures become excessive relative to the biosolids ash fusion point, "clinkers" form as masses of fused ash that can block drop holes and foul or jam the ash conveyor. This problem becomes excessive at temperatures above 1000°C (1800°F).

### 4.4.1.2. Fluidized Bed Systems

Drawing on the experience in the petroleum and ore processing industries, fluid bed furnaces were introduced to the combustion of sewage biosolids in 1962. Since then, they have seen rapid growth (when fuel was relatively cheap) and severe cutbacks (when, in the late 1970s, fuel costs began to rise rapidly). More recently, the introduction of improved energy-conserving designs, their favorable capital cost, and their favorable environmental characteristics have led to strengthening of their use in biosolids service. The fluid bed (FB) incinerator is well suited to the drying and combustion of a wide variety of biosolids waste. At present, over 85 FB systems are operating in North America and many more in Europe.

Figure 10.24 (53) shows a cross-section of a typical FB installation. The system in Figure 10.24 is designed for high temperature operation such that a refractory arch is indicated as the distribution plate located above the windbox into which combustion air is introduced. Other designs use alloy metal distribution plates. In operation, air at 0.2 to 0.35 bar (3 to 5 psig) is forced into the windbox and passes into the cylindrical furnace through an array of openings (tuyeres) in the distribution plate. Resting on the plate is a mass of graded sand (usually 20 to 80 mesh) from 0.75 to 1 m (2.5 to 3 feet) deep.

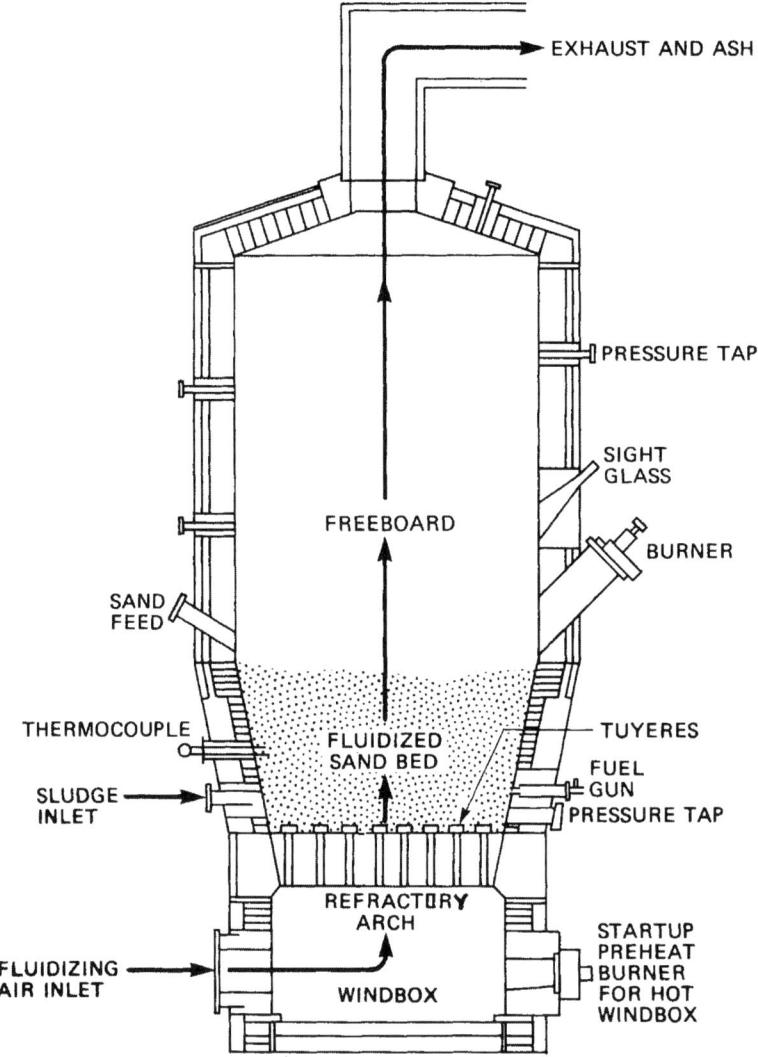

**Fig. 10.24.** Cross section of fluidized bed biosolids incinerator (53).

The tuyeres are designed to pass air into the bed but inhibit sand from draining into the windbox.

As the rate of air flow increases, the sand bed expands to about twice its original volume—sufficient growth to expand the bed to a working density high enough that the biosolids will not float to the top of the bed, yet insufficient air flow to blow the sand out of the reactor. Typically, the superficial air velocity at the bottom of the bed (based on the bed diameter) is 0.5 to 1 m/s (1.5 to 3 ft/s). The sand in the bed is in a state of violent recirculating motion that maintains a remarkable degree of uniformity in bed temperature. The large mass of the sand, heated to bed temperature, provides a large thermal flywheel effect that protects against rapid fluctuations in bed temperature if the feed rate or net heating value changes. For example, a 4.5-m (15-foot) diameter freeboard FB reactor, during normal operation, contains about 2 million kcal (8.5 million BTU) in the bed.

Other than the bed itself, key equipment includes the following:

- The fluidizing air turbine blower servicing air to the system
- The freeboard zone (above the bed) wherein combustion is completed; sometimes overfire air jets are mounted in the freeboard to assist the radial mixing process
- An (optional) heat exchanger mounted after the freeboard to preheat fluidizing air countercurrent to the hot gases leaving the furnace
- The APC equipment, which can include the Venturi scrubbers common in the U.S. or electrostatic precipitators common in Europe and Canada

*4.4.2. Pyrolysis Mode Systems*

4.4.2.1. MULTIPLE HEARTH SYSTEMS

In conventional MHFs, the system is allowed to operate with full burning. Thus, the gases leaving the furnace are primarily carbon dioxide, water vapor, and the nitrogen and oxygen from the air. In the pyrolysis mode, with a deficiency of air relative to the stoichiometric requirement, the off-gas contains a spectrum of hydrocarbons, hydrogen, and considerable carbon monoxide. Thus, in the pyrolysis system, a second afterburner chamber is required wherein a burner is mounted (to ensure ignition), and additional air (to, perhaps, 40% excess overall) is added. In some configurations, the "afterburner" function is accomplished on the top hearth and the furnace is configured to bypass the top hearth with the sludge feed. This so-called zero hearth afterburner is simpler in design and lower in cost than configurations using a second chamber as the afterburner. However, the difficulty in achieving good mixing and reliable energy exchange from an afterburner burner to the incompletely burned furnace gases often leads to higher than desired emissions of carbon monoxide, hydrocarbons, and odorous organic compounds.

In the starved air mode, perhaps 80% of theoretical air is supplied to the pyrolysis zones of the furnace. This releases approximately 80% of the heat of combustion in the biosolids. About 5% of the biosolids fuel value can be lost in unburned carbon char in the ash unless the lower hearths are operated above the stoichiometric level to ensure burnout. Looking at the furnace as a whole (including the air addition in the afterburner), the system operates at a net of about 40% excess air. This is in comparison with a conventional MHF system that uses 100% to 150% excess. Since most biosolids incinerators are energy deficient (biosolids energy is insufficient to dry the biosolids), the saving in fuel by reducing air use can be substantial.

Pyrolysis reactions of biosolid organic matter begin at about 200°C (400°F). Generally, pyrolysis reactions are initiated not by indirect heating but by careful air control (the starved air approach) such that the partial combustion of the biosolids material provides the energy for pyrolysis. Although not rigorous, the degree of pyrolysis that is affected at a given temperature T (°C) may be roughly estimated by:

$$Fraction\ pyrolyzed = 1.0663 \left(1 - \exp\left[\frac{(200 - T)}{360}\right]\right) \qquad (38)$$

The products of pyrolysis range broadly and include simple, low molecular weight hydrocarbons, complex polyaromatics, and a wide variety of partially oxidized alcohols, aldehydes, ketones, and the like. Higher temperatures favor the simpler, low molecular weight compounds, and lower temperatures favor the tarry, heavy oils. Importantly, however, the process involves gasification; the solid organic matter has been moved to the gas phase where combustion can be readily affected and the mixing processes of fuel vapors with oxygen are under the control of the designer. Further, because the temperatures where pyrolysis occurs are several hundred degrees lower than for full combustion conditions, heavy metals remain in the ash. Low emissions are also favored by the low gas velocities (discouraging entrainment) occasioned by the small air flow needed for partial combustion.

In the pyrolysis mode, there is considerably less experience than in the full combustion mode. Data from the Contra Costa County California plant (55) and from extensive US EPA-sponsored pilot plant studies (56) at Nichols' Research and Engineering (Bell Meade, New Jersey) suggest that the pyrolysis mode is easy to control, free of the clinkers that can form under full combustion conditions if hearth temperatures rise to levels much above 950°C (1750°F).

There are insufficient data to authoritatively document the environmental implications of operating in the starved air mode. However, the limited data available suggests lower particulate emissions, good control of hydrocarbon emissions in the afterburner, and considerably reduced heavy metal enrichment of the particulate. It is speculated that the reducing condition in the furnace inhibits the oxidation of chromium from the relatively nontoxic +3 form to the toxic and carcinogenic +6 form. This can be a significant environmental problem that is observed in fully combusting systems when burning high-chromium biosolids sometimes found in communities with substantial metal plating or leather tanning operations.

The primary beneficial features of the MHF system in the starved air mode relate to energy conservation where the lower excess air operation significantly (20% to 40% overall) significantly reduces fuel use. Clinker problems are reduced to a minimum, and it would be anticipated that refractory maintenance would be reduced due to the less severe operating temperatures.

Pyrolysis mode MHF operation is a combination of equipment configuration and operating methodology. The equipment for starved air mode is essentially the same as for conventional MHF operation and is well proven in biosolids incineration service. However, other than the limited Contra Costa and Nichols work mentioned above, there are only limited data on the pyrolysis operating configuration. Two plants (Alexandria, Virginia, and Cranston, Rhode Island) were originally constructed to operate in the pyrolysis mode but have not. Several other plants were run in starved air mode for a short time when "hot" Zimpro conditioned biosolids were being processed and temperatures became excessive.

Unfortunately, many owners and operators of MHF systems have a concern (although not realized in the limited operational experience with biosolids nor in the hundreds of MHF furnaces used for activated charcoal manufacture) about explosions if furnace doors are inadvertently opened when the furnace volume is filled with (combustible) pyrolysis gases. No such events were experienced in any of the furnace tests at Contra

Costa or Nichols. Nonetheless, the absence of a substantial body of working experience requires one to consider the pyrolysis mode as experimental. As noted above, however, it is fortunate that these systems can be readily designed with both conventional and starved air mode capability such that an inherent process backup exists so that experimentation is not costly.

#### 4.4.2.2. Fluidized Bed Systems

As with the MHF, the predominant application of fluid bed furnaces has been in the fully oxidizing mode. However, in the general case, the operation of FB units under sub-stoichiometric conditions is not without experience. Such conditions are used for the fluid bed processing of iron ores, for iron reduction, and for activated carbon regeneration. Installations at the Hyperion POTW in California showed some interesting applications of pyrolysis mode: a three-step, starved air system. The bed was operated at 30% theoretical air receiving a dried sludge from the Carver-Greenfield multiple effect evaporator technology. Published data on the Hyperion plant performance have been limited. However, it is reported that the performance of the FB portion of the plant was very satisfactory. It should be noted, however, with the fully dried biosolids feed, that the Hyperion FB behavior may be atypical. At Hyperion, pneumatic feeder problems were noted with the dry (>95% solids) feed. Air was added from sidewall jets just above the sand bed so that the freeboard operated at 80% theoretical air. The first (of two) afterburners operated under stoichiometric conditions. This was followed by a fully combusting afterburner with, overall, 135% theoretical air at the stack.

With steady biosolids flow, the excellent mixing in the fluid bed unit is indeed conducive to the attainment of a uniform product gas with hydrogen and carbon monoxide (plus small quantities of low molecular weight hydrocarbons) being the predominant forms of unreleased fuel value. The secondary (afterburner) chamber can involve high turbulence with air jets and chamber design combining to ensure complete burnout. In the case of the Hyperion plant, an intermediate boiler was installed to withdraw heat before completing combustion. Intermediate cooling at the stoichiometric point limited $NO_x$ formation (from oxidation of biosolids nitrogenous matter as well as thermal $NO_x$). Also, the reducing environment would be expected to minimize chromium oxidation.

With conventional biosolids incineration, overall excess air for the process can be limited to about 25% thus reducing fuel requirements somewhat in comparison to fuel use with conventional 40 + % excess air fully oxidizing fluid bed operations. Note, however, that the environmental and operational benefits indicated above are obtained at a significant increase in capital investment and the slightly lower excess air levels allow only a limited recovery of cost due to lower fuel requirements. Problem areas in maintenance are uncertain since so little data is available.

### 4.4.3. Full Combustion Mode Systems

Full combustion systems are the most common, and their operating characteristics and design features are presented elsewhere (1, 53). In this operational mode, sufficient air is supplied to meet and generally exceed stoichiometric requirements such that temperatures exceed 800°C (1500°F) in most cases.

4.4.3.1. MULTIPLE HEARTH SYSTEMS

The MHF system, operated in the full combustion mode, is the most common means of biosolids incineration in use today. Although not without its problems, the system has shown itself capable of responding to wide variations in feed rate and biosolids characteristics and to provide the flexibility to adapt to seasonal changes and industrialization.

The design features of the MHF system for the full combustion mode are essentially identical to those of the pyrolysis mode. Since combustion of gasified combustible is ideally completed within the unit, a separate afterburner is not provided. However, when biosolids with high grease content are burned, there is often excessive smoking and hydrocarbon carryover. When this problem is anticipated, the system is often configured to bypass the top (renamed the zeroth) hearth, which is then strongly fired with fuel to provide an afterburner function that is relatively inefficient. Retrofit of existing units with a hydrocarbon emission problem is also practiced with this reconfiguration, although usually with a loss in processing capacity.

The key process feature of the MHF is the regenerative aspect wherein the wet, incoming biosolids are dried on the top hearths and then pass through three oxidation zones followed by ash cooling. After drying, the biosolids temperature increases until, in the first of the three zones, thermal breakdown (pyrolysis) takes place with ignition and burning of the pyrolysis gases above the surface. Heat transfer to the dry solids is partially convective (from the hot gases rising from the fully burning zones below), combined with radiation from the hot refractory walls, roof, and pyrolysate flame. The second oxidation zone involves conventional oxidation of biosolids on the hearth. Because diffusion of oxygen from the gas phase to the burning mass is slow, a substantial fraction of the biosolids' organic matter remains as a high-carbon char. The char oxidizes slowly in the third combustion zone. Since there is essentially no liquid water to moderate the temperature, the solids often approach 1000°C (1800°F) in this zone. Finally, cooling of the mineral ash residue takes place on the lowest hearths. Typical temperature levels corresponding to this chain of processes are shown in Table 10.23.

The MHF systems are commonly operated at more than 100% excess air; levels as high as 200% excess are not uncommon. Ideally, a well-operated furnace can operate successfully at, say, 75% excess air, although stable feed composition and rate are essential to achieve these levels. With a stable feed rate (a risky assumption for many plants), the hearth-to-hearth temperatures hold steady hour after hour and, with trim of the system using the variable center shaft speed control, excellent ash burnout and heat recovery from the ash are achieved.

To fully realize the energy economy of the regenerative operating mode, the top hearth temperature gas must be allowed to fall considerably below levels where combustion

Table 10.23
Temperature distribution in MHF systems

|  | Drying zone | Pyrolysis/burning zone | Cooling zone |
| --- | --- | --- | --- |
| Biosolids | 70°C | 730°C+ | 200°C |
| Flue gases | 425°C+ | 830°C+ | 175°C+ |

takes place 350°C (650°F). If the biosolids contains large amounts of grease, this may lead to aerosol formation. These grease aerosols may crack or partially burn, leading to emission of fine particle carbonaceous soot (opacity) and odor. Also, raw biosolids can be carried off by entrainment in the gas flow rising through the drop holes. Heavy metal enrichment of the hard-to-collect fine particle fraction may also be a problem if the char combustion hearth temperatures exceed 900°C (1650°F), a condition that is not uncommon.

Several states require that gases in contact with raw biosolids must be reheated. For example, New Jersey requires reheat of the gases leaving the top hearth to 815°C (1500°F). Such a reheat requirement totally loses the regenerative mode advantage, and fuel use becomes comparable to plug flow fluid bed units but without the potential for "energy recycling" with the hot windbox design recuperative mode.

Perhaps the greatest benefit of the MHF is its energy economy (when environmental regulations permit). Also, the inherent inertia of the system provides a flywheel effect that is tolerant of fluctuations in the character of the feed. The problems of the system, aside from the environmental emissions and clinkering problems noted above, result from the considerable complexity and structural frailty of the design. The complexity requires considerable operator skill, and often results in the simple expedient of turning up the air supply to wash out the need for deft operating style and paying close attention to process changes. Clearly, overuse of air greatly increases fuel use per unit mass processed. The structural problem arises due to the flatness of the hearths such that careful control over the rate of temperature rise and fall must be observed to avoid catastrophic hearth failures.

#### 4.4.3.2. FLUIDIZED BED SYSTEMS

Although its application has been dwarfed by MHF installations, the economic and performance advantages of the fluid bed system have made it the dominant choice in new installations. This reflects the greater degree of control available with fluid beds, superior energy economy (with the hot windbox design concept discussed below), and improved air emissions.

The process description of the full combustion mode fluid bed is identical to the pyrolysis mode. The only difference is the proportions of combustible matter and air. Feed systems involve screw or diaphragm pumps discharging into either the bed or dropping through the freeboard. For large furnaces, a "flinger" (a rotating metal vane much like the mechanical spreader stoker discussed for burning RDF from municipal solid waste) mounted at the top of the bed (which is maintained with the draft in balance with atmospheric pressure) can be used to improve distribution.

Ideally, combustion of biosolids organic matter is completed within the sand bed such that a maximum of energy from biosolids combustion is available to dry the incoming biosolids. With high grease-content biosolids or when the biosolids is dropped through the freeboard space, a fraction of the burning takes place in the freeboard. This can starve the bed for drying energy and may lead to overheating (slag formation and buildup) in the freeboard and outlet flues.

The FB biosolids incineration at 750°C (1400°F) and 40% excess air is autogenous at an energy parameter of 225 to 250. As the feed becomes "hotter" (lower EP), the temperature will rise. The maximum bed temperature is set partially by materials of construction, design features (especially the ability of a refractory air distribution plate to accommodate the thermal expansion or, for alloy metal plates, to survive at the working temperature), and the desire to keep the bed cool enough to both prevent stickiness (incipient fusion) of the sand (rapidly leading to bed defluidization) or excessive volatilization of heavy metals. Bed temperatures at or below 875°C (1600°F) are common.

To improve energy economy, almost all recent designs include a combustion air preheater. In this design, air from the combustion air blower is passed on the shell side of a stainless steel shell and tube heat exchanger to preheat it prior to introduction to the windbox—the "hot windbox" design concept. Flue gases from the furnace are passed through the tube side. With proper distribution plate design (to accommodate the large thermal expansion from the cold state), this technique allows preheating of combustion air to as high as 650°C (1200°F). The combination of modern mechanical biosolids dewatering (to, say, 28% solids) and these levels of air preheat allows essentially autogenous operation (no fuel use).

Because of the good biosolids burnout affected by the FB, boilers can be added to recover heat from the furnace off-gases without undue risk of fouling. In some plants, boilers are used in addition to the hot windbox design to bring the gases to approximately 250°C (500°F), where they can be passed to an electrostatic precipitator for particulate control. Steam can be used with a turbine drive to reduce electrical costs for the large, high-horsepower combustion air blower as well as for building heat and process hot water.

The FB furnace is stable in operation with, commonly, oxygen instrumentation adjusting the fuel feed to maintain a relatively constant excess air level. Because of the simple refractory design, furnace maintenance costs are low. Maintenance problems with the hot windbox design have been experienced from thermal stresses in the expansion joints of the heat exchanger and some fouling can occur. In general, however, these problems can be dealt with by proper design and operation. One should note in this regard that the on-line availability of the biosolids incinerator is dependent on the availability of the heat exchanger (the most vulnerable part of the system), so prudent vendor and equipment selection has a particularly good payoff.

A unique feature of the FB system, brought about by the high sensible heat content (thermal inertia) of the sand bed and the simple mechanical design, is the ability of the systems to be rapidly shut down (simply turn off the air and the biosolids supply) and held without significant heat loss for extended periods. Thus, a system can be shut down for as much as a day or two and brought back into service in less than a half-hour.

Because of the excellent mixing in the bed and the long residence times of gases in the freeboard, organic emissions from the FB are small. If volatiles do become a problem, addition of air jets in the freeboard is often effective in realizing acceptable burnout of gas phase pollutants. The low bed temperature of the FB allows the operator to avoid excessive heavy metal volatilization.

Although most U.S. systems utilize a Venturi followed by a perforated tray scrubber system, many European plants, in combination with energy recovery in a boiler, use an electrostatic precipitator for particulate control. Excellent particulate resistivity properties lead to ESP collection efficiency that meets or exceeds U.S. standards. When heavy metals are a concern, several U.S. plants have installed wet electrostatic precipitators with excellent heavy metal collection performance.

The FB units offer a broad range of economic and performance benefits in comparison to MHF systems: simplicity of construction and operation, lower first and operating cost, the potential for energy recovery, and superior environmental emissions. The ability to shut down and restart (say, over a weekend) is also advantageous for smaller POTWs with modest biosolids incineration requirements. The fluid bed, either with a cold or hot windbox, and with energy recovery, is well developed and commercially available from several firms.

4.4.3.3. INFRARED SYSTEMS

The first electric furnace (infrared) furnace was installed in Richardson, Texas, in 1975. Since then, only a few units (less than 10) have been installed for biosolids incineration, importantly due to equipment problems and high electricity cost. The electric furnace system is composed of a horizontal woven stainless steel endless belt that is drawn through a long, refractory-lined chamber. Biosolids are fed and leveled onto one end of the belt and move down the furnace. Electrically heated glow-bars radiate heat to dry and heat the biosolid mass until it bursts into flame. The hot combustion gases pass countercurrent to the entering biosolids to assist in the drying process. Due to the formation of an insulating crust on the top of the biosolids cake, breaker bars are positioned at intervals along the belt to disrupt the crust and expose new surface. Ash is discharged from the end of the belt to an ash-handling system. Combustion air is introduced at the ash discharge end.

The combustion air rate for the electric furnace ranges from 20% to 70% excess, thus offering reasonable thermal efficiency from that point of view. Electrical heating requirements are similar to that of the MHF, but typically the cost of energy as electricity is considerably higher than when it is obtained by burning fossil fuels. When the biosolids are autogenous (e.g., as for some Zimpro biosolids), the energy economy of the electric furnace is superior to the MHF.

The electric furnace operating problems relate, importantly, to the highly mechanical nature of the device. Belt and glow-bar replacement have been costly and outages may be frequent and prolonged. However, the attractive capital cost and modular construction makes the electric furnace potentially attractive for small treatment plant systems with biosolids disposal problems.

The nature of the electric furnace process has shown excessive hydrocarbon and CO emissions. This comes from the poor mixing and stratification within the furnace. In states such as New Jersey, the high fuel costs associated with the state's requirement for an afterburner eliminate the other energy or capital advantages of the electric furnace system. Although data are not available, one could anticipate that excessive heavy metal volatilization may occur in the combustion zone.

The primary benefit of the electric furnace arises from its low capital cost. In general, system operational problems and high energy costs greatly overshadow this benefit. The electric furnace is still in the advanced state of development. Limited sales and experience in biosolids management service has not supported extensive experimentation to optimize the configuration of air systems, glowbar design, belt materials, and support/drive concepts.

### 4.5. High-Range (>1100°C or >2000°F) Combustion Processes

Increasing concern regarding the leaching of heavy metals from biosolids ash has led to the development of several combustion systems wherein the biosolids ash is heated beyond the point of fusion. Then, a glassy slag is formed. Leaching tests on the slag show essentially no solvation of metals. Several companies have plants, and plants based on this operating concept are in planning for use in Japan, where limited land area focuses great value on the minimization of ultimate residue (ash) disposal. The Kubota Company, drawing on Woetchke furnace designs developed in the 1960s in West Germany, offers one such design.

The primary furnace used by Kubota is a slowly rotating, refractory lined cylindrical chamber with an outlet tap in the center (Figure 10.25). The chamber has a relatively flat, refractory-lined, fixed, reverse-conical roof equipped with down-firing burners. There is a clearance (the annulus between the roof and lower chamber) through which a stream of predried biosolids is moved by the rotation of the lower body into the cavity between the roof and the lower chamber. There, the burners, augmented with preheated air, reduce the biosolids to a molten slag. The combustion heat contained in the biosolids solids is released by furnishing excess air through the burner ports. The size of the cavity may be adjusted to increase (larger cavity) or decrease (smaller cavity) the processing rate.

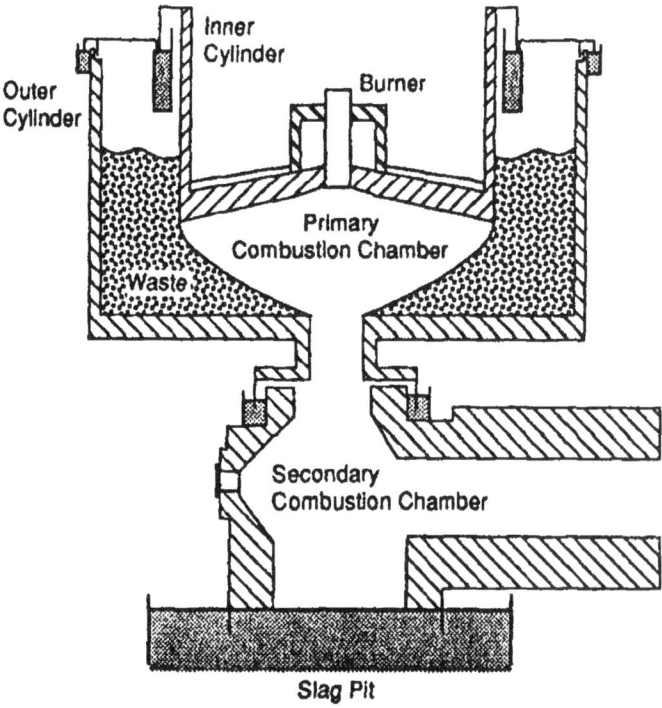

**Fig. 10.25.** Kubota melting furnace.

The primary furnace cavity is maintained slightly sub-stoichiometric. This produces both maximum temperatures and minimizes the gas flow in order to minimize entrainment.

The molten slag and off-gas from the primary chamber flow down through a water-cooled "monkey" discharge opening into a secondary combustion chamber. The slag continues to fall to a solidification area or directly into a water quench. Slow cooling favors crystal growth and forms a dense, strong, obsidian-like black glass. Faster cooling results in a granular black fractured glass with poorer structural characteristics than the slow-cooled product.

Addition of secondary air completes the combustion in a tunnel-type afterburner. The hot gases from the secondary are used to preheat the combustion air for the primary furnace (using a heat exchanger similar to the hot windbox fluid bed) and then are tempered and used to dry the incoming biosolids. Off-gas from the dryer is cleansed with an electrostatic precipitator, condenser, and scrubber. The use of fossil fuel is limited by the low overall air quantity supplied to the unit (10% to 30% excess). The temperature in the primary furnace is maintained at approximately 1450°C for most biosolids to yield acceptable biosolids fluidity. The furnace operation is stable as seen from strip charts of temperature and gas composition.

The operation of the primary chamber sub-stoichiometric greatly limits the formation of fuel-nitrogen $NO_x$. Net $NO_x$ in the flue gases from several plants averaged 100 to 150 ppm. Although one might expect the high primary furnace temperatures to result in emission of heavy metals, data from the manufacturer indicate that the majority of the metals are bound into the fluid slag that blankets the incoming biosolids and do not show up as heightened metal emissions.

The key environmental characteristic of the Kubota furnace is the ash, which exhibits essentially no leaching (below the detection limit by atomic absorption for cadmium, chromium +6 and +3, arsenic, mercury, and lead. There was also no detection of PCBs or cyanide ion, although these species appear in the raw biosolids. The slag formed has a specific gravity of from 2.4 to 2.7 and appears useful as a clean fill or road-bed material.

The primary benefits of the Kubota furnace relate to the potential to have no net waste for disposal (assuming that the excellent leaching characteristics and strength characteristics of the glassy ash make it disposable as a fill or usable as an aggregate with cement in patio blocks or other items of commerce rather than a waste). Fuel requirements are modest, but the plant has a high capital cost. Operations are complex including dryers, several air cleansing systems, and a mechanically sophisticated, high-temperature furnace and heat-recovery system.

Data indicate that acceptable long-term operation is possible. Three or more plants are now operating in Japan using the Kubota melting process, with the earliest plant coming on-line in 1975. The system must be considered to be in an advanced state of development.

## 4.6. Discussion

There are many alternatives open to the biosolids management professional for the processing of wastewater biosolids using thermal methods. These processes range from simple and mechanically complex drying systems to high-temperature slagging furnaces. The environmental implications of these plants are generally controllable and focus on air emissions (including odor). Although the capital costs are intensive and often there is the burden of continuing high operating expense, the significant reduction in weight or volume effected by these processing methods is increasingly justified by spiraling hauling and dumping fees.

The selection of a biosolids management system requires many factors (technical, economic, political, etc.) to be weighed and balanced. These decisions are rightly the responsibility of the ultimate owner or operator of the system. Often, cost factors dominate in the selection decision. Capital and operating and maintenance costs are basic and important criteria. In comparing alternatives, however, one must be sure that the capital cost estimate includes all elements (spare parts, provision for extra-rugged construction and materials of construction, etc.) that adequately protect continued operability and on-line availability. Operating cost estimates should include consideration of training expenses, higher than normal operator wage scales (to ensure a sophistication of operator compatible with sophisticated equipment and other reasonable and prudent contributions to keeping the costly and critical equipment operating at peak. Finally, maintenance cost should include the special elements that recognize the more sophisticated and maintenance-prone working environment of high-temperature incineration systems. These include contract maintenance of instrumentation and control systems, regular refractory replacement and repair, and so forth. Flexibility and expandability are important criteria for wastewater systems with changing processes, dewatering methods or polymers, growth of the system, etc., which may result in a continuously changing operating mode for the system.

Environmental impact must be evaluated in depth and carefully considered. Often, a risk assessment type document must be prepared as part of that evaluation. Also, community and regulatory perspectives change; the high capitalization systems used for most thermal processing alternatives cannot be readily modified to adapt. Fit with the community labor forces and willingness to spend are particularly important for small communities where sophisticated, high-cost systems, however appealing, may be inappropriate selections in view of competition with high-tech industry (with higher wages and benefits) for high-quality operators; the imperative for high-quality maintenance and associated costs for thermal systems; and other factors reflecting problematical local conditions. It should be said, however, that I have seen many small communities with excellent records in the operation of very costly and sophisticated sludge management systems.

## 5. ECONOMICS OF INCINERATION

Incineration is often the highest-cost approach to waste management. Consequently, an understanding of the economics of incineration is an important part of engineering and management decision making. Unfortunately, providing definitive guides to economic analysis is as difficult as it is for design.

The costs of incineration have increased rapidly since the 1970s. The reasons underlying the cost increases parallel those that have gripped the overall U.S. economy during this period: rising interest rates, equipment and construction costs, and labor and energy rates. Beyond these background inflationary factors, the increasing equipment

**Table 10.24**
**Elements of capital costs for incineration systems**

I. Incineration system
   A. Waste conveyance
      1. Open or compaction vehicles, commercial containers
      2. Special design containers
      3. Piping, ducting, conveyors
   B. Waste storage and handling at incinerator
      1. Waste receipt and weighing
      2. Pit and crane, floor dump and front-end loader
      3. Holding tanks, pumps, piping
   C. Incinerator
      1. Outer shell
      2. Refractory
      3. Incinerator internals (grates, catalyst)
      4. Burners
      5. Fans and ducting (forced and induced draft)
      6. Flue gas conditioning (water systems, boiler systems)
      7. Air pollution control and continuous emission monitoring (CEMs)
      8. Stacks and flues
      9. Residue handling
      10. Automatic control and indicating instrumentation
      11. Worker sanitary, locker, and office space

II. Auxiliary systems
   A. Land preparation, grading, clearing and grubbing, etc.
   B. Buildings, roadways, parking areas
   C. Foundations
   D. Special maintenance facilities
   E. Steam, electrical, water, fuel, and compressed air supply
   F. Secondary pollution control
      1. Residue disposal (landfill, etc.)
      2. Scrubber wastewater treatment

III. Nonequipment expenses
   A. Engineering fees
   B. Land costs
   C. Permits
   D. Interest during construction
   E. Spare parts inventory (working capital)
   F. Investments in operator training
   G. Start-up expenses (engineering, testing, laboratories, etc.)
   H. Technology fees to engineers, vendors

investment associated with the sophisticated air pollution control systems required by toughening emission codes and the investment involved in the generation of electrical power (turbines, condensers, cooling towers, switchgear, etc.) add capital expense and increase the needed skills and thus cost of the labor force. For refractory incinerators and

**Table 10.25**
**Elements of operating cost for incineration systems**

I. Fixed costs (credits)
   A. Repayment of debt capital
   B. Payment of interest on outstanding capital
   C. Tax credits for depreciation

II. Semivariable costs
   A. Labor (including supervision) with overhead
   B. Insurance
   D. Operating supplies
   D. Maintenance and maintenance supplies

III. Variable costs (credits)
   A. Steam usage (or credits)
   B. Electricity usage (or credits)
   C. Water supply and sewerage fees
   D. Oil or natural gas fuels
   E. Compressed air
   F. Chemicals (catalysts, water treatment)
   G. By-product credits
   H. Disposal fees

particularly for upgraded older units where operation at high excess-air levels yielded large flue-gas flow rates, the requirement for sophisticated air-pollution control systems was a major factor leading to shutdown.

## 5.1. General

It would be a gross simplification, in most cases, to suggest that incineration system capital costs could be reduced to a simple table or nomograph. Usually, the designs are highly customized, reflecting unique waste handling, ducting, local regulatory requirements, degree of automatic control, enclosure aesthetics, etc., which greatly affect the final system cost, even if the basic incinerator itself is of predictable cost. Similarly, operating costs reflect staffing practices, localized labor and utility costs, localized unit values for byproduct materials or energy, etc. Further, the bonded capital investment borne by the host community (assuming a design, build, and operate [DBO] facility development by a system vendor) depends importantly on the facility profit objectives of the vendor. In many cases, the vendor sees the long-term waste disposal service as the primary money-maker, and the construction carries only a nominal profit margin. Others seek a substantial profit "up front" on the capital investment as well as the longer term tipping fee income stream that is collected over the operating life of the service contract.

Consequently, the following subsections emphasize the elements of cost analysis as applied to incineration rather than simplified general numbers. The experienced engineer will recognize this not as an evasion but rather as an explicit recognition of the dangers of oversimplification.

Table 10.26
Solid waste weight percent characterization data (1988, 1990) (58)

| Waste category | Indianapolis, IN | Kauai, HI | King Co., WA | Bergen Co., NJ | Monroe Co., NY | Ann Arbor, MI | Portland, OR | San Diego, CA | Santa Cruz, Co., CA | National Estimate |
|---|---|---|---|---|---|---|---|---|---|---|
| Paper | 38 | 26 | 27 | 44 | 42 | 29 | 29 | 26 | 33 | 34 |
| Plastics | 8 | 7 | 7 | 9 | 10 | 8 | 7 | 7 | 8 | 9 |
| Yard debris | 13 | 20 | 19 | 9 | 7 | 8 | 11 | 21 | 15 | 20 |
| Misc. organics | 22 | 24 | 31 | 17 | 24 | 39 | 33 | 23 | 18 | 20 |
| Glass | 7 | 5 | 4 | 7 | 10 | 4 | 3 | 4 | 7 | 7 |
| Aluminum | 1 | 1 | 1 | 1 | 1 | 1 | 1 | 1 | 1 | 1 |
| Ferrous metal | 4 | 6 | 3 | 5 | 5 | 5 | 7 | 3 | 5 | 7 |
| Nonferrous metal | 0 | 1 | 1 | 4 | 0 | 2 | 0 | 0 | 0 | 0 |
| Misc. Inorganics | 0 | 3 | 7 | 5 | 0 | 2 | 9 | 6 | 6 | 2 |
| Other | 0 | 9 | 0 | 3 | 0 | 2 | 1 | 10 | 7 | 0 |

Co., county.

Table 10.27
Estimated average percent moisture in refuse on an "as-discarded" and "as-fired" basis (59)

| Component | As-fired (typical) | As-discarded |
|---|---|---|
| Food wastes | 63.6 | 70.0 |
| Yard wastes | 37.9 | 55.3 |
| Miscellaneous | 3.0 | 2.0 |
| Glass | 3.0 | 2.0 |
| Metal | 6.6 | 2.0 |
| Paper | 24.3 | 8.0 |
| Plastics | 13.8 | 2.0 |
| Leather and rubber | 13.8 | 2.0 |
| Textiles | 23.8 | 10.0 |
| Wood | 15.4 | 15.0 |

### 5.2. Capital Investment

For many prospective incinerator owners, the initial capital investment is the crucial issue. This is particularly true for industry, where the return on invested capital is often the prime measure of business performance. Table 10.24 shows the major capital expenditures. Note also that both purchased equipment and installation cost (the latter can be over 200% of the equipment cost alone) should be evaluated.

### 5.3. Operating Costs

Although capital investment is an important aspect, the actual *total unit cost*, allowing for the cost of capital, but including all operating costs, is a more incisive measure of economic impact. For example, contract hauling, typically, requires little or no investment by the owner, but may represent an unacceptably high unit cost for disposal or a long-term liability of considerable worth. Similarly, high-energy scrubbers are lower in capital cost than electrostatic precipitators, but at equivalent efficiency consume so much power that their cost per unit of gas cleaned is much higher. Typical elements of operating cost for incineration systems are shown in Table 10.25.

## 6. AN APPROACH TO DESIGN

How then do you design an incinerator? I wish the answer were straightforward, with the underlying principles uncluttered with contradiction and free of the need to apply both technical and value judgments. This section, however, can only scratch the surface of the challenge of system design; we will attempt generally to structure, if not guide in detail, the design process.

### 6.1. Characterize the Waste

Obtain the best practical characterization of the quantity and composition of the waste. Keep in mind future growth and the impact of seasonality and changes in technology and economics on operational patterns and decision making.

**Table 10.28**
**Estimated final analysis of refuse categories, percent, dry basis**

| Category | C | H | O | N | Ash | S | Fe | Al | Cu | Zn | Pb | Sn | P* | Cl | Se | Fixed carbon |
|---|---|---|---|---|---|---|---|---|---|---|---|---|---|---|---|---|
| Metal | 4.5 | 0.6 | 4.3 | 0.05 | 90.5 | 0.01 | 77.3 | 20.1 | 2.0 | – | 0.02 | 0.6 | 0.03 | – | – | 0.5 |
| Paper | 45.4 | 88 | 42.1 | 0.3 | 6.0 | 0.12 | – | – | – | – | – | – | – | – | Trace | 11.3 |
| Plastics | 59.8 | 8.3 | 19.0 | 1.0 | 11.6 | 0.3 | – | – | – | – | – | – | 0.01 | 6.0 | – | 5.1 |
| Leather/rubber | 59.8 | 8.3 | 19.0 | 1.0 | 11.6 | 0.3 | – | – | – | 2.0 | – | – | – | – | – | 6.4 |
| Textiles | 46.2 | 6.4 | 41.8 | 2.2 | 3.2 | 0.2 | – | – | – | – | – | – | 0.03 | – | – | 3.9 |
| Wood | 48.3 | 6.0 | 42.4 | 0.3 | 2.9 | 0.11 | – | – | – | – | – | – | 0.05 | – | – | 14.1 |
| Food waste | 41.7 | 5.8 | 27.6 | 2.8 | 21.9 | 0.25 | – | – | – | – | – | – | 0.24 | – | – | 5.3 |
| Yard waste | 49.2 | 6.5 | 36.1 | 2.9 | 5.0 | 0.35 | – | – | – | – | – | – | 0.04 | – | – | 19.3 |
| Glass | 0.52 | 0.07 | 0.36 | 0.36 | 99.0 | – | – | – | – | – | – | – | – | – | – | 0.4 |
| Misc. | 13.0[a] | 2.0[a] | 12.0[a] | 12.0[a] | 70.0 | – | – | – | – | – | – | – | – | – | – | 7.5 |

* Estimated (varies widely), excludes phosphorus in $CaPO_4$.
[a] From ref. 7.

Table 10.29
Ultimate analysis of annul average 1970 mixed municipal refuse

| Category | Wt% As fired | Wt% As discarded | % Moisture as discarded | Composition of average refuse, kg/100 kg dry solids | | | | | | Total Wt, kg |
|---|---|---|---|---|---|---|---|---|---|---|
| | | | | Ash | C | $H_2$ | $O_2$ | S | $N_2$ | |
| Metal | 8.7 | 8.2 | 2.0 | 10.13 | 0.50 | 0.067 | 0.481 | 0.0011 | 0.0056 | 11.19 |
| Paper | 44.2 | 35.6 | 8.0 | 2.74 | 20.70 | 2.781 | 19.193 | 0.0547 | 0.1368 | 45.59 |
| Plastics | 1.2 | 1.1 | 2.0 | 0.17 | 0.90 | 0.125 | 0.285 | 0.0045 | 0.0150 | 1.50 |
| Leather and rubber | 1.7 | 1.5 | 2.0 | 0.24 | 1.23 | 0.170 | 0.390 | 0.0062 | 0.0205 | 2.05 |
| Textiles | 2.3 | 1.9 | 10.0 | 0.08 | 1.10 | 0.152 | 0.995 | 0.0048 | 0.0523 | 2.38 |
| Wood | 2.5 | 2.5 | 15.0 | 0.09 | 1.43 | 0.178 | 1.260 | 0.0033 | 0.0089 | 2.96 |
| Food waste | 16.6 | 23.7 | 70.0 | 2.17 | 4.13 | 0.574 | 2.730 | 0.0248 | 0.2772 | 9.90 |
| Yard waste | 12.6 | 15.5 | 50.0 | 0.54 | 5.31 | 0.701 | 3.890 | 0.0378 | 0.3129 | 10.79 |
| Glass | 8.5 | 8.3 | 2.0 | 11.21 | 0.06 | 0.008 | 0.041 | – | 0.0034 | 11.32 |
| Miscellaneous | 1.7 | 1.7 | 2.0 | 1.62 | 0.30 | 0.046 | 0.278 | – | 0.0696 | 2.32 |
| | 100.0 | 100.0 | | 28.99 | 35.66 | 4.802 | 29.543 | 0.1372 | 0.9022 | 100.00 |

From ref. 7.

Combustion and Incineration Engineering

**Table 10.30**
**Average refuse summary, as-fired basis: 100-kg average refuse**

| Component | Weight percent (dry basis) | Weight percent (wet basis) | Mols (wet basis) |
|---|---|---|---|
| Moisture ($H_2O$) | – | 28.16 | 1.564 |
| Carbon, C | 35.66 | 25.62 | 2.135 |
| Hydrogen, $H_2$ = bound | 3.69 | 2.65 | 1.326 |
| Oxygen, O = bound | 29.52 | 21.21 | 1.326 |
| Hydrogen, $H_2$ | 1.11 | 0.80 | 0.399 |
| Sulfur, S | 0.1372 | 0.10 | 0.003 |
| Nitrogen, $N_2$ | 0.0396 | 0.64 | 0.023 |
| Ash | 28.99 | 20.82 | – |
| Total | 100.00 | 100.00 | 6.770 |

From ref. 59.

### 6.2. Lay Out the System in Blocks

Too often, incineration facilities are developed in pieces, with insufficient attention being given to the mating of interfaces between various elements. Remember the concept of system.

### 6.3. Establish Performance Objectives

Review present and prospective regulatory requirements. Evaluate the needs for volume reduction, residue burnout, or detoxification. Apply these to appropriate points in the facility layout.

### 6.4. Develop Heat and Material Balances

Using the techniques developed early in this chapter, determine the flows of material and energy in the waste, combustion air, and flue gases. Take into consideration probable materials of construction and establish reasonable limits on temperatures. Explore the impact of variations from the average waste feed composition and quantity. In practice, these off-average characteristics generally better characterize the day-to-day operating conditions.

### 6.5. Develop Incinerator Envelope

Using heat release rates per unit area and per unit volume, the overall size of the system can be established. Using burning rate, flame length and shape, kinetic expressions, and other analysis tools (1), establish the basic incinerator envelope. The final shape depends on judgment, as well as on these calculations. Draw on the literature and the personal experience of others. Interact with other engineers, vendors, operators, and designers of other combustion systems with similar operating goals or physical arrangements. Attempt to find the balance between being overly conservative at high cost and the unfortunate fact that few of the answers are tractable to definitive analysis

## Table 10.31
## Ultimate analyses and heating value of waste components

| Waste component | Weight %, dry | | | | | | HHV, kcal/kg | |
|---|---|---|---|---|---|---|---|---|
| | C | H | O | N | S | Noncombustible | As received | Dry |
| Paper and paper products | | | | | | | | |
|   Paper, mixed | 43.41 | 5.82 | 44.32 | 0.25 | 0.20 | 6.00 | 3778 | 4207 |
|   Newsprint | 49.14 | 6.10 | 43.03 | 0.05 | 0.16 | 1.52 | 4430 | 4711 |
|   Brown paper | 44.90 | 6.08 | 47.34 | 0.00 | 0.11 | 1.07 | 4031 | 4281 |
|   Trade magazines | 32.91 | 4.95 | 38.55 | 0.07 | 0.09 | 23.43 | 2919 | 3044 |
|   Corrugated boxes | 43.73 | 5.70 | 44.93 | 0.09 | 0.21 | 5.34 | 3913 | 4127 |
|   Plastic-coated paper | 45.30 | 6.17 | 45.50 | 0.18 | 0.08 | 2.77 | 4078 | 4279 |
|   Paper food cartons | 44.74 | 6.10 | 41.92 | 0.15 | 0.16 | 6.93 | 4032 | 4294 |
| Food and food waste | | | | | | | | |
|   Vegetable food waste | 49.06 | 6.62 | 37.55 | 1.68 | 0.20 | 4.89 | 997 | 4594 |
|   Fried fats | 73.14 | 11.54 | 14.82 | 0.43 | 0.07 | 0.00 | 9148 | 9148 |
|   Mixed garbage I | 44.99 | 6.43 | 28.76 | 3.30 | 0.52 | 16.00 | 1317 | 4713 |
|   Mixed garbage II | 41.72 | 5.75 | 27.62 | 2.97 | 0.25 | 21.87 | – | 4026 |
| Trees, wood, brush, plants | | | | | | | | |
|   Green logs | 50.12 | 6.40 | 42.26 | 0.14 | 0.08 | 1.00 | 1168 | 2336 |
|   Demolition softwood | 51.0 | 6.2 | 41.8 | 0.1 | <0.1 | 0.8 | 4056 | 4398 |
|   Waste hardwood | 49.4 | 6.1 | 43.7 | 0.1 | <0.1 | 0.6 | 3572 | 4056 |
|   Lawn grass | 46.18 | 5.96 | 36.43 | 4.46 | 0.42 | 6.55 | 1143 | 4618 |
|   Ripe leaves | 52.15 | 6.11 | 30.34 | 6.99 | 0.16 | 4.25 | 4436 | 4927 |
|   Wood and bark | 50.46 | 5.97 | 42.37 | 0.15 | 0.05 | 1.00 | 3833 | 4785 |
|   Brush | 42.52 | 5.90 | 41.20 | 2.00 | 0.05 | 8.33 | 2636 | 4389 |
|   Grass, dirt, leaves | 36.20 | 4.75 | 26.61 | 2.10 | 0.26 | 30.08 | – | 3491 |
| Domestic wastes | | | | | | | | |
|   Upholstery | 47.1 | 6.1 | 43.6 | 0.3 | 0.1 | 2.8 | 3867 | 4155 |
|   Tires | 79.1 | 6.8 | 5.9 | 0.1 | 1.5 | 6.6 | 7667 | 7726 |
|   Leather | 60.00 | 8.00 | 11.50 | 10.00 | 0.40 | 10.10 | 4422 | 4917 |
|   Leather shoes | 42.01 | 5.32 | 22.83 | 5.98 | 1.00 | 22.86 | 4024 | 4348 |
|   Rubber | 77.65 | 10.35 | – | – | 2.00 | 10.00 | 6222 | 6294 |
|   Mixed plastics | 66.00 | 7.20 | 22.60 | – | – | 10.20 | 7833 | 7982 |
|   Polyethylene | 84.54 | 14.18 | 0.00 | 0.06 | 0.03 | 1.19 | 10,932 | 10,961 |
|   Polystyrene | 87.10 | 8.45 | 3.96 | 0.21 | 0.02 | 0.45 | 9122 | 9139 |
|   Polyvinyl chloride | 45.14 | 5.61 | 1.56 | 0.08 | 0.14 | 2.06* | 5419 | 5431 |
|   Linoleum | 48.06 | 5.34 | 18.70 | 0.10 | 0.40 | 27.40 | 4528 | 4617 |
|   Rags | 55.00 | 6.60 | 31.20 | 4.12 | 0.13 | 2.45 | 3833 | 4251 |
| Municipal wastes | | | | | | | | |
|   Street sweepings | 34.70 | 4.76 | 35.20 | 0.14 | 0.20 | 25.00 | 2667 | 3333 |

\* Remaining 45.41% is chlorine.
From ref. 2.

**Table 10.32**
Range of total metal concentration in various ash types—mg/kg (18)

| Ash type | No. samples | Cadmium | Copper | Lead | Zinc |
|---|---|---|---|---|---|
| Bottom ash | | | | | |
|   Grate | 4 | <1.0–48.2 | 420–12,600 | 300–2750 | 903–2420 |
|   Siftings | 60 | <0.68–67.6 | 122–21,200 | 738–103,000 | 412–46,100 |
| Fly ash | | | | | |
|   Boiler tube | 7 | 130–389 | 534–988 | 4280–16,100 | 11,100–24,300 |
|   Spray dryer | 5 | 38.0–59.4 | 312–880 | 1060–1710 | 2830–9630 |
|   Bag house | 56 | 40–578 | 142–4,399 | 1100–10,340 | 280–92,356 |
| Combined | 54 | 7.7–120 | 445–17,355 | 561–5,100 | 733–53,800 |

and computation. Particularly, talk to operators of systems. Too often the designers speak only to one another, and the valuable insights of direct personal experience go unheard and, worse, unasked for.

### 6.6. Evaluate Incinerator Dynamics

Apply the jet evaluation methodology, buoyancy calculations, empirical relationships, and conventional furnace draft and pressure drop evaluation tools to grasp, however inadequately, the dynamics of the system.

### 6.7. Develop the Design of Auxiliary Equipment

Determine the sizes and requirements of burners, fans, grates, materials-handling systems, pumps, air compressors, air pollution control systems, and the many other auxiliary types of equipment comprising the system. Here, again, the caution is to be generous, protective, and rugged. The cardinal rule is to prepare for *when* "it" happens, not to argue about *whether* "it" will happen.

### 6.8. Review Heat and Material Balances

This self-explanatory step helps to reinforce the systems perspective by tracking the flows through one after another component part.

### 6.9. Build and Operate

In many cases, fortunately, nature is kind; reasonable engineering designs will function, though perhaps not to expectations or with maintenance expense or equipment availability different from what might be wished. At times, plants built using the most detailed analysis and care result in failure. Such is the lot of workers in the complex but fascinating field of combustion and incineration.

## APPENDIX: WASTE THERMOCHEMICAL DATA

The chemical composition and heat of combustion of waste materials is a matter of concern to the design engineer, to the operator of an incinerator, and to the engineer involved in incinerator troubleshooting.

Waste composition and heat content affect combustion air requirements, flame temperatures, and bed burning processes, and impact on corrosion, air pollution, and other important incinerator operating characteristics. Ideally, refuse composition is determined empirically in a comprehensive and scientifically planned and executed program of waste sampling and analysis. Often, however, an engineer requires compositional data for use in preliminary studies. In such cases, data are needed that might be called typical or average. The data presented below may serve this need. A guideline, however, is offered: There is no typical municipal refuse. Indeed, I recommend that the word *average* be redefined, as follows: *average* is the value assigned to an often critical design parameter that is *never* observed in practice. While tongue-in-cheek, this redefinition is intended to encourage the analyst to extensively use "what-if" calculation methods to assess possible system requirements and needs.

A more comprehensive assembly of data on the properties and generation rates of municipal refuse and other wastes is available elsewhere (1, 2).

## A.1. Refuse Composition

The seasonal and annual average compositions shown in Table 10.26 were derived from an analysis of data sets from municipalities throughout the U.S. (59). These data illustrate the wide variation to be expected in waste sampling. In addition to these regional differences, one can expect seasonal variation in the composition and quantity of wastes. The average of 3 years of data from New Jersey (58) from a region where the waste was about 54% residential and 46% commercial and industrial showed 122% of the annual average volumetric monthly waste quantity was received in May; about the average from July through October; and then an almost linear decline to only 78% of the average in February, climbing back to the peak in May. The refuse density varied from $435 \, \text{kg/m}^3$ in the summer to $385 \, \text{kg/m}^3$ in the winter.

In reviewing waste composition data as input to engineering calculations, it is necessary to take account of the fact that waste materials often exchange moisture when they are mixed in the waste storage container. As an example, consider a sheet of newspaper with a typical moisture regain (the equilibrium moisture content achieved by exchange between the dry material and the normally humid atmosphere) corresponding to about 8% moisture. Wrapping the garbage from dinner (typically 60% to 70% moisture) makes the paper wet and dries the food material. Therefore, when one takes a sample of this mixed waste and weighs the newspaper fraction, the number of pounds in the "paper category" is higher than that which corresponds to the "standard" moisture 8% content, making it appear as though there is more "standard paper" in the waste than is really there. If, then, we take the chemical composition (percent carbon, hydrogen, etc.) and reference heat content of standard paper times its "weight" in the sample, we will calculate an incorrect air requirement and an incorrect heat release potential.

Consequently, when analyzing waste composition data, the results are more useful if the moisture levels of the components are adjusted, category by category from the moisture content found after moisture exchange (the "as-fired" basis), to a moisture content corresponding to the manufactured state of the materials (the "as-discarded" basis). The as-discarded basis is useful in indicating the true relative magnitude of waste generation for the various categories, as the appropriate basis for estimating salvage potential, and as the basis for forecasting refuse generation rates and chemical and physical properties for combustion calculations. Such basis adjustments can become critical for wastes where a substantial fraction of the waste is very moist and thus where profound effects of moisture transfer occur. (6, 8)

The moisture content values shown in Table 10.27 can be used to effect this basis shift. Note that it is generally necessary for the analyst to adjust the assumptions for the "as-fired" state such that the average total moisture percentage in the mixed waste is unchanged by the recalculation of the percentage composition. Spreadsheet models to facilitate such adjustments in composition are available elsewhere (1).

## A.2. Solid Waste Properties

The categorical composition is the starting point for the development of parameters of interest to the incinerator designer. Although the manipulation of gross categorical data to establish average chemical composition, heat content, and the like requires assumptions of questionable accuracy, it is a necessary compromise. Typically, between 1 and 3 tons of waste from a waste flow of 200 to 1000 ton/d are analyzed to produce a categorical composition. Then a still smaller sample is hammer-milled and mixed, and a 500-mg sample is taken. Clearly, a calorific value determination on this sample is, at best, a rough reflection of the energy content of the original waste.

### A.2.1. Thermochemical Analysis

Stepping from the categorical analysis to a mean thermochemical analysis provides the basis for stoichiometric calculations and energy balances:

- Mixed municipal refuse: Chemical data for average municipal refuse components and for the mixed refuse shown in Table 10.26 are presented in Tables 10.28 to 10.30. These data were developed with the typical refuse of 1970 in mind, but should be applicable to almost any First World waste material.
- Specific waste components: Data for specific waste components are given in Table 10.31. These data may be used when detailed categorical analyses are available or to explore the impact of refuse compositional changes. Once a weight-basis generation rate is established, data on the heating value of waste components are of great interest to the combustion engineer. Note that the "As received" column corresponds to the "As discarded" moisture category described above, a point that emphasizes the importance of considering the moisture transfer process between waste components in developing compositions and heating values for engineering analysis.

## A.3. Ash Composition

Table 10.32 shows the range of total metal concentration found in mass burn MWC systems.

## NOMENCLATURE

$c, u$ = concentration or velocity in a jet (g mol/L, m/s)
$c_m, u_m$ = concentration or velocity on the centerline of a jet (g mol/L, m/s)
$c_0, u_0$ = concentration or velocity in the nozzle fluid of a jet (g mol/L, m/s)
$d_0$ = initial diameter of carbon particle or nozzle diameter (m, ft)

$f_a$, $f_b$ = mol fraction of component a, b
$g_c$ = acceleration due to gravity (ft/s$^{-2}$)
HHV = higher heating value (kcal/kg, BTU/lb, kcal/kg mol, BTU/lb mol)
$k_d$ = diffusion rate constant for carbon consumption reaction
$K_p$ = equilibrium constant
$k_s$ = kinetic rate constant for carbon consumption reaction
$M$ = cross-flow parameter (see Equation 29)
$Mc_p^o$ = mean specific heat at zero pressure
$Mc_{p_{avg}}^o$ = average specific heat at zero pressure between two temperatures
$\dot{m}_o$ = integrated total jet flow at the nozzle (kg/sec)
$\dot{m}_x$ = integrated total jet flow at a distance $x$ (kg/sec)
$N$ = number of mols (gram mols, kilogram mols, pound mols)
$Nf_o$ = NO concentration for 100% conversion of fuel nitrogen ($N_f$) to NO
$N_{Re}$ = Reynolds number (dimensionless)
$P$ = pressure (atm, Kpa, psi, etc.)
$P_A$ $P_B$ = partial pressure of component A, B, etc
$PM_{2.5}$ = particulate matter with diameter less than 2.5 μm
$Q$ = rate of carbon consumption (g/cm$^{-2}$ s$^{-1}$)
$R$ = jet radius (m)
$R$, $R'$ = universal gas constant (see Table 10.1)
$S$ = mean residence time (sec)
$T$ = temperature (°C, °F, °K, °R)
$T$ = time (s)
$t_b$ = burning time of carbon particle
$T_o$ = temperature at reference condition (°C, °F, °K, °R)
$V$ = volume (m$^3$, ft$^3$)
$V$ = velocity
$W$ = emission factor of particulate matter (kg/ton)
$X$ = downstream distance in jet analysis (m)
$Y$ = fraction of fuel nitrogen converted to NO$_x$
$Y$ = distance perpendicular to the axis of a jet (m)
$y_o$ = generalized elevation above a reference (m)
$\Delta h$ = enthalpy change (kcal, BTU)
$\mu_o$ = viscosity (kg m$^{-1}$sec$^{-1}$)
$\rho$ = density (kg/m$^3$, lb/ft$^3$)
$\rho_a$ = density of ambient fluid (kg/m$^3$, lb/ft$^3$)
$\rho_o$ = density of nozzle fluid or initial value (kg/m$^3$, lb/ft$^3$)

## REFERENCES

1. W. R. Niessen, *Combustion and Incineration Processes*, 3rd ed., Dekker, New York (2002).
2. D. G. Wilson, *Handbook of Solid Waste Management*, Van Nostrand-Reinhold, New York (1977).

3. Y. C. Chang, Estimating heat of combustion for waste materials, *Pollution Engineering*, 29 (1979).
4. W. S. Selvig, and E. H. Gibson, in *Chemistry of Coal Utilization*, H. H. Lowry (ed.), John Wiley and Sons, New York (1945).
5. H. C. Hottel, G. C. Williams, N. M. Nerheim, and G. Schneider, Combustion of Carbon Monoxide and Propane, 10th International Symposium on Combustion, pp. 111–121, Combustion Institute, Pittsburgh, PA (1965).
6. M. A. Field, D. W. Gill, B. B. Morgan, and P. G. W. Hawksley, *Combustion of Pulverized Coal*, The British Coal Utilization Research Assn., Leatherhead, Surrey, England (1967).
7. W. R. Niessen, S. H. Chansky, E. L. Field, A. N. Dimetriou, C. P. La Mantia, R. E. Zinn, T. J. Lamb, and A. S. Sarofim, Systems Study of Air Pollution from Municipal Incineration, NAPCA, U.S. DHEW, Contract CPA-22-69-23, March (1970).
8. E. R. Kaiser, and S. B. Friedman, paper presented at 60th Annual Meeting, AIChE, November (1968).
9. D. A. Hoffman and R. A. Fritz, *Environmental Science and Technology*, 2 (11), 1023 (1968).
10. M. A. Kanury, *Combustion and Flame*, 18, 75–83 (1972).
11. U. K. Shivadev, and H. W. Emmons, *Combustion and Flame*, 22, 223–236 (1974).
12. S. Kim, J. K. Park, and H. Chun, Pyrolysis kinetics of scrap tire rubbers. I: using TGD and TGA, *Journal of Environmental Engineering*, p. 507, July (1995).
13. R. S. Burto, III, and R. C. Bailie, *Combustion*, 13–18, February (1974).
14. S. B. Alpert, and F. A. Ferguson, et al., Pyrolysis of Solid Waste: A Technical and Economic Assessment, NTIS Pb. 218–231, September (1972).
15. P. Nicholls, Underfeed Combustion, Effect of Preheat, and Distribution of Ash in Fuelbeds, U.S. Bureau of Mines, Bulletin 378 (1934).
16. R. H. Stevens et al., Incinerator Overfire Mixing Study—Demonstration of Overfire Jet Mixing, OAP, U.S. Contract 68020204 (1974).
17. E. R. Kaiser, personal communication to W. R. Niessen, C. M. Mohr, and A. F. Sarofim (1970).
18. J. L. Korn, and R.L. Huitrick, Commerce refuse-to-energy facility combined ash treatment process, 30th Annual International Solid Waste Exposition, Tampa FL, pp. 431–446, August 3–6 (1992).
19. H. Eberhardt and W. Mayer, Experiences with Refuse Incinerators in Europe, in Proceedings of the 1968 National Incineration Conference, pp. 142–153, ASME, New York (1968).
20. F. Nowak, *Brennstoft-Warme-Kraft*, 19(2), 71–76 (1967).
21. R. L. Stenburg, R. R. Horsley, R. A. Herrick, and A. H. Rose, Jr., *Journal of Air Pollution Control Association*, 10, 114–120 (1966).
22. A. B. Walker and F. W. Schmitz, Characteristics of Furnace Emissions from Large, Mechanically Stoked Municipal Incinerators, in *Proceedings of the 1964 National Incineration Conference*, pp. 64–73, ASME, New York (1964).
23. F. R. Rehm, *Journal of Air Pollution Control Association*, 6 (4), 199–204 (1957).
24. H. R. Johnstone, *University of Illinois Engineering Experiment Station Bulletin*, 228, 221 (1931).
25. National Renewable Energy Laboratory, Polyvinyl Chloride Plastics in Municipal Solid Waste Combustion, NREL/TP-430-5518, Golden Colorado, April (1993).
26. H. G. Rigo, A. J. Chandler, and S.W. Lanier, The Relationship Between Chlorine in Waste Streams and Dioxin Emissions from Waste Combustor Stacks, American Society of Mechanical Engineers Report CRTD-Volume 36, New York (1995).

27. B. Zeldovitch, P. Sadovnikov, D. Frank-Kamenetski, Oxidation of Nitrogen in Combustion, Academy of Sciences (USSR), Institute of Chemistry and Physics, Moscow-Leningrad (1947).
28. C. T. Bowman, Kinetics of pollutant formation and destruction in combustion, *Progress in Energy Combustion Sci*ence, 1, 33–45 (1975).
29. Y.-H. Kiang, The formation of nitrogen oxides in hazardous waste incinerators, 10th Biennial ASME Solid Waste Processing Conference, New York, pp. 169–176, May 2–5 (1984).
30. A. H. Rose, Jr., and H. R. Crabaugh, Research findings in standards of incinerator design, in *Air Pollution*, Reinhold, New York (1955).
31. R. G. Barton, W. R. Seeker, and H. E. Bostian, The behavior of metals in municipal sludge incinerators, *Transactions of the Institution of Chemical Engineers*, 69, Part B, 29–36, February (1991).
32. NATO CCMS, International Toxicity Equivalency Factors (I/TEF) Method of Risk Assessment for Complex Mixtures of Dioxins and Related Compounds, Report 178, December (1988).
33. D. G. Barnes, J. Bellin, and D. Cleverly, Interim procedure for estimating risks associated with exposures to mixtures of chlorinated dibenzodioxins and dibenzofurans (CDDs and CDFs), *C*hemosphere, 15, 1985 (1986).
34. Bundesgesunheitsamt (German Federal Health Office) (BGA), Sachstand Dioxine, *Bericht des Umweltbundesamtes*, 5/85, 264 (1985).
35. U. G. Ahlborg, Nordic Risk Assessment of PCDDs and PCDFs, *Chemosphere*, 19, 603, (1989).
36. Y. V. Ivanov, *Effective Combustion of Overfire Fuel Gases in Furnaces*, Estonian State Publishing House, Tallin, USSR (1959).
37. G. N. Abramovich, *The Theory of Turbulent Jets*, M.I.T. Press, Cambridge, MA (1963).
38. Layout and Application of Overfire Jets for Smoke Control in Coal Fired Furnaces, Section F-3, Fuel Engineering Data, National Coal Association, Washington, DC, December (1962).
39. M. A. Patrick, Experimental investigation of the mixing and penetration of a round turbulent jet injected perpendicularly into a transverse stream, *Transactions of the Institute of Chemical Engineering*, 45, T-16 to T-31 (1967).
40. W. R. Niessen, Municipal waste combustors: environmentally sound power plants, *Solid Waste and Power*, 7, 12–16, January/February (1993).
41. N. P. Getz, and R. W. Pease, Jr., Design considerations for dry scrubbers, in *Proceedings of the ASME National Solid Waste Processing Conference*, Philadelphia, pp. 113–118, May 1–4 (1988).
42. J. R. Donnelly, M. T. Quoch, and J. T. Moller, Joy/Niro spray dryer absorption flue gas cleaning system, Acid Gas and Dioxin Control Conference, Washington, DC (1985).
43. H. H. Krause, D. A. Vaughan, and W. K. Boyd, Corrosion and Deposits from Combustion of Solid Waste. Part IV. Combined Firing of Refuse and Coal, in *Proceedings of the ASME Winter Annual Meeting*, Houston, Texas (1975).
44. H. H. Krause, D. A. Vaughan, and W. K. Boyd, Corrosion and Deposits from Combustion of Solid Waste. Part III. Effects of Sulfur on Boiler Tube Metals, in *Proceedings of the ASME Winter Annual Meeting* (1974).
45. H. H. Krause, D. A. Vaughan, and P. D. Miller, *Journal of Engineering and Power Transmission ASME, A*, 95, 45–52 (1973).
46. H. H. Krause, D. A. Vaughan, and P. D. Miller, *Journal of Engineering and Power Transmission ASME, A*, 96, 216–222 (1974).

47. P. D. Miller and H. H. Krause, Corrosion of Carbon and Stainless Steels in Flue Gases from Municipal Incinerators, in *Proceedings of the 1972 ASME National Incineration Conference*, New York (1972).
48. P. D. Miller and H. H. Krause, *Corrosion*, 27, 31–45 (1971).
49. F. Nowak, Considerations in the Construction of Large Refuse Incinerators, in *Proceedings of the 1970 ASME National Incineration Conference*, pp. 86–92, New York (1970).
50. D. L. Klumb, Union electric facilities for burning municipal refuse at the Meramec power plant, paper presented at the Union Electric Co. Solid Waste Seminar, St. Louis, Mo., 26 October (1972).
51. A. J. Chandler, T. T. Eighmy, J. Hartlen, O. Hjelmar, D. S. Kosson, S. E. Sawell, H. A. van der Sloot, and J. Vehlow, *Municipal Solid Waste Incinerator Residues*, The International Ash Working Group, Elsevier Publishing (1997).
52. W. R. Niessen, C. H. Marks, and R. E. Sommerlad, Evaluation of Gasification and Novel Thermal Processes for the Treatment of Municipal Solid Waste, Report to the National Renewable Energy Laboratory, Contract YAR-5-15116-01, Golden, CO, July (1996).
53. Process Design Manual for Biosolids Treatment and Disposal, US EPA, Municipal Environmental Research laboratory, Office of Research and Development, Center for Environmental Research Information Technology Transfer, Cincinnati, OH (1979).
54. W. R. Niessen, Air emissions from thermal processing of wastewater sludges, conflicts in priorities, Sludge Management Conference, Boston, June (1987).
55. Solid waste resource recovery full scale test report, Report to Central Contra Costa Sanitary District, Brown and Caldwell, Inc., March (1977).
56. C. F. von Dreusche, and J. S. Netfa, Pyrolysis design alternatives and economic factors for pyrolyzing sewage sludge in multiple hearth furnaces, American Chemical Society Symposium Series No. 76, Solid Waste and Residues: Conversion by Advanced Thermal Processes (1978).
57. A. E. Gay, T. G. Beam, and B. W. Mar, Cost-effective solid waste characterization methodology, *Journal of Environmental Engineering*, ASCE, 119(4), 631, July/August (1993).
58. W. A. Sanders, II, and D. J. Birnesser, Use of solid waste quantification and characterization program to implement an integrated system in Mercer County, New Jersey," *Proceedings of the ASME National Solid Waste Processing Conference*, Long Beach, CA, pp. 221–227, June 3–6 (1990).
59. W. R. Niessen, and S. H. Chansky, The nature of refuse, *Proceedings of the 1970 ASME Incineration Conference*, New York, p. 1 (1970).

# 11
# Combustion and Incineration Management

## Mingming Lu and Yu-Ming Zheng

**CONTENTS**

INTRODUCTION
OPERATION AND MANAGEMENT OF THE MULTIPLE HEARTH FURNACE
OPERATION AND MANAGEMENT OF THE FLUIDIZED BED FURNACE
OTHER INCINERATION PROCESSES
NOMENCLATURE
REFERENCES

**Abstract** Biosolids incineration management is described in this chapter. The first section provides an overview of the dewatering process, the air pollution control equipment, and rules and regulations applicable to the biosolids incineration facilities. Details of the incineration process, operational and design parameters, and management and troubleshooting of the multiple hearth furnace and the fluidized bed furnace are provided. Other incineration technologies are briefly described.

**Key Words** Incineration • sewage sludge • biosolids • multiple hearth furnace • fluidized bed furnace • management • monitoring • troubleshooting.

## 1. INTRODUCTION

### 1.1. Overview of Biosolids Incineration

Biosolids (sewage sludge) incineration can reduce the volume of the sludge by up to 80% (1) and possibly even more than 90%, dependent on the incineration process. According to a U.S. Environmental Protection Agency (US EPA) estimate in 1993, there were 343 biosolids incinerators in the United States (2). The siting and new development of sludge incinerator has been limited in recently years, since the beneficial reuse of sludge, such as land application, has been encouraged by the US EPA. Currently, approximately 254 biosolids incinerators are in operation in the U.S., out of which 197 are multiple hearth furnaces (MHFs), 55 are fluidized bed furnaces (FBFs), and two are electronic furnaces (3). The earliest installations in the 1930s were mainly MHFs, and the newer installations are more likely to be FBFs. When the earlier MHFs are taken out of service now, they are mostly replaced with FBFs.

From: *Handbook of Environmental Engineering, Volume 7: Biosolids Engineering and Management*
Edited by: L. K. Wang, N. K. Shammas and Y. T. Hung © The Humana Press, Totowa, NJ

The selection of a sludge disposal method usually depends on many factors. The cost-effectiveness of sludge incineration is primarily decided by the plant capacity. Incineration is generally more cost-effective for larger plants; it is generally economically feasible when the plant is 20 MGD (million gallons per day) or larger for a multiple hearth furnace system and 10 MGD or larger for a fluidized bed furnace system (4). Other factors to consider include the location of the plant, the availability of land, the cost of other disposal methods, and the sludge characteristics, such as whether it is high in organic content.

The sludge incineration system usually contains a dewatering device, such as a vacuum filter, centrifuge, or belt press; an incinerator, such as a multiple hearth furnace or fluidized bed; a sludge feeding system, an ash handling system, an air pollution control, and other control devices (5). All of these are an integral part of the sludge incineration system, and the proper management of each part is essential to ensure successful operation. The air pollution control devices (APCDs) are used to remove particulate matter or the metals in the exhaust gas or to further decompose organic materials. Example APCDs for metal removal include wet scrubbers, dry and wet electrostatic precipitators (ESPs), and fabric filters. Afterburners are APCDs to ensure complete combustion, and are sometimes used for odor control (6). In recent years, regenerative thermal oxidizers (RTOs) have also been used for heat recovery and further reduction of by-products of incomplete combustion. The ash from incineration is virus and pathogen free and can be disposed into a landfill, in agricultural land application, or some other beneficial use, such as a sorbent material for a baseball field in Columbus, Ohio (6).

### 1.2. Overview of the Dewatering Process

The dewatering process has a major impact on the subsequent incineration process. Sludge high in water content requires more energy to dry off the moisture and can also result in low combustion efficiency. Wastewater sludge is usually dewatered to 15% to 35% of solids before incineration (2). When the solid content reaches approximately 30%, the incineration process can be self-sustained (autogenous) and needs no auxiliary fuels.

The often-used dewatering methods include vacuum filters, centrifuges, and belt press. The first two methods are sometimes associated with the problem of high water content. As an example, a wastewater treatment plant in Waldwick, New Jersey (operated under the Northwest Bergen County Utilities Authority) uses a belt press to dewater the sludge to 23% solid content (7).

Improvements in the dewatering process can result in lower operation costs. The Hartford Metropolitan District Commission in Hartford, Connecticut, is among the first in the U.S. to convert the sludge dewatering process from vacuum filters to continuous belt presses (8). This occurred between 1979 and 1982, and this conversion alone has resulted in a significant fuel saving of 65%.

Recently, indirect dryers have been used to further dewater the biosolids to reach its autogenous point. The indirect dryers transfer heat to the biosolids without direct contact with the solids. As a result, there is a lower volume of off-gases to treat comparing with direct drying. The Buffalo (New York) Sewer Authority has used an indirect dryer to further increase the sludge solid content to approximately 40% after it has been dewatered between 18 and 20% (9). The fuel usage was reduced; however, the drier was found to be of high maintenance.

The use of chemicals can result in higher solid content; however, some of the inorganic chemicals may lower the heating value of the sludge and thus affects combustion efficiency, and the chemicals may result in potential pollutant release during combustion.

## 1.3. Overview of Air Pollution Control Devices

The air pollution control devices are used to remove particulate matter or the metals in the exhaust gas or to further decompose organic materials. Examples APCDs for metal removal include wet scrubbers, dry and wet electrostatic precipitators, and fabric filters. Afterburners are APCDs to ensure complete combustion, and are sometimes used for odor control (6). More recently, regenerative thermal oxidizers or regenerative afterburners have been used to reduce the emission of hydrocarbons and carbon monoxide (10). The RTOs are afterburners with heat recovery; they can compensate for the high cost of fuel, and are mainly used with MHFs. Fabric filters and dry ESPs are not often used, as the high moisture content of the incinerator exhaust gas can result in caking of the fabric filter and short circuit the dry ESP.

The proper operation of these APCDs is essential for the plants to meet regulatory requirements, and the essential operating conditions governing the performance should be monitored. Based on the results of the performance test, the permitting authority will require operating parameters to be monitored. Examples of operating parameters to be monitored for the APCD and examples of measurement instruments are listed in Table 11.1. The monitoring frequency is specified in the air permit the incineration facility receives from the local authority.

Scrubbers are the most often used exhaust treatment device in sludge incineration systems; the Venturi scrubber with an impingement tray separator is the most common type. Figure 11.1 is an example of a Venturi scrubber with an impingement tray separator. The exhaust gas and water are fed separately into the inlet and both go through the Venturi section to increase flow rates. The resultant turbulence can increase the rate of particle collision with water. The turbulent water flow also avoids dust buildup and reduced abrasion of the Venturi section (11). The gas and water then go through a flooded elbow, the gas flow slows down due to the increase in the width, which allows for gas–water separation. The impingement trays are perforated, so that the particle-laden gas can flow through the holes to get into contact with the water flowing on top of the tray. This further removes the particles in the gas stream. This mist eliminator (demister) separates the water droplets from the exhaust gas and reduces moisture from the gas stream. The clean gas is then drawn with an induced draft (ID) fan and is released from the stack. The waste scrubbing water can be recycled and put back into the wastewater treatment processes or put into a separation lagoon.

A wet electrostatic precipitator (WESP) was installed in the sludge incinerator of a wastewater treatment plant in Waldwick, New Jersey, in 2000 under a state mandate, in

Table 11.1
Monitoring parameters of air pollution control devices (6)

| Operating parameter | Air pollution control device | Example measuring instrument |
|---|---|---|
| Pressure drop | Venturi scrubber, impingement scrubber, mist eliminator, fabric filter | Differential pressure gauge/transmitter |
| Liquid flow rate(s) | Venturi scrubber, impingement scrubber, mist eliminator, wet electrostatic, precipitator (ESP) | Orifice plate with differential pressure gauge/transmitter |
| Gas temperature (inlet and/or outlet) | Venturi scrubber, impingement scrubber, dry scrubber, fabric filter, wet ESP | Thermocouple/transmitter |
| Liquid/reagent flow rate to atomizer | Dry scrubber (spray dryer absorber) | Magnetic flow meter |
| pH of liquid/reagent to atomizer | Dry scrubber (spray dryer absorber) | pH meter/transmitter |
| Atomized motor power (for rotary atomizer) | Dry scrubber (spray dryer absorber) | Water meter |
| Compressed air pressure (for dual fluid flow) | Dry scrubber (spray dryer absorber) | Pressure gauge |
| Compressed airflow rate | Dry scrubber (spray dryer absorber) | Orifice plate with differential pressure gauge/transmitter |
| Opacity | Fabric filter | Transmissometer |
| Secondary voltage (for each transformer/rectifier) | Wet ESP | Kilovolt meters/transmitter |
| Secondary currents (for each transformer/rectifier) | Wet ESP | Milliammeters/transmitter |

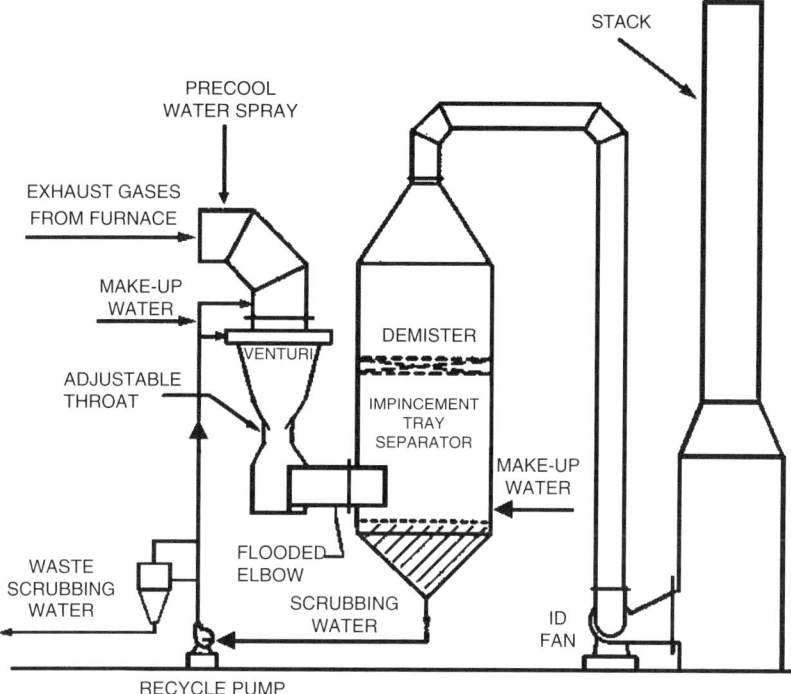

**Fig. 11.1.** Venturi scrubber (4).

addition to the Venturi scrubber (7). Pollutant emissions were compared between the two incinerator systems in the plant: an older unit installed in 1988 with a Venturi scrubber only and the new unit installed in 2000 with both a Venturi scrubber and a WESP. The WESP resulted in almost one order of magnitude of particulate removal. The particulate emission for the 1988 unit was 0.385 lb/h (0.011 grain/dscf (dry standard cubic feet) at 7% $O_2$) and the emission rate for the 2000 unit was only 0.07 lb/h (0.002 grain/dscf at 7% $O_2$). The emissions of carbon monoxide (CO), total hydrocarbons (THCs), and metals were close for the two systems.

### 1.4. Overview of the Ash-Handling System

There are two major ash-handling systems (4): the dry and the wet system. The dry system is also called the mechanical system, where the ash is discharged from the furnace and transported to a storage bin. The ash is then wetted or conditioned in the bin in order to reduce fugitive emissions and is normally disposed at a landfill by trucks. Figure 11.2 shows an example of a mechanical ash handling system, which consists of screw conveyors, a bucket elevator, a storage bin, and a rotary ash conditioner.

The wet system is also called the hydraulic system, where the ash is dropped into a storage tank filled with water. The resultant slurry is then pumped to a lagoon for further settling. Figure 11.3 shows an example of a hydraulic ash handling system, which consists of a steel ash hopper, a pump, water supply, and slurry discharge pipelines. The ash produced from the MHF can be handled in either of the two ways.

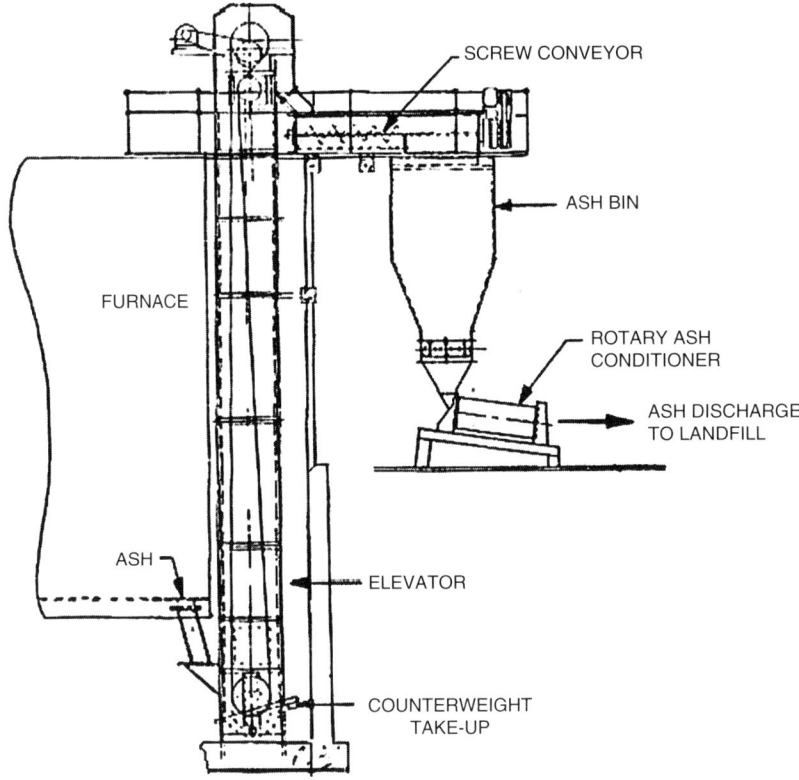

**Fig. 11.2.** Mechanical ash-handling system (4).

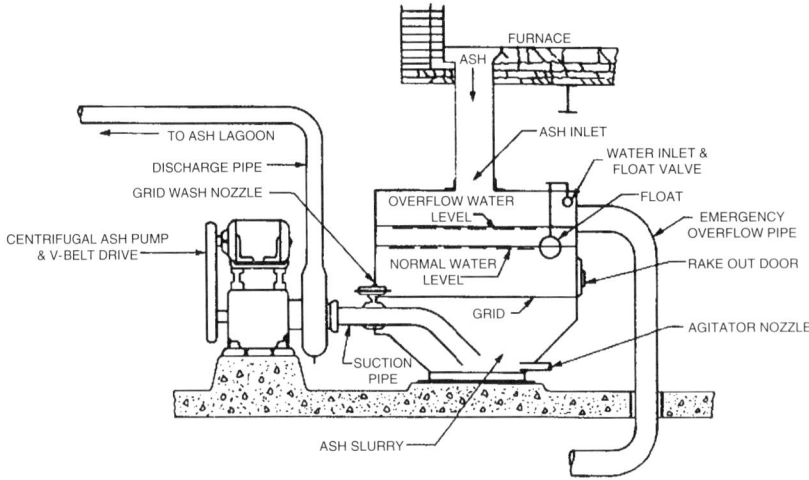

**Fig. 11.3.** Hydraulic ash-handling system (4).

## 1.5. U.S. Federal and State Regulations

### 1.5.1. Overview of Emission Regulations for Sludge Incineration

The sludge incineration facility must meet the pollutant emission requirements set forth by the federal and state regulatory agencies. The regulatory requirements typically include the following (2, 7):

1. Title 40 of the Code of Federal Regulations (CFR), Part 503 (Standards for the Use or Disposal of Sewage Sludge) subpart E (Incineration), which sets the minimum requirements for biosolids incinerators with respect to the emissions of CO, total hydrocarbons, total particulate matter, and seven heavy metals: arsenic, beryllium, cadmium, chromium, mercury, nickel, and lead. It also sets requirements for general management, reporting, record keeping, monitoring parameters, and frequency. Details of emission estimates and monitoring requirements are described in subsequent sections.
2. If the biosolid incinerators emit 9 Mg (10 tons) or more per year of any of the 189 pollutants or 23 Mg (25 tons) or more per year of any combination of the 189 pollutants, they are classified as major source of emissions. Incinerators that are classified as major source of emissions are regulated under Section 112 of the 1990 Clean Air Act Amendment (CAAA).
3. The new installations may be subject to the limit for particulate matter emissions under the New Source Performance Standards (NSPS).
4. The National Ambient Air Quality Standards: incinerators emitting more than 92 Mg (100 tons) per year must obtain Title V operating permits. Facilities emitting 23 to 92 Mg (25 to 100 tons) of $NO_x$ per year may be classified as a major source depending on the area attainment classification.
5. The state and local regulations: While some states adopt the federal regulation, other state and local governments have more stringent rules governing pollutant emissions based on their own air quality compliance status.

### 1.5.2. Overview of 40 CFR Part 503 Subpart E

Facilities/operations that are not governed by 40 CFR Part 503 subpart E includes the following:

1. The use of municipal solid waste (MSW) is greater than 30% (by dry weight) of the mixture of biosolids and as auxiliary fuel. The MSW is not considered as auxiliary fuel at this percentage, and the process is governed by 40 CFR Parts 60 and 61.
2. The nonhazardous ash generated in biosolids incineration is not applied to a MSW landfill.
3. An incinerator burns hazardous waste.

The limits set forth in subpart E of 40 CFR Part 503 protect human health from the reasonably anticipated harmful effects of these pollutants. The approaches for determination of the limits are listed in Table 11.2.

#### 1.5.2.1. Mercury and Beryllium

Mercury and beryllium emissions must meet the limits set forth by the National Emission Standards for Hazardous Air Pollutants (NESHAP, 40 CFR Part 61). The NESHAP requirement for beryllium and mercury are as follows: the total beryllium and mercury emission from each incinerator should not exceed 10 and 3200 g, respectively, during any 24-hour period. The NESHAP requirement for beryllium does not apply if

**Table 11.2**
**Determination of pollutant limits (6)**

| Pollutant | How to figure out pollutant limits | Determine dispersion factor (DF) | Determine control efficiency (CE) | Use National Ambient Air Quality Standard (NAAQS) | Use risk specific-concentration (RSC) | Use correction factor for oxygen | Use correction factor for moisture |
|---|---|---|---|---|---|---|---|
| Pollutant limits | | | | | | | |
| Arsenic | Use equation for arsenic | Yes | Yes | No | Yes | No | No |
| Beryllium | Use NESHAPs[a] | No | No | No | No | No | No |
| Cadmium | Use equation for cadmium | Yes | Yes | No | Yes | No | No |
| Chromium | Use equation for chromium | Yes | Yes | No | Yes | No | No |
| Lead | Use equation for lead | Yes | Yes | Yes | No | No | No |
| Mercury | Use NESHAPs[a] | No | No | No | No | No | No |
| Nickel | Use equation for nickel | Yes | Yes | No | Yes | No | No |
| Operational standard | | | | | | | |
| Total hydrocarbons or carbon monoxide [b] | Limit is 100 ppm$_v$ | No | No | No | No | Yes | Yes |

*Note*: Each of the requirements mentioned (e.g., dispersion factor, NAAQS) is explained in the text.
[a] National Emissions Standards for Hazardous Air Pollutant requirements are summarized in the text.
[b] THC or CO determinations are technology-based standards that in the judgment of the US EPA protect public health and the environment from the reasonably anticipated adverse effects of organic pollutants in the exit gas of biosolids incinerator.

(a) the ambient beryllium concentration in the proximity of the incinerator is less than 0.01 µg/m$^3$ averaged over a 30-day period, and (b) there is proof with historic data that the biosolids feed does not contain beryllium. A written approval has to be obtained from the regional administrator for the above two cases.

### 1.5.2.2. Lead

The allowable daily concentration limit of lead emission is estimated based on the following equation (6):

$$C_{lead} = \frac{0.1 \times NAAQS \times 86{,}400}{DF \times (1 - CE) \times SF} \quad (1)$$

where

$C$ = pollutant limit (allowable daily concentration of lead in biosolids, dry-weight basis, mg/kg

$0.1$ = allowable ground level concentration of lead from biosolids (10% of the NAAQS for lead).

$NAAQS$ = National Ambient Air Quality Standard for lead, µg/m$^3$ (currently this standard is 1.5 µg/m$^3$)

$DF$ = dispersion factor based on an air dispersion model, µg/m$^3$/g/s

$CE$ = biosolids incinerator control efficiency for lead based on a performance test, in hundredths

$SF$ = biosolids feed rate in dry metric tons per day, (T/d)

$86{,}400$ = time conversion factor, s/d

*Example*:

Given:
  Dispersion factor is 3.4 µg/m$^3$/g/s
  Control efficiency is 0.916
  Biosolids feed rate is 12.86 T/d
  NAAQS for lead is 1.5 µg/m$^3$

Solution:

$$C_{lead} = \frac{0.1 \times 1.5\,\mu g/m^3 \times 86{,}400}{3.4\,(\mu g/m^3/g/s) \times (1 - 0.916) \times 12.86\,(T/d)}$$

$$C_{lead} = 3529\,(mg/kg)$$

The DF factor can be determined in accordance with 40 CFR Part 503.42, and the CE factor can be determined in accordance with 40 CFR Part 503.43 through performance testing.

### 1.5.2.3. Arsenic, Cadmium, Chromium, and Nickel

The allowable daily concentration limits of arsenic, cadmium, chromium, and nickel emissions are estimated based on the following equation (6):

$$C = RSC \times \frac{86{,}400}{DF \times (1 - CE) \times SF} \quad (2)$$

where

$C$ = pollutant limit (allowable daily concentration of arsenic, cadmium, chromium, or nickel, mg/kg)

$RSC$ = risk-specific concentration (the allowable increase in the average daily ground-level ambient air concentration for pollutant at or beyond the property line of the site, $\mu g/m^3$)

$86,400$ = time conversion factor, s/d

$DF$ = dispersion factor (based on an air dispersion model), $\mu g/m^3/g/s$

$CE$ = biosolids incinerator control efficiency for lead (based on a performance test), in hundredths

$SF$ = biosolids feed rate, T/d

*Example for arsenic*:

If:

RSC is 0.023 $\mu g/m^3$
Dispersion factor is 3.4 $\mu g/m^3/g/s$
Control efficiency is 0.975
Biosolids feed rate is 12.86 T/d

Solution:

$$C_{arsenic} = \frac{0.023\,(\mu g/m^3) \times 86,400}{3.4\,(\mu g/m^3/g/s) \times (1 - 0.975) \times 12.86\,(T/d)}$$

$$C_{arsenic} = 1818\,(mg/kg)$$

If the dispersion factor were 0.6 instead of 3.4 $\mu g/m^3/g/s$, then the allowable concentration for arsenic would be 3.4/0.6, or 5.667 times greater at 10,300 mg/kg.

The RSC is based on human health concerns, provided in 40 CFR Part 503.43 and Table 11.3. The $RSC_s$ for arsenic, cadmium and nickel can be directly obtained from Table 11.3. However the RSC for chromium should be determined by either of the options listed in Table 11.3.

If any of the limits is exceeded, the incinerator would be "in violation" until adjustments are made to meet the requirements. Such adjustments include, but are not limited to, the improvement of sludge quality through pretreatment, reduction of the sludge feed rate, improvement of the furnace design, and installation of air pollution control devices. Performance tests must be repeated in the latter two cases. The operation conditions where the emission is in compliance must be maintained afterward.

#### 1.5.2.4. Total Hydrocarbons

Both CO and THC are indications of incomplete combustion in the incinerator and both can be used to represent all the organic compounds in the exhaust gas of the sludge incinerator. Either CO or THC should be continuously monitored in the stack gas to ensure that the monthly average is within limits. The Part 503 rule was recently amended for incinerators not exceeding 100 ppmv of CO (monthly average) in the stack gas. The US EPA will allow continuous CO monitoring as a surrogate for THC monitoring. The monthly average is the arithmetic mean of the hourly averages, and

## Table 11.3
### The RSC limits by metals and incinerator type (6)

| Pollutant | | Risk-specific concentration (RSC) (microgram per cubic meter) |
|---|---|---|
| Arsenic | | 0.023 |
| Cadmium | | 0.057 |
| Nickel | | 2.0 |
| Chromium | Option 1: Based on the type of incinerator and APCD | |
| | FBF with wet scrubber | 0.65 |
| | FBF with wet scrubber and wet electrostatic precipitator | 0.23 |
| | Other types with wet scrubber | 0.064 |
| | Other types with wet scrubber and wet electrostatic precipitator | 0.016 |
| | Option 2: Using equation | $RSC = 0.0085/r$ Where $r =$ the decimal fraction of the hexavalent chromium concentration in the total chromium concentration measured in the exit gas from the biosolids incinerator stack in hundredths |

FBF, Fluidized bed Furnace.

the hourly average must be based on at least two readings taken each hour when the incinerator is in operation.

The correction of 0% moisture and 7% oxygen are required for estimating THC or CO in stack gas to ensure that the concentration is evaluated on a standard basis. The correction equations are listed as following:

For the examples below, assume that the original THC (or CO monthly average) is 40 ppmv.

1. 0% moisture: The equation for correcting the THC (or CO) measurement for 0% moisture is as follows (6):

$$\text{Correction factor} = 1/(1 - X) \qquad (3)$$

   where
   $X =$ decimal fraction of the % moisture in the biosolids incinerator exit gas, in hundredths

   *Example*:

   Given:
   $X = 0.12$

   Solution:

   Correction factor for moisture $= 1/(1-0.12) = 1.14$
   Multiply the original THC (or CO) value (in this case, 40 ppmv) by the correction factor for moisture:

   $$40 \text{ ppmv} \times 1.14 = 45.6 \text{ ppmv}$$

   THC (or CO) concentration corrected for 0% moisture $= 45.6$ ppmv.

2. 7% oxygen: The equation for correcting the THC (or CO) measurement to 7% oxygen is as follows (6):

$$\text{Correction factor for oxygen} = 14/(21 - y) \quad (4)$$

where
- $14$ = the difference between the percent oxygen in air (21%) and 7% oxygen
- $21$ = percentage of oxygen in air
- $Y$ = percent oxygen concentration in the biosolids incinerator stack exit gas (dry volume/dry volume)

*Example*:

Given:
$Y = 10\%$

Solution:

$$\text{Correction factor for oxygen} = 14/(21 - 14) = 1.27$$

Multiply the THC or CO monthly average concentration (already corrected for moisture) by the correction factor for oxygen:

$$45.6 \text{ ppmv} \times 1.27 = 58 \text{ ppmv (rounded)}$$

Therefore the THC or CO monthly average concentration in this example corrected for 0% moisture and to 7% oxygen is 58 ppmv.

### 1.5.2.5. MANAGEMENT REQUIREMENT FOR BIOSOLIDS INCINERATION

The management requirements cover the operation and maintenance of instruments, temperature requirements, operation of APCDs, and protection of endangered species. Here is a summary of the management practice (6):

1. Instruments must be used that continuously measure and record THC (or CO) concentrations, oxygen levels, and the information needed to calculate moisture content in the stack exit gas and the combustion temperature in the furnace.
2. These instruments must be installed, calibrated, operated, and maintained according to guidance provided by the permitting authority. Calibration procedures are specified in the US EPA's new Continuous Emission Monitoring (CEM) guidance.
3. The instrument used for THC (or CO) measurements must
   a. use a flame ionization detector;
   b. have a sampling line heated to 150 °C or higher at all times; and
   c. be calibrated at least once every 24-hour operating period using propane.
4. The incinerator can be operated within the range of operating conditions set during the performance test and allowed in the permit, but it must not be operated above the maximum combustion temperature set by the permitting authority based on performance test conditions.
5. Conditions for operating the air pollution control devices must be followed; these conditions are also set by the permitting authority based on performance test conditions.
6. Biosolids may not be incinerated if incineration of biosolids is likely to negatively affect a threatened or endangered species or its critical habitat as listed in Section 4 of the Endangered Species Act. Critical habitat is any place where a threatened or endangered species lives and grows during any stage of its life cycle.

The following important parameters must be monitored:

1. Concentration of metals (arsenic, cadmium, chromium, lead, mercury, and nickel) in biosolids
2. Concentration of beryllium in the stack gas, unless the permitting authority approves a biosolids method
3. Concentration of THC (or CO) in stack exit gas
4. Oxygen concentration in stack exit gas
5. Information needed to determine moisture content in stack exit gas
6. Combustion temperature in the furnace
7. Operating conditions of air pollution control devices (conditions are set by the permitting authority based on performance test data)
8. Biosolids feed rate

The monitoring frequency is shown in Table 11.4. The monitoring frequency for beryllium and mercury is determined by the permitting authority, while the monitoring frequency for other metals is based on the incinerator capacity: the larger the capacity, the more often the metals need to be monitored.

The record-keeping requirements for the pollutant limits, monitoring and management requirements are as follows:

1. Records related to pollutant limits for metals:
   - Concentrations of arsenic, cadmium, chromium, lead, and nickel in biosolids fed to the incinerator
   - Information showing how the requirements for beryllium and mercury in the National Emission Standards for Hazardous Air Pollutants (NESHAPs) are being met, if applicable
   - Biosolids feed rate (for each incinerator, dry-weight basis), stack height, dispersion factor
   - Control efficiency for arsenic, cadmium, chromium, lead, and nickel (for each incinerator)
   - RSC for chromium
2. Records related to the THC (or CO) limit:
   - THC (or CO) monthly average concentration is the stack exit gas
   - Oxygen concentration in the stack exit gas
   - Information used to measure moisture content in the stack exit gas
3. Records related to management practices and monitoring requirements:
   - Combustion temperatures, including maximum daily combustion temperature, in the furnace
   - Measurements for required air pollution control device operating conditions
   - Calibration and maintenance log for instruments used to measure, including THC (or CO) level in stack exit gas, oxygen level in stack exit gas, moisture content in stack exit gas, and combustion temperature in furnace

The records must be kept for a minimum of 5 years.

All class 1 treatment works, that is, the works serving a population of more than 10,000 and with greater than 1 MGD design capacity, have to report annually (by February 19th) to the permitting authority the information collected as mentioned above.

Table 11.4
**Monitoring frequency of biosolids incinerators (6)**

| Pollutant/parameter | Amount of biosolids fired (metric tons per 365-day period, dry-weight basis) | Must monitor at least: |
|---|---|---|
| Arsenic, cadmium, chromium, lead, and nickel in biosolids | Greater than zero but less than 290 | Once per year |
| | Equal to or greater than 290 but less than 1500 | Once per quarter (four times per year) |
| | Equal to or greater than 15,000 | Once per month (12 times per year) |
| Beryllium and mercury in biosolids or stack exit gas | NA | As often as permitting authority requires |
| THC (or CO) concentration in stack exit gas | NA | Continuously; monthly average reported, which is the arithmetic mean of hourly average that include at least two readings per hours |
| Oxygen concentration in stack exit gas | NA | Continuously |
| Information needed to determine moisture content in stack exit gas | NA | Continuously |
| Combustion temperature in furnace | NA | Continuously |
| Air pollution control device conditions | NA | As often as permitting authority requires |

THC, total hydrocarbons.

# Combustion and Incineration Management

As mentioned in Table 11.4, biosolids incineration is not allowed if the "critical habitat" of the threatened or endangered species may be adversely affected. The Threatened and Endangered Species List should be reviewed prior to any incineration operations, and the state agencies governing fish and wildlife should also be consulted for state requirements.

### 1.5.3. Other Regulations

Due to its serious health effects, especially on newborns and pregnant women, the limits on mercury emission have been regulated the states and in other countries.

The European Union's directive on waste incineration sets the air emission limits for dioxins as $0.1\,\mu g/m^3$, cadmium (Cd) and thallium (Tl) as $0.05\,mg/m^3$, and mercury as $0.05\,mg/m^3$. This applies to almost all the waste incineration facilities including hazardous waste, and it became effective in 2000 (12).

In addition to 40 CFR 60, New York State sets it own limit on mercury under the 6 NYCRR (Regulations of the State of New York). The new subpart 219-7, Mercury Emission Limitations for Large Municipal Waste Combustors that are constructed on or before September 20, 1994, mandates that mercury emission for large municipal waste combustor plants must be in compliance with the emission limit of $28\,\mu g/dscm$ (corrected to 7% oxygen) or 85% removal, whichever is less stringent. This regulation was in effect in 2002 and is expected to reduce mercury emissions and its subsequent environmental loading in New York State and the northeast U.S. (13).

The National Ambient Air Quality Standards (NAAQS) mandates that incinerators emitting more than 92 Mg (100 tons) per year must obtain Title V operating permits. A new permit is required with major installation changes, such as furnace retrofit, replacement, or installation of a new APCD, as such processes tend to alter pollutant emissions (2). The air permits are usually issued by the state authority. Air permits were issued to both facilities by the Ohio US EPA, which mandates the emission limits and the standard methods to test these pollutants and the monitoring practice. The pollutant limits are summarized in Tables 11.5, 11.6, and recommended test methods in Table 11.7 (14, 15). Biosolids incineration is performed in two of the wastewater treatment facilities of the Metropolitan Sewage District (MSD) located in Cincinnati, Ohio. The Mill Creek facility handles 9.4 dry tons per hour (303 tons per day) of sludge in several MHFs. The Little Miami facility handles 72 dry tons per day of sludge in an FBF.

## 2. OPERATION AND MANAGEMENT OF THE MULTIPLE HEARTH FURNACE

### 2.1. Process Description

In the U.S., the first multiple hearth furnace for biosolids incineration was installed in Dearborn, Michigan, in 1935, and has been the dominant incinerator type until the 1960s. The MHFs are currently the most widely used in the incineration of municipal wastewater sludge, although most of the installations are aging now. The typical cross sectional structure of a MHF is shown in Figure 10.22 of Chapter 10. It consists of a

**Table 11.5**
**Air permit for the Little Miami Facility from Ohio, US EPA (14)**

| Pollutant | Tons/year |
|---|---|
| Organic compound | 36.1 |
| PM | 17.1 |
| $SO_2$ | 97.2 |
| $NO_x$ | 65.7 |
| CO | 124.8 |
| PM10 | 5.8 |
| Mercury (Hg) | 1.3 |
| Arsenic (As) | 0.3 |
| Beryllium (Be) | 0.004 |
| Cadmium | 0.7 |
| Chromium | 8.3 |
| Lead | 1.9 |
| Nickel | 25.8 |

**Table 11.6**
**Air permit for the Mill Creek Facility from Ohio US EPA (15)**

| Pollutant | Tons/year |
|---|---|
| PM | 60.8 |
| $SO_2$ | 44.2 |
| NOx | 276.5 |
| VOC | 115.02 |
| HCl | 16.6 |
| Mercury (Hg) | 1.3 |
| Beryllium (Be) | 10 g/24-hr period |

**Table 11.7**
**Test method recommended by the permit (15)**

| Pollutant | Test method |
|---|---|
| PM | 40 CFR 60.154 |
| $SO_2$ | US EPA Method 6 |
| $NO_x$ | US EPA Method 7 |
| CO | US EPA Method 10 |
| Metals | US EPA Method 29 |
| Mercury (Hg) | US EPA Method 29 |
| Beryllium (Be) | 101A, 103, or 104 |

steel shell surrounding a number of hearths made with refractory materials and a central shaft with rotating arms. Dewatered sludge enters the MHF from the top through a flap gate, and drops down to various stages of the MHF through the rabble arm. The MHFs can have four to 12 hearths, which are classified as the drying zone, the combustion zone, and the cooling zone from the top to the bottom of the furnace. The air enters counter-currently from the bottom of the MHF. The MHF is made of high heat and heavy-duty fire bricks. Since the operation temperature can reach 2000 °F, the central shaft and rabble arms are cooled with air supplied by a blower. The shaft is motor driven and the speed is adjustable from 0.5 to 1.5 rpm (5).

In each hearth, two or more rabble arms are connected to the shaft, which provide mixing and rotating the sludge downward. Sludge is also constantly broken into smaller sizes by the rabble arms, which can provide more contact with hot furnace gases, facilitates sludge drying, and results in better combustion of the sludge.

Each MHF usually has two doors for observation and sampling purposes. There is a central tube in each of the rabble arms for the cooling air, which can either be released to the atmosphere or returned to the bottom of the hearth as preheated air for combustion. Some MHFs are equipped with a heat exchanger, such as an RTO, to recover the heat of the exhaust gas prior to it release (10) to the atmosphere. The heat can be used to preheat the incoming combustion air, sludge conditioning, or elsewhere in the plant. The ash generated from a MHF can be handled with both a wet and a dry method, as described in Section 1.4.

## 2.2. Design and Operating Parameters

The capacity of an MHF is determined by the total furnace area. The diameters generally range from 54 inches to 21 feet and 6 inches, and the numbers of hearths range from four to 12 (Table 11.8). Capacities of MHF range from 200 to 8000 lb/h of dry sludge, and the operating temperature can be as high as 2000 °F.

Table 11.9 provides an example of the typical MHF loading rates for several types of sludge under a combustion temperature of 1400 °F, which is a temperature commonly assumed in many examples (16).

Some of the important operation parameters, such as the excess air, sludge loading rate, solids concentration, and moisture content can be estimated based on the operating conditions of the incinerator. The following are sample design calculations:

1. Excess air is the amount of air required beyond the theoretical air requirements for complete combustion: This parameter is expressed as a percentage of the theoretical air required. Sample calculation for excess air: Assume 1200 SCFM (Standard Cubic Foot per Minute) actual and 1000 SCFM theoretical.

$$\text{Excess air} = (\text{actual air rate} - \text{theoretical rate}) \times 100/\text{theoretical air rate}$$
$$= (1{,}200 - 1{,}000) \times 100/1{,}000 = 20\%$$

2. Sludge loading rate is the weight of wet sludge fed to the reactor (per square foot of reactor bed area per hour [lb/ft$^2$/h]).

**Table 11.8**
**Characteristics of multiple hearth furnaces**

| Effective hearth area, sq ft | Outer diameter, ft | Number hearths | Effective hearth area, sq ft | Outer diameter, ft | Number hearths |
|---|---|---|---|---|---|
| 85 | 6.75 | 6 | 988 | 16.75 | 7 |
| 98 | 6.75 | 7 | 1041 | 14.25 | 11 |
| 112 | 6.75 | 8 | 1068 | 18.75 | 6 |
| 125 | 7.75 | 6 | 1117 | 16.75 | 8 |
| 126 | 6.75 | 9 | 1128 | 14.25 | 12 |
| 140 | 6.75 | 10 | 1249 | 18.75 | 7 |
| 145 | 7.75 | 7 | 1260 | 16.75 | 9 |
| 166 | 7.75 | 8 | 1268 | 20.25 | 6 |
| 187 | 7.75 | 9 | 1400 | 16.75 | 10 |
| 193 | 9.25 | 6 | 1410 | 18.75 | 8 |
| 208 | 7.75 | 10 | 1483 | 20.25 | 7 |
| 225 | 9.25 | 7 | 1540 | 16.75 | 11 |
| 256 | 9.25 | 8 | 1580 | 22.25 | 6 |
| 276 | 10.75 | 6 | 1591 | 18.75 | 9 |
| 288 | 9.25 | 9 | 1660 | 20.25 | 8 |
| 319 | 9.25 | 10 | 1675 | 16.75 | 12 |
| 323 | 10.75 | 7 | 1752 | 18.75 | 10 |
| 351 | 9.25 | 11 | 1849 | 22.25 | 7 |
| 364 | 10.75 | 8 | 1875 | 20.25 | 9 |
| 383 | 9.25 | 12 | 1933 | 18.75 | 11 |
| 411 | 10.75 | 9 | 2060 | 20.25 | 10 |
| 452 | 10.75 | 11 | 2084 | 22.25 | 8 |
| 510 | 10.75 | 10 | 2090 | 18.75 | 12 |
| 560 | 10.75 | 12 | 2275 | 20.25 | 11 |
| 575 | 14.25 | 6 | 2350 | 22.25 | 9 |
| 672 | 14.25 | 7 | 2464 | 20.25 | 12 |
| 760 | 14.25 | 8 | 2600 | 22.25 | 10 |
| 845 | 16.75 | 6 | 2860 | 22.25 | 11 |
| 857 | 14.25 | 9 | 3120 | 22.25 | 12 |
| 944 | 14.25 | 10 | | | |

Sample loading rate: Assume 20 feet diameter reactor, 20% feed sludge moisture content, and 440 lb dry sludge/h.

$$\text{Loading rate} = \frac{(\text{lb dry sludge/h}) \times (100)}{(\% \text{ moisture content}) \times (\text{area})}$$

$$= \frac{440 \times 100}{20\% \times \frac{3.14(20)^2}{4}}$$

$$= 7.01 \text{ lb/ft}^2/\text{h}$$

3. Solids concentration is the weight of solids per unit weight of sludge. It is calculated as follows:

### Table 11.9
**Typical multiple hearth furnace loading with various sludge types (16)**

| Type of sludge | Percent solids | Percent VS | Chemical concentration* (mg/L) | Typical wet sludge loading rate** (lb/hr/sq ft) |
|---|---|---|---|---|
| 1. Primary | 30 | 60 | N/A | 7.0–12.0 |
| 2. Primary + $FeCl_3$ | 16 | 47 | 20 | 6.0–10.0 |
| 3. Primary + low lime | 35 | 45 | 298 | 8.0–12.0 |
| 4. Primary + WAS | 16 | 69 | N/A | 6.0–10.0 |
| 5. Primary + (WAS + $FeCl_3$) | 20 | 54 | 20 | 6.5–11.0 |
| 6. (Primary + $FeCl_3$) + WAS | 16 | 53 | 20 | 6.0–10.0 |
| 7. WAS | 16 | 80 | N/A | 6.0–10.0 |
| 8. WAS + $FeCl_3$ | 16 | 50 | 20 | 6.0–10.0 |
| 9. Digested Primary | 30 | 43 | N/A | 7.0–12.0 |

\* Assume no dewatering chemicals.
\** Low number is applicable to small plants, high number is applicable to large plants. WAS, waste-activated sludge.

Assume 120 lb wet sludge with 25 lb of dry solids.

$$\text{Concentration} = (\text{weight of dry sludge solids} \times 100)/(\text{weight of wet sludge})$$
$$= (25 \times 100)/120 = 20.8\%$$

4. Moisture content is the amount of water per unit weight of sludge. The moisture content is expressed as a percentage of the total weight of the wet sludge. This parameter is equal to 100 minus the percent solids concentration or can be computed as follows:
Same assumption as paragraph 3.

$$\text{Moisture content} = ((\text{weight of wet solids}) - (\text{weight of dry solids} \times 100))/(\text{weight of wet solids})$$
$$= ((120 - 25) \times 100)/120 = 79.2\%$$

As indicated in Figure 10.22 of Chapter 10, an MHF under normal loading generally can be divided into three zones: the drying zone, the combustion zone, and the cooling zone from the top to the bottom of the furnace. The temperatures in each zone are estimated in the range of 1000 °F, 1600 to 1800 °F, and 600 °F and lower, respectively (17). Some others further divide the combustion zone into a volatile combustion zone, where the volatile content of the sludge is combusted, and a fixed carbon combustion zone (4), with the typical temperatures provided in each zone (Figure 11.4).

The drying zone usually starts with two or more hearths from the top of a MHF, where most of the moisture is evaporated and the temperature is often at or below 140 °F. The odor is low at this stage, as most of the volatile content is not released at this temperature. For a sludge with 75% moisture, the evaporation of volatile content normally does not occur until 80% to 90% of the moisture has been removed. By this time, the sludge is usually in the lower stages of the MHF (in the combustion zone or the volatile combustion zone) and is in contact with hot gas, where both incineration and odor release can occur. The combustion zone usually includes two or more of the middle hearths where the sludge can burn at 1500 °F or higher. The cooling zone is at

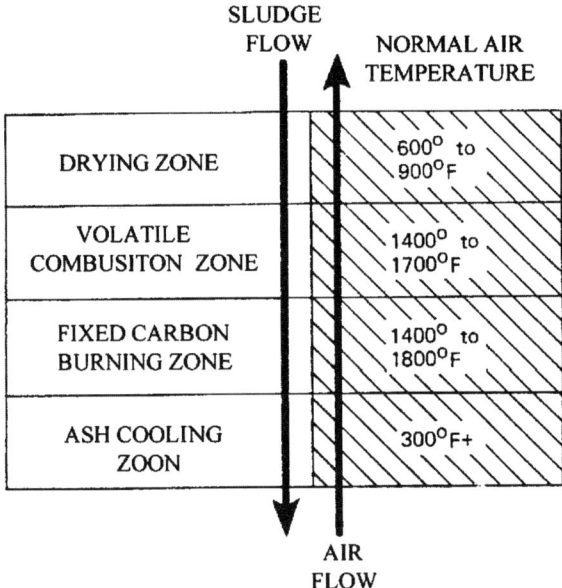

**Fig. 11.4.** MHF process zones (4).

the bottom hearth, where the ash is cooled and the incoming combustion air is heated. A temperature higher than 900 °F of the outlet gas can be an indication of fuel overuse.

### 2.3. Performance Evaluation, Management, and Troubleshooting of the Multiple Hearth Furnace

#### 2.3.1. Performance Evaluation

The incineration systems nowadays are mostly automatically controlled from a central control room. The temperatures of each hearth, the exhaust, and the inlet gases can be recorded, and the temperature of each hearth can be controlled to within ±40 °F. In case of a power or fuel failure, the furnace can be shut down automatically, and cooling air fan for the shaft can be run on backup power to prevent the shaft from melting.

The following items are usually checked for performance evaluation:

1. Hearth temperature check: check the temperature records to make sure that the temperatures of each hearth are uniform and within the specified range of the manufacturer.
2. CO check: the carbon monoxide of the stack gas should be measured, as it is an indication of incomplete combustion.
3. Record check: the maintenance record of the hearth and refractory should be checked, and the shutdown and start-up procedures should be reviewed if repairs are frequent.
4. Odor check: odor is an indication of MHF malfunctioning.
5. Furnace loading and fuel use check: fuel is typically required for the start-up process, for sludge with low solid content, and in some air pollution control devices. The heat needed for the MHF can be estimated based on different sludge solid contents and whether the combustion is autogenous.

The pollutants to be monitored and the monitoring frequency should follow the requirement of the Part 503 rule as described in Section 1.5. Continuous monitors are

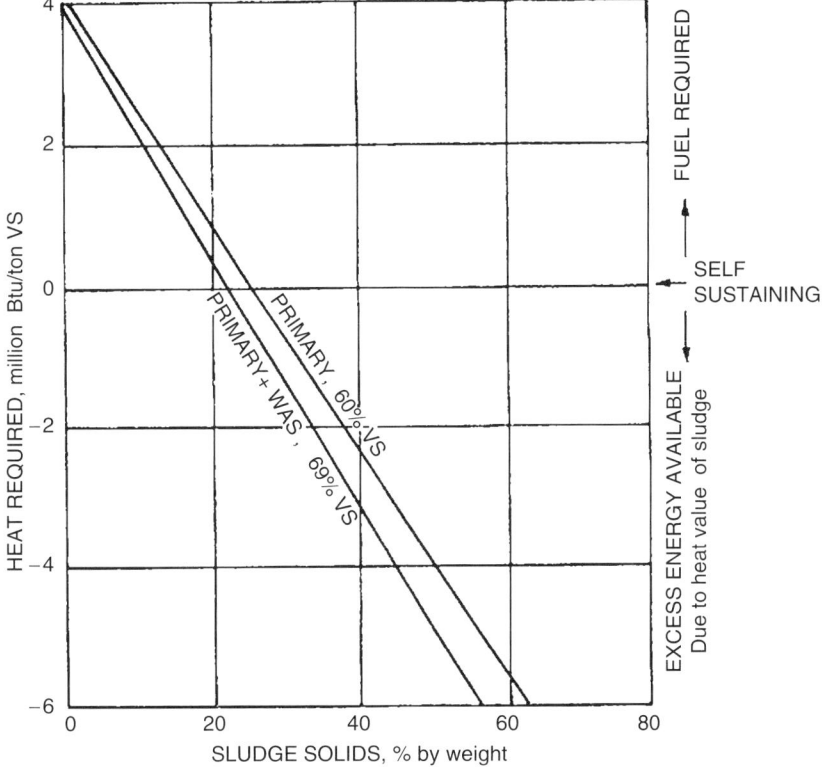

**Fig. 11.5.** Auxiliary heat required to sustain sludge combustion (5).

used for important parameters such as the CO and oxygen contents, and the sampling of many parameters nowadays are automatic. To ensure good incinerator operation, more air is provided than the stoichiometric value. Normally 75% to 100% of excess air is provided in an MHF incinerator, as too much excess air, such as more than 100%, results in fuel waste. When less excess air is provided, pyrolysis (incomplete combustion) can occur, and can result in the formation of CO, soot, and odorous hydrocarbons in the stack gas. Therefore, the stack gas composition is monitored to ensure the best thermal economy in the MHF. For example, a higher level of CO than the preset value is an indication of incomplete combustion, and more excess air is needed to correct this situation. However, if at the same time the oxygen level is at the preset value, this means the mixing of the sludge and air may be poor or the temperature is low due the higher sludge moisture content than normal.

Additional equipment may include a drying oven and a scale, to measure the sludge volatile content. The grab sample can be drawn from the sludge feedline or the preceding process, such as dewatering or storage. The sample should be capped to prevent moisture loss (16).

Figure 11.5 shows that the incineration process in the MHF can be self-sustaining if the sludge solid content is approximately 26% for primary sludge or 23% for primary sludge plus waste-activated sludge (WAS). When the solid content is high, excess energy

**Table 11.10**
**Heating value of various sludge types (5)**

| Type of sludge | Heating value (Btu/lb of dry solids) |
|---|---|
| Raw primary | 10,000–12,5000 |
| Activated sludge | 8500–10,000 |
| Anaerobically digested primary | 5500 |
| Raw primary (chemically precipitated) | 7000 |
| Biological filter | 8500–10,000 |
| Grease and scum | 16,7000 |
| Fine screenings | 7800 |
| Ground garbage | 8200 |
| High organic grit | 4000 |

can be generated. When the solid content is low, auxiliary fuel may be needed. The heating values of different types of sludge are listed in Table 11.10, which can be helpful in estimating auxiliary fuel usage.

The following steps can be implemented to increase the content of volatiles in sludge, in addition to further dewatering:

- Remove sludge inorganics such as grit.
- Avoid the use of inorganic chemicals such as ferric chloride and lime in the dewatering process.
- Avoid biological processes such as digestion before incineration.

### 2.3.2. Management and Maintenance

The general management, maintenance, monitoring, and record keeping should also meet the requirement of the Part 503 rule as summarized in Section 1.5. The management process requires thorough bookkeeping and regular attention of the major elements of the incineration system, such as drives and gear reducers, chains and sprockets, burners, air blowers, sludge conveying equipment, ash conveying equipment, furnace seals, draft controller, temperature controllers, standby engines or generators, and the air pollution control devices. Problems such as burner shutdown, furnace overheat, draft loss, and feed belt shutdown are also monitored. A good preventive maintenance program is essential to minimize incinerator shutdown, which can be costly to the plant and dangerous to the operators. An example of process checklist is presented in Table 11.11, which is qualitative and typically needs a yes or no response.

A more quantitative record is also kept, which includes the unit process of MHFs in Table 11.12 and historical data, such as the tonnage of sludge incinerated each year in the past, the fuel usage, and the maintenance record. The maintenance record can be used to predict and plan for the next repair or shutdown, which is extremely important.

Proper staff training is also invaluable to ensure the stable performance of the sludge incineration facilities. Consistent operation is essential to the stability of the incinerators and depends heavily on the experience of the engineers and the operators. The highly improved operation of the sludge incinerator at the Blue River Wastewater Treatment

## Table 11.11
### Incinerator process checklist for multiple hearth and fluidized bed furnaces (16)

| | | |
|---|---|---|
| 1. Are complete records kept on the following items? | | |
|    a. Temperature (each shift) | ( ) Yes | ( ) No |
|    b. Maintenance work accomplished | ( ) Yes | ( ) No |
|    c. Schedule of upcoming maintenance | ( ) Yes | ( ) No |
|    d. Fuel consumption (daily) | ( ) Yes | ( ) No |
|    e. Power consumption (daily) | ( ) Yes | ( ) No |
|    f. Sludge moisture content (each shift) | ( ) Yes | ( ) No |
|    g. Sludge volatile solids content (daily) | ( ) Yes | ( ) No |
| 2. Dose operator has a planned prodedure for slowly shutting down the process? | ( ) Yes | ( ) No |
| 3. Is there a plan for emergency operation for: | | |
|    a. Power outage? | ( ) Yes | ( ) No |
|    b. Fuel shortage? | ( ) Yes | ( ) No |
|    c. Other accidents? | ( ) Yes | ( ) No |
| 4. Is scum burned in the incinerator? | ( ) Yes | ( ) No |
| 5. If scum is burned, is the feed rate regulated? | ( ) Yes | ( ) No |
| 6. Is moisture content optimized for minimum total cost of dewatering and fuel consumption in incinerator? | ( ) Yes | ( ) No |

Plant (WWTP) in Kansas City, Missouri, clearly demonstrates the value of consistent staff training (18).

Prior to September 2003, the lack of consistent staff training was one of the key factors in the unsatisfactory performance of the MHF incinerators. The operators were trained only by oral instructions, but no written instructions followed. As a result, each operator developed his own way of operating. Consultants were hired during the furnace downtime in 2003 to develop written instructions and provide systematic training. A standard operation procedure has been developed and is now available in the control room. Together with structural and monitoring improvement, the performance of the sludge incinerator vastly improved despite budgetary constrains. The experience at the Blue River WWTP clearly demonstrates that adherence to consistent operating instructions can result in great improvement of furnace performance.

### 2.3.3. Troubleshooting

The proper maintenance of the three zones of an MHF is extremely important and can also be very challenging (18, 19). If sludge combustion occurs in the drying zone, the volatile component of the sludge will not be completely destroyed, and if sludge combustion occurs in the lower hearth, which is normally designed for ash burnout and cooling, the higher than normal temperature of the ash may damage the ash-handling equipment and result in a waste of fuels. Therefore, maintaining the fire at its desired location is essential to the stable operation of the MHF furnace.

The Blue River plant (105 MGD or 400,000 $m^3$/d) has several MHFs that were built in 1964; two were renovated in 1992. Each MHF has eight stages and is 22.25 ft (6.78 m) in diameter. The sludge is dried by a belt press, and the ash handling uses a wet system. In spite of the rehabilitation, the performance of the MHFs was unsatisfactory. The major

**Table 11.12**
**Incinerator record (16)**

| | | Plant size (MCD) | Test frequency | Location of sample | Method of sample | Reason for test |
|---|---|---|---|---|---|---|
| Suggested minimum | Temperature | | Mn | A | Mn | P |
| | Total volatile solids | | 1/D[1] | F<br>P | G | P |
| | Total solids | | 1/D[1] | F<br>P | G | P |
| | | | | | | |
| | | | | | | |
| | | | | | | |
| | | | | | | |
| | | | | | | |
| | | | | | | |
| | | | | | | |
| | | | | | | |
| | | | | | | |
| | | | | | | |
| | | | | | | |
| | | | | | | |
| Optional | | | | | | |
| | | | | | | |
| | | | | | | |
| | | | | | | |
| | | | | | | |
| | | | | | | |
| | | | | | | |

Estimated unit process sampling and testing needs

Incineration

(Multiple hearth)

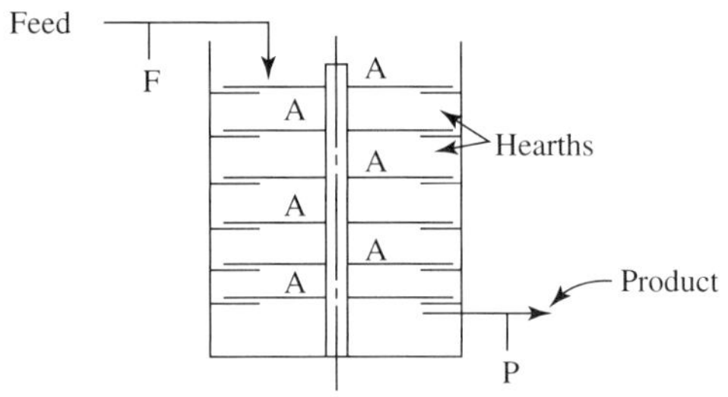


A. Test frequency
- H = Hour M = Month
- D = Day R = Record continuously
- W = Week Mn = Monitor continuously

B. Location of sample
- F = Feed
- P = Product
- A = Furnace atmosphere (at each hearth)

C. Method of sample
- 24C = 24-Hour composite
- G = Grab sample
- R = Record continuously
- Mn = Monitor continuously

D. Reason for test
- H = Historical knowledge
- P = Process control
- C = Cost control

E. Footnotes:
1. When furnace is operating

**Table 11.13**
**Troubleshooting guide for a multiple hearth furnace (5)**

| Indicators/observations | Probable cause | Check or monitor | Solutions |
|---|---|---|---|
| 1. Furnace temperature too high | 1a. Excessive fuel feed rate<br>1b. Greasy solids<br><br>1c. Thermocouple burned out | 1a. Fuel feed rate<br>1b. If fuel is off and temperature is rising, this may be the cause<br>1c. If temperature indicator is off scale, this is the likely cause | 1a. Decrease fuel feed rate<br>1b. Raise air feed rate or reduce sludge feed rate<br><br>1c. Replace thermocouple |
| 2. Furnace temperature too low | 2a. Moisture content of sludge has increased<br>2b. Fuel system malfunction | 2a. Moisture content and dewatering system operation<br>2b. Check fuel system (valve position, pressure, fuel supply, etc.) | 2a. Increase fuel feed rate until dewatering system operation is improved<br>2b. Establish proper fuel feed rate |
| 3. Oxygen content of stack gas is too high | 2c. Excessive air feed rate<br><br>2d. Flame out<br>3a. Sludge feed rate too low | 2c. If oxygen content of stack gas is high, this is the likely cause<br>2d. Visual check<br>3a. Check for blockage of sludge feed system and check feed rate | 2c. Reduce air feed rate or increase sludge feed rate<br>2d. Relight furnace<br>3a. Remove any blockages and establish proper feed rate |
| 4. Oxygen content of stack gas is too low | 3b. Air feed rate too high<br>4a. Volatile or grease content of sludge has increased<br>4b. Air feed rate too low | 3b. Air feed rate<br>4a. Sludge composition<br><br>4b. Check for malfunction of air supply and check feed rate | 3b. Decrease air feed rat<br>4a. Increase air feed rate or decrease sludge feed rate<br>4b. Increase air feed rate |
| 5. Furnace refractories have deteriorated | 5. Furnace has been started up and shut down too quickly | 5. Operating records | 5. Replace refractories and observe proper heating-up and cooling-down procedures in future |

**Table 11.13** (*Continued*)

| Indicators/observations | Probable cause | Check or monitor | Solutions |
|---|---|---|---|
| 6. Unusually high cooling effect from one hearth to another | 6. Air leak | 6. Hearth doors, discharge pipe, center shaft seal, air butterfly valves in inactive burners | 6. Stop leak |
| 7. Short hearth life | 7. Uneven firing | 7. Hearths have been operated with only one burner on | 7. Fire hearths equally on both sides |
| 8. Center shaft drive shear pin fails | 8. Rabble arm is dragging on hearth or foreign object is caught beneath arm | 8. Inspect each hearth | 8. Correct cause of failure and replace shear pin |
| 9. Furnace scrubber temperature too high | 9. Low water flow to scrubber | 9. Scrubber water flow | 9. Establish adequate scrubber water flow |
| 10. Stack gas temperature too low (500–600 °F) | 10. Inadequate fuel feed rate or excessive sludge feed rate | 10. Fuel and sludge feed rates | 10. Increase fuel or decrease sludge feed rates |
| 11. Stack gas temperature too high (1200–1600 °F) | 11. Excess heat value in sludge or excessive fuel feed rate | 11. Sludge characteristics and fuel rate | 11. Add more excess air or decrease fuel rate |
| 12. Furnace burners slagging up | 12. Burner design | | 12. Replace burners with newer designs that minimize slagging |
| 13. Rabble arms are drooping | 13. Excessive hearth temperatures or loss of cooling air | 13. Operating records: grease or scum is being injected into the hearth | 13. Maintain temperatures in proper range and maintain backup systems for cooling air in working condition; discontinue scum injection into hearth |
| 14. Stack gases contain excessive air pollutants | 14a. Incomplete combustion due to insufficient air | 14a. Monitor stack gases for unburned HC, NO$_x$, CO | 14a. Increase air fuel ratio to provide for more complete combustion |
| | 14b. Inoperative air pollution control equipment | | 14b. Repair or replace damaged equipment |
| 15. Flashing or explosion | 15. Scum or grease | 15. Sludge characteristics | 15. Provide for scum and grease removal before incineration |

problems include the following: the solids dewatering system cannot delivery sludge with consistent quality; the dewater rate is too low; the furnace is operated inconsistently by the staff; and the dysfunction of the in situ oxygen analyzer results in frequent occurrence of burnout and damage of the refractory. According to the plant, installing an on-the-fly air-fuel ratio control at each hearth (manually controlled) and fixing the oxygen analyzer greatly reduced refractory damage (18).

Typical problems observed in MHFs and problem solving procedures are listed in Table 11.13.

## 3. OPERATION AND MANAGEMENT OF THE FLUIDIZED BED FURNACE

### 3.1. Process Description

The first fluidized bed furnace in the U.S. was installed in Lynnwood, Washington, in 1962. Since then, FBF has becoming more widely used for wastewater sludge incineration especially with recent installations. When the MHF is taken out of service nowadays, it is likely to be replaced by a FBF. A typical schematic diagram of the fluidized bed furnace is shown in Figure 11.6. The FBF generally contains a vertical cylindrical vessel and a grid at the lower section to support a sand bed. The fluidized media are mainly from inert materials, with sand being the most often used. Dewatered sludge enters from the bottom of the incinerator above the grid, and pressurized air flows upward at approximately 3.5 to 5.0 psig to fluidize the sand and sludge. Unlike the MHF, the FBF is a single reactor system where sludge dewatering and burning occur in the same unit at temperatures between 1400 and 1700 °F. The gas residence time ranges from 5 to 10 seconds (4). Two zones generally exist in the fluidized bed incinerator: the bubbling zone at the bottom of the reactor, and the freeboard at the middle and upper section of the reactor. Both the sand and sludge are fluidized in the bubbling zone; the sludge is ground, dried, and combusted; and the uncombusted portion of the sludge continues to be burned to completion in the freeboard zone (20).

Auxiliary fuel can be supplied through the preheat burners, and exhaust gas is vented from the top of the incinerator. The fluidized bed furnace is also equipped with pressure taps, access doors, and thermocouples to monitor the incineration process. The air flow into the incinerator is controlled by an oxygen analyzer, and the feed rate of auxiliary fuel is controlled by a temperature recorder. The exhaust gas is usually scrubbed with effluent from the treatment plant, and the ash is separated from the liquid with a hydrocyclone.

Comparing with the MHF, the FBF promotes better dispersion of the sludge, better mixing, much quicker heating up, and hence more complete combustion (17). Since the sludge is in direct contact with the hot sand bed, the burning rate is much faster. The sand bed also serves as a heat reservoir, which allows for more rapid start-up when the unit is shut down for a short time (e.g., overnight). After start-up, a unit can operate 4 to 8 hours a day without reheating as a result of the heat preserved by the sand. Similar to that of an MHF, a scrubber is also used for an FBF. Since the scrubbing water contains ash, it is often treated and sent to the lagoon. In an FBF, the

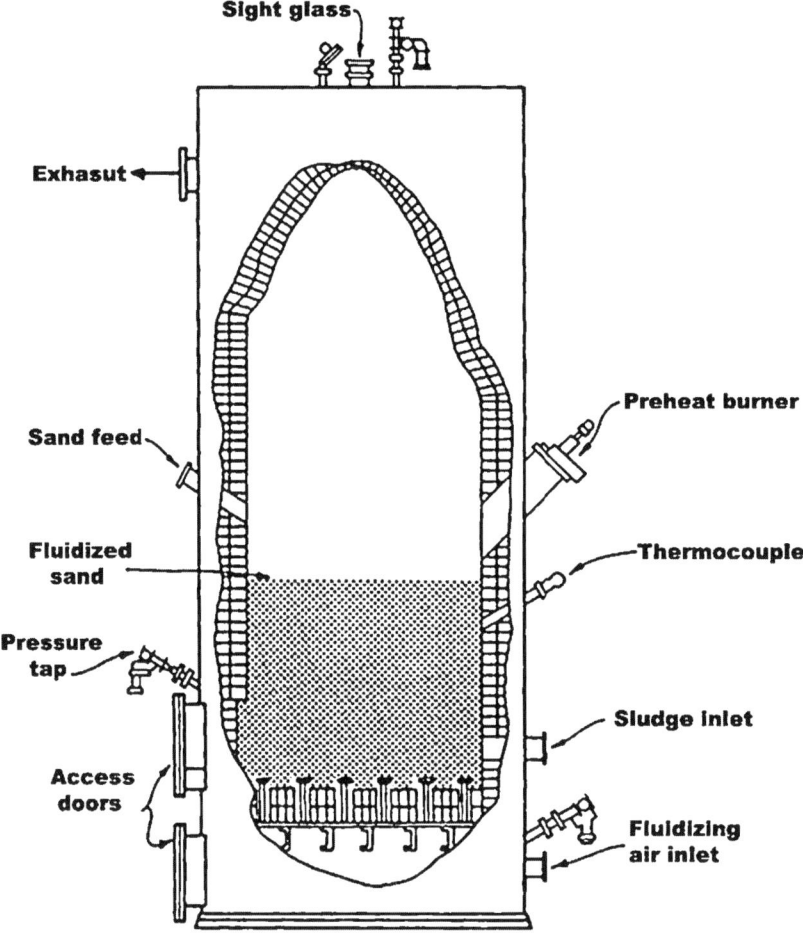

**Fig. 11.6.** Cross section of a fluidized bed furnace (5).

ash is usually handled with a hydraulic system, as the ash has already been wetted by a scrubber.

### 3.2. Design and Operating Parameters

The freeboard diameter of a typical FBF is 2.7 to 7.6 m (9 to 25 feet), and the thickness of the sand bed is 0.8 to 1.0 m (2.5 to 3 feet). When the sand expands, it can reach twice of its thickness. An FBF generally requires 20% to 50% of excess air to achieve complete combustion, which is much lower than in MHFs, and also results in less auxiliary fuel requirements. Sand loss of the FBF is at 5% per 300 hours of operation, since it can be carried out of the furnace by the gas stream (7, 21).

Air can be supplied to the FBF at ambient temperature (cold windbox) or at elevated temperatures (hot windbox). In the cold windbox design, there are no temperature resistance needs for the support rigs (grid) beneath the sand bed. In a hot windbox, air

# Combustion and Incineration Management

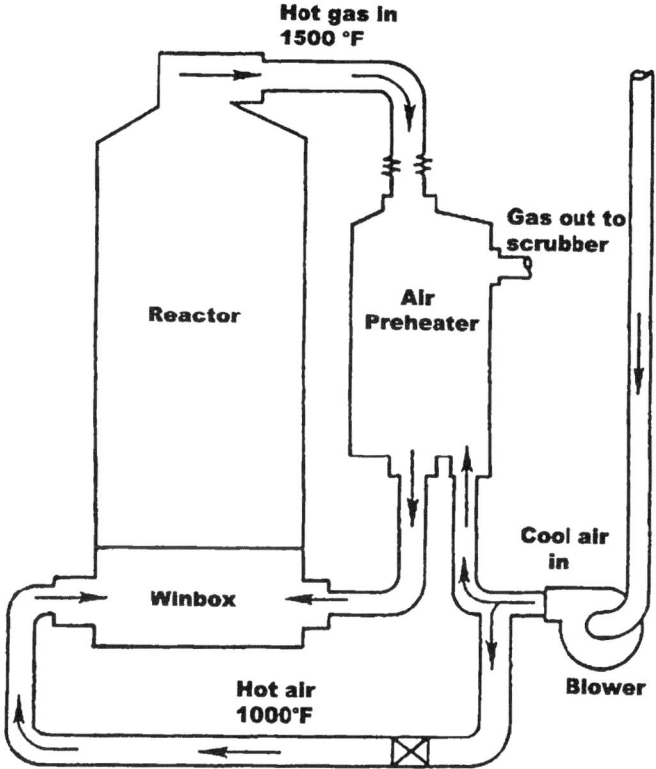

**Fig. 11.7.** Fluidized bed furnace with a heat exchanger (5).

can be heated by a heater within the windbox or with a separate air heater (Figure 11.7). The hot windbox can reduce the auxiliary fuel use and hence lower the operating cost, but a refractory dome is required due to the high temperatures.

Table 11.14 provides an example of the typical loading rates for several types of sludge in the conventional fluidized bed furnace, which are slightly higher than with the MHFs.

Similar to MHFs, the important operation parameters for an FBF include the excess air, sludge loading rate, solids concentration, and moisture content. These parameters can be estimated similar to those with the MHFs.

## 3.3. Performance Evaluation, Management, and Troubleshooting of the Fluidized Bed Furnace

The following operating parameters should be within the specified ranges in order to maintain normal operation:

1. The oxygen content of the stack gas should be between 3% and 6%.
2. The fluidized bed temperature should be between 1250 and 1300 °F.
3. The bed temperature record check demonstrates temperature consistency or any major changes.

### Table 11.14
**Typical loadings in a conventional fluidized bed furnace (16)**

| Type of sludge | Solids (%) | Volatile solids (%) | Chemical concentration,* (mg/L) | Wet sludge loading rate, (lb/h/sq ft) |
|---|---|---|---|---|
| 1. Primary | 30 | 60 | N/A | 14 |
| 2. Primary + FeCl$_3$ | 16 | 47 | 20 | 6.8 |
| 3. Primary + low lime | 35 | 45 | 298 | 18 |
| 4. Primary + WAS | 16 | 69 | N/A | 6.8 |
| 5. Primary + (WAS + FeCl$_3$) | 20 | 54 | 20 | 8.4 |
| 6. (Primary + FeCl$_3$) + WAS | 16 | 53 | 20 | 6.8 |
| 7. WAS | 16 | 80 | N/A | 6.8 |
| 8. WAS + FeCl$_3$ | 16 | 50 | 20 | 6.8 |
| 9. Digested primary | 30 | 43 | N/A | 14 |

*Assume no dewatering chemicals.

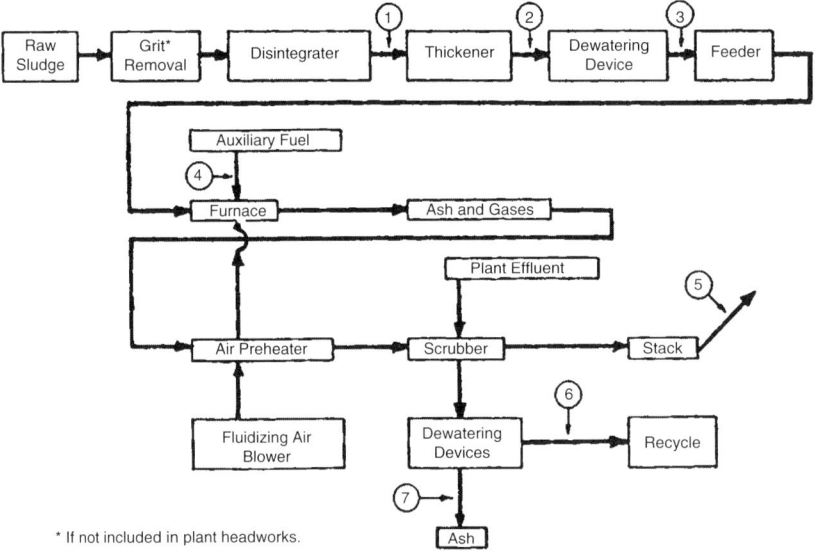

**Fig. 11.8.** FBF monitoring points (1).

4. The record for the depth of sand demonstrates if there is sand loss.
5. Check for sand leakage down to the windbox about once a month. When the reactor is not in operation, open the windbox manhole and rake for sand accumulation.
6. Clean the carbon deposit from the fuel burner. Carbon deposit near the tip of the burner can restrict fuel flow. The burner should be cleaned when the FBF is shutdown. If available, compressed air can be used to help inserting the gun back in the bed.

The loss of sand can occur as the grinding process can result into smaller sized sand particles, which can be carried out by the exhaust gas. When the bed level is determined to be low, add sand to the bed in the following order:

**Table 11.15**
**Emission monitoring points and monitoring frequency (16)**

| Monitoring point | Analysis | Frequency |
|---|---|---|
| 1 | Solids content | Weekly |
| 2 | Solids content | Weekly |
| 3 | Solids content | Weekly |
| 3 | Volatile solids | Weekly |
| 4 | Fuel quantity | Continuously |
| 5 | Oxygen content | Continuously |
| 5 | Particulate concentration | Weekly |
| 5 | Carbon monoxide | Monthly |
| 5 | Lead | Semiannual |
| 5 | Mercury | Semiannual |
| 5 | Hydrogen chloride | Semiannual |
| 5 | Sulfur dioxide | Semiannual |
| 5 | Oxides of nitrogen | Semiannual |
| 6 | $BOD_5$ | Weekly |
| 6 | Suspended solids | Weekly |
| 7 | Metals content | Semiannual* |
| 7 | Moisture content | Weekly |

*If ash used for soil conditioner.
BOD, biochemical oxygen demand.

- To avoid cooling of the sand bed below 1150 °F or to light the preheat burner, the temperature of the bed should be at least 1400 °F before adding any sand.
- The fluidizing blower should be completely stopped.
- Open the sand feed flange and attach the feed chute.
- Add sand in 10-bag (100 lb each) batches. If more than one batch is required, close the flange and reheat the sand bed to 1400 °F before adding the next batch. Repeat the process until the desired sand depth is reached.

The monitoring, maintenance, and record keeping of the FBF should follow the guidelines of the Part 503 rule, which was described in Sections 1.5 and 2.3. The record keeping is similar to that for MHF, which includes a yes or no checklist and a quantitative data sheet. Some of the operation manuals suggest the locations of the monitoring points as shown in Figure 11.8, and the analysis required and its frequency are listed in and Table 11.15. A detailed troubleshooting guide for FBFs is provided in Table 11.16.

### 3.4. Fluidized Bed Incinerator with Improved Design

Newer models of fluidized bed incinerators are also being used based on design improvements over the traditional configurations described earlier. A circulation fluidized bed incinerator has been developed by NKK Japan (20). The conventional two-zone design of the FBF is replaced with only one zone, where the quartz sand is circulated throughout the entire reactor. The sand layer becomes thinner as it approaches the top of the reactor. In this particular design, two-stage combustion is also implemented

### Table 11.16
### Troubleshooting guide for a fluidized bed furnace (5)

| Indicators/observations | Probable cause | Check or monitor | Solutions |
|---|---|---|---|
| 1. Bed temperature is falling | 1a. Inadequate fuel supply | 1a. Fuel system operation | 1a. Increase fuel feed rate or repair and fuel system malfunction |
| | 1b. Excessive rate of sludge feed | 1b. Sludge feed system | 1b. Decrease sludge feed rate |
| | 1c. Excessive sludge moisture | 1c. Dewatering system | 1c. Improve dewatering system operation |
| | 1d. Excessive air flow | 1d. Oxygen content of exhaust gas should not exceed 6% | 1d. Reduce air rate |
| 2. Low (<3%) oxygen in exhaust gas | 2a. Low air flow | 2a. Air flow rate | 2a. Increase air blower rate |
| | 2b. Fuel rate too high | 2b. Fuel rate too high | 2b. Decrease fuel rate |
| 3. Excessive (>6%) oxygen in exhaust gas | 3. Sludge feed rate too low | 3. Sludge feed rate | 3. Increase sludge feed rate and adjust fuel rate to maintain steady bed temperature |
| 4. Erratic bed depth readings on control panel | 4. Bed pressure taps plugged with solids | | 4a. Tap a metal rod into pressure tap pipe when reactor is not in operation |
| | | | 4b. Apply compressed air to pressure tap while the reactor is operation after reviewing manufacturer's safety instructions |
| 5. Preheat burner fails and alarm sounds | 5. Pilot flame not receiving fuel | 5. Fuel pressure and valves in fuel line | 5. Open appropriate valve and establish fuel supply |

**Table 11.17**
**The comparison of operating parameters between a multiple hearth furnace and a Fluidized bed furnace (10)**

| Parameters | Multiple hearth furnace | Fluidized bed furnace |
|---|---|---|
| Flow | Counter-current | Intense Back Mixing |
| Heat transfer | Poor | High |
| Biosolids detention time | 1/2–3 h | 1–5 min |
| Gas detention time at high temperature | 1–2 s | 6–8 s |
| Combustion temperature | 1500–1800 °F | 1400–1500 °F |
| Gas exit temperature | 800–1000 °F | 1500–1600 °F |
| Excess air | 75–100% | 40% |

where combustion air is introduced at two different locations. A cyclone is used to collect the sand at the exhaust gas exit of the incinerator. The circulation fluidized bed can reduce the space use significantly due to the combination of the two zones and the incinerator diameter is only two thirds or less of the conventional FBF. The staged combustion can improve the combustion efficiency and reduce $NO_x$ emissions (22). Other associated advantages include less fuel usage and less power consumption for the air blower.

### 3.5. Comparison Between Multiple Hearth and Fluidized Bed Furnaces

The FBFs have been reported to have fewer problems than the MHFs. The basic design differences result in several advantages of FBF, such as lower auxiliary fuel usage, lower emissions, and reduced maintenance and operating costs. The major design and operational parameters of a MHF and FBF are listed in Table 11.17 (10). In fact, the three-zone design of the MHF results in higher auxiliary fuel usage and potential odor emission than in the FBF; as the sludge is dried in the upper hearths, the loss of volatiles results in a lower heating value of the sludge and potential odor emissions (19).

The MHFs can be retrofitted with new technologies to meet more stringent environmental regulations, mainly for THC and odor reduction, and to reduce cost. Such retrofit technologies include zero-hearth burners, afterburners, and RTOs, with the latter two briefly discussed in Section 1.3.

A zero-hearth afterburner is usually installed at the top of the hearth, and the sludge feed needs to be rerouted to the second hearth. As a result, the sludge feed capacity is reduced. Such an installation in Manchester, New Hampshire, reduced the feed capacity by almost 28%. Due to other problems associated with the installation, the operation stopped after 3 months. In another application, 50% reduction of THC was reported, due to the higher temperature obtained in the zero-hearth afterburner.

These improvements result in less CO and THC emissions without much improvement in $NO_x$. However, considering the capital, operational, and maintenance cost, it is usually more cost-effective to install a new FBF than to retrofit the existing MHF.

## 4. OTHER INCINERATION PROCESSES

### 4.1. Electric Infrared Incinerators

The first electronic infrared incinerator was installed in 1975 in the U.S. and its use has been much less common than the MHF and FBF. A cross-sectional view of the electronic furnace can be found in Figure 11.9, which contains a horizontally oriented insulated furnace. A woven wire belt conveyor extends the length of the furnace, and infrared heating elements are located in the roof above the conveyor belt. The electronic incinerator can be assembled with several similar modules, which is determined by the length of the furnace. The sludge enters from the feeder and is spread on the conveyor belt at about 1-inch thickness. The sludge is first dried and then combusted as it moves along the continuous belt. The ash is discharged at the end of the furnace to a hopper. Combustion air is preheated by the flue gas, injected counter-currently from the sludge, and is further heated as it moves forward; 20% to 70% of excess air is typically applied. Exhaust gas leaves the furnace from the sludge feed side. The emission from this type of incinerator is mostly controlled with a scrubber, either a Venturi or other types of scrubbers (21).

Compared with the MHF and the FBF, the advantage of the electronic infrared furnace is the lower capital cost, especially for smaller wastewater treatment systems. The emission from electronic furnace is usually lower than from the MHF and FBF. However, the cost of electricity may vary with locations. The replacement of parts such as the woven belt and the infrared heater every 3 to 5 years can add additional cost.

### 4.2. Co-Incineration

Co-incineration refers to the incineration processes combining the wastewater sludge and another nonfuel combustible material, such as the municipal solid waste, wood waste, textile waste, bagasse, or farm waste. There are mainly two types of co-incinerators: biosolids incinerators, which accept solid waste, and solid waste incinerators, which accept biosolids (sludge) (23, 24).

The major advantage of co-incineration is to utilize the existing incinerator to achieve volume reduction of more than one waste material. The waste can serve as the auxiliary fuel needed for sludge incineration and thereby can result in better utilization of the incinerator and related cost benefits, and the benefits of reduced volume for disposal.

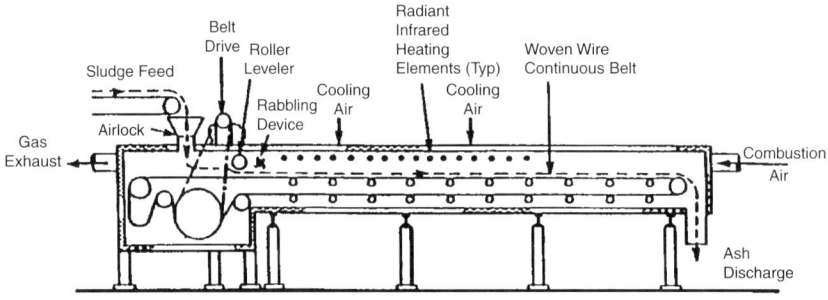

**Fig. 11.9.** Conceptual design of a circulating fluidized bed (20).

# Combustion and Incineration Management

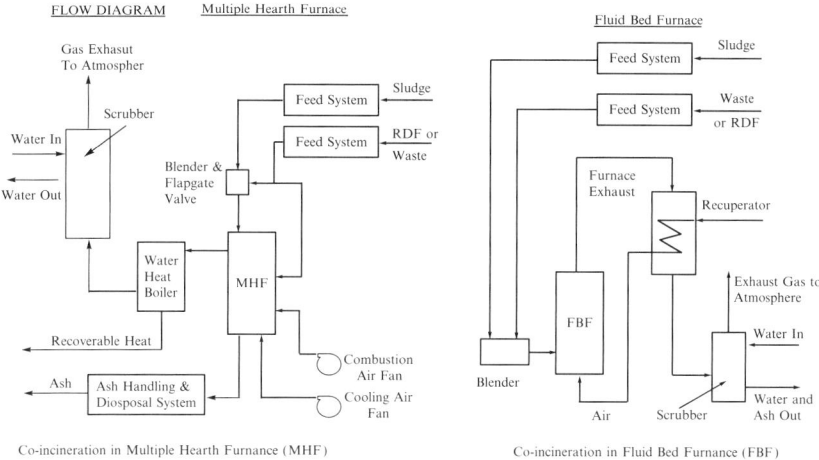

**Fig. 11.10.** Two-stage design of a circulating fluidized bed (20).

The limitation of co-incineration tends to result from institutional constraints, since the waste and sludge are oftentimes managed by separate government agencies. Therefore, proper coordination is necessary to ensure a stable supply of the feed materials. Additional devices are usually required to accommodate different feed streams or to dry the sludge to its autogenous point, and the excess air requirement is often larger than sludge combustion alone.

The environmental emission of this process strongly depends on the composition of feed materials: the sludge and the waste. The variation of the feed, especially of the waste, can result in different emissions. Pollutant emissions tend to be slightly different for the two co-incineration processes and will be discussed in each process, respectively.

### 4.2.1. Co-Incineration in Sludge Incinerators

Co-incineration using both MHF and FBF have been demonstrated in the U.S. and Europe. However, only three co-incineration plants are in operation in the U.S. since December 1996 (23). Figure 11.10 shows the schematic diagram of co-incineration in an MHF and an FBF sludge incinerator, respectively.

The limitation of this process is that a shredder is usually required to reduce the refuse size to approximately 1 inch to fit the feeding system of the sludge incinerators. The waste is usually fed into the incinerator through a separate feed system, and is blended with sludge prior to been fed into the incinerator.

Co-incineration in the U.S. was first demonstrated in Franklin, Ohio, using an FBF. The incinerator feed consisted of rejected organic waste from wood fiber recovery and wastewater sludge. The organic fiber waste consisted of approximately 20% solids, and the sludge consisted of 5% solids. The mixed waste was dewatered in a cone press to 45% solids prior to combustion. The FBF required about 3000 BTU/lb of energy to sustain the combustion process, and the mixed waste stream contained about 3600 BTU/lb of energy. An autogenous condition was achieved, but the potential of energy recovery was low, at only 600 BTU/lb.

Co-incineration using a MHF incinerator has been tested in the U.S. and Europe. The earlier tests in Europe using raw solid waste were not successful. After grinding the solid waste into smaller sizes, this process became feasible when tested in a wastewater treatment plant in Concord (central Contra Costa County), California, with support from the US EPA.

An existing six-hearth MHF (16-inch diameter) was modified to accommodate refuse-derived fuel (RDF). The RDF can be mixed with sludge, resulting in a mixture of 16% solid or fed separately into the third hearth, with the latter observed to be more efficient. The excess air used was between 70% and 100%. The system operated 8 hours a day for 2 months, with a combined wet feed rate of up to 10 ton/hour. Using the mixture with 16% solid, the combustion can be self-sustained with an RDF to sludge ratio of 1:2. The starved air combustion mode (excess air only added at the afterburner, oxygen deficient in MHF) was reported to be the preferred operation mode over the incineration mode (excess air added to the hearths).

The environmental emission of this process strongly depends on the composition of feed materials, the sludge, and the waste. The uncontrollable particulate matter emission from this process is about 10% higher than sludge incineration alone. It is reported that the air toxics can be destroyed, and most of the metals can be captured by particle collection devices, with the exception of mercury (Hg). Mercury emission should be evaluated against the designed value.

*4.2.2. Co-Incineration in Solid Waste Incinerators*

In this process, sludge is fed into a solid waste incinerator together with solid waste. The earlier trial of this type of co-incineration was not successful due to difficulties in materials handling, feeding, and firing. With the design improvement of solid waste incinerators, this process was tested again and became feasible.

A flow diagram of this process is shown in Figure 11.11. The sludge requires a separate feed system to reduce the high moisture content. In addition to the dewatering device, a dryer is also required. In some plants, the hot exhaust gas was used to dry the sludge to reach its autogenous point (self-sustained combustion).

Five plants worldwide have been reported to operate with this type of co-incineration and three of them are in the U.S. A plant in Ansonia, Connecticut, used hot flue gas to dry the sludge. The design capacity of the refractory incinerator is 200 ton/d and approximately 55 ton/d of mixed waste is disposed of in an 8-hour shift. The raw sludge from the wastewater treatment plant contains 4% solids and is dried in a high-speed disk co-current spray drier, where the hot flue gas from the secondary combustion chamber was introduced at 1200 °F. The vapor and dried sludge were both blown into the furnace above the second grate, where the sludge is combusted in suspension. Part of the dried sludge that was not incinerated was used as fertilizer instead by local residents (23).

Another small refractory incinerator with 50 ton/d throughput located in Holyoke, Massachusetts, also used hot flue gas to dry the sludge, after it was mechanically dewatered to 28% solids. The sludge is introduced into the incinerator about the refuse grate. A facility in Glen Cove, New York, used a slightly different approach, where the sludge was burned along with the solid waste. The sludge with 5% solids was sprayed

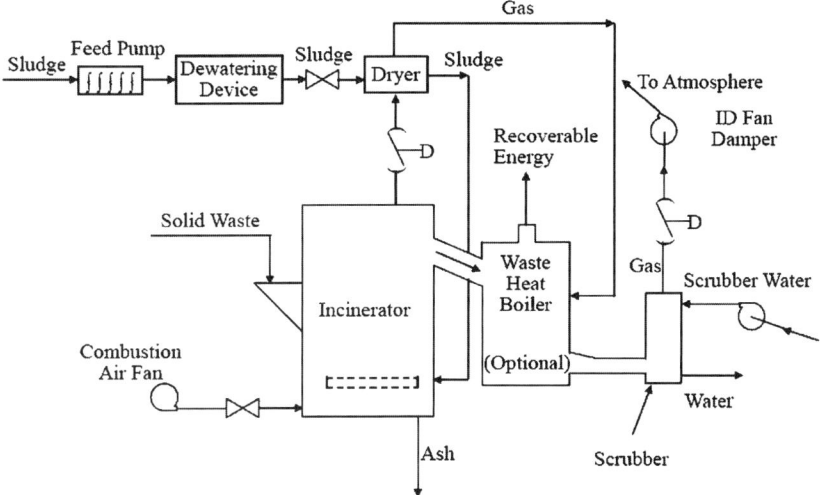

**Fig. 11.11.** Cross section of an electronic infrared incinerator (21).

into the charging chute, and a layer of sludge was formed on the solid waste. In the incinerator, the sludge was dried by the heat from the burning solid waste, and the dried sludge was then burned with the solid waste along the grate.

In Europe, one of the co-incinerator facilities was located in Dieppe, France, and another is in Krefeld, Germany. Both plants utilized a waterfall combustion unit for the incineration process.

One of the limitations of this process is that the incineration facility must be within pumping distance of the wastewater treatment plant. The process requires a minimum of 150% excess air and a minimum flue gas temperature of 1400°F for odor destruction.

### 4.3. Other Sludge Incineration Techniques

Other technologies used for the incineration of sewage sludge include cyclonic reactors, rotary kilns, and wet oxidation reactors. These processes are not commonly used in the U.S. and are described only briefly (21).

The cyclonic reactor is more suitable for small capacity plants. It consists of a vertical cylindrical chamber lined with refractory. The preheated combustion air is tangentially introduced into the incinerator at high velocities, similar to a cyclone. The sludge is sprayed radially toward the hot refractory walls, and the residence time of the sludge is on the order of 10 seconds.

Rotary kilns are also generally used in small-capacity applications. The kiln is inclined slightly from the horizontal plane, with both the sludge feed and the combustion air entering from the upper end. The burner is located at the lower end of the kiln. The kiln rotates at approximately 15 cm per second (6 inches per second).

The wet oxidation process operates at elevated temperature and pressure in the presence of water to achieve flameless combustion (oxidation). Thickened sludge with about 6% solids is first ground and mixed with a stoichiometric amount of compressed air. The mixture is then circulated through a series of heat exchangers before entering a pressurized reactor. The temperature of the reactor is held between 175 and 315 °C (350 and 600 °F). The pressure is normally 7000 to 12,500 kilopascals (1000 to 1800 psi grade). Steam is usually used to provide auxiliary heat to the reactor. The water and ash coming out of the reactor are separated in a lagoon. The liquid is recycled back to the treatment plant. The exhaust gas must be further treated using scrubbing, afterburning, or carbon absorption to remove odor.

## NOMENCLATURE

APCD = air pollution control devices
CAAA = Clean Air Act Amendment
$CE$ = control efficiency
CO = carbon monoxide
$DF$ = dispersion factor
ESP = electrostatic precipitator
FBF = fluidized bed furnace
ID = induced draft
MGD = million gallons per day
MHF = multiple hearth furnace
MSD = Metropolitan Sewage District
MSW = municipal solid waste
NAAQS = National Ambient Air Quality Standards
NESHAP = National Emission Standards for Hazardous Air Pollutants
NSPS = New Source Performance Standards
RDF = refuse-derived fuel
RSC = risk-specific concentration
RTO = regenerative thermal oxidizers
$SF$ = biosolids feed rate
T/d = dry metric tons per day
t/d = dry English tons per day
THC = total hydrocarbon
WAS = waste-activated sludge
WESP = wet electrostatic precipitator
WWTP = wastewater treatment plant

## REFERENCES

1. New England Interstate Water Pollution Control Commission, *Fact Sheet: Sewage Sludge Incineration*, Lowell, MA (2001).
2. U.S. Environmental Protection Agency, *Biosolids Technology Fact Sheet*, US EPA/832-F-03-013, Washington, DC (2003).
3. R. P. Dominak, Current practices and future direction for biosolids incinerators. In: *Proceedings of the Residuals and Biosolids Management Conference* (2001).
4. U.S. Environmental Protection Agency, *Multiple-Hearth and Fluid Bed Sludge Incinerators: Design and Operational Considerations*, US EPA/430/9-85-002, Washington, DC (1985).

5. U.S. Environmental Protection Agency, Field manual for performance evaluation and troubleshooting at municipal wastewater treatment facilities, US EPA/430/9-789-001, Washington, DC (1978).
6. U.S. Environmental Protection Agency, *A Plain English Guide to the US EPA Part 503 Biosolids Rule*, US EPA/832/R-93/003, Office of Wastewater Management, Washington, DC, September (1994).
7. J. W. Myer, J. F. Mullen, and K. Dangtran, Thirty Years of Satisfactory Fluid Bed Sludge Incineration, The Northwest Bergen County Utilities Authority Experience, http://www.infilcodegremont.com/images/pdf/Session_15A.pdf, May (2007).
8. U.S. Environmental Protection Agency, *Project summary: sewage sludge incinerator fuel reduction*, Hartford, Connecticut, US EPA/600/S2-84-146, Cincinnati, OH (1984).
9. National manual of good practice for biosolids, Chapter 15, *Biosolids Incineration Systems*, http://www.biosolids.org/docs/chapter15.pdf, April (2007).
10. K. Dangtran, J. Mullen, and D. Mayrose, A comparison of fluid bed and multiple hearth biosolids incineration, *14th Annual Residuals and Biosolids Management Conference*, February (2000).
11. Emission Factor Documentation for AP-42 Section 2.2, *Sewage Sludge Incineration*, http://www.epa.gov/ttn/chief/ap42/ch02/bgdocs/b02s02.pdf, April (2007).
12. Minutes of the tenth meeting of the Land-based Pollution Group, Helsinki Commission, Bonn, Germany, 31 May to 2 June (2005).
13. http://www.dec.ny.gov/regs/2492.html, May (2007).
14. Ohio EPA. *Air Quality Permit for Little Miami Facility*, Cincinnati, OH (1997).
15. Ohio EPA. *Air Quality Permit for Mill Creek Facility*, Cincinnati, OH (2002).
16. U.S. Environmental Protection Agency, Inspectors Guide for Evaluation of Municipal Wastewater Treatment Plants, US EPA/430/9-79-010, Washington, DC (1979).
17. J. Swithenbank, S. Basire, W. Y. Wong, Y. Lu, and V. Nasserzadeh, Sludge incineration in a spinning fluidized bed incinerator. In: *Proceedings of the 15$^{th}$ International Conference on Fluidized Bed Incineration*, May 16–19, Savanna, GA (1999).
18. S. O'Kelley, R. Williamson, and F. M. Lewis, Inspired incineration, *Water Environment & Technology*, February, 71–75 (2007).
19. P. A. Vesilind, and T. B. Ramsey, Effect of drying temperature on the fuel value of wastewater sludge, *Waste Management and Research*, 14, 2, 189–196 (1996).
20. http://www.gec.jp/JSIM_DATA/WASTE/WASTE_5/html/Doc_510.html, April (2007).
21. http://www.epa.gov/ttn/chief/ap42/ch02/final/c02s02.pdf, April (2007).
22. C. D. Cooper and F. C. Alley, *Air Pollution Control: A Design Approach*, 3rd ed., Waveland Press, Prospect Heights, IL, 507–511 (2002).
23. U.S. Environmental Protection Agency, *Innovative and Alternative Technology Assessment*, US EPA/430/9-78-009, Washington, DC (1980).
24. L. K. Wang, N. K. Shammas, and Y. T. Hung (eds.), *Biosolids Treatment Processes*. Humana Press, Totowa, NJ, 820 pp. (2007).

# 12
# Beneficial Utilization of Biosolids

## Nazih K. Shammas and Lawrence K. Wang

**CONTENTS**

INTRODUCTION
FEDERAL BIOSOLIDS REGULATIONS
LAND APPLICATION OF BIOSOLIDS
SURFACE DISPOSAL OF BIOSOLIDS
INCINERATION OF BIOSOLIDS AS AN ENERGY SOURCE
OTHER USES OF WASTEWATER SOLIDS AND SOLIDS BY-PRODUCTS
EXAMPLES
NOMENCLATURE
REFERENCES

**Abstract** Utilization refers to the beneficial use of biosolids or biosolids by-products. Biosolids may be used as a soil amendment and a source of nutrients, a source of heat and work, and a source of other useful products that include waste treatment chemicals, landfill toppings, industrial raw materials, animal feed, and materials of construction. The following aspects of utilization are discussed: biosolids regulations, land application, surface disposal, incineration of biosolids as an energy source, and other uses. Examples are given.

**Key Words** Utilization • biosolids • regulations • land application • surface disposal • energy source • design.

## 1. INTRODUCTION

Utilization refers to the beneficial use of biosolids or biosolids by-products. Biosolids may be used as the following (1–3):

1. *Soil amendment and source of nutrients*: Biosolids contain both crop nutrients and organic matter. Biosolids can be used as a fertilizer and in the reclamation of disturbed lands, such as construction sites, strip-mined lands, gravel pits, and clear-cut forests. They may be used to stabilize bank spoils and moving sand dunes.
2. *Source of heat and work*: Energy may be recovered from the gas produced during anaerobic stabilization, or partial or full pyrolysis of biosolids and from the direct burning of biosolids.

This energy may be converted to heat or work and put to a variety of in-plant uses or exported for uses outside the plant.
3. *Source of other useful products*: Other useful products include waste treatment chemicals, landfill toppings, industrial raw materials, animal feed, and materials of construction.

The thrust of recent legislation has been to encourage beneficial reuse. The Federal Water Pollution Control Act of 1972 stated, "The Administrator shall encourage waste treatment management which results in the construction of revenue producing facilities for the recycling of potential sewage pollutants through the production of agriculture, silviculture, or agricultural products, or any combination thereof." The Clean Water Act (CWA) of 1977 offered further incentives for projects that involved innovative and alternative technology (for example, biosolids utilization, energy recovery). In addition, the CWA required the establishment of industrial waste pretreatment programs with the objective of reducing toxic pollutant loadings to municipal treatment facilities. Implementation of pretreatment programs has made more municipal solids suitable for reuse.

The pretreatment program supplements programs established by the Toxic Substance Control Act which authorized the U.S. Environmental Protection Agency (US EPA) to obtain production and test data from industry on selected chemical substances and regulate them where they pose an unreasonable risk to the environment. Steps toward the goal of furthering biosolids utilization were taken by the Resources Conservation and Recovery Act (RCRA), which authorized the US EPA to develop treatment and application rate criteria for biosolids to be applied to land growing food-chain crops, as well as to nonagricultural areas. The RCRA also authorized funds for research, demonstrations, training, and other activities related to development of other resource recovery schemes.

At the same time, it is recognized that there are potential hazards associated with wastewater biosolids utilization, and that utilization without careful planning, management, and operation could present a danger to human health and to the environment. The most recent US EPA Part 503 rule (see details in the next section) provides the comprehensive requirements for the management of biosolids to safeguard public health and the environment and at the same time creates incentives for beneficial use of biosolids. The US EPA believes that biosolids are an important resource that can and should be safely used, for example, to condition soils and provide nutrients for agricultural, horticultural, and forest crops and vegetation, and for reclaiming and revegetating areas disturbed by mining, construction, and waste disposal facilities (4).

There are many advantages and disadvantages of biosolids utilization (3, 5–9). The advantages are as follows:

1. Improve soil properties for optimum plant growth, including texture, fertility, and water holding capacity.
2. Improve drainage of wet clay.
3. Reduce need for commercial fertilizers.
4. Slow the release of nutrients.
5. Reclaim strip-mined lands.
6. Enrich forest land.
7. Provide topsoil for recreational use.
8. Enhance conditions for vegetative growth.

9. Decrease the need for pesticide use by suppression of pathogenic soil organisms such as nematodes that damage plant roots as well as specific plant root diseases that otherwise cause damage to potted plants.
10. Decrease erosion.
11. Conserve landfill space and less leachate production.
12. Provide economic incentives and "green" grants and awards.

The disadvantages are as follows:

1. Lots of paper work
2. Monitoring and control demands
3. Increased analytical work
4. Increased time needed for application and storage
5. Weather limitations, such as rain or snow
6. Labor intensive
7. Odors (10)
8. Health issues (10, 11)
9. Long-term detrimental effect on soil (12)
10. The public's negative perception and opposition (13–16)

## 2. FEDERAL BIOSOLIDS REGULATIONS

### 2.1. Background

As required by the Clean Water Act Amendments of 1987, the US EPA developed a new regulation to protect public health and the environment from any reasonably anticipated adverse effects of certain pollutants that might be present in biosolids. This regulation, the Standards for the Use or Disposal of Sewage Sludge (Biosolids) (Title 40 of the Code of Federal Regulations [CFR], Part 503), was published in the Federal Register in February 1993, and became effective on March 22, 1993. This regulation is usually referred to as Part 503 or the Part 503 rule (4).

The Part 503 rule establishes requirements for the final use or disposal of biosolids when biosolids are

1. applied to land to condition the soil or fertilize crops or other vegetation grown in the soil,
2. placed on a surface disposal site for final disposal, or
3. fired in a biosolids incinerator.

The rule also indicates that if biosolids are placed in a municipal solid waste landfill, the biosolids must meet the provisions of 40 CFR Part 258.

The Part 503 rule was amended in February 1994. The amendment made two changes. It deleted pollutant limits for molybdenum in biosolids applied to land but retained the molybdenum ceiling limits, and in certain situations it permitted carbon monoxide (CO) monitoring in place of total hydrocarbon (THC) monitoring for biosolids incinerators.

The Part 503 rule is designed to protect public health and the environment from any reasonably anticipated adverse effects of certain pollutants and contaminants that may be present in biosolids. The provisions of the Part 503 rule are consistent with the US EPA's policy of promoting beneficial uses of biosolids. Land application takes advantage of the soil conditioning and fertilizing properties of biosolids. It is important to note that state rules also apply to biosolids use or disposal. Persons using or disposing of biosolids are subject to state and local regulations as well. Furthermore, these state and other regulations may be more stringent generally than the federal Part 503 rule, may define biosolids differently, or may regulate certain types of biosolids more stringently than the Part 503 rule.

## 2.2. Risk Assessment Basis of Part 503

Many of the requirements of the Part 503 rule are based on the results of an extensive multimedia risk assessment. This risk assessment was more comprehensive than for any previous federal biosolids rulemaking effort, the earliest of which began in the mid-1970s. Research results and operating experience over the past 30 years have greatly expanded the US EPA's understanding of the risks and benefits of using or disposing of biosolids.

Development of the Part 503 rule began in 1984. During this extensive effort, the US EPA addressed 25 pollutants using 14 exposure pathways in the risk assessment. In this assessment, the US EPA also developed a new methodology that provided for the protection of the environment and public health. The new method for conducting the multimedia risk assessment was reviewed and approved by rthe US EPA's Science Advisory Board.

The US EPA proposed the Part 503 rule in February 1989. During the 4 years between the publication of the proposed and final rule, the data, models, and assumptions used in the risk assessment process were reviewed and revised in an effort involving internationally recognized experts working closely with the US EPA. The US EPA feels this process has resulted in the establishment of state-of-the-art risk-based standards for controlling the use or disposal of biosolids. Part 503 did not consider ocean disposal to be a viable option for biosolids disposal (4). The reason why Congress banned ocean dumping was not that biosolids were toxic to marine life. Rather, Congress recognized that the nutrients in biosolids could cause increased aquatic plant production, eventually leading to oxygen depletion at the site. Congress properly decided that it made much more sense and better policy to get biosolids out of the ocean and use the nutrients in biosolids more productively to provide crop nutrients and to improve soil quality (17).

In mid-1993, at the request of the US EPA, the Water Science and Technology Board of the National Research Council (NRC)/National Academy of Science (NAS) undertook an extensive review of the Part 503 rule, including an evaluation of public health concerns, current biosolids management practices and regulations, and implementation issues. Sponsors of the study included the US EPA, the U.S. Department of Agriculture, the Food and Drug Administration, the Bureau of Reclamation, the Water Environment Research Foundation, the Association of Metropolitan Sewerage Agencies, the National Water Research Institute, and the National Food Processors Association, as well as several water and wastewater authorities and private companies.

In 1996, the NAS released *Use of Reclaimed Water and Sludge in Food Crop Production* (18). This extensive peer review concluded that "the use of these materials in the production of crops for human consumption, when practiced in accordance with federal guidelines and regulations, presents negligible risk to the consumer, to crop production and to the environment."

# Beneficial Utilization of Biosolids

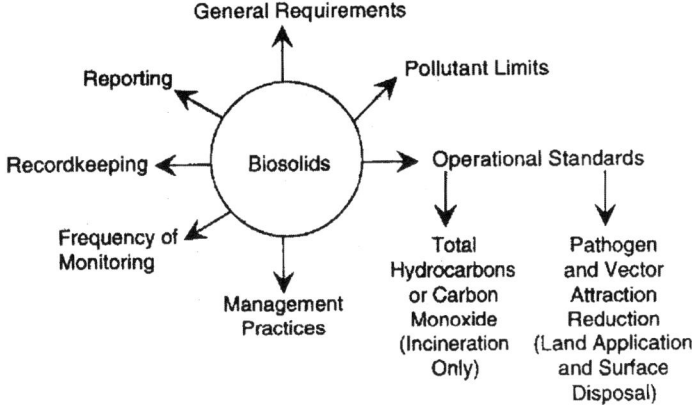

**Fig. 12.1.** Requirements of Part 503 standard. *Source*: US EPA.

In 2000, the US EPA again asked the NAS to review the science and methodology underlying the agency's current health and environmental standards for biosolids. On July 2, 2002, the panel released the results of its study, *Biosolids Applied to Land: Advancing Standards and Practices* (19), which confirms the findings of the 1996 NAS study. Its overarching finding is that "there is no documented scientific evidence that the Part 503 rule has failed to protect health," and it does not call for any restrictions on land application of biosolids. According to the report, "a causal association between biosolids exposure and adverse health outcomes has not been documented." It further states that the panel recognized that the land application of biosolids is a widely used, practical option for managing the large volume of sewage sludge generated at wastewater treatment plants.

Detailed information describing the risk assessment and technical basis of the Part 503 standards is contained in the preamble to the Part 503 rule and in a technical support document (20).

## 2.3. Overview of Part 503

The Part 503 rule includes five subparts: general provisions, requirements for land application, requirements for surface disposal, requirements for pathogen and vector attraction reduction, and requirements for incineration. For each of the regulated use or disposal practices, a Part 503 standard includes general requirements, pollutant limits, management practices, operational standards, and requirements for the frequency of monitoring, record keeping, and reporting, as shown in Figure 12.1. For the most part, the requirements of the Part 503 rule are self-implementing and must be followed even without the issuance of a permit.

Subpart A of the rule covers general provisions, such as the purpose and applicability of the rule, the compliance period, and exclusions from the rule. These general provisions apply to each of the three biosolids use or disposal practices.

## 2.4. Requirements for Land Application

Subpart B of the rule specifies requirements for biosolids applied to land. The word *apply* means to put biosolids on the land to take advantage of the nutrient content or soil conditioning properties of the biosolids. The requirements for land application also pertain to material derived from biosolids; that is, biosolids that have undergone a change in quality through treatment (e.g., composting) or by mixing with other materials (e.g., wood chips, municipal solid waste, yard waste).

The biosolids land application requirements are summarized below (21). There are several options for land applying biosolids under subpart B of the Part 503 rule, all of which are equally protective of human health and the environment:

1. *Exceptional quality (EQ) biosolids*: The term *exceptional quality* is used to characterize biosolids that meet low-pollutant and Class A pathogen reduction (virtual absence of pathogens) limits and that have a reduced level of degradable compounds that attract vectors. Once the requirements discussed in detail in Section 3, below, are met, EQ biosolids are considered a product that is virtually unregulated for use, whether used in bulk, or sold or given away in bags or other containers.
2. *Pollutant concentration (PC) biosolids*: The term *pollutant concentration* refers to biosolids that meet the same low-pollutant concentration limits as EQ biosolids, but only meet Class B pathogen reduction or are subjected to site management practices rather than treatment options to reduce vector attraction properties (see Section 3). Unlike EQ biosolids, PC biosolids may only be applied in bulk and are subject to general requirements and management practices; however, tracking of pollutant loadings to the land is not required.

    A majority of the biosolids currently generated in the United States are believed to be EQ or PC biosolids containing low levels of pollutants. The US EPA expects that many municipalities will strive to produce EQ or PC biosolids because of the reduced regulatory requirements and the anticipated improved public perception about using EQ and PC biosolids beneficially. Cumulative levels of pollutants added to land by EQ or PC biosolids do not have to be tracked because the risk assessment has shown that the life of a site would be at least 100 to 300 years under the conservative parameters assumed.
3. *Cumulative pollutant loading rate (CPLR) biosolids*: CPLR biosolids typically exceed at least one of the pollutant concentration limits for EQ and PC biosolids but meet the ceiling concentration limits (see Section 3). Such biosolids must be applied to land in bulk form. The cumulative levels of biosolids pollutants applied to each site must be tracked and cannot exceed the CPLR.
4. *Annual pollutant loading rate (APLR) biosolids*: APLR biosolids are biosolids that are sold or given away in a bag or other container for application to the land that exceed the pollutant limits for EQ biosolids but meet the ceiling concentration limits (see Section 3). These biosolids must meet APLR requirements and must be accompanied by specific biosolids application rate information on a label or handout that includes instructions on the material's proper use.

Each of the options for land applying biosolids are affected by the Part 503, 1994 amendment, which states that the US EPA is reconsidering appropriate land application and pollutant limits for molybdenum. During the period of reconsideration, only ceiling limits for molybdenum must be met. Molybdenum pollutant limits for EQ, PC, CPLR, or APLR biosolids have been deleted.

The options for using or disposing of domestic septage under subpart B are as follows: If domestic septage is applied to land with a high potential for contact by the public (e.g., public parks, ball fields, cemeteries, plant nurseries, and golf courses), the Part 503 land application requirements apply. However, when domestic septage is applied to nonpublic contact sites (e.g., agricultural land, forests, and reclamation sites), less burdensome

## 2.5. Requirements for Biosolids Placed on a Surface Disposal Site

Subpart C of the rule covers requirements for biosolids—including domestic septage—placed on a surface disposal site. Placement refers to the act of putting biosolids on a parcel of land at high rates for final disposal rather than using the organic content in the biosolids to condition the soil or using the nutrients in the biosolids to fertilize crops. Placing biosolids in a monofill, in a surface impoundment, on a waste pile, or on a dedicated site is considered surface disposal.

Certain materials derived from biosolids, the quality of which has been changed by treating the biosolids or by mixing them with other materials (e.g., wood chips), are subject to the surface disposal requirements in Part 503 with one exception. If biosolids are mixed with nonhazardous solid wastes, the mixture and the land onto which the mixture is placed are subject to the solid waste regulations (40 CFR Part 258) instead of Part 503.

## 2.6. Requirements for Pathogen and Vector Attraction Reduction

Subpart D of the Part 503 rule covers requirements for the control of disease-causing organisms, called pathogens, in biosolids and the reduction of the attractiveness of biosolids to vectors, such as flies, mosquitoes, and other potential disease-carrying organisms. Pathogen and vector attraction reduction requirements are described for biosolids applied to land or placed on a surface disposal site in (see Sections 3 and 4 of this chapter). More detailed guidance on meeting pathogen and vector attraction reduction requirements is provided in an US EPA publication (23).

## 2.7. Requirements for Biosolids Fired in Incinerators

Subpart E of the rule covers the requirements for biosolids fired in a biosolids incinerator. The firing of biosolids with auxiliary fuels also is covered by the Part 503 incineration requirements. Auxiliary fuel materials include gas, oil, coal, and other materials that serve as a fuel source.

The co-firing of biosolids in an incinerator with other wastes is generally not regulated under Part 503. It should be noted, however, that wastes either in auxiliary fuel or mixed and co-fired with biosolids are considered to be auxiliary fuel when the weight is less than or equal to 30% (by dry weight) of the total biosolids and auxiliary fuel mixture. The requirements in subpart E for biosolids incineration are discussed in Section 5.

The February 1994 amendment to the Part 503 rule states that under certain conditions the US EPA will allow continuous monitoring of carbon monoxide emissions from biosolids incinerators as an alternate to continuous monitoring of total hydrocarbons in emissions. The details of the amendment are also discussed in Section 5.

Part 503 specifies certain exclusions from the rule. These exclusions are listed in Table 12.1. Also listed in Table 12.1 are the federal regulations that apply to biosolids-related activities not covered by the Part 503 rule.

**Table 12.1**
**Exclusions from the Part 503 rule**

| Part 503 does not include requirements for: | Applicable federal regulation |
|---|---|
| *Treatment of biosolids*<br>Processes used to treat sewage sludge prior to final use or disposal (e.g., thickening, dewatering, storage, heat drying) | None (except for operational parameters used to meet the Part 503 pathogen and vector attraction reduction requirements) |
| *Selection of use or disposal practice*<br>The selection of a biosolids use or disposal practice | None (the determination of the biosolids use or the disposal practice is a local decision) |
| *Incineration of biosolids with other wastes*<br>Biosolids co-fired in an incinerator with other wastes (other than as an auxiliary fuel) | 40 CFR Parts 60, 61 |
| *Storage of biosolids*<br>Placement of biosolids on land for 2 years or less (or longer when demonstrated not to be a surface disposal site but rather, based on practices, constitutes treatment or temporary storage) | None |
| *Industrial sludge*<br>Sludge generated at an industrial facility during the treatment of industrial wastewater with or without combined domestic sewage | 40 CFR Part 257 if land applied<br>40 CFR Part 258 if placed in a municipal solid waste landfill |
| *Hazardous sewage sludge*<br>Sewage sludge determined to be hazardous in accordance with 40 CFR Part 261, *Identification and Listing of Hazardous Waste* | 40 CFR Parts 261–268 |
| *Sewage sludge containing PCBs $\geq 50\,mg/kg$*<br>Sewage sludge with a concentration of polychlorinated biphenyls (PCBs) equal to or greater than 50 milligrams per kilogram of total solids (dry-weight basis) | 40 CFR Part 761 |
| *Incinerator ash*<br>Ash generated during the firing of biosolids in a biosolid incinerator | 40 CFR Part 257 if land applied<br>40 CFR Part 258 if placed in a municipal solid waste landfill<br>or<br>40 CFR Parts 261–268 if hazardous |
| *Grit and screenings*<br>Grit (e.g., sand, gravel, cinders) or screenings (e.g., relatively large materials such as rags) generated during preliminary treatment of domestic sewage in a treatment works | 40 CFR Part 257 if land applied<br>40 CFR Part 258 if placed in a municipal solid waste landfill |
| *Drinking water sludge*<br>Sludge generated during the treatment of either surface water or ground water used for drinking water | 40 CFR Part 257 if land applied<br>40 CFR Part 258 if placed in a municipal solid waste landfill |
| *Certain nondomestic septage*<br>Septage that contains industrial or commercial septage, including grease-trap pumpings | 40 CFR Part 257 if land applied<br>40 CFR Part 258 if placed in a municipal solid waste landfill |

*Source*: US EPA (4).

## 2.8. Enforcement of Part 503 and Reporting Requirements

To ensure compliance with Part 503, regulatory authorities have the right to inspect operations involved in the use or disposal of biosolids, review and evaluate required reports and records, sample biosolids at regulated facilities, and respond to complaints from persons affected by an alleged improper use or disposal of biosolids. If records are not kept or other Part 503 requirements are not met, the US EPA can initiate enforcement actions.

Violations of the Part 503 requirements are subject to the same sanctions as wastewater effluent discharge violations. The US EPA can sue in civil court and seek remediation and penalties, and it can prosecute willful or negligent violations as criminal acts. If a problem occurred (e.g., ground-water contamination), the government could seek to have the offending party correct the situation. The US EPA can pursue civil fines of up to $25,000 per day per violation (a single violation that occurs over a 1-year period could result in a fine of over $9 million). Filing a false report carries a fine of up to $10,000 and up to 2 years in prison. Negligent violations carry a criminal fine of $2500 to $25,000 per day of violation and up to 1 year in prison. Willful violations carry a criminal fine of $5000 to $50,000 per day of violation and up to 3 years in prison.

Where the US EPA is unable to take an enforcement action, Section 505 of the CWA authorizes any citizen (e.g., a landowner, neighbor, lending institution) to bring a civil action against the violator for corrective action and the same penalties that the US EPA could have sought (i.e., $25,000 per violation per day).

The Part 503 rule includes reporting requirements for the following types of facilities:

1. Publicly owned treatment works (POTWs) with a design flow rate equal to or greater than 1 million gallon per day (MGD)
2. POTWs that serve a population of 10,000 or greater
3. Class 1 biosolids management facilities that are POTWs required to have an approved pretreatment program (5 MGD or greater as per 40 CFR Part 403.3a) and POTWs located in states that have elected to assume local program responsibilities for pretreatment (140 CFR 403.10e), and treatment works processing domestic sewage (TWTDS) that the US EPA or the state have classified as class 1 because of the potential to negatively affect public health and the environment

## 2.9. Relationship of the Federal Requirements to State Requirements

Part 503 does not replace any existing state regulations; rather, it sets minimum national standards for the use or disposal of biosolids. In some cases, the state requirements may be more restrictive or administered in a manner different from the federal regulation.

States can change their regulations to meet the minimum federal standards. The US EPA works with states to encourage them to gain approval for administering the Part 503 rule. States can apply to the US EPA for approval of a biosolids program at any time, but they are under no obligation to do so.

Knowing exactly which state or federal rules to follow can sometimes be complicated. Users or disposers of biosolids should keep the following situations in mind when considering the applicability of requirements (4):

1. In all cases, users or disposers of biosolids must comply with all applicable requirements of the new federal rule (Part 503).
2. If a state has its own rules governing the use or disposal of biosolids and has not yet adopted the federal rule, the owner/operator will have to follow the most restrictive portions of both the federal and state rules. For a summary of the 50 states' biosolids rules, the reader is referred to the Water Environment Federation's Manual of Practice FD-15 (24).

## 3. LAND APPLICATION OF BIOSOLIDS

In 2000 an estimated 7.1 million dry tons, or 63% of all biosolids generated, were recycled. By 2005 that amount was expected to reach 7.6 million ton (66% of the total) and 8.2 million tons (70%) by 2010 (5). The biosolids quantities generated in the coming years will depend in great part on the extent to which municipalities adopt land treatment of wastewater. Land treatment, which is an alternative to conventional forms of wastewater treatment, reduces substantially the amount of biosolids produced.

Biosolids are used to fertilize fields for raising crops. Agricultural uses of biosolids that meet strict quality criteria and application rates have been shown to produce significant improvements in crop growth and yield (25, 26). Nutrients found in biosolids, such as nitrogen, phosphorus, and potassium, and trace elements such as calcium, copper, iron, magnesium, manganese, sulfur, and zinc, are necessary for crop production and growth (25). The use of biosolids reduces the farmer's production costs and replenishes the organic matter that has been depleted over time. The organic matter improves soil structure by increasing the soil's ability to absorb and store moisture. The organic nitrogen and phosphorus found in biosolids are used very efficiently by crops because these plant nutrients are released slowly throughout the growing season. This enables the crop to absorb these nutrients as the crop grows. This efficiency lessens the likelihood of groundwater pollution of nitrogen and phosphorus.

### 3.1. Perspective

Biosolids provide farmers with $60 to $160 per acre worth of fertilizer, including many essential nutrients that the farmer may not normally replenish in the soil (17). Biosolids also contain valuable organic matter that improves the health, quality, and structure of the soil. Table 12.2 shows the elemental composition of biosolids collected from various municipalities in the U.S. (24).

While the values of nutrients in biosolids are small relative to the dollar values of commercial fertilizers, they are by no means insignificant to those who would benefit monetarily. For example, wastewater treatment plants could reduce operating costs by biosolids sales or by elimination of more expensive treatment and disposal methods. Biosolids users, for example private citizens, can obtain nutrients for farms, lawns, and gardens at low cost (Table 12.3).

It was estimated back in 1990 that annual savings in treatment costs have reached $100 to $500 million when biosolids utilization was about 50%. In the following decade utilization increased, in part from the incentives for innovative and alternative technologies provided by the 1977 CWA and the 1993 Part 503 regulations. As shown in Table 12.4, the US EPA reported that beneficial use of biosolids has actually increased to 60% in 1998 and to 66% in 2005. The US EPA estimated an increase of up to 70% in 2010.

**Table 12.2**
**Elemental composition of biosolids from various municipalities in the United States**

| Component | Concentration (dry basis) | | |
|---|---|---|---|
| | Minimum | Maximum | Medium |
| | Percent | | |
| Organic C | 6.5 | 48 | 30.4 |
| Inorganic C | 0.3 | 54.3 | 1.4 |
| Total N | <0.1 | 17.6 | 3.3 |
| $NH_3$-N | <0.1 | 6.7 | 1 |
| $NO_3$-N | <0.1 | 0.5 | <0.1 |
| Total P | <0.1 | 14.3 | 2.3 |
| Inorganic P | <0.1 | 2.4 | 1.6 |
| Total S | 0.6 | 1.5 | 1.1 |
| Ca | 0.1 | 25 | 3.9 |
| Fe | <0.1 | 15.3 | 1.1 |
| Al | 0.1 | 13.5 | 0.4 |
| Na | 0.1 | 3.1 | 0.2 |
| K | 0.2 | 2.6 | 0.3 |
| Mg | 0.3 | 2 | 0.4 |
| | mg/L | | |
| Zn | 101 | 27,800 | 1740 |
| Cu | 84 | 10,400 | 850 |
| Ni | 2 | 3515 | 82 |
| Cr | 10 | 99,000 | 890 |
| Mn | 18 | 7000 | 260 |
| Cd | 3 | 3410 | 16 |
| Pb | 13 | 19,730 | 500 |
| Hg | <1 | 10,600 | 5 |
| Co | 1 | 18 | 4 |
| Mo | 5 | 39 | 30 |
| Ba | 21 | 8980 | 162 |
| As | 6 | 230 | 10 |
| B | 4 | 757 | 33 |

*Source*: WEF (24).

Of the biosolids sent for disposal in 1998 and 2005, 17% and 13% went to surface disposal/landfill, respectively. The US EPA expects that percentage to decrease to just 10% in 2010, with corresponding increases in land application and other beneficial uses (5).

A large number of locations and case studies of sites where various biosolids utilization options are currently being employed are listed and discussed in the US EPA Process Design Manual (2), the Water Environment Federation (WEF) Manual of Practice FD-15 (24), and in the WEF's special publication (27). Some of these operations have only recently started up, while others have been in operation for as long as 75 years (for example, Los Angeles County, California).

**Table 12.3**
**Value of 5 to 10 dry ton/acre of a typical anaerobically digested dewatered biosolids applied to soil**

| Nutrient | lb/acre applied | Value/acre |
|---|---|---|
| Nitrogen | 150 | $30.00 |
| Phosphorus ($P_2O_5$) | 150 | $30.00 |
| Potassium ($K_2O$) | 10 | $1.00 |
| Copper (Cu) | 7 | $14.00 |
| Zinc (Zn) | 10 | $12.50 |
| Sulfur | 20 | $10.00 |
| Lime | 1 ton | $28.00 |
| Spreading | | $15.00 |
| Total value* | | $140.50 |

* Value of organic matter is in addition to this total.
Source: US EPA (3).

**Table 12.4**
**Projection of beneficial use and disposal in 2000, 2005, and 2010**

| | Beneficial use | | | | Disposal | | | |
|---|---|---|---|---|---|---|---|---|
| Year | Land application (%) | Advanced treatment (%) | Other beneficial use (%) | Total (%) | Surface disposal/ landfill (%) | Incineration (%) | Other (%) | Total (%) |
| 1998 | 41 | 12 | 7 | 60 | 17 | 22 | 1 | 40 |
| 2000 | 43 | 12.5 | 7.5 | 63 | 14 | 22 | 1 | 37 |
| 2005 | 45 | 13 | 8 | 66 | 13 | 20 | 1 | 34 |
| 2010 | 48 | 13.5 | 8.5 | 70 | 10 | 19 | 1 | 30 |

Source: US EPA (5).

### 3.2. Principles and Design Criteria

Certain basic elements are common to all land application projects, no matter how or where the biosolids are to be applied. These elements include preliminary planning, site selection, process design (which includes determination of biosolids application rates; see the example in Section 7.1, below), facilities design, and facility management and operation (2). Full and complete discussions of each of these elements are too lengthy to be included in this chapter. Therefore, this section will provide only a brief outline. Full details are available elsewhere (2, 4, 28–30).

#### 3.2.1. Preliminary Planning

Preliminary planning consists of the following steps (2, 28):

1. A planning team is formed of individuals who are interested in the proposed program and whose expertise and support are required. A major activity of the planning team is to solicit and obtain public support for the program, particularly the support of potential

# Beneficial Utilization of Biosolids

biosolids users and local government. The importance of obtaining public support cannot be overemphasized. Many utilization projects have failed because planners have failed to recognize this necessity.
2. Basic data are collected, including biosolids quantities and characteristics, climatic conditions, and local, state, and federal regulations including the Part 503 rule.

### 3.2.2. Site Selection

Site selection consists of the following (2, 31):

1. *Preliminary screening*: A rough estimate of total acreage required is obtained by dividing the total biosolids quantity by the calculated application rate. Land that might be available within about 30 miles is identified; obviously unsuitable sites are immediately eliminated. If this rough analysis indicates that sufficient land is available, a more detailed study of potential sites is initiated.
2. *Site identification*: Potentially available sites remaining after preliminary screening are characterized as to topography, land use, soil characteristics, geology, and distance from treatment plant. The characterization at first is general, taken from published and readily available sources of information, such as soils surveys and topographical maps. The least suitable sites are eliminated by an objective ranking procedure, similar to the analysis described in the Chapter 13. The procedure is reiterated, with more detailed and site-specific information in each iteration, until finally the best site or sites are determined.
3. *Site acquisition*: Sites are acquired either by outright purchase or by the municipality obtaining a contract for the right to use private land for biosolids utilization.

### 3.2.3. Part 503 Criteria for Determination of Design Application Rates

1. All biosolids applied to the land must meet the ceiling concentrations for pollutants, listed in the first column of Table 12.5. The ceiling concentrations are the maximum concentration limits for 10 heavy metal pollutants in biosolids, specifically arsenic, cadmium, chromium, copper, lead, mercury, molybdenum, nickel, selenium, and zinc. If a limit for any one of the pollutants is exceeded, the biosolids cannot be applied to the land until such time as the ceiling concentration limits are no longer exceeded. The ceiling concentrations for pollutants are included in Part 503 to prevent the land application of biosolids with the highest levels of pollutants and to encourage pretreatment efforts that will result in lower levels of pollutants (Table 12.5).
2. Biosolids applied to the land must also meet pollutant concentration limits, cumulative pollutant loading rate limits, or annual pollutant loading rate limits for these same heavy metals.
3. Either Class A or Class B pathogen requirements (summarized in Table 12.6) must be met before the biosolids can be land applied; the two classes differ depending on the level of pathogen reduction that has been obtained (Tables 12.7 and 12.8).
4. Finally, one of 10 options specified in Part 503 and summarized in Table 12.9 to achieve vector attraction reduction must be met when biosolids are applied to the land.

## 3.3. Options for Meeting Land Application Requirements

The Part 503 requirements can be grouped into four options for meeting pollutant limits and pathogen and vector attraction reduction operational standards when biosolids are applied to the land (4):

1. The exceptional quality (EQ) option
2. The pollutant concentration (PC) option

**Table 12.5**
**Pollutant limits (dry-weight basis)**

| Pollutant | Ceiling concentration limits for all biosolids applied to land (mg/kg) | Pollutant concentration limits for EQ and PC biosolids (mg/kg) | Cumulative pollutant loading rate limits for CPLR biosolids (kg/ha) | Annual pollutant loading rate limits for APLR biosolids (kg/ha/yr) |
|---|---|---|---|---|
| Arsenic | 75 | 41 | 41 | 2.0 |
| Cadmium | 85 | 39 | 39 | 1.9 |
| Chromium | 3000 | 1200 | 3000 | 150 |
| Copper | 4300 | 1500 | 1500 | 75 |
| Lead | 840 | 300 | 300 | 15 |
| Mercury | 57 | 17 | 17 | 0.85 |
| Molybdenum | 75 | — | — | — |
| Nickel | 420 | 420 | 420 | 21 |
| Selenium | 100 | 36 | 100 | 5.0 |
| Zinc | 7500 | 2800 | 2800 | 140 |
| Applies to: | All biosolids that are land applied | Bulk biosolids and bagged biosolids | Bulk biosolids | Bagged biosolids |

*Source*: US EPA (4).

3. The cumulative pollutant loading rate (CPLR) option
4. The annual pollutant loading rate (APLR) option

It is very important to realize that each option is equally protective of public health and the environment; that is, EQ, PC, CPLR, and APLR biosolids used in accordance with the Part 503 rule are equally safe. This safety is ensured by the combination of pollutant limits and management practices imposed by each option.

Whichever option is chosen, at a minimum, the ceiling concentrations for pollutants (listed in Table 12.5) and the frequency of monitoring (Table 12.10), reporting, and record-keeping requirements must be met. The four options are summarized in Table 12.11, illustrated in Figure 12.2, and discussed in detail below.

Depending on the land application option under consideration, site restrictions, general requirements, and management practices also apply. These additional restrictions, requirements, and practices are summarized in Tables 12.12 and 12.13. Table 12.12 graphically displays the level of required regulatory control for each option. The types of land onto which these different biosolids may be applied are listed in Table 12.13.

### 3.3.1. Option 1: Exceptional Quality (EQ) Biosolids

For biosolids to qualify under the EQ option, the following requirements must be met:

1. The ceiling concentrations for pollutants in Table 12.5 may not be exceeded.
2. The pollutant concentration limits in Table 12.5 may not be exceeded.
3. One of the Class A pathogen requirements in Table 12.6 must be met.
4. One of the first eight vector attraction reduction options in Table 12.9 must be achieved.

## Table 12.6
## Summary of Class A and Class B pathogen reduction requirements

Class A
In addition to meeting the requirements in one of the six alternatives listed below, fecal coliform or *Salmonella* sp. bacteria levels must meet specific density requirements at the time of biosolids use or disposal or when prepared for sale or give-away

*Alternative 1: Thermally treated biosolids*
Use one of four time-temperature regimens

*Alternative 2: Biosolids treated in a high pH–high temperature process*
Specifies pH, temperature, and air-drying requirements

*Alternative 3: For biosolids treated in other processes*
Demonstrate that the process can reduce enteric viruses and viable helminth ova; maintain operating conditions used in the demonstration

*Alternative 4: Biosolids treated in unknown processes*
Demonstration of the process is unnecessary; instead, test for pathogens—*Salmonella* sp. or fecal coliform bacteria, enteric viruses, and viable helminth ova—at the time the biosolids are used or disposed of or are prepared for sale or giveaway

*Alternative 5: Use of PFRP*
Biosolids are treated in one of the processes to further reduce pathogens (PFRP)

*Alternative 6: Use of a process equivalent to PFRP*
Biosolids are treated in a process equivalent to one of the PFRPs, as determined by the permitting authority

Class B
The requirements in one of the three alternatives below must be met

*Alternative 1: Monitoring of indicator organisms*
Test for fecal coliform density as an indicator for all pathogens at the time of biosolids use or disposal

*Alternative 2: Use of PSRP*
Biosolids are treated in one of the processes to significantly reduce pathogens (PSRP)

*Alternative 3: Use of processes equivalent to PSRP*
Biosolids are treated in a process equivalent to one of the PSRPs, as determined by the permitting authority

*Source*: US EPA (4).

Methods that typically achieve the pathogen and vector attraction reduction requirements and allow biosolids to meet EQ requirements include alkaline stabilization, composting, and heat drying. The Part 503 frequency of monitoring (Table 12.10), record-keeping, and reporting requirements also must be met for EQ biosolids.

Once biosolids meet EQ requirements, they are not subject to the land application general requirements and management practices in Part 503, with one possible exception: if the regional administrator or the state director determines, on a case-by-case basis, that such requirements are necessary to protect public health and the environment (this exception applies only to bulk biosolids). Once biosolids have been established as meeting EQ requirements, whether in bulk form or in bags or other containers, they can

**Table 12.7**
**Processes to further reduce pathogens (PFRPs)**

1. Composting
   Using either the within-vessel composting method or the static aerated pile composting method, the temperature of the biosolids is maintained at 55°C or higher for 3 days.

   Using the windrow composting method, the temperature of the biosolids is maintained at 55°C or higher for 15 days or longer. During the period when the compost is maintained at 55°C or higher, the windrow is turned a minimum of five times.

2. Heat drying
   Biosolids are dried by direct or indirect contact with hot gases to reduce the moisture content of the biosolids to 10% or lower. Either the temperature of the biosolids particles exceeds 80°C or the wet bulb temperature of the gas in contact with the biosolids as the biosolids leave the dryer exceeds 80°C.

3. Heat treatment
   Liquid biosolids are heated to a temperature of 180°C or higher for 30 minutes.

4. Thermophilic aerobic digestion
   Liquid biosolids are agitated with air or oxygen to maintain aerobic conditions, and the mean cell residence time of the biosolids is 10 days at 55° to 60°C.

5. Beta-ray irradiation
   Biosolids are irradiated with beta rays from an accelerator at dosages of at least 1.0 megarad at room temperature (ca. 20°C).

6. Gamma-ray irradiation
   Biosolids are irradiated with gamma rays from certain isotopes, such as cobalt 60 and cesium 137, at room temperature (ca. 20°C).

7. Pasteurization
   The temperature of the biosolids is maintained at 70°C or higher for 30 minutes or longer.

*Source*: US EPA (4).

generally be applied as freely as any other fertilizer or soil amendment to any type of land. While not required by the Part 503 rule, EQ biosolids should be applied at a rate that does not exceed the agronomic rate that supplies the nitrogen needs of the plants being grown, just as for any other commercial fertilizer or soil amending material that contains nitrogen.

*3.3.2. Option 2: Pollutant Concentration (PC) Biosolids*

To qualify under the PC option, biosolids must meet several requirements:

1. The ceiling concentration for pollutants in Table 12.5 may not be exceeded.
2. The pollutant concentration limits in Table 12.5 may not be exceeded (same requirement as for EQ biosolids, discussed above).
3. One of three Class B pathogen requirements must be met (Table 12.6), as well as Class B site restrictions.
4. One of 10 vector attraction reduction options must be achieved (Table 12.9).
5. Frequency of monitoring (Table 12.10), as well as record-keeping and reporting requirements must be met.
6. Applicable site restrictions, general requirements, and management practices must be met (summarized in Tables 12.12 and 12.13).

## Table 12.8
## Processes to significantly reduce pathogens (PSRPs)

1. Aerobic digestion
   Biosolids are agitated with air or oxygen to maintain aerobic conditions for a specific mean cell residence time at a specific temperature. Values for the mean cell residence time and temperature shall be between 40 days at 20°C and 60 days at 15°C.

2. Air drying
   Biosolids are dried on sand beds or on paved or unpaved basins. The biosolids dry for a minimum of 3 months. During 2 of the 3 months, the ambient average daily temperature is above 0°C.

3. Anaerobic digestion
   Biosolids are treated in the absence of air for a specific mean cell residence time at a specific temperature. Values for the mean cell residence time and temperature shall be between 15 days at 35°C to 55°C and 60 days at 20°C.

4. Composting
   Using either the within-vessel, static aerated pile, or windrow composting methods, the temperature of the biosolids is raised to 40°C or higher and maintained for 5 days. For 4 hours during the 5-day period, the temperature in the compost pile exceeds 55°C.

5. Lime stabilization
   Sufficient lime is added to the biosolids to raise the pH of the biosolids to 12 after 2 hours of contact.

*Source*: US EPA (4).

## Table 12.9
## Summary of vector attraction reduction options

Requirements in one of the following options must be met:

*Option 1*: Reduce the mass of volatile solids by a minimum of 38%
*Option 2*: Demonstrate vector attraction reduction with additional anaerobic digestion in a bench-scale unit
*Option 3*: Demonstrate vector attraction reduction with additional aerobic digestion in a bench-scale unit
*Option 4*: Meet a specific oxygen uptake rate for aerobically treated biosolids
*Option 5*: Use aerobic processes at greater than 40°C (average temperatures 45°C) for 14 days or longer (e.g., during biosolids composting)
*Option 6*: Add alkaline materials to raise the pH under specified conditions
*Option 7*: Reduce moisture content of biosolids that do not contain unstabilized solids from other than primary treatment to at least 75% solids
*Option 8*: Reduce moisture content of biosolids with unstabilized solids to at least 90%
*Option 9*: Inject biosolids beneath the soil surface within a specified time, depending on the level of pathogen treatment
*Option 10*: Incorporate biosolids applied to or placed on the land surface within specified time periods after application to or placement on the land surface

*Source*: US EPA (4).

**Table 12.10**
Frequency of monitoring for pollutants, pathogen densities, and vector attraction reduction

| | Amounts of biosolids* | | |
|---|---|---|---|
| Ton/yr | ton/d | ton/yr | Frequency |
| Greater than zero but less than 290 | >0 to <0.85 | >0 to <320 | Once per year |
| Equal to or greater than 290 but less than 1500 | 0.85 to <4.5 | 320 to <1,650 | Once per quarter (4 times per year) |
| Equal to or greater than 1500 but less than 15,000 | 4.5 to <45 | 1,650 to <16,500 | Once per 60 days (6 times per year) |
| Equal to or greater than 15,000 | ≥45 | ≥16,500 | Once per month (12 times per year) |

* Either the amount of bulk biosolids applied to the land or the amount of biosolids received by a person who prepares biosolids for sale or giveaway in a bag or other container for application to the land (dry-weight basis).
*Source*: US EPA (4).


**Table 12.11**
Options for meeting pollutant limits and pathogen and vector attraction reduction requirements for land application

| Option* | Pollutant limits | Pathogen requirements | Vector attraction reduction requirements |
|---|---|---|---|
| EQ biosolids | Bulk or bagged biosolids meet pollutant concentration limits in Table 12.5 | Any 1 of the Class A requirements in Table 12.6 | Any 1 of the requirements in options 1 through 8 in Table 12.9 |
| PC biosolids | Bulk biosolids meet pollutant concentration limits in Table 12.5 | Any 1 of the Class B requirements in Table 12.6 Any 1 of the Class A requirements in Table 12.6 | Any 1 of the 10 requirements in Table 12.9 Requirements 9 or 10 in Table 12.9 |
| CPLR biosolids | Bulk biosolids applied subject to cumulative pollutant loading rate (CPLR) limits in Table 12.5 | Any 1 of the Class A or Class B requirements in Table 12.6 | Any 1 of the 10 requirements in Table 12.9 |
| APLR biosolids | Bagged biosolids applied subject to annual pollutant loading rate (APLR) limits in Table 12.5 | Any 1 of the Class A requirements in Table 12.6 | Any 1 of the first 8 requirements in Table 12.9 |

* Each option requires that the biosolids meet the ceiling concentrations for pollutants listed in Table 12.5.
*Source*: US EPA (4).

# Beneficial Utilization of Biosolids

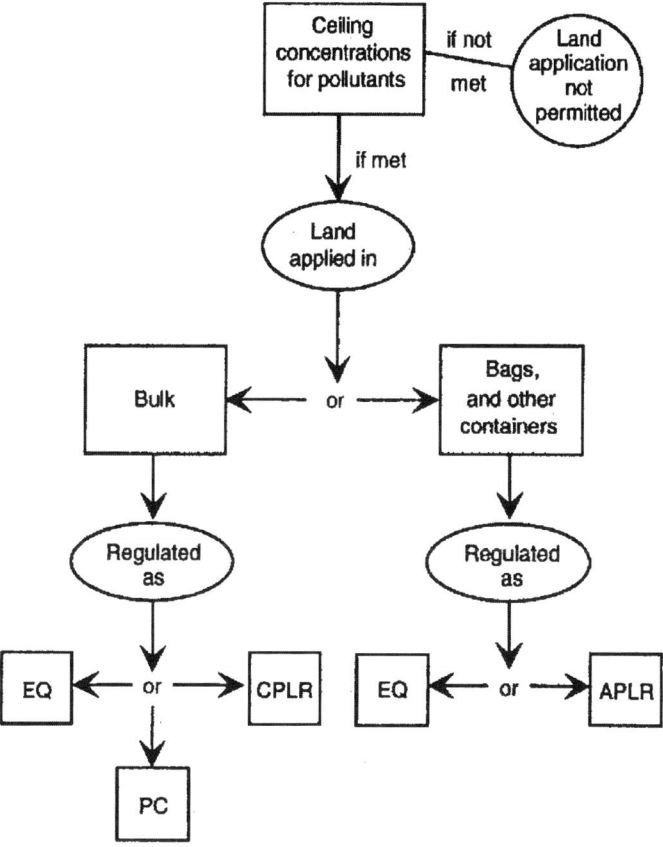

**Fig. 12.2.** Options for meeting Part 503 land application. *Source*: US EPA (4).

Class A biosolids meeting vector attraction reduction requirements 9 and 10 in Table 12.9 are another type of biosolids material that would fit in the PC category.

Thus, PC biosolids must meet more requirements than EQ biosolids, but are subject to fewer requirements than CPLR biosolids. Currently, the majority of biosolids in the U.S. could be characterized as PC biosolids.

### 3.3.3. Option 3: Cumulative Pollutant Loading Rate (CPLR) Biosolids

The third option for meeting land application requirements allows bulk biosolids that do not meet the pollutant concentration limits in Table 12.5 to be land applied as safely as EQ and PC biosolids. To qualify as CPLR biosolids, the following requirements must be met:

1. The ceiling concentrations for pollutants in Table 12.5 may not be exceeded.
2. Cumulative pollutant loading rates (CPLRs) listed in Table 12.5 may be not be exceeded.
3. Either the Class A or Class B pathogen requirements in Table 12.6 must be met.
4. One of the 10 vector attraction reduction options in Table 12.9 must be met.
5. Frequency of monitoring (Table 12.10), as well as record-keeping and reporting requirements must be met.
6. Applicable site restrictions, general requirements, and management practices must be met (summarized in Tables 12.12 and 12.13).

**Table 12.12**
**Summary of regulatory requirements for different types of biosolids**

| Type of biosolids and class of pathogens | Meet ceiling concentration for pollutants | Meet pollutant concentration limits | Site restrictions | General requirements and management practices | Track added pollutants |
|---|---|---|---|---|---|
| EQ Bag or bulk Class A | Yes | Yes | No | No | No |
| PC Bulk only Class A[a] | Yes | Yes | No | Yes | No |
| PC Bulk only Class B | Yes | Yes | Yes | Yes | No |
| CPLR Bulk only Class A | Yes | No | No | Yes | Yes |
| CPLR Bulk only Class B | Yes | No | Yes | Yes | Yes |
| APLR Bag only Class A | Yes | No | No | Yes[b] | Yes[c] |

[a] Biosolids meeting Class A pathogen reduction requirements but following options 9 or 10 vector attraction reduction requirements are also considered PC biosolids.

[b] The only general and management practice requirement that must be met is a labeling requirement.

[c] The amount of biosolids that can be applied to a site during the year must be consistent with the annual whole sludge application rate (AWSAR) for the biosolids that does not cause any of the ALPRs to be exceeded.

*Source*: US EPA (4).

The CPLR is the maximum amount of regulated pollutants in biosolids that can be applied to a site. When the CPLR for any one of the 10 heavy metals listed in Table 12.5 is reached at a site, no additional bulk biosolids, subject to the CPLR limits, may be applied to the site.

*3.3.4. Option 4: Annual Pollutant Loading Rate (APLR) Biosolids*

The fourth option only applies to biosolids that are sold or given away in a bag or other container for application to land. Under this option, the following requirements must be met:

1. The ceiling concentrations for pollutants in Table 12.5 may not be exceeded.
2. The annual pollutant loading rates (APLRs) listed in Table 12.5 may not be exceeded.
3. The Class A pathogen requirements in Table 12.6 must be met.
4. One of the first eight vector attraction reduction options in Table 12.9 must be met.

**Table 12.13**
**Types of land onto which different types of biosolids may be applied**

| Biosolids option | Pathogen class | VAR[a] options | Type of land | Other restrictions |
|---|---|---|---|---|
| EQ | A | 1–8 | All[b] | None |
| PC | A | 9 or 10 | All except lawn and home gardens[c] | Management practices |
|  | B | 1–10 | All except lawn and home gardens[c] | Management practices and site restrictions |
| CPLR | A | 1–10 | All except lawn and home garden[d] | Management practices |
|  | B | 1–10 | All except lawn and home garden[c,d] | Management practices and site restrictions |
| APLR | A | 1–8 | All, but most likely lawns and home gardens | Labeling management practice |

[a] VAR means vector attraction reduction.
[b] Agricultural land, forest, reclamation sites, and lawns and home gardens.
[c] It is not possible to impose site restrictions on lawns and home gardens.
[d] It is not possible to track cumulative additions of pollutants on lawns and home gardens.
*Source*: US EPA (4).

5. The frequency of monitoring (Table 12.10) as well as record-keeping and reporting requirements must be met.
6. Applicable site restrictions, general requirements, and management practices must be met (summarized in Tables 12.12 and 12.13).

An APLR is the maximum amount of regulated pollutants in biosolids that can be applied to a site in any 1 year. APLRs rather than CPLRs are used for biosolids sold or given away in a bag or other container because tracking the amount of pollutants applied in biosolids is not feasible in this situation.

A labeling requirement for bagged or containerized APLR biosolids is required. To meet the labeling requirement, the preparer of biosolids must calculate the amount of biosolids that can be applied to a site during the year so that none of the APLRs are exceeded. This amount of biosolids is referred to as the annual whole sludge (biosolids) application rate (AWSAR). The AWSAR can be determined once the pollutant concentrations in the biosolids are known. The procedure for determining the AWSAR is explained in the example in Section 7.1, below. The AWSAR must be calculated for each of the 10 metals listed in Table 12.5, and the lowest AWSAR for the 10 metals is the allowable AWSAR for the biosolids. The AWSAR on the required label or information sheet has to be equal to or less than the AWSAR calculated using the procedure in Section 7.1.

While not required by the Part 503 rule, it would also be good practice to provide information about the nitrogen content of the biosolids as well as the AWSAR on the label or information sheet that accompanies the biosolids. The example in Section 7.2, below, shows calculations that can be useful for determining how much nitrogen is being applied to land relative to the AWSAR and the nitrogen requirements of the plants being grown.

## 3.4. Site Restrictions, General Requirements, and Management Practices

The Part 503 site restrictions, general requirements, and management practices must be met for all but EQ biosolids. The types of restrictions, specific general requirements, and kinds of management practices that apply to each type of biosolids are listed and discussed in detail in the guide to Part 503 (4).

Site restrictions refer to the time requirement for harvesting of crops, grazing of animals, and public access on sites where Class B biosolids are applied.

The general requirements specify the responsibilities of the biosolids preparer and applier in record keeping, notification, and responsibilities.

The management practices refer to labeling of biosolids and other requirements during application as not to harm endangered species, not to apply biosolids within 10 m or less from U.S. waters, not to apply on wetlands, or on flooded or snow-covered areas.

## 3.5. Process Design

Process design involves selecting suitable crops and determining appropriate biosolids application rates as well as application methods. Although basic design goals (maximization of crop yield and quality, and minimization of environmental damage) remain constant regardless of projected land use, design procedures differ for applications on agricultural, forested, and reclaimed lands:

1. *Application on agricultural land*: Biosolids should be applied to agricultural land at a rate equal to the nitrogen uptake rate of the crop unless lesser application rates are required because of pollutants limitations (see example in Section 7.2). Annual loading rates for pollutants on soils have been set by Part 503 as shown in Table 12.5. The basis for the nitrogen criterion is to minimize nitrate leaching to groundwater. Site lifetime limits are established on the basis of maximum cumulative loadings of lead, zinc, copper, nickel, and cadmium (32–40).
2. *Application on forested land*: Biosolids improve forest productivity, increase growth of hybrid poplars, and enhance the aesthetic value of Christmas trees. Where biosolids have been used, the trees grow faster than those living in unfertilized soils. Wildlife populations often increase in these areas because the understory vegetation is more abundant. Biosolids have been found to promote rapid timber growth, allowing quicker and more efficient harvest of an important natural resource. As with agronomic crops, the harvesting of a forest stand removes the nutrients accumulated during growth. However, the amounts removed in forest harvesting annually are significantly lower than in agronomic crop harvesting. Therefore, forest systems rely primarily on soil processes (denitrification) to minimize nitrate leaching into groundwater. As a result, nutrient loadings on forested lands must generally be less than those on agricultural sites. Lifetime metals limits used for agricultural sites are suggested for forested land; this would minimize metal toxicity to trees and allow growth of other crops if the area were cleared at a future date (2, 41–45).
3. *Application on reclaimed land*: Biosolids are usually applied to impoverished lands at rates sufficient to satisfy the nutrient requirements of the cover crop. Severely disturbed soils can be reclaimed through the addition of biosolids to replace lost topsoil. Biosolids have been used successfully to reclaim surface strip mines, large construction sites, parks, wetlands, and landfills. Biosolids improve soil fertility and stability, aiding revegetation and decreasing erosion. Biosolids have been used successfully at mine sites to establish sustainable vegetation. The organic matter, inorganic matrix, and nutrients present in the biosolids not only reduce the bioavailability of toxic substances often found in highly

disturbed mine soils, but also regenerate the soil layer. This regeneration is very important for reclaiming abandoned mine sites with little or no topsoil. The biosolids application rate for mine reclamation is generally higher than the agronomic rate, which cannot be exceeded for use of agricultural soils (2, 44, 45).

4. *Application to public contact sites*: The biosolids application to public parks, plant nurseries, roadsides, golf courses, lawns, and home gardens continues to increase. Treatment such as heat-drying, composting, and treatment with alkaline materials convert biosolids into useful products that can be considered exceptional quality (EQ) if pollutant concentrations in the biosolids do not exceed the minimum levels specified in Table 12.5. These products are safe for unrestricted use by the general public. Generators of these products are required to have an ongoing monitoring program to ensure that the biosolids continually meet the exceptional quality requirements (39, 40, 44).

Products of this nature have sold in bulk for as for as much as $190/ton if high in nitrogen content and aesthetically pleasing. Kellogg Supply Company, a private firm in California, has been producing and marketing composted biosolids products (e.g., Nitrohumus, Topper, Gro-Mulch), mostly in California, Arizona, and Nevada, for a long time. Their products include composted biosolids that have come predominantly from Los Angeles County wastewater treatment facilities. Both Milorganite and Nitrohumus have been used to establish and maintain grass playing fields in sports stadiums across the country, including the Rose Bowl in Pasadena, California (3).

A composted biosolids product from Philadelphia called Earthgro has been used with great success for growing container plants and chrysanthemums. Even the White House has used composted biosolids to reestablish its lawns. Several years ago, 825 tons of composted biosolids (Compro) were used in this highly successful project. Similarly, the lawns at Mount Vernon, the Washington Monument grounds, and the governor's mansion in Annapolis, Maryland, were renewed with Compro. The first use of composted biosolids on the Washington, DC, Mall (nearly 6000 tons) was in 1976 to establish the Constitution Gardens in time for the U.S. bicentennial birthday celebration. Compro is currently being sold for $10 to $50 per cubic yard in bulk depending on quantity delivery. The cost of their bagged product is $5 to $6/ft$^3$ (3).

## 3.6. Facilities Design

Once the site has been chosen and crops and approximate biosolids application rates have been decided upon, the project can proceed to the facility design stage. This phase of the project is site-specific and consists of the following (2):

1. *Detailed site investigations*: On-site soil analyses are conducted to determine such factors as available phosphorus and potassium, soil pH and lime requirements, cation exchange capacity, and organic matter. Such information will allow for finalizing biosolids application rates determined in the process design phase. Soil should be characterized to provide baseline data against which subsequent analyses can be compared. This will allow documentation of changes in the physical and chemical properties of the soil due to biosolids application.
2. *Determining preapplication treatment*: This refers to upstream biosolids treatment, including thickening, stabilization, disinfection, conditioning, dewatering, and drying (46, 61–65). For new plants, the method of biosolids disposal or utilization may dictate the preapplication processing configuration. For existing plants, preapplication treatment influences biosolids form and composition, and thus affects application rate, the method of spreading, and the mode of biosolids transportation.
3. *Determining biosolids application mode*: The application mode depends on the biosolids form. Liquid biosolids can be spread by tank truck, sprayed, injected, or applied by

the ridge-and-furrow technique. Dewatered biosolids are usually applied by conventional fertilizer spreading equipment.
4. *Determining biosolids storage requirements*: Storage should be provided when biosolids cannot be spread (for example, during inclement weather). Storage can also provide additional stabilization and disinfection.

### 3.7. Facility Management, Operations, and Monitoring

Once the system has been constructed, it must be made to run smoothly and efficiently:

1. Operations must be scheduled. Spreading must be timed to satisfy farming requirements. If the municipality grows its own crops, tilling, planting, and harvesting operations must also be scheduled.
2. Operations must be managed to reduce off-site impacts, such as odors (66), and contamination of groundwater and surface water.
3. Operations must be monitored to ensure that the system is operating as intended. Biosolids must be analyzed to ensure their acceptability to the user and to provide a record of nutrient and metal additions to the soil. Soil, crops, groundwater, and surface water also need to be monitored to detect any possible contamination. Table 12.10 shows the required frequency of monitoring by Part 503 Rule for pollutants, pathogen densities, and vector attraction reduction (4).

## 4. SURFACE DISPOSAL OF BIOSOLIDS

### 4.1. Perspective

An activity is defined as surface disposal if biosolids are placed on an area of land for final disposal. Some surface disposal sites may be used for beneficial purposes as well as for final disposal. Owners and operators of surface disposal sites and anyone who prepares biosolids for final disposal of only biosolids on a surface disposal site must meet the requirements in subpart C of the Part 503 rule. Surface disposal sites include monofills, surface impoundments, lagoons, waste piles, dedicated disposal sites, and dedicated beneficial use sites (4).

*Monofills* are landfills in which only biosolids are disposed. Monofills include trenches and area fills. In trenches, biosolids are placed in an excavated area that can be a wide, shallow trench or a narrow, deep trench. In area fills, biosolids are placed on the original ground surface in mounds, layers, or diked containments. With area fills, excavation is not required (as it is with trenches) because biosolids are not placed below the ground surface. Area fills often are used when shallow bedrock or ground water is present.

*Surface impoundments and lagoons* are disposal sites in which biosolids with high water content are placed in an open, excavated area. If lagoons are used for treatment, they are not considered surface disposal sites.

*Waste piles* are mounds of dewatered biosolids placed on the soil surface for final disposal.

*Dedicated disposal sites* receive repeated applications of biosolids for the sole purpose of final disposal. Such sites often are located at POTW sites.

*Dedicated beneficial use sites* are surface disposal sites where biosolids are placed on the land at higher rates or with higher pollutant concentrations than are allowed when biosolids are land applied for farming or reclamation. Such sites might receive repeated applications of biosolids. In contrast to dedicated disposal sites, dedicated beneficial use sites are used to grow crops for beneficial purposes. For such sites, the permitting authority issues a permit that specifies appropriate management practices that ensure the protection of public health and the environment from any reasonably anticipated adverse effects of certain pollutants that may be present in biosolids if crops are grown or animals are grazed.

### 4.2. Differentiation Among Surface Disposal, Storage, and Land Application

An activity is considered storage if biosolids are placed and remain on land for 2 years or less. If biosolids remain on land for longer than 2 years, this land is considered an active biosolids unit and the surface disposal requirements in Part 503 have to be met. An active biosolids unit is the area, trench, waste pile, or lagoon in which biosolids are currently being placed. Note, however, that biosolids can remain on the land for longer than 2 years, but the person who prepares the biosolids must demonstrate that the site is not an active biosolids unit (4).

Any practice in which biosolids that meet pollutant concentrations, CPLRs, or APLRs, as well as ceiling limits are applied to land at agronomic rates to condition soil or to fertilize crops or vegetation is considered land application, not surface disposal.

The surface disposal provisions of the Part 503 rule do not apply when biosolids are treated on the land, such as in a treatment lagoon or stabilization pond, and treatment could be for an indefinite period. Placement of biosolids on the land in a municipal solid waste landfill also is not considered surface disposal under Part 503.

### 4.3. Pollutant Limits for Biosolids

For surface disposal, a pollutant limit is the amount of pollutant allowed per unit amount of biosolids. Subpart C of Part 503 sets pollutant limits for arsenic, chromium, and nickel in biosolids (4). These limits apply only to active biosolids units without liners and leachate collection systems.

A liner is a layer of relatively impervious soil, such as clay, or a layer of synthetic material that covers the bottom of an active biosolids unit and has a hydraulic conductivity of $1 \times 10^{-9}$ m/s or less. The liner slows the seeping of liquid on the surface disposal site into the ground water below. A leachate collection system is a system or device installed immediately above a liner that collects and removes leachate as it seeps through the disposal site. Biosolids placed on an active biosolids unit with a liner and leachate collection system do not have to meet pollutant limits, based on the assumption that these systems prevent pollutants from migrating to ground water.

There are two options for meeting the pollutant limits for arsenic, chromium, and nickel in active biosolids units without using liners and leachate collection systems. The first option is to ensure that the levels of arsenic, chromium, and nickel are below the pollutant limits listed in Table 12.14. These limits are based on how far the boundary of each active biosolids unit is from the surface disposal site property line. For example, the limits are 73 mg/kg for arsenic, 600 mg/kg for chromium, and 420 mg/kg for nickel

**Table 12.14**
**Pollutant limits for surface disposal of biosolids**

| Location in the Part 503 Rule | Distance from the boundary of active biosolids unit to surface disposal site property line (m) | Pollutant concentration* | | |
|---|---|---|---|---|
| | | Arsenic (mg/kg) | Chromium (mg/kg) | Nickel (mg/kg) |
| Table 2 of Section 503.23 | 0 to less than 25 | 30 | 200 | 210 |
| | 25 to less than 50 | 34 | 220 | 240 |
| | 50 to less than 75 | 39 | 260 | 270 |
| | 75 to less than 100 | 46 | 300 | 320 |
| | 100 to less than 125 | 53 | 360 | 390 |
| | 125 to less than 150 | 62 | 450 | 420 |
| Table 1 of Section 503.23 | Equal to or greater than 150 | 73 | 600 | 420 |

*Dry-weight basis (basically, 100% solids content).
*Source*: US EPA (4).

if the boundary of the active biosolids unit closest to the site's property line is greater than 150 m away (4).

There may be more than one active biosolids unit at a surface disposal site. If the boundary of a second active biosolids unit on the same site is 75 m from the property line, then the arsenic limit for that second unit would be 46 mg/kg. Thus different active biosolids units on the same site can have different pollutant limits, based on the closest distance between the active biosolids unit boundaries and the property line of the surface disposal site.

The second option for meeting pollutant limits is to meet "site-specific" limits set by the permitting authority, which would determine the limits after evaluating site data. The owner/operator of the surface disposal site must request site-specific limits when applying for a permit. The permitting authority then must determine whether site-specific pollutant limits are appropriate for the particular site (4).

Site-specific limits may be justified if the site conditions vary significantly from those assumed in the risk assessment used to derive the Part 503 pollutant limits. In general, if the depth to ground water is considerable or a natural clay layer underlies the site, the permittee may consider requesting site-specific pollutant limits.

### 4.4. Pathogens and Vector Attraction Reduction Requirements

Pathogens are disease-causing organisms, such as certain bacteria and viruses that might be present in biosolids. Vectors are animals, such as rats or insects, that might be attracted to biosolids and can spread disease after coming into contact with the biosolids. The Part 503 rule includes requirements concerning the control of pathogens and the reduction of vector attraction for biosolids placed on a surface disposal site. Biosolids can be placed on an active biosolids unit only if the pathogen and vector attraction reduction requirements are met (Table 12.15).

**Table 12.15**
**Pathogen and vector attraction reduction requirements for surface disposal sites**

Pathogen reduction requirements
  *Options (must meet one of these)*:
  - Place a daily cover on the active biosolids unit
  - Meet one of six Class A pathogen reduction requirements (see Table 12.6)
  - Meet one of three Class B pathogen reduction requirements, except site restrictions (see Table 12.6)

Vector attraction reduction requirements
  *Options (must meet one of these)*:
  - Place a daily cover on the active biosolids unit
  - Reduce volatile solids content by a minimum of 38% or less under specific laboratory test conditions with anaerobically and aerobically digested biosolids
  - Meet a specific oxygen uptake rate (SOUR)
  - Treat the biosolids in an aerobic process for a specified number of days at a specified temperature
  - Raise the pH of the biosolids with an alkaline material to a specified level for a specified time
  - Meet a minimum percent solids content
  - Inject or incorporate the biosolids into soil

*Source*: US EPA (4).

For pathogen reduction, the biosolids placed on an active biosolids unit must meet either Class A or Class B pathogen requirements, or a cover (soil or other material) must be placed on the active biosolids unit at the end of each day (47). If a daily cover is placed on the active biosolids unit, no other pathogen reduction requirements apply. If the biosolids meet Class B requirements, the "site restrictions" that apply to Class B do not have to be followed because the management practices for surface disposal already include these site restrictions.

For vector attraction reduction, one of several options listed in Table 12.15 must be met. Representative samples of biosolids must be collected and analyzed to demonstrate that the pathogen and vector attraction reduction requirements have been met. In most cases, owners or operators of surface disposal sites place a daily cover on the biosolids unit to meet pathogen and vector attraction reduction requirements (4).

### 4.5. Frequency of Monitoring Requirements

The monitoring of several different parameters is required at surface disposal sites. Monitoring is required for surface disposal sites without liners to determine levels of arsenic, chromium, and nickel in biosolids. Monitoring is required in both lined and unlined sites to show that the chosen pathogen and vector attraction reduction requirement is being met and to measure the amount of methane gas in air at a covered surface disposal site. How frequently biosolids must be monitored is determined according to the amount of biosolids placed on an active biosolids unit, as shown in Table 12.10. The permitting authority may require more frequent monitoring, for example, if the pollutant and pathogen levels in the biosolids are highly variable (4).

After biosolids have been monitored for 2 years at the frequency specified in Table 12.10, the permitting authority may reduce the frequency of monitoring for arsenic, chromium, nickel, and, under limited circumstances, pathogens in biosolids placed on an active biosolids unit. The frequency may be reduced, for example, if the pollutant levels in biosolids do not vary greatly or if pathogens are never detected when using Class A, Alternative 3, to meet pathogen reduction requirements.

Methane gas in air must be monitored continuously, both at the property line of the surface disposal site and within each structure at the site, if an active biosolids unit is covered. Methane gas monitors can be installed permanently to continuously test the air and provide readings of methane.

### 4.6. Regulatory Requirements for Surface Disposal of Domestic Septage

The regulatory requirements for the surface disposal of septage are not as extensive as those for biosolids. The requirements for surface disposal of domestic septage include meeting the same management practices that are required for the surface disposal of biosolids and one of the vector attraction reduction alternatives 9 to 12 (listed in Table 12.16). Note that vector attraction reduction 12 would require a determination that the pH of the septage had been raised to 12 for a period of 30 minutes. The person who places the domestic septage on the surface disposal site must certify that vector attraction reduction has been achieved and develop a description of how it was achieved. The certification and description must be kept on file for 5 years (4, 22).

There are no pathogen requirements for the surface disposal of domestic septage.

**Table 12.16**
**Summary of options for meeting vector attraction reduction**

| | |
|---|---|
| Option 1: | Meet 38% reduction in volatile solids content. |
| Option 2: | Demonstrate vector attraction reduction with additional anaerobic digestion in a bench-scale unit. |
| Option 3: | Demonstrate vector attraction reduction with additional aerobic digestion in a bench-scale unit. |
| Option 4: | Meet a specific oxygen uptake rate for aerobically digested biosolids. |
| Option 5: | Use aerobic processes at greater than 40°C for 14 days or longer. |
| Option 6: | Alkali addition under specified conditions. |
| Option 7: | Dry biosolids with no unstabilized solids to at least 75% solids. |
| Option 8: | Dry biosolids with unstabilized solids to at least 90% solids. |
| Option 9: | Inject biosolids beneath the soil surface. |
| Option 10: | Incorporate biosolids into the soil within 6 hours of application to or placement on the land. |
| Option 11: | Cover biosolids placed on a surface disposal site with soil or other material at the end of each operating day. (Note: only for surface disposal.) |
| Option 12: | Alkaline treatment of domestic septage to pH 12 or above for 30 minutes without adding more alkaline material. |

*Source*: US EPA (4).

## 5. INCINERATION OF BIOSOLIDS AS AN ENERGY SOURCE

Whether produced from direct burning of biosolids or from the combustion of biosolids-derived fuels such as digester gas or pyrolysis gas, the end product is energy (48, 49). Heat can be made to perform a variety of useful functions.

### 5.1. Perspective

The precipitous rise in energy prices during the 1970s has generated intense interest in the conservation and recovery of this precious commodity. For example, the U.S. Department of Energy has proposed one seventh of the U.S. energy requirements be produced by bioconversion processes (for example anaerobic digestion) by the year 2020 (50). Clearly, however, this awesome quantity of energy will not be generated from municipal wastewater biosolids; there are simply insufficient biosolids. Very large external organic sources (for example, manure from feed lots or municipal refuse) and external processing systems (energy farms) will be required to effect such production. As with utilization of biosolids on land, the impact of energy recovery from municipal biosolids will be largely local, that is, it will be felt most strongly at the treatment plant and in its immediate vicinity. Here, the effects can be significant.

The energy value of methane generated from the anaerobic digestion process exceeds the energy requirements of the digestion process. The excess can be used to supply the energy needs of other plant processes. In some instances, the gas generated is sufficient to supply the energy needs of the entire wastewater treatment plant, with excess gas available for sale. Notable examples are the British Southern and Mogden plants, the county sanitation districts of the Los Angeles County Joint Disposal Plant (51), and the Tampa, Florida, treatment works (3).

The Tampa, Florida, treatment works recovers about $700,000 worth of electricity each year from the methane it produces during anaerobic digestion. This is equivalent to approximately $65 worth of net electricity being produced per dry ton of volatile biosolids removed from the digester. Tampa also uses the heat removed from the electrical generators to provide more than 95% of the warmth needed for the digesters. All but 10% to 15% of Tampa's anaerobically digested biosolids are being heat-dried and marketed for $85 to $120 per dry ton. The balance is being land applied in dewatered form. Tampa was recognized for this highly efficient operation in the US EPA's 1992 Beneficial Use of Biosolids Awards Program (3).

Heat recovery is possible even if digestion is not used; for example, heat recovery from co-incineration of biosolids and municipal refuse was planned to provide all the energy needs of the Central Contra Costa Sanitary District (CCCSD) plant in Concord, California (52).

In January 1978, the State of California Public Utilities Commission (PUC) passed a resolution directing all state utilities to augment cogeneration projects by setting up new rate schedules covering interruptible electric service; by creating new specific rates to encourage cogeneration, including revisions to standby rates; and by developing guidelines covering the price and conditions for the purchase of energy and capacity from cogeneration facilities owned by others (53). The term *cogeneration* in this context means the production of power by utilization of waste heat; it also covers power produced through the burning of alternative fuels, such as municipal waste. The resolution significantly changed the economics of power generation at California wastewater treatment plants and encouraged the use of in-plant energy recovery.

On June 27, 1979, the Federal Energy Regulatory Commission issued regulations providing for the qualification of small power production and cogeneration facilities under Section 201 of the Public Utility Regulatory Policies Act of 1978 (54). The regulations were set up to ensure opportunities for small power producers (<80 MW) to sell electricity to electric utilities when such electricity is generated through the use of renewable energy sources (such as biosolids) or recovered process heat.

## 5.2. Recovery of Energy from Biosolids

Special consideration must be given when designing processes to recover energy from wastewater biosolids. Some of these considerations are discussed in the following subsections (2, 55).

### 5.2.1. Treatment of Digester Gas

The treatment required depends on the anticipated use of the digester gas. Treatment is minimal if the gas is burned in a boiler or in a high-temperature internal combustion engine. Conversely, if it is sold for utilities as a natural gas substitute, it must be upgraded to natural gas quality. This involves treatment to remove particulates, $H_2S$, $CO_2$, and water. As a general rule, gas treatment should be avoided to as great a degree as possible. It is preferable to set up recovery systems that can be operated with untreated digester gas.

Particulates are carried over with the gas as it leaves the digester. They may be removed in large sedimentation traps and cyclonic separators.

$H_2S$ is most commonly removed by iron-sponge scrubbers. The "sponge" consists of wood shavings impregnated with iron oxide. $H_2S$ reacts with iron oxide to form nonvolatile ferric sulfide. The sponge can be regenerated with air. Sponge capacity is about 0.6 lb of sulfur/lb of iron oxide (0.6 kg/kg). Problems have been experienced with fouling of the iron-sponge by oils and greases entrained in the digester gas. Iron-sponge scrubbers are commercially available. Other $H_2S$ scrubbing processes are less commonly used and are proprietary.

$CO_2$ removal processes can be divided into three broad categories: absorption (both physical and chemical), adsorption, and cryogenic processing (2). Many $CO_2$ removal processes also remove $H_2S$. The only process that has received much use in wastewater treatment plants is absorption in water; this process has been tested at Modesto, California, and Los Angeles County.

Gas leaves the digestion system at approximately 95°F (35°C) and is saturated with water vapor. During transport the gas is cooled. Condensate formed must be removed to protect downstream equipment. Water traps should be installed at low spots in the gas pipe and at frequent intervals. If moisture must be reduced substantially, adsorption drying or glycol dehydration can be used.

### 5.2.2. Gas-Burning Equipment
#### 5.2.2.1. Corrosion Factors

One of the major problems associated with recovering heat from digester gas is corrosion caused by $SO_2$ and $SO_3$, the combustion products of $H_2S$. If the exhaust gas temperature is allowed to drop below its dewpoint, the condensate that forms is acidic as the result of absorbing $SO_2$ and $SO_3$. The acidic condensate is corrosive to metallic elements of the exhaust-carrying system. There are two alternatives to alleviate the problem. The first is scrubbing of $H_2S$ from the gas before combustion. The second is maintaining the exhaust gas at temperatures considerably greater than its dewpoint, to prevent condensation. This generally requires that the water temperature of any boiler or engine using unscrubbed gas be at least 212°F (100°C). Also, stack gas temperatures should not be allowed to drop below 350° to 400°F (177° to 204°C). Use of unscrubbed digester gas is preferred. Equipment fueled by unscrubbed digester gas should not be used in intermittent service, since condensation will occur each time the unit is shut down. Shutdowns should be minimized. Similarly, the equipment should be designed so that even when operated at its lowest loadings, exhaust gas temperatures are sufficiently high to prevent condensation (2).

#### 5.2.2.2. Boilers

Scotch-type tube boilers and cast iron sectionalized boilers have both worked well with untreated digester gas as long as the water or steam temperatures are maintained above 212°F (100°C). The heat source (the boiler) and heat demands are not directly tied together, but separated by a condenser. The condenser is mounted directly above the boiler. The specific gravity of the steam/water mixture produced in the boiler tubes is less than that of the water returning to the boiler. The mixture is displaced upward into the condenser, gives up its heat, and then flows by gravity back to the boiler. A natural circulation pattern is thus set up.

If heat supply exceeds the heat demand, the excess heat is released by venting steam from the condensers. Temperature control is automatic, being set by the vent pressure. Advantages of this system are simplicity, elimination of costs associated with pumping, automatic temperature control, and independent operation of the boiler from other heat sources and heat demands. Independent operation is particularly important; it allows the boiler to operate at its own best conditions, without being affected by the operations of other components of the system.

#### 5.2.2.3. Prime Movers

Digester gas can be used to fuel reciprocating engines and gas turbines. Prime movers convert part of the fuel's energy to work, rejecting the remainder as waste heat. Thermal efficiency can be dramatically improved if portions of the rejected heat can be recovered and used for process or building heating. Waste heat recovery is more efficient if prime movers are run hot, since heat rejected at higher temperatures can be put to a greater variety of uses than heat rejected at low temperature. Also, exhaust systems last longer because $SO_2$-$SO_3$ corrosion is reduced.

5.2.2.4. RECIPROCATING ENGINES

Engines may be cooled using either a forced circulation system, in which water is pumped through the engine, or a natural draft system. The equipment configuration for natural circulation cooling is similar to that described for boiler natural circulation systems, except that the engine replaces the boiler in the flow diagram. The advantages of natural circulation cooling are the same as those discussed for natural circulation boiling. Cooling system pressures are limited to about 10 psig (69 kN/m$^2$); if operated at higher pressures, cooling water could leak past the cylinder liner seals and into the cylinder. The maximum cooling water temperature is thus about 240°F (116°C), corresponding to the temperature of saturated steam at 10 psig (69 kN/m$^2$). Engines using natural circulation cooling are relatively small, typically developing less than 1500 horsepower (1120 kW). Flow rates developed by natural circulation cooling may be insufficient to cool larger engines. Flow rates may be increased by installing a booster pump in the circulating loop near the entrance to the engine jacket. There are reciprocating engines on the market designed to operate at temperatures in the 160° to 180°F (71° to 82°C) range. However, they are not recommended for services with unscrubbed digester gas because of potential problems with $SO_2$-$SO_3$ corrosion. Heat recovered from the engine jacket is typically used to sustain the digestion process and for space heating (2).

Reciprocating engines commonly employed in wastewater treatment plants fall into two categories: dual-fuel (compression ignited) and spark ignited engines. Dual-fuel engines use a blend of diesel fuel and digester gas; the fraction of diesel fuel can be varied from a minimum of 4% all the way to 100% of the mixture. Dual-fuel engines are typically used if there is insufficient digester gas to satisfy power demands. Dual-fuel engines have been specified for new plants where digester gas production is expected to lag behind power demands for several years.

Spark-ignited engines are generally used when there is sufficient digester gas to satisfy power demands. Spark-ignited engines can operate on several different types of fuel (for example, digester gas and natural gas). Special carburetors are provided to blend digester gas with an air-diluted backup fuel (for example, natural gas) during infrequent periods when not enough digester gas is available to satisfy power requirements. Spark-ignited engines are less complex than dual-fuel engines, are available in smaller sizes, and are less costly to operate (56).

Naturally aspirated feed systems are preferred to turbocharge systems for spark-ignited engines. Turbocharged systems require that gas be delivered at high pressure, which means the gas must be first compressed, and then delivered through a fuel metering system with restricted openings. Gas impurities (oils, greases, and water) are condensed when the gas is compressed and cooled; these impurities often clog the fuel metering system. Naturally aspirated systems operate at low pressures (<0.5 psig or 3.4 kN/m$^2$). With careful design of the gas transport systems, compression of the feed gas is not required. Low-pressure fuel metering systems also have relatively large openings compared to metering systems used with turbocharged units. For these reasons, naturally aspirated fuel systems are therefore less susceptible to clogging than systems with turbocharged units.

Engines represent a large capital investment and should be conservatively designed to protect that investment. For four-stroke engines it is recommended that brake mean effective pressure (BMEP) not exceed 80 to 85 psig (550 to 590 kN/m$^2$) to minimize strain on the equipment. Engine speeds in the 700 to 1000 rpm range are preferred as are average piston speeds in the range of 1200 to 1500 ft/min (370 to 460 m/min). Heavy-duty industrial engines should be specified, not automotive engines (2, 56).

5.2.2.5. GAS TURBINES

Gas turbines have had relatively limited use to date. Where used, there have been fouling problems, which are inherent with compressing a dirty gas through fuel metering systems with small clearances. However, new developments in the turbine field and the fact that less nitric oxide (NO) is produced by turbines than by reciprocating engines has led to a second look at turbines, particularly in nonattainment air quality areas. A new system that uses a relatively low (4:1) pressure ratio turbine with recuperation has the potential to solve many of the problems that plagued earlier installations (57). The normally low efficiency of the low-pressure ratio turbine is boosted by preheating the compressed air with heat recovered from the exhaust gas. Ignition for this turbine can be staged to minimize NO$_x$ generation. Emissions control is particularly important in nonattainment areas where new stationary sources must use the best available control technology (BACT). The BACT for reciprocating engines is considered to be catalytic denitrification, while the BACT for low-pressure ratio turbines can be staged ignition.

### 5.2.3. Generators

Generators may be synchronous or induction types. Synchronous generators are by far the most common. However, in smaller sizes (below 5 or 10 MW) induction units are generally less expensive than synchronous units. They are also easier to maintain since they require no governor or synchronizing equipment. Induction generators have the disadvantage of being unable to operate unless paralleled with synchronous generation, either utility or in-plant. Thus an induction generator by itself cannot be used to provide emergency power.

### 5.3. Factors Affecting Heat Recovery

Factors a designer must consider in conducting a heat recovery analysis include the following (2, 56, 58, 59):

1. The full range of conditions expected at the plant must be evaluated, not just average conditions. Energy supply and energy demand schedules must be established. Heat recovery equipment must be sized to handle peak demands. Storage requirements for primary and backup fuels must be determined.
2. A source of backup energy must be available in the event that plant energy recovery systems experience partial or total failure.
3. The physical and chemical nature of flue gases generated must be considered (for example, temperature, corrosiveness, particulate concentration, and moisture content).
4. The equipment must be designed to withstand the conditions to which it will be subjected. Appropriate materials of construction must be used.
5. Any solid, liquid, or gaseous residual from the heat recovery operation must be collected and disposed of in a safe and environmentally sound manner.
6. Chemical and physical treatments for makeup and circulating water or steam must be established.

7. Manpower to operate the heat recovery system must be determined. Specialists may be required for certain equipment, for example, stationary engineers for high pressure boilers and engine specialists for internal combustion engines.
8. Control strategies must be decided upon, and instrumentation to carry them out must be provided.
9. Economic analyses must be performed to determine if the system can be economically justified. As a rule of thumb, the larger the plant, the more sophisticated the heat recovery system that can be justified.

### 5.4. Pollutant Limits for Biosolids Fired in Incinerators

A pollutant limit is the amount of pollutant allowed per unit amount of biosolids before incineration. Subpart E of the Part 503 rule regulates seven metals: arsenic, beryllium, cadmium, chromium, lead, mercury, and nickel (4). The limits protect human health from the reasonably anticipated harmful effects of these pollutants when biosolids are incinerated. The approaches for determining the limit for each pollutant and for total hydrocarbons are summarized in Table 12.17.

#### 5.4.1. Beryllium and Mercury Pollutant Limits

Levels of beryllium and mercury emitted from a biosolids incinerator must meet the National Emission Standards for Hazardous Air Pollutants (NESHAPs, 40 CFR Part 61).

The NESHAP for beryllium requires that the total quantity of beryllium emitted from each incinerator not exceed 10 g during any 24-hour period. The NESHAP for beryllium does not apply if written approval has been obtained (a) when the ambient concentration of beryllium in the proximity of the biosolids incinerator does not exceed $0.01\ \mu g/m^3$ when averaged over a 30-day period, or (b) if the biosolids incinerator operator can prove (with historical data) that the biosolids fired in the incinerator do not contain beryllium.

The NESHAP for mercury requires that the total quantity of mercury emitted into the atmosphere from all incinerators at a given site does not exceed 3200 g during any 24-hour period.

#### 5.4.2. Lead, Arsenic, Cadmium, Chromium, and Nickel Pollutant Limits

The pollutant limits for lead, arsenic, cadmium, chromium, and nickel in biosolids fired in a biosolids incinerator are calculated using the equations presented in Tables 12.18 and 12.19. Also, incinerators control efficiencies, dispersion factors, and biosolids feed rates must be determined to calculate the limits for these pollutants. For an explanation of control efficiencies, dispersion factors, and feed rates, see the guide for Part 503 (4).

Instead of using a National Ambient Air Quality Standard (NAAQS) as is done for lead, risk-specific concentrations (RSCs) are used to calculate limits for arsenic, cadmium, chromium, and nickel. The RSCs, which are based on human health concerns, represent the allowable increase in the average daily ground-level ambient air concentrations of pollutants at or beyond the property line of the site where the biosolids incinerator is located. The RSCs for arsenic, cadmium, and nickel are 0.023, 0.057, and $2.0\ \mu g/m^3$, respectively.

Table 12.17
Summary of pollutant limits for biosolids incineration

| Pollutant | Figure out pollutant limit | Determine dispersion factor (DF) | Determine control efficiency (CE) | National Ambient Air Quality Standard (NAAQS) | Risk specific concentration (RSC) |
|---|---|---|---|---|---|
| Pollutant limits | | | | | |
| Arsenic | Use equation for arsenic | Yes | Yes | No | Yes |
| Beryllium | Use NESHAPs | No | No | No | No |
| Cadmium | Use equation for cadmium | Yes | Yes | No | Yes |
| Chromium | Use equation for chromium | Yes | Yes | No | Yes |
| Lead | Use equation for lead | Yes | Yes | Yes | No |
| Mercury | Use NESHAPs | No | No | No | No |
| Nickel | Use equation for nickel | Yes | Yes | No | Yes |
| Operational Standard | | | | | |
| THC or CO | Limit is 100 ppm$_v$ | No | No | No | No |

NESHAPs, National Emission Standards for Hazardous Air Pollutants.
*Source*: US EPA (4).

In contrast, the RSC for chromium is based on the type of incinerator used. Operators can use the following values, which are based on incinerator type:

| | |
|---|---|
| Fluidized bed with wet scrubber | 0.65 |
| Fluidized bed with wet scrubber and wet electrostatic precipitator | 0.23 |
| Other types with wet scrubber | 0.064 |
| Other types with wet scrubber and wet electrostatic precipitator | 0.016 |

### 5.4.3. Total Hydrocarbons

Organic compounds that are emitted as a result of incomplete combustion or the generation of combustion by-products (e.g., benzene, phenol, vinyl chloride) can be present in biosolids incinerator emissions. Because these compounds can be harmful to the public health, the Part 503 rule regulates the emission of organic pollutants from biosolids incinerators through an operational standard that limits the amount of THC

**Table 12.18**
**Determination of the pollutant limit for lead**

Equation for determining the pollutant limit for lead:

$$C_{lead} = \frac{0.1 \times NAAQS \times 86,400}{DF \times (1-CE) \times SF}$$

where
- $C$ = The pollutant limit (allowable daily concentration of lead in biosolids, in mg/kg of total solids, dry-weight basis)
- $0.1$ = The allowable ground level concentration of lead from biosolids is 10 percent of the NAAQS for lead
- $NAAQS$ = National Ambient Air Quality Standard for lead in µg/m$^3$ (currently this standard is 1.5 µg/m$^3$)
- $DF$ = Dispersion factor in µg/m$^3$/g/s based on an air dispersion model
- $CE$ = Biosolids incinerator control efficiency for lead (in hundredths; based on a performance test)
- $SF$ = Biosolids feed rate (in dry Ton/d)
- $86,400$ = Time conversion factor (number of seconds per day)

**Example:**
If:
the dispersion factor is 3.4 µg/m$^3$/g/s
the control efficiency is 0.916
the biosolids feed rate is 12.86 Ton/d
the NAAQS for lead is 1.5 µg/m$^3$
Then:

$$C_{lead} = \frac{0.1 \times 1.5\,\mu g/m^3 \times 86,400}{3.4\,(\mu g/m^3/g/s) \times (1-0.916) \times 12.86\,\text{Ton/d}}$$

$$C_{lead} = 3529\,mg/kg$$

*Source*: US EPA (4).

or CO allowed in stack gas. According to a recent amendment of the Part 503 rule, for incinerators not exceeding 100 ppmv (parts per million, volume basis) of CO in the exhaust gas, the US EPA will allow continuous CO monitoring as a surrogate for THC monitoring.

The THC concentration (or CO) is used to represent all organic compounds in the exit gas covered by the Part 503 rule. The US EPA does not require that biosolids themselves be monitored for THC (or CO), as is required for metals. Instead, the stack gas must be monitored for THC (or CO) because organic pollutants could be present due to incomplete combustion of organic compounds or the generation of by-products of combustion.

The Part 503 rule allows a monthly average concentration of up to 100 ppmv of THC (or CO) in the stack gas. Thus, an incineration facility operator firing biosolids must continuously monitor THC (or CO) levels in the stack gas to ensure that the monthly average concentration of THC (or CO) is at or below the limit. The monthly average THC (or CO) concentration is the arithmetic mean of the hourly averages; the hourly

## Table 12.19
### Determination of the pollutant limits for arsenic, cadmium, chromium, and nickel

Equation for determining the pollutant limits for arsenic, cadmium, chromium, and nickel:

$$C = RSC \times \frac{86{,}400}{DF \times (1 - CE) \times SF}$$

where
- $C$ = The pollutant limit (allowable daily concentration of arsenic, cadmium, chromium, or nickel in mg/kg of total solids, dry-weight basis)
- $RSC$ = Risk-specific concentration (the allowable increase in the average daily ground-level ambient air concentration for a pollutant at or beyond the property line of the site in μg/m³)
- $86{,}400$ = Conversion factor (seconds per day)
- $DF$ = Dispersion factor μg/m³/g/s based on an air dispersion model
- $CE$ = Control efficiency for arsenic, cadmium, chromium, or nickel (in hundredths; based on a performance test)
- $SF$ = Biosolids feed rate in dry Ton/d

**Example for arsenic:**
If:
- the RSC is 0.023 μg/m³
- the dispersion factor is 3.4 μg/m³/g/s
- the control efficiency is 0.975
- the biosolids feed rate is 12.86 Ton/d

Then:

$$C_{arsenic} = \frac{0.023\,\mu g/m^3 \times 86{,}400}{3.4\,(\mu g/m^3/g/s) \times (1 - 0.975) \times 12.86\,\text{Ton/d}}$$

$C_{arsenic} = 1818\,mg/kg$

If the dispersion factor were 0.6 instead of 3.4 μg/m³/g/s, then the allowable concentration for arsenic would be 3.4/0.6, or 5.667 times greater at 10,300 mg/kg.

*Source*: US EPA (4).

averages must be calculated based on at least two readings taken each hour that the incinerator operates in a day (i.e., in a 24-hour period).

### 5.4.4. Frequency of Monitoring Requirements for Biosolids Incineration

The person firing biosolids in a biosolids incinerator must monitor at specified intervals for certain metals in the biosolids; for the THC (or CO) concentration, oxygen content, and information needed to determine moisture in the stack exit gas; for combustion temperature in the furnace; and for certain conditions of air pollution control device operation.

The minimum frequency for monitoring is based on the amount of biosolids incinerated (Table 12.20). The greater the amount of biosolids incinerated, the more frequently metals must be monitored. Continuous monitoring is required for THC (or CO) concentrations, oxygen levels, and information used to calculate moisture content in the stack exit gas. Continuous monitoring also is required for combustion temperature in the furnace.

**Table 12.20**
**Monitoring frequency for biosolids incinerators**

| Pollutant/Parameter | Amount of biosolids fired Ton/yr, dry-weight basis | Must monitor at least |
|---|---|---|
| Arsenic, cadium, chromium, lead, and nickel in biosolids | Greater than zero but less than 290 | Once per year |
| | Equal to or greater than 290 but less than 1500 | Once per quarter (four times per year) |
| | Equal to or greater than 1500 but less than 15,000 | Once per 60 days (six times per year) |
| | Equal to or greater than 15,000 | Once per month (12 times per year) |
| Berylium and mercury in biosolids or stack exit gas | NA | As often as permitting authority requires |
| THC (or CO) in stack exit gas | NA | Continuously; monthly averages reported, which is the arithmetic mean of hourly averages that include at least two readings per hour |
| Oxygen in stack exit gas | NA | Continuously |
| Information needed to determine moisture content in stack exit gas | NA | Continuously |
| Combustion temperature in furnace | NA | Continuously |
| Air pollution control device condition | NA | As often as permitting authority requires |

*Source*: US EPA (4).

## 6. OTHER USES OF WASTEWATER SOLIDS AND SOLIDS BY-PRODUCTS

Wastewater solids may sometimes be used beneficially in ways other than as a soil amendment or as a source of recoverable energy. Lime and activated carbon have been recovered from biosolids for many years at plant scale. Stabilized biosolids, when mixed with soil, are used as interim or final cover over completed areas of refuse landfills (47). Wastewater scum has been collected (sometimes purchased) by renderers at several treatment plants for use as a raw material in the manufacturing of cosmetics and other products. Grit, particularly incinerated grit, may be used as an aggregate (60), for example, as a road sub-base. Other beneficial uses of wastewater solids have been considered (2):

1. Recovery of ammonia from the filtrate or centrate following anaerobic digestion and dewatering of biosolids. Ammonia is stripped from the liquor, absorbed in sulfuric acid, and crystallized as ammonium sulfate.

# Beneficial Utilization of Biosolids

2. Recovery of ammonia and phosphates by precipitation of $MgNH_4PO_4$ from digester supernatants. The precipitate is used as a fertilizer.
3. Additions of biosolids to processes designed to compost or anaerobically digest municipal refuse. In such situations, biosolids serve principally as a nutrient source.
4. Recycling of wastewater biosolids for use as a foodstuff for livestock (cattle, sheep, goats, poultry, and fish). Note that solids used for this purpose have generally not originated from municipal wastewater treatment plants, but from systems treating purely industrial or animal wastes. However, the use of dried municipal biosolids disinfected by gamma irradiation has been investigated as a food source for grazing animals.
5. Use of wastewater biosolids as an organic substrate in worm farming.
6. Use of biosolids as a raw material for the production of powdered activated carbon.

## 7. EXAMPLES

### 7.1. Example 1: Determination of the Annual Whole Sludge (Biosolids) Application Rate (AWSAR)

1. Analyze a sample of the biosolids to determine the concentration of each of the 10 regulated metals in the biosolids.
2. Using the pollutant concentrations from step 1 and the APLRs from Table 12.1, calculate an AWSAR for each pollutant using Equation 1:

$$AWSAR = \frac{APLR}{0.001C} \quad (1)$$

where
   $AWSAR$ = annual whole sludge (biosolids) application rate (dry Ton/ha/yr)
   $APLR$ = annual pollutant loading rate (from Table 12.1) (kg/ha/yr)
   $C$ = pollutant concentration (mg/kg of biosolids, dry weight)
   $0.001$ = a conversion factor
3. The AWSAR for the biosolids is the lowest AWSAR calculated for each pollutant in step 2.

The following calculations illustrate the above procedure:

1. Biosolids to be applied to land are analyzed for each of the 10 metals regulated in Part 503. Analysis of the biosolids indicates the pollutant concentration in the second column of Table 12.21.
2. Using these test results and the APLR for each pollutant from Table 12.5, the AWSAR for all the pollutants are calculated as shown in the fourth column of Table 12.21.
3. The AWSAR for the biosolids is the lowest AWSAR calculated for all 10 metals. In our example, the lowest AWSAR is for copper at 20 Ton of biosolids/ha/yr. Therefore, the controlling AWSAR to be used for the biosolids is 20 Ton/ha/yr. The 20 Tons of biosolids/ha is the same as 410 lb biosolids/1000 ft$^2$ (20 Ton × 2205 lb/Ton/107,600 ft$^2$/ha). The AWSAR on the label or information sheet would have to be equal to or less than 410 lb/1000 ft$^2$.

### 7.2. Example 2: Determination of the Amount of Nitrogen Provided by the AWSAR Relative to the Agronomic Rate

In Example 1, the AWSAR for the biosolids in the example calculation was determined to be 410 lb of biosolids per 1000 ft$^2$ of land. If biosolids were to be placed on

**Table 12.21**
**Calculation of AWSAR for example 1**

| Metal | Biosolids concentrations (mg/kg) | APLR (kg/ha/yr) | $AWSAR = \dfrac{APLR}{Conc.\ in\ biosolids(0.001)} = Ton/ha$ |
|---|---|---|---|
| Arsenic | 10 | 2.0 | $2/(10 \times 0.001) = 200$ |
| Cadmium | 10 | 1.9 | $1.9/(10 \times 0.001) = 190$ |
| Chromium | 1000 | 150 | $150/(1000 \times 0.001) = 150$ |
| Copper | 3750 | 75 | $75/(3750 \times 0.001) = 20$ |
| Lead | 150 | 15 | $15/(150 \times 0.001) = 100$ |
| Nitrogen | 2 | 0.85 | $0.85/(2 \times 0.001) = 425$ |
| Nickel | 100 | 21 | $21/(100 \times 0.001) = 210$ |
| Selenium | 15 | 5.0 | $5/(15 \times 0.001) = 333$ |
| Zinc | 2000 | 140 | $140/(2000 \times 0.001) = 70$ |

*Source*: US EPA (4).

a lawn that has a nitrogen requirement of about 200 lb of available nitrogen per acre per year, the following steps would determine the amount of nitrogen provided by the AWSAR relative to the agronomic rate if the AWSAR was used:

1. The nitrogen content of the biosolids indicated on the label is 1% total nitrogen and 0.4% available nitrogen the first year.
2. The AWSAR is 410 lb of biosolids per 1000 ft$^2$, which is 17,860 lb of biosolids per acre:

   $$410\ lb/1000\ ft^2 \times 43,560\ ft^2/acre \times 0.001 = 17,860\ lb/acre$$

3. The available nitrogen from the biosolids is 71 lb/acre:

   $$17,860\ lb\ biosolids/acre \times 0.004 = 71\ lb/acre$$

4. Since the biosolids application will only provide 71 lb of the total 200 lb of nitrogen required, in this case the AWSAR for the biosolids will not cause the agronomic rate for nitrogen to be exceeded and an additional 129 lb/acre of nitrogen would be needed from some other source to supply the total nitrogen requirement of the lawn.

## NOMENCLATURE

APLR = annual pollutant loading rate, kg/ha/yr
AWSAR = annual whole sludge (biosolids) application rate, dry Ton/ha/yr
$C$ = pollutant concentration, mg/kg of biosolids, dry weight
CPLR = cumulative pollutant loading rate biosolids
CFR = United States Code of Federal Regulations
EQ = exceptional quality biosolids
PC = pollutant concentration biosolids
PFRPs = processes to further reduce pathogens
PSRPs = processes to significantly reduce pathogens
THC = total hydrocarbon

## REFERENCES

1. U.S. Environmental Protection Agency, *Biosolids Recycling: Benefit Technology for a Better Environment*, National Small Flows Clearing House, WWBLGN59, Morgantown, WV, June (1994).
2. U.S. Environmental Protection Agency, *Process Design Manual: Sludge Treatment and Disposal*, US EPA 625/1-79-011, Center for Environmental Research Information Technology Transfer, Municipal Environmental Research Laboratory, Cincinnati, OH, September (1979).
3. J. Bartlett, and E. Kilillea, The characterization, treatment and sustainable reuse of biosolids in Ireland, *Water Science Technology*, 44, 10, 35–40 (2001).
4. U.S. Environmental Protection Agency, A Plain English Guide to the US EPA Part 503 Biosolids Rule, US EPA/832/R-93/003, Office of Wastewater Management, Washington, DC, September (1994).
5. U.S. Environmental Protection Agency, *Biosolids Generation, Use and Disposal in the United States*, US EPA530-R-99-009, Solid Waste and Emergency Response, Washington, DC, September (1999).
6. M. Payn, Biosolids and Contract Bean and Grain Production, OMAFRA, Ontario Ministry of Agriculture, Food and Rural Affairs, http://www.gov.on.ca/OMAFRA/english/crops/field/news/croptalk/2004/ct_0604a9.htm (2008).
7. M. Payn, Land application of sewage biosolids for crop production, *Environmental Science & Engineering*, www.esemag.com, September (2001).
8. J. Slagle, Biosolids management with a utilization core, *Biocycle*, 35, 10, 30–33 (1994).
9. L. H. Moss, E. Epstein, and T. Logan, Comparing the characteristics, risks and benefits of soil amendments and fertilizers used in agriculture, *16th Annual Residuals and Biosolids Management Conference, Privatization, Innovation and Optimization—How to Do More for Less*, March (2002).
10. E. Epstein, Land application of biosolids and residuals: public and worker health issues, *Joint Residuals and Biosolids Management*, February (2003).
11. E. Epstein, Health issues related to beneficial use of biosolids, *16th Annual Residuals and Biosolids Management Conference 2002 Privatization, Innovation and Optimization—How to Do More for Less*, March (2002).
12. R. C. Stehouwer, and A. L. Shober, Effects of agronomic biosolids utilization on soil quality, *Joint Residuals and Biosolids Management*, February (2003).
13. T. K. O'Connor, R. Lanyon, and J. T. Zurad, Beneficial utilization of biosolids at the metropolitan water reclamation district of greater Chicago, *WEFTEC 2002 Conference Proceedings*, September (2002).
14. T. A. Angelo, Sales and marketing of alkaline stabilized EQ biosolids: the nature's blend program, *Biosolids 2001: Building Public Support*, February (2001).
15. N. C. Egigian, M. Moore, J. Burror, L. Baroldi, and F. Soroushian, Assessment of viable product markets in Orange County and Southern California, *Residuals and Biosolids Management Conference and Exhibition*, February (2004).
16. R. Wallin, and C. Drill, Marketing N-viro soil for specialty crops in Ontario Canada, *Joint Residuals and Biosolids Management*, February (2003).
17. *Biosolids* Web site: http://www.biosolids.com (2008).
18. National Academy of Science, *Use of Reclaimed Water and Sludge in Food Crop Production* (1996).
19. National Academy of Science, *Biosolids Applied to Land: Advancing Standards and Practices*, July (2002).

20. U.S. Environmental Protection Agency, *A Guide to the Biosolids Risk Assessment Methodology for the US EPA 503 Rule*, US EPA/832–R-93–009, Office of Wastewater Management, September (1993).
21. U.S. Environmental Protection Agency, *Process Design Manual: Land Application of Sewage Sludge and Domestic Septage*, Center for Environmental Research Information, Cincinnati, OH (1995).
22. U.S. Environmental Protection Agency, *Domestic Septage Regulatory Guidance: A Guide to the US EPA 503 Rule*, US EPA/832–B-92–005, Office of Wasewater Management, September (1993)
23. U.S. Environmental Protection Agency, *Environmental Regulations and Technology: Control of Pathogens and Vector Attraction in Sewage Sludge*, US EPA/625–R-92–013, Office of Research and Development, December (1992).
24. Water Environment Federation (formerly, Water Pollution Control Federation), *Beneficial Use of Waste Solids*, Manual of Practice FD-15, Alexandria, VA (1989).
25. D. Sullivan, *Fertilizing with Biosolids*, Oregon State University Extension Publications, Corvallis, OR, June (1998).
26. H. G. Adegbidi, R. D. Briggs, D. J. Robison, T. A. Volk, L. P. Abrahamson, and E. H. White, Biomass Power for Rural Development, Technical Report: Use of Biosolids as Organic Soil Amendment in Willow Bioenergy Plantations, Interim Program Report Prepared for the United States Department of Energy Under Cooperative Agreement No. DE-FC36–96GO10132, http://www.esf.edu/willow/PDFs/2000%20biosolids.pdf, July (2000).
27. Water Environment Federation, *Beneficial Use Programs for Biosolids Management*, Alexandria, VA (1994).
28. U.S. Environmental Protection Agency, *Principals and Design Criteria for Sewage Sludge Application on Land*, US EPA-625/4–78–012, Sludge Treatment and Disposal, Part 2, Technology Transfer, Cincinnati, October (1978).
29. J. Duggan, Studying land-applied biosolids: an integration of research and teaching in an environmental engineering curriculum, 2002 ASEE Annual Conference & Exposition, Vive L'ingenieur!, Montreal, Canada, 16–19 June (2002).
30. L. Spinosa, and P. A. Vesilind, *Sludge into Biosolids—Processing, Disposal, Utilization*, IWA Publishing, December (2001).
31. J. H. Huddleston, and M. P. Ronayne, *Guide to Soil Suitability and Site Selection for beneficial Use of Domestic Wastewater biosolids*, Manual 8, Oregon State University, Corvallis, OR (1995).
32. J. W. Greer, Lessons learned in seven successful years of biosolids land application, *Joint Residuals and Biosolids Management*, February (2003).
33. M. Payne, Biosolids Utilization on Agricultural Land—Phosphorus, Ministry of Agriculture, Food and Rural Affairs, http://www.gov.on.ca/OMAFRA/english/crops/field/news/croptalk/2003 (2008).
34. D. L. Binder, A. Dobermann, and D. H. Sander, Potential benefits of land applying biosolids in Eastern Nebraska, Biosolids 2001: Building Public Support, February (2001).
35. R. G. Duckworth, and A. Jerry, Beneficial use of solids, *Industrial Wastewater*, March/April (1999).
36. K. Blanton, D. Kowalski, C. Crosby, and G. Misterly, Benefits of a progressive centralized biosolids management system, in *WEFTEC 2003 Conference Proceedings*, October (2003).
37. R. F. Anderson, Biosolids land application in the 21st century: will the regulators base public policy on fact or fiction and fear? Biosolids 2001: Building Public Support, February (2001).

38. N. Beecher, Cultivating New England biosolids recycling, 14th Annual Residuals and Biosolids Management Conference, February/March (2000).
39. S. Worley, R. Bates, G. Swanson, A. Hogge, and J. Pavoni, Development of a biosolids marketing strategy: "a cultural change," WEFTEC 2004, October (2004).
40. S. Worley, D. Carty, R. Bates, and G. J. Swanson, Biosolids as a product rather than waste illustrates importance of EMS, Joint Residuals and Biosolids Management Conference, April (2005).
41. G. Velema, Domtar communication papers' land application...Experience at Cornwall 1995–1999, 14th Annual Residuals and Biosolids Management Conference, February/March (2000).
42. Department of Environmental Quality, Michigan Biosolids Program, http://www.michigan.gov/deq/0,1607,7-135-3313_3683_3720—,00.html (2008).
43. J. Glass, W. Steven, B. David, and R. E. Sosebee, Reclamation of semi-arid rangeland with biosolids, 16th Annual Residuals and Biosolids Management Conferenc, Privatization, Innovation and Optimization—How to Do More for Less, March (2002).
44. South Dakota Department of Environment and Natural Resources, Beneficial Uses of Biosolids, http://www.state.sd.us/denr/DES/Surfacewater/biosolids.htm (2007).
45. R. B. Stallings, Beneficial reuse of US EPA Class B biosolids—lessons learned, *Joint Residuals and Biosolids Management*, February (2003).
46. L. K. Wang, N. K. Shammas, and Y. T. Hung (eds.) *Biosolids Treatment Processes*, Humana Press, Inc., Totowa, NJ, 820 pp. (2007).
47. K. Fanfoni, and L. M. Naylor, Beneficial reuse of biosolids in landfill closure, Biosolids 2001: Building Public Support, February (2001).
48. G. Mininni, Options for biosolids utilization and sludge disposal: incineration with energy recovery, *Water Intelligence Online*, IWA Publishing (2002).
49. U.S. Environmental Protection Agency, *Technical Support Document for Incineration of Sewage Sludge*, Office of Water, NTIS: PB93–110617 (1993)
50. U.S. bares solar energy program to year 2020, *Chicago Sun-Times*, p. 29, August 14 (1975).
51. P. S. Ward, Digester gas helps most energy needs, *Journal of Water Pollution Control Federation*, 46, 620 (1974).
52. Brown and Caldwell, Solid Waste Resource Recovery Study, Prepared for the Central Contra Costa Sanitary District, Walnut Creek, CA, August (1974).
53. California Public Utilities Commission, Staff Report on California Cogeneration Activities, Utilities Division, San Francisco, January 17 (1978).
54. U.S. Department of Energy, *Proposed Regulations Providing for Qualification of Small Power Production and Cogeneration Facilities Under Section 201 of the Public Utility Regulatory Policies Act of 1978*, Federal Energy Regulatory Commission, Washington, DC, June (1979).
55. N. Cumiskey, and D. P. Capon, Recovering energy from biogas—a comparison of available technology, Joint Residuals and Biosolids Management Conference, April (2005).
56. J. L. Strehler, and D. L. Parry, Systems approach to integrating biosolids and energy management, Joint Residuals and Biosolids Management Conference, April (2005).
57. Alpha National Inc., Solid Waste and Biomass Low BTU Gas Conversion System Program, El Segundo, CA, April (1978).
58. U.S. Environmental Protection Agency, *Energy Conservation in Municipal Wastewater Treatment*, US EPA 430/9–77–001, Office of Water Program Operations, Washington, DC, March (1978).
59. K. Bolin, The SlurryCarb Process: Turning Municipal Wastewater Solids into a Profitable Renewable Fuel, Biosolids 2001: Building Public Support, February (2001).

60. H. K. Moo-Young, and C. E. Ochola, The characterization of paper mill biosolids and its application in civil engineering construction, 14th Annual Residuals and Biosolids Management Conference, February/March (2000).
61. L. K. Wang, Quaternary ammonium thickening of sewage sludge in magnetic field, *Industrial and Engineering Chemistry*, 16, 4, 311–315 (1977).
62. M. Krofta, and L. K. Wang, Sludge thickening and dewatering by dissolved air flotation: Floatpress, *Drying*, 2, 765–771, Harper & Row, New York (1986).
63. M. Krofta, and L. K. Wang, Sludge thickening and dewatering by dissolved air flotation: process design, *Drying*, 2, 772–780, Harper & Row, New York (1986).
64. L. K. Wang, *Guidelines for Disposal of Solid Wastes and Hazardous Wastes*, Volumes I to VI, U.S. Department of Commerce, National Technical Information Service, Springfield, VA, Technical Report Nos. PB88–178066/AS, PB88–178074/AS, PB88–178082/AS, PB88–178090/AS, PB88–178108/AS and PB89–158596/AS (1989).
65. L. K. Wang, *Sludge Treatment Apparatus*, U.S. Patent and Trade Marks Office, Washington, DC, Patent No. 5068031, November (1991).
66. L. K. Wang, *Analysis and Identification of Odorous Compounds in Emissions from Springfield Regional Wastewater Treatment Plant*, Massachusetts, Lenox Institute of Water Technology, Lenox, MA, Technical Report LIR/05–85/137 (1985).

# 13
# Process Selection of Biosolids Management Systems

## Nazih K. Shammas and Lawrence K. Wang

**CONTENTS**

INTRODUCTION
THE LOGIC OF PROCESS SELECTION
SIZING OF EQUIPMENT
APPROACHES TO SIDESTREAM MANAGEMENT
CONTINGENCY PLANNING
SITE VARIATIONS
ENERGY CONSERVATION
COST-EFFECTIVE ANALYSES
CHECKLISTS
U.S. PRACTICES IN MANAGING BIOSOLIDS
REFERENCES

**Abstract** This chapter presents a methodology for process selection and design of biosolids management systems. Topics discussed include systems approach, process selection logic, approaches to sidestream management, the concept of sizing equipment, contingency planning, and other general design considerations such as energy conservation and cost-effective analysis.

**Key Words** Biosolids • management • process selection • selection logic • sizing equipment • planning • design.

## 1. INTRODUCTION

This chapter presents a methodology for process selection and design of biosolids management systems. Topics discussed include systems approach, process selection logic, approaches to sidestream management, the concept of sizing equipment, contingency planning, and other general design considerations such as energy conservation and cost-effective analysis.

Overall wastewater treatment plant performance is the sum of the combined performances of the plant's linked components. The actions of one component affect the performance of all the others. As examples (1–3):

1. Materials not captured in solids treatment processes are returned in the sidestreams to the wastewater treatment system as a recirculating load. This load may cause degradation in effluent quality, an increase in wastewater treatment costs, and process upsets.
2. Failure to remove and to treat solids at the same rate as they are produced within the wastewater treatment system eventually causes effluent degradation and may increase wastewater treatment operating costs.
3. Hydraulic overloads resulting from inadequate solids thickening can cause downstream solids treatment processes (such as anaerobic digestion) to operate less effectively.
4. The addition of chemicals to the wastewater treatment process for purposes of nutrient and suspended solids removal increases the quantity and alters the characteristics of solids that must be treated and disposed.

It is important to understand the relationship between process parameters and the performance of processes, for example, how thickener feed rate affects thickener performance. It is equally important to understand how individual processes affect one another when combined into a system, for instance, how the performance of the thickener affects digestion and dewatering. Interactions between the processes in a system are described in this chapter.

## 2. THE LOGIC OF PROCESS SELECTION

Wastewater treatment and wastewater solids and disposal systems must be put together so as to ensure the most efficient utilization of resources such as money, materials, energy, and labor in meeting treatment requirements. Logic dictates what the process elements must be and the order in which they go together.

A methodical process of selection must be followed in choosing a resource-efficient and environmentally sound system from the myriad of treatment and disposal options available (4, 5). The basic selection mechanism is the "principle of successive elimination," an iterative procedure in which less effective options are progressively deleted from the list of candidate systems until only the most suitable system or systems for the particular site remain (3, 6).

The concept of a "treatment train" has been propounded as a result of a systems approach to problem solving. However, this concept is useful only if all components of the train are considered. This includes not only biosolids treatment and disposal components but also wastewater treatment options and other critical linkages such as biosolids transportation, storage, and sidestream treatment. The successful development of a treatment train from a collection of individual components depends on a rigorous system selection procedure or logic. For large plants, system selection is complex and a methodical approach is required. Progressive and concurrent documentation of the procedure is mandatory in that it prevents a cursory dismissal of options. For smaller plants (that is, less than one million gallons per day [1 MGD]), the system choices are often necessarily more obvious and the selection procedure is usually shorter and less complex.

The general sequence of events in system selection is as follows (3, 6):

1. Select the relevant criteria.
2. Identify the options.

3. Narrow the list of candidate systems.
4. Select a system.

## 2.1. Identification of Relevant Criteria

Criteria for system selection must be pinpointed prior to system synthesis. A listing of potential criteria for consideration is shown in Figure 13.1. The list is not necessarily complete and planners may find other criteria that they wish to include. The relative importance of each criterion varies from site to site. For example, reliability may be the most important at one site, whereas minimizing costs may be the most important at another. Criteria deemed relevant for each site in question are subsequently used in the system selection procedure.

## 2.2. Identification of System Options

Candidate systems are synthesized from an array of components. Wastewater and solids management components (7–10) are listed as a reminder that all components of the train must be considered. Process streams can be drawn on copies of the master drawing. Relevant information such as solids concentrations and mass flow rates can be entered directly on the flow sheet, if desired. The advantages of using arrays are that nearly all potential options are identified and process streams are clearly displayed.

## 2.3. System Selection Procedure

The process selection procedure consists of (a) developing treatment/beneficial use/disposal systems that are compatible with one another and appear to satisfy local relevant criteria, and (b) choosing the best system or systems by progressive elimination of weaker candidates. Related to these are the concepts of base and secondary alternatives.

### 2.3.1. Base and Secondary Alternatives

A base alternative is defined as a wastewater biosolids management system that, during evaluation, appears able to provide reliable treatment and beneficial use/disposal at all times under all circumstances for the biosolids. It therefore meets the prime criterion of reliability.

It must also satisfy the following seven conditions (3, 11, 12):

1. It must be legally acceptable.
2. Sites for processing and beneficial use/disposal operations must be readily available.
3. Environmental and health risks must be sufficiently low to satisfy the public and all agencies having jurisdiction.
4. The cost must be competitive with the costs of other alternatives on a first-round analysis.
5. The necessary equipment and material must be readily available.
6. The contractor must be able to begin construction immediately following design and have the system operational almost immediately after construction.
7. Financing of the system must be straightforward and ensured.

A secondary alternative is defined as a wastewater solids management system that does not meet the prime criterion of reliability, that is, the system cannot accept all of the biosolids under all circumstances all of the time. This does not mean secondary

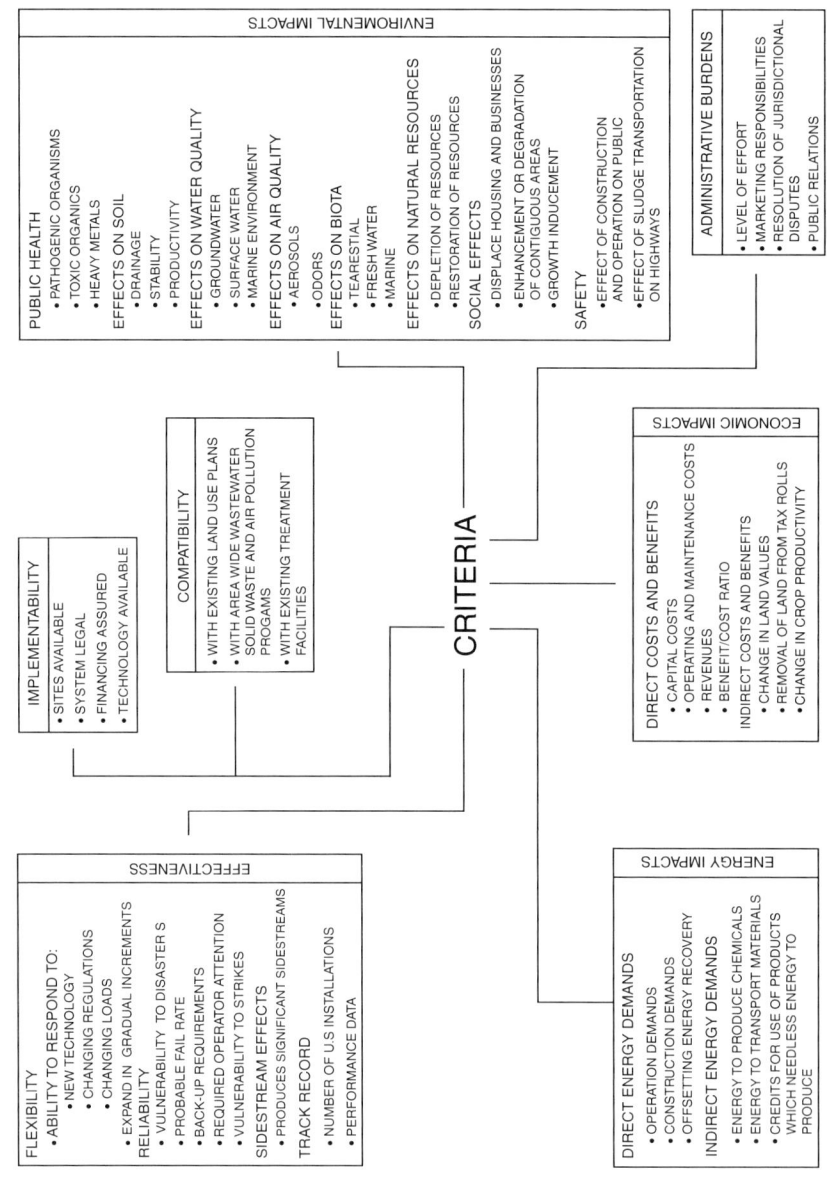

**Fig. 13.1.** Criteria for system selection. *Source*: US EPA.

alternatives are without value; they may in fact be used to great advantage in tandem with base alternatives and may in fact accept a greater quantity of biosolids than the base alternative. As an example, a city's horticultural market may be insufficiently developed to accept all of the city's biosolids all of the time; therefore, horticulture cannot be considered a base disposal alternative. However, it may cost less to release the biosolids to horticulture than to dispose of it by means of city's base disposal alternative, for example, landfilling (13). The city, therefore, should make every effort to dispose of as many solids as possible via horticulture, the secondary alternative. However, should the secondary alternative fail or be interrupted for any reason, the biosolids going to the secondary system must be readily and quickly diverted back to the base alternative, which must remain fully operational and thus immediately capable of receiving the entire biosolids flow.

*2.3.2. Choosing a Base Alternative: First Stage*

The purpose of the first stage is to rapidly and with minimum effort produce a list of candidate base alternatives that are technically feasible and reasonably cost-effective. The alternatives must be environmentally acceptable and implementable in the time frame of the project. Analyses are qualitative at this stage. The first stage involves determining the following (3, 14, 15):

1. Practical base utilization/disposal options
2. Practical base solids treatment systems
3. Practical treatment/utilization/disposal combinations

2.3.2.1. DETERMINATION OF PRACTICAL BASE
         UTILIZATION/DISPOSAL OPTIONS

The method of solids utilization/disposal usually controls the selection of solids treatment systems and not vice versa. Thus, the system selection procedure normally begins when the solids utilization/disposal option is specified.

In the first stage, feasible base utilization/disposal alternatives and relevant criteria are set up in matrix form. An example is shown in Table 13.1. Feasible alternatives are those that appear to be suitable for the situation at hand. Obviously inapplicable alternatives would not be included in this matrix. For example, we are at a point where marine dispersal is banned, incinerators are no longer popular, and landfill is diminishing in opportunity and cost-effectiveness (16). The United States Congress recognized that the nutrients in biosolids could cause increased algae production, eventually leading to oxygen depletion at the site. Thus Congress properly decided that it made much more sense and better policy to get biosolids out of the ocean and use the nutrients in biosolids more productively to provide crop nutrients and to improve soil quality (17).

Only those criteria that the planner sees as critical for the site at hand should be considered in this first stage. Other, less critical criteria can be considered in subsequent iterations, where more in-depth investigation is needed for each of the candidate processes.

For the hypothetical situation described in Table 13.1, nine utilization/disposal options are considered feasible and are set up for evaluation (3). The criteria most important to the site are judged to be reliability, environmental impacts, site availability, and cost.

**Table 13.1**
Example of initial screening matrix for base biosolids disposal options

| Utilization/disposal options | Relevant criteria | | | | |
| --- | --- | --- | --- | --- | --- |
| | Reliability | Environmental impacts | Site availability | Cost | Acceptable for base alternative |
| Bag-market as fertilizer | o[a] | x[b] | x | x | o |
| Agricultural land (private) | o | x | x | x | o |
| Agricultural land (public) | x | x | x | x | x |
| Forested land (private) | o | x | o | o | o |
| Forested land (public) | x | x | o | o | o |
| Give to citizens (horticulture) | o | x | x | x | o |
| Combine with commercial topsoil | o | x | x | x | o |
| Dedicated land disposal | x | x | x | x | x |
| Landfill | x | x | x | x | x |

[a] o, unacceptable.
[b] x, acceptable.
*Source*: US EPA.

Base utilization/disposal alternatives are judged to be practical only if they satisfy all the relevant criteria. In Table 13.1, utilization of biosolids on private agricultural land is an unacceptable base disposal alternative. Reasons for this might be insufficient acreage or a lack of assurance that the farmers would accept all of the biosolids. Alternatives that would seem to satisfy relevant criteria for base utilization/disposal alternatives are utilization on public agricultural land, landfill, and dedicated land disposal. Before considering these, however, one must determine what combinations of solids treatment processes make sense for the site in question.

2.3.2.2. DETERMINE PRACTICAL BASE TREATMENT SYSTEMS

Table 13.2 illustrates a process compatibility matrix for treatment alternatives. Incompatible processes and processes that are not applicable in given locations are eliminated. The combination of drying beds and mechanical dewatering, for example, is considered incompatible because both dewatering and drying take place on the drying bed; mechanical dewatering is not needed. On the other hand, the combination of incineration and mechanical dewatering of unstabilized biosolids is generally compatible, but for the hypothetical case investigated it is ruled out because of air pollution considerations. After first-stage analysis, seven base treatment options are considered feasible and are further evaluated.

2.3.2.3. DETERMINE PRACTICAL BASE TREATMENT/UTILIZATION/ DISPOSAL COMBINATIONS

Practical base treatment and utilization/disposal combinations are then combined in a matrix, which is subjected to further elimination. Table 13.3 shows the matrix of base treatment/utilization/disposal combinations made by bringing forward the base utilization/disposal and treatment options from Tables 13.1 and 13.2. Incompatible combinations and systems ruled out by local constraints are then eliminated. For example, undewatered wastewater biosolids are not generally disposed of in landfills. An example of local constraints is the ruling out of applying lime stabilized biosolids on agricultural lands because of already high soil pH.

The number of candidate base treatment/utilization/disposal systems is thus reduced. For the hypothetical case of Table 13.3, sixteen systems remain for further evaluation.

### 2.3.3. Choosing a Base Alternative: Second Stage

The purpose of second-stage analyses is to further reduce the list of candidate systems. Analyses are more quantitative than in the first stage, but the level of effort used to investigate each option is not yet intensive. Information used in the second stage is general and readily available, for instance, equipment cost curves that are not site-specific; area-wide evaluation of soils, geology, hydrology, topography, and land use; and general energy costs.

One approach is to set up a numerical rating system for the remaining candidate systems, such as that shown in Table 13.4. The list of criteria to be considered may be expanded beyond those critical criteria used in first-stage analyses to encompass the full range of criteria listed in Figure 13.1, or any fraction of it. This follows the principle that as the list of candidate process narrows each will be analyzed in greater detail.

In the second stage, subjective judgments are combined with technical measurements. Numerical values are assigned to all criteria for all alternative systems. The planner's perception of the relative importance of each criterion is indicated on a rating scale, say, of 0 to 5, with highest ratings given to criteria the planner considers to be of greatest importance and the lowest to those of least important. For example, if reliability is highly valued for the site in question, reliability may be assigned a relative weight of 5.

Next, each alternative system is rated according to its anticipated performance with respect to the various criteria, again by using a rating scale, say, 0 to 10. An alternative that rates favorably is given high scores; one that rates less favorably is given lesser scores. For example, an alternative that is not dependable may be rated 2 with respect to reliability.

The relative weight is then multiplied by the alternative rating to produce a weighted rating for each criteria/alternative combination. For the examples described in the previous two paragraphs, the weighted rating for the alternative in question with respect to reliability is $5 \times 2 = 10$.

Finally, the weighted ratings are summed for each alternative to produce a total or overall rating. Systems with lowest overall ratings are eliminated, with higher rated systems carried forward for further evaluations. In the example shown in Table 13.4, alternatives 3 and 4 are eliminated and alternatives 1, 2, and $n$ are carried forward.

### 2.3.4. Third Stage

The third stage uses the same methodology as the second, but the number of alternatives remaining is more limited, typically to a maximum of three to five, and the analysis is more detailed. Information may include the following (3, 18–22):

Table 13.2
Example of process compatibility matrix

| Final processing step | Digestion options | | Undigested biosolids options | | | | |
|---|---|---|---|---|---|---|---|
| | Anaerobically or aerobically digested | | Not stabilized | | Lime stabilized | Thermally conditioned | Wet air oxidation |
| | Mechanically dewatered | Not dewatered | Mechanically dewatered | Not dewatered | Mechanically dewatered | Mechanically dewatered | Mechanically dewatered |
| No further processing | x[a] | x | o[b] | o | x | ø[c] | ø |
| Drying beds | o | x | o | o | o | o | o |
| Heat dry | x | o | o | o | o | o | ø |
| Pyrolysis | o | o | ø | o | o | ø | o |
| Incineration | o | o | ø | o | o | ø | o |
| Compost | x | o | x | o | o | o | o |

[a] x, generally compatible.
[b] o, generally not compatible.
[c] ø, generally compatible, but ruled out by local constraints.
Source: US EPA.

**Table 13.3**
**Example of treatment/disposal compatibility matrix**

| Viable local disposal options | Treatment options | | | | | |
|---|---|---|---|---|---|---|
| | Digested biosolids options | | | | Undigested biosolids options | |
| | Mechanically dewatered | Mechanically dewatered, heat dry | Mechanically dewatered, compost | Not mechanically dewatered | Not mechanically dewatered, drying beds | Mechanically dewatered, composted | Lime stabilized, mechanically dewatered |
| Agricultural land (public) | x[a] | x | x | – | x | x | ø[b] |
| Landfill | x | x | ø[c] | o | x | x | x |
| Dedicated land disposal | x | x | o | x | x | x | ø |

[a] x, generally compatible.
[b] ø, generally compatible, but ruled out by local considerations.
[c] o, generally not compatible.
*Source*: US EPA.

### Table 13.4
### Example of numerical rating system for alternatives analysis

| Categories and criteria | Relative weight[a] | Alternative 1 AR[b] | Alternative 1 WR[c] | Alternative 2 AR | Alternative 2 WR | Alternative 3 AR | Alternative 3 WR | Alternative 4 AR | Alternative 4 WR | ... | Alternative n AR | Alternative n WR |
|---|---|---|---|---|---|---|---|---|---|---|---|---|
| Effectiveness | | | | | | | | | | | | |
| Flexibility | 3 | 4 | 12 | 6 | 18 | 9 | 27 | 5 | 15 | ... | 6 | 18 |
| Reliability | 5 | 3 | 15 | 5 | 25 | 5 | 25 | 2 | 10 | ... | 2 | 10 |
| Sidestream affects | 3 | 10 | 30 | 9 | 27 | 5 | 15 | 6 | 18 | ... | 7 | 21 |
| Track record | 2 | 5 | 10 | 7 | 14 | 4 | 8 | 9 | 18 | ... | 6 | 12 |
| Compatibility | | | | | | | | | | | | |
| With existing land use plans | 2 | 8 | 16 | 8 | 16 | 8 | 16 | 7 | 14 | ... | 4 | 8 |
| With areawide wastewater, solid waste and air pollution programs | 3 | 3 | 9 | 6 | 18 | 3 | 9 | 5 | 15 | ... | 7 | 21 |
| With existing treatment facilities | 4 | 5 | 20 | 5 | 20 | 6 | 24 | 8 | 32 | ... | 3 | 12 |
| Economic impacts | | | | | | | | | | | | |
| Net direct costs | 4 | 7 | 28 | 8 | 32 | 8 | 32 | 9 | 36 | ... | 7 | 28 |
| Net indirect costs | 1 | 8 | 8 | 9 | 9 | 6 | 6 | 3 | 33 | ... | 8 | 8 |
| Environmental impacts | | | | | | | | | | | | |
| Public health | 5 | 7 | 35 | 6 | 30 | 4 | 20 | 6 | 30 | ... | 7 | 35 |
| . | . | . | . | . | . | . | . | . | . | ... | . | . |
| . | . | . | . | . | . | . | . | . | . | ... | . | . |
| . | . | . | . | . | . | . | . | . | . | ... | . | . |
| Administrative burdens | | | | | | | | | | | | |
| Level of effort | 1 | 4 | 4 | 6 | 6 | 5 | 5 | 7 | 7 | ... | 4 | 4 |
| Marketing responsibilities | 2 | 5 | 10 | 5 | 10 | 4 | 8 | 7 | 14 | ... | 9 | 18 |
| Resolution of jurisdictional disputes | 1 | 3 | 3 | 4 | 4 | 4 | 4 | 4 | 4 | ... | 2 | 2 |
| Public relations | 2 | 4 | 8 | 2 | 4 | 5 | 10 | 5 | 10 | ... | 3 | 6 |
| Total weighted alternative rating[d] | – | – | 1,576 | – | 1,430 | – | 963 | – | 840 | ... | – | 1,317 |

[a] Relative importance of criteria as perceived by reviewer scale, 0 to 5, no importance rated zero, most important rated 5.
[b] Alternative rating. Rates the alternatives according to their anticipated performance with respect to the various criteria; scale 0 to 10, least favorable rated zero, most favorable rated 10.
[c] Weighted rating. Relative weight for each criteria multiplied by alternative rating.
[d] Sum of weighted ratings for each alternative.

*Source*: US EPA.

1. Analyses of potential biosolids utilization/disposal sites (soils, geology, and groundwater)
2. Local surveys to determine marketability of biosolids and biosolids by-products
3. Possible effects of industrial source control/pretreatment programs on process viability and quality of biosolids for disposal
4. Data-oriented literature search
5. Detailed analysis of effect of candidate systems on the environment (air, water, land)
6. Information developed from site-specific pilot work
7. Mass balances
8. Energy analyses
9. Detailed cost analyses

### 2.3.5. Subsequent Stages

Subsequent stages are even more detailed. Analyses are repeated until the optimum base treatment/utilization/disposal alternative is defined.

A simple example to illustrate the process at this stage is the determination of the optimum percent solids to attain for a combination of dewatering and disposal/utilization alternative solutions.

For ultimate biosolids disposal/agricultural application, the total cost consists of several elements:

1. Costs of disposal in landfill/agricultural utilization
2. Costs of transportation
3. Costs of dewatering

The key to each of these elements is the degree of processing and the percent solids in the dewatered biosolids. To establish the most cost-effective biosolids management system, representative costs of dewatering and costs of transport and disposal in landfill/agricultural utilization were developed for processing stabilized combined primary and secondary biosolids. These costs and the total costs are shown graphically in Figures 13.2 and 13.3. The optimum system costs are shown as the minimum cost per ton of dry solids on the total cost curve. For biosolids disposal in landfill the optimum system requires a biosolids cake of 36% solids; for agricultural utilization, an optimum biosolids cake of approximately 27% solids is required.

## 2.4. Parallel Elements

By means of the procedure discussed above, a base alternative is selected. However, the optimum system may include more than just this base alternative. A number of parallel elements may be involved that provide flexibility, reliability, and operating advantages. For example, the base alternative for the system depicted in Figure 13.4 is thickening, anaerobic digestion, storage in facultative biosolids lagoons, and spreading of liquid biosolids on agricultural land. Parallel elements consist of the application of liquid biosolids on forest land and drying beds followed by distribution for horticultural purposes. If horticultural and forest land outlets were each large enough to accept all of the biosolids under all circumstances and at all times, three base alternatives are then available. If not, the forest land and drying beds/horticulture applications would be considered secondary alternatives.

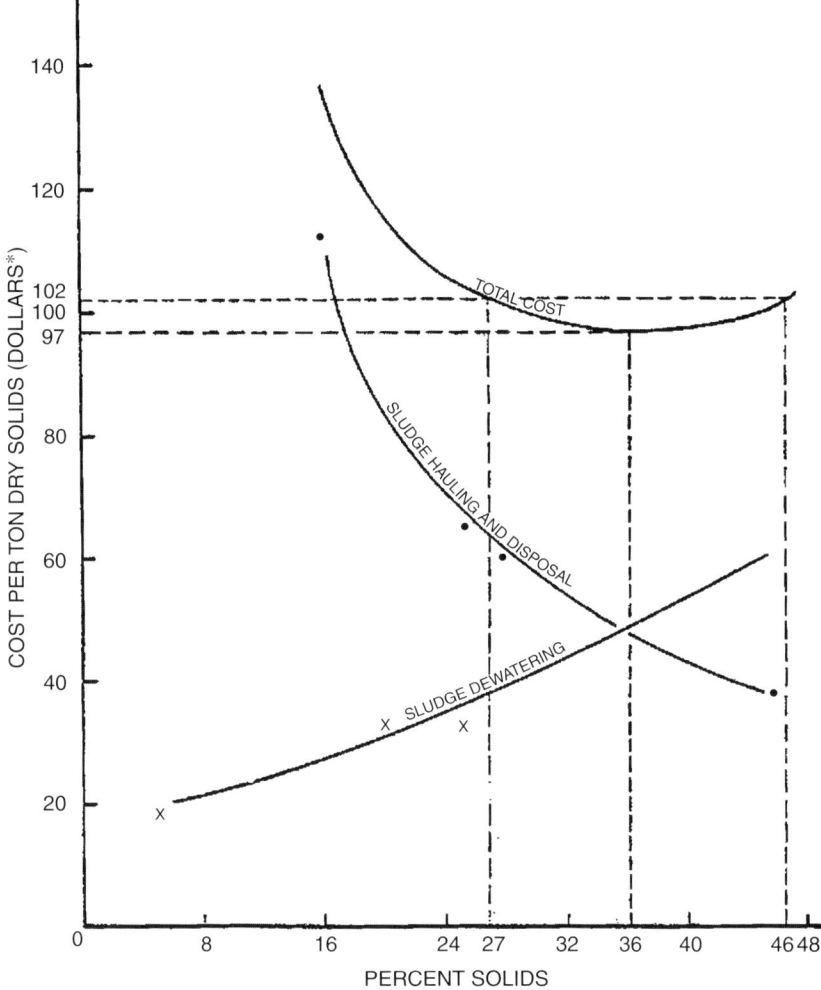

**Fig. 13.2.** Dewatering and landfill disposal of combined primary and secondary biosolids. *Source*: US EPA.
* Multiply cost by 2.1 to obtain cost in 2007 U.S. Dollars.

The concept of providing for more than one base alternative may at first seem contradictory, but a given base alternative might not always be reliable because unpredictable events might occur. For example, new owners of farmland may decide they do not wish to accept biosolids, or a disaster or strike could interrupt one method of transporting biosolids to its ultimate destination. To minimize risks, therefore, municipalities may wish to provide more than one base alternative. The selection procedure presented here has the advantage of clearly depicting which is the second or even third most desirable base alternative.

Parallel base alternatives are more common in large systems, which are generally located in urban areas where land is scarce than in small plants, which are usually located in rural areas where land is more plentiful and temporary storage and disposal options

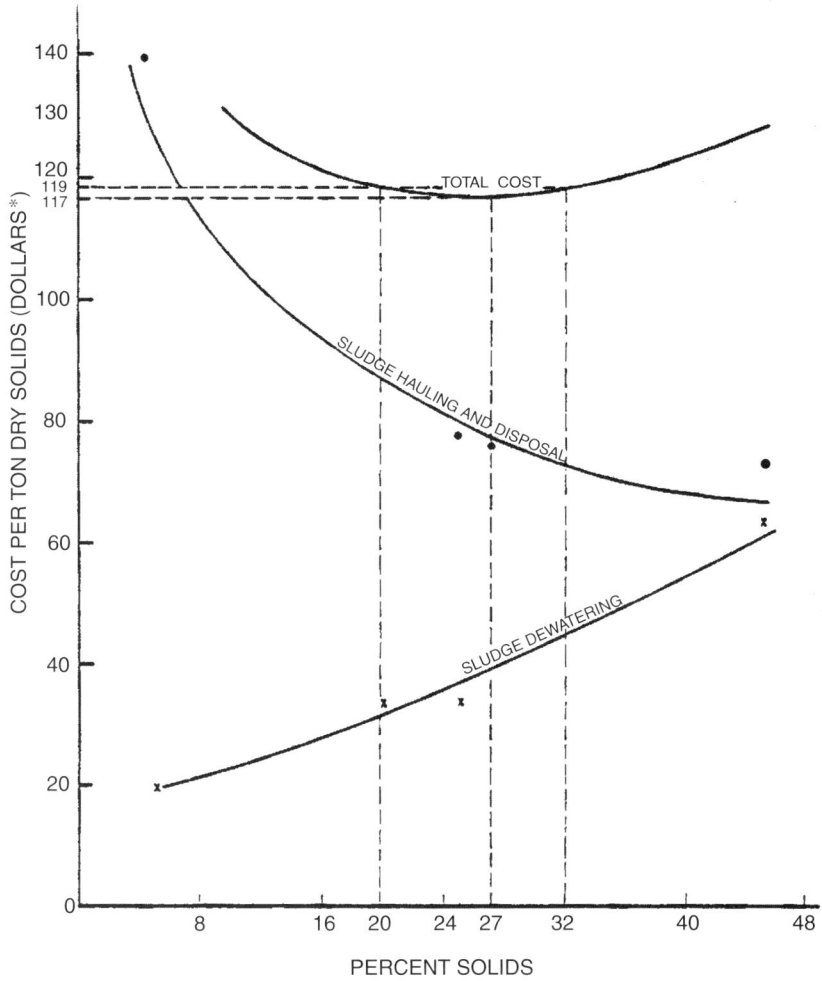

**Fig. 13.3.** Dewatering and agricultural utilization of combined primary and secondary biosolids. *Source*: US EPA.
* Multiply cost by 2.1 to obtain cost in 2007 U.S. Dollars.

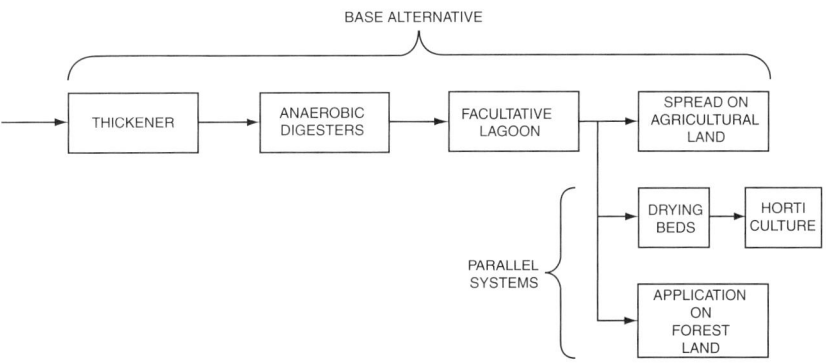

**Fig. 13.4.** Parallel elements. *Source*: US EPA.

therefore more numerous. Large plants may maintain two or three base alternatives to ensure biosolids disposal. Since this may increase the cost of operation, it leads to the observation that very large systems do not necessarily benefit from economies of scale when it comes to wastewater biosolids disposal.

### 2.5. Example of Process Selection at Eugene, Oregon

Eugene, a city of 100,000 people, is located at the southern end of the agricultural Willamette Valley in western Oregon. The Metropolitan Wastewater Management Commission (MWMC) was formed to implement the findings of a facility planning effort that called for the construction of a regional wastewater treatment plant (3). The plant, to be constructed on the site of the existing Eugene plant, will serve the whole metropolitan area. This area is composed of the cities of Eugene, Springfield, and urbanized portions of Lane County.

Regionalization and upgrading of the plant to meet a 10/10 summer effluent standard for 5 days biochemical oxygen demand ($BOD_5$) and suspended solids prior to discharge to the Willamette River, means that biosolids quantities are dramatically increased. The plant is to serve a population of 277,000 by the end of its life design period. Design average dry weather flow is 49 MGD ($2.15\,m^3/s$), wet weather flow is 70 MGD ($3.07\,m^3/s$), and peak wet weather flow is 175 MGD ($7.67\,m^3/s$).

The plant will use an activated sludge process, with flexibility for operation in plug, step, contact stabilization, or complete mix modes. Provision is also made for the addition of mechanical flocculators in the secondary clarifiers and tertiary filtration if either or both prove desirable at a later date.

It was decided early that biosolids thickening would be economical, regardless of the biosolids management system that would eventually be used. Consequently, two existing thickeners, one gravity and one flotation, will be retained for thickening primary solids, waste-activated sludge, or a combination of the two.

A key provision in the selection of a suitable biosolids management system was that the system be fully operational by the time the wastewater treatment system is started up. This seemingly straightforward condition was complicated by the fact that planning for the biosolids system did not start until design of the wastewater treatment plant was already under way. This meant that the biosolids management system would be forced to fit into an already developed plan for the wastewater treatment facility (which is by no means unusual).

As a first stage, biosolids utilization/disposal options were immediately developed and screened for acceptability as part of a base alternative, using a matrix similar to that developed in Table 13.1. Practical treatment systems were identified from a process compatibility matrix similar to that in Table 13.2. Practical utilization/disposal/processing combinations were then developed in a matrix form (as in Table 13.3). Physically incompatible or otherwise unsuitable combinations were eliminated in this matrix. A flowsheet was then prepared for the remaining options, with necessary intermediate storage and transport requirements added in. The flowsheet of alternatives for Eugene second stage analysis is shown in Figure 13.5.

It is worth noting that utilization on agricultural land could not be considered as a base alternative despite the large agricultural acreage north of Eugene and the fact that the new

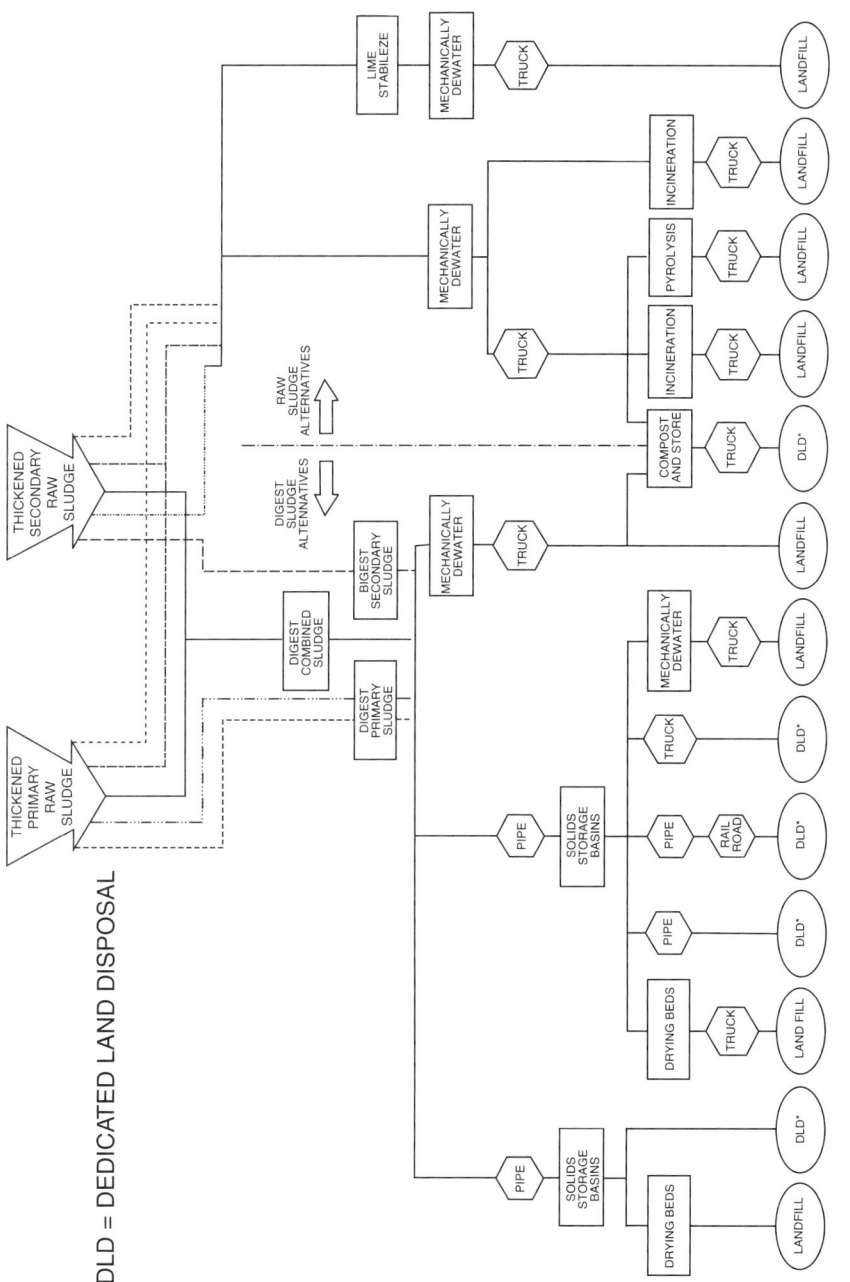

**Fig. 13.5.** Candidate base alternatives for Eugene-Springfield. *Source*: US EPA.

regional plant is on the north side of the city. It would have been a requisite that MWMC own sufficient farmland (2000 to 3000 acres) to accept all of the biosolids generated. The cost of purchasing such acreage was deemed unacceptably high; furthermore, there was opposition to converting private land to public land. Thus agricultural utilization was not considered further in the search for a base alternative.

The second-stage analysis was more quantitative. Information used was general and readily available. For example, costs were taken from current cost curves, and certain environmental impacts were assessed from projects with similar disposal systems and soil/groundwater conditions. With numerical data established for each criterion, a rating table was produced similar to that of Table 13.4. The data were developed by the project engineers, but the ratings were analyzed extensively by the Citizens Participation Committee (CPC) on biosolids management, which had been recruited from the population at large at the very beginning of the project. The committee was composed of various vested interest groups, representatives of government agencies, and private unaffiliated citizens who were interested in the project.

Systems with the lowest total ratings were then eliminated. Incineration was found to be unacceptable primarily because it would impact the already limited dilution capacity available during the summer in the trapped valley air shed of Eugene; pyrolysis was eliminated primarily because of its perceived inability to meet the construction deadline for plant start-up; and lime stabilization with disposal to landfill was eliminated primarily on a cost-effective basis. At the end of the second-stage analysis, all alternatives that could accommodate raw biosolids were eliminated, since, as indicated, most raw biosolids options (incineration, pyrolysis, lime stabilization) were not viable and there was a strong desire to make use of existing digesters. A decision was made to combine primary and secondary biosolids in order to avoid the cost and problems of constructing and operating separate systems for each.

The same methodology used in the second stage was used in the third; however, data used in the analysis were more site specific, so that economic and environmental comparisons could be better refined. As examples (3):

1. Actual routes were selected to off-site facilities; river crossings were defined, and decisions were made on routing pipes under bridges or jacking under freeways.
2. For disposal at the local sanitary landfill, estimates were made of (a) the contribution of the biosolids to landfill leachate production and subsequent marginal leachate treatment costs to be passed back from the Lane County Solid Waste Division to MWMC, and (b) the actual net volume of landfill required for biosolids disposal, allowing for biosolids consolidation.
3. For dedicated land disposal, seasonal water tables and detailed groundwater migration patterns, as well as private well locations and depths were determined.
4. Estimates were made of comparative nitrate loadings that would eventually reach the Willamette River from treated landfill leachate discharge; from groundwater migration from dedicated land disposal; and from filtrates from mechanical biosolids dewatering (which is subsequently discharged with the effluent).
5. Transportation modes were analyzed in detail and costed for various biosolids concentrations and transport routes and distances.

These detailed analyses still left a number of viable base alternatives. At this point, other less tangible factors were considered. These were (a) that the chosen base alternative(s) be compatible with desired secondary alternatives, and (b) that flexibility and reliability be provided through the use of parallel systems. After intensive screening, it was decided that two base alternatives would be used: spreading of liquid biosolids on dedicated land and open-air drying followed by landfill disposal. Both alternatives included force main transport of digested biosolids from the regional treatment plant to a remote biosolids management site, where the biosolids were to be stored in facultative lagoons. Liquid biosolids would be spread on dedicated land at the biosolids management site. Dried biosolids would be trucked to landfill. Operations associated with disposal (spreading, drying, and landfilling) would be carried out during dry weather. These systems provide the desired flexibility and reliability and are compatible with preferred secondary alternatives.

Several variations of biosolids utilization on land were adopted as secondary alternatives, since there was a strong feeling that biosolids should be used beneficially. The alternatives of particular interest to the Eugene-Springfield area included agricultural use on private farm land, use for ornamental horticulture, in nurseries and public parks, and use in a mixture with commercial topsoils in landscaping. Biosolids would be provided to these outlets as the market demands. Variable demand is particularly important in Oregon's Willamette Valley, where prolonged winter rainfall and summer harvesting schedules control the timing of agricultural biosolids use.

The flowsheet for the Eugene system is shown in Figure 13.6. The ability to use base facilities and equipment for desired secondary alternatives was a major consideration in selecting the base system. In Eugene, the force main, biosolids lagoons, and application equipment to be used for dedicated land disposal of the liquid biosolids are also required for agricultural use. Trucks to transport liquid biosolids from the biosolids management site to agricultural sites are an additional expense for the secondary alternatives.

It is hoped that eventually all biosolids can be utilized on land. As indicated, however, on Table 13.5, full agricultural utilization of biosolids is estimated to be more costly than either of the pure disposal options. This is because more equipment is needed to transport biosolids to, and spread it on, the agricultural sites than is needed for the pure disposal options. Thus, any system that even partially incorporates agricultural utilization will be more costly than pure disposal options. This could change if the farmers can be persuaded to pay for the biosolids.

## 3. SIZING OF EQUIPMENT

Components should be sized to handle the most rigorous loading conditions they are expected to encounter. These loadings are usually not determined by applying steady-state models to peak plant loads. Because of storage and plant scheduling considerations, the rate of solids reaching any particular piece of equipment does not usually rise and fall in direct proportion to the rate of solids arriving at the plant headworks. If maximum solids loads at the headworks are twice the average value, it does not necessarily follow that at that instant maximum dewatering loads are twice the average dewatering load (3).

To pursue this further, consider the design of a centrifuge intended to dewater anaerobically digested primary and secondary biosolids at a small treatment plant. The plant is staffed on only one shift per day, 7 days per week. The digesters are complete-mix

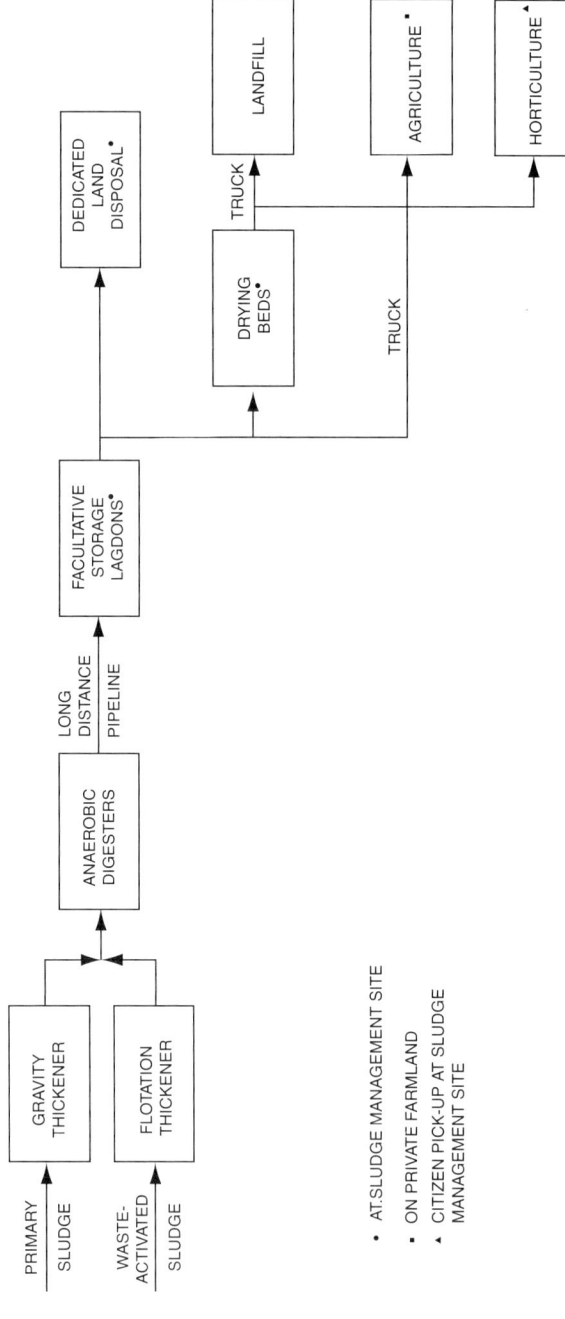

**Fig. 13.6.** Flowsheet for the Eugene-Springfield biosolids management system. *Source*: US EPA.

Table 13.5
Estimated costs of alternatives for Eugene-Springfield in 2007 US Dollars

| Biosolids form | Alternative | Total annual cost, million dollars |
|---|---|---|
| Liquid | Dedicated land disposal only | 2.16 |
|  | Agricultural utilization only | 3.21 |
| Dried | Landfill only | 2.39 |
|  | Agricultural utilization only | 2.77 |

Source: US EPA.

units equipped with floating covers. Because of the floating covers, digester volume can vary. Secondary biosolids are wasted from the activated sludge systems to a dissolved air flotation thickener prior to digestion whenever operators are available to operate the thickener.

Computation of the necessary centrifuge capacity requires analysis of both the load-dampening effect of the storage in the digesters and the plant operating schedule. During periods of peak plant solids loadings, loads to the dewatering units may be attenuated by storing portions of the peak loadings within the digester. This can be done by either of the following two mechanisms, acting either singly or in concert:

1. Digester volume is increased by allowing the digester floating cover to rise.
2. Solids are allowed to concentrate and thus accumulate within the digester.

The effect of both mechanisms 1 and 2 is storage within the digester of part of the load that would otherwise go to the centrifuge. Thus peak dewatering loads will not be 2.5 times the average when the peak solids mass withdrawn from primary and secondary sedimentation tanks is 2.5 times the average, but something less, for example, only 1.4 times the average value. The degree of load dampening is a direct function of the size and operating configuration of the digester (3).

Since the centrifuge only operates when attended, the "design" loading must account for this factor. Either the centrifuge must be capable of processing, during one shift, all the biosolids that must be extracted from the digester during the peak day (for example, 1.4 times average quantity), or the operators must dewater biosolids for longer than one shift per day. A judgment would be needed at this point whether to pay for increased equipment capacity or operator overtime to handle the peak loads. With no operator overtime, the "design" centrifuge capacity would have to be $1.4 \times 24/8 = 4.2$ times the average daily digested biosolids production to account for the effect of biosolids peaking, storage volume, and only one operation shift per day.

Note that the dissolved air flotation thickener (23) would need to be designed for $24/8 \times 2.5 = 7.5$ times the average daily rate of waste-activated sludge production if it is assumed no upstream storage is available for dampening thickener loadings, the thickener itself has no storage capacity, and the thickener is operated only one shift per day.

The foregoing example shows the influence of solids peaking, storage volume, and operating strategy on the selection of design loadings for a particular biosolids handling process. Several other factors are important in selecting the capacity a unit must have (3, 24):

1. Uncertainties: When systems are designed without the benefits of pilot or full-scale testing, actual biosolids quantities and characteristics as well as efficiencies of the biosolids handling system components may not be known with certainty. The degree and potential significance of the uncertainties must be considered when developing design criteria. This usually has the effect of introducing a safety factor into the design so that reliable performance can be obtained no matter what conditions are encountered in the full-scale application. The magnitude of the safety factor must be determined by the designer, based on his judgment and experience.
2. Equipment reliability: Greater capacity or parallel units must be specified if there is reason to believe that downtime for any particular units will be high.
3. Sensitivity of downstream components: If losses in efficiency of a particular biosolids handling component at peak loading conditions would cause problems for downstream processes, this upstream process should be designed conservatively. Conversely, if reduced efficiency could be tolerated, design need not be so conservative.

## 4. APPROACHES TO SIDESTREAM MANAGEMENT

Sidestreams are a major reason why solids treatment and disposal facilities often become trouble spots at wastewater treatment plants. Failure to account for these biosolids processing liquors in the wastewater treatment design can result in overloading of the treatment facility. It has been conventional practice to return biosolids sidestreams to the treatment plant at a convenient point, usually at the headworks, with no pretreatment and with little concern for its pollutant loadings. These sidestreams can increase the organic loading by 5% to 50%, depending on the type and number of solids treatment processes used (3).

The major objectives of this section are to describe the sidestreams produced by biosolids treatment processes, factors that affect sidestream quality, and options available to designers in managing the sidestreams.

### 4.1. Sidestream Production

Sidestreams are produced when wastewater solids are concentrated, and when water, usually plant effluent, is used to remove odors or particulate matter from flue gases, or to wash and transport debris from structures and equipment. Some sidestreams require special attention because of their impact on a wastewater treatment plant's efficiency (1).

Usually several sidestreams are produced at a particular plant. Figure 13.7 is a flow diagram showing eight wastewater solids sidestreams:

1. Screenings centrate
2. Grit separator overflow
3. Gravity thickener supernatant
4. Dissolved air flotation subnatant
5. Decantate following heat treatment

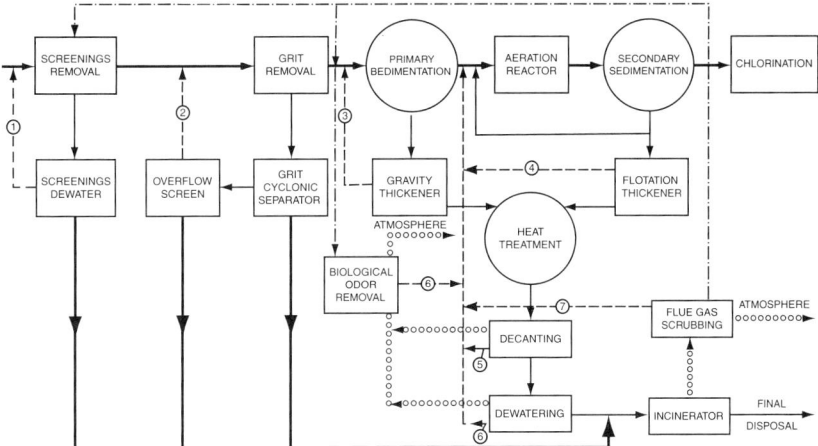

**Fig. 13.7.** Example of sidestream production. *Source*: US EPA.

6. Vacuum filter filtrate and wash water
7. Scrubber water from furnace flue gas cleanup
8. Overflow from biological odor removal system

This section devotes special attention to the most pronounced examples of the problem: anaerobic digester supernatant and thermal conditioning liquor. For additional information on production and treatment of wastewater solids sidestreams, several publications are available. *Municipal Wastewater Treatment Plant Sludge and Liquid Sidestreams* (25) deals with sidestreams from several solids handling and treatment processes. *Effects of Thermal Treatment of Sludge on Municipal Wastewater Treatment Costs* (26) describes the increased wastewater treatment capacity required by use of thermal conditioning.

### 4.2. Sidestream Quality and Potential Problems

The interrelationship between a wastewater treatment plant's effluent quality requirements and the processes used for solids treatment and disposal must be carefully scrutinized during planning and design to avoid problems caused by sidestreams. Generally, more sophisticated wastewater treatment plants produce greater quantities of more difficult to manage biosolids and chemical residues. When processed, these biosolids and residues may indirectly cause the production of sidestreams containing large quantities of soluble and colloidal materials including nutrients.

Sidestream quality from a specific process is strongly affected by upstream solids handling processes. Vacuum filter filtrate and washwater quality, for example, are determined by the upstream conditioning or stabilization process.

Sidestreams should be returned to points in the wastewater treatment process that will result in treatment of the sidestream and prevent nuisances and operational problems. The return points shown in Figure 13.7 comply with this requirement.

Runoff from biosolids composting areas and leachate from landfilling areas may pose a unique problem, since it may be difficult and costly to return them to the treatment plant if the landfill or composting site is remote from the treatment facility. Chian and DeWalle (27) extensively investigated the composition and treatment of sanitary landfill leachate, including anaerobic biological filtration, chemical precipitation, chemical oxidation, and activated carbon treatment. Data are also available on groundwater monitoring near wastewater biosolids or combined solid waste biosolids landfills (28–30). In addition, the U.S. Environmental Protection Agency's (US EPA) Process Design Manual, *Municipal Sludge Landfills*, discusses methods of handling leachate (31). In dry climates, leachate can often be recycled to the landfill site.

At Beltsville, Maryland, runoff from a composting site is stored in a pond and periodically sprayed on a forest floor. Monitoring wells have been installed, and no groundwater contamination has been detected. At Durham, New Hampshire, and Bangor, Maine, runoff is recycled back to the treatment works without pretreatment. At Sacramento, California, runoff has been returned from a dedicated land disposal site to the plant headworks and has been monitored for several parameters (32). It was found that runoff is polluted with constituents, particularly the first runoff following the spreading of biosolids. The concentrations of insoluble constituents such as heavy metals, however, were low.

### 4.3. General Approaches to Sidestream Problems

Several general approaches to preventing or solving problems that may result from sidestreams can be identified:

1. Modification of solids treatment and disposal systems to eliminate particular sidestreams
2. Modification of previous solids processing steps to improve sidestream quality from a particular solids treatment process
3. Changing the timing, return rate, or return point for reintroducing sidestreams into the wastewater treatment process
4. Modification of wastewater treatment facilities to accommodate sidestream loadings
5. Provision of separate sidestream treatment prior to return (33)

### 4.4. Elimination of Sidestream

Although not generally practical, specific situations arise in which it is possible to modify the solids treatment and disposal system and eliminate a troublesome sidestream. A particular case involves anaerobic digester supernatant, which has often been identified as a source of problems when a mixture of primary and waste-activated sludges is digested. Mignone (34–36) has pointed out that where mechanical dewatering follows anaerobic digestion, it would be beneficial to eliminate the secondary (unmixed) digester by converting it to a primary mode. There would be no variable supernatant stream, only a predictable filtrate or centrate stream of low solids content, which would be amenable to biological treatment.

### 4.5. Modification of Upstream Solids Processing Steps

Thickening of biosolids prior to anaerobic digestion by the use of gravity, flotation, or centrifugal thickeners can improve the quality and reduce the quantity of digester supernatant (37). Residence time in the digesters is increased, and smaller digesters can

Process Selection of Biosolids Management Systems

**Table 13.6**
**Effect of polymer on elutriation**

| Parameter | Before polymer use | After polymer use |
|---|---|---|
| Elutriate suspended solids, mg/L | 3,385 | 365 |
| Solids capture, percent | 65.1 | 95.3 |
| Underflow solids concentration, percent | 3.5 | 4.3 |

*Source*: US EPA.

be constructed. Liquid that would otherwise be produced by the secondary digester as supernatant is produced instead in the thickening step. Its quality will be better, and it will have a lesser impact when returned to the wastewater treatment facility.

Other digester operating parameters such as organic loading and temperature also affect supernatant quality. An increase in organic loading will generally result in poorer supernatant quality (38). Thermophilic digestion produces poorer supernatant quality than mesophilic digestion.

Substitution of an equivalent solids treatment process for another may also reduce sidestream problems. For example, substitution of chemical conditioning for elutriation or heat treatment can reduce the level of contaminants in sidestreams from subsequent dewatering steps.

The high colloidal content of elutriate has been successfully reduced in several instances by addition of chemicals, particularly polymer, to the elutriation process. The biosolids treatment system at the District of Columbia's Blue Plains plant (a 253-MGD, 11.1-m$^3$/s facility) consisted of gravity thickening, single-stage anaerobic digestion, elutriation of digested biosolids, chemical biosolids conditioning, and vacuum filtration. Large quantities of fines and activated sludge solids were recycled with the elutriate, and the primary clarifiers and aeration process could not accommodate them. Solids accumulated in the plant; upsets occurred in both the wastewater and biosolids treatment systems and it became necessary to temporarily discharge elutriate directly to the plant effluent. Eventually, addition of polyelectrolyte to the elutriation process, coupled with intensive effort on the part of plant staff to improve elutriation and vacuum filtration performance, resulted in a 90% solids capture through the two processes.

The Metropolitan Toronto main plant and the Richmond, California, facility experienced the same results as the Blue Plains plant. An example of successful use of polymer to improve elutriation is shown in Table 13.6 (39).

### *4.6. Change in Timing, Return Rate, or Return Point*

Sidestreams are normally returned to the wastewater treatment facilities at the plant headworks. In general, return of sidestreams to plant headworks should be at a low, steady rate rather than in slugs, since these are likely to cause upsets and overloads. In instances where there are high diurnal load fluctuations and the plant is approaching capacity, consideration should be given to returning sidestreams during off-peak hours, thus equalizing wastewater loadings. Adverse effects on primary treatment facilities, such as septicity, odors, and floating sludge, can be avoided by returning sidestreams to the biological treatment process influent. Alternatively, mixing supernatant with

**Table 13.7**
**Effect of supernatant return**

|  | Suspended solids, lb/d | | Phosphorus, lb/d | |
| --- | --- | --- | --- | --- |
| Point of measurement | With supernatant return[a] | Without supernatant return | With supernatant return[a] | Without supernatant return |
| Raw wastewater | 10,520 | 16,035 | 756 | 857 |
| To primary clarifiers | 36,801 | 15,969 | 1,304 | 914 |
| To secondary clarifiers | 15,306 | 9,501 | 991 | 803 |
| Final effluent | 3,467 | 2,836 | 435 | 500 |
| Primary sludge | 19,626 | 13,249 | 299 | 156 |
| Waste-activated sludge | 14,645 | 9,593 | 453 | 287 |

[a] Returned ahead of primary clarifiers.
*Source*: US EPA.

**Table 13.8**
**Increase in wastewater stream biological treatment capacity required to handle sidestreams from various solids treatment processes**

| Treatment process | Required capacity increase, % |
| --- | --- |
| Liquid biosolids to land | 0 |
| Raw biosolids to drying beds | 7 |
| Chemical conditioning and filter pressing | 6–11 |
| Rotoplug dewaterer | 10–30 |
| Digestion and drying beds | 0.6 |
| Digestion, chemical conditioning, and filter pressing | 5 |
| Digestion, chemical conditioning, and vacuum filtration | 4 |
| Heat treatment of raw biosolids | 30 |
| Heat treatment of digested biosolids | 7 |

*Source*: US EPA.

waste-activated sludge before returning it to the headworks may also aid in reducing odors because of the adsorptive nature of the activated sludge particles (85).

### 4.7. Modification of Wastewater Treatment Facilities

Liquid treatment facilities should be designed with the capacity to treat recycled sidestreams whenever the sidestream will contain significant concentrations of pollutants or have a large hydraulic impact. Table 13.7 shows an example of the effect of supernatant return on suspended solids and phosphorus loadings at an activated sludge plant using two-stage anaerobic digestion (40). Table 13.8 shows estimated increases in $BOD_5$ treatment capacity required by sidestreams from several biosolids treatment processes (41).

The Central Contra Costa Sanitary District Water Reclamation Plant, an advanced waste treatment facility, removes nutrients through chemical-primary treatment and nitrification-denitrification. Recycled sidestreams were taken into account in plant design by allowing for additional loads of 12% for $BOD_5$ and 21% for suspended solids. Recycled streams include gravity thickener overflow, centrate from a two-stage dewatering centrifuge, and drainwater from a wet scrubber.

Sidestreams may contain compounds that are difficult to remove in wastewater treatment facilities. For example, the nonbiodegradable chemical oxygen demand (COD) in heat treatment liquor will pass through normal secondary treatment unchanged. Digester and biosolids lagoon supernatant may contain high concentrations of nutrients. In some instances separate treatment may be appropriate. The Metropolitan Sanitary District of Greater Chicago has conducted several investigations involving nitrification and nitrogen removal from biosolids lagoon supernatant, using both attached growth and suspended growth biological processes (42–44).

In evaluating solids treatment and disposal processes, both the direct costs of the solids treatment and disposal systems and the indirect costs associated with return of sidestreams to the wastewater treatment facilities should be included in the cost-effectiveness analysis. The cost of handling the increased sidestream flows may or may not be negligible, but capital and operating expenses will surely increase as a result of the $BOD_5$ and suspended solids load of the returned stream. Major components of such indirect costs include increased aeration tank size and blower capacity (for diffused air-activated sludge systems), increased biosolids treatment capacity, increased power requirements for blowers, and increased labor for operating and maintaining more heavily loaded secondary treatment facilities. Additional costs will also be incurred if odor control facilities are required (85).

Indirect solids treatment costs for handling sidestreams will vary significantly. The indirect costs associated with heat treatment have been estimated as 20% of the direct thermal treatment costs. A report has been prepared describing the effects of biosolids heat treatment on overall wastewater treatment costs (26).

### 4.8. Separate Treatment of Sidestreams

Most sidestreams from properly operating solids treatment and disposal systems can be recycled to the wastewater treatment facilities without significant problems. In many cases, two-stage anaerobic digester supernatant return to the wastewater treatment facilities causes operating difficulties. Heat treatment is less widely used, but it results in conversion of some of the COD to the soluble form. Furthermore, a portion of the COD can be nonbiodegradable.

#### 4.8.1. Anaerobic Digester Supernatant

In most cases, $BOD_5$ and suspended solids are of concern, although under certain circumstances nitrogen and phosphorus removal may also be desirable. Anaerobic digester supernatant typical values are given as a part of the example in Figure 13.8. Table 13.9 lists possible treatment processes for each major constituent (45). Chemical treatment of digester supernatant has been studied for many years (46–48). Rudolfs and Fontenelli (46) studied coagulation using ferric chloride, lime, caustic soda, sulfuric acid, chlorine,

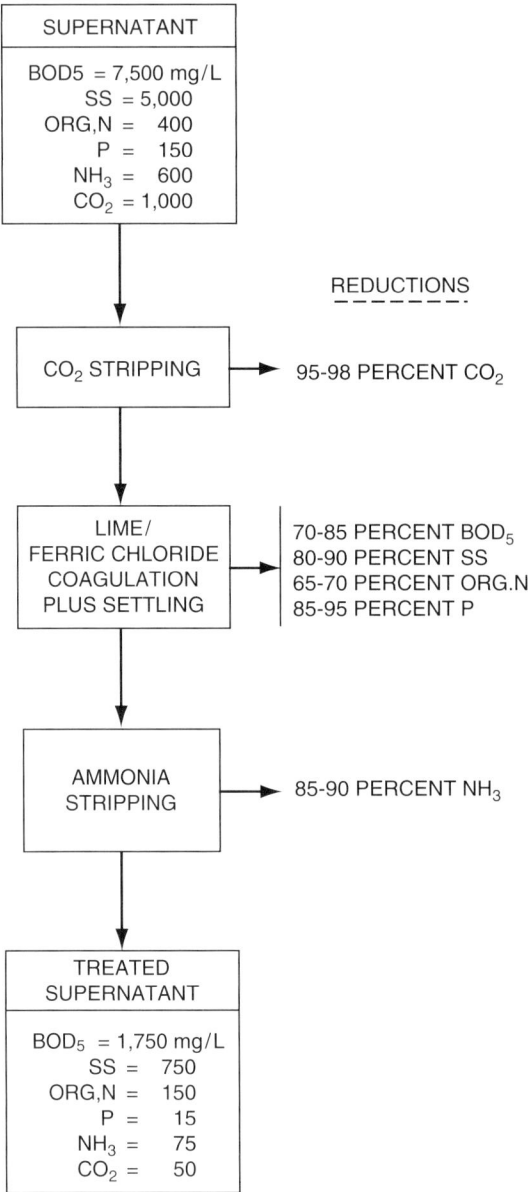

**Fig. 13.8.** Possible treatment scheme for anaerobic digester supernatant. *Source*: US EPA.

bentonite clay, and zeolite. It was found that a lime/ferric chloride combination gave the best results and 150 mg/L ferric chloride and 1200 mg/L lime reduced turbidity from 420 to 110 units.

The carbon dioxide in digester supernatant will react with the lime to form calcium carbonate precipitate. Lime requirements and the quantity of lime sludge produced can be reduced significantly by first air stripping carbon dioxide from the supernatant. This may also release odors, and for this reason its use should be approached with caution.

**Table 13.9**
**Possible digester supernatant treatment processes**

| Constituent | Processes |
|---|---|
| Suspended solids | Coagulation, settling, microstraining |
| $BOD_5$ | Removal with suspended solids, stripping of volatile acids, biological treatment, adsorption on activated carbon |
| Phosphorus | Removal with suspended solids, chemical precipitation, ion exchange |
| Nitrogen | Removal with suspended solids (limited), ammonia stripping, ion exchange |
| $CO_2$ | Lime addition, air stripping |

*Source*: US EPA.

Because lime raises the pH of the supernatant and under high pH conditions the ammonia molecule tends to be in the nondissociated form, ammonia stripping can be affected after coagulation. The relatively high temperature of digester supernatant also aids ammonia stripping for the same reason.

Figure 13.8 shows a possible treatment scheme for digester supernatant based principally on chemical coagulation (45). Also shown are probable removals and common influent and expected effluent concentrations. Straight aeration of digester supernatant at plant scale has also been attempted (37–50). Even where the supernatant after aeration was not settled prior to return and no discernible improvement in quality resulted, it was found that wastewater treatment operation improved, probably as a result of better settling in the primary clarifiers.

Biological filters, either aerobic or anaerobic, appear to be feasible methods of biologically treating digester supernatant. The Greater London Council studied aerobic biofilter treatment of supernatant liquor using coke as the filter medium (51). At a 1:1 dilution with clarified plant effluent, 85% to 90% $BOD_5$ removal and 60% ammonia removal were obtained.

Howe (47) suggested storage of digester supernatant in lagoons for long periods to reduce contaminant levels. In one experiment, a detention time of 60 days reduced $BOD_5$, suspended solids, color, and ammonia by about 85%; hydrogen sulfide was reduced by approximately 95%. Facultative biosolids lagoons designed for long-term storage have been found to reduce levels of all contaminants except ammonia.

*4.8.2. Thermal Conditioning Liquor*

Heat treated biosolids liquor, which is received as decantate and filtrate or centrate, contains high levels of soluble pollutants and a significant fraction of nonbiodegradable COD. The color level of the liquor may also be high, affecting the color of the final effluent (52). Furthermore, chlorination of effluent containing recycled heat treatment liquor may cause taste and odor problems if the receiving stream is used for drinking water (53).

Loll (54) has cited average $BOD_5$ loading increases of 7% to 15% and COD increases of 10% to 20% at wastewater facilities recycling untreated liquor (54). Recycle of heat treatment liquor at Colorado Springs, Colorado, caused the $BOD_5$ loading to be increased by 20% and the suspended solids load by 30% (52).

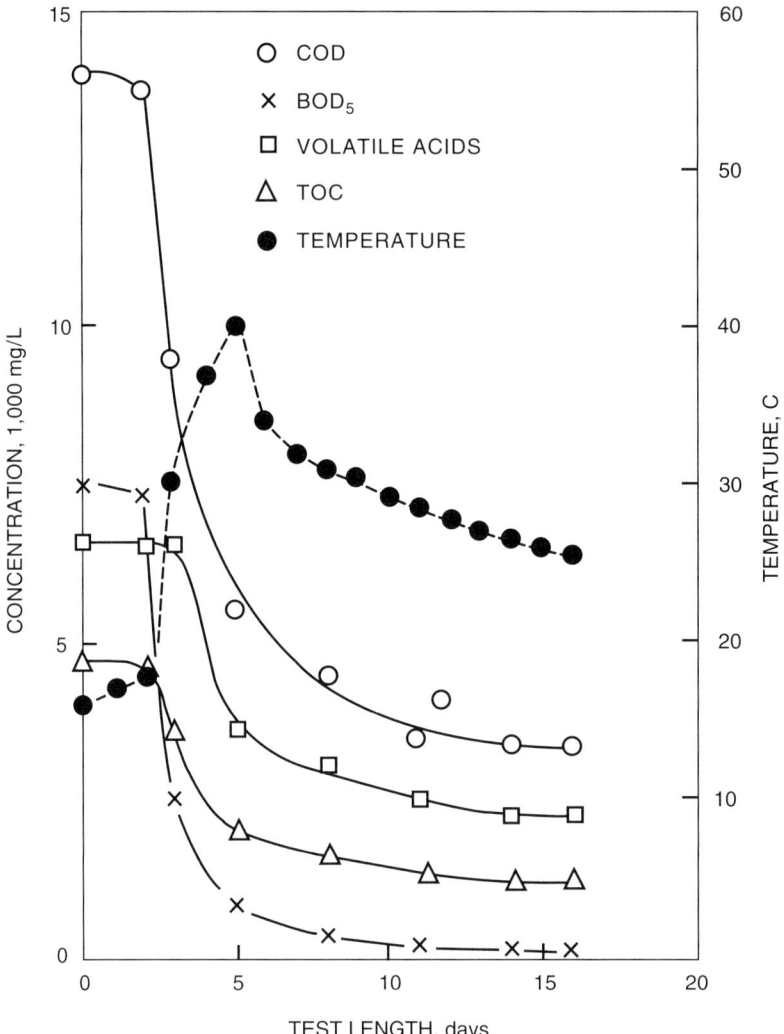

**Fig. 13.9.** Aerobic digestion of heat treatment, batch tests. *Source*: US EPA.

Trickling filters, activated sludge process, anaerobic biological filtration, and aerobic digestion have been used to treat the liquor. To reduce the nonbiodegradable COD, activated carbon has been used. Ozonation or chlorination can also be used to reduce COD levels.

Loll (54) has described experiments using autothermal thermophilic aerobic digestion of heat treatment liquors. Because the reactions are exothermic, the process is thermally self-supporting. Presented in Figure 13.9 are the results of batch aerobic digestion tests. Note that the temperature rose during the period of most rapid degradation. The results of continuous flow tests are presented in Table 13.10 at residence times of 5 and 10 days. The COD reduction is significantly less than the $BOD_5$ reduction, reflecting the nonbiodegradable character of a portion of the waste.

### Table 13.10
### Aerobic digestion of heat treatment liquor

| Parameter | Residence time, d | |
|---|---|---|
| | 5 | 10 |
| Temperature, °C | 38 | 34 |
| COD | | |
|   Influent, mg/L | 13,500 | 12,400 |
|   Effluent, mg/L | 4,100 | 3,800 |
|   Reduction, % | 66 | 71 |
| $BOD_5$ | | |
|   Influent, mg/L | 6,900 | 6,100 |
|   Effluent, mg/L | 510 | 250 |
|   Reduction, % | 94 | 96 |

*Source*: US EPA.

### Table 13.11
### Activated sludge treatment of thermal conditioning liquor

| Parameter | Aeration time, h | |
|---|---|---|
| | 21.8 | 40.9 |
| Temperature, °C | 33.4 | 31.7 |
| COD | | |
|   Influent, mg/L | 10,600 | 11,900 |
|   Effluent, mg/L | 4,300 | 2,000 |
|   Reduction, % | 59 | 83 |
| $BOD_5$ | | |
|   Influent, mg/L | 4,700 | 5,900 |
|   Effluent, mg/L | 400 | 110 |
|   Reduction, % | 91 | 98 |

*Source*: US EPA.

Erickson and Knopp (55) used the activated sludge process for heat treatment liquor. They reported a COD reduction of 83% and a $BOD_5$ reduction of 98% with an aeration time of 41 hours. Results are shown in Table 13.11.

Anaerobic biological filtration of heat treatment liquor has been tested for use at the City of Los Angeles Hyperion treatment plant (56). The waste-activated sludge treatment scheme is shown in Figure 13.10. The anaerobic filter, originally developed by Young and McCarty (57) is similar to the conventional aerobic trickling filter in that organisms are attached to the media surface and a short hydraulic detention time results. Advantages are that the production of methane can result in energy recovery and that no power is required for oxygen addition. Care must be taken, however, to avoid any plugging from periodic high suspended solids loadings. Results of a 2-month test are shown in

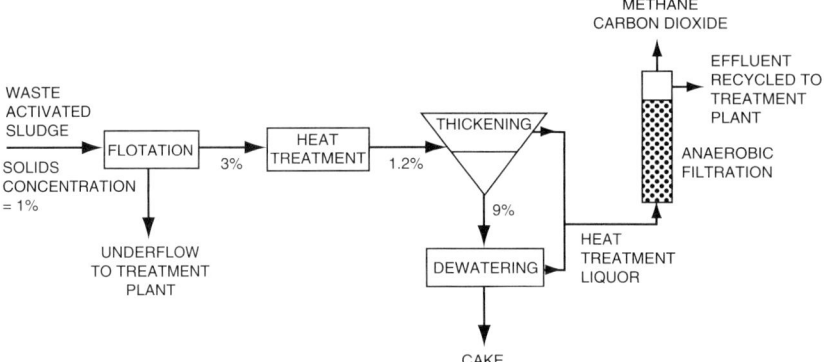

**Fig. 13.10.** Flow diagram, anaerobic filtration of heat treatment liquor. *Source*: US EPA.

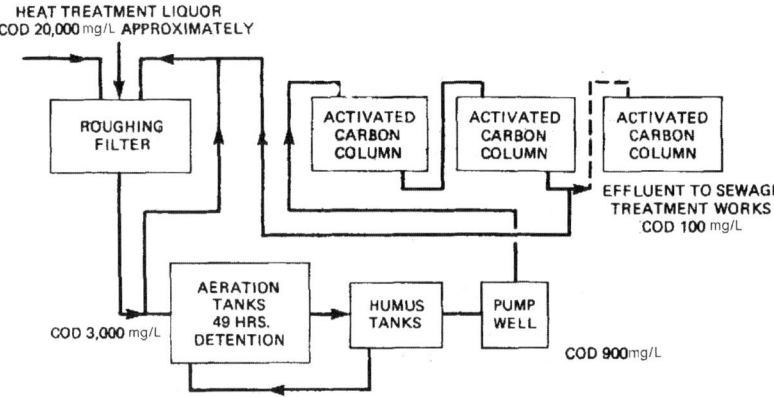

**Fig. 13.11.** Schematic diagram of plant for processing heat treatment liquor. *Source*: US EPA.

Table 13.12. At a hydraulic detention time of 2 days, $BOD_5$ and COD removals averaged 85% and 76%, respectively. This study concluded that detention time could be reduced to about 0.5 to 1.0 days without significant deterioration in performance. Other pilot scale tests on anaerobic filtration of heat treatment liquor have been conducted. One study reported COD removals of approximately 65% at detention times of 3.5 days and organic loadings of 125 lb COD/1000 ft$^3$/d (2.0 kg/m$^3$/d).

Figure 13.11 illustrates the treatment scheme used in a pilot study in Great Britain (53). The purpose of the study was to reduce the quantity of refractory organics entering the Thames River from treatment plants conditioning biosolids with heat treatment. The study was prompted by the fact that the Thames is used for water supply, and possible taste and odor problems would result from chlorinating the water; in addition, there was uncertainty about the exact composition and effects of the organics in the liquor. The process can reduce COD from 20,000 mg/L to about 100 mg/L or by approximately 99.5%.

**Table 13.12**
**Aerobic biological filtration of thermal condition liquor**

| Parameter | Value |
|---|---|
| Hydraulic detention time, d | 2.0 |
| Temperature, °C | 32 |
| COD | |
|   Influent, mg/L[a] | 9500 |
|   Effluent, mg/L | 2300 |
|   Reduction, % | 76 |
| $BOD_5$ | |
|   Influent, mg/L[a] | 3000 |
|   Effluent, mg/L | 450 |
|   Reduction, % | 85 |
| Suspended solids | |
|   Influent, mg/L[a] | 110 |
|   Effluent, mg/L | 100 |
| Total solids | |
|   Influent, mg/L[a] | 8800 |
|   Effluent, mg/L | 4900 |
| Volatile acids | |
|   Influent, mg/L[a] | 520 |
|   Effluent, mg/L | 300 |
| Alkalinity, as $CaCO_3$ | |
|   Influent, mg/L[a] | 2200[b] |
|   Effluent, mg/L | 3500 |
| pH | |
|   Influent[a] | 7.1[b] |
|   Effluent | 7.1 |

[a] Decant liquor.
[b] pH following thermal conditioning was approximately 5.5; 1,600 mg/L alkalinity added to influent for pH adjustment.
*Source*: US EPA.

## 5. CONTINGENCY PLANNING

### 5.1. Contingency Problems and Their Solutions

As indicated previously, flexibility to cope with unforeseen problems is highly desirable in any wastewater solids management system. Such problems and possible solutions include the following (3, 58, 59):

1. Equipment breakdowns: Downtime may be minimized by having maintenance people on call, by advance purchase of key spare parts, by providing parallel processing units, and by making use of storage.

2. Solids disposal problems: These may include closures of landfills, unwillingness of current users to further utilize biosolids, failure of a process to provide biosolids suitable for utilization, strikes by biosolids transporters, and inability to dispose of biosolids due to inclement weather. Disposal problems can be reduced by providing long-term storage and/or more than one disposal alternative.
3. Biosolids production greater than expected: In some instances this may be dealt with by operating for more hours per week than normal or by using chemicals to modify biosolids characteristics, thus increasing solids processing capacity.

Because of these factors, it is desirable to have more than one process for biosolids treatment and disposal. Often it is possible to add considerable flexibility with modest investment. Backup or alternative wastewater solids treatment processes often have higher operating costs per ton of biosolids processed than the primary processes. This is acceptable if the alternative process is not frequently needed and can be provided at minimum capital cost.

### 5.2. Example of Contingency Planning for Breakdowns

Assume the plant is a 10-MGD activated sludge facility with biosolids thickening, anaerobic digestion, and digested biosolids dewatering as shown in Figure 13.12. Pertinent design details include the following (3):

1. The waste-activated sludge (WAS) thickener can be operated with or without polymers. If polymers are used, more concentrated biosolids can be produced. The WAS can be diverted to the headworks if the WAS thickener is removed from service.
2. Two complete-mix digesters with floating covers are provided. Each digester has a net volume of 610,000 gal (2310 m$^3$) at minimum cover height. Net volume at maximum cover height is 740,000 gal (2803 m$^3$), thus total digester storage volume is 2(740,000 − 610,000) = 260,000 gal (984 m$^3$).
3. Two dewatering units are provided. Each unit, when fed at 90 gallons per minute (gpm) (40.8 m$^3$/h) can produce a 22% solids cake. When the dewatering units are fed at 110 gpm (49.9 m$^3$/h), a 17% solids cake is produced. The units are fed at 90 gpm (40.8 m$^3$/h) unless conditions dictate otherwise. The bulk density of each cake is 65.5 lb/ft$^3$ (1050 kg/m$^3$).
4. The cake is trucked to ultimate disposal. Each truck holds 16 yd$^3$ (12 m$^3$) of cake.
5. A dewatered biosolids storage area of capacity 750 yd$^3$ (574 m$^3$) is available.
6. Weekends are 2.7 days long (from 5 p.m. Friday to 8 a.m. Monday).

#### 5.2.1. Case A: All Units Available
1. Digester detention time $\frac{2(610,000) \text{ gal}}{(24,000+27,000) \text{ gpd}} = 24$ days.
2. Dewatering operation:
   a. Weekly biosolids feed = 7 (24,000 + 27,000 gpd) = 357,000 gal (1350 m$^3$).
   b. Hourly throughput = 2 × (90 gpm) (60 min /h) = 10,800 gal/h (40.8 m$^3$/h).
   c. Hours of operation per week = $\frac{357,000 \text{ gal}}{10,800 \text{ gal/hr}}$ = 33 h/week.
   d. 26.2 yd$^3$ (20.0 m$^3$) of 22% solids cake is produced each day.
3. If dewatering is not operated over the weekend, then 51,000 gpd (2.7 d) = 138,000 gal (522 m$^3$) of digested biosolids must be stored in the digesters during this period. Available storage that can be obtained by letting the floating cover rise is 260,000 gal (983 m$^3$). Therefore, digester storage capacity is adequate for weekend storage, including long (3.7 d) weekends.

## Process Selection of Biosolids Management Systems

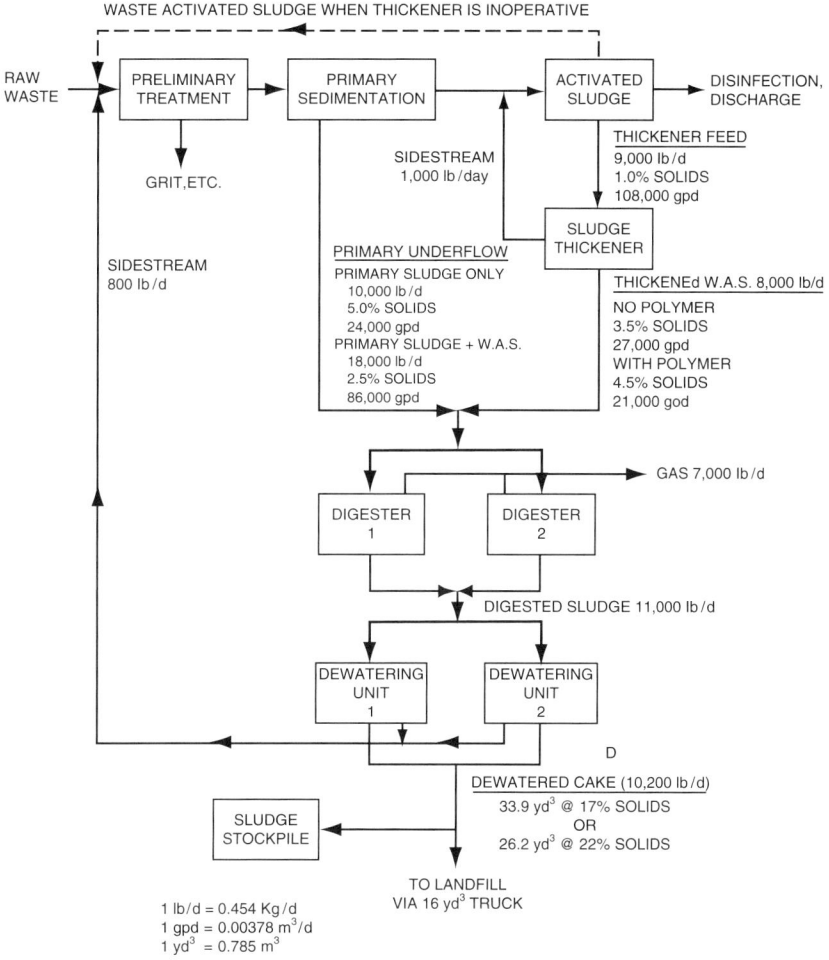

**Fig. 13.12.** Contingency planning example. *Source*: US EPA.

4. Truckloads required to haul dewatered cake $= 26.2\,\text{yd}^3/\text{d}$ divided by $16\,\text{yd}^3/\text{truck} = 1.6$ truckloads/d (11 per week).

In summary, the dewatering operation can be carried out in a normal 5-day, 8-h/d week. Time is available for start-up and shutdown and for providing good supervision. Digester detention time is more than adequate for good digestion.

### 5.2.2. Case B: Thickener Is Out of Service

All other units are available. Waste-activated sludge is diverted to the plant headworks and is subsequently removed in the primary sedimentation tank.

1. Digester detention time $= \frac{2(610,000)\,\text{gal}}{86,000\,\text{gpd}} = 14$ days; short, but tolerable.
2. Dewatering operation:
   a. Weekly biosolids feed $= 7\,(86,000\,\text{gpd}) = 602,000\,\text{gal}\,(2280\,\text{m}^3)$.

b. Hourly throughput. At 90 gal/min, throughput is 10,800 gal/h (40.8 m$^3$/h). At 110 gal/min, throughput is 13,200 gal/h (49.9 m$^3$/h).
   c. Operating hours required. At 90 gal/min (40.8 m$^3$/h), required hours per week = $\frac{602,000 \text{ gal}}{10,800 \text{ gph}}$ = 56 h/week. This requires substantial overtime or a second shift. At 110 gal/min (49.9 m$^3$/h), required operating hours = 602,000 gal/13,200 gph = 46 h/week. This reduces the amount of overtime required.
   d. If the dewatering units operate at 90 gal/min (40.8 m$^3$/h), 26.2 yd$^3$/d (20.0 m$^3$/d) of 22% cake is produced. Operation at 110 gal/min (49.9 m$^3$/h) produces 33.9 yd$^3$/d (25.9 m$^3$/d) of a 17% solids cake.
3. If dewatering units are not run on weekends, 86,000 gal/d × 2.7 d = 232,000 gal (878 m$^3$) must be stored in the digesters. Digester storage capacity is adequate for normal weekends, but not long weekends.
4. For 22% cake, 11 truckloads/week are required. For 17% cake, 15 truckloads/week are required.

In summary, loss of the thickener reduces digester detention time, increases required dewatering unit operating time and the amount of trucking required for disposal of cake. The operation can be managed, but with more difficulty. This example also illustrates the value of the thickener.

### 5.2.3. Case C: One Digester Is Out of Service

All other units are operating.

1. Digester detention time $\frac{610,000 \text{ gal}}{24,000+27,000 \text{ gpd}}$ = 12 days. This is only marginally adequate. By using polymers in the thickener, assume waste activated sludge thickness is increased from 3.5% to 4.5%. Detention time is increased to $\frac{610,000 \text{ gal}}{24,000+21,000 \text{ gpd}}$ = 14 days, still short, but an improvement.
2. Dewatering operation. This is not greatly affected by loss of the digester. It can still be operated with a single shift and a 22% cake can be produced.
3. Weekend storage. Without polymer addition to the thickener, required storage volume = 2.7 d × 51,000 gpd = 138,000 gal (522 m$^3$). One digester (130,000 gal or 492 m$^3$) has inadequate storage and a dewatering machine must be run part of the weekend. If polymer is used, required storage = 2.7 × 45,000 = 122,000 gal (462 m$^3$). One digester is marginally adequate for storage.
4. Eleven truckloads/week are required to transport the cake.

In summary loss of a digester can be compensated for by using polymer in the thickener.

### 5.2.4. Case D: One Dewatering Machine Is Out of Service

All other units are available.

1. Digestion is not affected.
2. Dewatering operation. Try the following alternatives:
   a. Feed rate 90 gal/min (40.8 m$^3$/h). Required operating time = $\frac{51,000 \text{ gpd}}{90 \text{ gpm} (60 \text{ min /hr})}$ = 9.4 h/d, every day, excluding start-up and shutdown time.
   b. Feed rate is 110 gal/min. Required operating time = $\frac{51,000 \text{ gpd}}{110 \text{ gpm} (60 \text{ min /hr})}$ = 7.8 h/d, every day, excluding start-up and shutdown time.

c. Try adding polymers to thickener and maintaining 110 gal/min feed rate to the dewatering units. Required operating time $= \frac{45,000 \text{ gpd}}{110 \text{ gpm} (60 \text{ min/hr})} = 6.8$ h/d, every day, excluding start-up and shutdown times.
3. Weekend digester storage is not an issue as dewatering units must be run 7 days a week.
4. Eleven truckloads are required to transport 22% cake, and 15 truckloads are required for 17% cake.

In summary, loss of one dewatering unit will require operation of the remaining unit for 7 days a week. Overtime costs will be high.

### 5.2.5. Case E: Truck Strike Lasting a Month

Assuming 22% cake, biosolids accumulate at about 25 yd$^3$/d (19 m$^3$/d). The biosolids storage area stockpile must, therefore, be able to store about 25 (30) = 750 yd$^3$ (570 m$^3$) of biosolids to avoid major problems due to the strike. Odors from the stockpile could be a problem.

Conclusion: The system as designed should be able to handle contingencies.

## 6. SITE VARIATIONS

Characteristics such as size and location of the plant and solids disposal sites strongly affect the nature and cost of treatment and disposal systems (3, 60, 61):

1. Disposal may often be accomplished on land, thus eliminating expensive dewatering, provided adequately sized sites are within reasonable distances from the treatment plant. However, dewatering is usually required if the amount of land available for biosolids disposal is limited or if the biosolids must be trucked long distances for disposal. Sufficient land also permits long-term storage in facultative lagoons, which can also provide some inexpensive disinfection.
2. Zoning regulations are different for different sites.
3. Locations near waterways and railroads provide opportunities for barge and rail transportation of biosolids and supplies.
4. Structures are less costly if foundation conditions are good. Quite often, however, wastewater treatment plants are located in valley bottoms, tidelands, or reclaimed landfills where expensive foundations are required.
5. Costs for labor, electricity, freight on chemicals, and trucking can vary markedly from one region to another.

Because of these variations, the best alternative for one site is often not the best at another site. Also, reported capital and operating costs from one site must be carefully adjusted before being used at another site.

## 7. ENERGY CONSERVATION

As fossil fuel supplies become scarcer and more expensive, energy conservation becomes increasingly important. The designer should employ energy-efficient processes and recover energy from biosolids and biosolids by-products, where practical. The following points should be considered in the design of energy-utilization processes (3, 62–65):

1. Energy from high-temperature sources is generally more useful than energy derived from low-temperature sources, since it can be put to a wider variety of uses.
2. The evaporation of water in dryers and furnaces consumes large amounts of energy. Such processes should therefore be provided with well-dewatered biosolids. Inert materials such as chemicals or ash used to condition biosolids for dewatering are also energy consumers.
3. Energy required for digestion and thermal conditioning is minimized where thickening is used to reduce the water content of process feed biosolids.
4. Trucking energy can be reduced if haul distances are short and the biosolids are well dewatered.
5. Energy is required for the manufacture and transportation of chemicals. Therefore, chemicals should be added in minimum amounts that are consistent with good operation. Whenever possible chemicals should be employed that require the least energy to produce and transport.
6. Costs saved by reducing peak energy demands can be substantial. In some instances, a treatment plant's electrical bills are largely determined by peak energy loadings, as opposed to total energy consumed. The designer should actively seek solutions to reduce peak energy demand. Energy recovered from biosolids and biosolids-derived fuel can be used for this purpose.
7. Motors should be accurately sized. Motors are most efficient when operated near capacity. However, motors in wastewater treatment plants are frequently operated far below capacity.
8. Where anaerobic digester gas is utilized, gas storage should be provided to minimize wastage.
9. Recycle loads from solids treatment processes should be minimized. Recycled loads increase the power and chemical requirements of wastewater treatment processes.

The designer should always keep in mind, however, that true economy is not found by minimizing specific uses of energy but by minimizing overall costs (65).

Energy costs for many of the biosolids treatment and disposal options are contained in the chapters describing those options (4). An US EPA publication contains more detailed guidance on making energy-effective analyses as well as a great deal of information on primary energy consumption, the electricity and fuel consumed directly at the treatment plant and secondary energy consumption, and the energy required to manufacture chemicals used in biosolids treatment (66).

## 8. COST-EFFECTIVE ANALYSES

One of the decisive factors in process selection is cost. Cost analyses must be carried out so that all alternatives are evaluated on an equivalent basis (67, 68). The US EPA has issued guidelines for cost-effective analyses (69). Monetary costs may be calculated in terms of present worth values or equivalent annual values over a defined planning period. Capital and operating and maintenance costs must be considered in the evaluation. Indirect costs should be included such as loss of property taxes when private land is acquired and incremental costs that the wastewater treatment facility must bear when sidestreams are returned to them. Credits for such items as crops and recovered energy should be taken where appropriate. The discount rate to be used in the analysis is established annually by the U.S. Water Resources Council. All construction cost data are referenced to a specific location and year using cost indices such as the Engineering News-Record construction cost index, the U.S. Army of Engineers Civil Works construction cost index (70), the US EPA sewage treatment plant index, or the US EPA sewer construction cost index. Inflation in costs and wages are not considered in the analyses, since it is assumed all prices will tend to change over time by the same percentage.

Cost-effective analysis for biosolids treatment and disposal systems has been discussed in somewhat greater detail in an US EPA publication (71). Present worth and equivalent annual value calculations are discussed elsewhere (72, 73).

## 9. CHECKLISTS

The following three checklists provide information a designer must have to design wastewater biosolids treatment and disposal systems:

1. Table 13.13 is a solids properties checklist that summarizes the required information concerning raw solids entering the solids treatment system and solids produced in the various processes and operations.
2. Table 13.14 is a process design checklist that describes information necessary to select and design biosolids treatment and disposal processes.
3. Table 13.15 is a public health and environmental impact checklist that summarizes key interactions that must be resolved between the proposed process and the surrounding environment.

Designers should refer frequently to these checklists to ensure that all relevant topics are given proper consideration during planning stages and system design. An extensive series of checklists dealing with wastewater solids management has also been prepared for the US EPA (71). The checklists are intended to serve as aids for the review of facility plans, for preparation of designs and specifications, and for the writing of operations and maintenance manuals.

**Table 13.13**
**Biosolids properties checklist**

| Item | Parameter |
|---|---|
| 1. | Origin and type |
| 2. | Quantity |
| 3. | Concentrations |
| 4. | Chemical composition and biological properties including biodegradability |
| 5. | Specific gravity |
| 6. | Rheological properties (e.g., viscosity) |
| 7. | Settling properties |
| 8. | Dewatering properties |
| 9. | Fuel value |
| 10. | Suitability for utilization or disposal without further processing |

*Source*: US EPA.

**Table 13.14**
**Process design checklist**

| Item | Parameter |
|---|---|
| 1. | Description of process |
| | Details of works, schematic drawing, logical location in overall sludge treatment flowsheet |
| 2. | Process theory |
| 3. | Current status |
| | Number of suppliers; usage in U.S.; good and bad experience and potential for avoiding problems; advantages and disadvantages with respect to competing processes |
| 4. | Design criteria |
| | Process loading (solids and hydraulic); pilot scale investigations (when to make them, methods, costs, limitations); special considerations (solids origin) |
| 5. | Instrumentation specific to the process |
| 6. | Operational considerations: flexibility |
| 7. | Energy impacts |
| | Primary and secondary requirements |
| | Potential for energy recovery |
| 8. | Actual performance data and case histories |
| 9. | Public health and environmental impacts |
| 10. | Solids production and properties |
| 11. | Sidestream production and properties |
| 12. | Cost information |
| | Construction/operation (tie to ENR and US EPA construction cost indexes); constraints (site-specific); break down costs by category (labor, electricity, etc.) so that adjustments can be made for different conditions |

ENR, Engineering News Record.
*Source*: US EPA.

**Table 13.15**
**Public health and environmental impact checklist**

| Item | Parameter |
|---|---|
| 1. | Control of vectors (bacteria, parasites, virus, flies, rats) |
| 2. | Odor |
| 3. | Air pollution |
| 4. | Groundwater contamination |
| 5. | Surface water contamination (by run-off) |
| 6. | Soils contamination |
| 7. | Land use |
| 8. | Socioeconomic |
| 9. | Utilization (biosolids or by-products used beneficially) |
| 10. | Occupational safety |
| 11. | Risk of accidents involving the public |
| 12. | Control of potentially hazardous substances |
| 13. | Effects on biota including transfer and accumulation of pollutants in the food chain |
| 14. | Use of material resources |

*Source*: US EPA.

## 10. U.S. PRACTICES IN MANAGING BIOSOLIDS

This section illustrates how the different wastewater treatment plants in the U.S. process their biosolids (74). Biosolids management has become the center of attention in the wastewater treatment industry. In examining the operation of many treatment plants, biosolids management is often found to command a major portion of the budget, whether it be for capital, labor, materials, or energy (75–79). The biosolids management problem is becoming more complicated as some of the more economical and conventional practices such as ocean disposal and incineration become environmentally unacceptable (16, 17, 74).

Usable data were obtained from 98 plants serving a total population of over 54 million people, or roughly one third of the sewered population of the U.S. The treatment plants are located throughout the country, but nearly 50% of the flow treated occurs in the heavily populated areas of the East and West Coasts. The sample group included a wide range of treatment processes, from virtually no treatment to tertiary treatment with chemical precipitation. Plant sizes range from $15\,m^3/d$ (4000 gpd) to over $3,000,000\,m^3/d$ (800 MGD), and provide a wide sampling of plant types.

The 98 plants for which usable data were available were placed into one or two of three classifications, as follows (3):

1. *Plants processing primary biosolids*: This category includes 40 plants that process only primary biosolids and one plant that processes primary biosolids independent of secondary biosolids (also included in the secondary category).
2. *Plants processing secondary biosolids*: Of the 21 plants in this category 14 preprocess secondary biosolids before combining with primary biosolids for further dewatering (also included in the combined category), six plants process only secondary biosolids, and one plant processes secondary biosolids independent of primary biosolids (also included in the primary category).
3. *Plants processing a combined mixture of primary and secondary biosolids*: This category includes 51 plants with 37 mixing primary and secondary biosolids before processing and 14 plants preprocessing secondary biosolids before combining with primary for final processing (also included in the secondary category).

This distinction between plant types allowed the development of Figures 13.13, 13.14, and 13.15, which graphically show the use of the different types of equipment for handling the different types of biosolids. None of the plants handles inorganic biosolids from chemical physical treatment. The amount of biosolids processed by each method is partially indicated by the size of the process representations included in the figures. The number of plants using each train of processes is also included on these diagrams. Since some plants truck partially processed biosolids to other plants, or use more than one method to process biosolids, the number of plants may change along a process branch line.

### 10.1. Primary Biosolids Processing Trains

Forty-one plants process primary biosolids only or process it separately from secondary biosolids. These are shown in Figure 13.13. The plants range in size from $295\,m^3/d$ (0.078 MGD) to over $3 \times 10^6\,m^3/d$ (800 MGD). The total average wastewater flow treated by this group is $12.3 \times 106\,m^3/d$ (3260 MGD). The reported

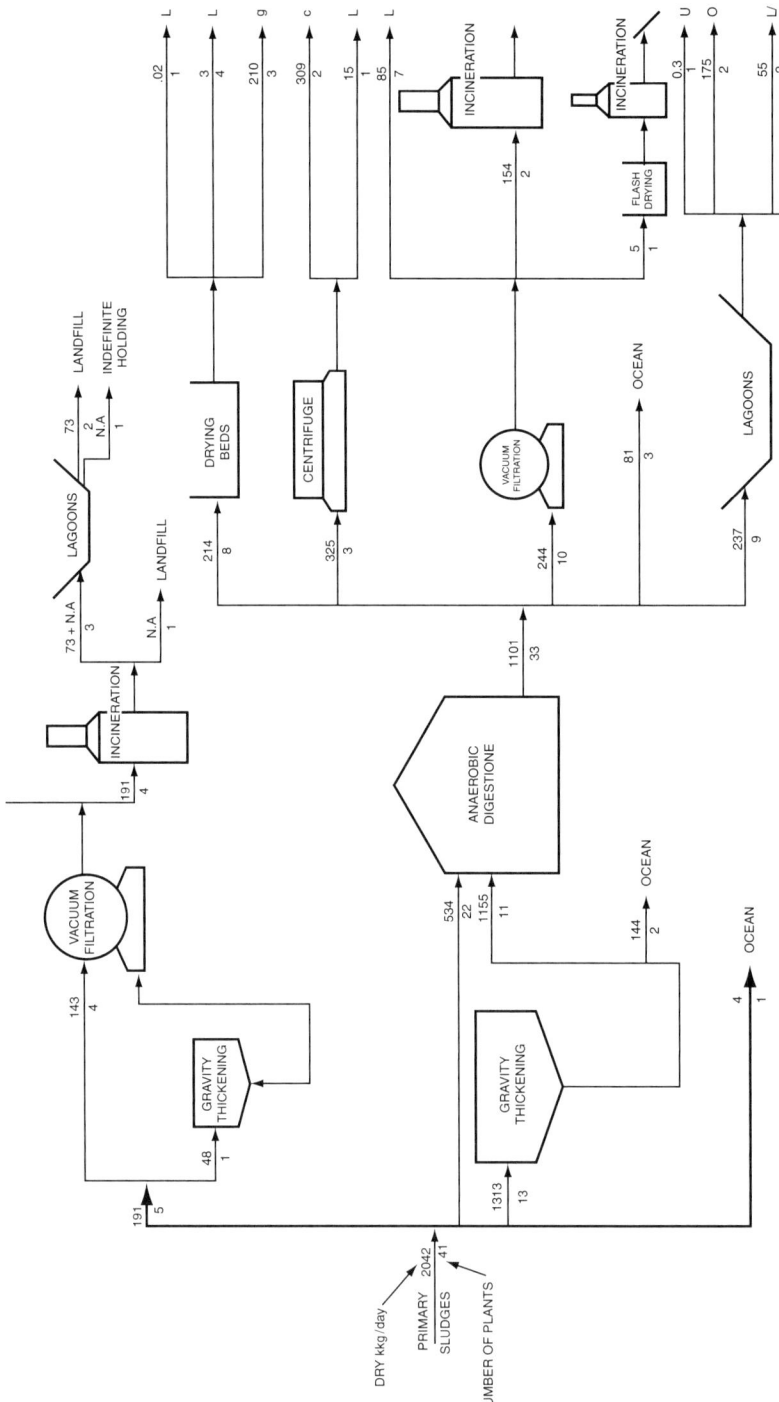

**Fig. 13.13.** Use of processes for primary biosolids management. *Source*: US EPA.

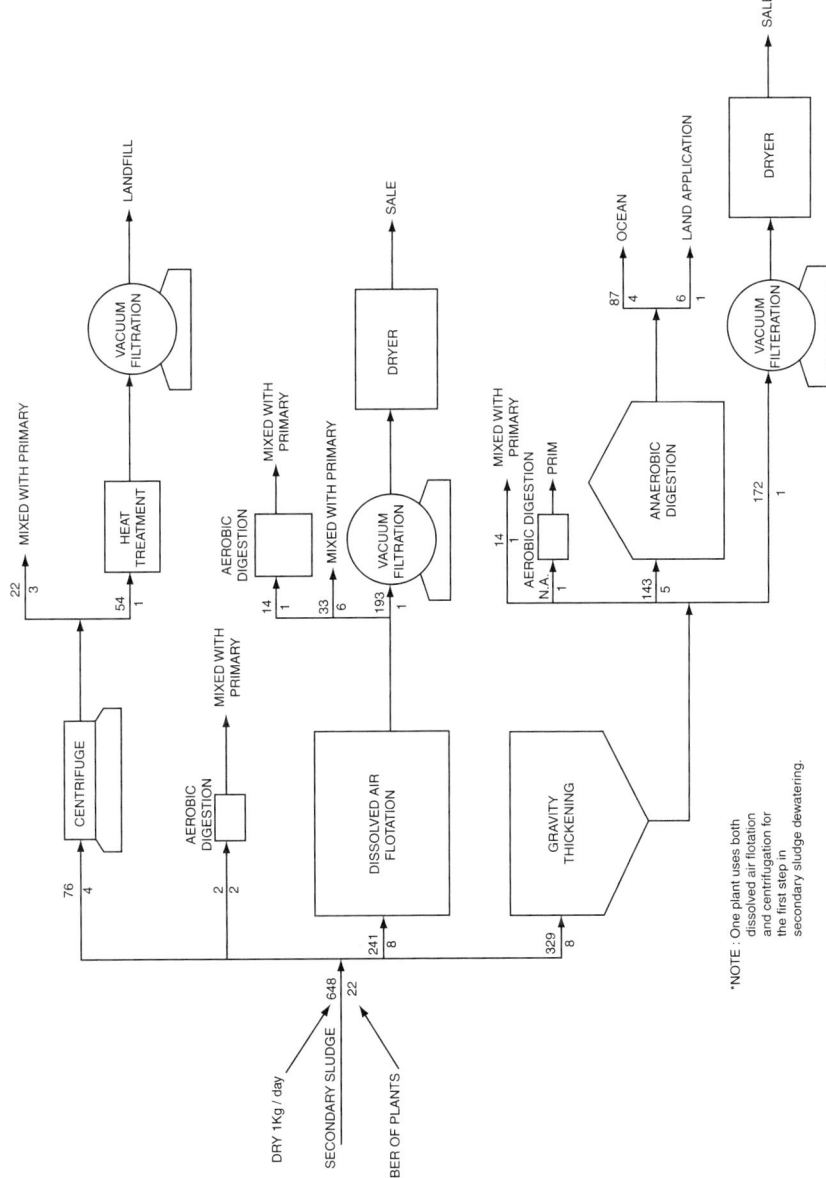

**Fig. 13.14.** Use of processes for secondary biosolids management. *Source*: US EPA.

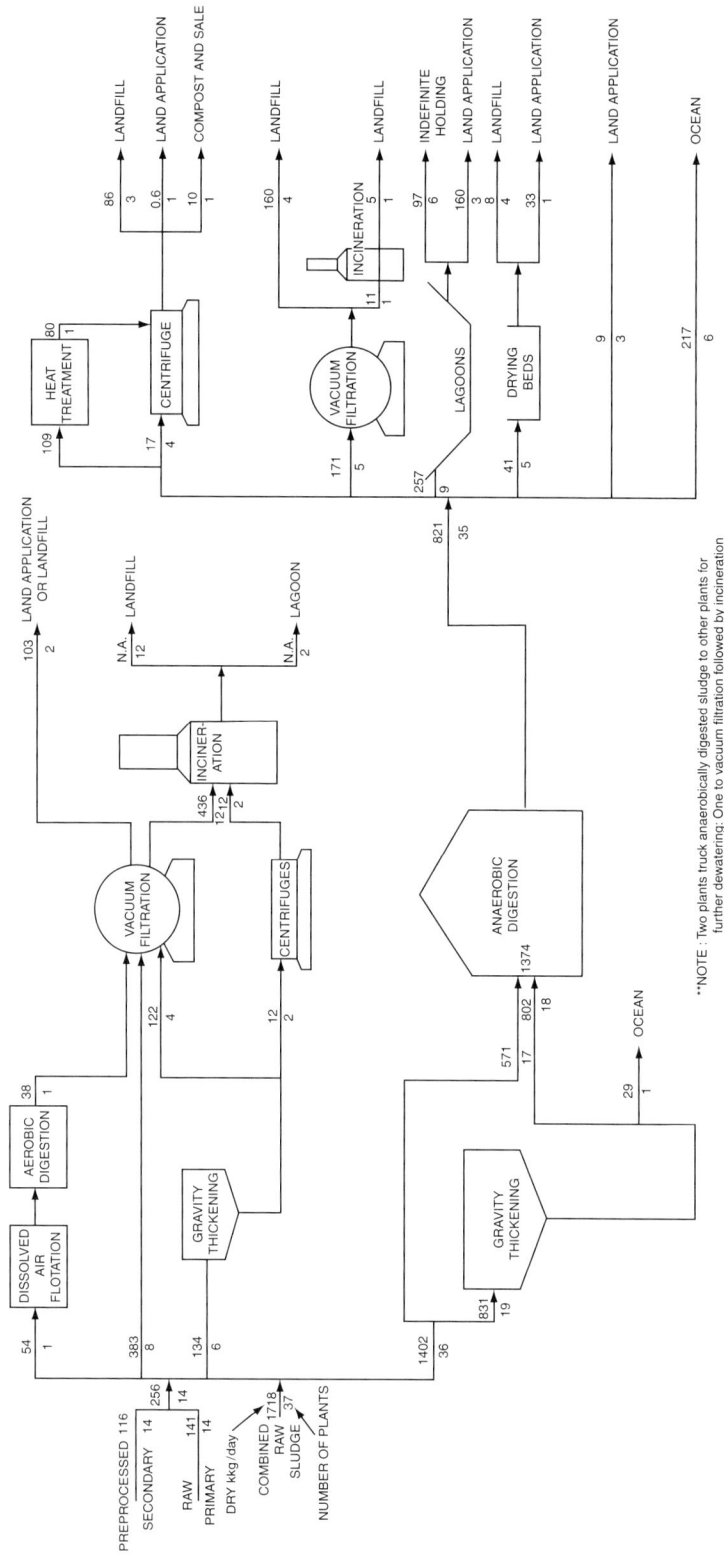

**Fig. 13.15.** Use of processes for combined biosolids management. *Source*: US EPA.

### Table 13.16
### Primary biosolids processing methods

| Biosolids treatment process | No. of plants | %* | Ton/d | ton/d | %* |
|---|---|---|---|---|---|
| Anaerobic digestion | 22** | 54 | 534 | 589 | 26 |
| 1. Drying beds | 7 | 17 | 5 | 6 | 1 |
| 2. Vacuum filtration | 6 | 15 | 66 | 73 | 3 |
|    a. Disposal | 4 | 10 | 47 | 52 | 2 |
|    b. Incineration | 2 | 5 | 19 | 21 | 1 |
| 3. Lagoons | 4 | 10 | 8 | 9 | 1 |
| 4. Centrifuge | 1 | 2 | 227 | 250 | 11 |
| 5. Ocean | 1 | 2 | 18 | 20 | 1 |
| Gravity thickening | 14 | 34 | 1367 | 1507 | 67 |
| 1. Anaerobic digestion | 11*** | 27 | 1155 | 1273 | 56 |
|    a. Vacuum filters | 4 | 10 | 178 | 196 | 9 |
|       Disposal | 3 | 7 | 38 | 42 | 2 |
|       Incineration | 1 | 2 | 140 | 154 | 7 |
|    b. Lagoons | 3 | 7 | 229 | 252 | 11 |
|    c. Centrifuge | 2 | 5 | 98 | 108 | 5 |
|    d. Ocean | 2 | 5 | 63 | 69 | 3 |
|    e. Drying beds | 1 | 2 | 209 | 230 | 10 |
| 2. Ocean | 2 | 5 | 164 | 180 | 8 |
| 3. Vacuum filters—Incineration | 1 | 2 | 48 | 53 | 2 |
| Vacuum filters | 4 | 10 | 143 | 157 | 7 |
| 1. Incineration | 3 | 7 | 142 | 156 | 7 |
| 2. Disposal | 1 | 2 | 1 | 1 | 1 |
| Ocean discharge | 1 | 1 | 4 | 4 | 1 |

   * As a percentage of primary plants or primary sludge.
  ** Three plants truck anaerobically digested sludge to a fourth plant for vacuum filtering.
*** One plant uses both vacuum filters and centrifuges to dewater anaerobically digested sludge.
*Source*: US EPA.

quantities of raw primary biosolids processed daily ranges from 0.01 to 364 Ton/d (0.01 to 400 ton/d), totaling 2048 Ton/d (2260 ton/d). This represents over 160 mg/L of suspended solids removed from the flow of $12.3 \times 10^6 \, m^3/d$ (3260 MGD).

Table 13.16 is a condensation of much of the information in Figure 13.13, which tabulates the dewatering methods used and indicates the popularity of each flow scheme. Figure 13.13 and Table 13.16 both show the popularity of gravity thickening and anaerobic digestion for primary biosolids, but no particular preference was evident for a dewatering method. Filters, centrifuges, drying beds, and lagoons take similar quantities of primary biosolids. Of the four categories, the largest dry weight quantity of biosolids is dewatered by centrifuge, but at only three plants. This indicates a greater use of centrifuges among larger plants. Three plants truck anaerobically digested biosolids to a fourth plant for filtering, and one plant uses both filters and centrifuges before landfilling.

Within the primary biosolids group, biosolids that are neither anaerobically digested nor discharged raw are dewatered using filters. Five plants process 191 Ton/d (211 tons/d) in this manner. Only one such installation uses a separate gravity thickener before filtration. Four of the five plants use incineration after filtration to reduce the bulk before landfilling, in some cases preceded by interim ash lagooning. All five of these plants are located in large midwestern cities, four of which have populations of over 300,000.

Anaerobic digestion, the process most frequently used to handle primary biosolids, is used at 33 of the plants. Eighty-two percent of the separately processed primary biosolids are handled in this manner. The average total solids reduction through this process is 35% for those plants reporting solids values both into and out of their digesters. Gravity thickeners are used to reduce the volumetric load to some of the digesters, most notably those at larger installations.

Many different processes were reported for dewatering anaerobically digested biosolids before disposal or use. Many of the smaller plants use drying beds. Filters are used at 10 plants, with three of these incinerating the resultant filter cake. The other plants truck the dewatered cake to landfill. Seven of the plants store the stabilized biosolids in lagoons either temporarily or indefinitely. Those plants that must periodically remove these biosolids do so to land or landfill.

## *10.2. Secondary Biosolids Processing Trains*

Those plants processing secondary biosolids alone or preprocessing it before combining it with primary biosolids are in the secondary plant category. Twenty-one of the 98 plants belong to this category. One plant processes secondary biosolids independently of the primary biosolids, and six are contact stabilization plants, with only secondary biosolids. The average daily flows within this category ranged from $6400 \, m^3/d$ (1.7 MGD) to $3.2 \times 10^6 \, m^3/d$ (847 MGD) with a total flow of $7 \times 10^6 \, m^3/d$ (1866 MGD) among all 21 plants. The quantity of secondary biosolids processed at each plant ranged from 0.39 Ton/d (0.43 ton/d) to 164 Ton/d (181 ton/d) for a total of 648 Ton/d (714 ton/d) for all of the plants. This total figure represents over 90 mg/L of secondary biosolids removed from the wastewater treated.

The manner in which secondary biosolids are processed varies widely. For this reason, it was difficult to arrange the flow diagram for these plants around a few central processes. Instead, Figure 13.14 has been arranged around that process by which the secondary biosolids is first handled. Table 13.17 is a summary of the information contained in Figure 13.14.

The greater use of thickening processes before dewatering or stabilization in this category reflects the more dilute nature of secondary biosolids. Four plants use centrifuges to thicken secondary biosolids before combining it with primary biosolids or further dewatering the biosolids. Dissolved air flotation is also used as a preliminary thickening process. Six of the eight plants using flotation mix the thickened secondary biosolids with primary biosolids following flotation. One large municipality uses a filter to dewater flotation thickened biosolids, dryers to further dewater the cake, and then sells the dried material. Gravity thickeners are used by eight plants, most frequently before anaerobic digestion for contact stabilization biosolids. In a process arrangement similar to the one

**Table 13.17**
**Secondary biosolids processing methods**

| Biosolids treatment process | No. of plants | %* | Ton/d | ton/d | %* |
|---|---|---|---|---|---|
| Gravity thickening | 8 | 38 | 329 | 363 | 51 |
| 1. Anaerobic digestion | 5 | 24 | 143 | 158 | 22 |
| 2. Vacuum filter-dryer-sale | 1 | 5 | 172 | 190 | 27 |
| 3. Mixed with primary | 1 | 5 | 14 | 15 | 2 |
| 4. Aerobic dig.-mixed w/primary | 1 | 5 | n.a. | – | – |
| Dissolved air flotation | 8** | 38 | 241 | 266 | 37 |
| 1. Mixed with primary | 6 | 29 | 63 | 69 | 10 |
| 2. Vacuum filter-dryer-sale | 1 | 5 | 163 | 180 | 25 |
| 3. Aerobic dig.-mixed w/primary | 1 | 5 | 14 | 15 | 2 |
| Centrifuge | 4** | 19 | 76 | 84 | 12 |
| 1. Mixed with primary | 3 | 14 | 22 | 24 | 3 |
| 2. Heat treatment—vacuum filter | 1 | 5 | 54 | 60 | 8 |
| Aerobic digestion | 2 | 10 | 2 | 2 | 1 |

\* As a percentage of secondary plants or secondary sludge.
\*\* One plant uses dissolved air flotation and centrifugation for the first step in secondary sludge dewatering.
*Source*: US EPA.

mentioned above, gravity thickening is used rather than dissolved air flotation, followed by vacuum filtration, drying, and sale as fertilizer/soil conditioner. One plant uses both dissolved air flotation and centrifugation for the first step in secondary biosolids dewatering.

### *10.3. Combined Biosolids Processing Trains*

The majority of the plants within the sample group have both primary and secondary biosolids to dispose of, and combine the two at some point within their processing scheme. Fifty-one plants fit this category, ranging in average daily flow from $3800 \, m^3/d$ (1.0 MGD) to $1.32 \times 10^6 \, m^3/d$ (348 MGD), and treat a total average daily flow of $10.7 \times 10^6 \, m^3/d$ (2831 MGD). Raw biosolids production ranges from 0.5 to 250 Ton/d (0.5 to 274 ton/d), totaling 1972 Ton/d (2174 ton/d) for all 51 of these plants. This represents an average removal of over 180 mg/L of solids for these plants.

As with uncombined primary biosolids, the most common process within the combined category is anaerobic digestion as shown in Figure 13.15 and Table 13.18. A total of 35 plants digest 1374 Ton/d (1515 ton/d) of combined biosolids. Gravity thickening preceded digestion in more than half of these plants, indicating a preference among larger plants for using separate gravity thickeners for combined biosolids. The total solids reduction for plants reporting influent and effluent solids from the digesters was 40%.

As shown on Figure 13.15, many methods are used to dewater and dispose of anaerobically digested combined biosolids. Five installations reported using centrifuges before land application, landfilling, or composting the dewatered biosolids. An equal number of plants use filters, and one plant incinerates the filtered cake. Fourteen plants use lagoons

**Table 13.18**
**Combined biosolids processing methods**

| Biosolids treatment process | No. of plants | %* | Ton/d | ton/d | %* |
|---|---|---|---|---|---|
| Gravity thickening | 25 | 49 | 965 | 1064 | 49 |
|   1. Anaerobic digestion | 18 | 35 | 803 | 885 | 41 |
|     a. Lagoons | 6 | 12 | 208 | 229 | 11 |
|     b. Disposal | 5 | 10 | 59 | 66 | 3 |
|     c. Vacuum filter | 3 | 6 | 78 | 86 | 4 |
|       Disposal | 2 | 4 | 67 | 74 | 3 |
|       Incineration | 1 | 2 | 11 | 12 | 1 |
|     d. Drying beds | 3 | 6 | 37 | 41 | 2 |
|     e. Heat treat—centrifuge | 1 | 2 | 109 | 120 | 6 |
|   2. Vacuum filters | 4*** | 8 | 122 | 135 | 6 |
|     a. Incineration | 4 | 8 | 112 | 123 | 6 |
|     b. Disposal | 1 | 2 | 9 | 10 | 1 |
|   3. Centrifuges—incineration | 2 | 4 | 13 | 14 | 1 |
|   4. Ocean | 1 | 2 | 28 | 31 | 1 |
| Anaerobic digestion | 17** | 33 | 571 | 630 | 29 |
|   1. Disposal | 4 | 8 | 166 | 183 | 8 |
|   2. Centrifuge | 4 | 8 | 17 | 19 | 1 |
|   3. Lagoons | 3 | 6 | 49 | 54 | 2 |
|   4. Vacuum filters | 2 | 4 | 93 | 103 | 5 |
|   5. Drying beds | 2 | 4 | 3 | 3 | 1 |
| Vacuum filters | 8 | 16 | 383 | 422 | 19 |
|   1. Incineration | 7 | 14 | 288 | 318 | 15 |
|   2. Disposal | 1 | 2 | 94 | 104 | 5 |
| Flotation–aerobic digestion–vacuum filters–incinceration | 1 | 2 | 54 | 60 | 3 |

   * As a percentage of combined plants or combined sludges.
   ** Two plants truck anaerobically digested sludge to other plants for further dewatering: one to vacuum filtration followed by incineration and the other to a plant not shown.
   *** One plant disposes of sludge by landfilling and incineration.
   *Source*: US EPA.

or drying beds to concentrate a total of almost 300 Ton/d (331 ton/d) before landfill or land application. Three of the smaller plants use land application as a means of directly disposing of such digested biosolids.

Filtration is the second most commonly used process to handle raw combined biosolids. Four of the 13 plants that handle biosolids in this manner use gravity thickeners to increase feed solids to the filters. One plant uses flotation thickening followed by aerobic digestion to increase dewaterability and decrease the loading on the filters. Of the 13 plants using filters, 12 incinerate the filter cake, and either lagoon or landfill the resultant ash.

Two plants use centrifuges to dewater raw combined biosolids. Both of these plants use gravity thickening before centrifuging, then incinerate the centrifuged cake.

**Table 13.19**
**Summary of types of treatment processes**

| Type of equipment | No. of plants | % of 98 plants | Ton/d | ton/d |
|---|---|---|---|---|
| Anaerobic digestion | 73 | 74 | 3206 | 3535 |
| Gravity thickening | 47 | 48 | 2661 | 2934 |
| Vacuum filtration | 36 | 37 | 1538 | 1695 |
| Incineration | 22 | 22 | 812 | 895 |
| Lagoons | 22 | 22 | 567 | 625 |
| Centrifuge | 14 | 14 | 540 | 595 |
| Dryers | 3 | 3 | 340 | 375 |
| Dissolved air flotation | 9 | 9 | 295 | 325 |
| Drying beds | 13 | 13 | 255 | 281 |
| Heat treatment | 2 | 2 | 163 | 180 |
| Aerobic digestion | 5 | 5 | 70 | 77 |

*Source*: US EPA.

### 10.4. Types of Unit Processes

The preceding sections summarized the information about the different trains used by wastewater treatment plants to process various types of biosolids. This section summarizes the data on the types of processes used within these systems (4, 75–85). The numbers of plants using each of these unit processes are tabulated in Table 13.19. The most commonly used types of processes in decreasing order of frequency, are anaerobic digestion, gravity thickening, and filtration (3).

#### 10.4.1. Anaerobic Digestion

Within the sample group, 73 plants use anaerobic digestion for stabilization and volume reduction of wastewater biosolids. The breakdown of biosolids types so processed is as follows:

| | |
|---|---|
| Combined: | 35 plants |
| Primary: | 33 plants |
| Secondary: | 5 plants |

The total feed of solids was 3206 Ton/d (3535 ton/d). The average solids content of the feed biosolids was 5.5%, which indicates that over 58,000 m$^3$/d (15 MGD) of biosolids are anaerobically digested within this sample group alone.

The 12 plants using anaerobic digestion provided detailed information about the operation of their digesters. The group was evenly divided between single- and multiple-stage digesters. All use internal gas mixers, and seven use external heat exchangers. The remainder use steam injection or internal hot water circulation pipes for heating. The average volatile suspended solids reduction through the digesters was reported as 50%. The detention times reported ranged from 15 to 65 days, and averaged 30 days. The cleaning schedules for digesters ranged from none (8-year-old digesters) to every 2 to 3 years. The addition of pretreatment by major industrial users was cited by some members as a great aid in extending the time between cleanings.

Of the 73 plants using anaerobic digestion, data on gas production, utilization, and wastage were available from 52. The quantity of gas produced per pound of volatile matter destroyed could not be determined from the data; 152 m$^3$/Ton (4880 ft$^3$/ton) were produced, however. The total quantity of digester gas produced was indicated as 178 × 10$^6$ m$^3$/yr (6.3 × 10$^9$ ft$^3$/yr). Within the group, the number of plants practicing digester gas energy recovery is as follows:

| | |
|---|---|
| Plants recovering energy as usable heat | 30 |
| Plants recovering energy as electricity | 12 |
| Plants recovering energy as mechanical energy | 4 |

One facility with extensive digester operations indicated that over half of the daily digester gas production is sold. Other facilities indicated plans to sell digester gas in the future. Two agencies specifically indicated that digester gas is not wasted.

### 10.4.2. Filtration

Thirty-six plants use filters to dewater raw and digested biosolids. Of these plants, 19 process raw biosolids, 16 process digested biosolids, and one processes heat-treated biosolids with vacuum filters. Twenty incinerate the cake and 17 landfill it or dispose of it in some other manner. A total of 1538 Ton/d (1695 ton/d) are dewatered on filters within this group.

Information on the common operational characteristics of filters was provided by 16 plants. Within this group, eight plants are equipped with belt type filters, and an equal number have coil spring media filters. A preference for either type for either raw or digested biosolids was not apparent. The average solids content for raw biosolids cake was 29%. Digested biosolids yielded a wetter cake with 21% solids. Six plants rely solely on polymer conditioners, three on lime and ferric chloride, and three reported using a combination of all three chemicals. Personnel at one plant stated that polymer works well in winter months, but becomes very difficult to control in the warmer months.

### 10.4.3. Centrifuges

Fourteen plants use centrifuges for dewatering or thickening, eight for digested and six for raw biosolids. Most of the centrifuges are the solid bowl type, except for two plants that use disk-nozzle type machines. Both facilities using the disk-nozzle centrifuges reported concentration to 4% to 4.5% solids with waste activated sludge. Nozzle wear and plugging were cited as difficulties with the machines, and one plant noted that fine particles buildup was a potential problem.

Seven plants centrifuging primary biosolids provided information on solid bowl centrifuge operations. The average cake dryness was 22%, with feed solids averaging 5.7%. Four of these plants use polymers. Most of the authorities were pleased with their centrifuges, but stated that maintenance was a high cost item for the process.

### 10.4.4. Incineration

Twenty-two plants incinerate 812 Ton/d (853 ton/d) of primary and combined solids. The resultant ash quantities were not reported in many cases. Twenty of the plants use filters before incineration, with two using centrifuges. Only four plants incinerate anaerobically digested biosolids.

*10.4.5. Other Processes*

Few process or operating details were available for the other biosolids handling methods and equipment, such as heat treatment and dryers. As would be expected, there is little operational information available for simpler operations such as lagooning and sand bed drying.

**REFERENCES**

1. L. McGovern, P. Shoenberger, and W. Won, Case study: management of wastewater recycle streams at the west basin water recycling plant, *Biosolids 2001: Building Public Support*, February (2001).
2. M. Mickley, Directions in management of membrane side streams, *Joint Residuals and Biosolids Management*, February (2003).
3. U.S. Environmental Protection Agency, Process Design Manual: Sludge Treatment and Disposal, US EPA 625/1-79-011, Center for Environmental Research Information Technology Transfer, Municipal Environmental Research Laboratory, Cincinnati, OH, September (1979).
4. L. K. Wang, N. K. Shammas, and Y. T. Hung (eds.), *Biosolids Treatment Processes*, Humana Press, Inc., Totowa, NJ, 820 pp. (2007).
5. L. Spinosa, and P. A. Vesilind, *Sludge into Biosolids—Processing, Disposal, Utilization*, IWA Publishing, December (2001).
6. R. J. Bjork, Planning for biosolids reuse a comprehensive assessment of local and regional alternatives, *WEFTEC 2003 Conference Proceedings*, October (2003).
7. Australia US EPA, *Draft Guidelines for Biosolids Management, Best Practice Environmental Management Series*, Environment Protection Authority Melbourne, Australia (1999).
8. R. E. Kilian, A. C. Todd, and A. K. Wason, How to put one egg in multiple baskets? EMWD's regional biosolids management approach makes sense, *Joint Residuals and Biosolids Management*, February (2003).
9. R. G. O'Dette, National issues impacting biosolids management options, *WEFTEC 2001 Conference Proceedings*, October (2001).
10. P. E. Heath, and M. C. Jeffrey, An integrated waste management approach to biosolids: a public/private partnership, *14th Annual Residuals and Biosolids Management Conference*, February/March (2000).
11. R. M. Pickett, Adventures in biosolids management: the Toronto experience, *16th Annual Residuals and Biosolids Management Conference, Privatization, Innovation and Optimization—How to Do More for Less*, March (2002).
12. D. Tenenbaum, The beauty of biosolids, *Environmental Health Perspectives*, 105, 1, January (1997).
13. M. Bethel, Best practice management of effluent and biosolids, *1999 Conference Paper*, *62$^{nd}$ Annual Water Industry Engineers and Operators Conference* Water Industry Operators Association (WIOA) Australia, Http://Www.Wioa.Org.Au/Conf_Papers/99/Paper14.Htm, September (2002).
14. S. Laha, and W. Parker, Biosolids and Sludge Management, *Water Environment Research Literature Review*, September/October (2003).
15. W. Parker, and S. Laha, Biosolids and sludge management, *Water Environment Research Literature Review*, September/October (2004).
16. P. Matthews, A millennium perspective on biosolids and sludge management, *14th Annual Residuals and Biosolids Management Conference*, February/March (2000).

17. *Biosolids*, http://www.biosolids.com (2008).
18. D. T. Hull, J. Russell, and J. Madigan, A new horizon: private biosolids management for the City of Atlanta, *16th Annual Residuals and Biosolids Management Conference, Privatization, Innovation and Optimization—How to Do More for Less*, March (2002).
19. F. Munsey, and P. Schlecht, Management of multiple biosolids options Milwaukee's experience, *WEFTEC 2000 Conference Proceedings*, October (2000).
20. BHE and LW Diamond Environmental, *Beneficial Utilization Initiatives for Waste stream Organics Sludge/Biosolids*, http://www.wastenot.mb.ca/sludgebiosolids.htm (2008).
21. W. P. Dennis, and C. Stanley, 15 years of successful and cost-effective biosolids management in new jersey, *Residuals and Biosolids Management Conference and Exhibition*, February (2004).
22. C. Peot, S.A. Gabriel, S. Vilalai, M. Ramirez, and S. Murthy, A proactive approach to managing biosolids: using statistical modeling to optimize cost, product quality, and public relations, *WEFTEC 2004*, October (2004).
23. L. K. Wang, N. K. Shammas, W. A. Selke, and D. B. Aulenbach, Flotation thickening. In *Biosolids Treatment Processes*, L. K. Wang, N. K. Shammas, and Y. T. Hung (eds.), The Humana Press, Inc., Totowa, NJ, 820 pp. (2007).
24. L. Slattery, P. Loar, and P. Sanghavi, Biosolids management alternatives study for Arlington water pollution control plant, *Joint Residuals and Biosolids Management 2003*, February (2003).
25. U.S. Environmental Protection Agency, *Municipal Wastewater Treatment Plant Sludge and Liquid Sidestreams*, US EPA 430/9-76-007, Office of Water Program Operations, June (1976).
26. L. J. Ewing, Jr., H. H. Almgren, and R. L. Culp, *Effects of Thermal Treatment of Sludge on Municipal Wastewater Treatment Costs*, US EPA-600/2-78-073, June (1978).
27. E. S. K. Chian, and F. B. DeWalle, Sanitary landfill leachates and their treatment, *Proceedings ASCE, Journal of the Environmental Engineering Division*, 102, 411 (1976).
28. R. J. Lofy, H. T. Phung, R. P. Stearns, and J. J. Walsh, *Subsurface Disposal of Municipal Wastewater Treatment Sludge, Environmental Assessment*, US EPA 68-01-4166, Office of Solid Waste (1978).
29. L. J. Sikora, C. M. Murray, N. H. Frankos, and J. M. Walker, Water quality at a sludge entrenchment Site, *Groundwater*, 16 (1978).
30. J. M. Walker, L. Ely, P. Hundenmann, N. Frankos, and A. Kaminski, *Sewage Sludge Entrenchment System for Use by Small Communities*, U.S. Environmental Protection Agency, US EPA-600/2-78-018, February (1978).
31. U.S. Environmental Protection Agency, Technology Transfer, *Process Design Manual, Municipal Sludge Landfills*, October (1978).
32. Sacramento Area Consultants, *Dedicated Land Disposal Study*, Sewage Sludge Management Program, Final Report, Volume 5, Sacramento Regional County Sanitation District, Sacramento, CA, September (1979).
33. N. R. Mishalani, and J. A. Husband, Nitrogen removal from dewatering sidestreams, *Biosolids 2001: Building Public Support*, February (2001).
34. N. A. Mignone, Digester supernatant does not have to be a problem, *Water & Sewage Works*, p. 57, December (1976).
35. N. A. Mignone, Survey of anaerobic digestion supernatant treatment alternatives, *Water & Sewage Works*, p. 42, January (1977).
36. N. A. Mignone, Elimination of Anaerobic Digester Supernatant, *Water & Sewage Works*, p. 48, February (1977).

37. S. E. Kappe, Digester supernatant: problems, characteristics and treatment, *Sewage and Industrial Wastes*, 30, 937 (1958).
38. L. Mueller, E. Hindin, J. V. Lundsford, and G. H. Dunstan, Some characteristics of anaerobic sludge digestion—I, effect of loading, *Sewage and Industrial Wastes*, 31, 669 (1959).
39. R. S. Burd, Use of new polyelectrolytes in sewage sludge conditioning, *Proceedings of the 2nd Vanderbilt Sanitary Engineering Conference*, May (1963).
40. A. Geinopolos, and F. I. Vilen, Process evaluation—phosphorus removal, *Journal Water Pollution Control Federation*, 43, 1975 (1971).
41. G. F. G. Clough, The effect of sludge treatment processes on the design and operation of sewage treatment plants, *Water Pollution Control*, 76, 452 (1977).
42. C. Lue-Hing, A. W. Obayashi, D. R. Zenz, B. Washington, and B. M. Sawyer, Nitrification of a High Ammonia Content Sludge Supernatant by Use of Rotating Discs, Presented at the *29th Annual Purdue Industrial Waste Conference*, West Lafayette, Indiana, p. 245, May (1974).
43. C. Lue-Hing, A. W. Obayashi, D. R. Zenz, B. Washington, and B. M. Sawyer, Biological nitrification of a high ammonia content sludge supernatant under ambient winter and summer conditions by use of rotating discs, Presented at the *47th Annual New York Water Pollution Control Conference*, January (1975).
44. T. B. S. Prakasam, W. E. Robinson, and C. Lue-Hing, Nitrogen removal from digested sludge supernatant liquor using attached and suspended growth systems, Presented at the *32nd Annual Purdue Industrial Waste Conference*, West Lafayette, Indiana, p. 745, May (1977).
45. J. F. Malina, and J. DiFilippo, Treatment of supernatant and liquids associated with sludge treatment, *Water & Sewage Works*, p. R-30, (1971).
46. W. Rudolfs, and L. S. Fontenelli, Supernatant liquor treatment with chemicals, *Sewage Works Journal*, 17, 538 (1945).
47. R. H. Howe, What to do with supernatant, *Wastes Engineering*, 30, 12 (1959).
48. C. E. Keefer, and H. Kratz, Jr., Treatment of supernatant sludge liquor by coagulation and sedimentation, *Sewage Works Journal*, 12, 738 (1940).
49. C. V. Erickson, Treatment and disposal of digestion tank supernatant liquor, *Sewage Works Journal*, 17, 889 (1945).
50. The PFT supernatant liquor treater, *Sewage Works Journal*, 15, 1018 (1943).
51. B. R. Brown, L. B. Wood, and H. J. Finch, Experiments on the dewatering of digested and activated sludge, *Water Pollution Control*, 71, 61 (1972).
52. J. D. Boyce, and D. D. Gruenwald, Recycle of liquor from heat treatment of sludge, *Journal Water Pollution Control Federation*, 47, 2482–2489 (1975).
53. K. D. Corrie, Use of activated carbon in the treatment of heat treatment plant liquor, *Water Pollution Control*, 71, 629 (1972).
54. U. Loll, Treatment of thermally conditioned sludge liquors, *Water Research*, 11, 869–872 (1977).
55. A. H. Erickson, and P. V. Knopp, Biological treatment of thermally conditioned sludge liquors, *Proceedings of the 5th International Water Pollution Control Research Conference*, San Francisco, Vol. II, p. 30 (1970).
56. R. T. Haug, S. K. Raksit and G. G. Wong, Anaerobic filter treats waste activated sludge, *Water & Sewage Works*, p. 40, February (1977).
57. J. C. Young, and P. L. McCarty, The anaerobic filter for waste treatment, *Journal Water Pollution Control Federation*, 41, research supplement, p. R-l60 (1969).
58. M. VanderMarck, E. McCormick, and R. Gillette, An evaluation of biosolids management strategies: selecting an effective approach to an uncertain future, *WEFTEC 2002 Conference Proceedings*, September (2002).

59. R. H. Forbes, G. Brubaker, and L. Kellam, Have you prepared your biosolids management contingency plan? (An example from Camp Lejeune Marine Corps Base), *16th Annual Residuals and Biosolids Management Conference, Privatization, Innovation and Optimization—How to Do More for Less*, March (2002).
60. O. E. Tyler, Incorporating marketing and public education programs into your composting operation to increase product demand, *16th Annual Residuals and Biosolids Management Conference, Privatization, Innovation and Optimization—How to Do More for Less*, March (2002).
61. B. James, Long-Range Biosolids Management Plan Project Summary, *Joint Residuals and Biosolids Management Conference*, April (2005).
62. V. M. Grace, M. W. Thayer, and R. S. Hogan, A biosolids management approach that emphasizes environmental sustainability, *Joint Residuals and Biosolids Management*, February (2003).
63. C. Egigian-Nichols, M. Moore, J. Burror, L. Baroldi, and F. Soroushian, Assessment of viable product markets in Orange County and Southern California, *Residuals and Biosolids Management Conference and Exhibition*, February (2004).
64. L. P. David, M. T. Hogan, P. P. Berge and M. L. Van Cleave, Comprehensive operations management optimizes bioenergy, *Residuals and Biosolids Management Conference and Exhibition*, February (2004).
65. L. Jennifer, D. Strehler, and L. Parry, Systems approach to integrating biosolids and energy management, *Joint Residuals and Biosolids Management Conference*, April (2005).
66. U.S. Environmental Protection Agency, Energy Conservation in Municipal Wastewater Treatment, US EPA 4-30/9-77-011, Office of Water Program Operations, Washington, DC, March (1978).
67. B. L. Pastushenko, A. G. Rodieck, L. D. Tortorici, and M. S. Norris, A diverse biosolids management program for Oxnard, CA, *WEFTEC 2003 Conference Proceedings*, October (2003).
68. S. Jeyanayagam, P. Sandstrom, and A. Turk, A model for developing technical and cost-effective solutions for biosolids processing, *Residuals and Biosolids Management Conference and Exhibition*, February (2004).
69. Federal Register, *Cost-Effectiveness Analyses*, 40 CFR 35, Appendix A, September (1975).
70. U.S. Army Corps of Engineers, *Civil Works Construction Cost Index System Manual*, 110-2-1304, 2000 Tables, Washington, DC, http://www.nww.usace.army.mil/cost (2007).
71. U.S. Environmental Protection Agency, Evaluation of Sludge Management Systems: Evaluation Checklist and Supporting Commentary, Office of Water Program Operations, Washington, DC, August (1978).
72. E. L. Grant, and W. G. Ireson, *Principles of Engineering Economy*, 4th ed., Ronald Publishing Co., New York (1964).
73. M. S. Peters, and K. D. Timmerhaus, *Plant Design and Economics for Chemical Engineers*, McGraw-Hill Book Co. New York (1962).
74. U.S. Environmental Protection Agency, Sludge Handling And Disposal Practices At Selected Municipal Wastewater Treatment Plants, US EPA-430/9-77-007, Office of Water Program Operations, Municipal Construction Division, Washington, DC, April (1977).
75. P. E. Brubaker, F. Gregory, G. H. Davis, and K. Kellum, Camp Lejeune's biosolids application program—a model for environmentally sustainable biosolids management, *Biosolids Technical Bulletin*, February/March (2000).

76. G. F. Shimp, B. Childress, and M. Sweeney, A solid finish, a Kentucky municipality reigns in an innovative biosolids-management solution, *Water Environment & Technology*, February (2002).
77. L. S. Pringle, J. Ollerenshaw, and P. Burnet, Aligning environmental management systems with organizational priorities—biosolids and beyond, *WEFTEC 2003 Conference Proceedings*, October (2003).
78. G. Hyare, D. C. Scholler, and T. Battenfield, Flexibility: Houston's approach to regional biosolids handling, *16th Annual Residuals and Biosolids Management Conference 2002 Privatization, Innovation and Optimization—How to Do More for Less*, March (2002).
79. B. S. Ronald, and B. Michael, Solids master planning for smaller communities a case study in Snyderville Utah, *Residuals and Biosolids Management Conference and Exhibition*, February (2004).
80. L. K. Wang, Quaternary ammonium thickening of sewage sludge in magnetic field, *Industrial and Engineering Chemistry*, 16, 4, 311–315 (1977).
81. M. Krofta, and L. K. Wang, Sludge thickening and dewatering by dissolved air flotation: Floatpress, *Drying*, 2, 765–771, Harper & Row Publishers, New York (1986).
82. M. Krofta, and L. K. Wang, Sludge thickening and dewatering by dissolved air flotation: process design, *Drying*, 2, 772–780, Harper & Row Publishers, New York (1986).
83. L. K. Wang, *Guidelines for Disposal of Solid Wastes and Hazardous Wastes*, Volumes I to VI, U.S. Department of Commerce, National Technical Information Service, Springfield, VA, Technical Report Nos. PB88-178066/AS, PB88-178074/AS, PB88-178082/AS, PB88-178090/AS, PB88-178108/AS, and PB89-158596/AS (1989).
84. L. K. Wang, *Sludge Treatment Apparatus*, U.S. Patent and Trade Marks Office, Washington, DC. Patent No. 5068031, November (1991).
85. L. K. Wang, *Analysis and Identification of Odorous Compounds in Emissions from Springfield Regional Wastewater Treatment Plant, Massachusetts*, Lenox Institute of Water Technology, Lenox, MA, Technical Report LIR/05-85/137 (1985).

# Appendix: Conversion Factors for Environmental Engineers

## Lawrence K. Wang

**CONTENTS**

CONSTANTS AND CONVERSION FACTORS
BASIC AND SUPPLEMENTARY UNITS
DERIVED UNITS AND QUANTITIES
PHYSICAL CONSTANTS
PROPERTIES OF WATER
PERIODIC TABLE OF THE ELEMENTS

**Abstract** With the current trend toward metrication, the question of using a consistent system of units has been a problem. Wherever possible, the authors of this *Handbook of Environmental Engineering* series have used the British system (fps) along with the metric equivalent (mks, cgs, or SIU) or vice versa. For the convenience of the readers around the world, this book provides a detailed Conversion Factors for Environmental Engineers. In addition, the basic and supplementary units, the derived units and quantities, important physical constants, the properties of water, and the Periodic Table of the elements, are also presented in this document.

**Key Words** Conversion factors • British units • metric units • physical constants • water properties • periodic table of the elements • environmental engineers • Lenox Institute of Water Technology • mks (meter-kilogram-second) • cgs (centimeter-gram-second) • SIU (Système international d'unités; International System of Units) • fps (foot-pound-second).

# 1. CONSTANTS AND CONVERSION FACTORS

| Multiply | by | to obtain |
|---|---|---|
| abamperes | 10 | amperes |
| abamperes | $2.99796 \times 10^{10}$ | statamperes |
| abampere-turns | 12.566 | gilberts |
| abcoulombs | 10 | coulombs (abs) |
| abcoulombs | $2.99796 \times 10^{10}$ | statcoulombs |
| abcoulombs/kg | 30,577 | statcoulombs/dyne |
| abfarads | $1 \times 10^{9}$ | farads (abs) |
| abfarads | $8.98776 \times 10^{20}$ | statfarads |
| abhenries | $1 \times 10^{-9}$ | henries (abs) |
| abhenries | $1.11263 \times 10^{-21}$ | stathenries |
| abohms | $1 \times 10^{-9}$ | ohms (abs) |
| abohms | $1.11263 \times 10^{-21}$ | statohms |
| abvolts | $3.33560 \times 10^{-11}$ | statvolts |
| abvolts | $1 \times 10^{-8}$ | volts (abs) |
| abvolts/centimeters | $2.540005 \times 10^{-8}$ | volts (abs)/inch |
| acres | 0.4046 | ha |
| acres | 43,560 | square feet |
| acres | 4047 | square meters |
| acres | $1.562 \times 10^{-3}$ | square miles |
| acres | 4840 | square yards |
| acre-feet | 43,560 | cubic feet |
| acre-feet | 1233.5 | cubic meters |
| acre-feet | 325,850 | gallons (U.S.) |
| amperes (abs) | 0.1 | abamperes |
| amperes (abs) | $1.036 \times 10^{-5}$ | faradays/second |
| amperes (abs) | $2.9980 \times 10^{9}$ | statamperes |
| ampere-hours (abs) | 3600 | coulombs (abs) |
| ampere-hours | 0.03731 | faradays |
| amperes/sq cm | 6.452 | amps/sq in |
| amperes/sq cm | $10^{4}$ | amps/sq meter |
| amperes/sq in | 0.1550 | amps/sq cm |
| amperes/sq in | 1550.0 | amps/sq meter |
| amperes/sq meter | $10^{-4}$ | amps/sq cm |
| amperes/sq meter | $6.452 \times 10^{-4}$ | amps/sq in |
| ampere-turns | 1.257 | gilberts |
| ampere-turns/cm | 2.540 | amp-turns/in |
| ampere-turns/cm | 100.0 | amp-turns/meter |
| ampere-turns/cm | 1.257 | gilberts/cm |
| ampere-turns/in | 0.3937 | amp-turns/cm |
| ampere-turns/in | 39.37 | amp-turns/meter |
| ampere-turns/in | 0.4950 | gilberts/cm |
| ampere-turns/meter | 0.01 | amp-turns/cm |
| ampere-turns/meter | 0.0254 | amp-turns/in |
| ampere-turns/meter | 0.01257 | gilberts/cm |

## Conversion Factors

| Multiply | by | to obtain |
|---|---|---|
| angstrom units | $1 \times 10^{-8}$ | centimeters |
| angstrom units | $3.937 \times 10^{-9}$ | inches |
| angstrom unit | $1 \times 10^{-10}$ | meter |
| angstrom unit | $1 \times 10^{-4}$ | micron or μm |
| ares | 0.02471 | acre (U.S.) |
| ares | 1076 | square feet |
| ares | 100 | square meters |
| ares | 119.60 | sq yards |
| assay tons | 29.17 | grams |
| astronomical unit | $1.495 \times 10^{8}$ | kilometers |
| atmospheres (atm) | 0.007348 | tons/sq inch |
| atmospheres | 76.0 | cms of mercury |
| atmospheres | $1.01325 \times 10^{6}$ | dynes/square centimeter |
| atmospheres | 33.90 | ft of water (at 4°C) |
| atmospheres | 29.92 | inches of mercury (at 0°C) |
| atmospheres | 1.033228 | kg/sq cm |
| atmospheres | 10,332 | kg/sq meter |
| atmospheres | 760.0 | millimeters of mercury |
| atmospheres | 14.696 | pounds/square inch |
| atmospheres | 1.058 | tons/sq foot |
| avograms | $1.66036 \times 10^{-24}$ | grams |
| bags, cement | 94 | pounds of cement |
| barleycorns (British) | 1/3 | inches |
| barleycorns (British) | $8.467 \times 10^{-3}$ | meters |
| barrels (British, dry) | 5.780 | cubic feet |
| barrels (British, dry) | 0.1637 | cubic meters |
| barrels (British, dry) | 36 | gallons (British) |
| barrels, cement | 170.6 | kilograms |
| barrels, cement | 376 | pounds of cement |
| barrels, cranberry | 3.371 | cubic feet |
| barrels, cranberry | 0.09547 | cubic meters |
| barrels, oil | 5.615 | cubic feet |
| barrels, oil | 0.1590 | cubic meters |
| barrels, oil | 42 | gallons (U.S.) |
| barrels, (U.S., dry) | 4.083 | cubic feet |
| barrels (U.S., dry) | 7056 | cubic inches |
| barrels (U.S., dry) | 0.11562 | cubic meters |
| barrels (U.S., dry) | 105.0 | quarts (dry) |
| barrels (U.S., liquid) | 4.211 | cubic feet |
| barrels (U.S., liquid) | 0.1192 | cubic meters |
| barrels (U.S., liquid) | 31.5 | gallons (U.S.) |
| bars | 0.98692 | atmospheres |
| bars | $10^{6}$ | dynes/sq cm |
| bars | $1.0197 \times 10^{4}$ | kg/sq meter |
| bars | 1000 | millibar |
| bars | 750.06 | mm of Hg (0°C) |

| Multiply | by | to obtain |
| --- | --- | --- |
| bars | 2089 | pounds/sq ft |
| bars | 14.504 | pounds/sq in |
| barye | 1.000 | dynes/sq cm |
| board feet | 1/12 | cubic feet |
| board feet | 144 sq.in. × 1 in. | cubic inches |
| boiler horsepower | 33,475 | BTU (mean)/hour |
| boiler horsepower | 34.5 | pounds of water evaporated from and at 212°F (per hour) |
| bolts (U.S., cloth) | 120 | linear feet |
| bolts (U.S., cloth) | 36.576 | meters |
| bougie decimales | 1 | candles (int) |
| BTU (mean) | 251.98 | calories, gram (g. cal) |
| BTU (mean) | 0.55556 | centigrade heat units (chu) |
| BTU (mean) | $1.0548 \times 10^{10}$ | ergs |
| BTU (mean) | 777.98 | foot-pounds |
| BTU (mean) | $3.931 \times 10^{-4}$ | horsepower-hrs (hp-hr) |
| BTU (mean) | 1055 | joules (abs) |
| BTU (mean) | 0.25198 | kilograms, cal (kg cal) |
| BTU (mean) | 107.565 | kilogram-meters |
| BTU (mean) | $2.928 \times 10^{-4}$ | kilowatt-hr (Kwh) |
| BTU (mean) | 10.409 | liter-atm |
| BTU (mean) | $6.876 \times 10^{-5}$ | pounds of carbon to $CO_2$ |
| BTU (mean) | 0.29305 | watt-hours |
| BTU (mean)/cu ft | 37.30 | joule/liter |
| BTU/hour | 0.2162 | foot-pound/sec |
| BTU/hour | 0.0700 | gram-cal/sec |
| BTU/hour | $3.929 \times 10^{-4}$ | horsepower-hours (hp-hr) |
| BTU/hour | 0.2930711 | watt (w) |
| BTU/hour (feet)°F | 1.730735 | joule/sec (m)°k |
| BTU/hour (feet$^2$) | 3.15459 | joule/m$^2$-sec |
| BTU (mean)/hour(feet$^2$)°F | $1.3562 \times 10^{-4}$ | gram-calorie/second (cm$^2$)°C |
| BTU (mean)/hour(feet$^2$)°F | $3.94 \times 10^{-4}$ | horsepower/(ft$^2$)°F |
| BTU (mean)/hour(feet$^2$)°F | 5.678264 | joule/sec (m$^2$)°k |
| BTU (mean)/hour(feet$^2$)°F | 4.882 | kilogram-calorie/hr (m$^2$)°C |
| BTU (mean)/hour(feet$^2$)°F | $5.682 \times 10^{-4}$ | watts/(cm$^2$)°C |
| BTU (mean)/hour(feet$^2$)°F | $2.035 \times 10^{-3}$ | watts/(in$^2$)°C |
| BTU (mean)/(hour)(feet$^2$) (°F/inch) | $3.4448 \times 10^{-4}$ | calories, gram (15°C)/sec (cm$^2$) (°C/cm) |
| BTU (mean)/(hour)(feet$^2$) (°F/in.) | 1 | chu/(hr)(ft$^2$)(°C/in) |
| BTU (mean)/(hour)(feet$^2$) (°F/inch) | $1.442 \times 10^{-3}$ | joules (abs)/(sec)(cm$^2$) (°C/cm) |
| BTU (mean)/(hour)(feet$^2$) (°F/inch) | $1.442 \times 10^{-3}$ | watts/(cm$^2$) (°C/cm) |
| BTU/min | 12.96 | ft lb/sec |
| BTU/min | 0.02356 | hp |
| BTU/min | 0.01757 | kw |
| BTU/min | 17.57 | watts |

## Conversion Factors

| Multiply | by | to obtain |
|---|---|---|
| BTU/min/ft$^2$ | 0.1221 | watts/sq inch |
| BTU/pound | 0.5556 | calories-gram(mean)/gram |
| BTU/pound | 0.555 | kg-cal/kg |
| BTU/pound/°F | 1 | calories, gram/gram/°C |
| BTU/pound/°F | 4186.8 | joule/kg/°k |
| BTU/second | 1054.350 | watt (W) |
| buckets (British, dry) | $1.818 \times 10^4$ | cubic cm |
| buckets (British, dry) | 4 | gallons (British) |
| bushels (British) | 1.03205 | bushels (U.S.) |
| bushels (British) | 1.2843 | cubic feet |
| bushels (British) | 0.03637 | cubic meters |
| bushels (U.S.) | 1.2444 | cubic feet |
| bushels (U.S.) | 2150.4 | cubic inch |
| bushels (U.S.) | 0.035239 | cubic meters |
| bushels (U.S.) | 35.24 | liters (L) |
| bushels (U.S.) | 4 | pecks (U.S.) |
| bushels (U.S.) | 64 | pints (dry) |
| bushels (U.S.) | 32 | quarts (dry) |
| butts (British) | 20.2285 | cubic feet |
| butts (British) | 126 | gallons (British) |
| cable lengths | 720 | feet |
| cable lengths | 219.46 | meters |
| calories (thermochemical) | 0.999346 | calories (Int. Steam Tables) |
| calories, gram (g. cal or simply cal.) | $3.9685 \times 10^{-3}$ | BTU (mean) |
| calories, gram (mean) | 0.001459 | cubic feet atmospheres |
| calories, gram (mean) | $4.186 \times 10^7$ | ergs |
| calories, gram (mean) | 3.0874 | foot-pounds |
| calories, gram (mean) | 4.186 | joules (abs) |
| calories, gram (mean) | 0.001 | kg cal (calories, kilogram) |
| calories, gram (mean) | 0.42685 | kilograms-meters |
| calories, gram (mean) | 0.0011628 | watt-hours |
| calories, gram (mean)/gram | 1.8 | BTU (mean)/pound |
| cal/gram-°C | 4186.8 | joule/kg-°k |
| candle power (spherical) | 12.566 | lumens |
| candles (int) | 0.104 | carcel units |
| candles (int) | 1.11 | hefner units |
| candles (int) | 1 | lumens (int)/steradian |
| candles (int)/square centimeter | 2919 | foot-lamberts |
| candles (int)/square centimeter | 3.1416 | lamberts |
| candles (int)/square foot | 3.1416 | foot-lamberts |
| candles (int)/square foot | $3.382 \times 10^{-3}$ | lamberts |
| candles (int)/square inch | 452.4 | foot-lamberts |
| candles (int)/square inch | 0.4870 | lamberts |
| candles (int)/square inch | 0.155 | stilb |
| carats (metric) | 3.0865 | grains |
| carats (metric) | 0.2 | grams |
| centals | 100 | pounds |

| Multiply | by | to obtain |
|---|---|---|
| centares (centiares) | 1.0 | sq meters |
| centigrade heat units (chu) | 1.8 | BTU |
| centigrade heat units (chu) | 453.6 | calories, gram (15°C) |
| centigrade heat units (chu) | 1897.8 | joules (abs) |
| centigrams | 0.01 | grams |
| centiliters | 0.01 | liters |
| centimeters | 0.0328083 | feet (U.S.) |
| centimeters | 0.3937 | inches (U.S.) |
| centimeters | 0.01 | meters |
| centimeters | $6.214 \times 10^{-6}$ | miles |
| centimeters | 10 | millimeters |
| centimeters | 393.7 | mils |
| centimeters | 0.01094 | yards |
| cm of mercury | 0.01316 | atm |
| cm of mercury | 0.4461 | ft of water |
| cm of mercury | 136.0 | kg/square meter |
| cm of mercury | 1333.22 | newton/meter$^2$ (N/m$^2$) |
| cm of mercury | 27.85 | psf |
| cm of mercury | 0.1934 | psi |
| cm of water (4°C) | 98.0638 | newton/meter$^2$ (N/m$^2$) |
| centimeters-dynes | $1.020 \times 10^{-3}$ | centimeter-grams |
| centimeter-dynes | $1.020 \times 10^{-8}$ | meter-kilograms |
| centimeter-dynes | $7.376 \times 10^{-8}$ | pound-feet |
| centimeter-grams | 980.7 | centimeter-dynes |
| centimeter-grams | $10^{-5}$ | meter-kilograms |
| centimeter-grams | $7.233 \times 10^{-5}$ | pound-feet |
| centimeters/second | 1.969 | fpm (ft/min) |
| centimeters/second | 0.0328 | fps (ft/sec) |
| centimeters/second | 0.036 | kilometers/hour |
| centimeters/second | 0.1943 | knots |
| centimeters/second | 0.6 | m/min |
| centimeters/second | 0.02237 | miles/hour |
| centimeters/second | $3.728 \times 10^{-4}$ | miles/minute |
| cms/sec./sec. | 0.03281 | feet/sec/sec |
| cms/sec./sec. | 0.036 | kms/hour/sec |
| cms/sec./sec. | 0.02237 | miles/hour/sec |
| centipoises | 3.60 | kilograms/meter hour |
| centipoises | $10^{-3}$ | kilograms/meter second |
| centipoises | 0.001 | newton-sec/m$^2$ |
| centipoises | $2.089 \times 10^{-5}$ | pound force second/square foot |
| centipoises | 2.42 | pounds/foot hour |
| centipoises | $6.72 \times 10^{-4}$ | pounds/foot second |
| centistoke | $1.0 \times 10^{-6}$ | meter$^2$/sec |
| chains (engineers' or Ramden's) | 100 | feet |
| chains (engineers' or Ramden's) | 30.48 | meters |
| chains (surveyors' or Gunter's) | 66 | feet |
| chains (surveyors' or Gunter's) | 20.12 | meters |

## Conversion Factors

| Multiply | by | to obtain |
| --- | --- | --- |
| chaldrons (British) | 32 | bushels (British) |
| chaldrons (U.S.) | 36 | bushels (U.S.) |
| cheval-vapours | 0.9863 | horsepower |
| cheval-vapours | 735.5 | watts (abs) |
| cheval-vapours heures | $2.648 \times 10^6$ | joules (abs) |
| chu/(hr)(ft$^2$)(°C/in.) | 1 | BTU/(hr)(ft$^2$)(°F/in.) |
| circular inches | 0.7854 | square inches |
| circular millimeters | $7.854 \times 10^{-7}$ | square meters |
| circular mils | $5.067 \times 10^{-6}$ | square centimeters |
| circular mils | $7.854 \times 10^{-7}$ | square inches |
| circular mils | 0.7854 | square mils |
| circumferences | 360 | degrees |
| circumferences | 400 | grades |
| circumferences | 6.283 | radians |
| cloves | 8 | pounds |
| coombs (British) | 4 | bushels (British) |
| cords | 8 | cord feet |
| cords | $8' \times 4' \times 4'$ | cubic feet |
| cords | 128 | cubic feet |
| cords | 3.625 | cubic meters |
| cord-feet | $4' \times 4' \times 1'$ | cubic feet |
| coulombs (abs) | 0.1 | abcoulombs |
| coulombs (abs) | $6.281 \times 10^{18}$ | electronic charges |
| coulombs (abs) | $2.998 \times 10^9$ | statcoulombs |
| coulombs (abs) | $1.036 \times 10^{-5}$ | faradays |
| coulombs/sq cm | 64.52 | coulombs/sq in |
| coulombs/sq cm | $10^4$ | coulombs/sq meter |
| coulombs/sq in | 0.1550 | coulombs/sq cm |
| coulombs/sq in | 1550 | coulombs/sq meter |
| coulombs/sq meter | $10^{-4}$ | coulombs/sq cm |
| coulombs/sq meter | $6.452 \times 10^{-4}$ | coulombs/sq in |
| cubic centimeters | $3.531445 \times 10^{-5}$ | cubic feet (U.S.) |
| cubic centimeters | $6.102 \times 10^{-2}$ | cubic inches |
| cubic centimeters | $10^{-6}$ | cubic meters |
| cubic centimeters | $1.308 \times 10^{-6}$ | cubic yards |
| cubic centimeters | $2.6417 \times 10^{-4}$ | gallons (U.S.) |
| cubic centimeters | 0.001 | liters |
| cubic centimeters | 0.033814 | ounces (U.S., fluid) |
| cubic centimeters | $2.113 \times 10^{-3}$ | pints (liq.) |
| cubic centimeters | $1.057 \times 10^{-3}$ | quarts (liq.) |
| cubic feet (British) | 0.9999916 | cubic feet (U.S.) |
| cubic feet (U.S.) | 0.8036 | bushels (dry) |
| cubic feet (U.S.) | 28317.016 | cubic centimeters |
| cubic feet (U.S.) | 1728 | cubic inches |
| cubic feet (U.S.) | 0.02832 | cubic meters |
| cubic feet (U.S.) | 0.0370 | cubic yard |

| Multiply | by | to obtain |
| --- | --- | --- |
| cubic feet (U.S.) | 7.48052 | gallons (U.S.) |
| cubic feet (U.S.) | 28.31625 | liters |
| cubic feet (U.S.) | 59.84 | pints (liq.) |
| cubic feet (U.S.) | 29.92 | quarts (liq.) |
| cubic feet of common brick | 120 | pounds |
| cubic feet of water (60°F) | 62.37 | pounds |
| cubic foot-atmospheres | 2.7203 | BTU (mean) |
| cubic foot-atmospheres | 680.74 | calories, gram (mean) |
| cubic foot-atmospheres | 2116 | foot-pounds |
| cubic foot-atmospheres | 2869 | joules (abs) |
| cubic foot-atmospheres | 292.6 | kilogram-meters |
| cubic foot-atmospheres | $7.968 \times 10^{-4}$ | kilowatt-hours |
| cubic feet/hr | 0.02832 | $m^3$/hr |
| cubic feet/minute | 472.0 | cubic cm/sec |
| cubic feet/minute | 1.6992 | cu m/hr |
| cubic feet/minute | 0.0283 | cu m/min |
| cubic feet/minute | 0.1247 | gallons/sec |
| cubic feet/minute | 0.472 | liter/sec |
| cubic feet/minute | 62.4 | lbs of water/min |
| cubic feet/min/1000 cu ft | 0.01667 | liter/sec/cu m |
| cubic feet/second | 1.9834 | acre-feet/day |
| cubic feet/second | 1.7 | cu m/min |
| cubic feet/second | 0.02832 | $m^3$/sec |
| cubic feet/second | 448.83 | gallons/minute |
| cubic feet/second | 1699 | liter/min |
| cubic feet/second | 28.32 | liters/sec |
| cubic feet/second (cfs) | 0.64632 | million gallons/day (MGD) |
| cfs/acre | 0.07 | $m^3$/sec-ha |
| cfs/acre | 4.2 | cu m/min/ha |
| cfs/sq mile | 0.657 | cu m/min/sq km |
| cubic inches (U.S.) | 16.387162 | cubic centimeters |
| cubic inches (U.S.) | $5.787 \times 10^{-4}$ | cubic feet |
| cubic inches (U.S.) | 1.0000084 | cubic inches (British) |
| cubic inches (U.S.) | $1.639 \times 10^{-5}$ | cubic meters |
| cubic inches (U.S.) | $2.143 \times 10^{-5}$ | cubic yards |
| cubic inches (U.S.) | $4.329 \times 10^{-3}$ | gallons (U.S.) |
| cubic inches (U.S.) | $1.639 \times 10^{-2}$ | liters |
| cubic inches (U.S.) | 16.39 | mL |
| cubic inches (U.S.) | 0.55411 | ounces (U.S., fluid) |
| cubic inches (U.S.) | 0.03463 | pints (liq.) |
| cubic inches (U.S.) | 0.01732 | quarts (liq.) |
| cubic meters | $8.1074 \times 10^{-4}$ | acre-feet |
| cubic meters | 8.387 | barrels (U.S., liquid) |
| cubic meters | 28.38 | bushels (dry) |
| cubic meters | $10^6$ | cubic centimeters |
| cubic meters | 35.314 | cubic feet (U.S.) |

## Conversion Factors

| Multiply | by | to obtain |
|---|---|---|
| cubic meters | 61,023 | cubic inches (U.S.) |
| cubic meters | 1.308 | cubic yards (U.S.) |
| cubic meters | 264.17 | gallons (U.S.) |
| cubic meters | 1000 | liters |
| cubic meters | 2113 | pints (liq.) |
| cubic meters ($m^3$) | 1057 | quarts (liq.) |
| cubic meters/day | 0.183 | gallons/min |
| cubic meters/ha | 106.9 | gallons/acre |
| cubic meters/hour | 0.2272 | gallons/minute |
| cubic meters/meter-day | 80.53 | gpd/ft |
| cubic meters/minute | 35.314 | cubic ft/minute |
| cubic meters/second | 35.314 | cubic ft/sec |
| cubic meters/second | 22.82 | MGD |
| cubic meters/sec-ha | 14.29 | cu ft/sec-acre |
| cubic meters/meters$^2$-day | 24.54 | gpd/ft$^2$ |
| cubic yards (British) | 0.9999916 | cubic yards (U.S.) |
| cubic yards (British) | 0.76455 | cubic meters |
| cubic yards (U.S.) | $7.646 \times 10^5$ | cubic centimeters |
| cubic yards (U.S.) | 27 | cubic feet (U.S.) |
| cubic yards (U.S.) | 46,656 | cubic inches |
| cubic yards (U.S.) | 0.76456 | cubic meters |
| cubic yards (U.S.) | 202.0 | gallons (U.S.) |
| cubic yards (U.S.) | 764.6 | liters |
| cubic yards (U.S.) | 1616 | pints (liq.) |
| cubic yards (U.S.) | 807.9 | quarts (liq.) |
| cubic yards of sand | 2700 | pounds |
| cubic yards/minute | 0.45 | cubic feet/second |
| cubic yards/minute | 3.367 | gallons/second |
| cubic yards/minute | 12.74 | liters/second |
| cubits | 45.720 | centimeters |
| cubits | 1.5 | feet |
| dalton | $1.65 \times 10^{-24}$ | gram |
| days | 1440 | minutes |
| days | 86,400 | seconds |
| days (sidereal) | 86164 | seconds (mean solar) |
| debye units (dipole moment) | $10^{18}$ | electrostatic units |
| decigrams | 0.1 | grams |
| deciliters | 0.1 | liters |
| decimeters | 0.1 | meters |
| degrees (angle) | 60 | minutes |
| degrees (angle) | 0.01111 | quadrants |
| degrees (angle) | 0.01745 | radians |
| degrees (angle) | 3600 | seconds |
| degrees/second | 0.01745 | radians/seconds |
| degrees/second | 0.1667 | revolutions/min |
| degrees/second | 0.002778 | revoltuions/sec |

| Multiply | by | to obtain |
|---|---|---|
| degree Celsius | °F = (°C × 9/5) + 32 | Fahrenheit |
| degree Celsius | °K = °C + 273.15 | Kelvin |
| degree Fahrenheit | °C = (°F − 32) × 5/9 | Celsius |
| degree Fahrenheit | °K = (°F + 459.67)/1.8 | Kelvin |
| degree Rankine | °K = °R/1.8 | Kelvin |
| dekagrams | 10 | grams |
| dekaliters | 10 | liters |
| dekameters | 10 | meters |
| drachms (British, fluid) | $3.5516 \times 10^{-6}$ | cubic meters |
| drachms (British, fluid) | 0.125 | ounces (British, fluid) |
| drams (apothecaries' or troy) | 0.1371429 | ounces (avoirdupois) |
| drams (apothecaries' or troy) | 0.125 | ounces (troy) |
| drams (U.S., fluid or apoth.) | 3.6967 | cubic cm |
| drams (avoirdupois) | 1.771845 | grams |
| drams (avoirdupois) | 27.3437 | grains |
| drams (avoirdupois) | 0.0625 | ounces |
| drams (avoirdupois) | 0.00390625 | pounds (avoirdupois) |
| drams (troy) | 2.1943 | drams (avoirdupois) |
| drams (troy) | 60 | grains |
| drams (troy) | 3.8879351 | grams |
| drams (troy) | 0.125 | ounces (troy) |
| drams (U.S., fluid) | $3.6967 \times 10^{-6}$ | cubic meters |
| drams (U.S., fluid) | 0.125 | ounces (fluid) |
| dynes | 0.00101972 | grams |
| dynes | $10^{-7}$ | joules/cm |
| dynes | $10^{-5}$ | joules/meter (newtons) |
| dynes | $1.020 \times 10^{-6}$ | kilograms |
| dynes | $1 \times 10^{-5}$ | newton (N) |
| dynes | $7.233 \times 10^{-5}$ | poundals |
| dynes | $2.24809 \times 10^{-6}$ | pounds |
| dyne-centimeters (torque) | $7.3756 \times 10^{-8}$ | pound-feet |
| dynes/centimeter | 1 | ergs/square centimeter |
| dynes/centimeter | 0.01 | ergs/square millimeter |
| dynes/square centimeter | $9.8692 \times 10^{-7}$ | atmospheres |
| dynes/square centimeter | $10^{-6}$ | bars |
| dynes/square centimeter | $2.953 \times 10^{-5}$ | inch of mercury at 0°C |
| dynes/square centimeter | $4.015 \times 10^{-4}$ | inch of water at 4°C |
| dynes/square centimeter | 0.01020 | kilograms/square meter |
| dynes/square centimeter | 0.1 | newtons/square meter |
| dynes/square centimeter | $1.450 \times 10^{-5}$ | pounds/square inch |
| electromagnetic fps units of magnetic permeability | 0.0010764 | electromagnetic cgs units of magnetic permeability |
| electromagnetic fps units of magnetic permeability | $1.03382 \times 10^{-18}$ | electrostatic cgs units of magnetic permeability |

## Conversion Factors

| Multiply | by | to obtain |
|---|---|---|
| electromagnetic cgs units, of magnetic permeability | $1.1128 \times 10^{-21}$ | electrostatic cgs units of magnetic permeability |
| electromagnetic cgs units of mass resistance | $9.9948 \times 10^{-6}$ | ohms (int)-meter-gram |
| electronic charges | $1.5921 \times 10^{-19}$ | coulombs (abs) |
| electron-volts | $1.6020 \times 10^{-12}$ | ergs |
| electron-volts | $1.0737 \times 10^{-9}$ | mass units |
| electron-volts | 0.07386 | rydberg units of energy |
| electronstatic cgs units of Hall effect | $2.6962 \times 10^{31}$ | electromagnetic cgs units of Hall effect |
| electrostatic fps units of charge | $1.1952 \times 10^{-6}$ | coulombs (abs) |
| electrostatic fps units of magnetic permeability | 929.03 | electrostatic cgs units of magnetic permeability |
| ells | 114.30 | centimeters |
| ells | 45 | inches |
| ems, pica (printing) | 0.42333 | centimeters |
| ems, pica (printing) | 1/6 | inches |
| ergs | $9.4805 \times 10^{-11}$ | BTU (mean) |
| ergs | $2.3889 \times 10^{-8}$ | calories, gram (mean) |
| ergs | 1 | dyne-centimeters |
| ergs | $7.3756 \times 10^{-8}$ | foot-pounds |
| ergs | $0.2389 \times 10^{-7}$ | gram-calories |
| ergs | $1.020 \times 10^{-3}$ | gram-centimeters |
| ergs | $3.7250 \times 10^{-14}$ | horsepower-hrs |
| ergs | $10^{-7}$ | joules (abs) |
| ergs | $2.390 \times 10^{-11}$ | kilogram-calories (kg cal) |
| ergs | $1.01972 \times 10^{-8}$ | kilogram-meters |
| ergs | $0.2778 \times 10^{-13}$ | kilowatt-hrs |
| ergs | $0.2778 \times 10^{-10}$ | watt-hours |
| ergs/second | $5.692 \times 10^{-9}$ | BTU/min |
| ergs/second | $4.426 \times 10^{-6}$ | foot-pounds/min |
| ergs/second | $7.376 \times 10^{-8}$ | foot-pounds/sec |
| ergs/second | $1.341 \times 10^{-10}$ | horsepower |
| ergs/second | $1.434 \times 10^{-9}$ | kg-calories/min |
| ergs/second | $10^{-10}$ | kilowatts |
| farad (international of 1948) | 0.9995 | farad (F) |
| faradays | 26.80 | ampere-hours |
| faradays | 96,500 | coulombs (abs) |
| faradays/second | 96,500 | amperes (abs) |
| farads (abs) | $10^{-9}$ | abfarads |
| farads (abs) | $10^{6}$ | microfarads |
| farads (abs) | $8.9877 \times 10^{11}$ | statfarads |
| fathoms | 6 | feet |
| fathom | 1.829 | meter |
| feet (U.S.) | 1.0000028 | feet (British) |
| feet (U.S.) | 30.4801 | centimeters |

| Multiply | by | to obtain |
| --- | --- | --- |
| feet (U.S.) | 12 | inches |
| feet (U.S.) | $3.048 \times 10^{-4}$ | kilometers |
| feet (U.S.) | 0.30480 | meters |
| feet (U.S.) | $1.645 \times 10^{-4}$ | miles (naut.) |
| feet (U.S.) | $1.893939 \times 10^{-4}$ | miles (statute) |
| feet (U.S.) | 304.8 | millimeters |
| feet (U.S.) | $1.2 \times 10^{4}$ | mils |
| feet (U.S.) | 1/3 | yards |
| feet of air (1 atmosphere, 60°F) | $5.30 \times 10^{-4}$ | pounds/square inch |
| feet of water | 0.02950 | atm |
| feet of water | 0.8826 | inches of mercury |
| feet of water at 39.2°F | 0.030479 | kilograms/square centimeter |
| feet of water at 39.2°F | 2988.98 | newton/meter$^2$ (N/m$^2$) |
| feet of water at 39.2°F | 304.79 | kilograms/square meter |
| feet of water | 62.43 | pounds/square feet (psf) |
| feet of water at 39.2°F | 0.43352 | pounds/square inch (psi) |
| feet/hour | 0.08467 | mm/sec |
| feet/min | 0.5080 | cms/sec |
| feet/min | 0.01667 | feet/sec |
| feet/min | 0.01829 | km/hr |
| feet/min | 0.3048 | meters/min |
| feet/min | 0.01136 | miles/hr |
| feet/sec | 30.48 | cm/sec |
| feet/sec | 1.097 | km/hr |
| feet/sec | 0.5921 | knots |
| feet/sec | 18.29 | meters/min |
| feet/sec | 0.6818 | miles/hr |
| feet/sec | 0.01136 | miles/min |
| feet/sec/sec | 30.48 | cm/sec/sec |
| feet/sec/sec | 1.097 | km/hr/sec |
| feet/sec/sec | 0.3048 | meters/sec/sec |
| feet/sec/sec | 0.6818 | miles/hr/sec |
| feet/100 feet | 1.0 | percent grade |
| firkins (British) | 9 | gallons (British) |
| firkins (U.S.) | 9 | gallons (U.S.) |
| foot-candle (ft-c) | 10.764 | lumen/sq m |
| foot-poundals | $3.9951 \times 10^{-5}$ | BTU (mean) |
| foot-poundals | 0.0421420 | joules (abs) |
| foot-pounds | 0.0012854 | BTU (mean) |
| foot-pounds | 0.32389 | calories, gram (mean) |
| foot-pounds | $1.13558 \times 10^{7}$ | ergs |
| foot-pounds | 32.174 | foot-poundals |
| foot-pounds | $5.050 \times 10^{-7}$ | hp-hr |
| foot-pounds | 1.35582 | joules (abs) |
| foot-pounds | $3.241 \times 10^{-4}$ | kilogram-calories |
| foot-pounds | 0.138255 | kilogram-meters |

## Conversion Factors

| Multiply | by | to obtain |
| --- | --- | --- |
| foot-pounds | $3.766 \times 10^{-7}$ | kwh |
| foot-pounds | 0.013381 | liter-atmospheres |
| foot-pounds | $3.7662 \times 10^{-4}$ | watt-hours (abs) |
| foot-pounds/minute | $1.286 \times 10^{-3}$ | BTU/minute |
| foot-pounds/minute | 0.01667 | foot-pounds/sec |
| foot-pounds/minute | $3.030 \times 10^{-5}$ | hp |
| foot-pounds/minute | $3.241 \times 10^{-4}$ | kg-calories/min |
| foot-pounds/minute | $2.260 \times 10^{-5}$ | kw |
| foot-pounds/second | 4.6275 | BTU (mean)/hour |
| foot-pounds/second | 0.07717 | BTU/minute |
| foot-pounds/second | 0.0018182 | horsepower |
| foot-pounds/second | 0.01945 | kg-calories/min |
| foot-pounds/second | 0.001356 | kilowatts |
| foot-pounds/second | 1.35582 | watts (abs) |
| furlongs | 660.0 | feet |
| furlongs | 201.17 | meters |
| furlongs | 0.125 | miles (U.S.) |
| furlongs | 40.0 | rods |
| gallons (Br.) | $3.8125 \times 10^{-2}$ | barrels (U.S.) |
| gallons (Br.) | 4516.086 | cubic centimeters |
| gallons (Br.) | 0.16053 | cu ft |
| gallons (Br.) | 277.4 | cu inches |
| gallons (Br.) | 1230 | drams (U.S. fluid) |
| gallons (Br.) | 4.54596 | liters |
| gallons (Br.) | $7.9620 \times 10^{4}$ | minims (Br.) |
| gallons (Br.) | $7.3783 \times 10^{4}$ | minims (U.S.) |
| gallons (Br.) | 4545.96 | mL |
| gallons (Br.) | 1.20094 | gallons (U.S.) |
| gallons (Br.) | 160 | ounces (Br., fl.) |
| gallons (Br.) | 153.72 | ounces (U.S., fl.) |
| gallons (Br.) | 10 | pounds (avoirdupois) of water at 62°F |
| gallons (U.S.) | $3.068 \times 10^{-4}$ | acre-ft |
| gallons (U.S.) | 0.031746 | barrels (U.S.) |
| gallons (U.S.) | 3785.434 | cubic centimeters |
| gallons (U.S.) | 0.13368 | cubic feet (U.S.) |
| gallons (U.S.) | 231 | cubic inches |
| gallons (U.S.) | $3.785 \times 10^{-3}$ | cubic meters |
| gallons (U.S.) | $4.951 \times 10^{-3}$ | cubic yards |
| gallons (U.S.) | 1024 | drams (U.S., fluid) |
| gallons (U.S.) | 0.83268 | gallons (Br.) |
| gallons (U.S.) | 0.83267 | imperial gal |
| gallons (U.S.) | 3.78533 | liters |
| gallons (U.S.) | $6.3950 \times 10^{4}$ | minims (Br.) |
| gallons (U.S.) | $6.1440 \times 10^{4}$ | minims (U.S.) |
| gallons (U.S.) | 3785 | mL |

| Multiply | by | to obtain |
|---|---|---|
| gallons (U.S.) | 133.23 | ounces (Br., fluid) |
| gallons (U.S.) | 128 | ounces (U.S., fluid) |
| gallons | 8 | pints (liq.) |
| gallons | 4 | quarts (liq.) |
| gal water (U.S.) | 8.345 | lb of water |
| gallons/acre | 0.00935 | cu m/ha |
| gallons/day | $4.381 \times 10^{-5}$ | liters/sec |
| gpd/acre | 0.00935 | cu m/day/ha |
| gpd/acre | 9.353 | liter/day/ha |
| gallons/capita/day | 3.785 | liters/capita/day |
| gpd/cu yd | 5.0 | L/day/cu m |
| gpd/ft | 0.01242 | cu m/day/m |
| gpd/sq ft | 0.0408 | cu m/day/sq m |
| gpd/sq ft | $1.698 \times 10^{-5}$ | cubic meters/hour/sq meter |
| gpd/sq ft | 0.283 | cu meter/minute/ha |
| gpm (gal/min) | 8.0208 | cfh (cu ft/hr) |
| gpm | $2.228 \times 10^{-3}$ | cfs (cu ft/sec) |
| gpm | 4.4021 | cubic meters/hr |
| gpm | 0.00144 | MGD |
| gpm | 0.0631 | liters/sec |
| gpm/sq ft | 2.445 | cu meters/hour/sq meter |
| gpm/sq ft | 40.7 | L/min/sq meter |
| gpm/sq ft | 0.679 | liter/sec/sq meter |
| gallons/sq ft | 40.743 | liters/sq meter |
| gausses (abs) | $3.3358 \times 10^{-4}$ | electrostatic cgs units of magnetic flux density |
| gausses (abs) | 0.99966 | gausses (int) |
| gausses (abs) | 1 | lines/square centimeter |
| gausses (abs) | 6.452 | lines/sq in |
| gausses (abs) | 1 | maxwells (abs)/square centimeters |
| gausses (abs) | 6.4516 | maxwells (abs)/square inch |
| gausses (abs) | $10^{-8}$ | webers/sq cm |
| gausses (abs) | $6.452 \times 10^{-8}$ | webers/sq in |
| gausses (abs) | $10^{-4}$ | webers/sq meter |
| gilberts (abs) | 0.07958 | abampere turns |
| gilberts (abs) | 0.7958 | ampere turns |
| gilberts (abs) | $2.998 \times 10^{10}$ | electrostatic cgs units of magneto motive force |
| gilberts/cm | 0.7958 | amp-turns/cm |
| gilberts/cm | 2.021 | amp-turns/in |
| gilberts/cm | 79.58 | amp-turns/meter |
| gills (Br.) | 142.07 | cubic cm |
| gills (Br.) | 5 | ounces (British, fluid) |
| gills (U.S.) | 32 | drams (fluid) |
| gills | 0.1183 | liters |

## Conversion Factors

| Multiply | by | to obtain |
|---|---|---|
| gills | 0.25 | pints (liq.) |
| grade | 0.01571 | radian |
| grains | 0.036571 | drams (avoirdupois) |
| grains | 0.01667 | drams (troy) |
| grains (troy) | 1.216 | grains (avdp) |
| grains (troy) | 0.06480 | grams |
| grains (troy) | $6.480 \times 10^{-5}$ | kilograms |
| grains (troy) | 64.799 | milligrams |
| grains (troy) | $2.286 \times 10^{-3}$ | ounces (avdp) |
| grains (troy) | $2.0833 \times 10^{-3}$ | ounces (troy) |
| grains (troy) | 0.04167 | pennyweights (troy) |
| grains | 1/7000 | pounds (avoirdupois) |
| grains | $1.736 \times 10^{-4}$ | pounds (troy) |
| grains | $6.377 \times 10^{-8}$ | tons (long) |
| grains | $7.142 \times 10^{-8}$ | tons (short) |
| grains/imp gal | 14.254 | mg/L |
| grains/imp. gal | 14.254 | parts/million (ppm) |
| grains/U.S. gal | 17.118 | mg/L |
| grains/U.S. gal | 17.118 | parts/million (ppm) |
| grains/U.S. gal | 142.86 | lb/mil gal |
| grams | 0.5611 | drams (avdp) |
| grams | 0.25721 | drams (troy) |
| grams | 980.7 | dynes |
| grams | 15.43 | grains |
| grams | $9.807 \times 10^{-5}$ | joules/cm |
| grams | $9.807 \times 10^{-3}$ | joules/meter (newtons) |
| grams | $10^{-3}$ | kilograms |
| grams | $10^{3}$ | milligrams |
| grams | 0.0353 | ounces (avdp) |
| grams | 0.03215 | ounces (troy) |
| grams | 0.07093 | poundals |
| grams | $2.205 \times 10^{-3}$ | pounds |
| grams | $2.679 \times 10^{-3}$ | pounds (troy) |
| grams | $9.842 \times 10^{-7}$ | tons (long) |
| grams | $1.102 \times 10^{-6}$ | tons (short) |
| grams-calories | $4.1868 \times 10^{7}$ | ergs |
| gram-calories | 3.0880 | foot-pounds |
| gram-calories | $1.5597 \times 10^{-6}$ | horsepower-hr |
| gram-calories | $1.1630 \times 10^{-6}$ | kilowatt-hr |
| gram-calories | $1.1630 \times 10^{-3}$ | watt-hr |
| gram-calories | $3.968 \times 10^{-3}$ | British Thermal Units (BTU) |
| gram-calories/sec | 14.286 | BTU/hr |
| gram-centimeters | $9.2967 \times 10^{-8}$ | BTU (mean) |
| gram-centimeters | $2.3427 \times 10^{-5}$ | calories, gram (mean) |
| gram-centimeters | 980.7 | ergs |
| gram-centimeters | $7.2330 \times 10^{-5}$ | foot-pounds |

| Multiply | by | to obtain |
| --- | --- | --- |
| gram-centimeters | $9.8067 \times 10^{-5}$ | joules (abs) |
| gram-centimeters | $2.344 \times 10^{-8}$ | kilogram-calories |
| gram-centimeters | $10^{-5}$ | kilogram-meters |
| gram-centimeters | $2.7241 \times 10^{-8}$ | watt-hours |
| grams-centimeters$^2$ (moment of inertia) | $2.37305 \times 10^{-6}$ | pounds-feet$^2$ |
| grams-centimeters$^2$ (moment of inertia) | $3.4172 \times 10^{-4}$ | pounds-inch$^2$ |
| gram-centimeters/second | $1.3151 \times 10^{-7}$ | hp |
| gram-centimeters/second | $9.8067 \times 10^{-8}$ | kilowatts |
| gram-centimeters/second | 0.065552 | lumens |
| gram-centimeters/second | $9.80665 \times 10^{-5}$ | watt (abs) |
| grams/cm | $5.600 \times 10^{-3}$ | pounds/inch |
| grams/cu cm | 62.428 | pounds/cubic foot |
| grams/cu cm | 0.03613 | pounds/cubic inch |
| grams/cu cm | 8.3454 | pounds/gallon (U.S.) |
| grams/cu cm | $3.405 \times 10^{-7}$ | pounds/mil-foot |
| grams/cu ft | 35.314 | grams/cu meter |
| grams/cu ft | $10^6$ | micrograms/cu ft |
| grams/cu ft | $35.314 \times 10^6$ | micrograms/cu meter |
| grams/cu ft | $35.3145 \times 10^3$ | milligrams/cu meter |
| grams/cu ft | 2.2046 | pounds/1000 cu ft |
| grams/cu m | 0.43700 | grains/cubic foot |
| grams/cu m | 0.02832 | grams/cu ft |
| grams/cu m | $28.317 \times 10^3$ | micrograms/cu ft |
| grams/cu m | 0.06243 | pounds/cu ft |
| grams/liter | 58.417 | grains/gallon (U.S.) |
| grams/liter | $9.99973 \times 10^{-4}$ | grams/cubic centimeter |
| grams/liter | 1000 | mg/L |
| grams/liter | 1000 | parts per million (ppm) |
| grams/liter | 0.06243 | pounds/cubic foot |
| grams/liter | 8.345 | lb/1000 gal |
| grams/sq centimeter | 2.0481 | pounds/sq ft |
| grams/sq centimeter | 0.0142234 | pounds/square inch |
| grams/sq ft | 10.764 | grams/sq meter |
| grams/sq ft | $10.764 \times 10^3$ | kilograms/sq km |
| grams/sq ft | 1.0764 | milligrams/sq cm |
| grams/sq ft | $10.764 \times 10^3$ | milligrams/sq meter |
| grams/sq ft | 96.154 | pounds/acre |
| grams/sq ft | 2.204 | pounds/1000 sq ft |
| grams/sq ft | 30.73 | tons/sq mile |
| grams/sq meter | 0.0929 | grams/sq ft |
| grams/sq meter | 1000 | kilograms/sq km |
| grams/sq meter | 0.1 | milligrams/square cm |
| grams/sq meter | 1000 | milligrams/sq meter |
| grams/sq meter | 8.921 | pounds/acre |
| grams/sq meter | 0.2048 | pounds/1000 sq ft |
| grams/sq meter | 2.855 | tons/sq mile |

## Conversion Factors

| Multiply | by | to obtain |
| --- | --- | --- |
| g (gravity) | 9.80665 | meters/sec$^2$ |
| g (gravity) | 32.174 | ft/sec$^2$ |
| hand | 10.16 | cm |
| hands | 4 | inches |
| hectare (ha) | 2.471 | acre |
| hectares | $1.076 \times 10^5$ | sq feet |
| hectograms | 100 | grams |
| hectoliters | 100 | liters |
| hectometers | 100 | meters |
| hectowatts | 100 | watts |
| hemispheres | 0.5 | spheres |
| hemispheres | 4 | spherical right angles |
| hemispheres | 6.2832 | steradians |
| henries (abs) | $10^9$ | abhenries |
| henries | 1000.0 | millihenries |
| henries (abs) | $1.1126 \times 10^{-12}$ | stathenries |
| hogsheads (British) | 63 | gallons (British) |
| hogsheads (British) | 10.114 | cubic feet |
| hogsheads (U.S.) | 8.422 | cubic feet |
| hogsheads (U.S.) | 0.2385 | cubic meters |
| hogsheads (U.S.) | 63 | gallons (U.S.) |
| horsepower | 2545.08 | BTU (mean)/hour |
| horsepower | 42.44 | BTU/min |
| horsepower | $7.457 \times 10^9$ | erg/sec |
| horsepower | 33,000 | ft lb/min |
| horsepower | 550 | foot-pounds/second |
| horsepower | $7.6042 \times 10^6$ | g cm/sec |
| horsepower, electrical | 1.0004 | horsepower |
| horsepower | 10.70 | kg.-calories/min |
| horsepower | 0.74570 | kilowatts (g = 980.665) |
| horsepower | 498129 | lumens |
| horsepower, continental | 736 | watts (abs) |
| horsepower, electrical | 746 | watts (abs) |
| horsepower (boiler) | 9.803 | kw |
| horsepower (boiler) | 33.479 | BTU/hr |
| horsepower-hours | 2545 | BTU (mean) |
| horsepower-hours | $2.6845 \times 10^{13}$ | ergs |
| horsepower-hours | $6.3705 \times 10^7$ | ft poundals |
| horsepower-hours | $1.98 \times 10^6$ | foot-pounds |
| horsepower-hours | 641,190 | gram-calories |
| horsepower-hours | $2.684 \times 10^6$ | joules |
| horsepower-hours | 641.7 | kilogram-calories |
| horsepower-hours | $2.737 \times 10^5$ | kilogram-meters |
| horsepower-hours | 0.7457 | kilowatt-hours (abs) |
| horsepower-hours | 26,494 | liter atmospheres (normal) |
| horsepower-hours | 745.7 | watt-hours |

| Multiply | by | to obtain |
|---|---|---|
| hours | $4.167 \times 10^{-2}$ | days |
| hours | 60 | minutes |
| hours | 3600 | seconds |
| hours | $5.952 \times 10^{-3}$ | weeks |
| hundredweights (long) | 112 | pounds |
| hundredweights (long) | 0.05 | tons (long) |
| hundredweights (short) | 1600 | ounces (avoirdupois) |
| hundredweights (short) | 100 | pounds |
| hundredweights (short) | 0.0453592 | tons (metric) |
| hundredweights (short) | 0.0446429 | tons (long) |
| inches (British) | 2.540 | centimeters |
| inches (U.S.) | 2.54000508 | centimeters |
| inches (British) | 0.9999972 | inches (U.S.) |
| inches | $2.540 \times 10^{-2}$ | meters |
| inches | $1.578 \times 10^{-5}$ | miles |
| inches | 25.40 | millimeters |
| inches | $10^3$ | mils |
| inches | $2.778 \times 10^{-2}$ | yards |
| inches$^2$ | $6.4516 \times 10^{-4}$ | meter$^2$ |
| inches$^3$ | $1.6387 \times 10^{-5}$ | meter$^3$ |
| in. of mercury | 0.0334 | atm |
| in. of mercury | 1.133 | ft of water |
| in. of mercury (0°C) | 13.609 | inches of water (60°F) |
| in. of mercury | 0.0345 | kgs/square cm |
| in. of mercury at 32°F | 345.31 | kilograms/square meter |
| in. of mercury | 33.35 | millibars |
| in. of mercury | 25.40 | millimeters of mercury |
| in. of mercury (60°F) | 3376.85 | newton/meter$^2$ |
| in. of mercury | 70.73 | pounds/square ft |
| in. of mercury at 32°F | 0.4912 | pounds/square inch |
| in. of water | 0.002458 | atmospheres |
| in. of water | 0.0736 | in. of mercury |
| in. of water (at 4°C) | $2.540 \times 10^{-3}$ | kgs/sq cm |
| in. of water | 25.40 | kgs/square meter |
| in. of water (60°F) | 1.8663 | millimeters of mercury (0°C) |
| in. of water (60°F) | 248.84 | newton/meter$^2$ |
| in. of water | 0.5781 | ounces/square in |
| in. of water | 5.204 | pounds/square ft |
| in. of water | 0.0361 | psi |
| inches/hour | 2.54 | cm/hr |
| international ampere | .9998 | ampere (absolute) |
| international volt | 1.0003 | volts (absolute) |
| international volt | $1.593 \times 10^{-19}$ | joules (absolute) |
| international volt | $9.654 \times 10^4$ | joules |
| joules | $9.480 \times 10^{-4}$ | BTU |
| joules (abs) | $10^7$ | ergs |

## Conversion Factors

| Multiply | by | to obtain |
| --- | --- | --- |
| joules | 23.730 | foot poundals |
| joules (abs) | 0.73756 | foot-pounds |
| joules | $3.7251 \times 10^{-7}$ | horsepower hours |
| joules | $2.389 \times 10^{-4}$ | kg-calories |
| joules (abs) | 0.101972 | kilogram-meters |
| joules | $9.8689 \times 10^{-3}$ | liter atmospheres (normal) |
| joules | $2.778 \times 10^{-4}$ | watt-hrs |
| joules-sec | $1.5258 \times 10^{33}$ | quanta |
| joules/cm | $1.020 \times 10^4$ | grams |
| joules/cm | $10^7$ | dynes |
| joules/cm | 100.0 | joules/meter (newtons) |
| joules/cm | 723.3 | poundals |
| joules/cm | 22.48 | pounds |
| joules/liter | 0.02681 | BTU/cu ft |
| joules/$m^2$-sec | 0.3167 | BTU/$ft^2$-hr |
| joules/sec | 3.41304 | BTU/hr |
| joules/sec | 0.056884 | BTU/min |
| joules/sec | $1 \times 10^7$ | erg/sec |
| joules/sec | 44.254 | ft lb/min |
| joules/sec | 0.73756 | ft lb/sec |
| joules/sec | $1.0197 \times 10^4$ | g cm/sec |
| joules/sec | $1.341 \times 10^{-3}$ | hp |
| joules/sec | 0.01433 | kg cal/min |
| joules/sec | 0.001 | kilowatts |
| joules/sec | 668 | lumens |
| joules/sec | 1 | watts |
| kilograms | 564.38 | drams (avdp) |
| kilograms | 257.21 | drams (troy) |
| kilograms | 980,665 | dynes |
| kilograms | 15,432 | grains |
| kilograms | 1000 | grams |
| kilograms | 0.09807 | joules/cm |
| kilograms | 9.807 | joules/meter (newtons) |
| kilograms | $1 \times 10^6$ | milligrams |
| kilograms | 35.274 | ounces (avdp) |
| kilograms | 32.151 | ounces (troy) |
| kilograms | 70.93 | poundals |
| kilograms | 2.20462 | pounds (avdp) |
| kilograms | 2.6792 | pounds (troy) |
| kilograms | $9.84207 \times 10^{-4}$ | tons (long) |
| kilograms | 0.001 | tons (metric) |
| kilograms | 0.0011023 | tons (short) |
| kilogram-calories | 3.968 | British Thermal Units (BTU) |
| kilogram-calories | 3086 | foot-pounds |
| kilogram-calories | $1.558 \times 10^{-3}$ | horsepower-hours |
| kilogram-calories | 4186 | joules |

| Multiply | by | to obtain |
|---|---|---|
| kilogram-calories | 426.6 | kilogram-meters |
| kilogram-calories | 4.186 | kilojoules |
| kilogram-calories | $1.162 \times 10^{-3}$ | kilowatt-hours |
| kg-cal/min | 238.11 | BTU/hr |
| kg-cal/min | 3.9685 | BTU/min |
| kg-cal/min | $6.9770 \times 10^8$ | erg/sec |
| kg-cal/min | 3087.4 | ft-lb/min |
| kg-cal/min | 51.457 | ft-lb/sec |
| kg-cal/min | $7.1146 \times 10^5$ | g cm/sec |
| kg-cal/min | 0.0936 | hp |
| kg-cal/min | 69.769 | joules/sec |
| kg-cal/min | 0.0698 | kw |
| kg-cal/min | 46636 | lumens |
| kg-cal/min | 69.767 | watts |
| kgs-cms. squared | $2.373 \times 10^{-3}$ | pounds-feet squared |
| kgs-cms. squared | 0.3417 | pounds-inches squared |
| kilogram-force (kgf) | 9.80665 | newton |
| kilogram-meters | 0.0092967 | BTU (mean) |
| kilogram-meters | 2.3427 | calories, gram (mean) |
| kilogram-meters | $9.80665 \times 10^7$ | ergs |
| kilogram-meters | 232.71 | ft poundals |
| kilogram-meters | 7.2330 | foot-pounds |
| kilogram-meters | $3.6529 \times 10^{-6}$ | horsepower-hours |
| kilogram-meters | 9.80665 | joules (abs) |
| kilogram-meters | $2.344 \times 10^{-3}$ | kilogram-calories |
| kilogram-meters | $2.52407 \times 10^{-6}$ | kilowatt-hours (abs) |
| kilogram-meters | $2.7241 \times 10^{-6}$ | kilowatt-hours |
| kilogram-meters | 0.096781 | liter atmospheres (normal) |
| kilogram-meters | $6.392 \times 10^{-7}$ | pounds carbon to $CO_2$ |
| kilogram-meters | $9.579 \times 10^{-6}$ | pounds water evap. at 212°F |
| kilograms/cubic meter | $10^{-3}$ | grams/cubic cm |
| kilograms/cubic meter | 0.06243 | pounds/cubic foot |
| kilograms/cubic meter | $3.613 \times 10^{-5}$ | pounds/cubic inch |
| kilograms/cubic meter | $3.405 \times 10^{-10}$ | pounds/mil. foot |
| kilograms/$m^3$-day | 0.0624 | lb/cu ft-day |
| kilograms/cu meter-day | 62.43 | pounds/1000 cu ft-day |
| kilograms/ha | 0.8921 | pounds/acre |
| kilograms/meter | 0.6720 | pounds/foot |
| kilograms/sq cm | 980,665 | dynes |
| kilograms/sq cm | 0.96784 | atmosphere |
| kilograms/sq cm | 32.81 | feet of water |
| kilograms/sq cm | 28.96 | inches of mercury |
| kilograms/sq cm | 735.56 | mm of mercury |
| kilograms/sq cm | 2048 | pounds/sq ft |
| kilograms/sq cm | 14.22 | pounds/square inch |
| kilograms/sq km | $92.9 \times 10^{-6}$ | grams/sq ft |

## Conversion Factors

| Multiply | by | to obtain |
|---|---|---|
| kilograms/sq km | 0.001 | grams/sq meter |
| kilograms/sq km | 0.0001 | milligrams/sq cm |
| kilograms/sq km | 1.0 | milligrams/sq meter |
| kilograms/sq km | $8.921 \times 10^{-3}$ | pounds/acre |
| kilograms/sq km | $204.8 \times 10^{-6}$ | pounds/1000 sq ft |
| kilograms/sq km | $2.855 \times 10^{-3}$ | tons/sq mile |
| kilograms/sq meter | $9.6784 \times 10^{-5}$ | atmospheres |
| kilograms/sq meter | $98.07 \times 10^{-6}$ | bars |
| kilograms/sq meter | 98.0665 | dynes/sq centimeters |
| kilograms/sq meter | $3.281 \times 10^{-3}$ | feet of water at 39.2°F |
| kilograms/sq meter | 0.1 | grams/sq centimeters |
| kilograms/sq meter | $2.896 \times 10^{-3}$ | inches of mercury at 32°F |
| kilograms/sq meter | 0.07356 | mm of mercury at 0°C |
| kilograms/sq meter | 0.2048 | pounds/square foot |
| kilograms/sq meter | 0.00142234 | pounds/square inch |
| kilograms/sq mm. | $10^6$ | kg/square meter |
| kilojoule | 0.947 | BTU |
| kilojoules/kilogram | 0.4295 | BTU/pound |
| kilolines | 1000.0 | maxwells |
| kiloliters | $10^3$ | liters |
| kilometers | $10^5$ | centimeters |
| kilometers | 3281 | feet |
| kilometers | $3.937 \times 10^4$ | inches |
| kilometers | $10^3$ | meters |
| kilometers | 0.53961 | miles (nautical) |
| kilometers | 0.6214 | miles (statute) |
| kilometers | $10^6$ | millimeters |
| kilometers | 1093.6 | yards |
| kilometers/hr | 27.78 | cm/sec |
| kilometers/hr | 54.68 | feet/minute |
| kilometers/hr | 0.9113 | ft/sec |
| kilometers/hr | 0.5396 | knot |
| kilometers/hr | 16.67 | meters/minute |
| kilometers/hr | 0.2778 | meters/sec |
| kilometers/hr | 0.6214 | miles/hour |
| kilometers/hour/sec | 27.78 | cms/sec/sec |
| kilometers/hour/sec | 0.9113 | ft/sec/sec |
| kilometers/hour/sec | 0.2778 | meters/sec/sec |
| kilometers/hour/sec | 0.6214 | miles/hr/sec |
| kilometers/min | 60 | kilometers/hour |
| kilonewtons/sq m | 0.145 | psi |
| kilowatts | 56.88 | BTU/min |
| kilowatts | $4.425 \times 10^4$ | foot-pounds/min |
| kilowatts | 737.6 | ft-lb/sec |
| kilowatts | 1.341 | horsepower |
| kilowatts | 14.34 | kg-cal/min |

| Multiply | by | to obtain |
| --- | --- | --- |
| kilowatts | $10^3$ | watts |
| kilowatt-hrs | 3413 | BTU (mean) |
| kilowatt-hrs | $3.600 \times 10^{13}$ | ergs |
| kilowatt-hrs | $2.6552 \times 10^6$ | foot-pounds |
| kilowatt-hrs | 859,850 | gram-calories |
| kilowatt-hrs | 1.341 | horsepower hours |
| kilowatt-hrs | $3.6 \times 10^6$ | joules |
| kilowatt-hrs | 860.5 | kg-calories |
| kilowatt-hrs | $3.6709 \times 10^5$ | kilogram-meters |
| kilowatt-hrs | 3.53 | pounds of water evaporated from and at 212°F |
| kilowatt-hrs | 22.75 | pounds of water raised from 62° to 212°F |
| knots | 6080 | feet/hr |
| knots | 1.689 | feet/sec |
| knots | 1.8532 | kilometers/hr |
| knots | 0.5144 | meters/sec |
| knots | 1.0 | miles (nautical)/hour |
| knots | 1.151 | miles (statute)/hour |
| knots | 2,027 | yards/hr |
| lambert | 2.054 | candle/in$^2$ |
| lambert | 929 | footlambert |
| lambert | 0.3183 | stilb |
| langley | 1 | 15° gram-calorie/cm$^2$ |
| langley | 3.6855 | BTU/ft$^2$ |
| langley | 0.011624 | Int. kw-hr/m$^2$ |
| langley | 4.1855 | joules (abs)/cm$^2$ |
| leagues (nautical) | 3 | miles (nautical) |
| leagues (statute) | 3 | miles (statute) |
| light years | 63,274 | astronomical units |
| light years | $9.4599 \times 10^{12}$ | kilometers |
| light years | $5.8781 \times 10^{12}$ | miles |
| lignes (Paris lines) | 1/12 | ponces (Paris inches) |
| lines/sq cm | 1.0 | gausses |
| lines/sq in | 0.1550 | gausses |
| lines/sq in | $1.550 \times 10^{-9}$ | webers/sq cm |
| lines/sq in | $10^{-8}$ | webers/sq in |
| lines/sq in | $1.550 \times 10^{-5}$ | webers/sq meter |
| links (engineer's) | 12.0 | inches |
| links (Gunter's) | 0.01 | chains (Gunter's) |
| links (Gunter's) | 0.66 | feet |
| links (Ramden's) | 0.01 | chains (Ramden's) |
| links (Ramden's) | 1 | feet |
| links (surveyor's) | 7.92 | inches |
| liters | $8.387 \times 10^{-3}$ | barrels (U.S.) |
| liters | 0.02838 | bushels (U.S. dry) |

## Conversion Factors

| Multiply | by | to obtain |
|---|---|---|
| liters | 1000.028 | cubic centimeters |
| liters | 0.035316 | cubic feet |
| liters | 61.025 | cu inches |
| liters | $10^{-3}$ | cubic meters |
| liters | $1.308 \times 10^{-3}$ | cubic yards |
| liters | 270.5179 | drams (U.S. fl) |
| liters | 0.21998 | gallons (Br.) |
| liters | 0.26417762 | gallons (U.S.) |
| liters | 16,894 | minims (Br.) |
| liters | 16,231 | minims (U.S.) |
| liters | 35.196 | ounces (Br. fl) |
| liters | 33.8147 | ounces (U.S. fl) |
| liters | 2.113 | pints (liq.) |
| liters | 1.0566828 | quarts (U.S. liq.) |
| liter-atmospheres (normal) | 0.096064 | BTU (mean) |
| liter-atmospheres (normal) | 24.206 | calories, gram (mean) |
| liter-atmospheres (normal) | $1.0133 \times 10^9$ | ergs |
| liter-atmospheres (normal) | 74.735 | foot-pounds |
| liter-atmospheres (normal) | $3.7745 \times 10^{-5}$ | horsepower hours |
| liter-atmospheres (normal) | 101.33 | joules (abs) |
| liter-atmospheres (normal) | 10.33 | kilogram-meters |
| liter-atmospheres (normal) | $2.4206 \times 10^{-2}$ | kilogram calories |
| liter-atmospheres (normal) | $2.815 \times 10^{-5}$ | kilowatt-hours |
| liter/cu m-sec | 60.0 | cfm/1000 cu ft |
| liters/minute | $5.885 \times 10^{-4}$ | cubic feet/sec |
| liters/minute | $4.403 \times 10^{-3}$ | gallons/sec |
| liter/person-day | 0.264 | gpcd |
| liters/sec | 2.119 | cu ft /min |
| liters/sec | $3.5316 \times 10^{-2}$ | cu ft /sec |
| liters/sec | 15.85 | gallons/minute |
| liters/sec | 0.02282 | MGD |
| $\log_{10} N$ | 2.303 | $\log_e N$ or $\ln N$ |
| $\log_e N$ or $\ln N$ | 0.4343 | $\log_{10} N$ |
| lumens | 0.07958 | candle-power (spherical) |
| lumens | 0.00147 | watts of maximum visibility radiation |
| lumens/sq. centimeters | 1 | lamberts |
| lumens/sq cm/steradian | 3.1416 | lamberts |
| lumens/sq ft | 1 | foot-candles |
| lumens/sq ft | 10.764 | lumens/sq meter |
| lumens/sq ft/steradian | 3.3816 | millilamberts |
| lumens/sq meter | 0.09290 | foot-candles or lumens/sq |
| lumens/sq meter | $10^{-4}$ | phots |
| lux | 0.09290 | foot-candles |
| lux | 1 | lumens/sq meter |
| lux | $10^{-4}$ | phots |

| Multiply | by | to obtain |
| --- | --- | --- |
| maxwells | 0.001 | kilolines |
| maxwells | $10^{-8}$ | webers |
| megajoule | 0.3725 | horsepower-hour |
| megalines | $10^6$ | maxwells |
| megohms | $10^{12}$ | microhms |
| megohms | $10^6$ | ohms |
| meters | $10^{10}$ | angstrom units |
| meters | 100 | centimeters |
| meters | 0.5467 | fathoms |
| meters | 3.280833 | feet (U.S.) |
| meters | 39.37 | inches |
| meters | $10^{-3}$ | kilometers |
| meters | $5.396 \times 10^{-4}$ | miles (naut.) |
| meters | $6.2137 \times 10^{-4}$ | miles (statute) |
| meters | $10^3$ | millimeters |
| meters | $10^9$ | millimicrons |
| meters | 1.09361 | yards (U.S.) |
| meters | 1.179 | varas |
| meter-candles | 1 | lumens/sq meter |
| meter-kilograms | $9.807 \times 10^7$ | centimeter-dynes |
| meter-kilograms | $10^5$ | centimeter-grams |
| meter-kilograms | 7.233 | pound-feet |
| meters/minute | 1.667 | centimeters/sec |
| meters/minute | 3.281 | feet/minute |
| meters/minute | 0.05468 | feet/second |
| meters/minute | 0.06 | kilograms/hour |
| meters/minute | 0.03238 | knots |
| meters/minute | 0.03728 | miles/hour |
| meters/second | 196.8 | feet/minute |
| meters/second | 3.281 | feet/second |
| meters/second | 3.6 | kilometers/hour |
| meters/second | 0.06 | kilometers/min |
| meters/second | 1.944 | knots |
| meters/second | 2.23693 | miles/hour |
| meters/second | 0.03728 | miles/minute |
| meters/sec/sec | 100.0 | cm/sec/sec |
| meters/sec/sec | 3.281 | feet/sec/sec |
| meters/sec/sec | 3.6 | km/hour/sec |
| meters/sec/sec | 2.237 | miles/hour/sec |
| microfarad | $10^{-6}$ | farads |
| micrograms | $10^{-6}$ | grams |
| micrograms/cu ft | $10^{-6}$ | grams/cu ft |
| micrograms/cu ft | $35.314 \times 10^{-6}$ | grams/cu m |
| micrograms/cu ft | 35.314 | microgram/cu m |
| micrograms/cu ft | $35.314 \times 10^{-3}$ | milligrams/cu m |
| micrograms/cu ft | $2.2046 \times 10^{-6}$ | pounds/1000 cu ft |

## Conversion Factors

| Multiply | by | to obtain |
|---|---|---|
| micrograms/cu m | $28.317 \times 10^{-9}$ | grams/cu ft |
| micrograms/cu m | $10^{-6}$ | grams/ cu m |
| micrograms/cu m | 0.02832 | micrograms/cu ft |
| micrograms/cu m | 0.001 | milligrams/cu m |
| micrograms/cu m | $62.43 \times 10^{-9}$ | pounds/1000 cu ft |
| micrograms/cu m | $\dfrac{0.02404}{\text{molecular weight of gas}}$ | ppm by volume (20°C) |
| micrograms/cu m | $834.7 \times 10^{-6}$ | ppm by weight |
| micrograms/liter | 1000.0 | micrograms/cu m |
| micrograms/liter | 1.0 | milligrams/cu m |
| micrograms/liter | $62.43 \times 10^{-9}$ | pounds/cu ft |
| micrograms/liter | $\dfrac{24.04}{\text{molecular weight of gas}}$ | ppm by volume (20°C) |
| micrograms/liter | 0.834.7 | ppm by weight |
| microhms | $10^{-12}$ | megohms |
| microhms | $10^{-6}$ | ohms |
| microliters | $10^{-6}$ | liters |
| microns | $10^4$ | angstrom units |
| microns | $1 \times 10^{-4}$ | centimeters |
| microns | $3.9370 \times 10^{-5}$ | inches |
| microns | $10^{-6}$ | meters |
| miles (naut.) | 6080.27 | feet |
| miles (naut.) | 1.853 | kilometers |
| miles (naut.) | 1.853 | meters |
| miles (naut.) | 1.1516 | miles (statute) |
| miles (naut.) | 2027 | yards |
| miles (statute) | $1.609 \times 10^5$ | centimeters |
| miles (statute) | 5280 | feet |
| miles (statute) | $6.336 \times 10^4$ | inches |
| miles (statute) | 1.609 | kilometers |
| miles (statute) | 1609 | meters |
| miles (statute) | 0.8684 | miles (naut.) |
| miles (statute) | 320 | rods |
| miles (statute) | 1760 | yards |
| miles/hour | 44.7041 | centimeter/second |
| miles/hour | 88 | feet/min |
| miles/hour | 1.4667 | feet/sec |
| miles/hour | 1.6093 | kilometers/hour |
| miles/hour | 0.02682 | km/min |
| miles/hour | 0.86839 | knots |
| miles/hour | 26.82 | meters/min |
| miles/hour | 0.447 | meters/sec |
| miles/hour | 0.1667 | miles/min |
| miles/hour/sec | 44.70 | cms/sec/sec |
| miles/hour/sec | 1.4667 | ft/sec/sec |
| miles/hour/sec | 1.6093 | km/hour/sec |

| Multiply | by | to obtain |
| --- | --- | --- |
| miles/hour/sec | 0.4470 | m/sec/sec |
| miles/min | 2682 | centimeters/sec |
| miles/min | 88 | ft/sec |
| miles/min | 1.609 | km/min |
| miles/min | 0.8684 | knots/min |
| miles/min | 60 | miles/hour |
| miles-feet | $9.425 \times 10^{-6}$ | cu inches |
| millibars | 0.00987 | atmospheres |
| millibars | 0.30 | inches of mercury |
| millibars | 0.75 | millimeters of mercury |
| milliers | $10^3$ | kilograms |
| millimicrons | $1 \times 10^{-9}$ | meters |
| milligrams | 0.01543236 | grains |
| milligrams | $10^{-3}$ | grams |
| milligrams | $10^{-6}$ | kilograms |
| milligrams | $3.5274 \times 10^{-5}$ | ounces (avdp) |
| milligrams | $2.2046 \times 10^{-6}$ | pounds (avdp) |
| milligrams/assay ton | 1 | ounces (troy)/ton (short) |
| milligrams/cu m | $283.2 \times 10^{-6}$ | grams/cu ft |
| milligrams/cu m | 0.001 | grams/cu m |
| milligrams/cu m | 1000.0 | micrograms/cu m |
| milligrams/cu m | 28.32 | micrograms/cu ft |
| milligrams/cu m | 1.0 | micrograms/liter |
| milligrams/cu m | $62.43 \times 10^{-6}$ | pounds/1000 cu ft |
| milligrams/cu m | $\dfrac{24.04}{\text{molecular weight of gas}}$ | ppm by volume (20°C) |
| milligrams/cu m | 0.8347 | ppm by weight |
| milligrams/joule | 5.918 | pounds/horsepower-hour |
| milligrams/liter | 0.05841 | grains/gallon |
| milligrams/liter | 0.07016 | grains/imp. gal |
| milligrams/liter | 0.0584 | grains/U.S. gal |
| milligrams/liter | 1.0 | parts/million |
| milligrams/liter | 8.345 | lb/mil gal |
| milligrams/sq cm | 0.929 | grams/sq ft |
| milligrams/sq cm | 10.0 | grams/sq meter |
| milligrams/sq cm | $10^4$ | kilograms/sq km |
| milligrams/sq cm | $10^4$ | milligrams/sq meter |
| milligrams/sq cm | 2.048 | pounds/1000 sq ft |
| milligrams/sq cm | 89.21 | pounds/acre |
| milligrams/sq cm | 28.55 | tons/sq mile |
| milligrams/sq meter | $92.9 \times 10^{-6}$ | grams/sq ft |
| milligrams/sq meter | 0.001 | grams/sq meter |
| milligrams/sq meter | 1.0 | kilograms/sq km |
| milligrams/sq meter | 0.0001 | milligrams/sq cm |
| milligrams/sq meter | $8.921 \times 10^{-3}$ | pounds/acre |
| milligrams/sq meter | $204.8 \times 10^{-6}$ | pounds/1000 sq ft |

## Conversion Factors

| Multiply | by | to obtain |
|---|---|---|
| milligrams/sq meter | $2.855 \times 10^{-3}$ | tons/sq mile |
| millihenries | 0.001 | henries |
| milliliters | 1 | cubic centimeters |
| milliliters | $3.531 \times 10^{-5}$ | cu ft |
| milliliters | $6.102 \times 10^{-2}$ | cu in |
| milliliters | $10^{-6}$ | cu m |
| milliliters | $2.642 \times 10^{-4}$ | gal (U.S.) |
| milliliters | $10^{-3}$ | liters |
| milliliters | 0.03381 | ounces (U.S. fl) |
| millimeters | 0.1 | centimeters |
| millimeters | $3.281 \times 10^{-3}$ | feet |
| millimeters | 0.03937 | inches |
| millimeters | $10^{-6}$ | kilometers |
| millimeters | 0.001 | meters |
| millimeters | $6.214 \times 10^{-7}$ | miles |
| millimeters | 39.37 | mils |
| millimeters | $1.094 \times 10^{-3}$ | yards |
| millimeters of mercury | $1.316 \times 10^{-3}$ | atmospheres |
| millimeters of mercury | 0.0394 | inches of mercury |
| millimeters of mercury (0°C) | 0.5358 | inches of water (60°F) |
| millimeters of mercury | $1.3595 \times 10^{-3}$ | kg/sq cm |
| millimeter of mercury (0°C) | 133.3224 | newton/meter$^2$ |
| millimeters of mercury | 0.01934 | pounds/sq in |
| millimeters/sec | 11.81 | feet/hour |
| million gallons | 306.89 | acre-ft |
| million gallons | 3785.0 | cubic meters |
| million gallons | 3.785 | mega liters ($1 \times 10^6$) |
| million gallons/day (MGD) | 1.547 | cu ft/sec |
| MGD | 3785 | cu m/day |
| MGD | 0.0438 | cubic meters/sec |
| MGD | 43.808 | liters/sec |
| MGD/acre | 9360 | cu m/day/ha |
| MGD/acre | 0.039 | cu meters/hour/sq meter |
| mils | 0.002540 | centimeters |
| mils | $8.333 \times 10^{-5}$ | feet |
| mils | 0.001 | inches |
| mils | $2.540 \times 10^{-8}$ | kilometers |
| mils | 25.40 | microns |
| mils | $2.778 \times 10^{-5}$ | yards |
| miner's in. | 1.5 | cu ft/min |
| miner's inches (Ariz., Calif. Mont., and Ore.) | 0.025 | cubic feet/second |
| miner's in. (Colorado) | 0.02604 | cubic feet/second |
| miner's inches (Idaho, Kan., Neb., Nev., N. Mex., N. Dak., S. Dak. and Utah) | 0.020 | cubic feet/second |

| Multiply | by | to obtain |
| --- | --- | --- |
| minims (British) | 0.05919 | cubic centimeter |
| minims (U.S.) | 0.06161 | cubic centimeters |
| minutes (angles) | 0.01667 | degrees |
| minutes (angles) | $1.852 \times 10^{-4}$ | quadrants |
| minutes (angles) | $2.909 \times 10^{-4}$ | radians |
| minutes (angle) | 60 | seconds (angle) |
| months (mean calendar) | 30.4202 | days |
| months (mean calendar) | 730.1 | hours |
| months (mean calendar) | 43805 | minutes |
| months (mean calendar) | $2.6283 \times 10^6$ | seconds |
| myriagrams | 10 | kilograms |
| myriameters | 10 | kilometers |
| myriawatts | 10 | kilowatts |
| nepers | 8.686 | decibels |
| newtons | $10^5$ | dynes |
| newtons | 0.10197 | kilograms |
| newtons | 0.22481 | pounds |
| newtons/sq meter | 1.00 | pascals (Pa) |
| noggins (British) | 1/32 | gallons (British) |
| No./cu.cm. | $28.316 \times 10^3$ | No./cu ft |
| No./cu.cm. | $10^6$ | No./cu meter |
| No./cu.cm. | 1000.0 | No./liter |
| No./cu.ft. | $35.314 \times 10^{-6}$ | No./cu cm |
| No./cu.ft. | 35.314 | No./cu meter |
| No./cu.ft. | $35.314 \times 10^{-3}$ | No./liter |
| No./cu. meter | $10^{-6}$ | No./cu cm |
| No./cu. meter | $28.317 \times 10^{-3}$ | No./cu ft |
| No./cu. meter | 0.001 | No./liter |
| No./liter | 0.001 | No./cu cm |
| No./liter | 28.316 | No./cu ft |
| No./liter | 1000.0 | No./cu meter |
| oersteds (abs) | 1 | electromagnetic cgs units of magnetizing force |
| oersteds (abs) | $2.9978 \times 10^{10}$ | electrostatic cgs units of magnetizing force |
| ohms | $10^9$ | abohms |
| ohms | $1.1126 \times 10^{-12}$ | statohms |
| ohms | $10^{-6}$ | megohms |
| ohms | $10^6$ | microhms |
| ohms (International) | 1.0005 | ohms (absolute) |
| ounces (avdp) | 16 | drams (avoirdupois) |
| ounces (avdp) | 7.2917 | drams (troy) |
| ounces (avdp) | 437.5 | grains |
| ounces (avdp) | 28.349527 | grams |
| ounces (avdp) | 0.028350 | kilograms |
| ounces (avdp) | $2.8350 \times 10^4$ | milligrams |

# Conversion Factors

| Multiply | by | to obtain |
|---|---|---|
| ounces (avdp) | 0.9114583 | ounces (troy) |
| ounces (avdp) | 0.0625 | pounds (avoirdupois) |
| ounces (avdp) | 0.075955 | pounds (troy) |
| ounces (avdp) | $2.790 \times 10^{-5}$ | tons (long) |
| ounces (avdp) | $2.835 \times 10^{-5}$ | tons (metric) |
| ounces (avdp) | $3.125 \times 10^{-5}$ | tons (short) |
| ounces (Br. fl) | $2.3828 \times 10^{-4}$ | barrels (U.S.) |
| ounces (Br. fl) | $1.0033 \times 10^{-3}$ | cubic feet |
| ounces (Br. fl) | 1.73457 | cubic inches |
| ounces (Br. fl) | 7.6860 | drams (U.S. fl) |
| ounces (Br. fl) | $6.250 \times 10^{-3}$ | gallons (Br.) |
| ounces (Br. fl) | 0.07506 | gallons (U.S.) |
| ounces (Br. fl) | $2.84121 \times 10^{-2}$ | liters |
| ounces (Br. fl) | 480 | minims (Br.) |
| ounces (Br. fl) | 461.160 | minims (U.S.) |
| ounces (Br. fl) | 28.4121 | mL |
| ounces (Br. fl) | 0.9607 | ounces (U.S. fl) |
| ounces (troy) | 17.554 | drams (avdp) |
| ounces (troy) | 8 | drams (troy) |
| ounces (troy) | 480 | grains (troy) |
| ounces (troy) | 31.103481 | grams |
| ounces (troy) | 0.03110 | kilograms |
| ounces (troy) | 1.09714 | ounces (avoirdupois) |
| ounces (troy) | 20 | pennyweights (troy) |
| ounces (troy) | 0.068571 | pounds (avdp) |
| ounces (troy) | 0.08333 | pounds (troy) |
| ounces (troy) | $3.061 \times 10^{-5}$ | tons (long) |
| ounces (troy) | $3.429 \times 10^{-5}$ | tons (short) |
| ounces (U.S. fl) | $2.48 \times 10^{-4}$ | barrels (U.S.) |
| ounces (U.S. fl) | 29.5737 | cubic centimeters |
| ounces (U.S. fl) | $1.0443 \times 10^{-3}$ | cubic feet |
| ounces (U.S. fl) | 1.80469 | cubic inches |
| ounces (U.S. fl) | 8 | drams (fluid) |
| ounces (U.S. fl) | $6.5053 \times 10^{-3}$ | gallons (Br.) |
| ounces (U.S. fl) | $7.8125 \times 10^{-3}$ | gallons (U.S.) |
| ounces (U.S. fl) | 29.5729 | milliliters |
| ounces (U.S. fl) | 499.61 | minims (Br.) |
| ounces (U.S. fl) | 480 | minims (U.S.) |
| ounces (U.S. fl) | 1.0409 | ounces (Br. fl) |
| ounces/sq inch | 4309 | dynes/sq cm |
| ounces/sq. inch | 0.0625 | pounds/sq inch |
| paces | 30 | inches |
| palms (British) | 3 | inches |
| parsecs | 3.260 | light years |
| parsecs | $3.084 \times 10^{13}$ | kilometers |
| parsecs | $3.084 \times 10^{16}$ | meters |

| Multiply | by | to obtain |
|---|---|---|
| parsec | $19 \times 10^{12}$ | miles |
| parts/billion (ppb) | $10^{-3}$ | mg/L |
| parts/million (ppm) | 0.07016 | grains/imp. gal. |
| parts/million | 0.058417 | grains/gallon (U.S.) |
| parts/million | 1.0 | mg/liter |
| parts/million | 8.345 | lbs/million gallons |
| ppm by volume (20°C) | $\dfrac{\text{molecular weight of gas}}{24.04}$ | micrograms/liter |
| ppm by volume (20°C) | $\dfrac{\text{molecular weight of gas}}{0.02404}$ | micrograms/cu meter |
| ppm by volume (20°C) | $\dfrac{\text{molecular weight of gas}}{24.04}$ | milligrams/cu meter |
| ppm by volume (20°C) | $\dfrac{\text{molecular weight of gas}}{28.8}$ | ppm by weight |
| ppm by volume (20°C) | $\dfrac{\text{molecular weight of gas}}{385.1 \times 10^6}$ | pounds/cu ft |
| ppm by weight | $1.198 \times 10^{-3}$ | micrograms/cu meter |
| ppm by weight | 1.198 | micrograms/liter |
| ppm by weight | 1.198 | milligrams/cu meter |
| ppm by weight | $\dfrac{28.8}{\text{molecular weight of gas}}$ | ppm by volume (20°C) |
| ppm by weight | $7.48 \times 10^{-6}$ | pounds/cu ft |
| pecks (British) | 0.25 | bushels (British) |
| pecks (British) | 554.6 | cubic inches |
| pecks (British) | 9.091901 | liters |
| pecks (U.S.) | 0.25 | bushels (U.S.) |
| pecks (U.S.) | 537.605 | cubic inches |
| pecks (U.S.) | 8.809582 | liters |
| pecks (U.S.) | 8 | quarts (dry) |
| pennyweights | 24 | grains |
| pennyweights | 1.555174 | grams |
| pennyweights | 0.05 | ounces (troy) |
| pennyweights (troy) | $4.1667 \times 10^{-3}$ | pounds (troy) |
| perches (masonry) | 24.75 | cubic feet |
| phots | 929.0 | foot-candles |
| phots | 1 | lumen incident/sq cm |
| phots | $10^4$ | lux |
| picas (printers') | 1/6 | inches |
| pieds (French feet) | 0.3249 | meters |
| pints (dry) | 33.6003 | cubic inches |
| pints (liq.) | 473.179 | cubic centimeters |
| pints (liq.) | 0.01671 | cubic feet |
| pints (liq.) | $4.732 \times 10^{-4}$ | cubic meters |
| pints (liq.) | $6.189 \times 10^{-4}$ | cubic yards |
| pints (liq.) | 0.125 | gallons |
| pints (liq.) | 0.4732 | liters |

## Conversion Factors

| Multiply | by | to obtain |
|---|---|---|
| pints (liq.) | 16 | ounces (U.S. fluid) |
| pints (liq.) | 0.5 | quarts (liq.) |
| planck's constant | $6.6256 \times 10^{-27}$ | erg-seconds |
| poise | 1.00 | gram/cm sec |
| poise | 0.1 | newton-second/meter$^2$ |
| population equivalent (PE) | 0.17 | pounds BOD |
| pottles (British) | 0.5 | gallons (British) |
| pouces (Paris inches) | 0.02707 | meters |
| pouces (Paris inches) | 0.08333 | pieds (Paris feet) |
| poundals | 13,826 | dynes |
| poundals | 14.0981 | grams |
| poundals | $1.383 \times 10^{-3}$ | joules/cm |
| poundals | 0.1383 | joules/meter (newton) |
| poundals | 0.01410 | kilograms |
| poundals | 0.031081 | pounds |
| pounds (avdp) | 256 | drams (avdp) |
| pounds (avdp) | 116.67 | drams (troy) |
| pounds (avdp) | 444,823 | dynes |
| pounds (avdp) | 7000 | grains |
| pounds (avdp) | 453.5924 | grams |
| pounds (avdp) | 0.04448 | joules/cm |
| pounds (avdp) | 4.448 | joules/meter (newtons) |
| pounds (avdp) | 0.454 | kilograms |
| pounds (avdp) | $4.5359 \times 10^5$ | milligrams |
| pounds (avdp) | 16 | ounces (avdp) |
| pounds (avdp) | 14.5833 | ounces (troy) |
| pounds (avdp) | 32.17 | poundals |
| pounds (avdp) | 1.2152778 | pounds (troy) |
| pounds (avdp) | $4.464 \times 10^{-4}$ | tons (long) |
| pounds (avdp) | 0.0005 | tons (short) |
| pounds (troy) | 210.65 | drams (avdp) |
| pounds (troy) | 96 | drams (troy) |
| pounds (troy) | 5760 | grains |
| pounds (troy) | 373.2418 | grams |
| pounds (troy) | 0.37324 | kilograms |
| pounds (troy) | $3.7324 \times 10^5$ | milligrams |
| pounds (troy) | 13.1657 | ounces (avdp) |
| pounds (troy) | 12.0 | ounces (troy) |
| pounds (troy) | 240.0 | pennyweights (troy) |
| pounds (troy) | 0.8229 | pounds (avdp) |
| pounds (troy) | $3.6735 \times 10^{-4}$ | tons (long) |
| pounds (troy) | $3.7324 \times 10^{-4}$ | tons (metric) |
| pounds (troy) | $4.1143 \times 10^{-4}$ | tons (short) |
| pounds (avdp)-force | 4.448 | newtons |
| pounds-force-sec/ft$^2$ | 47.88026 | newton-sec/meter$^2$ |
| pounds (avdp)-mass | 0.4536 | kilograms |

| Multiply | by | to obtain |
| --- | --- | --- |
| pounds-mass/ft$^3$ | 16.0185 | kilogram/meter$^3$ |
| pounds-mass/ft-sec | 1.4882 | mewton-sec/meter$^2$ |
| pounds of BOD | 5.882 | population equivalent (PE) |
| pounds of carbon to $CO_2$ | 14,544 | BTU (mean) |
| pounds of water | 0.0160 | cu ft |
| pounds of water | 27.68 | cu in |
| pounds of water | 0.1198 | gallons |
| pounds of water evaporated at 212°F | 970.3 | BTU |
| pounds of water per min | $2.699 \times 10^{-4}$ | cubic feet/sec |
| pound-feet | 13,825 | centimeter-grams |
| pound-feet (torque) | $1.3558 \times 10^7$ | dyne-centimeters |
| pound-feet | 0.1383 | meter-kilograms |
| pounds-feet squared | 421.3 | kg-cm squared |
| pounds-feet squared | 144 | pounds-inches squared |
| pounds-inches squared | 2926 | kg-cm squared |
| pounds-inches squared | $6.945 \times 10^{-3}$ | pounds-feet squared |
| pounds/acre | 0.0104 | grams/sq ft |
| pounds/acre | 0.1121 | grams/sq meter |
| pounds/acre | 1.121 | kg/ha |
| pounds/acre | 112.1 | kilograms/sq km |
| pounds/acre | 0.01121 | milligrams/sq cm |
| pounds/acre | 112.1 | milligrams/sq meter |
| pounds/acre | 0.023 | pounds/1000 sq ft |
| pounds/acre | 0.32 | tons/sq mile |
| pounds/acre/day | 0.112 | g/day/sq m |
| pounds/cu ft | 0.0160 | g/mL |
| pounds/cu ft | 16.02 | kg/cu m |
| pounds/cu ft | $16.018 \times 10^9$ | micrograms/cu meter |
| pounds/cu ft | $16.018 \times 10^6$ | micrograms/liter |
| pounds/cu ft | $16.018 \times 10^6$ | milligrams/cu meter |
| pounds/cu ft | $\dfrac{385.1 \times 10^6}{\text{molecular weight of gas}}$ | ppm by volume (20°C) |
| pounds/cu ft | $133.7 \times 10^3$ | ppm by weight |
| pounds/cu ft | $5.787 \times 10^{-4}$ | lb/cu in |
| pounds/cu ft | $5.456 \times 10^{-9}$ | pounds/mil-foot |
| pounds/1000 cu ft | 0.35314 | grams/cu ft |
| pounds/1000 cu ft | 16.018 | grams/cu m |
| pounds/1000 cu ft | $353.14 \times 10^3$ | micrograms/cu ft |
| pounds/1000 cu ft | $16.018 \times 10^6$ | microgram/cu m |
| pounds/1000 cu ft | $16.018 \times 10^3$ | milligrams/cu m |
| pounds/cubic inch | 27.68 | grams/cubic cm |
| pounds/cubic inch | $2.768 \times 10^4$ | kgs/cubic meter |
| pounds/cubic inch | 1728 | pounds/cubic foot |
| pounds/cubic inch | $9.425 \times 10^{-6}$ | pounds/mil foot |
| pounds/day/acre-ft | 3.68 | g/day/cu m |
| pounds/day/cu ft | 16 | kg/day/cu m |

## Conversion Factors

| Multiply | by | to obtain |
|---|---|---|
| pounds/day/cu yd | 0.6 | kg/day/cu m |
| pounds/day/sq ft | 4,880 | g/day/sq m |
| pounds/ft | 1.488 | kg/m |
| pounds/gal | 454 g/3.7851L = 119.947 | g/liter |
| pounds/1000-gal | 120 | g/1000-liters |
| pounds/horsepower-hour | 0.169 | mg/joule |
| pounds/in | 178.6 | g/cm |
| pounds/mil-foot | $2.306 \times 10^6$ | gms/cu cm |
| pounds/mil gal | 0.12 | g/cu m |
| pounds/sq ft | $4.725 \times 10^{-4}$ | atmospheres |
| pounds/sq ft | 0.01602 | ft of water |
| pounds/sq ft | 0.01414 | inches of mercury |
| pounds/sq ft | $4.8824 \times 10^{-4}$ | kgs/sq cm |
| pounds/sq ft | 4.88241 | kilograms/square meter |
| pounds/sq ft | 47.9 | newtons/sq m |
| pounds/sq ft | $6.944 \times 10^{-3}$ | pounds/sq inch |
| pounds/1000 sq ft | 0.4536 | grams/sq ft |
| pounds/1000 sq ft | 4.882 | grams/sq meter |
| pounds/1000 sq ft | 4882.4 | kilograms/sq km |
| pounds/1000 sq ft | 0.4882 | milligrams/sq cm |
| pounds/1000 sq ft | 4882.4 | milligrams/sq meter |
| pounds/1000 sq ft | 43.56 | pounds/acre |
| pounds/1000 sq ft | 13.94 | tons/sq mile |
| pounds/sq in | 0.068046 | atmospheres |
| pounds/sq in | 2.307 | ft of water |
| pounds/sq in | 70.307 | grams/square centimeter |
| pounds/sq in | 2.036 | in of mercury |
| pounds/sq in | 0.0703 | kgs/square cm |
| pounds/sq in | 703.07 | kilograms/square meter |
| pounds/sq in | 51.715 | millimeters of mercury |
| pounds/sq in | 6894.76 | newton/meter$^2$ |
| pounds/sq in | 51.715 | millimeters of mercury at 0°C |
| pounds/sq in | 144 | pounds/sq foot |
| pounds/sq in (abs) | 1 | pound/sq in (gage) + 14.696 |
| proof (U.S.) | 0.5 | percent alcohol by volume |
| puncheons (British) | 70 | gallons (British) |
| quadrants (angle) | 90 | degrees |
| quadrants (angle) | 5400 | minutes |
| quadrants (angle) | $3.24 \times 10^5$ | seconds |
| quadrants (angle) | 1.571 | radians |
| quarts (dry) | 67.20 | cubic inches |
| quarts (liq.) | 946.4 | cubic centimeters |
| quarts (liq.) | 0.033420 | cubic feet |
| quarts (liq.) | 57.75 | cubic inches |
| quarts (liq.) | $9.464 \times 10^{-4}$ | cubic meters |
| quarts (liq.) | $1.238 \times 10^{-3}$ | cubic yards |

| Multiply | by | to obtain |
| --- | --- | --- |
| quarts (liq.) | 0.25 | gallons |
| quarts (liq.) | 0.9463 | liters |
| quarts (liq.) | 32 | ounces (U.S., fl) |
| quarts (liq.) | 0.832674 | quarts (British) |
| quintals (long) | 112 | pounds |
| quintals (metric) | 100 | kilograms |
| quintals (short) | 100 | pounds |
| quires | 24 | sheets |
| radians | 57.29578 | degrees |
| radians | 3438 | minutes |
| radians | 0.637 | quadrants |
| radians | $2.063 \times 10^5$ | seconds |
| radians/second | 57.30 | degrees/second |
| radians/second | 9.549 | revolutions/min |
| radians/second | 0.1592 | revolutions/sec |
| radians/sec/sec | 573.0 | revs/min/min |
| radians/sec/sec | 9.549 | revs/min/sec |
| radians/sec/sec | 0.1592 | revs/sec/sec |
| reams | 500 | sheets |
| register tons (British) | 100 | cubic feet |
| revolutions | 360 | degrees |
| revolutions | 4 | quadrants |
| revolutions | 6.283 | radians |
| revolutions/minute | 6 | degrees/second |
| revolutions/minute | 0.10472 | radians/second |
| revolutions/minute | 0.01667 | revolutions/sec |
| revolutions/minute$^2$ | 0.0017453 | radians/sec/sec |
| revs/min/min | 0.01667 | revs/min/sec |
| revs/min/min | $2.778 \times 10^{-4}$ | revs/sec/sec |
| revolutions/second | 360 | degrees/second |
| revolutions/second | 6.283 | radians/second |
| revolutions/second | 60 | revs/minute |
| revs/sec/sec | 6.283 | rads/sec/sec |
| revs/sec/sec | 3600 | revs/min/min |
| revs/sec/sec | 60 | revs/min/sec |
| reyns | $6.8948 \times 10^6$ | centipoises |
| rod | .25 | chain (gunters) |
| rods | 16.5 | feet |
| rods | 5.0292 | meters |
| rods | $3.125 \times 10^{-3}$ | miles |
| rods (surveyors' means) | 5.5 | yards |
| roods (British) | 0.25 | acres |
| scruples | 1/3 | drams (troy) |
| scruples | 20 | grains |
| sections | 1 | square miles |
| seconds (mean solar) | $1.1574 \times 10^{-5}$ | days |

## Conversion Factors

| Multiply | by | to obtain |
|---|---|---|
| seconds (angle) | $2.778 \times 10^{-4}$ | degrees |
| seconds (mean solar) | $2.7778 \times 10^{-4}$ | hours |
| seconds (angle) | 0.01667 | minutes |
| seconds (angle) | $3.087 \times 10^{-6}$ | quadrants |
| seconds (angle) | $4.848 \times 10^{-6}$ | radians |
| slugs | 14.59 | kilogram |
| slugs | 32.174 | pounds |
| space, entire (solid angle) | 12.566 | steradians |
| spans | 9 | inches |
| spheres (solid angle) | 12.57 | steradians |
| spherical right angles | 0.25 | hemispheres |
| spherical right angles | 0.125 | spheres |
| spherical right angles | 1.571 | steradians |
| square centimeters | $1.973 \times 10^5$ | circular mils |
| square centimeters | $1.07639 \times 10^{-3}$ | square feet (U.S.) |
| square centimeters | 0.15499969 | square inches (U.S.) |
| square centimeters | $10^{-4}$ | square meters |
| square centimeters | $3.861 \times 10^{-11}$ | square miles |
| square centimeters | 100 | square millimeters |
| square centimeters | $1.196 \times 10^{-4}$ | square yards |
| square centimeters-square centimeter (moment of area) | 0.024025 | square inch-square inch |
| square chains (gunter's) | 0.1 | acres |
| square chains (gunter's) | 404.7 | square meters |
| square chains (Ramden's) | 0.22956 | acres |
| square chains (Ramden's) | 10000 | square feet |
| square feet | $2.29 \times 10^{-5}$ | acres |
| square feet | $1.833 \times 10^8$ | circular mils |
| square feet | 144 | square inches |
| square feet | 0.092903 | square meters |
| square feet | 929.0341 | square centimeters |
| square feet | $3.587 \times 10^{-8}$ | square miles |
| square feet | 1/9 | square yards |
| square feet/cu ft | 3.29 | sq m/cu m |
| square foot-square foot (moment of area) | 20,736 | square inch-square inch |
| square inches | $1.273 \times 10^6$ | circular mils |
| square inches | 6.4516258 | square centimeters |
| square inches | $6.944 \times 10^{-3}$ | square feet |
| square inches | 645.2 | square millimeters |
| square inches | $10^6$ | square mils |
| square inches | $7.71605 \times 10^{-4}$ | square yards |
| square inches-inches sqd. | 41.62 | sq cm-cm sqd |
| square inches-inches sqd. | $4.823 \times 10^{-5}$ | sq feet-feet sqd |
| square kilometers | 247.1 | acres |
| square kilometers | $10^{10}$ | square centimeters |

| Multiply | by | to obtain |
|---|---|---|
| square kilometers | $10.76 \times 10^6$ | square feet |
| square kilometers | $1.550 \times 10^9$ | square inches |
| square kilometers | $10^6$ | square meters |
| square kilometers | 0.3861006 | square miles (U.S.) |
| square kilometers | $1.196 \times 10^6$ | square yards |
| square links (Gunter's) | $10^{-5}$ | acres (U.S.) |
| square links (Gunter's) | 0.04047 | square meters |
| square meters | $2.471 \times 10^{-4}$ | acres (U.S.) |
| square meters | $10^4$ | square centimeters |
| square meters | 10.76387 | square feet (U.S.) |
| square meters | 1550 | square inches |
| square meters | $3.8610 \times 10^{-7}$ | square miles (statute) |
| square meters | $10^6$ | square millimeters |
| square meters | 1.196 | square yards (U.S.) |
| square miles | 640 | acres |
| square miles | $2.78784 \times 10^7$ | square feet |
| square miles | 2.590 | sq km |
| square miles | $2.5900 \times 10^6$ | square meters |
| square miles | $3.098 \times 10^6$ | square yards |
| square millimeters | $1.973 \times 10^3$ | circular mils |
| square millimeters | 0.01 | square centimeters |
| square millimeters | $1.076 \times 10^{-5}$ | square feet |
| square millimeters | $1.550 \times 10^{-3}$ | square inches |
| square mils | 1.273 | circular mils |
| square mils | $6.452 \times 10^{-6}$ | square centimeters |
| square mils | $10^{-6}$ | square inches |
| square rods | 272.3 | square feet |
| square yard | $2.1 \times 10^{-4}$ | acres |
| square yards | 8361 | square centimeters |
| square yards | 9 | square feet |
| square yards | 1296 | square inches |
| square yards | 0.8361 | square meters |
| square yards | $3.228 \times 10^{-7}$ | square miles |
| square yards | $8.361 \times 10^5$ | square millimeters |
| statamperes | $3.33560 \times 10^{-10}$ | amperes (abs) |
| statcoulombs | $3.33560 \times 10^{-10}$ | coulombs (abs) |
| statcoulombs/kilogram | $1.0197 \times 10^{-6}$ | statcoulombs/dyne |
| statfarads | $1.11263 \times 10^{-12}$ | farads (abs) |
| stathenries | $8.98776 \times 10^{11}$ | henries (abs) |
| statohms | $8.98776 \times 10^{11}$ | ohms (abs) |
| statvolts | 299.796 | volts (abs) |
| statvolts/inch | 118.05 | volts (abs)/centimeter |
| statwebers | $2.99796 \times 10^{10}$ | electromagnetic cgs units of magnetic flux |
| statwebers | 1 | electrostatic cgs units of magnetic flux |

## Conversion Factors

| Multiply | by | to obtain |
| --- | --- | --- |
| stilb | 2919 | footlambert |
| stilb | 1 | int. candle cm$^{-2}$ |
| stilb | 3.142 | lambert |
| stoke (kinematic viscosity) | $10^{-4}$ | meter$^2$/second |
| stones (British) | 6.350 | kilograms |
| stones (British) | 14 | pounds |
| temp. (degs. C.) + 273 | 1 | abs. temp. (degs. K.) |
| temps (degs. C.) + 17.8 | 1.8 | temp. (degs. Fahr.) |
| temps. (degs. F.) + 460 | 1 | abs. temp. (degs. R.) |
| temps. (degs. F.) − 32 | 5/9 | temp. (degs. Cent.) |
| toises (French) | 6 | paris feet (pieds) |
| tons (long) | $5.734 \times 10^5$ | drams (avdp) |
| tons (long) | $2.613 \times 10^5$ | drams (troy) |
| tons (long) | $1.568 \times 10^7$ | grains |
| tons (long) | $1.016 \times 10^6$ | grams |
| tons (long) | 1016 | kilograms |
| tons (long) | $3.584 \times 10^4$ | ounces (avdp) |
| tons (long) | $3.267 \times 10^4$ | ounces (troy) |
| tons (long) | 2240 | pounds (avdp) |
| tons (long) | 2722.2 | pounds (troy) |
| tons (long) | 1.12 | tons (short) |
| Tons (metric) (T) | 1000 | kilograms |
| Tons (metric) (T) | 2204.6 | pounds |
| Tons (metric) (T) | 1.1025 | tons (short) |
| tons (short) | $5.120 \times 10^5$ | drams (avdp) |
| tons (short) | $2.334 \times 10^5$ | drams (troy) |
| tons (short) | $1.4 \times 10^7$ | grains |
| tons (short) | $9.072 \times 10^5$ | grams |
| tons (short) | 907.2 | kilograms |
| tons (short) | 32,000 | ounces (avdp) |
| tons (short) | 29,166.66 | ounces (troy) |
| tons (short) | 2000 | pounds (avdp) |
| tons (short) | 2.430.56 | pounds (troy) |
| tons (short) | 0.89287 | tons (long) |
| tons (short) | 0.9078 | Tons (metric) (T) |
| tons (short)/sq ft | 9765 | kg/sq meter |
| tons (short)/sq ft | 13.89 | pounds/sq inch |
| tons (short)/sq in | $1.406 \times 10^6$ | kg/sq meter |
| tons (short)/sq in | 2000 | pounds/sq inch |
| tons/sq mile | 3.125 | pounds/acre |
| tons/sq mile | 0.07174 | pounds/1000 sq ft |
| tons/sq mile | 0.3503 | grams/sq meter |
| tons/sq mile | 350.3 | kilograms/sq km |
| tons/sq mile | 350.3 | milligrams/sq meter |
| tons/sq mile | 0.03503 | milligrams/sq cm |
| tons/sq mile | 0.03254 | grams/sq ft |

| Multiply | by | to obtain |
|---|---|---|
| tons of water/24 hours | 83.333 | pounds of water/hr |
| tons of water/24 hours | 0.16643 | gallons/min |
| tons of water/24 hours | 1.3349 | cu ft/hr |
| torr (mm Hg, 0°C) | 133.322 | newton/meter$^2$ |
| townships (U.S.) | 23040 | acres |
| townships (U.S.) | 36 | square miles |
| tuns | 252 | gallons |
| volts (abs) | $10^8$ | abvolts |
| volts (abs) | $3.336 \times 10^{-3}$ | statvolts |
| volts (international of 1948) | 1.00033 | volts (abs) |
| volt/inch | .39370 | volt/cm |
| watts (abs) | 3.41304 | BTU (mean)/hour |
| watts (abs) | 0.0569 | BTU (mean)/min |
| watts (abs) | 0.01433 | calories, kilogram (mean)/minute |
| watts (abs) | $10^7$ | ergs/second |
| watts (abs) | 44.26 | foot-pounds/minute |
| watts (abs) | 0.7376 | foot-pounds/second |
| watts (abs) | 0.0013405 | horsepower (electrical) |
| watts (abs) | $1.360 \times 10^{-3}$ | horsepower (metric) |
| watts (abs) | 1 | joules/sec |
| watts (abs) | 0.10197 | kilogram-meters/second |
| watts (abs) | $10^{-3}$ | kilowatts |
| watt-hours | 3.415 | British Thermal Units |
| watt-hours | $3.60 \times 10^{10}$ | ergs |
| watt-hours | 2655 | foot-pounds |
| watt-hours | 859.85 | gram-calories |
| watt-hours | $1.34 \times 10^{-3}$ | horsepower-hours |
| watt-hours | $3.6 \times 10^3$ | joule |
| watt-hours | 0.8605 | kilogram-calories |
| watt-hours | 367.1 | kilogram-meters |
| watt-hours | $10^{-3}$ | kilowatt-hours |
| watt (international) | 1.0002 | watt (absolute) |
| watt/(cm$^2$)(°C/cm) | 693.6 | BTU/(hr)(ft$^2$)(°F/in) |
| wave length of the red line of cadmium | $6.43847 \times 10^{-7}$ | meters |
| webers | $10^3$ | electromagnetic cgs units |
| webers | $3.336 \times 10^{-3}$ | electrostatic cgs units |
| webers | $10^5$ | kilolines |
| webers | $10^8$ | lines |
| webers | $10^8$ | maxwells |
| webers | $3.336 \times 10^{-3}$ | statwebers |
| webers/sq in | $1.550 \times 10^7$ | gausses |
| webers/sq in | $10^8$ | lines/sq in |
| webers/sq in | 0.1550 | webers/sq cm |
| webers/sq in | 1,550 | webers/sq meter |

## Conversion Factors

| Multiply | by | to obtain |
|---|---|---|
| webers/sq meter | $10^4$ | gausses |
| webers/sq meter | $6.452 \times 10^4$ | lines/sq in |
| webers/sq meter | $10^{-4}$ | webers/sq cm |
| webers/sq meter | $6.452 \times 10^{-4}$ | webers/sq in |
| weeks | 168 | hours |
| weeks | 10,080 | minutes |
| weeks | 604,800 | seconds |
| yards | 91.44 | centimeters |
| yards | 3 | feet |
| yards | 36 | inches |
| yards | $9.144 \times 10^{-4}$ | kilometers |
| yards | 0.91440 | meters |
| yards | $4.934 \times 10^{-4}$ | miles (naut.) |
| yards | $5.682 \times 10^{-4}$ | miles (stat.) |
| yards | 914.4 | millimeters |
| years (sidereal) | 365.2564 | days (mean solar) |
| years (sidereal) | 366.2564 | days (sidereal) |
| years (tropical, mean solar) | 365.2422 | days (mean solar) |
| years (common) | 8760 | hours |
| years (tropical, mean solar) | 8765.8128 | hours (mean solar) |
| years (leap) | 366 | days |
| years (leap) | 8784 | hours |
| years (tropical, mean solar) | $3.155693 \times 10^7$ | seconds (mean solar) |
| years (tropical, mean solar) | 1.00273780 | years (sidereal) |

## 2. BASIC AND SUPPLEMENTARY UNITS

A *meter (m)* is 1,650,763.73 wavelengths in vacuo of the radiation corresponding to the transition between the energy levels $2p_{10}$ and $5d_5$ of the krypton 86 atom.

A *kilogram (kg)* is the mass of the international prototype in the custody of the Bureau International des Poids et Mesures at Sevres in France.

A *second (sec)* is the interval occupied by 9,192,631,770 cycles of the radiation corresponding to the transition of the cesium-133 atom when unperturbed by exterior fields.

An *ampere* is the constant current that if maintained in two parallel rectilinear conductors of infinite length of negligible circular cross section and placed at a distance of one meter apart in vacuo would produce between these conductors a force equal to $2 \times 10^{-7}$ newton per meter length.

A *kelvin* ($°K$) is the degree interval of the thermodynamic scale on which the temperature of the triple point of water is 273.16 degrees.

A *candle* is such that the luminance of a full radiator at the temperature of solidification of platinum is 60 units of luminous intensity per square centimeter.

A *mole (mol)* is the amount of substance which contains as many elementary units as there are atoms in 0.012 kg of carbon-12. The elementary unit must be specified and may be an atom, an ion, an electron, a photon, etc., or a given group of such entities.

A *radian* is the angle subtended at the center of a circle by an arc of the circle equal in length to the radius of the circle.

A *steradian* is the solid angle that, having its vertex at the center of a sphere, cuts off an area of the surface of the sphere equal to that of a square with sides of length equal to the radius of the sphere.

## 3. DERIVED UNITS AND QUANTITIES

The *liter* was defined in 1901 as the volume of 1 kilogram of pure water at normal atmospheric pressure and maximum density equal therefore to 1.000028 dm$^3$. This 1901 definition applied for the purpose of the 1963 Weights and Measures Acts.

By a resolution of the 12th Conference General des Poids et Mesures (CGPM) in 1964 the word *liter* is now recognized as a special name for the dm$^3$, but is not used to express high precision measurements. It is used widely in engineering and the retail business, where the discrepancy of 28 parts in 1 million is of negligible significance.

A *newton (N)* is the force that, when applied to a body of mass of one kilogram, gives it an acceleration of one meter per second per second.

*Stress* is defined as the resultant internal force per unit area resisting change in the shape or size of a body acted on by external forces, and is therefore measured in *newtons per square meter* (N/m$^2$).

A *bar* is a pressure equivalent to 100,000 newtons acting on an area of one square metor.

A *joule (J)* is the work done when the point of application of a force of one newton is displaced through a distance of one meter in the direction of the force.

A *watt* is equal to one joule per second.

*Dynamic viscosity* is the property of a fluid whereby it tends to resist relative motion within itself. It is the shear stress, i.e., the tangential force on unit area, between two infinite horizontal planes at unit distance apart, one of which is fixed while the other moves with unit velocity. In other words, it is the shear stress divided by the velocity gradient, i.e., (N/m$^2$) ÷ (m/sec/m) = N sec/m$^2$.

*Kinematic viscosity* is the dynamic viscosity of a fluid divided by its density, i.e., (N sec/m$^2$)/(kg/m$^3$) = m$^2$/sec.

*Density of heat flow rate* (or heat flux) is the heat flow rate (W) per unit area, i.e., W/m$^2$.

*Coefficient of heat transfer* is the heat flow rate (W) per unit area per unit temperature difference, i.e., W/m$^2$°C.

*Thermal conductivity* is the quantity of heat that will be conducted in unit time through unit area of a slab of material of unit thickness with a unit difference of temperature between the faces; in other words, the heat flow rate (W) per unit area per unit temperature gradient, i.e., W/[m$^2$(°C/m)] = W/m°C.

The *heat capacity* of a substance is the quantity of heat gained or lost by the substance per unit temperature change, i.e., J/°C.

*Specific heat capacity* is the heat capacity per unit mass of the substance, i.e., J/kg°C.

*Internal energy* is the kinetic energy possessed by the molecules of a substance due to temperature and is measured in joules (J).

*Specific internal energy* (u) is the internal energy per unit mass of the substance, i.e., J/kg. When a small amount of heat is added at constant volume the increase in specific internal energy is given by: $du = c_v \, dT$, where $c_v$ is the specific heat capacity at constant volume, and $dT$ is the increase in absolute temperature.

*Specific enthalpy* (h) is defined by the equation: $h = u + pv$, where $p$ is the pressure and $v$ is the specific volume. Specific enthalpy is measured in J/kg. When a small amount of heat is added to a substance at constant pressure, the increase in specific enthalpy is given by: $-dh = cp \, dT$, where $cp$ is the specific heat capacity at constant pressure.

The *specific latent heat* of a substance is the heat gained per unit mass without an accompanying rise in temperature during a change of state at constant pressure. It is measured in J/kg.

The *entropy* (S) of a substance is such that when a small amount of heat is added, the increase in entropy is equal to the quantity of heat added ($dQ$) divided by the absolute temperature ($T$) at which the heat is absorbed; i.e., $dS = dQ/T$, measured in J/°K.

The *specific entropy* (s) of a substance is the entropy per unit mass, i.e., J/kg°K.

A *volt* is the difference of electric potential between two points of a conductor carrying a constant current of one ampere when the power dissipated is one watt.

A *weber* (Wb) is the magnetic flux through a conductor with a resistance of one ohm when reversal of the direction of the magnetic flux causes the transfer of one coulomb in the conductor loop.

*Tesla*: The magnetic flux density is the normal magnetic flux per unit area and is measured in *teslas*.

A *lumen*, the unit of luminous flux, is the flux emitted within unit solid angle of one steradian by a point source having a uniform intensity of one candle.

A *lux* is an illumination of one lumen per square meter.

*Luminance* is the luminous intensity per unit area of a source of light or of an illumination. It is measured in candles per square meter.

## 4. PHYSICAL CONSTANTS

Standard temperature and pressure (S.T.P.) $\begin{cases} = 273.15°\text{K and } 1.013 \times 10^5 \text{ N/m}^2 \\ = 0°\text{C and } 1.013 \text{ bar} \\ = 0°\text{C and } 760 \text{ mm Hg} \end{cases}$

Molecular volume of ideal gas at S.T.P. $= 22.41 \text{ liters/mol}$
Gas constant (R) $= 8.314 \text{ J/mol°K}$
$^{RT}(273.15°\text{K})$ $= 2.271 \times 10^3 \text{ J/mol}$
Avogadro constant $= 6.023 \times 10^{23}/\text{mol}$
Boltzmann constant $= 1.3805 \times 10^{-23} \text{ J/K}$
Faraday constant $= 9.6487 \times 10^4 °\text{C/mol } (= \text{A s/mol})$
Planck constant $= 6.626 \times 10^{-34} \text{ J sec}$
Stefan-Boltzman constant $= 5.6697 \times 10^{-8} \text{ W/m}^2 \text{ K}^4$
Ice point of water $= 273.15°\text{K } (0°\text{C})$
Triple point of water $= 273.16°\text{K } (0.01°\text{C})$
Speed of light $= 2.998 \times 10^8 \text{ m/sec}$

Acceleration of gravity (standard) (Greenwich) $\begin{cases} = 9.80665 \text{ m/s}^2 \\ = 9.81188 \text{ m/s}^2 \end{cases} \begin{bmatrix} \text{take g as} \\ 9.81 \text{ m/s}^2 \end{bmatrix}$

Universal constant of gravitation $= 6.670 \times 10^{-11} \text{ Newton m}^2/\text{kg}^2$
Mass of hydrogen atom $= 1.6734 \times 10^{-27} \text{ kg}$

## 5. PROPERTIES OF WATER

| Temperature (°F) | Specific weight, $\gamma$ (lb/ft$^3$) | Mass density, $\rho$ (lb-sec$^2$/ft$^4$) | Dynamic viscosity, $\mu \times 10^5$ (lb-sec/ft$^2$) | Kinematic viscosity, $\nu \times 10^5$ (ft$^2$/sec) | Surface energy, $\sigma \times 10^3$ (lb/ft) | Vapor pressure, $\rho$ (lb/in.$^2$) | Bulk modulus, $E \times 10^{-3}$ (lb/in.$^2$) |
|---|---|---|---|---|---|---|---|
| 32  | 62.42 | 1.940 | 3.746 | 1.931 | 5.18 | 0.09  | 290 |
| 40  | 62.43 | 1.938 | 3.229 | 1.664 | 5.14 | 0.12  | 295 |
| 50  | 62.41 | 1.936 | 2.735 | 1.410 | 5.09 | 0.18  | 300 |
| 60  | 62.37 | 1.934 | 2.359 | 1.217 | 5.04 | 0.26  | 312 |
| 70  | 62.30 | 1.931 | 2.050 | 1.059 | 5.00 | 0.36  | 320 |
| 80  | 62.22 | 1.927 | 1.799 | 0.930 | 4.92 | 0.51  | 323 |
| 90  | 62.11 | 1.923 | 1.595 | 0.826 | 4.86 | 0.70  | 326 |
| 100 | 62.00 | 1.918 | 1.424 | 0.739 | 4.80 | 0.95  | 329 |
| 110 | 61.86 | 1.913 | 1.284 | 0.667 | 4.73 | 1.24  | 331 |
| 120 | 61.71 | 1.908 | 1.168 | 0.609 | 4.65 | 1.69  | 333 |
| 130 | 61.55 | 1.902 | 1.069 | 0.558 | 4.60 | 2.22  | 332 |
| 140 | 61.38 | 1.896 | 0.981 | 0.514 | 4.54 | 2.89  | 330 |
| 150 | 61.20 | 1.890 | 0.905 | 0.476 | 4.47 | 3.72  | 328 |
| 160 | 61.00 | 1.896 | 0.838 | 0.442 | 4.41 | 4.74  | 326 |
| 170 | 60.80 | 1.890 | 0.780 | 0.413 | 4.33 | 5.99  | 322 |
| 180 | 60.58 | 1.883 | 0.726 | 0.385 | 4.26 | 7.51  | 318 |
| 190 | 60.36 | 1.876 | 0.678 | 0.362 | 4.19 | 9.34  | 313 |
| 200 | 60.12 | 1.868 | 0.637 | 0.341 | 4.12 | 11.52 | 308 |
| 212 | 59.83 | 1.860 | 0.593 | 0.319 | 4.04 | 14.7  | 300 |

# 6. PERIODIC TABLE OF THE ELEMENTS (COMPLIMENTS OF THE LENOX INSTITUTE OF WATER TECHNOLOGY)

| Groups & Periods & sub-shells | 1 IA | 2 IIA | 3 IIIB | 4 IVB | 5 VB | 6 VIB | 7 VIIB | 8 VIII | 9 VIII | 10 VIII | 11 IB | 12 IIB | 13 IIIA | 14 IVA | 15 VA | 16 VIA | 17 VIIA | 18 O |
|---|---|---|---|---|---|---|---|---|---|---|---|---|---|---|---|---|---|---|
| 1 1s | 1 H 1.00794 Hydrogen | | | | | | | | | | | | | | | | 1 H 1.00794 Hydrogen | 2 He 4.00260 Helium |
| 2 2s2p | 3 Li 6.941 Lithium | 4 Be 9.01218 Beryllium | | | | | | | | | | | 5 B 10.811 Boron | 6 C 12.011 Boron | 7 N 14.0067 Nitrogen | 8 O 15.9994 Oxygen | 9 F 18.9984 Fluorine | 10 Ne 20.179 Neon |
| 3 3s3p | 11 Na 22.9897 Sodium | 12 Mg 24.305 Magnesium | | | | | | | | | | | 13 Al 26.9815 Aluminum | 14 Si 28.0855 Silicon | 15 P 30.9738 Phosphorus | 16 S 32.066 Sulfur | 17 Cl 35.4527 Chlorine | 18 Ar 39.948 Argon |
| 4 4s3d4p | 19 K 39.098 Potassium | 20 Ca 40.078 Calcium | 21 Sc 44.9559 Scandium | 22 Ti 47.88 Titanium | 23 V 50.9415 Vanadium | 24 Cr 51.996 Chromium | 25 Mn 54.938 Manganese | 26 Fe 55.847 Iron | 27 Co 58.933 Cobalt | 28 Ni 58.69 Nickel | 29 Cu 63.546 Copper | 30 Zn 65.39 Zinc | 31 Ga 69.723 Gallium | 32 Ge 69.561 Germanium | 33 As 74.9216 Arsenic | 34 Se 78.96 Selenium | 35 Br 79.904 Bromine | 36 Kr 83.80 Krypton |
| 5 5s4d5p | 37 Rb 85.468 Rubidium | 38 Sr 87.62 Strontium | 39 Y 88.9059 Yttrium | 40 Zr 91.224 Zirconium | 41 Nb 92.9064 Niobium | 42 Mo 95.94 Molybdenum | 43 Tc (98) Technetium | 44 Ru 101.07 Ruthenium | 45 Rh 102.906 Rhodium | 46 Pd 106.42 Palladium | 47 Ag 107.868 Silver | 48 Cd 112.411 Cadmium | 49 In 114.82 Indium | 50 Sn 118.710 Tin | 51 Sb 121.75 Antimony | 52 Te 127.60 Tellurium | 53 I 126.90 Iodine | 54 Xe 131.29 Xenon |
| 6 6s4f5d6p | 55 Cs 132.905 Cesium | 56 Ba 137.327 Barium | 57 La 138.906 Lanthanum | 72 Hf 178.49 Hafnium | 73 Ta 180.948 Tantalum | 74 W 183.85 Tungsten | 75 Re 186.207 Rhenium | 76 Os 190.2 Osmium | 77 Ir 192.22 Iridium | 78 Pt 195.08 Platinum | 79 Au 196.97 Gold | 80 Hg 200.59 Mercury | 81 Tl 204.383 Thallium | 82 Pb 207.2 Lead | 83 Bi 208.98 Bismuth | 84 Po (209) Polonium | 85 At (210) Astatine | 86 Rn (222) Radon |
| 7 7s5f6d | 87 Fr (223) Francium | 88 Ra (226) Radium | 89 Ac (227) Actinium | 104 Rf (261) Rutherfordium | 105 Ha (262) Dubnium | 106 Sg (263) Seaborgium | 107 Ns (262) Bohrium | 108 Hs (265) Hassium | 109 Mt (266) Meitnerium | 110 | | 112 | | | | | | |

| | | | | | | | | | | | | | | |
|---|---|---|---|---|---|---|---|---|---|---|---|---|---|---|
| 6 4f | 58 Ce 140.116 Cerium | 59 Pr 140.91 Praseodymium | 60 Nd 144.24 Neodymium | 61 Pm (145) Promethium | 62 Sm 150.35 Samarium | 63 Eu 107.26 Europium | 64 Gd 157.25 Gadolinium | 65 Tb 158.925 Terbium | 66 Dy 162.50 Dysprosium | 67 Ho 104.930 Holmium | 68 Er 167.26 Erbium | 69 Tm 168.934 Thulium | 70 Yb 173.04 Ytterbium | 71 Lu 174.967 Lutetium |
| 7 5f | 90 Th 232.038 Thorium | 91 Pa (231) Protactinium | 92 U 238.029 Uranium | 93 Np (237) Neptunium | 94 Pu (244) Plutonium | 95 Am (243) Americium | 96 Cm (247) Curium | 97 Bk (247) Berkelium | 98 Cf (251) Californium | 99 Es (252) Einsteinium | 100 Fm (257) Fermium | 101 Md (258) Mendelevium | 102 No (259) Nobelium | 103 Lr (262) Lawrencium |

# Subject Index

**A**

Abbottstown, soil, 358
Acid gases, 507
Acidification, 83
Acoustic geophysical survey, 459
Activated sludge, 358–360
    dewatered, 358
    liquid, 360
    treatment of thermal conditioning liquor, 719
Advanced alkaline stabilization, 77
Aerated sludge basin, 232
Aeration reactor, 202
Aerobic digester, 207
Aerobic digestion, of heat treatment liquor, 719
Aerobically digested biosolids, 359
Agricultural land, 365, 400
    application
        design, 364
        biosolids, 343–413
        management practice, 360
Agronomic rate, 370, 378, 380, 391, 394, 400
Air pollution
    control devices, 317, 533, 609
    incineration, 502
Air toxics, 508
Albrights, soil, 358
Alfalfa, 360, 364, 367, 372
Alkali stabilization, 429
Alkaline pretreatment, 76
Alkaline stabilization, 75–77
Allenwood, soil, 358
Alternatives for meeting Class A pathogen requirements, 321
Ammonia nitrogen, 357
Ammonium nitrate, 364
Ammonium nitrogen volatilization factor, 370
Ammonium polyphosphate, 364
Ammonium sulfate, 364

Ampere, 784
Amphipod, 453–454
Anaerobic digester, 206, 715–717
Anaerobic digestion, 77, 737–738
    with ozone treatment, 83
Anaerobic liquid sludge lagoon, 229
Anaerobically digested biosolids, 359
Anaerobic-baffled reactor (ABR), 80
Anhydrous ammonia, 364
Annual Pollutant Loading Rate (APLR)
    biosolids, 289–290, 400, 666–667
Annual whole sludge application rate (AWSAR), 289–290, 400
Anoxic aerobic digestion, 75
Anoxic gas flotation thickening, 97
APLR. *See* Annual Pollutant Loading Rate
Apples, 367
Application frequency, biosolids, 365
Application land use restrictions, biosolids, 351
Application rate
    biosolids, 359, 365
    determination, 366
Area of cropland, 400
Arsenic pollutant limits, 312–313, 680–683
Ash composition, 601
Ash conveyance, 545
Ash discharge, 545
Ash-handling system, 611
ATAD. *See* Autothermal thermophilic aerobic digestion
Autothermal thermophilic aerobic digestion (ATAD), 75
Auxiliary heat, sludge combustion, 627
Availability factor
    phosphorus in biosolids, 371
    potassium in biosolids, 372
Available mineralized organic nitrogen, 378
Available phosphoric acid, 364
Avogadro constant, 787
AWSAR. *See* Annual whole sludge application rate

## B

Bagged biosolids, 400
Bar, 785
Barge transportation, 42–51
Barley, crop, 367, 372
Basher, soil, 358
Belt conveyors, 29–32
Belt filter press, 100
Beneficial use, biosolids, 347
Beneficial utilization, of biosolids, 647–690
Bepex dryer, 573–574
Beryllium pollutant limits, 306–308, 680
Best available technology, land application, 382
Biological flotation, 140
Biological testing, ocean disposal, 450
Biosolids mineralization factor, 359
Biosolids
   aerobically digested, 359
   anaerobically digested, 359
   application, vegetation sites, 70
   ceiling concentration limit (CCL), 349, 352
   characteristics, 256, 345, 418, 421
   chemical fixation, 470
   chlorination, 152
   class A, class B, 257–260
   composted, 359
   culmulative pollutant loading rate (CPLR), 349, 352
   dewatering, 121–140, 563
      effectiveness levels, 563
   environmental management system, 444, 469
   exceptional quality, EQ, 351
   field storage, 242
   fired in incinerators, 653
   frequency of monitoring for land application, 292
   incineration, 607, 618
   injection equipment, 355–356
   land application restrictions, 351
   landfill, 470
   lime stabilized, 359
   management practices for land application, 290–292
   maximum metal concentration, 71
   mineralization rate factor, 359
   monofill, 433
   nutrient concentration, 347
   ocean disposal, 470
   pathogen class, 352
   pathogen reduction, 350
   physical and chemical characteristics, 425
   placed on a surface disposal site, 653
   pollutant concentration (PC), 352
   pre-treatment, 344
   production, 344
   properties checklist, 727
   quality categories, 351
   record-keeping for land application, 292–293
   regulations, 70, 649–690
   reporting requirements for land application, 292–293
   siting restrictions, 352
   stabilization, 429–430
   storage, 193
      design, 261–262
   surface disposal, 294–305
   thickening, 121–140
   treatment for landfilling, 427
   vector attraction reduction (VAR), 352
   volatilization factor, 360
Biosolids application rate, 364
   based on
      heavy metal contents, 371
      nitrogen contents, 394–395
      phosphorus contents, 371, 394
      potassium contents, 372, 397
   determination of, 366
   simplified, 372
Biosolids crop year organic nitrogen mineralization factor, 369
Biosolids disposal
   cost, 470
   land application, 470
   on-site energy recovery, 470
Biosolids land application
   forest land, 356
   incorporation, 354
   site suitability, 353
   surface application, 354
   time, 359

# Subject Index

Biosolids landfilling, 415
  case studies, 438
  design, 432, 436, 439
  design criteria, 436
  future trend, 438
  gas collection, 437
  methods of, 434
  pollutant concentration, 437
  regulations, 416
Biosolids requirement, landfilling, 421
Black Point Disposal Site, 462
Bluegrass, 367
Boilers, 677
Boltzmann constant, 787
Bulk biosolids, 400
Bulk modulus, water, 787

## C

Cabbage, 367
Cadmium pollutant limits, 312–313, 680–683
Calorific value, pyrolysis gases, 498
Candle, 784
Carbon monoxide measurement, 314
CBFT3. See Columbus biosolids flow-through thermophilic treatment
CCC. See Chlorine contact chamber
CCL. See Ceiling concentration limit
Ceiling concentration limit (CCL), 349, 352, 400
Cell destruction, 89
Centrifugal pumps, 3–4
Centrifugation thickening, 95
Centrifuges, 103, 738
Centripress, 127–128
Cgs system, centimeter-gram-second system, 745
Change in return rate, 713–714
Characteristics, multiple hearth furnaces, 624
Checklists, 727–728
Chemical and biological sampling, ocean, 460
Chemical conditioning, 87
Chemical fixation, biosolids disposal, 470

Chemical screening, ocean disposal, 450
Chemical stabilization, lime requirement, 77
Chlorination, 151
Chlorine and ammonia nitrogen breakpoint ratio, 156
Chlorine contact chamber (CCC), 153, 167
Chlorine contact time, 154
Chlorine dosage, 152, 155
Chlorine stabilization, 152, 167, 169
Composted biosolids, 359
Chromium pollutant limits, 312–313, 680–683
Chutes and inclined planes, 36
Circulating fluidized bed, 2 stage design, 641
Class A pathogen reduction requirements, 282, 321–327, 421
Class B pathogen reduction requirements, 282, 327–328, 423
Class I sludge, 400
Clover grass, 372
CND. See Crop nitrogen deficit
CNFR. See Crop nitrogen fertilizer rate
Coastal Bermuda grass, 367
Coefficient of heat transfer, 785
Co-incineration, 640–642
  sludge incinerators, 641
  solid waste incinerators, 642
Collection
  of leachate, 300
  runoff, 299–290
Columbus biosolids flow-through thermophilic treatment (CBFT3), 81
Combined biosolids
  processing methods, 736
  processing trains, 735–737
Combustible gases, 505
Combustible liquids, 505
Combustible solids, 505
Combustion air, 530
Combustion, 479
Combustor, 521
Community septic tank, 203
Composition, pyrolysis gases, 499
Compost, biosolids, 470

Composting, 84–86, 207, 431
  aerated static pile, 85
  in-vessel, 86
  windrow, 85
Conditioning, 87, 428
Contingency planning, for breakdowns, 721–725
Contingency problems, 721–722
Control efficiency for lead, 308
Control of pumping, 14–15
Convention on the Prevention of Marine Pollution by Dumping of Wastes and Other Matter (London Convention), 446
Conventional fluidized bed furnace, loading, 636
Conventional nitrogen fertilizer materials, 364
Conversion factors, 745
Cooling
  by heat withdrawal, 533
  by water evaporation, 532
Corn grain, 358, 360, 367, 372
Corn silage, 358, 360, 372
Corn, crop, 367, 372
Corrosion issues, incineration, 540
Corrosion regimes in incinerator boiler systems, 542
Cost relationships, 336
Cost-effective analyses, 726–727
Costs
  of biosolids disposal and reuse, 273
  evaporation lagoon, 205–206
  facultative sludge lagoon, 229
  land application, 376, 398–399
  for sludge disposal alternatives, 332–337
  sludge holding tank, 210
Cotton, 367
Coxsackie Sewage Treatment Plant, 165–167, 178–183
CPLR. See Cumulative pollutant loading rate
Crop
  fruit, 367
  vegetable, 367
Crop group, 357, 367, 400
Crop nitrogen deficit (CND), 366, 368, 370, 393, 401

Crop nitrogen fertilizer rate (CNFR), 366, 368, 393, 401
Crop productivity group, 358
Crop year, 357, 401
  biosolids PAN applied, 369
  nonbiosolids PAN applied, 369
  organic nitrogen, 369
Crop yield capability, 358
Cumulative pollutant loading rate (CPLR), 349, 352, 383–384, 401, 665–666
Cumulative pollutant loading rate biosolids, 286–289
Currents and sediment transport survey, 460
Cylindrical furnace grates, 527

## D

DAMOS. See Disposal Area Monitoring System
Decay dilution, ocean, 466
Decomposition kinetics, 499
Dedicated beneficial use sites, 294
Dedicated disposal sites, 294
Deep-shaft digestion, 87
Density of heat flow rate, heat flux, 785
Design
  evaporation lagoon, 205
  land application, 376
  septage chlorination, 184
  of sludge/biosolids hauling, 51–55
  sludge storage, 209
Design application rates, to land, 659
Design criteria, for land application of biosolids, 658
Detroit stoker reciprocating grate furnace, 529
Dewatered sludge storage, 233
Dewatered solids, conveyance, 28–36
Dewatering and drying, 100
Dewatering process, 608
Diammonium phosphate, 364
Diaphragm pumps, 8–10
Diffuser, outfall design, 467
Digested biosolids, 359
Digester, 211
Digestion (anaerobic and aerobic), 430
Direct-fired rotary dryer, 571
Direct-fired systems, 568
Dispersion dilution, ocean, 466

Subject Index  793

Dispersion factor for lead, 308
Disposal Area Monitoring System
   (DAMOS), 461
Domestic septage, 293, 401
Dredged materials, 363, 450
Dry-basis yields, pyrolysis of refuse,
   501
Drying bed, 104, 207, 567
Drying lagoon, sludge, 235
Drying, 109
Dynamic viscosity, 785, 787

E

Effect of excess air, flame temperature,
   531
Effect of supernatant return, 714
Ejector pumps, 11–12
Electric infrared incinerators, 640
Electrocoagulation, 91
Electro-dewatering, 107
Electronic infrared incinerator, cross
   section, 643
Elemental composition of biosolids, 657
Elements of capital costs, incineration
   systems, 591
Elements of operating costs, incineration
   systems, 592
Elimination of sidestream, 712
Emergency
   operating procedures, land application,
      374
   operating procedures, sludge
      stabilization, 173
Emission monitoring points and monitoring
   frequency, 637
Energy conservation, 725–726
Energy content
   ash, 490
   flue gas, 490
Energy markets, 538
Energy recovery, incineration, 540
Enforcement of Part 503, 655
Entropy, 786
Environmental assessment, 450
Environmental impacts, land application,
   375
Enzyme conditioning, 92
EQ. *See* Exceptional Quality

Equilibrium constants of combustion
   reactions, 491
Evaporation lagoon, 204–205
Exceptional Quality (EQ) biosolids, 259,
   283–284, 351, 401, 660–662
Exclusions from the Part 503 rule, 654
Exposure pathways, 382
Expressor press, 123–125

F

Facilities design for land application,
   669–670
Facility management for land application,
   670
Factors affecting heat recovery, 679–680
Facultative sludge lagoon, 212–223
Faraday constant, 787
Farm field, 355, 401
Farm identification elements, 362
FBF monitoring points, 636
Feed crops, 401
Feed rate for lead, 308
Feeding solid waste, 521
FELL. *See* Focused electrode leak locator
Ferrate addition, 83
Fescue, crop, 367
Fiber crops, 401
Field storage, biosolids, 242
Filtration, 738
Fire tube boiler, 538
Flash dryer, 568
Flash pyrolysis, 548
Flotation clarification, 72
Flotation thickening, 97
Flue gas conditioning, 531
Fluid mechanics, furnace systems, 510
Fluidized bed furnace, 633–635
   cross section, 634
   operating parameters, 634
   operation and maintenance, 633
   performance evaluation, 635
   troubleshooting, 635
Fluidized bed incinerator, 579–580
   improved design, 637
Fluidized bed system, 583, 585
Focused electrode leak locator (FELL)
   electroscanning, 113
Food crops, 402

Forest land, 365
  biosolids land application, 356, 402
40 CFR Part 503 Subpart E, 613
Fps system, foot-pound-second system, 745
Freezing, 92
Frequency of monitoring, 303–305, 317–320, 664, 673–674, 684
Friction headloss factor, 21
Fruit crop, 367

## G

Gas constant, 787
Gas laws, 481
Gas lift pumps, 12
Gas turbines, 679
Gas-burning equipment, 677–679
Gasification, 548
Generators, 679
Geological stability, 298–299
Global Plan of Action for Protection of Marine Environment from Land-based Activities (GPA), 463
Glossary
  chlorination, 152
  land application, 400
GPA. See Global Plan of Action for Protection of Marine Environment from Land-based Activities
Gravity belt thickening, 97
Gravity thickener, 93, 206
Grit removal, 197
Groundwater impact, 224, 252

## H

Headloss calculations, 21–22
Health effect classifications, 509
Heat capacity, 785
Heat conditioning, 88
Heat drying, 431
Heat flux, 785
Heat recovery, 538
Heat transfer, 785
Heating values
  sludge, 628
  waste components, 598
  wastewater treatment solids, 564
Heavy metal, 368, 370, 402

Heavy metal limits, land application, biosolids, 349
High rate plug flow, 81
High-range combustion processes, 588
Holding tank, 208
Hollin Iron Works Screw Press, 128–132
Hot-water boiler, 538
Hydraulic ash-handling system, 612
Hygiene and sanitation, 58
Hyperion treatment plant, Los Angeles, California, 469

## I

Ideal gas, 787
Imhoff tank, 203
Immunization, 59
IMO. See International Maritime Organization
Incineration furnaces, grates and hearths, 524
Incineration of biosolids, 305–328
  as an energy source, 675–684
Incineration system
  process analysis, 480
  municipal solid waste, 515
Incinerator
  design approach, 594
  economics, 590
  firing diagram, 519
  process checklist, 629
Incorporation of biosolids, 354
Increase in wastewater stream biological treatment capacity, 714
Indicate organism, 402
Indirect fired systems, 573
Indirect heating, 547
Individual field unit, 357, 402
Infrared system, 587
Initial dilution, ocean, 466
Injection below surface, agricultural land application, 360
Installations, thermoselect process units, 555
Instruments operation and maintenance, 316
Internal energy, 785
International Convention for the Prevention of Pollution from Ships, marine pollution (MARPOL), 445

## Subject Index

International Maritime Organization (IMO), 463
Irradiation, 83

### J
Jet circulation patterns, 514
Jet cross-section, 514
Jet mixing analysis, 465

### K
Kelvin degree, 784
Kiln injection, 113
Kinematic parameters, scrap tire materials, 501
Kinematic viscosity, 785, 787
Kinetics, 493
    carbon oxidation, 493
    soot oxidation, 493
Korean screw press, 128
Kubota melting furnace, 588

### L
Lagoons, 294
Land application of biosolids, 277–294, 656–670
Land application, 402
    biosolids disposal, 343–413, 470
    design, 364
    heavy metal limits, 349
    process, 353
    regulations, 348
    requirements for, 651–653, 659–670
    site, 402
    siting restrictions, 352
Land requirements, land application, 394
Land-based waste discharge to the sea, 463
Landfill, biosolids disposal, 470
Landfilling, types of biosolids, 419
Lead Pollutant Limits, 680–683
Lenox Institute of Water Technology, 745, 788
Light speed, 787
Lime stabilization, 76–78
Lime stabilized biosolids
Logic of process selection, 692–701
London Convention, 446
Long-distance transport, 18–21
Long-distance wastewater solids hauling, 36–55
Los Angeles Joint Water Pollution Control Plant, 240–241
Low-range drying processes, 566
Luminance, 786
Lux, 786
Lystek thermal/chemical process, 113

### M
Management practices
    agricultural land application, 360
    for biosolids incineration, 316–320
    for surface disposal of biosolids, 297
Manure, 402
Marine pollution prevention, 445, 467, 469
Marine pollution, 444
Marine Protection Research and Sanctuaries Act (MPRSA, US), 448, 450
Marine waste characterization, 449
MARPOL. *See* International Convention for the Prevention of Pollution from Ships, marine pollution
Mass burning, 499
Mass density of water, 787
Maximum continuous rating (MCR), 518
MCR. *See* Maximum continuous rating
MCU. *See* Modular combustion unit
Mean yield, soil productivity, 360
Mechanical ash-handling system, 612
Mechanical grates
    batch operations, 527
    continuous operations, 528
Membrane clarification, 73
Membrane filter press, 109
Membrane schematic diagram, 73
Membrane thickening, 99
Mercury pollutant limits, 306–308, 680
Mercury, 449
Metal concentration, ash, incineration, 599
Metal screen filtration, 107
Metal screen thickening, 100
Metals, biosolids, 372
Methane gas concentrations, 300–301
MHFs. *See* Multiple hearth furnaces
MHF process zones, 626
MHF systems, temperature distribution, 584

Microtox, 455
Mid-range combustion processes, 576
Mineral particulate, 504
Mineralization factor, biosolids, 359
Mineralization, 369, 380–381, 392, 402
Minimum distance, land application, 354
Mks system, meter-kilogram-second system, 745
Modular combustion unit (MCU), 559, 561
Modular incineration systems, 558
Moisture, refuse, 594
Mole, 784
Molecular volume of ideal gas, 787
Monitoring, ocean disposal site, 457
Monofills, 294
Most probable number (MPN), 154–155
MPN. *See* Most probable number
MPRSA. *See* Marine Protection Research and Sanctuaries Act
Multibeam bathymetry, 459, 462
Multiple hearth and fluidized bed furnaces, comparison, 639
Multiple hearth furnace (MHFs), 577–578, 581, 621
   design parameters, 623
   loading, 625
   operation and maintenance, 621
   operation parameters, 623
   performance, 626
   trouble shooting guide 631
   troubleshooting, 626
Multiple hearth system, 584

**N**

National Biosolids Partnership, 469
National Sewage Sludge Survey (NSSS), 349
Net primary nutrient crop need, 387
Newton, 785
Nickel pollutant limits, 312–313, 680–683
Nitrate nitrogen, 357, 379
Nitrite nitrogen, 357, 379
Nitrogen, 370, 379, 385, 388
Nitrogen credit, soil productivity, 364
Nitrogen fertilizer rate, 368
Nitrogen removal, 358, 367
Nitrogen, percent content, 387

NSSS. *See* National Sewage Sludge Survey
Nutrient, 353, 402
   allocation, 363
   concentration, biosolids, 347
   management, land application, 361–363
Nutrient per acre, 386

**O**

Oats, 372
Ocean
   decay dilution, 466
   dispersion dilution, 466
   disposal site monitoring system, 458
Ocean disposal, 443–477
   permit, 449–450
   site, 457
Ocean initial dilution, 466
Ocean outfall
   design, 467–468
   system, 464
O'Connor combustor rotary kiln configuration, 527
Odor conditioning, 90
On-site energy recovery, biosolids disposal, 470
Operating procedures
   emergency, land application, 374
   emergency, sludge stabilization, 173
   normal, land application, 374
Optimal APC (air pollution control) technology, 535
Options for Meeting Land Application Requirements, 280–290
Options for Meeting vector attraction reduction, 329, 674
Orchard grass, 372
Organic chemical, biosolids, 350
Organic nitrogen, 357, 369
Outfall design, ocean, 467
Outfall diffuser system, 465

**P**

PAH. *See* Polynuclear aromatic hydrocarbon
PAN. *See* Plant available nitrogen
Part 503 Rule
   background 275–276
   exclusions, 278–279

## Subject Index

overview, 276–277
risk assessment, 276
Partial combustion, 550
Particulate control, 533
Pasteurization, 86
Pasture, crop, 360, 402
Pathogen, 224, 255–257, 402
    limits for land application, 280
    and vector attraction reduction, 302–303, 653, 672–673
Pathogen reduction
    alternatives, 320–328
    biosolids, 224, 252, 350
    guidelines, 422
    requirements, 282, 321–328, 421–423, 661
PC. See Pollutant concentration
PCB. See Polychlorinated biphenyls
PCL. See Pollutant concentration limits
Peaches, 367
Peanuts, 367
Pelletech dryer, 574–575
Percent organic nitrogen mineralized, 359
Periodic table, of elements, 788
Personal protective equipment, 58
PFRPs. See Processes to further reduce pathogens
pH, 403
Phosphate removal, 358
Phosphorus balance, 371
Pipe, fittings, and valves, 18
Pipelines, 18–28
    cost, 26–28
    design, 22–26
Piston pumps, 6–7
Planck constant, 787
Planning, agricultural land application, 361
Plant available nitrogen (PAN), 357, 369–370, 388, 403
Planting and harvesting periods, 357, 403
Planting season, 357, 403
Plunger pumps, 5–6
Pneumatic conveyors, 33–35
Pollutant concentration (PC) biosolids, 284–286, 352, 662–665
Pollutant concentration limits (PCL), 349, 351, 403

Pollutant limits
    for biosolids fired in an incinerator, 306, 680
    for biosolids incineration, 681
    calculations for lead, 308
    for land application, 280–281, 660
    for surface disposal of biosolids, 296–297, 672
Polychlorinated biphenyls (PCB), 449
Polynuclear aromatic hydrocarbon (PAH), 453
Porcupine dryer, 573
Positive-displacement–type Conveyors, 33
Potash removal, 358
Potassium balance, 371
Potassium nitrate, 364
Potato, 367
POTW. See Publicly owned treatment works
Preliminary treatment, 68
Prevention of Marine Pollution, 445, 469
Primary biosolids
    processing methods, 733
    processing trains, 729–734
Primary sedimentation, 202
Process design checklist, 728
Process design for land application, 668–669
Process selection for biosolids management, 691–743
Processes to further reduce pathogens (PFRPs), 326, 350, 424, 662
Processes to significantly reduce pathogens (PSRPs), 328, 350, 425, 663
Progressive cavity pumps, 7–8
Protection
    of ground water, 303
    of threatened or endangered species, 297, 317
    of wetlands, 299
    of workers, 57–59
PSRPs. See Processes to significantly reduce pathogens
Public contact site, 402
Public health and environmental impact checklist, 728
Publicly owned treatment works (POTW), 404

Pumping of sewage sludge and biosolids, 2–18
Pumping, 2–18
Pumps, trouble shooting, 16–17
Purifax process, 183, 185
Pyrolysis
　gasification systems, 547
　mode systems, 581
　products, 496
　time, 495

## R
Radian, 784
Rail transportation, 42–57
Range land, 365, 402
Raymond flash dryer, 569
Receipt, raw refuse, 519
Recessed-plate filter press, 101
Reciprocating engines, 678–679
Reciprocating grate, 528
Reclamation sites, 365, 402
Record-keeping, 305
Recovery of energy from biosolids, 676–684
Rectangular batch furnace grates, 528
Recuperative thickening, 100
Red clover, crop, 367
Reduction of vector attraction, 423
Reed bed, 137–139
Refuse composition, 600
Regions, jet flow, 512
Regulations of biosolids disposal and reuse, 273–342
REMOTS, 461
Reporting requirements, 305
Residue processing and disposal, 544
Residue properties, 546
Restriction(s)
　of base flood flow, 298
　on crop production, 301
　elements, nutrients, 363
　for harvesting of crops, 283
　on grazing, 301
　on public access, 302
Risk assessment
　basis of part, 650–651
　land application, 378, 382
Risk to biosolids exposure, 55–57

Rocking grate, 529
Rotary drum thickening, 97
Rotary pumps, 10–11
RSC limits, metals, incinerators, 617
Rye, 367

## S
Sacramento Central Wastewater Treatment Plant, 223, 228
Safety, land application, 373
San Jose-Santa Clara Water Pollution Control Plant, 234
Screening matrix, 696
Screw conveyors, 32–33
Sea urchin, 453
Secondary biosolids processing, 734–735
Secondary sedimentation, 202
Secondary treatment, 69
Sediment ammonia, 455
Sediment sulfate, 455
Sedimentaion clarification, 72
Separate treatment of sidestreams, 715
Septage, 183–185, 402
　chlorination, 183–185
　treatment, 140–47
　　composting, 145–146
　　lagoon disposal, 144–145
　　land application, 142–144
　　odor control, 146–147
　　receiving station, 141–142
Sewage
　clarification, 72
　composition, 66
　generation, 65
　treatment processes, 68–69
Sewage sludge, 193, 402
　and biosolids, 1–2
　stabilization, 73
　　aerobic digestion, 74
Side scan sonar, 459, 462
Sidestream
　management, 710–721
　production, 710
　quality and potential problems, 711–712
SilvaGas refuse gasification system, 552
Simultaneous sludge digestion and metal leaching (SSDML), 75
Site restrictions, 668

Site selection
    biosolids landfilling, 433
    for land application of biosolids, 659
Site variations, 725
Siting restrictions, biosolids land application, 352
SIU system, international system of units (French), 745
Sizing of equipment, 707–710
Sludge
    and biosolids pumps, types of, 2–12
    chlorination, 152, 167, 169, 178
    disposal cost curves, 334–336
    drying lagoon, 235
    equalization tank, 212
    generation, 65–66
    incineration techniques, 643
    stabilization, 167, 169, 178–183
    storage, design example, 261–262
    treatment, 68
    volume calculation, 67
Sludge-freezing bed, 139–140
Soil productivity
    mean yield, 360
    nitrogen credit, 364
Solid waste properties, 601
Soluble potash, 364
Som-A-System, 125–127
Sooting tendency of waste materials, 564
Soybeans, 358, 367, 372
Specific enthalpy, 786
Specific entropy, 786
Specific heat capacity, 785
Specific internal energy, 785
Specific latent heat, 786
Specific weight of water, 787
Spinach, 367
Spreader stoker, 542
SSDML. *See* Simultaneous sludge digestion and metal leaching
Standard pressure, 787
Standard temperature, 787
Stationary grates, 527
Stationary hearth, 524
Steradian, 784
Stoichiometry 480
Storage management, 243
Storage of sludge and biosolids, 193–197

Storage, biosolids, 242
Storage, dewatered sludge, 233
Stress, 785
Sub-bottom profiler, 460
Sun sludge system, 132–134
Surface application, biosolids land application, 354
Surface disposal of biosolids, 670–674
Surface disposal sites, requirements for, 295–296
Surface energy, water, 787, vapor pressure, water, 787
Surface impoundments, 294
Suspension burning, 502, 542–543
Syngas composition, 555
System selection procedure, 693–704

**T**

Tall-grass hay, 360
TBT. *See* Trubutyltin
Technology assessment, ocean disposal, 471
Temperature requirements, 317
Temperature-phased anaerobic digestion, 81
Terminal Island Treatment Plant, 469
Tesla, 786
Textile media filtration, 108
TFEs. *See* Toxicity equivalence factors
Thermal conditioning
    and dewatering, 109
    liquor, 717–721
Thermal conductivity, 785
Thermal decomposition (pyrolysis), 494
Thermal hydrolysis, 81
Thermal processing systems, biosolids, 560
Thermophilic anaerobic digestion fermentation, 82
Thermoselect refuse gasification system, 556
Thickening, 93, 428
Three-phase anaerobic digestion, 82
Timothy, crop, 367
TKN. *See* Total Kjeldahl nitrogen
TN. *See* Total nitrogen
Tobacco, 367
Tomatoes, 367
Torque flow pumps, 4–5
Total hydrocarbons, and measurement, 314–316, 616, 681–684

Total Kjeldahl nitrogen (TKN), 357, 402
Total nitrogen (TN), 357, 402
Total PAN applied, 369
Total solids, 404
Toxicity equivalence factors (TFEs), 511
TPS Termiska Processor–gasification, 550
Trace elements, 404
Training, 58
Transport of sewage sludge and biosolids, 1–64
Transportable configurations, 527
Traveling grate, 529
Treatment processes, types of, 737
Treatment works, 404
Trubutyltin (TBT), 450
Truck transportation, 36–42
Two-phase anaerobic digestion, 82
Two-stage digesters, 80

## U

U.S. Federal and State Regulations, incineration, 613
U.S. practices in managing biosolids, 729–743
UNFR. *See* Unit nitrogen fertilizer rate
Unit nitrogen fertilizer rate (UNFR), 368, 404
Universal constant of gravitation, 787
Unstabilized activated sludge, 359
Upstream solids processing, 712–713
Urea, 364
US EPA, Part 503 biosolids rule, 416
US EPA, Part 503 Rule, 70
US Marine Protection Research and Sanctuaries Act, MPRSA, 448, 450
Use elements, nutrients, 363
Uses of wastewater solids, 684–686

## V

Vacuum filtration, 106
Vacuum-assisted dewatering bed (VADB), 136–137
VADB. *See* Vacuum-assisted dewatering bed
Vector attraction reduction, 224, 249, 302–303, 328–332, 404, 653, 672–673
   for land application, 280
   options, 284, 663
   processes, 426
Vector, 224, 404
Vegetable crop, 367
Venturi scrubber, 611
Vermicomposting, 86
Viscosity, 785, 787
Volatility temperatures, air toxics metals, 510
Volatilization, 404
Volatilization factor
   ammonium nitrogen, 370, 379, 389–390
   biosolids, 360

## W

WAG. *See* Waste assessment guidance
WAS. *See* Waste activated sludge
Waste activated sludge (WAS), 359
Waste assessment audit, 447
Waste assessment guidance (WAG), 446–447
Waste heat boiler, 538
Waste piles, 294
Wastewater
   chlorination, 152–153, 158
      design example, 165
      troubleshooting, 160–164
   stabilization pond, 203
   treatment facilities, 714–715
   treatment sludge storage, 197
Water eductors, 12
Water property, 787
Water tube boiler, 538
Watt, 785
Weber, 786
Wedgewater bed, 134–136
West-Southwest wastewater treatment plant, 236
Wheat, 358, 360, 367, 372
Winter operation, biosolids land application, 354

## Y

Yield, 368, 404

## Z

Zoned partial combustion, 547

Site selection
   biosolids landfilling, 433
   for land application of biosolids, 659
Site variations, 725
Siting restrictions, biosolids land application, 352
SIU system, international system of units (French), 745
Sizing of equipment, 707–710
Sludge
   and biosolids pumps, types of, 2–12
   chlorination, 152, 167, 169, 178
   disposal cost curves, 334–336
   drying lagoon, 235
   equalization tank, 212
   generation, 65–66
   incineration techniques, 643
   stabilization, 167, 169, 178–183
   storage, design example, 261–262
   treatment, 68
   volume calculation, 67
Sludge-freezing bed, 139–140
Soil productivity
   mean yield, 360
   nitrogen credit, 364
Solid waste properties, 601
Soluble potash, 364
Som-A-System, 125–127
Sooting tendency of waste materials, 564
Soybeans, 358, 367, 372
Specific enthalpy, 786
Specific entropy, 786
Specific heat capacity, 785
Specific internal energy, 785
Specific latent heat, 786
Specific weight of water, 787
Spinach, 367
Spreader stoker, 542
SSDML. *See* Simultaneous sludge digestion and metal leaching
Standard pressure, 787
Standard temperature, 787
Stationary grates, 527
Stationary hearth, 524
Steradian, 784
Stoichiometry 480
Storage management, 243
Storage of sludge and biosolids, 193–197
Storage, biosolids, 242
Storage, dewatered sludge, 233
Stress, 785
Sub-bottom profiler, 460
Sun sludge system, 132–134
Surface application, biosolids land application, 354
Surface disposal of biosolids, 670–674
Surface disposal sites, requirements for, 295–296
Surface energy, water, 787, vapor pressure, water, 787
Surface impoundments, 294
Suspension burning, 502, 542–543
Syngas composition, 555
System selection procedure, 693–704

**T**

Tall-grass hay, 360
TBT. *See* Trubutyltin
Technology assessment, ocean disposal, 471
Temperature requirements, 317
Temperature-phased anaerobic digestion, 81
Terminal Island Treatment Plant, 469
Tesla, 786
Textile media filtration, 108
TFEs. *See* Toxicity equivalence factors
Thermal conditioning
   and dewatering, 109
   liquor, 717–721
Thermal conductivity, 785
Thermal decomposition (pyrolysis), 494
Thermal hydrolysis, 81
Thermal processing systems, biosolids, 560
Thermophilic anaerobic digestion fermentation, 82
Thermoselect refuse gasification system, 556
Thickening, 93, 428
Three-phase anaerobic digestion, 82
Timothy, crop, 367
TKN. *See* Total Kjeldahl nitrogen
TN. *See* Total nitrogen
Tobacco, 367
Tomatoes, 367
Torque flow pumps, 4–5
Total hydrocarbons, and measurement, 314–316, 616, 681–684

Total Kjeldahl nitrogen (TKN), 357, 402
Total nitrogen (TN), 357, 402
Total PAN applied, 369
Total solids, 404
Toxicity equivalence factors (TFEs), 511
TPS Termiska Processor–gasification, 550
Trace elements, 404
Training, 58
Transport of sewage sludge and biosolids, 1–64
Transportable configurations, 527
Traveling grate, 529
Treatment processes, types of, 737
Treatment works, 404
Trubutyltin (TBT), 450
Truck transportation, 36–42
Two-phase anaerobic digestion, 82
Two-stage digesters, 80

**U**
U.S. Federal and State Regulations, incineration, 613
U.S. practices in managing biosolids, 729–743
UNFR. See Unit nitrogen fertilizer rate
Unit nitrogen fertilizer rate (UNFR), 368, 404
Universal constant of gravitation, 787
Unstabilized activated sludge, 359
Upstream solids processing, 712–713
Urea, 364
US EPA, Part 503 biosolids rule, 416
US EPA, Part 503 Rule, 70
US Marine Protection Research and Sanctuaries Act, MPRSA, 448, 450
Use elements, nutrients, 363
Uses of wastewater solids, 684–686

**V**
Vacuum filtration, 106
Vacuum-assisted dewatering bed (VADB), 136–137
VADB. See Vacuum-assisted dewatering bed
Vector attraction reduction, 224, 249, 302–303, 328–332, 404, 653, 672–673
for land application, 280
options, 284, 663
processes, 426
Vector, 224, 404
Vegetable crop, 367
Venturi scrubber, 611
Vermicomposting, 86
Viscosity, 785, 787
Volatility temperatures, air toxics metals, 510
Volatilization, 404
Volatilization factor
ammonium nitrogen, 370, 379, 389–390
biosolids, 360

**W**
WAG. See Waste assessment guidance
WAS. See Waste activated sludge
Waste activated sludge (WAS), 359
Waste assessment audit, 447
Waste assessment guidance (WAG), 446–447
Waste heat boiler, 538
Waste piles, 294
Wastewater
chlorination, 152–153, 158
design example, 165
troubleshooting, 160–164
stabilization pond, 203
treatment facilities, 714–715
treatment sludge storage, 197
Water eductors, 12
Water property, 787
Water tube boiler, 538
Watt, 785
Weber, 786
Wedgewater bed, 134–136
West-Southwest wastewater treatment plant, 236
Wheat, 358, 360, 367, 372
Winter operation, biosolids land application, 354

**Y**
Yield, 368, 404

**Z**
Zoned partial combustion, 547

Printed in the United States of America